# SMALL MOLECULE MEDICINAL CHEMISTRY

# SMALL MOLECULE MEDICINAL CHEMISTRY

## Strategies and Technologies

Edited by

WERNGARD CZECHTIZKY
PETER HAMLEY

WILEY

Published by John Wiley & Sons, Inc., Hoboken, New Jersey

Published simultaneously in Canada



***Library of Congress Cataloging-in-Publication Data***

Small molecule medicinal chemistry : strategies and technologies / edited by Werngard Czechtizky, Peter Hamley.

pages cm

Includes bibliographical references and index.

ISBN 978-1-118-77160-0 (cloth)

1. Pharmaceutical chemistry. 2. Drug development. I. Czechtizky, Werngard, editor. II. Hamley, Peter, editor.

RS403.S62 2015

615.1′9–dc23

2015020801

Cover image courtesy of ademperçem/iStockphoto

Set in 10/12pt Times by SPi Global, Pondicherry, India

# CONTENTS

# LIST OF CONTRIBUTORS

**Muhammad Ayaz,** University of Arizona, Tucson, AZ, USA

**Karl-Heinz Baringhaus,** Sanofi-Aventis Deutschland GmbH, Frankfurt am Main, Germany

**Nina Bionda,** University of Rochester, Rochester, NY, USA

**Werngard Czechtizky,** Sanofi-Aventis Deutschland GmbH, Frankfurt am Main, Germany

**Hélène Y. Decornez,** Albany Molecular Research Inc. (AMRI), Albany, NY, USA

**Rudi Fasan,** University of Rochester, Rochester, NY, USA

**Warren R. J. D. Galloway,** University of Cambridge, Cambridge, UK

**Niels Griesang,** Sanofi R&D, Frankfurt am Main, Germany

**Peter Hamley,** Sanofi-Aventis Deutschland GmbH, Frankfurt am Main, Germany

**Lars Ole Haustedt,** AnalytiCon Discovery GmbH, Potsdam, Germany

**Gerhard Hessler,** Sanofi-Aventis Deutschland GmbH, Frankfurt am Main, Germany

**Christopher Hulme,** University of Arizona, Tucson, AZ, USA

**Jörg Hüser,** Bayer Pharma AG, Wuppertal, Germany

**Edgar Jacoby,** Janssen Research & Development, Beerse, Belgium

**Patrick Jimonet,** Sanofi-Aventis R&D, Vitry-sur-Seine, France

**Philip S. Jones,** European Screening Centre Newhouse, Lanarkshire, UK

**Christopher Kallus,** Sanofi R&D, Frankfurt am Main, Germany

**Douglas B. Kitchen,** Albany Molecular Research Inc. (AMRI), Albany, NY, USA

**Matthias Löhn,** Sanofi Deutschland GmbH, Frankfurt am Main, Germany

**Thomas C. Maier,** Sanofi-Aventis Deutschland GmbH, Frankfurt am Main, Germany

**Andres Mariscal,** Tucson Research Center, Sanofi US, Tucson, AZ, USA

**Alexander Marker,** Sanofi R&D, Frankfurt am Main, Germany

**Guillermo Martinez-Ariza,** University of Arizona, Tucson, AZ, USA

**Federico Medda,** University of Arizona, Tucson, AZ, USA

**Katharina Mertsch,** Sanofi R&D, Frankfurt am Main, Germany

**Adam Nelson,** Astbury Centre for Structural Molecular Biology, and School of Chemistry, University of Leeds, Leeds, UK

**Thomas Neumann,** NovAliX, BioParc, Illkirch Cedex, France

**Victor Nikolaev,** Tucson Research Center, Sanofi US, Tucson, AZ, USA

**Jacob Olsen,** Sanofi R&D, Frankfurt am Main, Germany

**Michelle Palmer,** Broad Institute of Harvard and MIT, Cambridge, MA, USA

**Marcel Patek,** Tucson Research Center, Sanofi US, Tucson, AZ, USA

**Oliver Plettenburg,** Sanofi Deutschland GmbH, Frankfurt am Main, Germany

**Jean-Paul Renaud,** NovAliX, BioParc, Illkirch Cedex, France

**Arthur Shaw,** University of Arizona, Tucson, AZ, USA

**Karsten Siems,** AnalytiCon Discovery GmbH, Potsdam, Germany

**Martin Smrcina,** Tucson Research Center, Sanofi US, Tucson, AZ, USA

**David R. Spring,** University of Cambridge, Cambridge, UK

**Jamie E. Stokes,** University of Cambridge, Cambridge, UK

**Peter ten Holte,** Janssen Research & Development, LLC, San Diego, CA, USA

**Luc Van Hijfte,** NovAliX, BioParc, Illkirch Cedex, France

**Eric Wegrzyniak,** Tucson Research Center, Sanofi US, Tucson, AZ, USA

**Martin Will,** Sanofi R&D, Frankfurt am Main, Germany

# INTRODUCTION

WERNGARD CZECHTIZKY and PETER HAMLEY
*Sanofi-Aventis Deutschland GmbH, Frankfurt am Main, Germany*

## I.1 MEDICINAL CHEMISTRY: A DEFINITION

The science of medicinal chemistry emerged in a recognizable form toward the end of the nineteenth century as a discipline exploring relationships between chemical structure and observed biological activity via chemical modification and structural mimicry of nature's materials. Its roots have been said to be in the fertile mix of ancient folk medicine and early awareness of the properties of natural products, hence the name [1]. A more recent definition is that it is a "traditional scientific discipline rooted in organic chemistry concerning the discovery, development, identification and interpretation of the mode of action of biologically active compounds at the molecular and cellular level" [2]. It has also been stated that "medicinal chemistry uses physical organic principles to understand the interaction of smaller molecular displays with the biological realm" [1].

## I.2 THE ROLE OF A MEDICINAL CHEMIST

Medicinal chemistry is pivotal to the process of discovering medicines. The goal is seemingly simple—the design and synthesis of new biologically active molecules with a new and useful medical advantage along with a safety profile good enough to obtain approval to reach the global pharmaceutical market. However, to achieve this is immensely challenging, and in order to have a chance of succeeding, a successful medicinal chemist must operate at the boundaries of many disciplines [3] to interact in and understand areas far outside organic chemistry and to analyze and understand a significant amount of data from various biological sources such as cell biology, molecular biology, and

*Small Molecule Medicinal Chemistry: Strategies and Technologies*, First Edition.
Edited by Werngard Czechtizky and Peter Hamley.

pharmacology. In addition, the medicinal chemist must constantly take the right decisions using analytical, creative, and teamworking skills to advance toward the goal.

Medicinal chemists are continuously working against the odds [4, 5]—the rate of molecules making it all the way to market approval is nowadays estimated to be 1:10,000 [6]—in iterations of compound design and synthesis, often referred to as design–make–test cycles. In order to increase the likelihood of success, what was once a process involving much trial and error has become more predictive over the last decade. Ideally, one would *only* consider the synthesis of molecules with a high chance of biological potency, a reasonable physicochemical and pharmacokinetic behavior, and an absence of properties predicted to lead to safety issues. To this end, medicinal chemists no longer rely on their own experience, but access new molecules in collaboration with biologists, chemoinformaticians [7] and drug designers [8], structural biologists, specialists for physicochemical and pharmacokinetic [9] profiling, and toxicologists. The creative forces within an individual medicinal chemistry project come together in a project team to give rise to a new chemical entity (NCE) [10] with a unique biological activity in a highly collaborative process; it requires a number of scientists to contribute their individual expertise and ideas. The investigation of the data associated with an emerging chemical series with computational models of drug–target interactions and the simulation and/or testing of the series' physicochemical and pharmacokinetic properties has become crucial for any drug discovery program.

The modern medicinal chemist must maintain an awareness of new developments in this constantly evolving field; otherwise, there is a risk of following unproductive paradigms and pathways that have been shown to be contributors to poor productivity of the pharmaceutical industry in the recent past [4, 5, 11]. We know now that successful, productive medicinal chemistry must go beyond "syntheses typically consisting of six steps, predominantly composed of amine deprotections to facilitate amide formation reactions and Suzuki couplings to produce insoluble biaryl derivatives, resulting in large, flat, achiral derivatives destined for screening cascades" [12]. New technologies and new strategies are continuously brought to bear to better enable the discovery of medicines. The landscape, the understanding, and the techniques involved in the chemistry aspects of drug discovery are very different now than they were even 10 years ago, and it is necessary to keep up to date with these new aspects in order to be effective and competitive when engaged in the field. That is the goal of this book.

## I.3 THE STATE OF THE ART

### I.3.1 The Drug Discovery Value Chain

The phases of drug discovery and development ordered by time are relatively distinct and universal [6, 13]. This is known as the value chain of research and development (R&D) (Fig. I.1).

The value chain consists of a series of individual steps that sum up a time period of normally between 10 and 15 years between the initial target hypothesis and the market launch of the drug [6]. Steps "target" to "preclinical" are parts of the typical research activities within a drug discovery program leading to a clinical candidate (see also Fig. I.2). Franz Hefti [14] nicely describes the properties of a clinical candidate as follows: "A drug candidate suitable for clinical testing is expected to bind selectively to

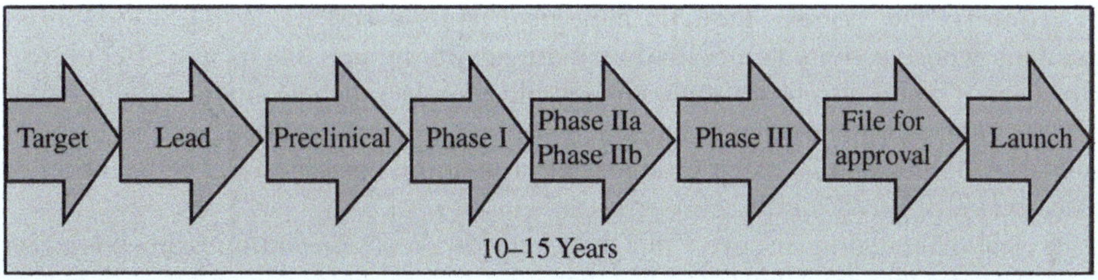


**FIGURE I.1** Sketch of the drug discovery and development value chain consisting of target hypothesis, lead identification and optimization to a clinical candidate, preclinical testing, phase I–III studies, approval, and launch.

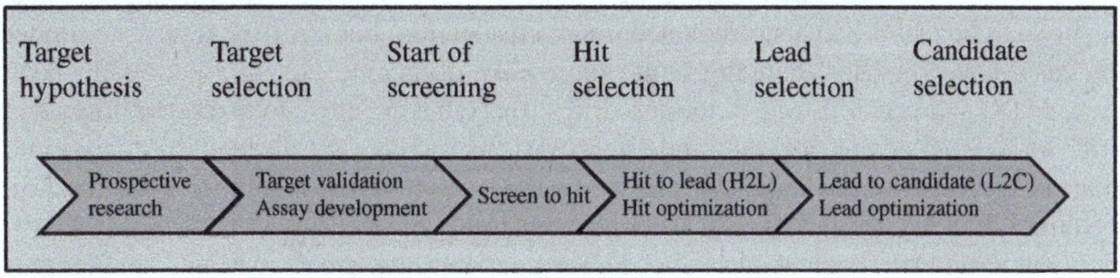


**FIGURE I.2** The value chain process focusing on the research phase, from target hypothesis to identification of a clinical candidate.

the receptor site on the target, to elicit the desired functional response of the target molecule, and to have adequate bioavailability and biodistribution to elicit the desired responses in animals and humans; it must also pass formal toxicity evaluation in animals."

Clinical phases I–III [15] comprise the phases of a clinical drug development program, culminating in the filing for approval followed (ideally) by market launch of a new drug (or NCE). In clinical phase I, researchers test a new drug or treatment in a small group of people for the first time to evaluate its safety, determine a safe dosage range, and identify side effects [15]. Normally, a small group of 20–100 healthy volunteers will be recruited. In phase II [15], the drug or treatment is given to a larger group of people to see if it is effective and to further evaluate its safety. Phase II trials are usually performed on larger groups (100–300) and are designed to assess how well the drug works. They are sometimes divided into phase IIA and phase IIB. Phase IIA is specifically designed to assess dosing requirements (how much drug should be given), while phase IIB is specifically designed to study efficacy (how well the drug works at the prescribed dose(s)). Drug development for a new drug often fails during phase II trials, when the drug is discovered not to work as planned or to have toxic effects. In phase III [15], the drug or treatment is given to even larger groups of patients (up to 10,000) to confirm its effectiveness, monitor side effects, compare it to commonly used treatments, and collect information that will allow the drug or treatment to be used safely.

### I.3.2 The Origin of a Drug Discovery Project

Drug discovery begins with a physiological or pharmacological hypothesis involving amplification or inhibition of a specific biological mechanism [1]. This is often a hypothesis involving a single protein target (Fig. I.2) along with its proposed mechanism of action (in this context, the term biological target describes the native protein in the body whose activity is modified by a drug resulting in a therapeutic effect [16]). However, it could also be a simple phenotypic response such as modulation of a biomarker [17]. A biomarker is a biological molecule found in the blood, other body fluids, or tissues and is a sign of a normal or abnormal process or of a condition or disease [17].

A clear trend in drug discovery pipelines today is a focus on portfolios around targets or phenotypes that are validated in the context of human disease, in an effort to reduce costly failure rates (attrition) at the proof-of-concept stage in humans, rather than the historic reliance on animal models of disease that are often artificially induced and have poor translatability to the species of interest, that is, human. Chemistry has a major role to play in the validation process by contributing chemical probes for target identification. Once, medicinal chemistry had a strong voice in target selection; but this is generally no longer the case since the "low-hanging fruit" of readily druggable targets has already been picked [18] and fast-follower or me-too drugs (ones that are close to marketed drugs and offer little or no advantage) are rarely approved these days [18]. Instead, biologists and pharmacologists select a target (or phenotype) that has a strong likelihood of efficacy in the clinic. Readily druggable targets (targets that are likely to be modulated with a small-molecule drug [19]) such as kinases, GPCRs, enzymes, etc. are becoming a smaller part of a modern portfolio—replaced by more challenging targets such as protein–protein interactions, transcription factors, or epigenetic targets. Because these target classes have proven more difficult to modulate with small molecules, the assessment of target "druggability" is becoming an important early step in delineating the likely challenges and hence approaches needed for a successful generation of useful hits [19].

The identification of biomarkers and the analysis of biological networks [20] and biochemical pathways [21] around the target of interest are nowadays further integral parts for the preparation of a drug discovery program. Deciphering biological signaling networks and the quantification of information flux through these networks has become one of the challenges of fundamental basic research for drug discovery. Systems biology, the computational and mathematical modeling of complex biological systems [22], is increasingly important for the development and detailed validation of highly selective tool compounds to perturb complex networks in order to discover nodes that can be targeted with innovative new drugs [2].

### I.3.3 Target Validation and Assay Development

Target selection is followed by target validation as the next crucial step before assay development and the start of the hit finding campaign. Target validation [23] is the process by which the predicted molecular target is verified. Target validation can include determining the structure–activity relationship (SAR) of analogues of the small molecule, generating a drug-resistant mutant of the presumed target, knockdown or overexpression of the presumed target, and monitoring the known signaling systems downstream of the presumed target [23]. However, in recent years, there has been more emphasis on using human patient data generated in the clinic or using epidemiological studies, and these sources are particularly powerful if this data is genetic in origin. In case the target validity is considered sufficient, assay development typically leads to the setup of biochemical and/or cellular assays to investigate the interaction of chemical compounds that amplify or attenuate the hypothesis-related biological target.

### I.3.4 The Generation of Hits

Once appropriate assays are in place, the discovery campaign can start. The initial challenge is to generate chemical matter that has some promising level of activity against the target or phenotype in question, although issues of selectivity and physical properties are at least as important.

There are many methods that can be used to generate these hit structures, and this subject is a central domain of medicinal chemistry. While once approaches such as modification of substrates or ligands were often used, the predominant form of lead generation technique in the last two decades has been high-throughput screening (HTS [24, 25]), whereby a large number of compounds are robotically screened in miniaturized assays. More recently, fragment screening [26] (using collections of compounds that have reduced complexity, typically with molecular weights under 300 Da) has become popular, and for targets for which structural information can be derived, the technique of virtual screening [27] in silico can be used. When resources are not an issue, these techniques are sometimes used in parallel to increase the chance of success. Alternative forms of screening, such as DNA-encoded library screening [28], have been introduced recently, and these can offer significant advantages in certain cases.

Screening nowadays utilizes screening collections from many sources. The classical big pharma screening collections, built up through many years of medicinal chemistry efforts and rounds of mergers and acquisitions and usually enriched with so-called rule-of-5-compliant compounds [29], are no longer the preserve of the major pharmaceutical companies. The advent of academic drug discovery and the proliferation of small biotech companies have led to the evolution of new models for access to quality collections such as risk sharing/partnership approaches or from international consortia.

Small molecules have intrinsic advantages such as oral bioavailability, accessibility of cellular compartments, simple manufacturing, and low cost of goods. However, they are also associated with high rates of attrition, despite the improvements in understanding of compound properties, and this has led to a revival of interest in peptides, peptidomimetics, oligonucleotides, novel protein formats, and natural products. In addition, the limits of chemical space exploration imposed by Lipinski's rule of 5 [29] have led to a greater emphasis on accessing more of the infinity of chemical space, resulting in new chemical collections using fundamentally different choices of chemical reactions (diversity-oriented synthesis (DOS) [30]), collections derived from multicomponent reactions (MCRs [31]), natural product-derived collections, or peptidomimetics and macrocycles. Such complex molecules are often richer in $sp^3$-configured carbons, which distinguish them from standard drug-like molecules from classical medicinal chemistry approaches [2].

### I.3.5 Hit to Lead

After screening, the prioritization of compounds from large hit lists derived from HTS (HTS triage [32]) for further follow-up is an especially challenging task for medicinal chemists. During this step of drug discovery and in addition to biological *in vitro* efficacy and drug-likeness [33], multiple parameters such as target specificity, physicochemical and ADME (absorption, distribution, metabolism, and excretion [34]) parameters must be considered simultaneously ("multiparameter optimization"). During the last 10 years, the industry has come to the realization that control of physicochemical and ADME properties is critical to improve success rates in delivering effective new drugs to patients. Most medicinal chemists nowadays have access to predictive ADME software and models that support compound design, but the accuracy of these models is still a limiting factor. Improving these models is an important challenge for medicinal chemists, experts in pharmacokinetics, and computational chemists and relies on access to experimental data available for model building.

From the filtered pool of most promising compounds, the medicinal chemist will select so-called hit series. These almost always must be further elaborated to generate a structure-activity relationship (SAR, [35])—the relationship between the chemical structure of the molecule and its biological activity—and an improved physicochemical and pharmacological profile. Parallel (or high-throughput) medicinal chemistry (either in solution or on solid phase) is routinely used as a tool allowing the medicinal chemist to prosecute multiple structurally distinct series concurrently and to develop rich SAR quickly. It allows the design team to draw conclusions based on data associated with a matrix of compounds instead of single compounds. The systems used are nowadays far more than just bench equipment tied together via robotics; an extensive infrastructure of databases and software has been built to facilitate interactive use of the systems, sometimes even remotely from around the world.

### I.3.6 Lead Optimization

The hit optimization *resp.* hit-to-lead (H2L) phase of the drug discovery program is crucial to select a lead, which usually has a suitable overall compound profile to show—for the first time—an *in vivo* efficacy of the compound series at the target of interest in animal disease models. After lead selection, an often resource-intensive lead optimization (or lead-to-candidate (L2C)) program is required to identify the endpoint of a discovery program, that is, a clinical candidate with suitable biological potency and physicochemical and pharmacological profile, which is then profiled in toxicity and dose-finding studies in animals during preclinical testing. This phase uses much of the same techniques as the H2L phase, but the number of compounds and series tends to decrease dramatically until just one candidate drug is identified. A more careful study of the properties of the reduced set of synthesized compounds needs to be made, for example, to assess behavior *in vivo*, both in animal models and in terms of pharmacokinetic properties (how quickly the drug is cleared from the body, how it is metabolized and distributed, etc.). These studies usually necessitate preparation of more material; therefore, efficient synthetic routes need to be devised, ideally in partnership with development (process) chemists. Closer to the clinic, the compounds of highest interest will be assessed for a suitable physical form to enable reproducible manufacture and often to increase solubility, typically by selecting an optimal salt form. If all results are acceptable, the final compound is tested for animal toxicity, usually in several species at ascending doses, and if there are no adverse effects, it is transferred into the clinic to be tested in humans.

## I.4 CURRENT AND FUTURE CHALLENGES FOR MEDICINAL CHEMISTRY

Drug discovery has undergone major strategic changes in the last decade, which affect both the setting and the practice of the discipline. The regulatory environment has become more stringent with safety requirements ever more challenging, while the industry faces substantial cost increases in tandem with declining R&D success rates, often due to lack of clinical efficacy in humans or unexpected toxicity [18, 5]. This has resulted in a productivity gap, and although there are many factors contributing to this, some techniques practiced by chemists in drug discovery in the past have been associated with this. Commonly cited examples include the advent of combinatorial chemistry and the associated inflation of molecular weight, the need for a large number of compounds to feed HTS leading to a lack of imagination in synthetic protocols and ultimately to "flat" molecules, the phasing out of natural product collections and skills associated with them, a race for potency rather than multidimensional optimization, and the list goes on [36]. What is exciting about recent developments in the field is that they are often at least in part answers to these particular criticisms—often associated with a greater awareness of chemical structure, the coverage of chemical space, and the properties required to make a successful drug.

Other challenges and insights remain to be satisfactorily tackled. Target occupancy and drug–target residence times are seen as crucial for a drug's final efficacy *in vivo* [37], but there is still a lack of understanding how they can be optimized, and even less is known about how they can be designed into a given chemical series. A better understanding of the energetic and kinetic aspects of protein–ligand interactions is likely to have a great impact in this area. Unexpected toxicities furthermore require an increase in drug selectivity and a shift of the equilibrium between the desired effect on target and unwanted side effects. However, the tendency to increase lipophilicity within H2L and L2C optimization to improve potency on the target of interest often counteracts selectivity, as nonpolar protein–ligand interactions are often less specific and lead to toxic side effects [36]. Since protein–protein interactions and other difficult targets are becoming more prevalent, the ability to optimize interactions while maintaining optimal levels of lipophilicity will become more important.

The identification of highly validated targets has become more difficult, and healthcare providers worldwide are trying to reduce costs and demanding more accountability. Medicinal chemists find themselves sandwiched between target discovery and the identification of clinical compounds; the need to focus more and more on target identification and validation has become critical for the success of many drug discovery programs [2]. Recent approaches toward more disease relevant mechanisms using polypharmacology [38]—tackling a disease with two or more compounds with different modes of action or with one compound showing different modes of action in parallel—will not lead to a reduction of complexity of the task.

The era of large pharmaceutical companies with huge internal and inward-looking departments of medicinal chemistry and expensive associated staff is over. So-called big pharma has made sustained efforts to reduce cost (often through layoffs and site closures), but in parallel, growing capabilities at many contract research organizations offer the opportunity to build an effective lower-cost global network while maintaining quality and efficiency. A notable globalization and outsourcing of research and innovation away from the traditional bastions of the United States, Europe, and Japan is another obvious sign of approaches toward cost reduction. At the same time, we see increased investment in lean, small biotechs and academia establishing their own efficient drug discovery facilities, often using highly innovative approaches to therapies and technologies.

It is the aim of the following chapters to cast light on these major challenges and to describe strategic and technological solutions that represent a panoramic snapshot of the status of the chemical aspects of drug discovery today.

## REFERENCES

[1] Erhardt, P. W.; *Pure Appl. Chem.* 2002, **74**(5), 703–785.

[2] Brenk, R.; Rauh, D.; *Bioorg. Med. Chem.* 2012, **20**, 3695–3697.

[3] Hart, T.; 2006, Medicinal chemistry: progress through innovation. Summer 2006. http://www.ddw-online.com/chemistry/p97059-medicinal-chemistry:-progress-through-innovation summer-06.html (accessed May 25, 2015).

[4] Munos, B.; *Nat. Rev. Drug Discov.* 2009, **8**, 959–968.

[5] Paul, S. M.; Mytelka, D. S.; Dunwiddie, C. T.; Persinger, C. C.; Munos, B. H.; Lindborg, S. R.; Schacht, A. L.; *Nat. Rev. Drug Discov.* 2010, **9**(3), 203–214.

[6] Castner, M.; Hayes, J.; Shankle, D.; 2007, Global value chains: shifts in the configuration of the industry from 1995 until present. The Global Pharmaceutical Industry. https://web.duke.edu/soc142/team2/shifts.html (accessed May 27, 2015).

[7] Brown, F. K.; *Annu. Rep. Med. Chem.* 1998, **33**, 375.

[8] Madsen, U.; Krogsgaard-Larsen, P.; Liljefors, T.; 2002, *Textbook of Drug Design and Discovery*. Washington, DC: Taylor & Francis.

[9] Ruiz-Garcia, A.; Bermejo, M.; Moss, A.; Casabo, V. G.; *J. Pharm. Sci.* 2008, **97**(2), 654–690.

[10] Branch, S. K.; Agranat, I.; *J. Med. Chem.* 2014, **57**(21), 8729–8765.

[11] Hann, M. M.; Keserü, G. M.; *Nat. Rev. Drug Discov.* 2012, **11**, 355–365.

[12] Roughley, S. D.; Jordan, A. M.; *J. Med. Chem.* 2011, **54**, 3451–3479.

[13] Walker, S. M.; Davies, B. J.; *Drug Discov. Today* 2011, **16**(11–12), 467–471.

[14] Hefti, F. F.; *BMC Neurosci.* 2008, **9**(Suppl 3), S7.

[15] NIH; 2008, FAQ ClinicalTrials.gov—Clinical Trial Phases. http://www.nlm.nih.gov/services/ctphases.html (accessed May 27, 2015).

[16] Rang, H. P.; Dale, M. M.; Ritter, J. M.; Flower, R. J.; Henderson, G. (eds); 2012, How drugs act: general principles. In *Rang and Dale's Pharmacology*. Edinburgh/New York: Elsevier/Churchill Livingstone, pp. 6–19.

[17] Strimbu, K.; Tavel, J. A.; *Curr. Opin. HIV AIDS* 2010, **5**(6), 463–466.

[18] Scanell, J. W.; Blanckley, A.; Boldon, H.; Warrington, B.; *Nat. Rev. Drug Discov.* 2012, **11**, 191–200.

[19] Cheng, A. C. et al.; *Nat. Biotechnol.* 2007, **25**, 71–75.

[20] Proulx, S. R.; Promislow, D. E. L.; Phillips, P. C.; *Trends Ecol. Evol.* 2005, **20**(6), 345–353.

[21] Krauss, G.; 2008, *Biochemistry of Signal Transduction and Regulation*. Weinheim/New York: Wiley-VCH, p. 15.

[22] Alberghina, L.; Westerhoff, H. V.; 2005, *Systems Biology: Definitions and Perspectives*. Topics in Current Genetics **13**. Berlin: Springer-Verlag, pp. 357–451.

[23] Fishman, M.; 2012, Target validation. Nature Publishing Group. http://www.nature.com/subjects/target-validation (accessed May 27, 2015).

[24] Mayr, L. M.; Bojanic, D.; *Curr. Opin. Pharmacol.* 2009, **9**, 580–588.

[25] Hertzberg, R. P.; Pope, A. J.; *Curr. Opin. Chem. Biol.* 2000, **4**, 445–451.

[26] Rees, D. C.; Congreve, M.; Murray, C. W.; Carr, R.; *Nat. Rev. Drug Discov.* 2004, **3**, 661–672.

[27] Drwal, M.; Griffith, R.; *Drug Discov. Today Technol.* 2013, **10**(3), 395–401.

[28] Clark, M. A. et al.; *Nat. Chem. Biol.* 2009, **5**, 647 – 654.

[29] Lipinski, C. A.; *Drug Discov. Today Technol.* 2004, **1**(4), 337–341.

[30] (a)Tan, D. S.; *Nat. Chem. Biol.* 2005, **1**, 74–84;(b)Spring, D. R.; *Org. Biomol. Chem.* 2003, **1**, 3867–3870.

[31] Ugi, I.; *Pure Appl. Chem.* 2001, **73**(1), 187–191.

[32] Cox, P. B.; Gregg, R. J.; Vasudevan, A.; *Bioorg. Med. Chem.* 2012, **20**(14), 4564–4573.

[33] Murcko, M. A.; Patrick Walters, W.; *Adv. Drug Deliv. Rev.* 2002, **54**(3), 255–271.

[34] (a)Cruciani, G.; Milletti, F.; Storchi, L.; Sforna, G.; Goracci, L.; *Chem. Biodivers.* 2009, **6**(11), 1812–1821;(b)Yu, H.; Adedoyin, A.; *Drug Discov. Today* 2003, **8**(18), 852–861.

[35] Cherkasov, A. et al.; *J. Med. Chem.* 2014, **57**, 4977–5010.

[36] (a)Leeson, P.; Springthorpe, B.; *Nat. Rev. Drug Discov.* 2007, **6**, 881–890;(b)Hann, M.; Keserü, G. M.; *Nat. Rev. Drug Discov.* 2012, **11**, 355–365.

[37] Copeland, R. A.; Pompliano, D. L.; Meek, T. D.; *Nat. Rev. Drug Discov.* 2006, **5**, 730–739.

[38] Anighoro, A.; Bajorath, J.; Rastelli, G.; *J. Med. Chem.* 2014, **57**, 7874–7887.

# PART I

# EXPLORING BIOLOGICAL SPACE: ACCESS TO NEW COLLECTIONS

# 1

# ELEMENTS FOR THE DEVELOPMENT OF STRATEGIES FOR COMPOUND LIBRARY ENHANCEMENT

EDGAR JACOBY
*Janssen Research & Development, Beerse, Belgium*

## 1.1 INTRODUCTION

The main purpose of a small molecule compound collection that is sometimes considered to constitute the crown jewels of a drug discovery organization is to supply the discovery pipeline with hit-to-lead compounds for today's and the future's portfolio of drug discovery programs and to provide tool compounds for the investigation of biological targets and pathways [1–7]. Independent of the followed discovery strategy relying on diversity or hypothesis-based screening, the automated access to high-quality compounds constitutes a key asset [8]. Accordingly, all major organizations, including the National Institutes of Health (NIH) and the European Union Innovative Medicines Initiative (EU IMI), have initiated over the last years dedicated compound collection enhancement projects [9]. In alignment with the general paradigm shift observed in drug discovery, going from quantity to quality, the fundamental principle aims to select both—at the chemical and the biological level—the best possible molecular starting points for lead discovery and development in the early drug discovery phases in order to reduce attrition at later preclinical and clinical stages.

To be successful on the long-term perspective, such design strategy addresses the known target space and tries to expand into nonprecedented areas of chemical and biological spaces using diversity principles [5, 6]. Directing the molecular properties toward the lead-like space is expected to improve overall success rates. The application of absorption, distribution, metabolism, excretion, and toxicity (ADMET) property models and rules of thumb aims to reduce the attrition risk and can be front-loaded into the design

*Small Molecule Medicinal Chemistry: Strategies and Technologies*, First Edition.
Edited by Werngard Czechtizky and Peter Hamley.

of the collection. On the other hand, a screening collection should allow for serendipitous discovery going in hand with diversity designs.

Drug discovery compound collections have evolved during recent history. Up to the early 1990s when drug discovery was mainly conduced in phenotypic *in vivo* screening of corporate medicinal chemistry compounds, the collections were limited to a few thousands of compounds that were carefully generated within the individual therapeutic programs. With the advances of molecular and cell biology and the advent of high-throughput chemistry and screening, the drug discovery world changed and compound collections were grown in the last 15 years to pass in a number of organizations beyond the one million number. Today, screening collections integrate design-focused and diversity-based compound sets from the synthetic and natural paradigms generated via corporate medicinal chemistry and combinatorial compound synthesis and external compound acquisition or merger projects [1–3]. The compound collections serve diverse screening paradigms, ranging from target-based to phenotypic-based screening, from biochemical to cell-based screening, and from focused hypothesis-based to diversity-based screening, opening a wide diversity of strategic choices for the future enhancement of the compound collection.

Herein, we review chemical, biological, and informatics elements for the development of strategies for compound library enhancement. The interdisciplinary nature of the library design activity is emphasized.

## 1.2 CHEMICAL SPACE FOR DRUG DISCOVERY

The chemical space is the ensemble of all possible molecules and comprises physically documented molecules available in the corporate and public databases as well as yet unknown, virtual molecules [10]. To delineate how many and which molecules populate unknown chemical space in total, Jean-Louis Reymond's group at the University of Berne performed a systematic computational enumeration and assembled the so-called chemical universe database—Figure 1.1 [10]. GDB-11 lists 26.4 million molecules of up to 11 atoms of C, N, O, and F, GDB-13 lists 977 million molecules up to 13 atoms of C, N, O, Cl, and S, and GDB-17 lists 166 billion molecules up to 17 atoms of C, N, O, S, and halogens [13]. The number of molecules enumerated in GDB increases exponentially with the number of atoms such that the database will become impracticably large as molecular size increases. For instance, extrapolation from the numbers in GDB-17 suggests that there would be approximately $10^{24}$ molecules up to 30 nonhydrogen atoms—typically, drug-sized molecules include up to 35 nonhydrogen atoms with molecular weight (MW) < 500 Da.

Within a drug discovery context, these astronomic numbers have to be placed in relation to the number of physically available chemicals and the actual number of around 1200 approved drugs satisfying stringent efficacy and safety criteria [14]. The Elsevier Medicinal Chemistry and Chemical Abstracts Service (CAS) Registry databases, which are up-to-date representatives of molecules described in the chemical literature, list, respectively, 5.5 and 74 million compounds [15, 16]. The eMolecules and ChemNavigator iResearch libraries, which are industry references for off-the-shelf compound acquisition, list, respectively, five and six million unique commercially available compounds [17, 18]. The screening collections of the major pharmaceutical companies include typically one to two million proprietary and nonproprietary compounds [7]. Given the practically infinite possibilities,

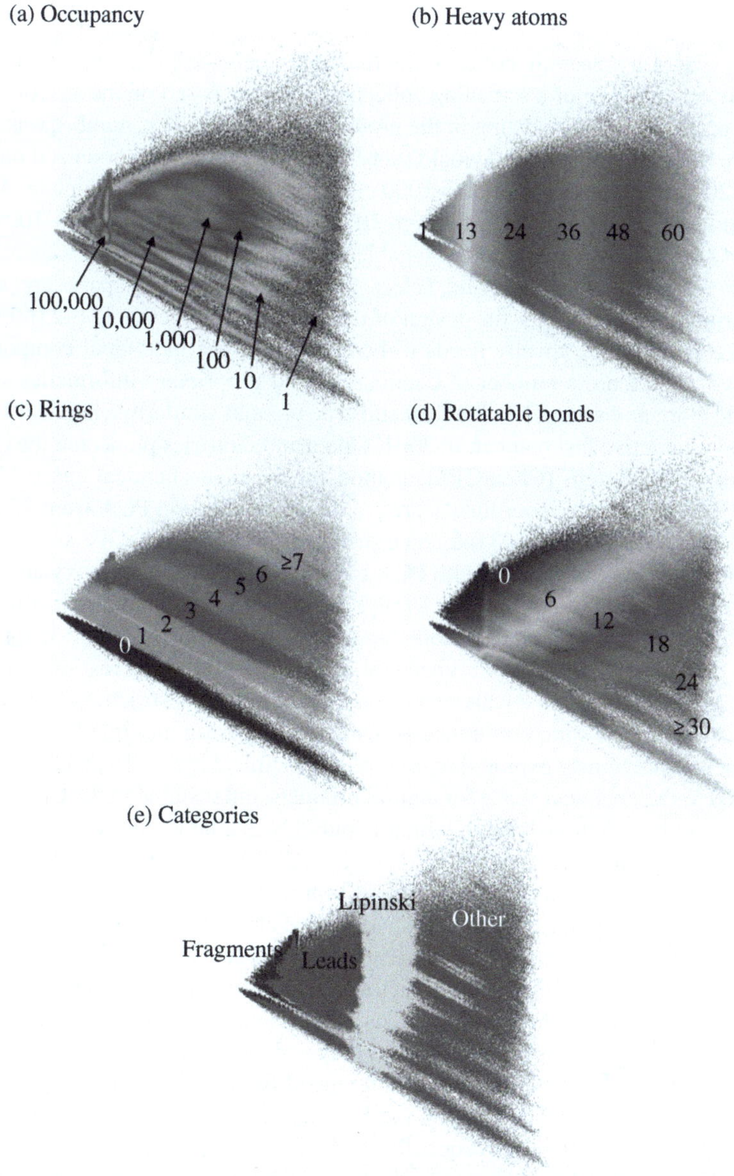


**FIGURE 1.1** Example of visualization of chemical space via principal component analysis (PCA) [10–12]. Color-coded molecular quantum number (MQN) maps of the chemical space of PubChem compounds up to 60 heavy atoms and a subset of GDB-13 compounds in the (PC1, PC2) plane (total: 66,647,914 molecules). (a) Occupancy map color coded by the number of molecules per pixel. (b–d) Descriptor value maps color coded by the average descriptor value in each pixel. Saturation to gray is used to show standard deviation. (e) Category map for blue, fragments (rule of 3 (*vide infra*), 32.5 million compounds); green, lead-like (Teague's NOT rule of 3 (*vide infra*), 2.7 million compounds) (note: in total 12.2 million structures follow Teague's lead-likeness criteria); and cyan, Lipinski (rule of 5 (*vide infra*) NOT leads or rule of 3, 31.4 million compounds); and red, not matching any rule (1.6 million compounds). Color coding according to the majority category in each pixel except for leads (green), which were given priority to make them visible. Reprinted with permission from Ref. 10. © 2014, Pan Stanford Publishing. (*See insert for color representation of the figure.*)

the optimal size of a screening collection is frequently debated [19]. One estimate of the theoretically optimal size of a screening collection could be based on the size of the finite number of protein domains existing in the protein universe [5]. This number was recently estimated to be 1,500 domains and would translate to 15,000,000 compounds if one would design 10–20 chemotypes each of 500–1000 compounds to target each domain. A similar number can be reached if one would design 10–20 chemotypes each of 500–1000 for the estimated 600–1500 disease relevant druggable protein targets [20].

Tools to visualize, navigate, and select within the chemical space are essential chemoinformatic objectives for the design of the screening collection [21–23]. For every newly added compound, novelty needs to be checked at the individual compound and scaffold level. There are a number of commercial and proprietary informatics solutions that allow to store and search by chemical substructure and similarity chemical spaces in a robust and interactive fast manner. In 2001, Oprea and Gottfries pioneered the chemical global positioning system (ChemGPS) method to visualize chemical space [24]. The ChemGPS drug space map coordinates are *t*-scores extracted via PCA from 72 descriptors that evaluate a total set of 423 reference structures. Global ChemGPS scores describe well the latent structures extracted with PCA for a set of 8599 monocarboxylates, a set of 45 heteroaromatic compounds, and for 87 alpha-amino acids. ChemGPS positions novel structures in drug space via PCA-score prediction, providing a unique mapping and prediction device for the drug-like chemical space. ChemGPS scores are comparable across a large number of chemicals and do not change as new structures are predicted, making this tool a well-suited reference system for comparing multiple libraries and for keeping track of previously explored regions of the chemical space. The method was later on expanded to the chemical space for natural products and resulted in the ChemGPS-NP visualization and prediction system, which is publicly available on the web ChemGPS-NP(Web) (http://chemgps.bmc.uu.se) [25, 26]. ChemGPS-NP(Web) can assist in compound selection and prioritization, property description and interpretation, cluster analysis and neighborhood mapping, as well as comparison and characterization of large compound data sets. Schuffenhauer et al. introduced scaffold tree to analyze the scaffold diversity of natural products [27]. The method is a hierarchical classification of chemical scaffolds that form the leaf nodes in the hierarchy trees. By an iterative removal of rings, scaffolds forming the higher levels in the hierarchy tree are obtained. Prioritization rules ensure that less characteristic, peripheral rings are removed first. All scaffolds in the hierarchy tree are well-defined chemical entities making the classification chemically intuitive. The scaffold tree classification procedure handles robustly synthetic structures and natural products. In the design of new screening collections, the scaffold tree method is invaluable. Integrated with a chemically aware visualization tool like Tibco Spotfire, it allows the immediate assessment of the abundance of a given chemical scaffold within the existing collection and the candidate collection to integrate [28]. Within a collaboration between the Max Planck Institute for Molecular Physiology and Novartis, the method was integrated in a structural classification of natural products (SCONP) to chart the known chemical space explored by nature [29]. SCONP arranges the scaffolds of the natural products in a treelike fashion and provides a viable analysis- and hypothesis-generating tool for the design of natural product-derived compound collections. The Waldmann group developed the method further into Scaffold Hunter, an interactive computer-based tool for navigation in chemical space that fosters intuitive recognition of complex structural relationships associated with bioactivity [30, 31]. The program reads compound structures and bioactivity data, generates compound scaffolds, correlates them in a hierarchical treelike arrangement, and annotates them with bioactivity.

In a need to enable navigation and selection with chemical space, researchers at Janssen presented library enhancement through the wisdom of crowds [32]. Compounds of interest are clustered together with the in-house collection using a fingerprint-based clustering algorithm that emphasizes common substructures and works with large data sets. Clusters populated exclusively by external compounds are identified as "diversity holes," and representative members of these clusters are presented to the global corporate medicinal chemistry community, who are asked to specify which ones they like, dislike, or are indifferent by using a simple point-and-click interface. The resulting votes are then used to rank the clusters from the most to the least desirable and to prioritize which ones should be targeted for acquisition.

Hypothesis-based selection in chemical space is supported by different types of virtual screening technologies depending on the size of the considered physical or virtual compound libraries. ChemNavigator offers, for instance, a comprehensive set of virtual screening services called 3DPL™ to select from their iResearch Library [18]. Researchers at Boehringer Ingelheim run virtual screening in a huge collection of virtual combinatorial libraries that led recently to the identification of two new structural classes of GPR119 agonists [33, 34]. Their virtual library called Boehringer Ingelheim Comprehensive Library of Accessible Innovative Molecules (BICLAIM) is based on combinatorial reactions and stored in a feature trees (FTrees) fragment space. The virtual chemical space contains about 1,600 scaffolds and 30,000 reagents encoding about $5 \times 10^{11}$ theoretically chemically accessible molecules. The chemical universe database GDB-17 of 166.4 billion molecules can be virtually screened using a hashed fingerprint derived from the 42 integer MQN molecular descriptors [12]. An MQN-searchable 50 million subset of GDB-17 is publicly available at http://www.gdb.unibe.ch.

## 1.3 MOLECULAR PROPERTIES FOR DRUG DISCOVERY

Given that the size of the chemical space is virtually infinite, the art of library design lies in parts in the selection of the appropriate molecular property spaces. Medicinal computational chemists developed over the last decade a number of statistical analyses and ADMET models that are easily applicable upfront compound synthesis and are intended to reduce attrition at various stages [35–39]. The simplest models include substructure filters for potentially problematic chemical functionalities. The rapid elimination of swill (REOS) filters published by Vertex flag false positives in screening due to assay interference and reactivity or compounds with poor ADMET properties [40]. The pan-assay interference compounds (PAINS) filters identify frequent hitter in HTS [41, 42]. The analysis of Thorne et al. on typical screening technology-related assay artifacts provides a further guide to eliminate undesirable compound classes [43]. Among the molecular properties that are essential to small molecules are the water solubility and membrane permeability that form the basis of the two-dimensional biopharmaceutics classification system for drug developability [44]. The two properties are dependent in the sense that for specific oral dosing regime, a minimum equilibrium solubility level is required given the compound permeability class. They are not only important for late drug developability but also for early drug discovery. A compound has to be sufficiently soluble to enable a dose–response-dependent readout. In addition, the compound has to have the appropriate permeability properties to reach its site of action within a cell or tissue. It is thus logical that a number of models focus on these properties.

Besides cheminformatics software like ACD/Labs Percepta Platform [45], Schrodinger's QikProp [46], or Simulation Plus' ADMET Predictor [47], which are based on advanced quantitative structure–property relationship (QSPR) modeling methods, there are a number of simpler heuristic-based models that have the advantage of being easily interpretable by the medicinal chemist. Chris Lipinski's pioneering work on the rule of 5, for instance, is derived from a quest for experimental and computational approaches to estimate solubility and permeability in drug discovery and development settings [48]. In the discovery setting, the rule of 5 predicts that poor absorption or permeation is more likely when there are more than 5 H-bond donors and 10 H-bond acceptors, the MW is greater than 500 Da, and the calculated LogP (CLogP) is greater than 5.

As the rule of 5 is, for instance, not able to statistically discriminate drugs from nondrugs, Lipinksi's thinking was initially highly controverted in an era where combinatorial chemistry aimed to deliver compounds that were easily synthesized and purified with having higher MW and LogP. Later, Lipinski's thinking influenced an entire school of thought around small molecules for drug discovery [49, 50].

The analysis provided by Wenlock showed that MW and ClogP distributions move through the phases of drug development and that the property distributions approach those of marketed oral drugs—Figure 1.2 [51]. Given that the early combinatorial chemistry was not successful, it is not surprising that over time a quantity to quality shift developed. The observation that medicinal chemists focus on potency during early lead optimization by making compounds even bigger and lipophilic led Oprea's group at AstraZeneca to introduce the concept of lead-likeness [52]. Larger size and lipophilicity drive also compound promiscuity and potential off-target effects [53–55]. Increasing MW and CLogP is an easy way to reach the common nanomolar potency. This tendency is, however, counterproductive for ADMET and moves the properties further away from historical drug space.

Over the time, Oprea refined his analysis of lead–drug pairs and recommended that lead-likeness libraries should have the following as characteristics: MW < 460 Da, $-4 <$ CLogP $< 4.2$, water solubility LogS $> -5$, number of rotatable bonds less than 10, number of rings less than 4, number of H-bond donors less than 5, and number of H-bond acceptors less than 9 [56]. These differences compared to drugs are thus subtle, and as concluded by Proudfoot, successful and timely drug discovery campaigns require high-quality lead structures, and these lead structures may need to be much more drug-like than is commonly accepted [57]. Similar conclusions were derived by Oprea when analyzing more recently chemical probes and leads [58]. The field of fragment-based drug discovery takes the concept of having small molecule starting points further. Following an analysis done at Astex, fragment libraries are often designed by applying a rule of 3 in which MW is less than 300 Da, the number of hydrogen-bond donors is less than or equal to 3, the number of hydrogen-bond acceptors is less than or equal to 3, and ClogP is less than or equal to 3. In addition, the analysis suggested that the number of rotatable bonds (NROT) (≤3) and topological polar surface area (tPSA) (≤60 Å$^2$) might also be useful criteria for fragment selection [59]. The Astex scientists argue that ADMET properties can be better controlled during optimization when starting with a fragment compound compared to a larger compound.

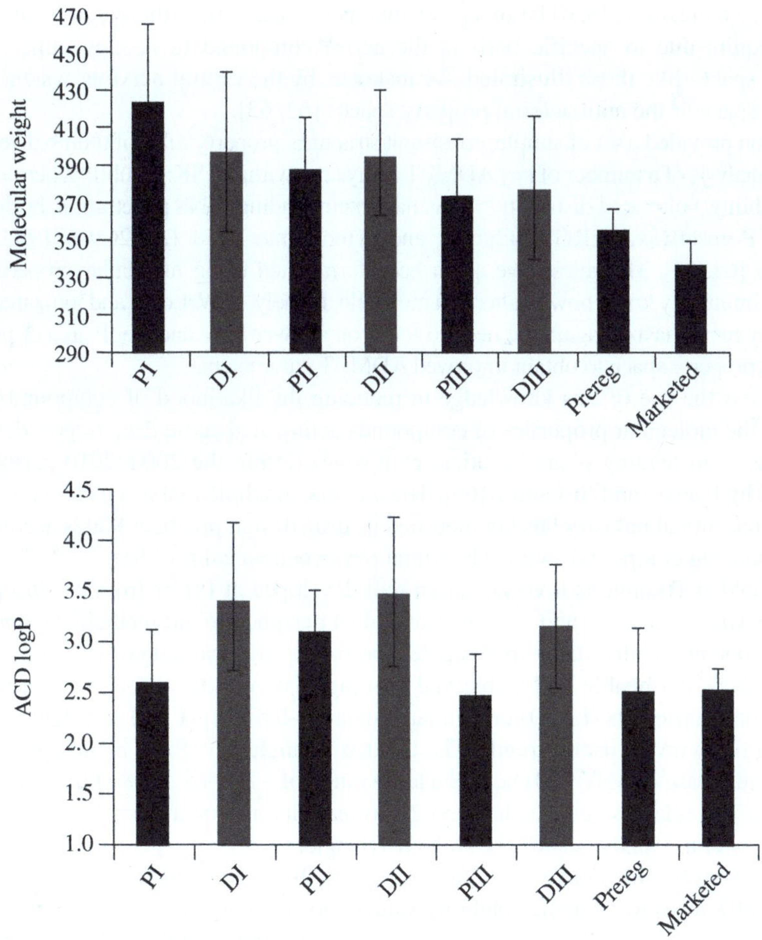


**FIGURE 1.2** Analysis by Wenlock et al. of the evolution of molecular property distributions with progressing through development stages [51]. Phase I (PI); discontinued phase I (DI); phase II (PII); discontinued phase II (DII); phase III (PIII); discontinued phase III; preregistration (Prereg); marketed oral drugs. The analysis shows that the mean molecular weight of orally administered drugs in development decreases on passing through each of the different clinical phases and gradually converges toward the mean molecular weight of marketed oral drugs. It is also clear that the most lipophilic compounds are being discontinued from development. Reprinted with permission from Ref. 51. © 2003, American Chemical Society.

The observation that smaller and less lipophilic starting points are better prompted researchers at GSK to propose the concept of lead-oriented synthesis (LOS), which aims for compounds with LogP values in the range −1 to 3 and MW in the range of 200–350 Da [60]. The authors emphasize the need to access to novel synthesis methodologies given that the current array chemistry has an unintentional bias toward the synthesis of less drug-like molecules.

Further analyses of computed and experimental physicochemical properties of drug compounds lead to the conclusion that the property spaces depend on the target class and therapeutic indication [35, 61]. For a given therapeutic indication, the site of *in vivo* action might require due to specific barriers the active compound to occupy quite specific property spaces like those illustrated, for instance, by the central nervous system (CNS) property space or the antibacterial property spaces [62, 63].

Gleeson provided a set of simple, consistent structure–property rules of thumb determined from an analysis of a number of key ADMET assays run within GSK: solubility, permeability, bioavailability, volume of distribution, plasma protein binding, CNS penetration, brain tissue binding, P-gp efflux, hERG inhibition, and cytochrome P450 1A2/2C9/2C19/2D6/3A4 inhibition [64, 65]. The rules have again been formulated using molecular properties that chemists intuitively know how to alter in a molecule, namely, MW, LogP, and ionization state. This study reemphasizes again the need to focus on a lower MW and LogP area of physicochemical property space to obtain improved ADMET parameters.

To assess the use of this knowledge in reducing the likelihood of compound-related attrition, the molecular properties of compounds acting at specific drug targets described in patents from leading pharmaceutical companies during the 2000–2010 period were analyzed by Leeson and St-Gallay [66]. The authors conclude that a substantial sector of the pharmaceutical industry has not modified its drug design practices and is according to them producing compounds with suboptimal physicochemical profiles.

The Golden Triangle is a visualization tool developed at Pfizer from *in vitro* permeability, *in vitro* clearance, and computational data designed to aid medicinal chemists in achieving metabolically stable, permeable, and potent drug candidates [67]. Classifying compounds as permeable and stable and plotting MW versus octanol–buffer (pH 7.4) distribution coefficients (LogD) or estimated octanol–buffer (pH 7.4) distribution coefficients (eLogD) reveal useful trends. The Golden Triangle is defined by an apex of MW 450 Da and a base of MW 200 Da, and a logD range of −2 to +5. 25% of the compounds in Golden Triangle has acceptable Caco-2 permeability and microsomal stability versus only 3% for compounds outside the Golden Triangle.

The analysis by Hill and Young of the relationship between hydrophobicity and approximately 100k measured kinetic solubility values showed that better solubility predictions are obtained by taking ACD clogD(pH 7.4) values together with the number of aromatic rings in a given molecule—Figure 1.3 [68]. The Solubility Forecast Index (SFI = clogD(pH 7.4) + #Ar) was proposed as a simple, yet effective, guide to predicting solubility.

Pfizer provided with the 3/75 rule an example of how physicochemical drug properties are associated with *in vivo* toxicity [69]. From a data set consisting of animal *in vivo* toleration studies on 245 preclinical Pfizer compounds across a broad swath of chemical space, an increased likelihood of toxic events across a wide range of types of toxicity is observed for less polar, more lipophilic compounds. Compounds with $\text{CLogP} < 3$ and a $\text{tPSA} > 75\,\text{Å}^2$ show a clear correlation of lower odds of promiscuity and toxicity.

Strict property-based assessment of drug-likeness has been recently criticized as being too blunt an instrument that affords only a yes–no answer. The quantitative estimate of drug-likeness (QED) has been introduced to overcome such limitations by characterizing how well physicochemical properties of a candidate compound match the property distributions of marketed oral drugs [70]. Ritchie and MacDonald showed that drugs with high QED scores exhibit higher absorption and bioavailability, are administered at lower doses, and have fewer drug–drug interaction warnings, P-glycoprotein interactions, and absorption issues due to a food effect. By contrast, the high-scoring drugs exhibit similar behavior to low-scoring drugs with respect to free fraction in plasma, extent of gut-wall

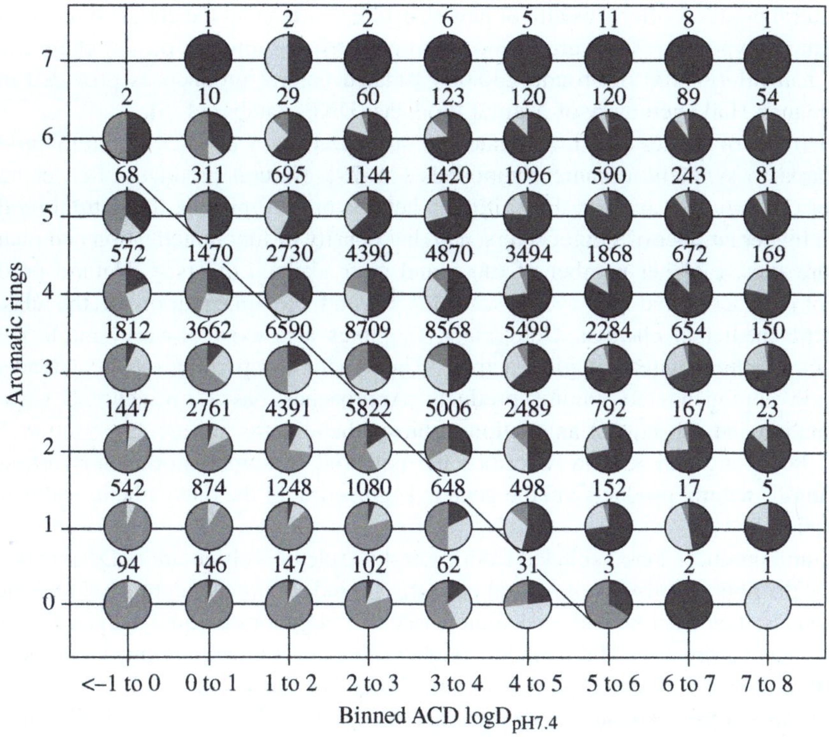


**FIGURE 1.3** Trelis plot of Hill and Young showing the distribution of water solubility as a function of computed LogD and # of aromatic rings [68]. Solubility classes—green, high (>200 μM); yellow, medium (30–200 μM); and red, low (<30 μM). The number above the pie charts corresponds to the number of compounds analyzed for each bin. Reprinted with permission from Ref. 68. © 2010, Elsevier. (*See insert for color representation of the figure.*)

metabolism, first-pass hepatic extraction, elimination half-life, clearance, volume of distribution, and frequency of dosing [71].

## 1.4 MAJOR COMPOUND CLASSES

Natural products, known bioactives, peptides, heterocycles, and DOS libraries, constitute the prevalent compound classes represented in screening collections and are reviewed in this section [4]. For obvious reasons, natural principles play a predominant role in the history of drug discovery. Diverse classes of natural products including carbohydrates, steroids, fatty acids, polyketides, peptides, terpenoids, flavonoids, alkaloids, and many other products were isolated initially from herbs and later from various micro and higher organisms for structure and activity characterization [72, 73]. Natural products are a major source of innovative tool compounds for the elucidation of signaling pathways and new medicines for most indications, such as lipid disorders, cancer, infectious diseases, and immunomodulation. Between 1981 and 2002, 5% of the around 870 new chemical entities approved by the US Food and Drug Administration (FDA) were natural products, and another 23% were molecules derived from natural products [74].

Natural products offer a wealth of new structures far beyond the classical repertoire of synthetic compounds. The current most comprehensive summary on the chemical and biological information of around 230,000 isolated natural products is provided in the Chapman & Hall Dictionary of Natural Products (DNP) database [75].

A number of studies have investigated the structural characteristics of natural products compared to synthetic organic compounds [76–79]. Natural products often contain a greater proportion of oxygen than nitrogen heteroatoms. Typically, the natural products have a higher number of stereocenters, a higher density of functionalization and pharmacophore sites, a higher number of rings, and more skeletal diversity. Natural products exemplify macro- and polycyclic scaffolds beyond the imagination of the classical synthetic medicinal chemist. Conversely, examples also exist of very simple natural product structures with biological activity. The structural repertoire can be extended by genomic approaches to natural products. Approaches based on genome sequence information and subsequent annotation of biosynthetic pathways are emerging technologies [80]. Tang and Khosla described the potential of combinatorial biosynthesis of "unnatural" natural products via the genetic engineering of the biosynthetic pathways of polyketides [81].

Natural products were excluded from Lipinski's rule of 5 observation. Despite the fact that the distribution profiles of natural products are indeed broader compared to synthetic compounds, their fraction with two or more rule of 5 violations is equal to that of synthetic drugs. One interpretation of this finding might be that evolutionary optimization has coded in addition to these essential properties other biocharacteristics that still need to be deciphered. An analysis by Ganesan showed that those natural products that violated the rule of 5 have higher MW, more rotatable bonds, and more stereocenters; however, they remain largely compliant in terms of logP and H-bond donors, highlighting the importance of these two metrics in predicting bioavailability [82]. Natural products have learned to maintain low hydrophobicity and intermolecular H-bond donating potential when it needs to make biologically active compounds with high MW and large numbers of rotatable bonds. In addition, natural products are more likely than purely synthetic compounds to resemble biosynthetic intermediates or endogenous metabolites and hence take advantage of active transport mechanisms. Conversely, a number of marketed natural product-based drugs are not orally available, but uniquely address a number of therapeutic applications.

One key dilemma for natural products drug discovery is that although the primary HTS hit rates in the micromolar affinity range are 5–10 times higher than the hit rates for synthetic compounds, the take-up rate of the compounds by chemists for follow-up lead optimization is significantly lower [1]. This finding is most probably due to the higher structural complexity and challenges related to the chemical structure elucidation and synthesis. A promising trend to broaden the scope of natural products is given by making small combinatorial libraries from natural products and natural product-like scaffolds. A systematic extension of such libraries based on protein structure similarity clustering (PSSC) was proposed by the Waldmann group [83]. The principles of this approach consider the domain organization and conservation of proteins and the corresponding needs for conservatisms of the architectures and interaction modes of their ligands.

Primary metabolites and marketed drugs form additional sets of biologically relevant and validated compounds that form an essential component of a comprehensive screening collection [4].

Primary metabolites, which are key intermediates of cellular metabolisms and which interact with key enzymes and cellular regulatory receptor systems, are systematically included in deorphanization libraries of orphan targets. The CheBI database organizes the relevant chemical and biological information [84]. Hits from such libraries allow the elucidation of the functional relevance of a new potential target protein.

Marketed drugs and derivative libraries are an important and invaluable compound source and provide the basis for the selective optimization of side activities (SOSA) approach [85]. The SOSA approach consists of testing old drugs on new pharmacological targets. The aim is to subject to pharmacological screening a limited number of drug molecules that are structurally and therapeutically very diverse and that have known safety and bioavailability in humans, thereby potentially shortening the time and the cost needed for hit optimization. Since bioavailability and toxicity studies have already been performed for those drugs and since they have proven their usefulness in human therapy, all hits are *per definition* drug-like. In the second stage, the hits are optimized by means of traditional, parallel, or combinatorial chemistry in order to increase the affinity for the new target and decrease the affinity for the other targets. The objective is to prepare analogues of the hit molecule in order to transform the observed “side activity” into the main effect and to strongly reduce or abolish the initial pharmacological activity.

Peptide–protein molecular interactions are the most ubiquitous mode for controlling and modulating cellular function, intercellular communication, and signal transduction pathways [86]. Peptides are key components of chemogenomics discovery libraries and are especially useful for the characterization of orphan targets. A number of successful deorphanizations, especially in the GPCR field, are based on peptides, resulting in new drug discovery projects. New peptides for such libraries are discovered using HPLC fractionations of tissue extracts together with random or designed peptide libraries based on the bioinformatics analysis of putatively secreted peptides and protein hormones defined in the genome [87].

Limiting factors of peptide-based drugs are directed by the number of amide bonds that determine properties like a high tPSA, a low membrane permeability, and a potentially high proteolytic degradation, resulting in quite poor ADME properties [88]. Mainly because of these reasons, robust strategies for the design of peptide mimetics have been successfully developed [89]. Oral delivery of therapeutic peptides is still a challenge. A number of factors including high proteolytic activity and low pH conditions of the gastrointestinal tract act as major barriers in the successful delivery of intact peptide to the targeted site. Low permeability of peptides across the intestinal barrier is also a factor adding to the low bioavailability. Nanocarrier-based delivery presents an appropriate choice of drug carriers owing to their property to protect proteins from degradation by the low pH conditions in the stomach or by the proteolytic enzymes in the gastrointestinal tract [90]. Recently, cell-penetrating peptides (CPPs) such as HIV-1 Tat, penetratin, and oligoarginine are considered as a useful tool for the intracellular delivery of therapeutic macromolecules [91]. CPPs are likely to become powerful tools for overcoming the low permeability of therapeutic peptides through the intestinal membrane, the major barrier to their oral delivery. Peptide-derived (natural and nonnatural amino acids) macrocycles are a relatively new trend in drug discovery [92–95]. Macrocycles are conformationally constrained molecules that can fix the bioactive conformation. Macrocycles come in different flavors and can’t be lumped into one class because they cover a wide range of different structural classes and different MW. A stapled peptide is very different from a large cyclic peptide, which is very different from a synthetic macrocycle, which again is very different from a natural product macrocycle [95]. Heterocycles form historically the most prevalent class of drug molecules. They cover a diverse set of ring systems with various types of heteroatoms and have been extensively patented. The quest for new rings was systematically investigated *in silico*. Researchers at UCB generated a complete list of 24,847 ring systems called virtual exploratory heterocyclic library (VEHICLe) [96]. Searching literature and compound databases, using this list as substructure queries, identified only 1701 as synthesized. Using a carefully validated machine learning approach, it was possible to estimate that the number of unpublished, but synthetically tractable, VEHICLe rings could be over 3000. This analysis also shows that the rate of publication of novel examples to be as low as 5–10 per year. Corroboratively, Ertl and coworkers at Novartis showed that bioactive molecules only contain a relatively limited number of unique ring types [97]. To identify those ring properties and structural characteristics that are necessary for biological activity, a large virtual library of nearly 600,000 heteroaromatic scaffolds was created and characterized by calculated properties. Using a self-organizing neural network, the scaffolds were clustered and showed that bioactivity is very sparsely distributed within the scaffold property and structural space, forming only several relatively small, well-defined "bioactivity islands." Such analyses provide a fresh stimulus to creative organic chemists by highlighting a small set of apparently simple ring systems that are predicted to be tractable but are unconquered. A recent trend in heterocyclic chemistry is to increase the ratio of $sp^3$-hybridized carbon atoms ($Fsp^3$) yielding more saturated ring systems. Lovering et al. showed that both complexity (as measured by $Fsp^3$) and the presence of chiral centers correlate with success as compounds transition from discovery, through clinical testing, to drugs. In an attempt to explain these observations, it was demonstrated that saturation correlates with solubility [98]. Within the same perspective, Ishikawa and Hashimoto provided examples how the breaking of molecular symmetry and planarity is effective to improve solubility despite increasing hydrophobicity [99]. The impact of carboaromatic, heteroaromatic, carboaliphatic, and heteroaliphatic ring counts and fused aromatic ring count on several developability measures (solubility, lipophilicity, protein binding, P450 inhibition, and hERG binding) was recently reviewed [100]. Increasing ring counts have detrimental effects on developability in the order carboaromatics $\gg$ heteroaromatics $>$ carboaliphatics $>$ heteroaliphatics, with heteroaliphatics exerting a beneficial effect in many cases. Increasing aromatic ring count exerts effects on several developability parameters that are lipophilicity and size independent, and fused aromatic systems have a beneficial effect relative to their nonfused counterparts.

The metabolism of heterocycles can result in challenges for the optimization of pharmacokinetics/pharmacodynamics (PK/PD) profiles of the compounds. Recently, systematic mitigating strategies for heterocycle metabolism have been established by St. Jean and Fotsch allowing the selection of improved building blocks for library design [101].

Diversity-oriented synthesis (DOS), as opposed to the traditional target-oriented synthesis (TOS) chemistry approach, was introduced by the Schreiber group for forward chemical genetic screening in order to mimic the structural complexity and the skeletal and stereochemical diversity of natural products [102]. Conversely to a convergent synthesis strategy resulting from the logic of retrosynthetic analysis of the target molecules, DOS, in the ideal state, allows the application of a diverse set of reagents and structural transformations on each synthesis intermediate; this results in diverging synthesis pathways that create a broad diversity of target molecules with different scaffolds. DOS compounds clearly share a number of characteristics with natural products including most notably the scaffold diversity and stereochemical complexity. The question remains, however, whether these products of pure chemist imagination capture the evolutionary advantages of natural products and natural product-based compounds. The DOS planning strategy allows, by enumeration over a larger number of steps, the genesis of truly novel structures that by itself is an innovative concept. In practice, DOS combinatorial libraries focus to leverage information about existing biologically active molecules in order to address the biologically relevant regions of chemical space. DOS libraries are not directed toward a single biological target and aim to provide diverse discovery libraries. DOS has increased the need for exceptionally efficient, stereoselective, and chemoselective reactions, including multicomponent reactions (MCR) that can be applied to a broad range of substrates.

A number of recent success stories prove that DOS compounds provide invaluable tools for target validation [103]. The validation of the ADMET and *in vivo* properties of these compounds and their value as therapeutics remains however to be proven. Comparable to natural products, as result of the structural complexity, a key challenge is expected in the lead optimization phase and for the industrial chemical development of the final compounds.

In a comparative analysis, Clemons et al. found that compounds from different sources (commercial, academic DOS, natural products) have different protein-binding behaviors against each of 100 diverse (sequence-unrelated) proteins [104]. These behaviors correlate with general trends in stereochemical and shape descriptors for these compound collections. Increasing the content of $sp^3$-hybridized and stereogenic atoms relative to compounds from commercial sources, which comprise the majority of current screening collections, improved binding selectivity and frequency.

## 1.5 CHEMICAL DESIGN APPROACHES TO EXPAND BIOACTIVE CHEMICAL SPACE

Systematic hypothesis-based expansion of the chemical space to reach a maximum of biological binding sites appears possible when conserved molecular recognition principles are the founding hypothesis for the design of the compounds. Such chemogenomics principles, including approaches focusing on target families, privileged scaffolds, protein secondary structure mimetics, cofactor mimetics, and BIOS libraries, were recently summarized by us [5]. To be broadly successful, these approaches are complemented by diversity-based principles like DOS, DNA-encoded libraries (DELs), and fragment-based approaches (FBS).

More than 50% of the marketed drugs target only four key gene families, including the rhodopsin-like GPCRs, nuclear receptors, ligand-gated ion channels, and voltage-gated ion channels [61, 105]. Historically, drug discovery has thus been focusing on a few "druggable" target families. The key design principles, focusing on similarities or differences in the physicochemistry of equivalent residues lining the binding site, can also rationalize the polypharmacology of many drugs. Because protein family-targeted library design requires extensive ligand-based or structure-based knowledge, it is not surprising that current design of chemical libraries directed to target classes focuses mainly on GPCRs, kinases, nuclear receptors, and more recently ion channels and epigenetic targets. Today, protein family-targeted libraries with a large diversity of chemotypes are specifically designed toward subfamilies with conserved molecular recognition [106]. Various strategies have been applied to design GPCR [107, 108] and ion channel libraries [109], mostly based on ligand information captured in the form of molecular descriptors, pharmacophores, and substructures extracted from active reference compounds. In addition to these methods, the large amount of structure-based information available from X-ray analysis of ligand-target complexes makes structure-based design approaches feasible in the protein kinase [110, 111], protease [112], and nuclear receptor [113] classes. Impressive progress was reached within the last 5 years on new epigenetic targets like histone methyltransferases and bromodomains, thanks to the initiative of the structural genomics consortium (SGC) for chemical probes and structural biology [114–116]. The typical size of a protein target-focused library is 100–1000 compounds if it is centered around one single chemotype; the library size can grow to 10,000–20,000 compounds if it is oriented around multiple chemotypes.

De novo design approaches for target families were recently reported [117]. Automated design of ligands to polypharmacological profiles was demonstrated by the Hopkins group by the evolution of an approved acetylcholinesterase inhibitor drug into brain-penetrable ligands with either specific polypharmacology or exquisite selectivity profiles for biogenic G-protein-coupled receptors [118]. Overall, 800 ligand-target predictions of prospectively designed ligands were tested experimentally, of which 75% were confirmed to be correct. The approach can be a useful source of leads when multitarget profiles are required to achieve either selectivity over other drug targets or a desired polypharmacology. Hartenfeller at Novartis published the libDOGS methods in extension to DOGS developed by the Schneider group at ETH [119, 120]. The principle is a reaction-driven de novo design method for automated design of target family-oriented libraries. Focusing on hypotheses (suggested molecules) that can easily be validated via straightforward syntheses expected to allow testing more hypotheses compared to scenarios where complex molecules require more synthetic effort.

The goal of a protein family-targeted library is not to target a specific target exclusively, but to address by different library members different members of the target subfamilies. This coverage of a targeted library could until recently only be addressed experimentally by the analysis of hit rates. Such analyses showed that some designed libraries hit not only the primary target family but also other a priori nonrelated target families [1].

Biology-oriented synthesis (BIOS) was introduced by the Waldmann group [121–123]. BIOS centers on the generation of small compound libraries based on scaffolds of proven biological relevance. Library generation is focused on compound classes from the "biological relevant space," that is, the natural products and drugs. BIOS unifies the aforementioned SCONP and PSSC concepts (*vide supra*) that, respectively, allow navigation in the chemical and biological spaces.

The use of privileged substructures or molecular master keys is an accepted concept in medicinal chemistry. The privileged structure approach emphasizes molecular scaffolds or selected substructures that are able to provide high-affinity ligands (agonist or antagonists) for diverse receptors and originates from work at Merck Research Laboratories on the design of benzodiazepine-based CCK (Cholecystokinin) antagonists, where the previously known κ-opioid tifluadom was identified as a lead structure [124]. A number of recent literature reviews provide impressive reference repertoires of empirically derived privileged structures for various target families [125–127].

Protein–protein molecular interactions are the most ubiquitous mode for controlling and modulating cellular function, intercellular communication, and signal transduction pathways. Peptide and protein mimetic libraries including β-turn and α-helix mimetics are recognized of central importance in chemogenomics. The design of drug-like active β-turn mimetics based on organic drug-like scaffolds advanced to a quite routine methodology and a variety of approaches have been validated as recently summarized by Marshall et al. [128], including α,β-dehydroamino acids, proline and heterochiral *N*-methyl amino acids, or cyclic α-peptides. The work of Garland and Dean [129, 130], defining a set of triangular distance constraints that the substitution points of a scaffold have to satisfy in order to mimic the specific $C_\alpha$ atoms of the peptide template, provided a generalized frame for the design of novel β-turn mimetic scaffolds and was in combination with database searches successfully applied for the design of CCK, SST, and $MC_4$ antagonists [131].

Other protein–protein interactions (PPI) are via key α-helix motifs. The work of the groups of Hamilton [132] has established a solid foundation for the rational design of α-helix mimetics. 3.2′.2″-substituted terphenyl-derived motifs were among the first designed motifs and were shown to be able to mimic the side-chain positions $i$, $i+3$ or $i+4$, and $i+7$, which are on the same face of a α-helix. Besides these rationally designed helix mimetics, diversity-based high-throughput screening (HTS) and virtual screening have identified a number of scaffolds for a variety of targets that allow the correct spatial orientation of substituents for interaction with the protein target [133]. Combinatorial libraries around such scaffolds are an essential component of a chemogenomics discovery library. Morelli et al. analyzed the PPI chemical space together with current structural knowledge regarding both protein–protein and protein–inhibitor complexes summarized in the 2P2I database [134]. The statistical analysis of the 39 inhibitors present in $2P2I_{DB}$ enabled them to calculate the general characteristics of the PPI chemical space. Average values for the MW (547±154 Da; thus, MW>400 Da), ALogP (3.99±2.37; thus, ALogP>4), number of rings (4.44±1.02; thus, #Rings>4), and number of hydrogen-bond acceptors (6.62±2.60; thus, #HBA>4) define the generic profile of a PPI inhibitor compound that could be further derived into a more specific inhibitor. These chemical rule of 4 properties can be used to filter "in-house" databases and accelerate the process of hit identification by lowering both cost and time.

With the successful design of MDM2-P53 inhibitors, the Dömling group introduced the ANCHOR method whereby computational and MCR chemistry converge [135]. Their general workflow for the rapid generation of low-molecular-weight (ant)agonists of PPIs relies on a sequence of steps. First, the presence of a highly buried amino acid is identified. In the next step, a molecular anchor is generated by fragmentation of the candidate ligand. Virtual chemistry employing the anchor and based on MCR is then used for the enumeration of further candidate compounds that are docked into the binding site using the anchor as a constraint. Finally, chemical synthesis and biological screening of the most promising candidates are completed.

As successfully exemplified by the protein kinase ATP-binding site, cofactor-binding sites are evolutionary highly conserved binding sites that provide rich opportunities for enzyme-based drug discovery. The detailed comparative structural analysis of substrate- and cofactor-binding sites shows indeed that cofactor analogues open a very wide target window. Ji et al. examined around 2200 well-defined small molecule ligands and thousands of protein domains derived from a database of druggable binding sites and found that cofactor molecules are the most prevalent bioligands in the protein fold universe and hence present a unique opportunity for chemogenomics systematization approaches [136].

Typical industrial screening libraries are of the size of approximately 1–2 Mio compounds, which is a drop in the ocean compared to the theoretical chemical space of drug-like molecules. FBS and DELs allow the access to a significantly larger chemical space compared to conventional screening libraries.

Considering the hypothetical case of a ligand comprising three fragments that bind to three subsites, if only 1000 fragments should be tested for each of the subsites, this would require the synthesis and testing of at least $1000 \times 1000 \times 1000 = 10^9$ molecules, without considering the number of possible linkers. In an FBS, it would only require the screening of 1000 compounds in this hypothetical case.

Key to the success of FBS, which can be performed by a variety of biophysical methods, is the quality of the FBS library, and different generations of designs have been introduced in the past [137–139]. More recently, the concepts of 3D fragments, fragments of known drugs, and fragments of natural products have been realized [140, 141]. Very important for the follow-up of the fragment hits is also the access to structural biology support providing high-resolution structures of the fragment-binding interactions and the three-dimensional building plan to expand the fragments into full-sized molecules [142, 143]. Typically, dedicated chemistry support is required, and a number of BioTech companies have focused on the FBS paradigm and delivered an impressive number of drug candidates [144, 145].

DELs provide access to $10^3$–$10^9$ diverse compounds in a single test tube. The scientific approach was first described by Lerner and Brenner and consists of small molecules covalently attached to unique DNA sequences that serve as PCR amplifiable identification barcodes [146]. Selection offers large advantages over screening in terms of numbers, flexibility, convenience, and costs. A number of protocols have been designed [147]. The approaches enable the deconvolution of SAR information and families of actives are observed. One potential limitation is the restriction of chemical reactions to aqueous and DNA-compatible conditions. DELs have started to yield novel modulators of biological targets [148–150].

## 1.6 CONCLUSION

The discovery of new innovative small molecule drugs will continue to rely on the systematic exploration of new chemotypes. The enhancement of the corporate screening collection with high-quality chemical matter is a long-term continuous process of capital interest to every drug discovery organization. Quality library design is one of the most important scientific drug discovery competencies and is recognized as critical to the future productivity [151, 152]. Disruptive versus incremental innovation can be promoted by moving on purpose into new areas of the chemical space. As outlined in this article, there are a number of strategic options that will need to be combined in an appropriate manner. The optimal strategy for one organization might be very different from that of another organization. Even within one same organization, there might be different needs from the different disease areas, for instance, which will call for a mixed-strategy approach. A very important aspect is the time horizon on which the yield of the screening collection will be measured. Diverse libraries will require a longer readout time than project-focused libraries. Compared to partial deck screening and screening deck turnover strategies, the full deck screening of a collection over a longer time period (~10 years) adds additional value in building a comprehensive experimental structure–activity relationship (SAR) matrix, which will help in making informed decisions relevant for many projects and also help to build the basis for future knowledge-based library design and virtual screening approaches [153–155].

## ACKNOWLEDGMENTS

Drs. D. Berthelot, C. Buyck, R. Desjarlais, C. Martinez-Lamenca, T. Mirzadegan, V. Pande, P. Ten Holte, G. Tresadern, and H. van Vlijmen (all Janssen associates) are acknowledged for various support and discussions. Figures 1.1, 1.2, and 1.3 were, respectively, reproduced with permission from references [10, 51, 68].

## REFERENCES

[1] Jacoby E, Schuffenhauer A, Popov M, Azzaoui K, Havill B, Schopfer U, Engeloch C, Stanek J, Acklin P, Rigollier P, Stoll F, Koch G, Meier P, Orain D, Giger R, Hinrichs J, Malagu K, Zimmermann J, Roth HJ (2005). Key aspects of the Novartis compound collection enhancement project for the compilation of a comprehensive chemogenomics drug discovery screening collection. *Curr. Top. Med. Chem.* **5**:397–411.

[2] Renner S, Popov M, Schuffenhauer A, Roth HJ, Breitenstein W, Marzinzik A, Lewis I, Krastel P, Nigsch F, Jenkins J, Jacoby E (2011). Recent trends and observations in the design of high-quality screening collections. *Future Med. Chem.* **3**:751–766.

[3] Djuric SW, Akritopoulou-Zanze I, Cox PB, Galasinski S (2010). Compound collection enhancement and paradigms for high-throughput screening-an update. *Ann. Rep. Med. Chem.* **45**:409–428.

[4] Jacoby E, Schuffenhauer A, Azzaoui K, Popov M, Dressler S, Glick M, Jenkins J, Davies J, Roggo S. Small molecules for chemogenomics-based drug discovery. In: Jacoby E, editor. *Chemogenomics—knowledge-based approaches to drug discovery*. London: Imperial College Press; 2006. p 1–38.

[5] Jacoby E, Mozzarelli A (2009). Chemogenomic strategies to expand the bioactive chemical space. *Curr. Med. Chem.* **16**:4374–4381.

[6] Drewry DH, Macarron R (2010). Enhancements of screening collections to address areas of unmet medical need: an industry perspective. *Curr. Opin. Chem. Biol.* **14**:289–298.

[7] Macarron R, Banks MN, Bojanic D, Burns DJ, Cirovic DA, Garyantes T, Green DV, Hertzberg RP, Janzen WP, Paslay JW, Schopfer U, Sittampalam GS (2011). Impact of high-throughput screening in biomedical research. *Nat. Rev. Drug Discov.* **10**:188–195.

[8] Wigglesworth M, Wood T, editors. *Management of chemical and biological samples for screening applications*. Weinheim: Wiley-VCH Verlag GmbH & co. KGaA; 2012.

[9] Mullard A (2013). European lead factory opens for business. *Nat. Rev. Drug Discov.* **12**:173–175.

[10] Reymond JL, Ruddigkeit L, Awale M. Considerations on the drug-like chemical space. In: Jacoby E, editor. *Computational chemogenomics*. Singapore: Pan Stanford; 2013.

[11] Deursen RV, Blum LC, Reymond JL (2010). A searchable map of PubChem. *J. Chem. Inf. Model.* **50**:1924–1934.

[12] Ruddigkeit L, Blum LC, Reymond JL (2013). Visualization and virtual screening of the chemical universe database GDB-17. *J. Chem. Inf. Model.* **53**:56–65.

[13] Ruddigkeit L, van Deursen R, Blum LC, Reymond JL (2012). Enumeration of 166 billion organic small molecules in the chemical universe database GDB-17. *J. Chem. Inf. Model.* **52**:2864–2875.

[14] Overington JP, Al-Lazikani B, Hopkins AL (2006). How many drug targets are there? *Nat. Rev. Drug Discov.* **5**:993–996.

[15] http://www.elsevier.com/online-tools/reaxys/reaxys-medicinal-chemistry

[16] http://www.cas.org/content/chemical-substances

[17] http://www.emolecules.com/

[18] http://www.chemnavigator.com/

[19] Lipkin MJ, Stevens AP, Livingstone DJ, Harris CJ (2008). How large does a compound screening collection need to be? *Comb. Chem. High Throughput Screen.* **11**:482–493.

[20] Hopkins AL, Groom CR (2002). The druggable genome. *Nat. Rev. Drug Discov.* **1**:727–730.

[21] Lajiness M, Watson I (2008). Dissimilarity-based approaches to compound acquisition. *Curr. Opin. Chem. Biol.* **12**:366–371.

[22] Medina-Franco JL. Cheminformatic characterization of the chemical space and molecular diversity of compound libraries. In: Trabocchi A, editor. *Diversity-oriented synthesis: basics and applications in organic synthesis, drug discovery, and chemical biology*. Hoboken: John Wiley & Sons, Inc.; 2013. p 325–352.

[23] Gibbs A, Agrafiotis DK. Chemical diversity: definition and quantification. In: Bartlett PA, Entzeroth M, editors. *Exploiting chemical diversity for drug discovery*. Cambridge: Royal Society of Chemistry; 2006. p 137–162.

[24] Oprea TI, Gottfries J (2001). Chemography: the art of navigating in chemical space. *J. Comb. Chem.* **3**:157–166.

[25] Larsson J, Gottfries J, Muresan S, Backlund A (2007). ChemGPS-NP: tuned for navigation in biologically relevant chemical space. *J. Nat. Prod.* **70**:789–794.

[26] Rosén J, Gottfries J, Muresan S, Backlund A, Oprea TI (2009). Novel chemical space exploration via natural products. *J. Med. Chem.* **52**:1953–1962.

[27] Schuffenhauer A, Ertl P, Roggo S, Wetzel S, Koch MA, Waldmann H (2007). The scaffold tree—visualization of the scaffold universe by hierarchical scaffold classification. *J. Chem. Inf. Model.* **47**:47–58.

[28] http://spotfire.tibco.com/

[29] Koch MA, Schuffenhauer A, Scheck M, Wetzel S, Casaulta M, Odermatt A, Ertl P, Waldmann H (2005). Charting biologically relevant chemical space: a structural classification of natural products (SCONP). *Proc. Natl. Acad. Sci. U. S. A.* **102**:17272–17277.

[30] Wetzel S, Klein K, Renner S, Rauh D, Oprea TI, Mutzel P, Waldmann H (2009). Interactive exploration of chemical space with Scaffold Hunter. *Nat. Chem. Biol.* **5**:581–583.

[31] Lachance H, Wetzel S, Kumar K, Waldmann H (2012). Charting, navigating, and populating natural product chemical space for drug discovery. *J. Med. Chem.* **55**:5989–6001.

[32] Hack MD, Rassokhin DN, Buyck C, Seierstad M, Skalkin A, ten Holte P, Jones TK, Mirzadegan T, Agrafiotis DK (2011). Library enhancement through the wisdom of crowds. *J. Chem. Inf. Model.* **51**:3275–3286.

[33] Lessel U, Wellenzohn B, Lilienthal M, Claussen H (2009). Searching fragment spaces with feature trees. *J. Chem. Inf. Model.* **49**:270–279.

[34] Wellenzohn B, Lessel U, Beller A, Isambert T, Hoenke C, Nosse B (2012). Identification of new potent GPR119 agonists by combining virtual screening and combinatorial chemistry. *J. Med. Chem.* **55**:11031–11041.

[35] Meanwell NA (2011). Improving drug candidates by design: a focus on physicochemical properties as a means of improving compound disposition and safety. *Chem. Res. Toxicol.* **24**:1420–1456.

[36] Gleeson MP, Montanari D (2012). Strategies for the generation, validation and application of in silico ADMET models in lead generation and optimization. *Expert Opin. Drug Metab. Toxicol.* **8**:1435–1446.

[37] Cox PB, Vasudevan A. Generating a high quality compound collection. In: Wigglesworth M, Wood T, editors. *Management of chemical and biological samples for screening applications.* Weinheim: Wiley-VCH Verlag GmbH & co. KGaA; 2012. p 9–22.

[38] Grime KH, Barton P, McGinnity DF (2013). Application of in silico, in vitro and preclinical pharmacokinetic data for the effective and efficient prediction of human pharmacokinetics. *Mol. Pharm.* **10**:1191–1206.

[39] Cumming JG, Davis AM, Muresan S, Haeberlein M, Chen H (2013). Chemical predictive modelling to improve compound quality. *Nat. Rev. Drug Discov.* **12**:948–962.

[40] Walters WP, Namchuk M (2003). Designing screens: how to make your hits a hit. *Nat. Rev. Drug Discov.* **2**:259–266.

[41] Baell JB, Holloway GA (2010). New substructure filters for removal of pan assay interference compounds (PAINS) from screening libraries and for their exclusion in bioassays. *J. Med. Chem.* **53**:2719–2740.

[42] Baell JB (2013). Broad coverage of commercially available lead-like screening space with fewer than 350,000 compounds. *J. Chem. Inf. Model.* **53**:39–55.

[43] Thorne N, Auld DS, Inglese J (2010). Apparent activity in high-throughput screening: origins of compound-dependent assay interference. *Curr. Opin. Chem. Biol.* **14**:315–324.

[44] Bergström CA, Andersson SB, Fagerberg JH, Ragnarsson G, Lindahl A (2013). Is the full potential of the biopharmaceutics classification system reached? *Eur. J. Pharm. Sci.* **57**:224–231.

[45] http://www.acdlabs.com/products/percepta/

[46] http://www.schrodinger.com/productpage/14/17/

[47] http://www.simulations-plus.com/Products.aspx?pID=13

[48] Lipinski CA, Lombardo F, Dominy BW, Feeney PJ (1997). Experimental and computational approaches to estimate solubility and permeability in drug discovery and development settings. *Adv. Drug Deliv. Rev.* **23**:3–25.

[49] Jarvis L (2007). Originator of 'the rule of five' reflects on 10 years of drug development under its influence. *Chem. Eng. News* **85**:18–18.

[50] Baell J, Congreve M, Leeson P, Abad-Zapatero C (2013). Ask the experts: past, present and future of the rule of five. *Future Med. Chem.* **5**:745–752.

[51] Wenlock MC, Austin RP, Barton P, Davis AM, Leeson PD (2003). A comparison of physiochemical property profiles of development and marketed oral drugs. *J. Med. Chem.* **46**:1250–1256.

[52] Teague SJ, Davis AM, Leeson PD, Oprea T (1999). The design of leadlike combinatorial libraries. *Angew. Chem. Int. Ed. Engl.* **111**:3962–3967.

[53] Leeson PD, Springthorpe B (2007). The influence of drug-like concepts on decision-making in medicinal chemistry. *Nat. Rev. Drug Discov.* **6**:881–890.

[54] Azzaoui K, Hamon J, Faller B, Whitebread S, Jacoby E, Bender A, Jenkins JL, Urban L (2007). Modeling promiscuity based on in vitro safety pharmacology profiling data. *ChemMedChem* **2**:874–880.

[55] Gleeson MP, Hersey A, Montanari D, Overington J (2011). Probing the links between in vitro potency, ADMET and physicochemical parameters. *Nat. Rev. Drug Discov.* **10**:197–208.

[56] Opera TI, editor. Chemoinformatics in lead discovery. In: *Chemoinformatics in drug discovery*. Weinheim: Wiley-VCH; 2005. p 25–42.

[57] Proudfoot JR (2005). The evolution of synthetic oral drug properties. *Bioorg. Med. Chem. Lett.* **15**:1087–1090.

[58] Oprea TI, Allu TK, Fara DC, Rad RF, Ostopovici L, Bologa CG (2007). Lead-like, drug-like or "Pub-like": how different are they? *J. Comput. Aided Mol. Des.* **21**:113–119.

[59] Congreve M, Carr R, Murray C, Jhoti H (2003). A 'rule of three' for fragment-based lead discovery? *Drug Discov. Today* **8**:876–877.

[60] Nadin A, Hattotuwagama C, Churcher I (2012). Lead-oriented synthesis: a new opportunity for synthetic chemistry. *Angew. Chem. Int. Ed. Engl.* **51**:1114–1122.

[61] Paolini GV, Shapland RH, van Hoorn WP, Mason JS, Hopkins AL (2006). Global mapping of pharmacological space. *Nat. Biotechnol.* **24**, 805–815.

[62] Hitchcock SA, Pennington LD (2006). Structure-brain exposure relationships. *J. Med. Chem.* **49**:7559–7583.

[63] O'Shea R, Moser HE (2008). Physicochemical properties of antibacterial compounds: implications for drug discovery. *J. Med. Chem.* **51**:2871–2878.

[64] Gleeson MP (2008). Generation of a set of simple, interpretable ADMET rules of thumb. *J. Med. Chem.* **51**:817–834.

[65] Gleeson MP, Bravi G, Modi S, Lowe D (2009). ADMET rules of thumb II: a comparison of the effects of common substituents on a range of ADMET parameters. *Bioorg. Med. Chem.* **17**:5906–5919.

[66] Leeson PD, St-Gallay SA (2011). The influence of the 'organizational factor' on compound quality in drug discovery. *Nat. Rev. Drug Discov.* **10**:749–765.

[67] Johnson TW, Dress KR, Edwards M (2009). Using the Golden Triangle to optimize clearance and oral absorption. *Bioorg. Med. Chem. Lett.* **19**:5560–5564.

[68] Hill AP, Young RJ (2010). Getting physical in drug discovery: a contemporary perspective on solubility and hydrophobicity. *Drug Discov. Today* **15**:648–655.

[69] Hughes JD, Blagg J, Price DA, Bailey S, DeCrescenzo GA, Devraj RV, Ellsworth E, Fobian YM, Gibbs ME, Gilles RW, Greene N, Huang E, Krieger-Burke T, Loesel J, Wager T, Whitely L, Zhang Y (2008). Physicochemical drug properties associated with in vivo toxicological outcomes. *Bioorg. Med. Chem. Lett.* **18**:4872–4875.

[70] Bickerton GR, Paolini GV, Besnard J, Muresan S, Hopkins AL (2012). Quantifying the chemical beauty of drugs. *Nat. Chem.* **4**:90–98.

[71] Ritchie TJ, Macdonald SJ (2014). How drug-like are 'ugly' drugs: do drug-likeness metrics predict ADME behaviour in humans? *Drug Discov. Today* **19**:489–495.

[72] Clardy J, Walsh C (2004). Lessons from natural molecules. *Nature* **432**:829–837.

[73] Koehn FE, Carter GT (2005). The evolving role of natural products in drug discovery. *Nat. Rev. Drug Discov.* **4**:206–220.

[74] Newman DJ, Cragg GM, Snader KM (2003). Natural products as sources of new drugs over the period 1981–2002. *J. Nat. Prod.* **66**:1022–1037.

[75] http://dnp.chemnetbase.com/intro/index.jsp

[76] Henkel T, Brunne R, Müller H, Reichel F (1999). Statistical investigation of structural complementarity of natural products and synthetic compounds. *Angew. Chem. Int. Ed. Engl.* **38**:643–647.

[77] Feher M, Scmidt JM (2003). Property distributions: differences between drugs, natural products, and molecules from combinatorial chemistry. *J. Chem. Inf. Comput. Sci.* **43**:218–227.

[78] Lee ML, Schneider G (2001). Scaffold architecture and pharmacophoric properties of natural products and trade drugs: applications in the design of natural products based combinatorial libraries. *J. Comb. Chem.* **3**:284–289.

[79] Vasilevich NI, Kombarov RV, Genis DV, Kirpichenok MA (2012). Lessons from natural products chemistry can offer novel approaches for synthetic chemistry in drug discovery. *J. Med. Chem.* **55**:7003–7009.

[80] Schmitt EK, Moore CM, Krastel P, Petersen F (2011). Natural products as catalysts for innovation: a pharmaceutical industry perspective. *Curr. Opin. Chem. Biol.* **15**:497–504.

[81] Tang Y, Khosla C. Biosynthesis of "unnatural" natural products. In: Bartlett PA, Entzeroth M, editors. *Exploiting chemical diversity for drug discovery*. Cambridge: Royal Society of Chemistry; 2006. p 137–162.

[82] Ganesan A (2008). The impact of natural products upon modern drug discovery. *Curr. Opin. Chem. Biol.* **12**:306–317.

[83] Koch MA, Wittenberg LO, Basu S, Jeyaraj DA, Gourzoulidou E, Reinecke K, Odermatt A, Waldmann H (2005). Compound library development guided by protein structure similarity clustering and natural product structure. *Proc. Natl. Acad. Sci. U. S. A.* **101**:16721–16726.

[84] Hastings J, de Matos P, Dekker A, Ennis M, Harsha B, Kale N, Muthukrishnan V, Owen G, Turner S, Williams M, Steinbeck C (2013). The ChEBI reference database and ontology for biologically relevant chemistry: enhancements for 2013. *Nucleic Acids Res.* **41**(Database issue):D456–D463.

[85] Wermuth CG (2006). Selective optimization of side activities: the SOSA approach. *Drug Discov. Today* **11**:160–164.

[86] Hruby VJ (2002). Designing peptide receptor agonists and antagonists. *Nat. Rev. Drug Discov.* **1**:847–858.

[87] Wise A, Jupe SC, Rees S (2004). The identification of ligands at orphan G-protein coupled receptors. *Annu. Rev. Pharmacol. Toxicol.* **44**:43–66.

[88] Craik DJ, Fairlie DP, Liras S, Price D (2013). The future of peptide-based drugs. *Chem. Biol. Drug Des.* **81**:136–147.

[89] Hummel G, Reineke U, Reimer U. Translating peptides into small molecules. In: Bartlett PA, Entzeroth M, editors. *Exploiting chemical diversity for drug discovery*. Cambridge: Royal Society of Chemistry; 2006. p 57–90.

[90] Gupta S, Jain A, Chakraborty M, Sahni JK, Ali J, Dang S (2013). Oral delivery of therapeutic proteins and peptides: a review on recent developments. *Drug Deliv.* **20**:237–246.

[91] Khafagy EL-S, Morishita M (2012). Oral biodrug delivery using cell-penetrating peptide. *Adv. Drug Deliv. Rev.* **64**:531–539.

[92] Driggers EM, Hale SP, Lee J, Terrett NK (2008). The exploration of macrocycles for drug discovery—an underexploited structural class. *Nat. Rev. Drug Discov.* **7**:608–624.

[93] Mallinson J, Collins I (2012). Macrocycles in new drug discovery. *Future Med. Chem.* **4**:1409–1438.

[94] Giordanetto F, Kihlberg J (2014). Macrocyclic drugs and clinical candidates: what can medicinal chemists learn from their properties? *J. Med. Chem.* **57**:278–295.

[95] http://www.nature.com/scibx/collections/macrocycles

[96] Pitt WR, Parry DM, Perry BG, Groom CR (2009). Heteroaromatic rings of the future. *J. Med. Chem.* **52**:2952–2963.

[97] Ertl P, Jelfs S, Mühlbacher J, Schuffenhauer A, Selzer P (2006). Quest for the rings. In silico exploration of ring universe to identify novel bioactive heteroaromatic scaffolds. *J. Med. Chem.* **49**:4568–4573.

[98] Lovering F, Bikker J, Humblet C (2009). Escape from flatland: increasing saturation as an approach to improving clinical success. *J. Med. Chem.* **52**:6752–6756.

[99] Ishikawa M, Hashimoto Y (2011). Improvement in aqueous solubility in small molecule drug discovery programs by disruption of molecular planarity and symmetry. *J. Med. Chem.* **54**:1539–1554.

[100] Ritchie TJ, Macdonald SJ, Young RJ, Pickett SD (2011). The impact of aromatic ring count on compound developability: further insights by examining carbo- and hetero-aromatic and -aliphatic ring types. *Drug Discov. Today* **16**:164–171.

[101] St Jean DJ Jr, Fotsch C (2012). Mitigating heterocycle metabolism in drug discovery. *J. Med. Chem.* **55**:6002–6020.

[102] Schreiber SL (2000). Target-oriented and diversity-oriented organic synthesis in drug discovery. *Science* **287**:1964–1969.

[103] Trabocchi A, editor. *Diversity-oriented synthesis: basics and applications in organic synthesis, drug discovery, and chemical biology*. Hoboken: John Wiley & Sons, Inc.; 2013.

[104] Clemons PA, Bodycombe NE, Carrinski HA, Wilson JA, Shamji AF, Wagner BK, Koehler AN, Schreiber SL (2010). Small molecules of different origins have distinct distributions of structural complexity that correlate with protein-binding profiles. *Proc. Natl. Acad. Sci. U. S. A.* **107**:18787–18792.

[105] Agarwal P, Sanseau P, Cardon LR (2013). Novelty in the target landscape of the pharmaceutical industry. *Nat. Rev. Drug Discov.* **12**:575–576.

[106] Lackey KE, editor. *Gene family targeted molecular design*. Hoboken: John Wiley & Sons, Inc.; 2008.

[107] Savchuk NP, Tkachenko SE, Balakin KV. Strategies for the design of pGPCR-targeted libraries. In: Rognan D, editor. *Methods and principles in medicinal chemistry, 30 (Ligand design for G protein-coupled receptors)*. Weinheim: Wiley-VCH; 2006. pp. 137–164.

[108] Jacoby E (2006). Designing compound libraries targeting GPCRs. *Ernst Schering Found. Symp. Proc.* **2**:93–103.

[109] Baringhaus KH, Hessler G. A chemical genomics approach for ion channel modulators. In: Kubinyi H, Müller G, editors. *Methods and principles in medicinal chemistry, 22 (Chemogenomics in drug discovery)*. Weinheim: Wiley-VCH; 2004. p 221–242.

[110] Harris CJ, Stevens AP (2006). Chemogenomics: structuring the drug discovery process to gene families. *Drug Discov. Today* **11**:880–888.

[111] Liao JJ (2007). Molecular recognition of protein kinase binding pockets for design of potent and selective kinase inhibitors. *J. Med. Chem.* **50**:409–424.

[112] Sedrani R, Hommel U, Eder J. Protease-directed drug discovery. In: Lackey KE, editor. *Gene-family targeted molecular design*. Hoboken: John Wiley & Sons, Inc.; 2008. p 159–197.

[113] Moore JT, Collins JL, Pearce KH. The nuclear receptor superfamily and drug discovery. In: Schreiber SL, Kapoor TM, Wess G, editors. *Chemical biology—from small molecules to systems biology and drug design*. Weinheim: Wiley-VCH; 2007. p 891–932.

[114] Edwards AM, Bountra C, Kerr DJ, Willson TM (2009). Open access chemical and clinical probes to support drug discovery. *Nat. Chem. Biol.* **5**:436–440.

[115] Sippl W, Jung M, editors. *Epigenetic targets in drug discovery*. Weinheim: Wiley-VCH Verlag GmbH & Co. KGaA; 2009.

[116] Helin K, Dhanak D (2013). Chromatin proteins and modifications as drug targets. *Nature* **502**:480–488.

[117] Schneider G, editor. *De novo molecular design*. Weinheim: Wiley-VCH Verlag GmbH & Co. KgaA; 2013.

[118] Besnard J, Ruda GF, Setola V, Abecassis K, Rodriguiz RM, Huang XP, Norval S, Sassano MF, Shin AI, Webster LA, Simeons FR, Stojanovski L, Prat A, Seidah NG, Constam DB, Bickerton GR, Read KD, Wetsel WC, Gilbert IH, Roth BL, Hopkins AL (2012). Automated design of ligands to polypharmacological profiles. *Nature* **492**:215–220.

[119] Hartenfeller M, Renner S, Jacoby E. Reaction-driven de novo design: a keystone for automated design of target family-oriented libraries. In: Schneider G, editor. *De novo molecular design*. Weinheim: Wiley-VCH Verlag GmbH & Co. KgaA; 2013. p 245–263.

[120] Hartenfeller M, Eberle M, Meier P, Nieto-Oberhuber C, Altmann KH, Schneider G, Jacoby E, Renner S (2012). Probing the bioactivity-relevant chemical space of robust reactions and common molecular building blocks. *J. Chem. Inf. Model.* **52**:1167–1178.

[121] Wilk W, Zimmermann TJ, Kaiser M, Waldmann H (2010). Principles, implementation, and application of biology-oriented synthesis (BIOS). *Biol. Chem.* **391**:491–497.

[122] Wetzel S, Bon RS, Kumar K, Waldmann H (2011). Biology-oriented synthesis. *Angew. Chem. Int. Ed. Engl.* **50**:10800–10826.

[123] Zimmermann TJ, Roy S, Martinez NE, Ziegler S, Hedberg C, Waldmann H (2013). Biology-oriented synthesis of a tetrahydroisoquinoline-based compound collection targeting microtubule polymerization. *Chembiochem* **14**:295–300.

[124] Evans BE, Rittle KE, Bock MG, DiPardo RM, Freidinger RM, Whitter WL, Lundell GF, Veber DF, Anderson PS, Chang RS (1988). Methods for drug discovery: development of potent, selective, orally effective cholecystokinin antagonists. *J. Med. Chem.* **31**:2235–2246.

[125] Müller G. Target family-directed masterkeys in chemogenomics. In: Kubinyi H, Müller G, editors. *Methods and principles in medicinal chemistry, 22 (Chemogenomics in drug discovery)*. Weinheim: Wiley-VCH; 2004. p 7–41.

[126] Hu Y, Wassermann AM, Lounkine E, Bajorath J (2010). Systematic analysis of public domain compound potency data identifies selective molecular scaffolds across druggable target families. *J. Med. Chem.* **53**:752–758.

[127] Kombarov R, Altieri A, Genis G, Kirpichenok M, Kochubey V, Rakitina N, Titarenko Z (2010). BioCores: identification of a drug/natural product-based privileged structural motif for small-molecule lead discovery. *Mol. Divers.* **14**:193–200.

[128] Marshall GR, Kuster DJ, Che Y (2009). Chemogenomics with protein secondary-structure mimetics. *Methods Mol. Biol.* **575**:123–158.

[129] Garland SL, Dean PM (1999). Design criteria for molecular mimics of fragments of the beta-turn. 1. C alpha atom analysis. *J. Comput. Aided Mol. Des.* **13**:469–483.

[130] Garland SL, Dean PM (1999). Design criteria for molecular mimics of fragments of the beta-turn. 2. C alpha-C beta bond vector analysis. *J. Comput. Aided Mol. Des.* **13**:485–498.

[131] Webb TR, Jiang L, Sviridov S, Venegas RE, Vlaskina AV, McGrath D, Tucker J, Wang J, Deschenes A, Li R (2007). Application of a novel design paradigm to generate general nonpeptide combinatorial templates mimicking beta-turns: synthesis of ligands for melanocortin receptors. *J. Comb. Chem.* **9**:704–710.

[132] Cummings CG, Hamilton AD (2010). Disrupting protein-protein interactions with non-peptidic, small molecule alpha-helix mimetics. *Curr. Opin. Chem. Biol.* **14**:341–346.

[133] Fry D, Huang KS, Di Lello P, Mohr P, Müller K, So SS, Harada T, Stahl M, Vu B, Mauser H (2013). Design of libraries targeting protein-protein interfaces. *ChemMedChem* **8**:726–732.

[134] Morelli X, Bourgeas R, Roche P (2011). Chemical and structural lessons from recent successes in protein-protein interaction inhibition (2P2I). *Curr. Opin. Chem. Biol.* **15**:475–481.

[135] Czarna A, Beck B, Srivastava S, Popowicz GM, Wolf S, Huang Y, Bista M, Holak TA, Dömling A (2010). Robust generation of lead compounds for protein-protein interactions by computational and MCR chemistry: p53/Hdm2 antagonists. *Angew. Chem. Int. Ed. Engl.* **49**:5352–5356.

[136] Ji HF, Kong DX, Shen L, Chen LL, Ma BG, Zhang HY (2007). Distribution patterns of small-molecule ligands in the protein universe and implications for origin of life and drug discovery. *Genome Biol.* **8**:R176.

[137] Schuffenhauer A, Ruedisser S, Marzinzik AL, Jahnke W, Blommers M, Selzer P, Jacoby E. (2005). Library design for fragment based screening. *Curr. Top. Med. Chem.* **5**:751–762.

[138] Siegal G, Ab E, Schultz J (2007). Integration of fragment screening and library design. *Drug Discov. Today* **12**:1032–1039.

[139] Chen IJ, Hubbard RE (2009). Lessons for fragment library design: analysis of output from multiple screening campaigns. *J. Comput. Aided Mol. Des.* **23**:603–629.

[140] Genis D, Kirpichenok M, Kombarov R (2012). A minimalist fragment approach for the design of natural-product-like synthetic scaffolds. *Drug Discov. Today* **17**:1170–1174.

[141] Over B, Wetzel S, Grütter C, Nakai Y, Renner S, Rauh D, Waldmann H (2013). Natural-product-derived fragments for fragment-based ligand discovery. *Nat. Chem.* **5**:21–28.

[142] Hubbard RE, Chen I, Davis B (2007). Informatics and modeling challenges in fragment-based drug discovery. *Curr. Opin. Drug Discov. Devel.* **10**:289–297.

[143] Murray CW, Blundell TL (2010). Structural biology in fragment-based drug design. *Curr. Opin. Struct. Biol.* **20**:497–507.

[144] Hajduk PJ, Greer J (2007). A decade of fragment-based drug design: strategic advances and lessons learned. *Nat. Rev. Drug Discov.* **6**:211–219.

[145] Baker M (2013). Fragment-based lead discovery grows up. *Nat. Rev. Drug Discov.* **12**:5–7.

[146] Brenner S, Lerner RA (1992). Encoded combinatorial chemistry. *Proc. Natl. Acad. Sci. U. S. A.* **89**:5381–5383.

[147] Clark MA (2010). Selecting chemicals: the emerging utility of DNA-encoded libraries. *Curr. Opin. Chem. Biol.* **14**:396–403.

[148] Mannocci L, Leimbacher M, Wichert M, Scheuermann J, Neri D (2011). 20 years of DNA-encoded chemical libraries. *Chem. Commun. (Camb.)* **47**:12747–12753.

[149] Kleiner RE, Dumelin CE, Liu DR (2011). Small-molecule discovery from DNA-encoded chemical libraries. *Chem. Soc. Rev.* **40**:5707–5717.

[150] Goodnow RA, editor. *A handbook for DNA-encoded chemistry: theory and applications for exploring chemical space and drug discovery.* Hoboken: John Wiley & Sons, Inc.; 2014.

[151] Hajduk PJ, Galloway WR, Spring DR (2011). Drug discovery: a question of library design. *Nature* **470**:42–43.

[152] Barker A, Kettle JG, Nowak T, Pease E (2013). Expanding medicinal chemistry space. *Drug Discov. Today* **18**:298–304.

[153] Zhou JZ. Chemical library design. In: Walker JM, editor. *Methods in molecular biology (Volume 685).* New York: Humana Press, Springer; 2011.

[154] Schneider G (2010). Virtual screening: an endless staircase? *Nat. Rev. Drug Discov.* **9**:273–276.

[155] Sotriffer C. Virtual screening: principles, challenges, and practical guidelines. In: Mannhold R, Kubinyi H, Folkers G, editors. *Methods and principles in medicinal chemistry (Volume 48).* Weinheim: VCH-Wiley; 2011.

# 2

# THE EUROPEAN LEAD FACTORY

Christopher Kallus[1], Jörg Hüser[2], Philip S. Jones[3], and Adam Nelson[4,5]

[1] *Sanofi R&D, Frankfurt am Main, Germany*
[2] *Bayer Pharma AG, Wuppertal, Germany*
[3] *European Screening Centre Newhouse, Lanarkshire, UK*
[4] *Astbury Centre for Structural Molecular Biology, University of Leeds, Leeds, UK*
[5] *School of Chemistry, University of Leeds, Leeds, UK*

## 2.1 INTRODUCTION

### 2.1.1 Background

Synthetic organic small molecules remain a major modality for therapeutics (in 2013, the vast majority of FDA-approved drugs is belonging to this class [1, 2]). Despite the impressive recent medical and commercial success stories of antibody-based drugs, small molecule drugs have a number of potential properties that render them irreplaceable in the doctor's tool box. The possibility to finely titrate and control the pharmacological effect through the dosing regime is frequently of great advantage when interfering with delicately balanced homeostatic processes. Many small molecules can penetrate cell membranes and address intracellular targets to reversibly manipulate cellular signaling pathways or regulate gene expression. In addition, the ability to penetrate the blood–brain barrier also allows the central nervous system (CNS) to be targeted. Finally, the option to generate orally available drugs greatly contributes to their widespread use outside the hospital or doctor's office. Obviously, none of these traits come as a matter of course but are typically the result of a resource-intensive multifactorial optimization in which medicinal chemistry takes a central role.

"Trial-and-error" screening of comprehensive chemical collections using appropriate bioassays, most efficiently realized in high-throughput screening (HTS), remains a major

*Small Molecule Medicinal Chemistry: Strategies and Technologies*, First Edition.
Edited by Werngard Czechtizky and Peter Hamley.

source of novel lead structures to drive the drug discovery process [3]. To justify the large investment in the research and development (R&D) process, the resulting therapeutics would need to provide clearly differentiated treatment options to patients with a significant benefit over existing medicines and therapies. This differentiation might originate from the properties of the small molecule itself leading, for example, to improved efficacy or safety. Alternatively, modulation of a novel molecular target might offer yet unprecedented or greatly improved treatment options for disease.

Altogether, the factors outlined previously render the discovery of a novel lead structure targeting a novel or yet intractable receptor a rare event. Moreover, future success critically depends on the continuous investments and improvements in chemical libraries, innovative drug targets, and improved screening and lead generation technologies [3]. In reality, however, low R&D output and increasing costs have resulted in industry's increasing risk aversion and diminished commitment to invest. Current research strategies do not include costly large-scale library enhancements or increased resources in discovery research [4–7]. Company mergers and the subsequent net loss of R&D resources have further aggravated this trend. The industry has increasingly abandoned research activities in selected large disease areas, very prominently, for example, in diseases of the CNS [8]. On the other hand, academic researchers working on innovative biological targets typically neither have access to high-quality drug-like compound collections nor to state-of-the-art HTS capacities to fully explore the potential of the novel approach. With its European Lead Factory (ELF) project, the Innovative Medicine Initiative (IMI) aims to address a number of the earlier described gaps by a means of a public–private partnership. Viewed from both sides as a large-scale "controlled experiment," it differs from basic research-driven academic initiatives in this area by fusing academia's innovation potential with industry's applied research processes and effectiveness of small- or medium-sized enterprises (SMEs) by goal settings emphasizing the generation of tangible values to justify additional downstream investments.

### 2.1.2 The European Lead Factory

The ELF is a large-scale public–private partnership in the field of small molecule drug discovery established in early 2013 under the umbrella of the IMI (http://www.imi.europa.eu) that promotes discoveries via open innovation and crowdsourcing. Funded for a 5-year period with a total budget of around €196 million, it brings together seven industry members of the European Federation of Pharmaceutical Industries and Associations (EFPIA) (http://www.efpia.eu) to join forces with originally two public consortia responding to IMI's fifth call for proposals published in spring 2012. The project budget is composed of €80 million from the European Commission and €91 million from EFPIA participants together with €25 million (to a great extent as "in-kind" contributions) coming from non-EFPIA participants. This joint effort consists of two major parts:

1. Assembly of unique and unprecedented *Joint European Compound Library* (*JECL*), which will consist of up to 500,000 high-quality and drug-like compounds (300,000 compounds selected from internal company collections that are otherwise not accessible and 200,000 compounds synthesized de novo)

2. Creation of a *European Screening Centre* (*ESC*), which is able to handle compound logistics and perform HTSs and hit follow-up in a highly professional manner

The consortium is responsible for the formation of the ESC and for the enhancement of the JECL.

In this section, we will briefly describe the concept of the ELF, the JECL will be the major subject of Section 2.2, and details on the ESC will be given in Section 2.3 of this chapter.

***2.1.2.1 A Consortium Combining Innovation with Value Generation*** The overall concept of the ELF provides a novel and unique setup to organize the early drug discovery activities around HTS of small molecule compound libraries. This initiative combines industry's operational expertise and effectiveness of SMEs with innovative ideas from academia. Its general ambition is to provide innovative impulses to the drug discovery process by means of unprecedented targets or novel chemical compounds not yet represented in compound collections. It also pioneers the idea of exploiting existing compound collections within pharmaceutical companies as part of a consortium. Chemical collections established using decades of medicinal chemistry expertise that have been highly safeguarded by pharmaceutical companies in the past will be tested against targets (and indications) that were not envisaged by the original compound owner. The ELF differs from the numerous academic initiatives in this field by emphasizing value generation. Thus, a single high-impact publication that triggers little or no follow-up in either the scientific community or in pharmaceutical and biotechnological companies is not viewed as sufficient success. More tangible value is seen in downstream alliances formed to progress the screening results along the pharmaceutical value chain, license agreements around intellectual property generated, or the foundation of spinouts based on assets generated by external contributors and ELF. The EFPIA consortium members will be available as primary contacts to create partnerships with industry. Publishing key results of target programs—preferentially in high-impact journals—will complement this activity and is encouraged to share the generated knowledge and further advertise the ELF concept. However, publication requires adherence to the consortium's approval process, which aims to balance knowledge sharing, value creation, and the interests of all participants.

Not all proposed targets will have sufficient supporting evidence to fulfill pharma's criteria for investing in lead generation. In many cases, useful pharmacological tools are lacking to further explore the target (or pathway) function or its potential disease relevance. Such target programs, provided a sufficiently justified rationale of the target's disease relevance can be formulated, address another bottleneck in the current drug research process. While industry increasingly avoids the risk to invest in nonvalidated targets, the identification of chemical tool compounds can greatly facilitate the process of target validation and benefits the industry in general. Therefore, the ELF and its partners also aim to generate quality tool compound(s) for subsequent target research.

Again, expertise relevant to the molecular target, corresponding assays, or disease biology together with validated hit molecules is a strong basis for the forming of partnerships between a public contributor and participating EFPIA companies or the participating SMEs.

Unlike many academic screening initiatives, the ELF will be different in restricting access to chemical structural data of compounds from the JECL even among consortium participants. In fact, an honest data broker system and process will regulate access to the structure database to a defined group of scientists driving the hit characterization and selection process in each target program. Preserving the IP integrity of the collection has been a crucial prerequisite to include high-quality assets of pharmaceutical industry in this project. Limiting the open distribution of screening data linked to structural information from the JECL will not only safeguard the future patentability of compounds in the collection. It is also a motivation to base the hit selection on rigorous criteria, including potency, specificity, and selectivity, which need to be agreed upon at the start of individual target programs.

Participants are grouped depending on their principal function within the ELF. The consortium formed by altogether thirty partners includes academic institutions and small- and medium-sized enterprises (SMEs) (Fig. 2.1).

| Participant | Principal role(s) |
|---|---|
| Bayer Pharma AG (coordinator) | EFPIA participant |
| AstraZeneca AB | |
| Lundbeck A/S | |
| Janssen Pharmaceutica NV | |
| Merck KGaA | |
| Sanofi SA | |
| UCB Pharma SA | |
| ChemAxon Kft, Hungary | Implementation and maintenance of web-based tool for the collation of library proposals |
| Leiden University, the Netherlands | Development and synthetic validation of library proposals |
| Max Planck Institute for Molecular Physiology, Germany | |
| Radboud University Nijmegen, the Netherlands | |
| Nederlands Kanker Instituut, the Netherlands | |
| Technical University of Denmark, Denmark | |
| Universität Duisburg-Essen, Germany | |
| University of Groningen, the Netherlands | |
| University of Leeds, United Kingdom | |
| University of Nottingham, United Kingdom | |
| VU University, Amsterdam, the Netherlands | |
| Lead Discovery Center GmbH, Germany | Nomination of libraries for production |
| Edelris SAS, France | Library production |
| Mercachem BV, the Netherlands | |
| Sygnature Discovery Ltd, United Kingdom | |
| Syncom BV, the Netherlands | |
| Taros Chemicals GmbH & Co KG, Germany | |
| Top Institute Pharma, the Netherlands | Scientific Programme Office |
| University of Dundee, United Kingdom | Hit follow-up and medicinal chemistry |
| University of Oxford, United Kingdom | Protein production and X-ray crystallography |
| BioAscent, United Kingdom | JECL storage and logistics |
| Pivot Park Screening Centre BV, the Netherlands | Assay development and uHTS |
| GABO:mi GmbH and Co KG, Germany | Project management |

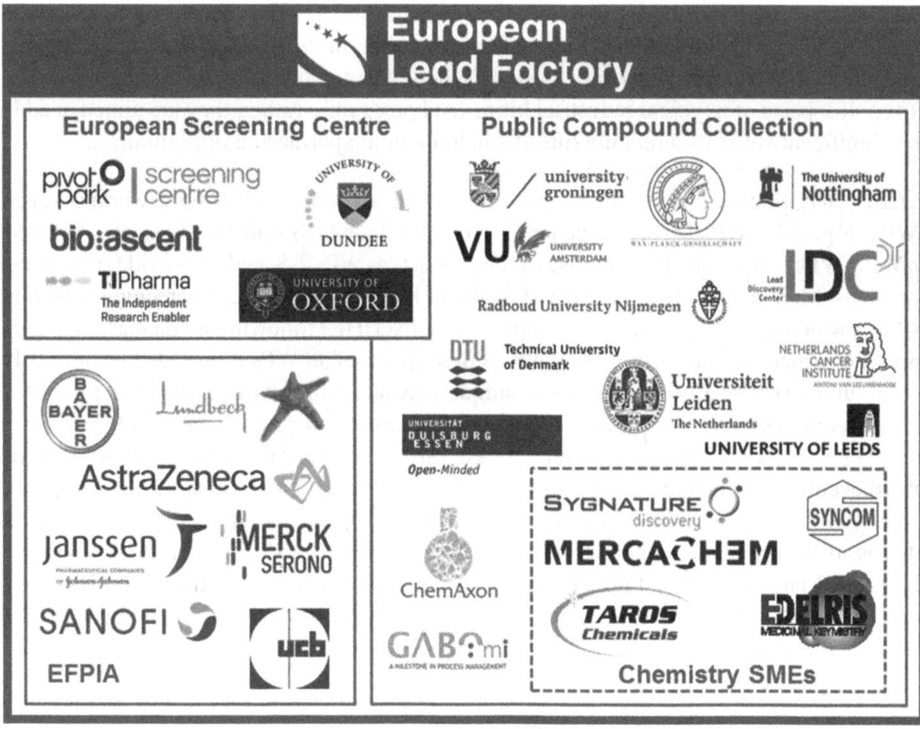


**FIGURE 2.1** Consortium members and their roles in the European Lead Factory. (*See insert for color representation of the figure.*)

In contrast to most other existing IMI-sponsored projects, it is uniquely organized to provide access to the ELF's offerings also to selected external partners, that is, academic institutions and SMEs. For this purpose, the broad public is invited via an open and competitive call to introduce into the ELF innovative ideas and projects. These ideas can include novel disease relevant molecular targets and associated bioassays for screening (referred to as "target programs") or proposals for chemical libraries (referred to "chemistry programs") to enhance the JECL. The ELF has established processes and infrastructure (see for further information: http://www.europeanleadfactory.eu) to manage the underlying crowdsourcing of these ideas. Public partners contributing selected proposals will either receive monetary rewards (for chemistry proposals) or cost-free access to all facets of early lead generation provided through the ESC (for target programs). The external participants will need to formally sign up to selected sections of the consortium's project agreement, that is, the contract regulating rights and obligations among participants, to become "contributing third parties." To learn about the details of these rights and obligations, the reader is again referred to the ELF's website (see earlier text). Importantly, the final selection process includes a peer review of the proposal by an expert panel comprised of scientists from participant organizations and independent external experts. As yet another differentiation to the many screening initiatives in the academic realm, the ESC also provides medicinal chemistry resources to selected public target programs. These chemistry activities aim toward exploring emerging structure–activity relationships (SAR) around singletons or improving the activity or selectivity of initial screening hits to facilitate their *in vitro* characterization in the hit qualification, that is, hit characterization phase. However, these activities are not intended to comprehensively cover all aspects of a typical industrial hit-to-lead program—rather, they are aimed at adding significant value to screening hits and help facilitate partnering opportunities.

***2.1.2.2 Organization*** The ELF consortium is organized in 13 distinct workpackages (WPs; Fig. 2.2). WPs 9–12 are generating the JECL, which will be physically handled in WP2. WPs 1–6 physically comprise the ESC, whereas WPs 7, 8, and 13 provide supporting functions. Main deliverables are a list of qualified hits from the primary screening activities (QHL), which can be extended to an improved hit list (IHL) following medicinal chemistry input for particularly promising programs. A description of all WPs is provided in Table 2.1.

In summary, the ELF represents a unique private–public partnership that promotes new discoveries via open innovation and crowdsourcing. It provides access to a unique 500,000 compound chemical library based on otherwise safeguarded libraries from pharmaceutical companies and entirely new chemical matter based on rewarded proposals from crowdsourcing. State-of-the-art ultra-HTS (uHTS) screening facilities at the ESC plus assay development capacities are a great opportunity for SMEs and academics to get their novel targets screened against this high-quality collection in a highly professional manner. As a further plus, ELF is an ideal forum to stimulate networking and partnering between academia, SMEs, and industrial partners.

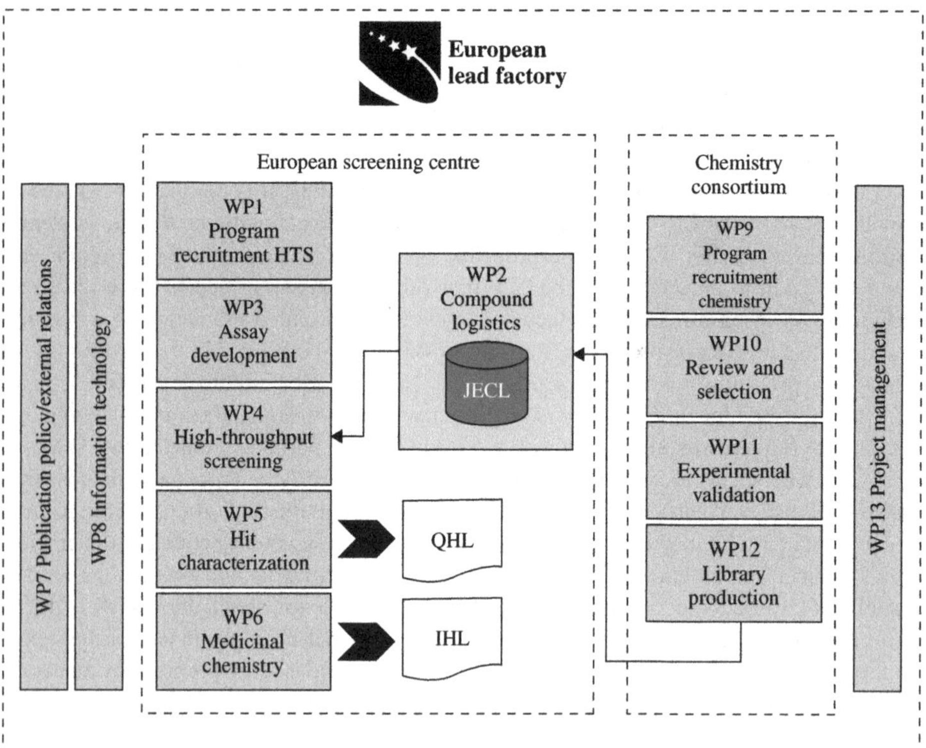


**FIGURE 2.2** Organization of the European Lead Factory in distinct workpackages (WPs).

**TABLE 2.1 Description of Workpackages of the ELF Project**

| WP # | WP Name | Short Description |
|---|---|---|
| 1 | Program recruitment HTS | Sourcing, selecting, and planning of screening programs |
| 2 | Compound logistics | Organization, storage, and compound supply of the JECL |
| 3 | Assay development | Ensure assay quality that is suitable for screening center's HTS technology |
| 4 | High-throughput screening | Perform screening programs, including confirmation and $IC_{50}/EC_{50}$ |
| 5 | Hit characterization | Perform specificity/selectivity and mode of action assays. Deliver a list of ≤50 qualified hits (QHL) |
| 6 | Medicinal chemistry | Perform analytics, extend the initial QHL by broadening of chemical series |
| 7 | Publication policy/external relations | Ensure dissemination in the scientific community, press releases, etc. |
| 8 | Information technology | Provide framework for managing and analyzing the data from assay results and compound logistics; guarantee security of compound structures |
| 9/10 | Program recruitment chemistry/ review and selection | Sourcing, selecting, and planning of chemical library proposals |
| 11 | Experimental validation | Validate selected chemistry proposals to ensure a low attrition during library production |
| 12 | Library production | Ensure a continuous flow of realized libraries |
| 13 | Project management | Governance |

## 2.2 BUILDING THE JOINT EUROPEAN COMPOUND LIBRARY

The JECL is at the heart of the ELF (Fig. 2.3). At the start of the project, the EFPIA partners have contributed a total of over 300,000 compounds to the JECL. The JECL will be enhanced through an estimated 200,000 additional compounds that will be produced as part of the project.

**Compounds Contributed by EFPIA Partners.** To ensure the value of compounds contributed by EFPIA, general rules for selecting compounds from in-house collections were agreed. They can be summarized as follows:

- Drug-likeness: Violate no more than two Lipinski rules, application of property filters developed in a participating company, and exclusion of compounds bearing unwanted motifs
- Originality: Excludes compounds under third-party rights, commercially available compounds, and singletons

An initial challenge was to ensure diversity between the compounds contributed by each company while minimizing disclosure of confidential compound details. This was

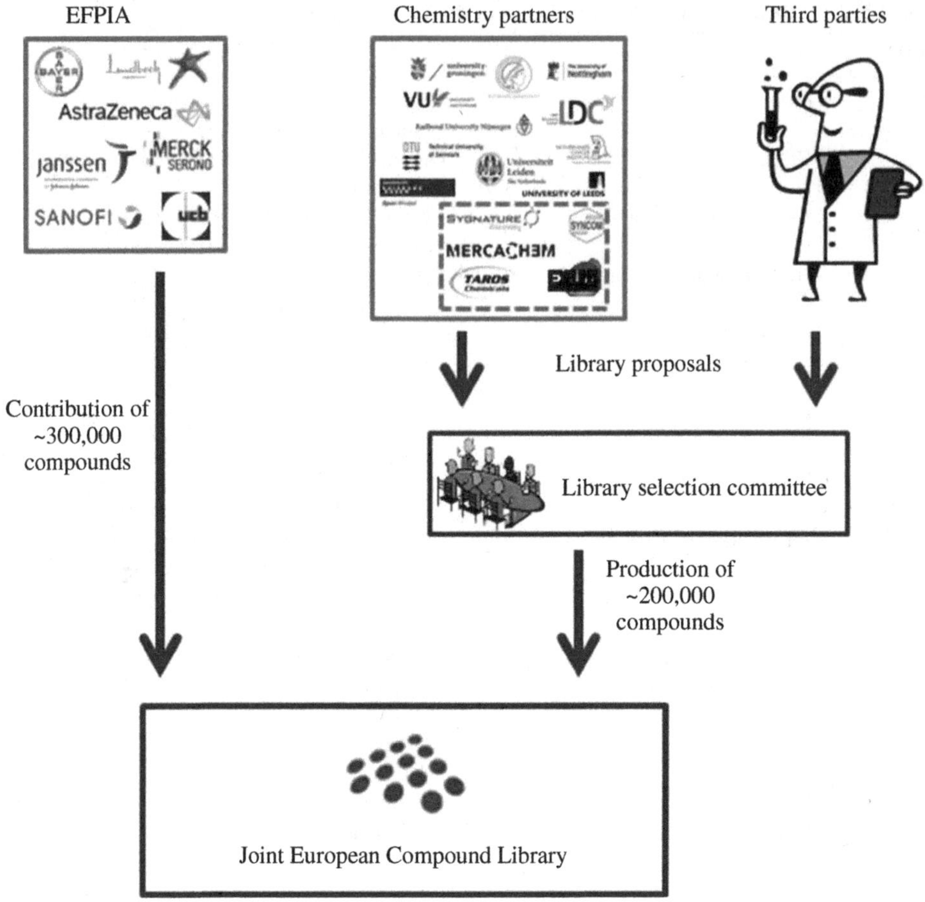


**FIGURE 2.3** Provenance of compounds in the Joint European Compound Library. The approximately 300,000 compounds contributed by the consortium's EPFIA partners will be complemented by approximately 200,000 additional compounds produced as part of the project.

achieved by developing a Pipeline Pilot protocol that was distributed to the EFPIA companies. The protocol assigned a unique identifier to each compound and calculated a wide range of physicochemical properties together with EFPF and FCFP circular fingerprints. These derived properties were returned to the University of Dundee where duplicate, or very similar, compounds were identified using fingerprint identity and removed. Replacement compounds to refill the deduplicated collection were resupplied by the companies to maintain the size of the collection. This procedure allowed a comparison of subsets from seven companies' databases to be compared without revealing structures or compromising intellectual property.

Figure 2.4 shows the analysis of the molecular properties of the combined approximately 300,000 compounds contributed by EFPIA partners.

It is easy to see that the careful selection of sublibraries resulted in satisfying library characteristics: the means of calculated logP (clogP) values and molecular weights are lead-like (3 and 350, respectively). Furthermore, the number of atoms per molecule and

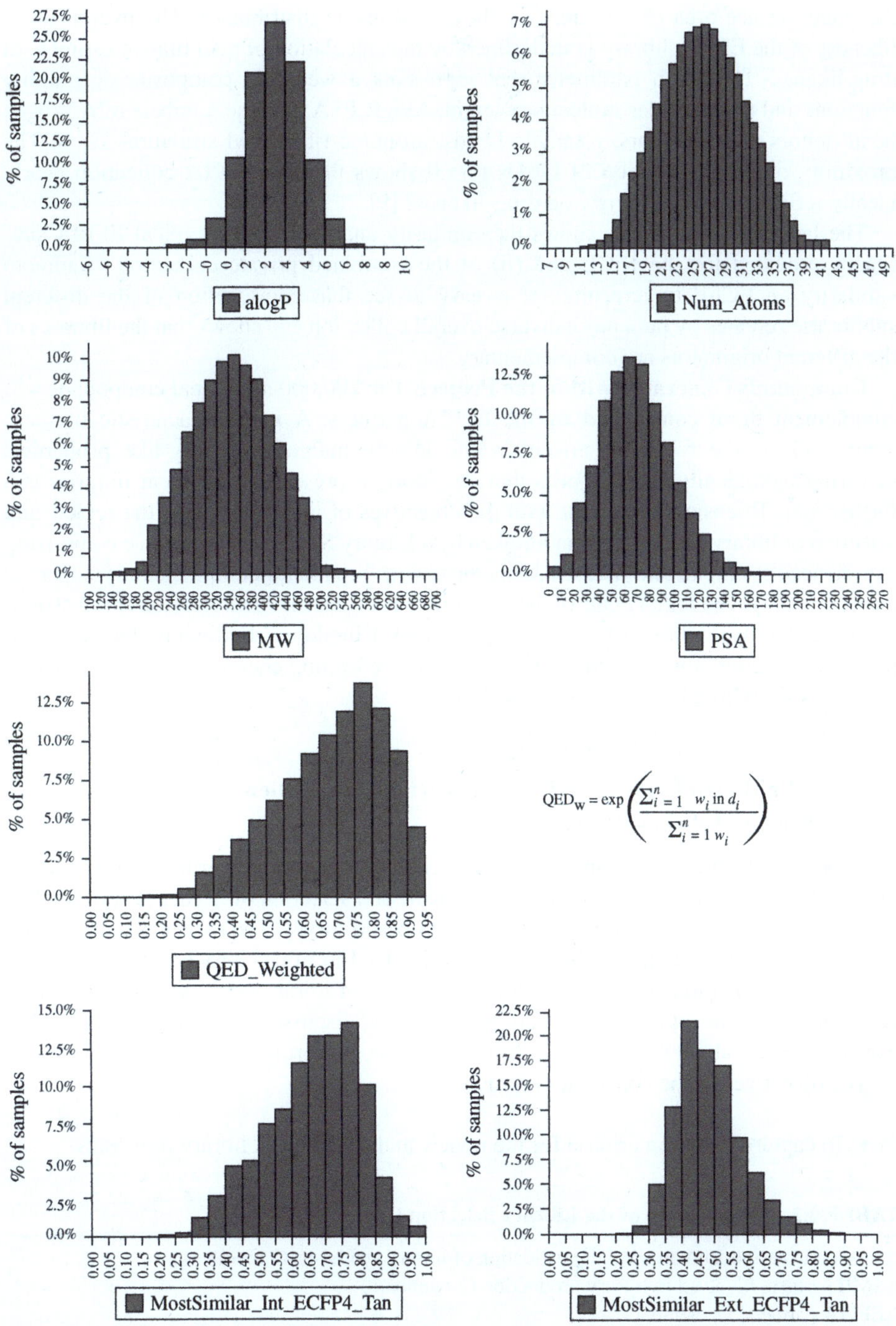


**FIGURE 2.4** Molecular properties of the approximately 300,000 compounds contributed by EFPIA partners to the Joint European Compound Library. d, desirability functions; QED, quantitative estimate of drug-likeness.

the polar surface area (PSA) analysis shows a sensible distribution. The overall drug-likeness of the EFPIA library is underlined by the calculation of quantitative estimate of drug-likeness (QED), a parameter that represents a weighted composite desirability functions and that includes molecular weight, AlogP, PSA, and the numbers of hydrogen bond donors and acceptors, rotatable bonds, aromatic rings, and structural alerts. The proximity of the distribution of QEDs to 1.0 shows that the EFPIA collection nearly ideally reflects the current view on drug-likeness [9].

The last pair of diagrams shows the similarity analysis of (i) a typical EFPIA sub-library provided for the JECL and (ii) of the combined private collection (Tanimoto similarity on ECFP fingerprints). It is easy to see that combination of the different sublibraries creates by far a more diverse overall collection and shows that the libraries of the different originators are complementary.

**Compounds Generated within the Project.** The 200,000 additional compounds will complement those contributed by the EFPIA partners. A key characteristic of these compounds is the focus on novelty while largely maintaining drug-like properties. Covering biologically active motifs that are poorly represented in current libraries is a further aim. To ensure alignment with the objectives of the consortium, the review and selection of library proposals is undertaken by a Library Selection Committee comprising experienced synthetic and medicinal chemists with broad and complementary expertise (see Table 2.2). The initial chair of the committee is Professor Adam Nelson (professor of chemical biology, University of Leeds, United Kingdom). Following the selection process, the actual libraries will be produced by five leading chemistry contract research organizations (CROs).

### 2.2.1 Definition of Criteria and an Approach for the Review and Selection of Library Proposals

At the start of the project (in January 2013), each of the consortium partners was consulted on possible criteria and approaches for the review and selection of library proposals. The partners were initially invited to complete a web-based survey. The survey was completed by the seven EFPIA partners, the five SMEs involved in library production, nine of the academic organizations responsible in the validation of library syntheses, and two other partners. The results of the survey were presented and discussed at a 1-day workshop that was attended by about 40 colleagues from across the consortium.

The objectives of the consultation process were:

- To capture views on criteria for the review and selection of library proposals

**TABLE 2.2 Membership of the Library Selection Committee**

| Membership |
|---|
| Academic partners involved in the validation of library syntheses (2 members) |
| SME partners involved in library production (2 members) |
| EFPIA partners[a] (2 members) |
| Partner from the European Screening Centre (1 member) |
| Lead Discovery Center (1 member) |

[a]The EFPIA partners may choose to appoint delegates to this committee—bound to confidentiality—to introduce additional industry expertise into this team.

- To capture views on chemoinformatic approaches that may assist the review and selection of library proposals
- To assist the Library Selection Committee to establish robust, yet efficient, processes for the review and selection of library proposals

Following this consultation process, the Library Selection Committee defined criteria and its approach for the review and selection of library proposals (which the committee may revise in the future in the light of their experience). It was agreed that each proposal would be scored against six criteria (see Table 2.3) using a simple scoring system:

- Exclude: Insufficient fit of the proposal to the criterion to allow selection
- 1: Some fit of the proposal to the criterion
- 2: Good fit of the proposal to the criterion
- 3: Excellent fit of the proposal to the criterion

It was agreed that a score of "exclude" against any criterion would necessarily lead to rejection of a library proposal. Crucially, proposals will be evaluated by tensioning all of the criteria against each other. This approach would, for example, allow the selection of a library of compounds of anomalously high molecular weight—such as macrocycles—provided that some other aspect, for example, the innovation in the library design, was ranked highly. The possible outcomes of library proposals are summarized in Table 2.4.

### 2.2.2 Collation, Review, and Selection of an Initial Wave of Library Proposals

In the first year of the project, an initial wave of library proposals was collated from consortium partners: the ten academic partners involved in the validation of library syntheses and the five SMEs involved in library production. This approach was undertaken to ensure that library validation and production could begin well before finalizing a functional web tool for crowdsourcing from the broader chemistry community.

**TABLE 2.3 Criteria for the Review and Selection of Library Proposals**

| Criterion | Comment |
|---|---|
| Molecular properties | Assessment for drug-relevant molecular properties[a] |
| Structural features | Assessment of features that may constitute a liability in screening compounds |
| Novelty | Assessment of novelty by comparison with commercially available and patented compounds |
| Diversity potential | Assessment of the potential to deliver a large library of diverse compounds |
| Synthetic tractability | Assessment of the likely synthetic accessibility of the library based on the precedent provided and the capabilities of the consortium's partners |
| Innovative library design | Assessment of the rationale that has been used in library design |

[a] Some information on drug-relevant molecular properties may be found in the Refs. 10–12. It was decided that a clogP of 4 was the preferred maximum lipophilicity of compounds.

**TABLE 2.4 Possible Outcomes Following the Review and Selection of Library Proposals**

| Outcome | Comment |
|---|---|
| Accepted for synthetic validation | The proposal aligns with all six criteria and is prioritized for synthetic validation |
| Refinement of the proposal required | Refinement of the proposal may allow fit with all six criteria. Refinement of the proposal will be undertaken by Lead Discovery Center before reconsideration by the Library Selection Committee |
| Proposal invited for resubmission | The proposal may have the potential to align with the criteria. The submitter is invited to submit a revised version that addresses the specific feedback provided by the Library Selection Committee |
| Proposal rejected | The proposal does not align with the criteria. It is felt that revision of the proposal would not enable the proposal to align with all six of the criteria |

**TABLE 2.5 Examples of Scores and Feedback from the Library Selection Committee**

| | Scores[a] | | | | | | | |
|---|---|---|---|---|---|---|---|---|
| # | MP | SF | N | DP | ST | ILD | Outcome | Feedback |
| A | 3 | 2 | 3 | 3 | 3 | 2 | Accept | It was noted that the proposal exploits efficient, well validated chemistry. There is good diversity potential and the potential to prepare a library with good molecular properties |
| B | 3 | 3 | 2 | 3 | 1 | 3 | Accept | An interesting natural product-like scaffold. The synthetic validation work will need to establish whether the diastereoselectivity can be controlled |
| C | Ex | Ex | 2 | 2 | 2 | 1 | Reject | The committee has poor experience of libraries based on similarly extended flat ring systems. The committee also considered that the molecular properties of the proposed library were poor |

[a]DP, diversity potential; Ex, exclude; ILD, innovative library design; MP, molecular properties; N, novelty; SF, structural features; ST, synthetic tractability.

In the first year, the Library Selection Committee met seven times and considered more than 175 proposals. About 125 library proposals were selected for synthetic validation (of which, in about 20 cases, a revised proposal had been invited by the committee). Some examples of scores and feedback on specific proposals are summarized in Table 2.5.

It is hoped that feedback from the committee—for both accepted and rejected proposals—will help submitters to align future proposals with the requirements of the consortium. For proposal B, for example, the diastereoselectivity of a key step had not been validated (leading to a score of "1" for synthetic tractability): this aspect was highlighted as a priority for the synthetic validation work. A highly innovative design idea, however, was the main argument to ultimately accept this proposal. For proposal C, it was explained why "Exclude" scores had been assigned for two of the criteria: Molecular Properties and Structural Features.

### 2.2.3 A Web-Based Tool to Support the Collation, Review, and Selection of Proposals

In parallel with the selection of an initial wave of proposals for synthetic validation, a web-based tool has been developed to support the collation, review, and selection of future proposals. The tool was operational from the start of the second year of the project. The capabilities of the tool were defined by the Library Selection Committee, and its development was managed by a small team of colleagues from the University of Leeds, ChemAxon, Taros, and AstraZeneca. The tool was built by, and is hosted by, ChemAxon (Budapest, Hungary). The tool enables library proposals to be submitted by both partners within the consortium and third parties external to the consortium. If necessary, the tool will be revised in the future to optimize the collation, review, and selection of proposals within the consortium.

The ELF seeks to encourage the submission of proposals based on innovative synthetic methodology. To encourage engagement, the web-based tool has been designed to be accessible to synthetic chemists who have no previous experience of library design (see http://www.europeanleadfactory.eu/proposals/chemical-scaffolds/).

Perhaps the easiest way to define a library proposal is to use the embedded tools for library enumeration (see Fig. 2.5).

Using this approach, the submitter sketches a scaffold (using Marvin Sketch), specifying the positions of variable substituents (as R1, R2, R3 etc.). The submitter can then select alternative possible substituents from predefined lists. It is also possible for submitters to define library proposals by uploading an appropriate file (e.g., in sd format).

Before submission of a library proposal to the consortium, the submitter is presented with a range of information about their proposal (Fig. 2.6). Crucially, after submission of the proposal, this information is used by the Library Selection Committee to inform their assessment. When satisfied that the proposal addresses the assessment criteria, the submitter may choose to submit the proposal. At this stage, the tool performs an assessment of similarity to a reference set of commercially available and patented compounds. An approach is also being developed to allow the comparison of library proposals with the compounds in the JECL. Diversity will be calculated using circular fingerprints to ensure consistency with the comparison to commercial compounds.

Submitted proposals are considered by the Library Selection Committee. An overview of how proposals are assessed against the six criteria is provided in Table 2.6. By consensus, the committee reaches a score against each of the criteria and decides on an outcome for each proposal. Following processing of the proposal, the scores, feedback, and outcome are visible to the submitter (Fig. 2.7).

### 2.2.4 Synthetic Validation of Library Proposals and Library Production

Following acceptance (and, for submitters external to the consortium, receipt of a signed form of accession), experimental validation of selected library proposals is undertaken (see Fig. 2.8).

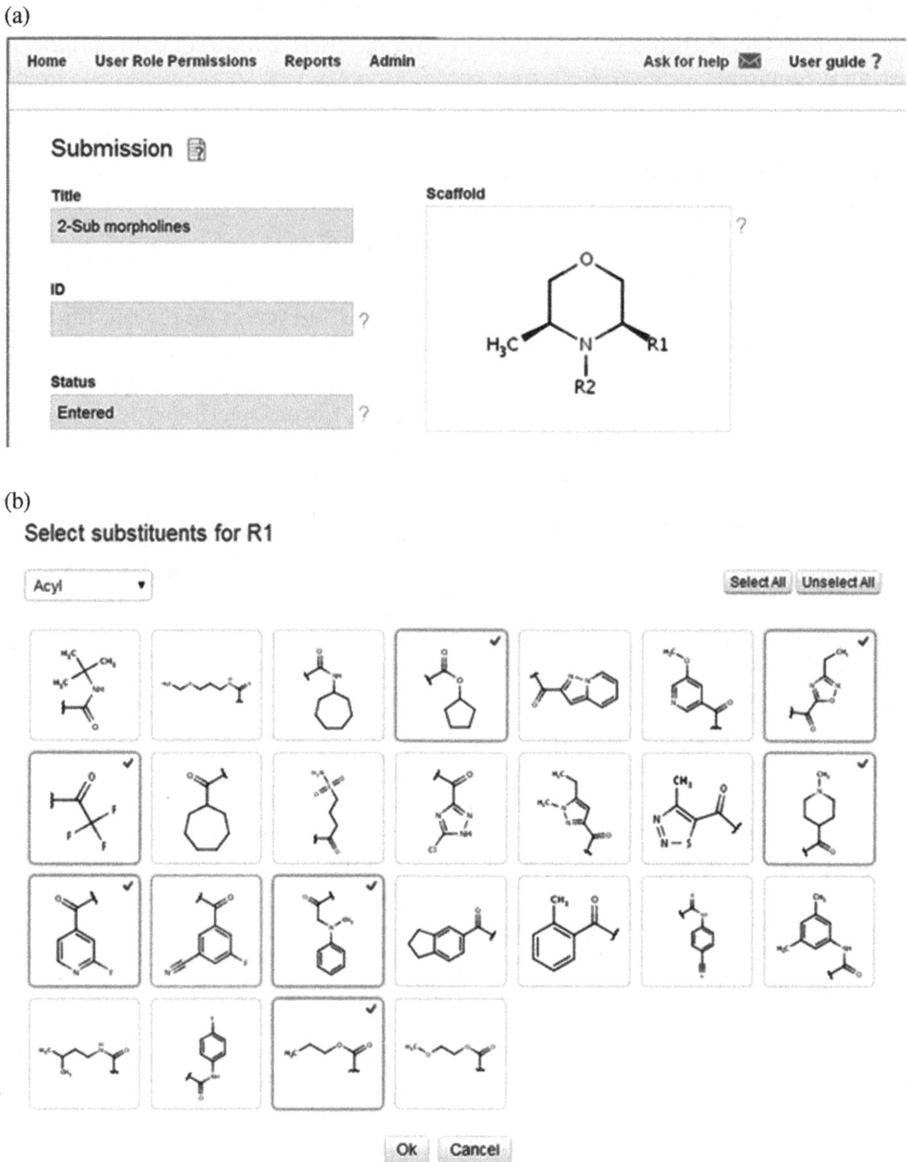


**FIGURE 2.5** Definition of a library proposal using the tools embedded within the consortium's web-based tool. (a) The submitter sketches a scaffold (using Marvin Sketch), specifying the positions of variable substituents (as R1, R2, R3 etc.). (b) The submitter can then select alternative possible substituents from predefined lists.

Synthetic validation work is carried out by academic partners within the consortium working in close cooperation with one of the SMEs who are responsible for library production. The objective of this work is to demonstrate that the scaffold can be prepared on a multigram scale by synthesizing and purifying an initial set of approximately

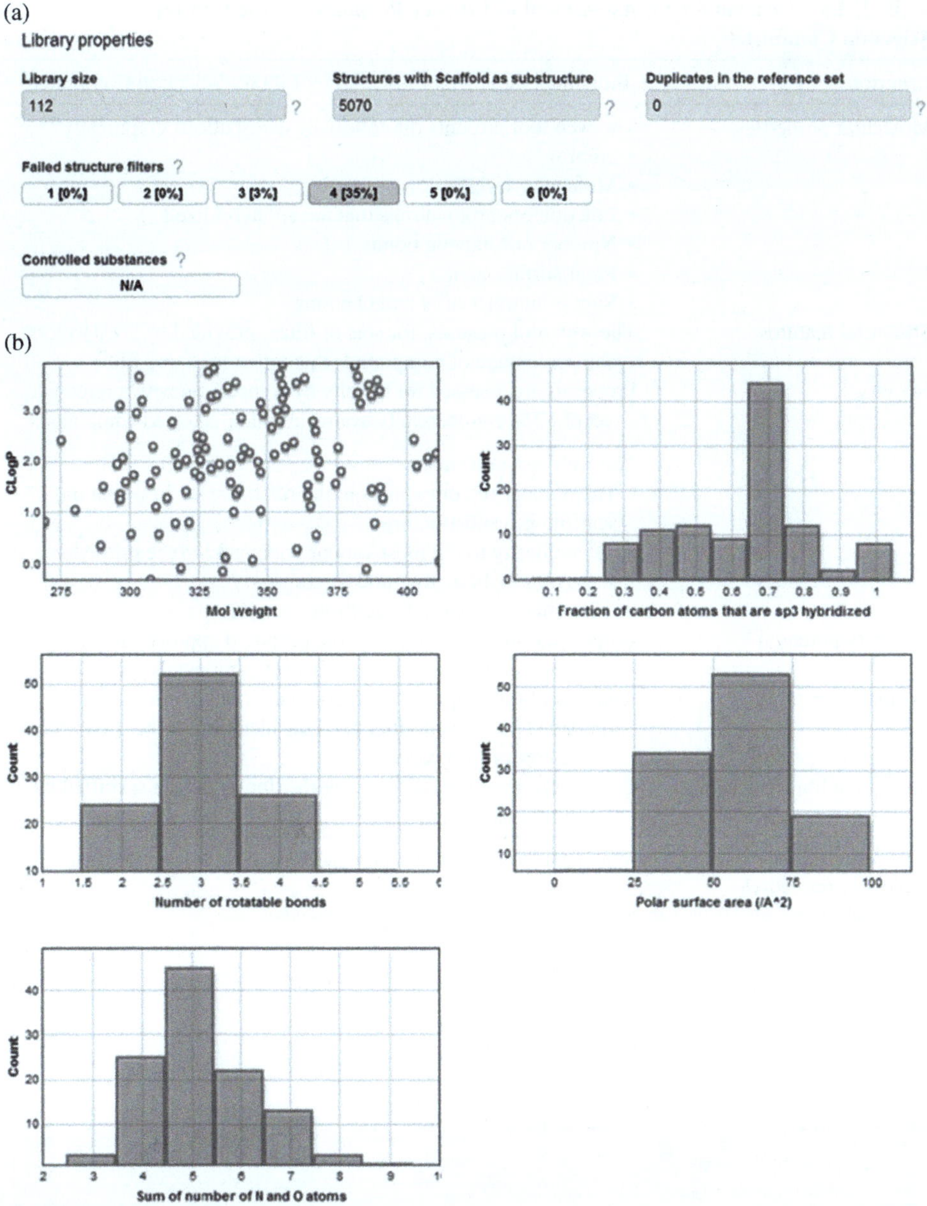


**FIGURE 2.6** Information presented to the submitter before submission of a library proposal (a) General information on proposed library e.g., size and fraction not passing structural filters. (b) Distribution of properties used in the assessment later on.

20 derivatives. For the proposed scaffolds, different routes toward final products need to be devised and evaluated on a 20- to 100-mg scale. In general, routes need to be optimized with respect to the required number of synthetic steps, potential for scale-up, orthogonality of decoration chemistry, and availability of required reagents.

**TABLE 2.6 Criteria for the Assessment of Library Proposals by the Library Selection Committee**

| Criterion | Individual Parameters assessed by Library Selection Committee |
|---|---|
| Molecular properties | The web tool presents the following distributions graphically:<br>• clogP<br>• Molecular weight<br>• Fraction of carbon atoms that are $sp^3$ hybridized<br>• Number of rotatable bonds<br>• Polar surface area<br>• Sum of number of N and O atoms |
| Structural features | The web tool presents, for sets of filters provided by six partners, the percentage of compounds that fail at least one filter |
| Novelty | Proposals are assessed for novelty by comparison with a reference set of 12 M commercially available and/or patented compounds[a]<br><br>The web tool presents:<br>• The number of compounds in the reference set in which the scaffold is a substructure<br>• 2D similarity to the most similar compound in the reference set (using ECFP4 structural fingerprints)[b]<br>• a 2D pharmacophore fingerprint comparison[b] |
| Diversity potential | Consideration of the fitness of the proposed approach for the synthesis of a large library of diverse compounds |
| Synthetic tractability | Analysis of the proposed synthetic route (and its precedent) uploaded by the submitter and consideration of the capabilities of the consortium partners |
| Innovative library design | Consideration of the rationale for the library design specified by the submitter |

[a] The reference set of commercially available and patented compounds was provided by eMolecules and SureChem, respectively.

[b] This analysis is visible to the submitter after submission of the library proposal.

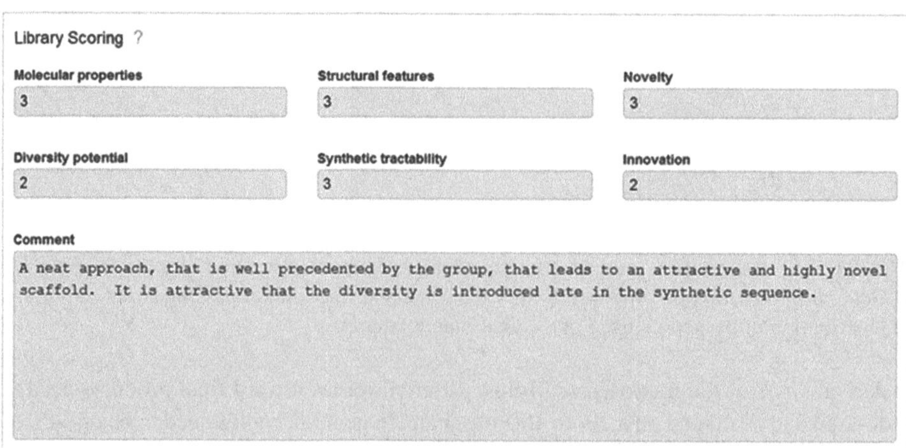


**FIGURE 2.7** An illustration of scores and feedback that might be provided to the submitter after consideration of the library proposal by the consortium's Library Selection Committee.

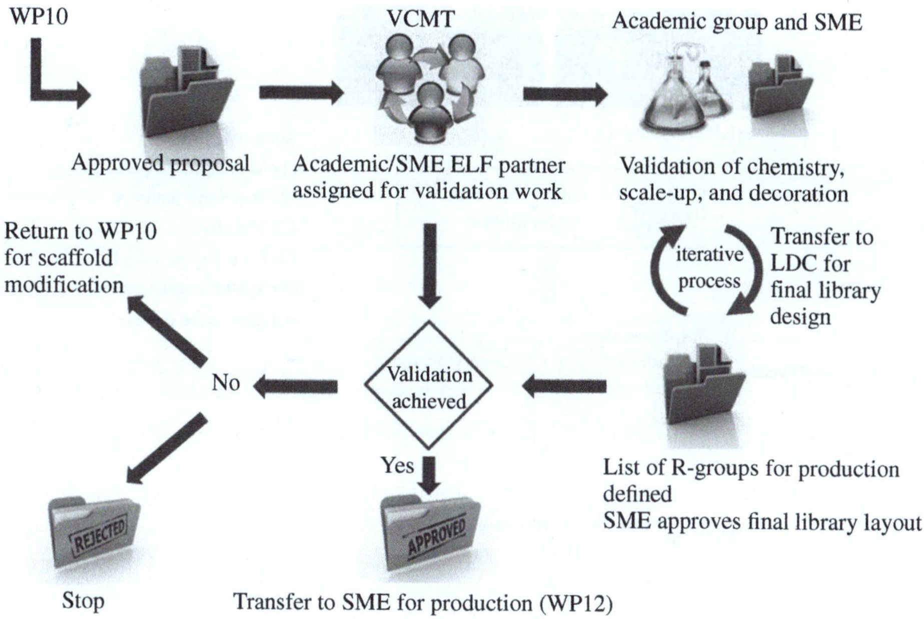


**FIGURE 2.8** Selected library proposals undergo chemical validation work. The validation chemistry management team (VCMT) assigns a consortium academic group to each proposal submitted. After library refinement in the light of validation work, the VCMT will review the validation work and decide on the further progress. After approval, library production is undertaken.

On the basis of the validation work, a management team of experts with complementary expertise then decides whether it is feasible to nominate a library for production. When difficulties with selected library proposals occur, a quick decision will be made by the management team on either continued prosecution with additional chemistry resources, modification of the original scaffold into a more feasible scaffold, or rejection of the scaffold due to chemistry-related problems.

Within the scope of the synthetic chemistry that has been validated, each library is optimized for diversity through library design at the Lead Discovery Center, Dortmund, Germany. Library production is then undertaken by one of the five SMEs within the consortium. The whole process is illustrated in Figure 2.9.

It is anticipated that from a large number of proposals collated, approximately 140 library proposals will undergo the experimental validation process each year. Assuming an attrition rate of about 30%, around 100 scaffolds from these proposals will proceed to library production by one of the chemistry SMEs (Taros, Mercachem, Edelris, Sygnature, and Syncom). An average library size of 500 compounds will lead to the intended size of ELF's Public Compound Collection after 5 years.

The experimental progress of the library validation and production is be monitored by TarosGate, software initially developed by Taros. A version of TarosGate was developed by colleagues at Taros in the first 9 months of the project specifically for use within the ELF. The software allows data to be shared between relevant members of the consortium (see Fig. 2.10) and enables the efficient transfer of data between WPs.

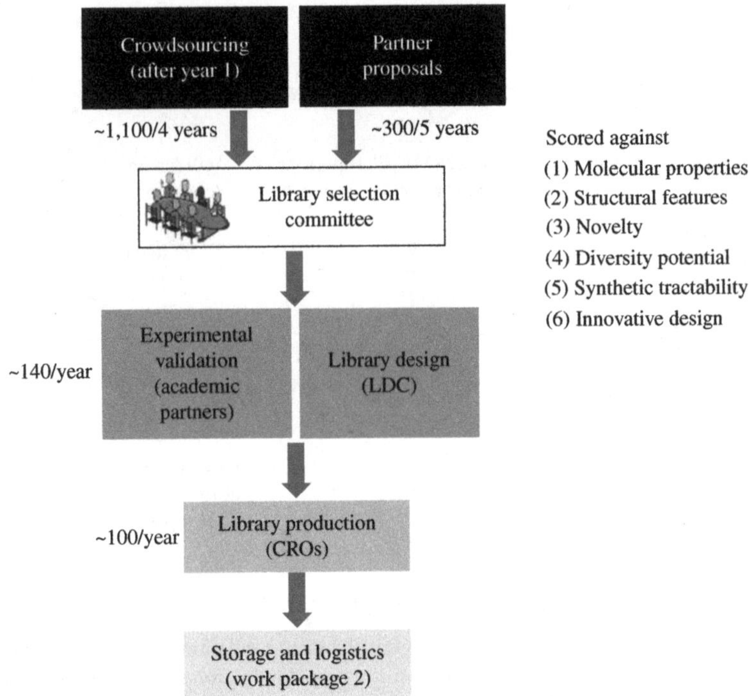


**FIGURE 2.9** Workflow of Library proposal, validation, design, and production.

## 2.3 QUALIFIED HIT GENERATION

### 2.3.1 Capabilities of the ESC

ELF target programs from the public domain is executed in the ESC located in Newhouse, Scotland, and in Oss, the Netherlands. The facilities in Scotland comprise a compound logistics centre and University of Dundee scientists engaged in hit characterization and medicinal chemistry (see the following). The logistics centre houses the JECL and distributes the compounds plates and samples for screening or hit characterization. The two partners, Pivot Park Screening Centre and BioAscent, emerged from the legacy Merck, Sharp & Dohme research sites and have a successful operational track record reaching back to their Organon heritage. Pivot Park Screening Centre is the location for the assay adaption to the automation platforms and uHTS activities. The ESC is completed by the University of Oxford, providing protein production and structural biology expertise to the ELF. The Universities of Dundee and Oxford are established centers of excellence in academic drug research and incorporate a great number of industry experienced scientist among their faculty and staff.

The compound storage is based on REMP® Storage μ Tube Technology including μ Tube Punching Technology. The large store available to the ESC allows compounds to be stored in single-use tubes at low temperature in a controlled atmosphere. In addition, the technology is able to "cherry-pick" any combination of samples and provide them in differing formats of output plates. Thus, the lifetime of samples is maximized firstly by storing compounds in conditions that minimizes decomposition, secondly by storing compounds in single-use tubes and hence minimizing freeze–thaw cycles, and thirdly by

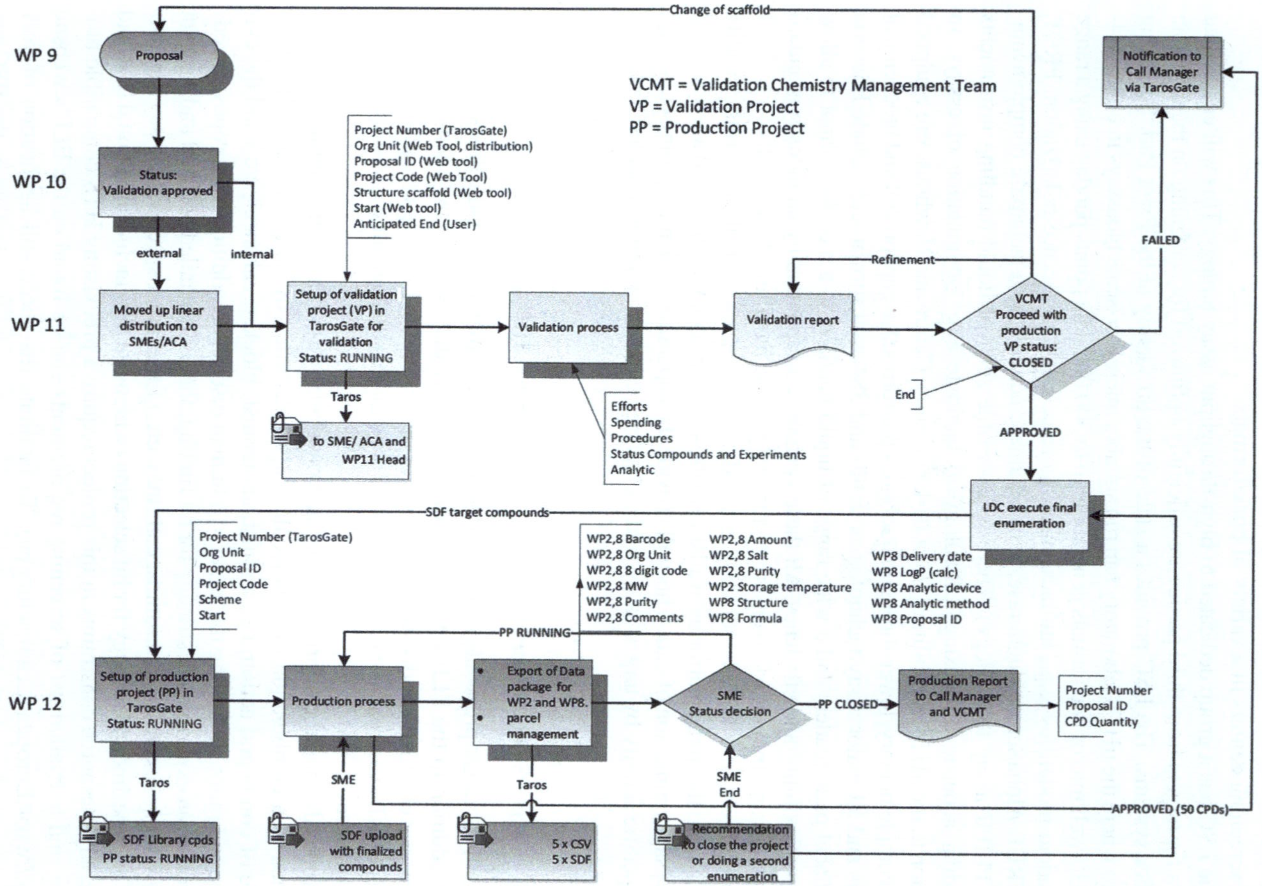


**FIGURE 2.10** Data workflow covered by TarosGate.

providing rapid access to precise libraries of follow-up samples. The logistics center is staffed with experts who have considerable experience carrying out these activities for major pharma. Further automation platforms are available to allow the rapid processing of samples into the store and fast and flexible delivery of compound sets to the ELF's eight screening centers in a variety of plate formats.

The ESC has a group dedicated to high-throughput assay biology. This will ensure that assays for targets from third parties are efficiently optimized for screening on the available robotics systems. The ESC provides a state-of-the-art facility at the Pivot Park Screening Centre where the uHTS laboratory can run suitable microliter-scale bioassays in either 384- or 1536-well microplate formats generating up to 300,000 data points per day using a range of readout modalities (e.g., luminescence, fluorescence, fluorescence polarization, FRET, TR-FRET, AlphaScreen, luciferase, beta-lactamase activity, beta-lactamase complementation, FLIPR assay technology). There is a versatile suite of liquid handling instruments including acoustic dispensing technology to facilitate easy optimization of assays. In addition to the uHTS capabilities, Pivot Park Screening Centre and Newhouse are equipped with complementary, lower-throughput robotics platforms that support cell and biochemical assays and all major assay technologies in 96- and 384-well format and include Envision multilabel plate readers and a wide range of liquid handlers in a modular format to allow process flexibility and efficiency. All these systems are supported by an industry-standard data processing framework to ensure secure and reliable handling of the results.

In addition, the ESC is adopting a standardized screening analysis format, readily adaptable to the assay platforms available to the screening centre, which will allow a broad pan-assay comparison of data. This will provide the capability to identify promiscuous and nonspecific activity by target, assay, and technology class using primary screening data.

### 2.3.2 Target Selection and Generation of Qualified Hits

The call for target proposals is not restricted to specific therapeutic indications.

In addition to the ELF's goal of value creation by discovering new chemical matter against novel targets for disease indications of interest to the EFPIA participants, a major aim is to achieve broad coverage of putative drug targets throughout the vast spectrum of human disease.

Proposals should comprise of a defined molecular target (a target program in ELF is defined as a combination of a molecular target identified by its unique gene ID and a mode of compound action, i.e., agonist, antagonist, blocker, enhancer, etc.) and the evidence and rationale linking the target to a human disease. In addition, the proposal needs to provide an experimental strategy for hit finding, that is, a detailed protocol and relevant material (e.g., expression plasmids, cell lines, etc.) for a target specific bioassay together with an experimental strategy for hit selection (which can also include additional tests and material). Resource constraints in this project require a preexisting HTS compatible bioassay, and a detailed list of screening requirements can be found on the ELF's website (http://www.EuropeanLeadFactory.eu). To facilitate the successful application of very novel targets for which HTS ready assays are not available, scientists in the ESC are happy to work together with applicants in a preselection process to address identified technical problems and enable—where possible—the assay to be implemented in a uHTS format. The recruitment and selection process together with the continuous monitoring of

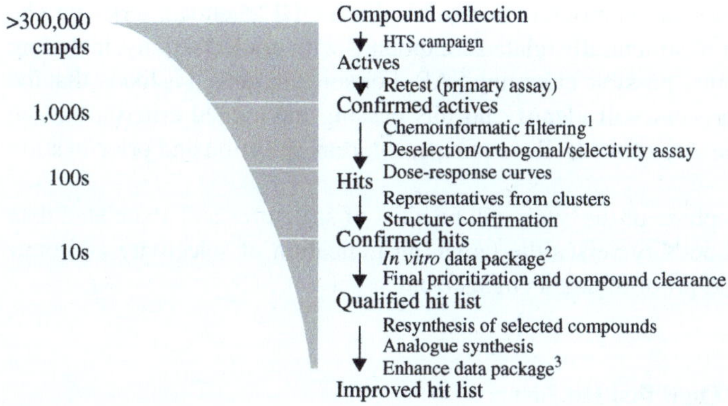


1. Bayesian model prediction of false negatives/positives, toxicophores, calculated physicochemical properties
2. Mechanistic assay (e.g., SPR, kinetics, mode of action)
3. Safety, ADME, physicochemical determinations, detailed selectivity studies, biostructural information.

**FIGURE 2.11** Process from HTS toward qualified hit lists and improved hit lists.

the progress and the appropriate involvement of the target program owner will be managed by a dedicated program office. This role is taken by an established not-for-profit research management organization, TI Pharma, based in Leiden, the Netherlands, whose function includes the setup and supervision of tools and processes to handle confidential information of participants and contributing external partners.

For accepted programs, a comprehensive program plan is agreed upon following a meeting between ESC representatives and the public target owner. This meeting is designed to be fully collaborative and bring together the expert target knowledge of the target owner and the screening expertise of the centre. The main aim of the program plan is to devise an agreed-upon screening cascade that has the highest likelihood of delivering high-quality hit molecules that will fulfill the target owner's major goals.

Following on from the actual uHTS campaign, hits will be characterized via the agreed-upon triaging cascade, which will be bespoke to the target and will include initial studies to confirm primary activity. Where possible, the list of confirmed actives will be filtered on the basis of results from a deselection or orthogonal assay, and then a subset selected for analyzing dose–response activity against the target. The full process of hit characterization will include a tailored set of experiments and can include characterization for specificity and selectivity of the hit compound–target interaction, exploring the mode of action together with a check of the samples purity and composition (Fig. 2.11).

All activities in this context will be embedded in, and guided by, chemoinformatic tools to enable triaging of the list of actives on an extensive range of parameters. These can include properties such as molecular weight and chirality; simple counts of hydrogen bonding potential, rotatable bonds, rings, and carbons with $sp^3$ hybridization; predicted values of lipophilicity, PSA, and QED drug-likeness; and a filter for CNS penetration depending upon the specified goals of the target owner. Compounds are preclustered, and Bayesian models of a broad range of potential selectivity or antitargets are included within the tools to allow further prioritization of the hit lists. The final selection will include review of the chemical structures by experienced medicinal chemists with the goal of compiling a "qualified hit list" (QHL) of no more than 50 different compounds that will be disclosed to the target program owner. Ideally, a QHL chemical series will be represented by a number of structurally related compounds with graded activity, including selected inactives to capture possible emerging SAR. For some targets, it is likely that the rigorous qualification process will identify no hits passing the agreed criteria, but the numeric limitation of the size of the QHL requires a rigorous selection and prioritization process does occur.

The output from this phase of the work will be a list of structures and associated data outlining the biological activity versus the target, an indication of selectivity and optimally mode of action and proof of target engagement.

### 2.3.3 Exploitation of Qualified Hit List

For public target program only, the ESC will also provide initial medicinal chemistry support to external partners. Some limited resources will be available for the synthesis of selected derivatives to probe potential emerging SAR or explore the chemical space around identified singletons, and these activities will be supported by computational chemistry. In addition, for some programs, it is envisaged that X-ray crystallographic information on protein–ligand complexes will be generated to support the ongoing studies and to serve as a source of future ideas. Overall, this chemistry input will serve to improve the quality of the screening results by adding to the hit qualification. In some cases, such activity might also contribute to disqualify identified series from further prosecution providing valuable clues for the target program owner.

More detailed biological evaluation will also be possible at this stage both using the resources available to the ESC and of the target owner. It will be possible to carry out detailed mechanism of action studies in addition to biophysical measurements. For some compounds, it will be possible to produce *in vitro* ADME data (e.g., lipophilicity, p$K_a$, solubility, hepatocyte/microsomal stability, plasma protein binding, permeability, plasma stability, cytochrome P450 inhibition) to further characterize the hit series and likely identify liabilities for resolution during later optimization. Clearly, it will only be possible to produce this information for a limited number of compounds and programs, and so studies will be prioritized in discussion between the program owner and scientists at the center. The ultimate output will be an IHL, which consists of no more than 50 further compounds for the target owner.

The ambition is to provide a high-quality package including a range of data and physical samples that will facilitate further exploitation of the structures by the target owner.

## 2.4 FUTURE PERSPECTIVES

The ELF has received a jump start from the EFPIA participants providing 300,000 different compounds to the JECL. This JECL together with its growing fraction of publicly sourced compounds will be available for the screening activities within selected public programs as well as in programs nominated by the EFPIA participants. Over the 5-year funding period of the project, the goal will be to identify and perform 120 public projects. During the peak activity period, in the years 2–4, the ESC thus will process up to around 30 screening programs per year. At the time of finalization of the manuscript, the ELF consortium has developed the technical and organizational setup to present its offerings to the broad public (see http://www.europeanleadfactory.eu). The first compound libraries generated in this project have been processed for integration in the screening process. In addition, initial target programs originating from participating public institutions have already entered the screening process after serving as test cases to check and improve all relevant process steps. With the opening to the general public, the ELF experiments enter its next and most exciting phase. There is little doubt that the screening activities of EFPIA participants exploiting the JECL will translate into tangible results, that is, novel lead structures active in suitable preclinical models. Given industry-standard probabilities of success together with the ambitious quantitative goal of performing 120 target programs during the funding period for EFPIA, it is not unrealistic to expect that the ELF will eventually provide the foundation for at least one drug product to reach the market. While this is likely to take another decade or more, initial results will become recognizable late in course of the project through patenting activities and other means of disclosures. On this foundation, all contributions from the public domain or SME/biotech companies, that is, innovative chemistry to enrich the JECL and unprecedented target programs to spark novel therapeutic approaches or explore cutting-edge biology, can add substantial additional value—as explained in this chapter.

## ACKNOWLEDGMENTS

The authors would like to thank Jérémy Besnard, Andrew Hopkins, and Andrew Pannifer (University of Dundee) for providing the analysis of the compounds from the EFPIA partners that will be published in more detail elsewhere in due course. We would like to thank Gerhard Müller (Mercachem), Alexander Piechot (Taros), Stuart McElroy (University of Dundee), and all other workpackage leaders for providing material and careful editing of the manuscript.

## REFERENCES

[1] Mullard, A.; 2013 FDA drug approvals, *Nat. Drug Discov.* **13**, 85–89 (2014).

[2] Kling, J.; Fresh from the pipeline—2013, *Nat. Biotech.* **32**, 121–124 (2014).

[3] Macarron, R.; Banks, M. N.; Bojanic, D.; Burns, D. J.; Cirovic, D. A.; Garyantes, T.; Green, D. V. S.; Hertzberg, R. P.; Janzen, W. P.; Paslay, J. W.; Schopfer, U.; Sittampalam, G. S.; Impact of high-throughput screening in biomedical research, *Nat. Drug Discov.* **10**, 188–195 (2011).

[4] Pammolli, F.; Magazzini, L.; Riccaboni, M.; The productivity crisis in R&D, *Nat. Drug Discov.* **10**, 428–438 (2011).

[5] Munos, B.; Lessons from 60 years of pharmaceutical innovation, *Nat. Rev. Drug Discov.* **8**, 959–968 (2009).

[6] Scannell, J. W.; Blanckley, A.; Boldon, H.; Warrington, B.; Diagnosing the decline in pharmaceutical R&D efficiency, *Nat. Rev. Drug Discov.* **11**, 191–200 (2012).

[7] Khanna, I.; Drugs discovery in pharmaceutical industry: productivity challenges and trends, *Drug Disc. Today* **17**, 1088–1102 (2012).

[8] Wegener, G.; Rujescu, D.; The current development of CNS drug research, *Int. J. Neuropsychopharmacol.* **16**, 1687–1693 (2013).

[9] Bickerton, G. R.; Paolini, G. V.; Besnard, J.; Muresan, S.; Hopkins, A. L.; Quantifying the chemical beauty of drugs, *Nat. Chem.* **4**, 90–98 (2012).

[10] Lipinski, C. A.; Lombardo, F.; Dominy, B. W.; Feeney, P. J.; Experimental and computational approaches to estimate solubility and permeability in drug discovery and development, *Adv. Drug Deliv. Rev.* **23**, 3–25 (1997).

[11] Teague, S. J.; Davis, A. M.; Leeson, P. D.; Oprea, T.; The design of leadlike combinatorial libraries, *Angew. Chem. Int. Ed.* **38**, 3743–3748 (1999).

[12] Walters, W. P.; Green, J.; Weiss, J. R.; Murcko, M. A.; What do medicinal chemists actually make? A 50 year retrospective, *J. Med. Chem.* **54**, 6405–6416 (2011).

# 3

# ACCESS TO COMPOUND COLLECTIONS: NEW BUSINESS MODELS FOR COMPOUND ACQUISITION AND SHARING

PETER TEN HOLTE
*Janssen Research & Development, LLC, San Diego, CA, USA*

## 3.1 INTRODUCTION

### 3.1.1 Vertical Disintegration and the Quest for Innovation

The pharmaceutical industry is increasingly aiming to broaden access to innovation [1]. While most companies once preferred an end-to-end value chain within their walls, performing drug discovery and development solely internally [2], they are now eager to partner with external research organizations, either of academic or commercial background, and this strategy is not just for cost reduction but to broaden access to the new skills and capabilities required to innovate. This process has been called "vertical disintegration." Materials, intellectual capital, and/or human resources essential to delivering the final product or service are accessed from independent, external sources [3] (Fig. 3.1).

Consequently, this leads to a virtual integration of external contributions with a significant impact on the operational aspects of drug discovery. "Because great new technologies are coming from people who want to do their own thing, and won't necessarily work for someone else, acquisition may be the only way for a large company to access them," says Mitchell Kertzman, partner at Hummer Winblad, a San Francisco-based venture capital firm [4]. The need for maximum input of novel ideas and scientific expertise is driven by the ever-increasing challenge to discover and develop truly novel, successful drugs [5]. From the 1960s to the 1990s, new starting points for drug discovery

*Small Molecule Medicinal Chemistry: Strategies and Technologies*, First Edition.
Edited by Werngard Czechtizky and Peter Hamley.

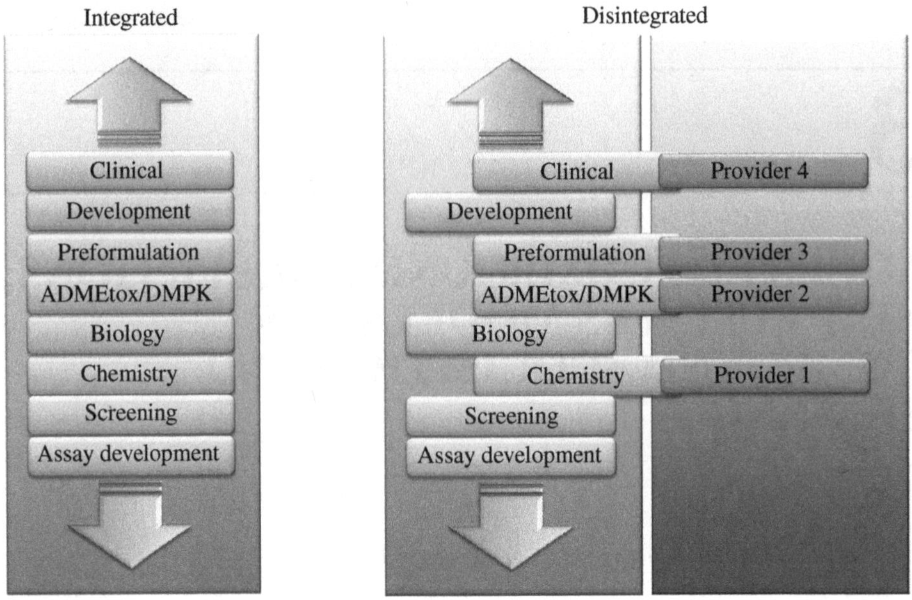


**FIGURE 3.1** Representation of a fully integrated and a partially disintegrated organization where four providers participate in different aspects of the R&D process.

projects were generated in-house, often as part of medicinal chemistry efforts focused on delivering new drug candidates for specific indications. Academic knowledge derived from literature provided medicinal chemists with ideas of biologically attractive chemical matter for chemical exploration, leading to novel drug discovery starting points. In the 1990s, medicinal chemistry teams—believing in the power of statistics and newly equipped with high-throughput screening (HTS) facilities—started to explore combinatorial chemistry as a source of massive numbers of new screening compounds. Across the industry, teams with specialized equipment, responsible for the production of libraries to support medicinal chemistry projects, were established. In parallel, new contract research organizations (CROs) focusing on library design and library production or compound purchasing were founded—examples of these early pioneers being Asinex and Specs and BioSpecs. In academia, many research groups developed new methodologies for parallel, combinatorial, and solid-phase synthesis, fueling medicinal chemistry efforts in the pharmaceutical field. Pharmaceutical companies therefore invested heavily in internal compound collection enhancement. However, inspired by external opportunities and focused on existing research programs, they also reserved considerable budgets for external compound acquisition to complement internal efforts. After a peak in the early 2000s, library development and production for compound library enhancement have now mostly been shifted to specialized library vendors. Internal, company-wide crowdsourcing initiatives have often been implemented to ensure optimal selection of preferred compound sets for acquisition [6]. Custom library production, where only library design is done internally while production is performed at a CRO, has become typical.

In the late 2000s, academia fully embarked on the setup of drug discovery facilities and expertise. This trend is expected to yield new ways of thinking and more readiness to embark on riskier areas of research—with universities and institutes typically covering more ground when it comes to basic, fundamental research [7]. This development has led to many new consortium-type collaborations and partnerships fueling medicinal chemistry with drug discovery innovation. Research institutes and pharmaceutical companies nowadays consider themselves as equal partners with complementary expertise and focus [8]. Under such circumstances, one can truly speak of a virtually integrated research model incorporating the potential to jointly produce solutions for unmet medical needs.

Vertical disintegration and subsequent virtual integration have had a significant impact on the practice of hit identification for novel biological targets. This chapter covers creative new ways to access novel chemical starting points for medicinal chemistry projects beyond pharmaceutical industry's legacy compound collections.

### 3.1.2 Innovative Chemistry

In order to bring new drugs to the market, the pharmaceutical industry must innovate. This innovation will come from biology, but a critical dimension is clearly also rooted in chemical innovation, a central domain of medicinal chemistry. This is vital for two purposes: (i) structures must interact with the biological target of interest, an increasingly difficult task as new types of biological targets arise that are no longer classical targets such as GPCRs, ion channels, kinases, and proteases; and (ii) a marketed drug requires a proprietary intellectual property (IP) position. Starting a medicinal chemistry program from a structurally novel starting point makes achieving this goal much easier. Therefore, the quest for access to novel chemical structures goes hand in hand with an innovative drug discovery strategy.

### 3.1.3 Access to Supplementary Compound Collections

Screening chemical compound libraries, most often by large-scale HTS, is still the default method for generating hits for biological targets of interest in most organizations. To obtain useful tool compounds, that is, compounds with some level of novelty that interact specifically with the target (often a new target where binders are hitherto unknown), the screening collection must contain compounds that cover a sufficiently large chemical space. Pharmaceutical companies with large "historical" collections of compounds derived from past medicinal chemistry campaigns still wish to augment these collections with new chemistries to cover new areas of chemical space and gain access to new IP. In contrast, smaller organizations often do not have the luxury of access to a vast historical repository.

Acquiring compounds "off the shelf," that is, directly from chemical suppliers, can become an expensive endeavor with major financial implications for small biotechs (often financed by start-up capital) and academic groups. Such entities are often considerably less well funded than industry, which is itself recently subject to a greater focus on cost control.

Obtaining innovative compound collections in an increasingly cost-constrained environment has had a serious impact on compound library providers. Pharmaceutical companies would ideally only like to pay compound library providers for compounds that show potency and are patentable. On the contrary, the compound library providers need to see a fair return on their investment (ROI) to operate sustainably. Therefore, they have become open to various diverse models for compound acquisition and sharing. Some providers have proactively built their business around the industry's preferred models—an example being BioFocus (A Charles River Company), who reshape their consortium model every year based on customers' feedback. Users and providers nowadays apply these business models to create low-hurdle and low-cost access to novel chemical starting points with the mutual hope for a greater payoff if the chemistry results in useful hits and leads. The challenge resides in the joint balancing of ownership and patentability with low-hurdle, ideally open, access.

In addition to establishing new business models between chemistry providers and drug discovery organizations, there is another increasingly popular way of dealing with the challenges by creating precompetitive consortia. In these, academia, chemistry providers, governments, and the pharmaceutical industry collaborate and share openly. The Academic Drug Discovery Consortium (ADDC), the EU-OPENSCREEN platform, the National Institutes of Health (NIH) Roadmap Molecular Libraries Screening Centers Network (MLSCN), and the IMI European Lead Factory creatively facilitate sharing of compound library collections between pharmaceutical companies and academia.

## 3.2 RISK-SHARING APPROACHES

### 3.2.1 Overview

Risk-sharing business models are models where a company provides some self-funding of a project with a partner in the hope of obtaining a (usually) greater downstream financial reward. In the context of chemistry innovation in drug discovery, a company provides novel chemical matter (the chemistry provider) and transfers this chemical matter to a partner (chemistry acquirer) who tests these compounds and *may* further use them for a drug discovery program. In contrary to classical compound acquisition models where the whole risk of investment in the compounds rests on the acquirer having paid the provider before knowing the true value of the acquired compounds, risk-sharing models make both parties collaborating partners. They share the risk of investment in new chemical matter but also the chances of success these compounds may bring. The challenge in all risk-sharing models therefore lies in a fair agreement between both partners; sufficient ROI must finally be generated for both parties involved. This is true for the monetary part of the deal, but can sometimes be an even bigger challenge for the IP portion generated in the collaboration.

Due to the fact that all partners share an avid interest in the success of the joint risk-sharing model, the level of commitment on both sides is usually much higher than in classical compound acquisition models. The chemistry provider's goal is to deliver successfully performing compounds generating future business, while the chemistry acquirer is fully driven to evaluate the compounds' value in a timely and high-quality manner to generate novel starting points for its drug discovery program and to save costs.

Risk-sharing models can be operative between two or more partners. There are usually two corporate partners, a drug discovery organization and a specialist compound provider, which may be complemented by academic groups or governmental organizations providing specific scientific expertise, for example, a special biological assay or a disease model.

The next sections describe various types of risk-sharing models in practice. All are based on an initial step of blinded screening, including the transfer of novel compounds from the provider to the acquirer for biological evaluation, but without transfer of structural information (see Section 3.2.2). Blinded screening can be followed up in different ways of collaborations; the "lease" model, the "appetizer" model, and the "subscription" model are described in Section 3.2.3.

### 3.2.2 Blinded Screening

Compounds are usually delivered to the partner in an assay-ready plated format sufficient for a onetime screen in order to prevent further distribution or analysis of the structures by the compound acquirer. The selection of the compounds is often done by the provider if he has expertise in the biological field in which the pharmaceutical partner is interested to screen or has state-of-the-art computational capabilities. Done this way, structures can remain unknown to the screening party and the provider's IP safeguarded for future screening partners. The prevention of so-called contamination can be another driver for blinded screening; contamination is present if structures, even under a confidentiality agreement, become known to the screening party (the acquirer), while in the same company, other researchers come up with a similar idea and cover the same chemical space. In that case, the acquirer needs to prove that the internal researchers haven't had any access to the provider's notes or structures and can demonstrate the originality of their idea. In many cases, though, the pharmaceutical company will have some level of input in the selection of the compounds, preferably via an independent (external) third-party consultant.

The biological evaluation of blinded structures is a true hurdle-lowering activity between partners that, if done under the proper agreement type, can yield a win–win situation for both partners. The acquirer gets confidence in the value of the structures of interest, whereas the chemistry provider usually gets important information concerning its chemistry offer, free of charge. In addition, the provider will be able to use that information to design new libraries or generate potential future deals with other pharmaceutical partners in case the original deal is ended. A reason for such a deal to end would be a lack of interest in the properties and activity of the compounds by the initial partner.

### 3.2.3 Follow-Up of Blinded Screening: Various Models

Once a blinded screen has been finished, there are several ways in which the risk-sharing partners can follow this up in a collaborative fashion.

***3.2.3.1 The "Lease" Model*** The "lease" or "license" model is the type of agreement with the biggest up-front commitment on the side of the chemistry provider. They invest into delivery of the blinded screening compounds, scale-up support, and the full transfer of IP. The chemistry acquirer pays for each step along the way (so-called milestone payments) and in return gets the compounds that are under evaluation at his sole disposal (Fig. 3.2).

This model is, for example, particularly interesting for natural products and/or natural product fractions as it helps the acquirer to take a stepwise approach and only acquire a natural product that really displays the desired activity profile and properties. The

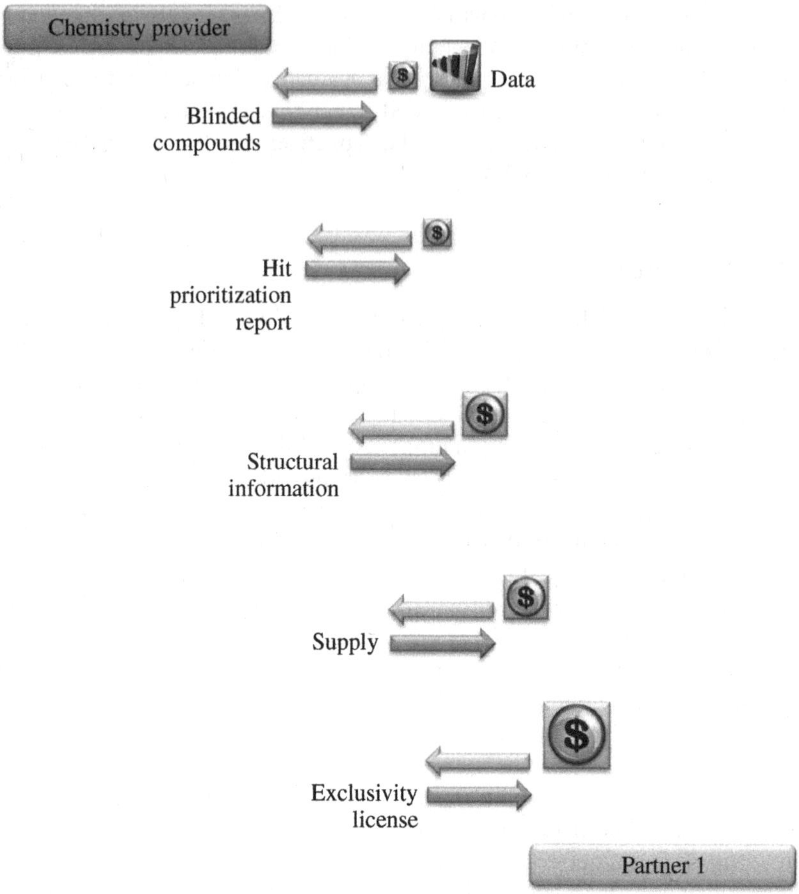


**FIGURE 3.2** Stepwise approach of a lease model, where there is full contractual commitment of the chemistry provider to support the partner in every step of the project.

providers, for example, AnalytiCon Discovery GmbH, are specialized companies that can help the client with all steps in discovering novel hits and leads from natural product screening samples.

A "lease" project starts with the blinded transfer of a set of screening compounds from the provider to the acquirer. The selection of the compound set can be done in different ways as described before. However, in the case of natural products, a specialized partner that knows the origin of the products (certain strains, plants, locations) and physicochemical parameters (e.g., polarity, molecular weight) is often in the best position to select products with the greatest chance of showing biological activity against a certain target or disease. After blinded screening, a hit prioritization report is supplied by the provider. The hit prioritization report provides blinded, Markush-type scaffolds and numbers of clusters and cluster analogues and describes whether there are any frequent hitters in the hit collection. A relatively small consideration is normally paid for these first two steps. If, based on the data and the hit prioritization report, the acquirer is interested in certain clusters or singletons, the structural information of those products can be purchased for a larger sum per compound. The compounds that are not of interest at this point can be put back into the collection for use with other clients. After disclosure of the structural information, supply of a larger amount of material along with short-term exclusivity (~3–6 months) may be acquired. The final and most costly step is then the transfer of IP of the product from the provider to the acquirer.

This step-by-step approach can be a very attractive alternative as it allows clients to enter a highly complex area of research while getting support from experts along the way and having full control over the spending of limited budgets.

With an additional refinement to their business model, AnalytiCon Discovery has recently launched another license-based model that allows their partner to tap into AnalytiCon's proprietary MacroEvolution technology for the identification of nature-inspired synthetic macrocycles. There are several phases with different exit options—the first phase giving access to the library, a macrocyclic screening set with high scaffold diversity. The partner acquires a nonexclusive license for screening. Phases II and III are then optional. Phase II includes the synthesis of exclusive or nonexclusive focused libraries, while Phase III comprises an optional medicinal chemistry hit-to-lead optimization project with AnalytiCon Discovery.

***3.2.3.2 The "Appetizer" Model*** Some chemistry providers utilize smaller, focused screening libraries with targeted chemotype design to attract business for their medicinal chemistry divisions. They do not make money by selling compound libraries, but rather offer them free of charge for pharmaceutical partners to screen. The chemistry provider is confident that hits will be picked up by the acquirer, making the small screening set an appetizer for a medicinal chemistry collaboration when biological activity surfaces. Besides having confidence in the biological target class and the chemotype design, the chemistry provider should be certain that the chemistry is not described in the literature, thus providing chemical novelty to a putative downstream project. Upon displaying biological activity, the "appetizer" compound set can jump-start a project with a ready-to-go patentable chemotype displaying some activity and selectivity.

An appetizer model usually starts with a blinded screening, including most of the steps described previously (Fig. 3.3).

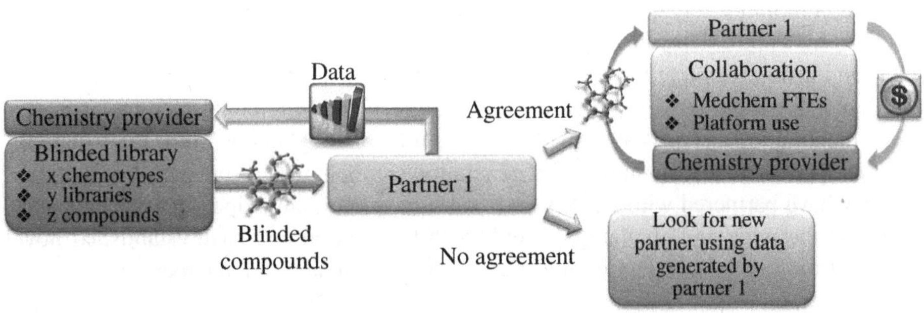


**FIGURE 3.3** The appetizer business model. Novel and blinded chemistry of a chemistry provider is exchanged for biological data from a chemistry acquirer in the first cycle. If the biological data suit the acquirer's needs, the two parties can come to an agreement about further collaboration. If this is not the case, the chemistry provider is free to use the generated data and look for an alternative partner interested in the compounds and data generated.

A material transfer agreement (MTA) is generally needed to gain access to the provider's screening compounds. This agreement type usually involves a negotiation period in which confidentiality of the data is maintained. If no further collaboration is successfully negotiated, the agreement is terminated and the chemistry provider is free to use the biological data that was generated during the term of the project. If an agreement for a follow-up contract research collaboration is successfully negotiated, the structures will be disclosed and the IP will be transferred to the partner kick-starting an optimization program.

In some cases, the chemistry service provider even prevalidates novel chemotypes in that a small set of prototype compounds is submitted to, for example, a target profiling CRO, delivering proof of concept for the underlying designs. This is a true win–win model with low risk for both parties, giving the pharmaceutical organization the opportunity to access external chemistry innovation while providing valuable biological data and potentially offering a research collaboration to the chemistry CRO.

The appetizer business model is characterized by the initially low degree of commitment needed from both parties. The client only needs to commit to screening a relatively small number of the provider's compounds in certain agreed-upon screening panels. The chemistry provider needs to be capable of shipping the small set of screening compounds to the client. There are no further commitments after the term of the agreement has expired, unless a collaboration is started.

A medicinal chemistry CRO actively pursuing this business model as part of its broader service offering is Mercachem BV. Mercachem's in-house compound library is based on privileged structures that are tailored toward pharmaceutically relevant gene families such as kinases, PDEs, epigenetic enzymes, and protein–protein interactions (PPIs). Although it is not available for purchase, under MTA, a partner can evaluate a focused subset in order to find innovative and patentable starting points for projects. Mercachem's design principles are backed by chemoinformatics-based chemical space and diversity analyses, using ChEMBL and scaffold identification and classification methods [9, 10], while otherwise tailored to the target. For example, a "retro design approach" has been used to identify deep pocket-directed scaffolds leading to several sets of novel kinase inhibitors with distinct binding kinetic signatures [11]. New HDAC targeting scaffolds were designed to achieve longer residence time by combining and extrapolating HDAC SAR described in the literature [12–14]. As a final example, in collaboration with ProQinase, an innovative approach combining screening of target family-centric compound libraries against a set of KMTs and multifactorial cellular phenotypic testing is followed to produce novel SAM- and histone-competitive inhibitors [15]. Mercachem, as a chemistry-centric CRO, considers it as essential to establish an inter-CRO network with reliable service providers complementing their own skill set with, for example, biology and pharmacology capabilities, therapeutic area know-how, or hit generation technologies, respectively. In that context, they have partnered with a DNA-encoded library company, Vipergen (Copenhagen), to address the wide and challenging field of PPI inhibition and have established novel scaffolds from the field of peptidomimetics, spiro compounds, and macrocycles.

***3.2.3.3 The "Subscription" Model*** A number of chemistry providers offer a subscription model to several acquirers to give clients full access to their screening compound collections (Fig. 3.4).

As a prerequisite for this model, there is usually a long-standing relationship between partners with sufficient faith in the quality and novelty of the products covered by the

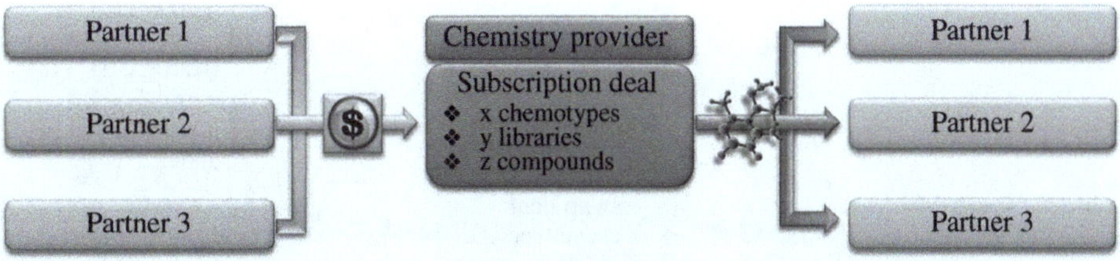


**FIGURE 3.4** Within a subscription model, several acquirers can sign up to get screening compound libraries from a chemistry provider. The up-front commitment from both the partner and the chemistry provider is high. The partner commits at the beginning of the deal to acquire the envisaged libraries, and the chemistry provider commits to the library design, synthesis, and timely delivery.

subscription deal. This is crucial because there is a financial commitment made by the client to develop the products during a set period of time.

The chemistry provider makes a commitment to deliver a predefined number of scaffolds, libraries, and compounds during the term of the agreement. Initially, the nature of the compounds produced is not disclosed. In return for commitment and trust, the acquirer is provided with unique, not widely accessible chemistry and a partner dedicated to providing high-level follow-up service and support. Typically, several clients join the same deal with a chemistry provider, but predefined limits are set to the number of participants.

A good example of library offering via a subscription model is that of BioFocus (A Charles River Company). The basic outline of their model is as described previously, but the reason to implement this model as well as its evolution over the past years is noteworthy. From 2000 until 2010, BioFocus offered nonexclusive, predesigned novel compound libraries. These were viewable and for sale to anyone via their website after agreeing to keep the information confidential. These SoftFocus libraries were mainly designed to target kinases, GPCRs, and ion channels and had an average size of approximately 1000 compounds, the libraries being considered rather large [16, 17]. In 2011, the subscription model was introduced with the main goal to enhance the chances of a limited number of customers (~10) to be able to secure IP. Based on feedback from their customers, BioFocus also started to design libraries targeting PPI and epigenetic targets and gradually reduced the size of the libraries while increasing the number of scaffolds, or modules [18]. These developments have created a more diverse and flexible offering that now covers five modules (kinases, GPCRs, ion channels, PPIs, and epigenetics), with each containing five to eight scaffolds with a maximum library size of 250 compounds. Design strategies include structure-based and ligand-based approaches as well as the use of chemogenomics, proprietary tools, and virtual (fragment) screening while using the PDB, ChEMBL, and patents to guide monomer selection.

## 3.3 LIBRARY EXCHANGE

In the field of compound libraries and HTS, pharmaceutical companies are—after a long period of keeping all secrets around setup and enhancement of screening compound decks in-house—willing to share their screening collections. There are several ways in which internal collection sharing can be realized—most often, two organizations with mutually interesting compound collections and either different or similar areas of research work together without payments involved (Fig. 3.5).

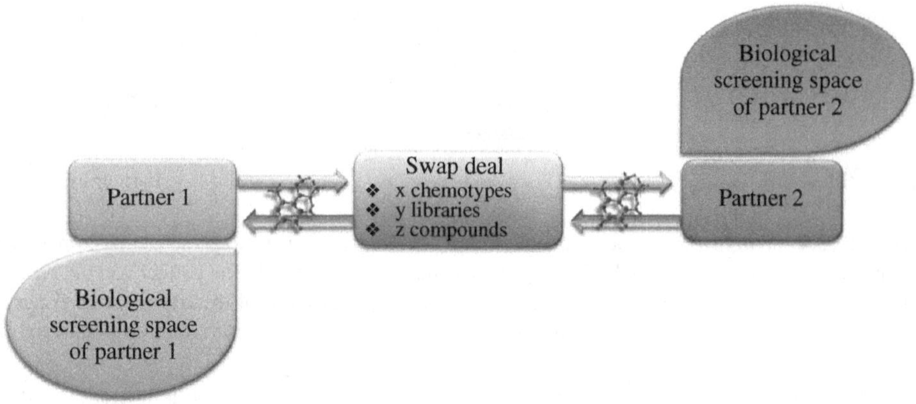


**FIGURE 3.5** Library exchange. The biological screening spaces are generally well defined and may or may not be overlapping (partners with different or similar scientific interest).

However, costs of compound testing, selection of biological targets of interest and conditions to screen them, logistics, and the preparation of the legal environment still need to be considered to set such a collaboration on a solid basis.

### 3.3.1 Partners with Different Scientific Interests

If there is different scientific interest in the two organizations, it is quite likely that compound collections, for example, of a pharmaceutical- and an agricultural-based organization contain subsets that are of mutual interest for both organization. Since the focus of one partner's collection and the way it was compiled served a totally different purpose in one company as compared to the other, it is highly likely that the subsets to be swapped fill unpopulated chemical space in both parties' existing collections.

Some prerequisites need to be fulfilled to make such deals efficient. It is of fundamental importance that the majority of the compounds are ideally proprietary; the likelihood of duplication (same type of chemical structure available in both collections) is strongly increased if the compounds were obtained through public vendors. The compound exchange partners should further define very clearly their field of interest for compound testing. Only when an unambiguous definition of two nonoverlapping fields of application can be articulated, such a deal can move forward without additional strings attached. This means that both parties agree on well-defined limitations on the use of the compounds in their specific field and for their specific purposes (e.g., the field of human health vs. animal health or pharmaceuticals vs. agrochemicals). Finally, the numbers of compounds to be exchanged, their desired purity, available amounts, formats, potential for resupply, calculated properties, etc. should be agreed.

### 3.3.2 Partners with Similar Scientific Interests

This type of partnering involves two parties with very similar research areas, for example, two pharmaceutical companies sharing their internal screening collections. The most prominent example of this approach contains a compound collection sharing agreement by AstraZeneca and Bayer presented to the public in 2013 [19]. This joint initiative, named Project Boomerang, was first established in 2010 and aimed at enabling both parties to seek chemical starting points not available in their internal collections. To estimate the potential value of a screening collaboration, the project started with a joint mutual assessment of each other's collection based on molecular data, similarity analysis, and overall comparison of the compositions of the individual compound collections. The agreed framework for this collaborative assessment included that no structures were exchanged or reviewed, that is, the partners only used each other's molecular fingerprints as blinded representation of structures. Furthermore, no reverse engineering of fingerprints was allowed to be done by the joint team. The outcome of this assessment showed a small 3.3% overlap between the two collections with 95% of this overlapping chemical space consisting of commercially available compounds. It was evident that the AstraZeneca and Bayer collections clearly access distinct parts of chemical space.

Instead of swapping the actual compound libraries, it was decided to swap biological targets (2–3 per year), sensible from a logistical point of view with both companies' collections numbering several million molecules. Each screening party provides the assay development, optimized assay conditions, and reagents necessary to execute a full screen-to-hit cascade. Once interesting hit series are found, the chemical starting points and related structures are transferred to the partner. Further progression of the discovery programs is then linked to milestone and royalty payments. This is an excellent example of two major competitors collaborating to achieve mutual benefits—something seeming unconceivable until then.

### 3.3.3 Compound Selection: Use and Potential Risks

The initial selection of compounds can be performed by exchanging and evaluating fingerprints instead of full molecular structures.

However, a big risk of selecting compounds for a library swap is the potential for IP contamination. The problem arises if the evaluating company views structures of the partnering organization that are not further pursued in the actual swap, but are later identified of interest for any other screening or optimization campaign in this company. If the evaluating party starts working on compounds that are related, similar, or even identical to those disclosed from the partnering organization, it is very difficult to prove that these were not inspired by the initial disclosure. There are several ways to mitigate this risk. For example, a noncontaminating confidentiality agreement can be signed, or a third-party expert can be hired to do the evaluation for the partner.

Another potential risk is that of patent genus overlap. This risk can be mitigated by agreement that regardless of who owns the patent, if a transferred compound falls within the scope of the partner's patent, the nonpatent holding partner will receive a nonexclusive license for its respective field of interest or purpose. The agreement, however, could also be to exchange the libraries as research tools, without transfer of patent rights and with the goal of entering new IP space quickly by preparing modified derivatives that diverge from the original starting point. Finally, it should be investigated if any of the compounds that are to be exchanged are already licensed to other partners. Sublicensing the partner's swapped compounds to a third party may no longer be possible.

## 3.4 SHARING COLLECTIONS FOR EXTERNAL SCREENING

### 3.4.1 Rationale

In addition to risk-sharing models and library exchange, the pharmaceutical industry has taken another step forward by sharing proprietary compound collections for external screening campaigns, either through precompetitive consortia models or with individual academic institutions and/or small biotech companies.

The underlying reason for exposing internal compound collections to external screening is the emerging need to identify novel biological targets for specific disease areas, where biological expertise is only found in the hands of an external partner.

The direct way for a pharmaceutical company to engage in such a model is to simply offer an exclusive in-house compound set (blinded and under MTA) to the external partner of interest with a unique biology platform or disease model. These external partners are often start-up biotechs or academic groups in need of high-quality leads lacking the resources to produce or acquire their own compound collections. If interesting biological activity is found, a follow-up partnership will be discussed according to the end point envisioned in the contract.

A related example, although unique in nature, is the announcement that Evotec, a major European CRO, will obtain Sanofi's entire compound library for screening as part of a larger collaboration between the two companies [20]. In this case, Evotec will be able to offer a huge collection of a major pharmaceutical company to third-party clients and will perform the screening. This is a unique opportunity for smaller companies or funded academic groups to obtain access to such a wide amount of chemical space. Sanofi benefits in the form of downstream royalties should a therapy reach the market.

Several consortia-like initiatives have been set up to take advantage of pharma's willingness to share chemistry. The result of these efforts is not an exclusive partnership, but rather the generation and publication of a wealth of information to advance science and drug discovery in general, often with the goal of generating tool compounds to cast light on new biological pathways. Three of these initiatives will be described below, while the IMI Lead Factory consortium is described in detail in this book in its own chapter.

### 3.4.2 Academic Drug Discovery Consortium (ADDC)

Academic institutions are responding to the paradigm shift in the pharmaceutical world, and a good example of this is the creation of the ADDC [21]. The founders of the ADDC realized that, more than ever, early drug discovery work, including target validation, assay development, HTS, and translational research, is done at academic institutions. However, the interconnection between these institutions is often not strong enough to lead to efficient drug discovery efforts.

In contrast, the ADDC creates a powerful network of academic groups with often complementary capabilities. It nurtures the individual centers by facilitating the exchange of expertise and the creation of collaborations between academic groups and pharmaceutical and biotech partners. The top three therapeutic areas represented in the ADDC are oncology, infectious diseases, and neurological disorders; research is both done on small molecules and biologics. Currently, about 80 academic research centers in the United States, Europe, and the Middle East have joined the ADDC. This initiative makes it very convenient and attractive for the pharmaceutical industry to start sharing their compound collections with the academic world and generate broad biological information around their lead compounds. AstraZeneca, for example, makes their compound library fully available to be screened by members of the consortium. This process will help the industry to identify future partners of choice in the academic world and could result in fruitful collaborations that would benefit drug discovery and potentially help patients worldwide.

### 3.4.3 EU-OPENSCREEN

EU-OPENSCREEN is a European infrastructure of open screening platforms for chemical biology [22]. The aim of EU-OPENSCREEN is to support academic research in chemical biology to develop new tool compounds for life sciences research and to ultimately identify new biological targets for drug discovery. The approach includes assay development, HTS, and optimization of an initially active molecule by medicinal chemistry. EU-OPENSCREEN has established a network of shared resources in the field of compound screening, medicinal chemistry resources, and a screening collection of commercial and proprietary compounds. This whole suite of capabilities is available to researchers in an open-access fashion. Chemists from academia and industry can submit their own compounds under an MTA in a blinded manner to enrich the open-access compound collection. All compounds are then screened against a panel of biological assays. Upon written agreement, both parties receive the unblinded results simultaneously, while publication of the data will only occur after 18 months. The data will then be accessible to the public via a central database. The initiative has already developed new library design principles, enabled the exchange of local compound libraries, and shaped joint screening projects.

The EU-OPENSCREEN framework is an ideal facilitator for a reversed compound acquisition model, in other words—the transfer of pharma screening sets to academic labs or biotechs. An internal compound set should, however, be selected with care. It should show activity in the relevant biological space, should be certain to never become a drug candidate, and should be IP protected or already published. Once those factors have been validated and a set of compounds has been selected, the pharmaceutical company can submit the compounds to EU-OPENSCREEN and obtain a wealth of chemical biology data from an international group of research centers.

### 3.4.4 NIH Roadmap

Like the EU-OPENSCREEN platform, the NIH Roadmap Molecular Libraries and Imaging Initiative aims to develop small molecule probe compounds for research purposes. More specifically, the probes are designed to open new avenues to studying the role of genes, cells, and biochemical pathways in health and disease. The NIH Roadmap Initiative governs the MLSCN consisting of ten labs offering biomedical researchers access to assay automation and HTS and medicinal chemistry expertise for the development of bioactive probe compounds. Moreover, information about the biological activity of compounds in the Small Molecule Repository (SMR), potentially useful as starting points for research projects, is provided. The SMR is the actual small molecule screening deck, containing both compounds of synthetic and biological (e.g., natural product) origin. Outside researchers, including pharmaceutical organizations, can submit small molecules to the SMR under an MTA. This transfer is generally at no cost, and the provider of the material agrees that the MLSCN Project Team will assume further management of the NIH-funded activities, executed by the MLSCN. These activities include the screening of the small molecules in HTS-ready assays solicited from the research community and follow-up hit optimization chemistry to develop useful probe compounds for the biological targets or phenotypes studied. The final step in this process is the data approval and data release through the public database PubChem to both the public and private scientific community for further use in studying biology and disease. The NIH Roadmap is therefore an effective tool for owners of unique small molecule compound collections to broadly profile and further chemically develop probe compounds in a cost-efficient manner while contributing to the knowledge of the general scientific community. It is particularly noteworthy for encouraging and sponsoring academic chemistry groups to submit libraries, thereby tapping into the chemical creativity of the academic sector.

More recently, the Molecular Libraries Probe Centers Network (MLPCN) was launched [23]. This network specifically focuses on probe discovery in an academic setting whereby a wide range of assay technologies will be applied to primary and secondary screening campaigns. Prestigious centers like the Broad Institute Comprehensive Screening Center and the Sanford-Burnham Center for Chemical Genomics are among the nine groups currently present in the network.

## 3.5 CONCLUSION

Innovative compound design will continue to fuel small molecule drug discovery in the future. In order to maintain a sustainable source of new chemistries to safeguard successful drug discovery, alternative and flexible business models for compound acquisition and sharing will be critical, not only for the benefit of the pharmaceutical industry but to allow universities and public institutions access to quality starting points. When an innovative chemistry provider obtains a fair ROI by getting compounds evaluated in as many assays as possible and is rewarded for success, the business will last and quality will improve. If low-hurdle access to innovative chemistries is enabled by implementing alternative business models, it will boost the discovery of new leads and thereby the whole discovery process. The reversed approach, where a pharmaceutical company shares proprietary compounds with the outside world for broad profiling, will greatly advance the science of drug discovery as a whole by making novel data available to the community. The new models that have sprung up in the last few years to meet these challenges amount to a revolution in the practice of drug discovery.

## ACKNOWLEDGMENTS

I thank colleagues Drs. Tara Mirzadegan, Renee Desjarlais, Edgar Jacoby, Christine Cully, Michelle Browner, Chris Flores, and Kristof Van Emelen (Janssen Research & Development) and external business partners Drs. Frank Leemhuis, Gerhard Müller (Mercachem BV), Zhenmin He (Wuxi AppTec), Richard Hill, Jackie Macritchie (BioFocus), Dietmar Wolf (AnalytiCon GmbH), Mark Spyvee (SAI Life Sciences, Ltd.),

Tom Smart and Peter Pallai (BioBlocks, Inc), Niko Gubernator (eMolecules, Inc.), and Sylvie Sakata (Pfizer, Inc.) for various fruitful discussions on the topic. The organizers of the International Conference series on Intelligent Compound Libraries and in particular Mr. Amir Hudson-Nahum (IQPC GmbH) are acknowledged for providing a platform to the screening compound library community, fostering creative thinking and collaboration. Finally, I thank Drs. Werngard Czechtizky and Peter Hamley for editing this chapter.

All figures are original and were created by the author using Microsoft PowerPoint 2010.

## REFERENCES

[1] Abou-Gharbia M, Childers WE (2014). Discovery of innovative therapeutics: today's realities and tomorrow's vision. 2. Pharma's challenges and their commitment to innovation. *J. Med. Chem.* **57**:5525–5553.

[2] Tapon F, Cadsby CB (1996). The optimal organization of research: evidence from eight case studies of pharmaceutical firms. *J. Econ. Behav. Organ.* **31**:381–399.

[3] Dixon J, Lawton G, Machin P (2009). Vertical disintegration: a strategy for pharmaceutical businesses in 2009? *Nat. Rev. Drug Discov.* **8**:435.

[4] Zachary GP (2008). Thinking outside the company's box. *The New York Times*, March 30, 2008. [Accessed May 26, 2015]. Available from: http://www.nytimes.com/2008/03/30/business/30ping.html

[5] Paul SM, Mytelka DS, Dunwiddie CT, Persinger CC, Munos BH, Lindborg SR, Schacht AL (2010). How to improve R&D productivity: the pharmaceutical industry's grand challenge. *Nat. Rev. Drug Discov.* **9**:203–214.

[6] Hack MD, Rassokhin DN, Buyck C, Seierstad M, Skalkin A, ten Holte P, Jones TK, Mirzadegan T, Agrafiotis DK (2011). Library enhancement through the wisdom of crowds. *J. Chem. Inf. Model.* **51**:3275–3286.

[7] Huryn DM (2013). Drug discovery in an academic setting: playing to the strengths. *ACS Med. Chem. Lett.* **4**:313–315.

[8] Schachter B (2012). Partnering with the professor. *Nat. Biotechnol.* **30**:944–952.

[9] Gaulton A, Bellis L , Bento AP, Chambers J, Davies M, Hersey A, Light Y, McGlinchey S, Michalovich D, Al-Lazikani B, Overington JP (2012). ChEMBL: a large-scale bioactivity database for drug discovery. *Nucleic Acids Res.* **40**:D1100–D1107.

[10] Hu Y, Bajorath J (2013). Systematic identification of scaffolds representing compounds active against individual targets and single or multiple target families. *J. Chem. Inf. Model.* **53**: 312–326.

[11] Müller G (2013). Charting new bioactive space for drug discovery: CROs as innovation source for pharma and biotech companies. 9th International Conference Compound Libraries 2013, Berlin, October 28–29, 2013.

[12] ten Holte P, Van Emelen K, Janicot M, Fong PC, de Bono JS, Arts J (2007). HDAC inhibition in cancer therapy: an increasingly intriguing tale of chemistry, biology and clinical benefit. *Topics Med. Chem.* **1**:293–331.

[13] Chou CJ, Herman D, Gottesfeld JM (2008). Pimelic diphenylamide 106 is a slow, tight-binding inhibitor of class I histone deacetylases. *J. Biol. Chem.* **283**:35402–35409.

[14] Thaler F (2012). Current trends in the development of histone deacetylase inhibitors: a review of recent patent applications. *Pharm. Pat. Anal.* **1**:75–90.

[15] Basavapathruni A, Jin L, Daigle SR, Majer CRA, Therkelsen CA, Wigle TJ, Kuntz KW, Chesworth R, Pollock RM, Scott MP, Moyer MP, Richon VM, Copeland RA, Olhava EJ (2012). Conformational adaptation drives potent, selective and durable inhibition of the human protein methyltransferase DOT1L. *Chem. Biol. Drug Des.* **80**:971–980.

[16] Harris CJ, Hill RD, Sheppard DW, Slater MJ, Stouten PFW (2011). The design and application of target-focused compound libraries. *Combin. Chem. High Throughput Screen.* **14**: 521–531.

[17] Crossley R, Macritchie JA, Slater MJ (2011). Thematic Analysis™: a chemogenomic approach to GPCR drug discovery. *Curr. Top. Med. Chem.* **11**:1925–1943.

[18] Sheppard DW, Lipkin MJ, Harris CJ, Catana C, Stouten PFW (2014). Strategies for small molecule library design. *Curr. Pharm. Des.* **20**:3314–3322.

[19] Rich K, Kogej T, Kalthof B, Schamberger J (2013). The AstraZeneca-Bayer alliance—evolution of a strategic partnership between pharma for a sustainable long-term relationship. 9th International Conference Compound Libraries 2013, Berlin, October 28–29, 2013.

[20] Evotec (2014). Evotec and Sanofi enter into exclusive negotiations for a major multi-component strategic collaboration [Press release]. [Accessed December 4, 2014]. Available from: http://www.evotec.com

[21] Slusher BS, Conn PJ, Frye S, Glicksman M, Arkin M (2013). Bringing together the academic drug discovery community. *Nat. Rev. Drug Discov.* **12**:811–812.

[22] Horvath D, Lisurek M, Rupp B, Kühne R, Specker E, von Kries J, Rognan D, Andersson CD, Almqvist F, Elofsson M, Enqvist P-A, Gustavsson A-L, Remez N, Mestres J, Marcou G, Varnek A, Hibert M, Quintana J, Frank R (2014). Design of a general-purpose European compound screening library for EU-OPENSCREEN. *ChemMedChem* **9**:2309–2326.

[23] Roy A, McDonald PR, Sittampalam S, Chaguturu R (2010). Open access high throughput drug discovery in the public domain: a Mount Everest in the making. *Curr. Pharm. Biotechnol.* **7**:764–778.

# PART II

# EXPLORING BIOLOGICAL SPACE: ACCESS TO NEW CHEMISTRIES

# 4

# NEW ADVANCES IN DIVERSITY-ORIENTED SYNTHESIS

WARREN R. J. D. GALLOWAY, JAMIE E. STOKES, and DAVID R. SPRING
*University of Cambridge, Cambridge, UK*

## 4.1 INTRODUCTION: SMALL MOLECULES AND BIOLOGY

Humanity has a significant dependence on small molecular mass chemical entities, so-called small molecules [1]. The use of small molecules to selectively modulate biological systems represents the basis for medicinal chemistry (wherein molecules are used to chemically modify disease states) and underpins the field of chemical biology (wherein molecules are used as "probes" to investigate biological processes) [1–3]. However, the discovery of new biologically active small molecules represents a significant challenge [4]. In cases where modulators of a specific biological molecule or family of molecules are sought, it is often possible to rationally design (or rationally select) small molecule binding partners using information about the structure of the target or a known native ligand [2, 4–6]. However, there are many situations in which a "rational" design or selection process is not appropriate, desired, or indeed possible. For example, it may be the case that the precise nature of the biological target is unknown (e.g., a phenotypic screen) or that no natural ligands have been identified or that a novel mode of binding to a particular target is sought [2, 4–6]. Such scenarios necessitate a different approach, namely, the screening of collections (libraries) of small molecules in order to identify those with the desired properties [4, 7]. Clearly, the molecular composition of such libraries (i.e., the nature of the compounds present) is a crucial consideration [2, 4, 8].

Broadly speaking, the molecules comprising such screening collections may be obtained from two main sources: natural (i.e., natural products) or nonnatural (chemical synthesis) [4]. Nature undoubtedly represents an extremely important source of biologically active

*Small Molecule Medicinal Chemistry: Strategies and Technologies*, First Edition.
Edited by Werngard Czechtizky and Peter Hamley.

small molecules [9, 10]. Natural products have been used medicinally for millennia and still provide many lead compounds and drugs today [4, 11]. However, there are well-documented drawbacks associated with the use of screening libraries consisting solely of natural products [2, 9, 10, 12]. Thus, deliberate chemical synthesis represents an important alternative source of small molecule screening libraries [2, 4].

## 4.2 THE NEED FOR STRUCTURAL DIVERSITY IN SYNTHETIC SMALL MOLECULE SCREENING COLLECTIONS

Between the late 1980s and mid-1990s, a "brute-force" approach toward small molecule library synthesis was typically adopted, enabled by the development of combinatorial chemistry-type methods that allow extremely large compound collections (literally millions of compounds in some cases) to be efficiently produced [6]. Such libraries could then be rapidly screened for biological activity using high-throughput methods. It was assumed that such collections would yield a plethora of drug leads, simply as a consequence of the sheer volume of molecules examined [6]. However, the expected surge in productivity did not materialize; indeed, libraries of this sort have had limited success in the discovery of new small molecules with useful biological activity [13]. This disappointing degree of productivity has been largely ascribed to the relative lack of structural diversity within the libraries, that is, a lack of structural variation between the compounds [6, 14]. Put simply, these libraries were comprised of very large numbers of molecules, but they all had very similar structures. Why should this be an issue? Biological macromolecules are, on a molecule scale, large three-dimensional (3D) environments with certain defined potential binding regions [4]. A given biological macromolecule will therefore only interact with those small molecules that have a complementary 3D binding surface; that is, nature "sees" small molecules as 3D surfaces of chemical information [2, 4, 15, 16]. Thus, the biological activity of any given small molecule is intrinsically related to its 3D shape [2, 17]. This in turn is directly controlled by its molecular structure [2, 6]. The more structural diversity within a small molecule library, the greater the range of distinct molecular shapes present; consequently, the library as a whole should display a broader range of biological activities [2, 17, 18]. It is now widely acknowledged that the structural diversity of a small molecule library plays a crucial role in determining its success (in terms of the hit frequency) in any screening process [2, 19, 20]. Indeed, the overall functional (biological) diversity of a library is directly correlated with its structural diversity, which in turn is proportional to the amount of chemical space that the library accesses [2, 13, 21]. Clearly, high functional diversity is of value in "unbiased" screens where the precise nature of the biological target is unknown (e.g., a phenotypic screen) [2]. Furthermore, the presence of multiple structural classes within a molecular library that is being screened against a single target also increases the probability of discovering a molecule capable of binding to that target in a novel (previously undefined) fashion [4, 22].

There are four principal components of structural diversity typically identified in the literature [2]: (i) appendage (building-block) diversity, variation in structural moieties around a common scaffold; (ii) functional group diversity, variation in the functional groups present; (iii) stereochemical diversity, variation in the orientation of potential macromolecule-interacting elements; and (iv) scaffold (skeletal) diversity, presence of many distinct molecular scaffolds (scaffold in this context can be taken to mean the core rigidifying structural feature of a molecule).

Of these, it is the scaffold diversity that has the largest impact upon library functional diversity [2]. The shape space coverage of any compound set stems mainly from the nature and geometries of the central scaffolds (with the peripheral substituents being of minor importance). Substantial shape space coverage (i.e., a high level of molecular shape diversity) is known to be correlated with broad biological activity [2, 17, 18]. Indeed, small multiple-scaffold libraries are widely regarded as being superior to large single-scaffold libraries in terms of biorelevant diversity [2, 17, 18]. The scaffold diversity of a library can be considered as a surrogate measure for shape diversity and thus overall functional diversity [2].

Structural complexity is also generally regarded as a desirable characteristic in the small molecules comprising screening collections. It has been argued that structurally complex molecules are more likely to exhibit selective interactions with biological macromolecules [2, 23].

The combinatorial-type libraries of the late 1980s to mid-1990s were typically generated by combining a number of building blocks in different ways, using the same synthetic methods [4]. This approach usually resulted in the variation of substituents around a common scaffold; thus, such libraries typically possessed low levels of scaffold diversity, which goes someway to explaining their aforementioned poor performances in many biological screens [4, 13]. Recent years have witnessed a growing realization of the importance of structural (scaffold) diversity in small molecule screening libraries [2]. However, a lack of scaffold diversity is still apparent in many of the synthetic collections that are commercially available or employed by pharmaceutical companies [24]. For example, it was recently reported that 83% of the core ring scaffolds found in natural products are unrepresented among commercially available compound collections [25]. The continuing decline in drug discovery successes can be attributed, at least in part, to this deficiency in current compound collections [2, 6, 13]. Overall, the universe of organic chemistry is dominated by an astonishingly small percentage of molecular scaffolds [6, 24]. An illustration of this point is provided by a recent study of all known cyclic molecules, which showed that a large percentage of the compounds were represented by only a small percentage of scaffolds (0.25% of the molecular scaffolds were found in 50% of known compounds) [26]. This scenario demonstrates a clear need for novel or atypical molecular scaffolds in small molecule screening libraries [24].

The lack of scaffold diversity, and thus functional diversity, in many typical screening sets can be partly attributed to the nature of medicinal chemistry research over the course of the last few decades, which has typically focused upon a limited set of biological targets [6]. Indeed, approximately 29% of all drugs are enzyme modulators, and nearly 36% act upon G-protein-coupled receptors [19, 27]. Consequently, typical commercially available and proprietary compound libraries are predominately comprised of compounds that satisfy predefined criteria for the modulation of such "traditional" targets (e.g., the Lipinski "rule of 5" criteria for orally bioavailable drugs) [2, 22, 28], that is, what can be regarded as traditional "drug-like" molecules, which are typically relatively "flat" and structurally simple, being based around aromatic core structures and containing a high proportion of $sp^2$-hybridized carbon atoms and few stereogenic centers [23, 29, 30]. Unsurprisingly, libraries of this sort are heavily weighted toward biologically active chemical space (i.e., the chemical space accessed by known biologically active molecules) [2]. By definition, this is a fruitful region of chemical space for the discovery of biologically useful molecules. However, there are large swathes of potentially biologically relevant chemical space underexplored by such libraries [2, 13]. Compounds in such areas may display activity against a chosen biological target through novel, even unexpected modes of action [2, 13]. Furthermore, the exploration of uncharted regions of chemical space is vitally important in the context of intellectual property considerations and patentability, as it is likely that the low-hanging fruit within the boundaries of known bioactive chemical space have already been picked [2, 31].

Another drawback associated with the aforementioned structural bias in typical synthetic small molecule screening sets is their lack of efficacy against "underexploited" drug targets. Overall, there are only approximately 500 distinct targets of the current pharmacopoeia [29, 32]. It is widely acknowledged that this represents only a tiny fraction of the potential targets that could impact on human health [2, 6, 29]. There are many biological molecules and processes that bear little resemblance to the molecular drug targets exploited in present-day drug therapy (examples include transcription factors and processes such as protein–protein interactions and protein–DNA interactions). These targets and processes have historically been considered "undruggable," that is, considered difficult, if not impossible, to modulate using small molecules [2, 6, 29, 33]. However, recent years have witnessed a growing realization that such targets and processes are indeed tractable to small molecule modulation; it is simply that the candidate molecules comprising typical commercially available and proprietary compound collections are not well suited to the task (i.e., they lack the necessary structural features) [6, 25, 29, 33]. Many of these classically "undruggable" biological macromolecules and processes are now validated targets for the treatment of a variety of human diseases, offering exciting (and largely untapped) therapeutic potential [2, 6, 29, 33]. However, in order to address these targets, new small molecule collections that span regions of bioactive chemical space not accessed by more "traditional" compound libraries are required [2, 6].

## 4.3 DIVERSITY-ORIENTED SYNTHESIS OF NEW STRUCTURALLY DIVERSE COMPOUND COLLECTIONS

### 4.3.1 General Principles of Diversity-Oriented Synthesis

Diversity-oriented synthesis (DOS) can be considered as a concept (or indeed a guiding principle [34]) in small molecule library synthesis that was born out of a desire to address the problems associated with the use of natural products and "traditional" commercially available compound collections in biological screens [2]. The aim of a DOS campaign is to deliberately generate a structurally, and thus functionally, diverse collection of structurally complex small molecules that efficiently interrogates wide regions of biologically active chemical space (i.e., with a relatively small number of total compounds per library), including known and previously "untapped" areas [2]. The molecules comprising such collections should ideally be nonnatural (i.e., structurally novel), but natural product-like in terms of their capability to interact with, and thus modulate biological systems (i.e., they should be structurally complex and rich in functionality and 3D structural features, with a higher level of saturation compared to more "typical" drug-like compounds) [2, 23, 35, 36]. DOS libraries are usually created without a particular target in mind. Instead, collections of small molecules with disparate biological properties are desired; in principle, the screening of such a collection should provide hits against a broad range of biological targets with increased frequency and decreased cost relative to less diverse libraries [8, 29, 34, 36, 37].

The term DOS was first used by Schreiber and coworkers in two seminal publications in 2000 [36, 37]. These reports highlighted the potential value of small molecule libraries with high levels of structural diversity and used the term "DOS" to describe the concept of synthesizing such collections in a deliberate fashion. However, it is reasonable to argue that many of the ideas behind DOS existed for some time before these reports [4, 8]. In 1997, Spaller et al. published a review article [38] in which they suggested that combinatorial libraries may be considered to fall into two classes: (i) "focused" libraries, where a number of similar compounds each based on the same privileged structure are generated with a specific, known biological target in mind; and (ii) "prospecting" libraries, where an entirely novel lead compound is desired, so the aim is to prepare and screen a large number of structurally varied molecules in an attempt to find a lead with a novel mode of action [4, 8]. The syntheses of these so-called "prospecting" libraries can conceivably be considered to be early examples of DOS, in that the goal was to achieve high levels of structural diversity and thus bioactive chemical space coverage, as is the case in contemporary DOS campaigns [4, 8]. However, it was not until the term "DOS" was first coined in 2000 that the strategies and ideas underpinning modern diversity-driven synthesis began to become more formalized, and the field of DOS has evolved rapidly since [4, 8].

### 4.3.2 Achieving Structural Diversity: The Importance of Scaffold Diversity

Traditionally, DOS campaigns have sought to address the four principal types of structural diversity mentioned previously [2]. As alluded to earlier, one can consider there to be a hierarchy within these diversity aspects that is based on both their relative perceived "value" and also synthetic tractability [4]. Appendage diversity in a compound set is generally accepted as the easiest to achieve but is of the least importance in terms of generating functional diversity across the compound set [4]. On the other hand, it is widely accepted that the incorporation of scaffold diversity into a compound collection is the most important in terms of producing functional diversity (vide supra) [2, 4]. Indeed, there are many examples of DOS strategies that focus almost entirely upon the generation of diverse molecular scaffolds [4, 39, 40]. The efficient generation of multiple molecular scaffolds in a DOS campaign is generally viewed as crucial in order to achieve high levels of library structural diversity (and thus large chemical space coverage) in an *efficient* manner (i.e., with the fewest number of molecules) [2]. However, this presents a formidable synthetic challenge.

### 4.3.3 Synthetic Principles in DOS

Broadly speaking, there are two basic approaches for the generation of scaffold diversity in a DOS context (Fig. 4.1). The first is a "branching" approach, where divergent reactions are carried out on a substrate to afford a number of different compounds with distinct molecular scaffolds. The second is a "folding" approach, where strategically positioned functional groups are reacted together ("paired") in an intramolecular fashion and so "fold" a substrate into distinct molecular scaffolds. This approach typically involves the

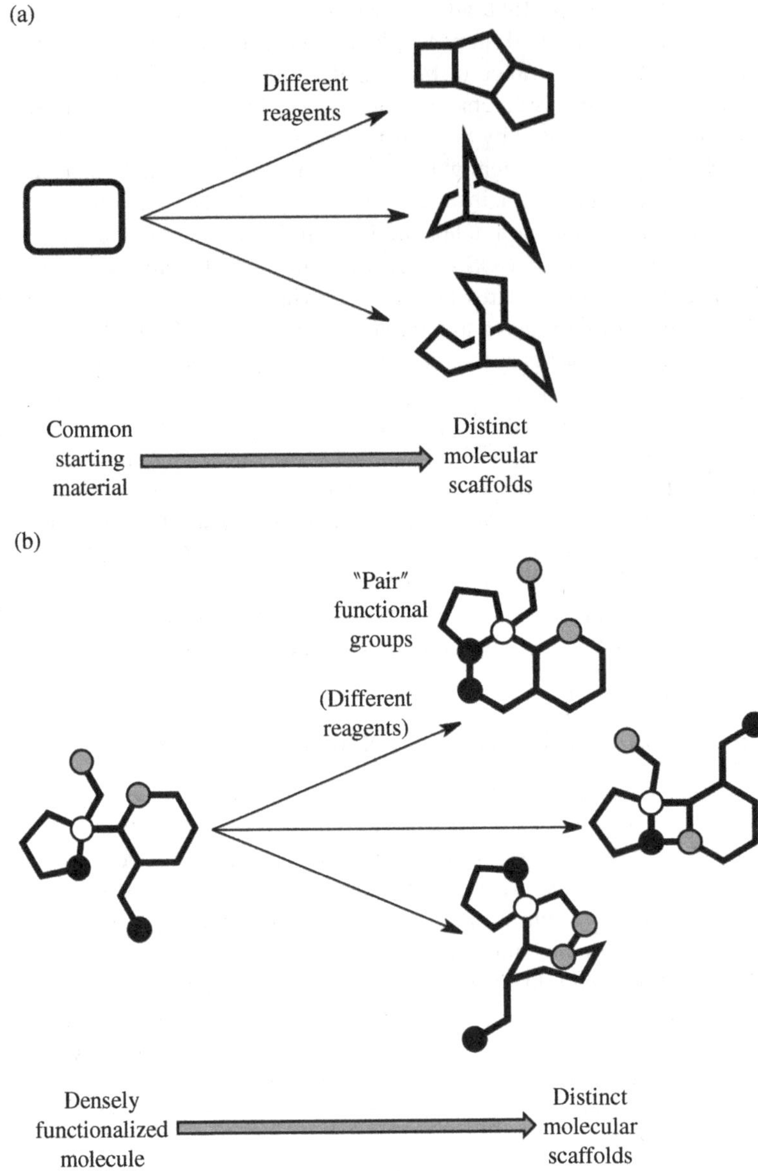


**FIGURE 4.1** General approaches for generating scaffold diversity in DOS. (a) The "branching" approach. (b) The "folding" approach.

use of either a densely functionalized molecule, where different functional groups can be reacted together under different reaction conditions to yield multiple scaffolds from one substrate, or different starting materials and common reaction conditions, such that each starting material will furnish a product based around a different scaffold. These two approaches toward scaffold diversification are not orthogonal, and most modern DOS strategies will incorporate aspects of both [2, 4, 19]. Other diversity aspects (appendage, functional group, and stereochemical) can be introduced into the compound libraries through variation in the starting materials and/or reagents used. Stereochemical diversity can also be incorporated if new stereocenters are created in any reactions [2, 35].

In 2008, Nielsen and Schreiber identified a common synthetic algorithm that was present in many DOS pathways, referred to as the build/couple/pair (B/C/P) three-phase strategy (Fig. 4.2) [31]. The "build" phase involves the synthesis of starting materials (or building blocks), which are ideally chiral. These building blocks are then coupled together in the "pair" phase to produce densely functionalized molecules; this process provides the basis for the introduction of stereochemical diversity. The "pair" stage is a folding-type process that provides the basis for the introduction of scaffold diversity; intramolecular reactions that join ("pair") functional groups incorporated in the "build" phase are performed, usually under reagent control, to generate diverse molecular scaffolds. The B/C/P algorithm subsequently found widespread application in a DOS context and has been used (often identified explicitly) as the basis for a large number of DOS pathways; indeed, it is arguably the most commonly used synthetic strategy in DOS.

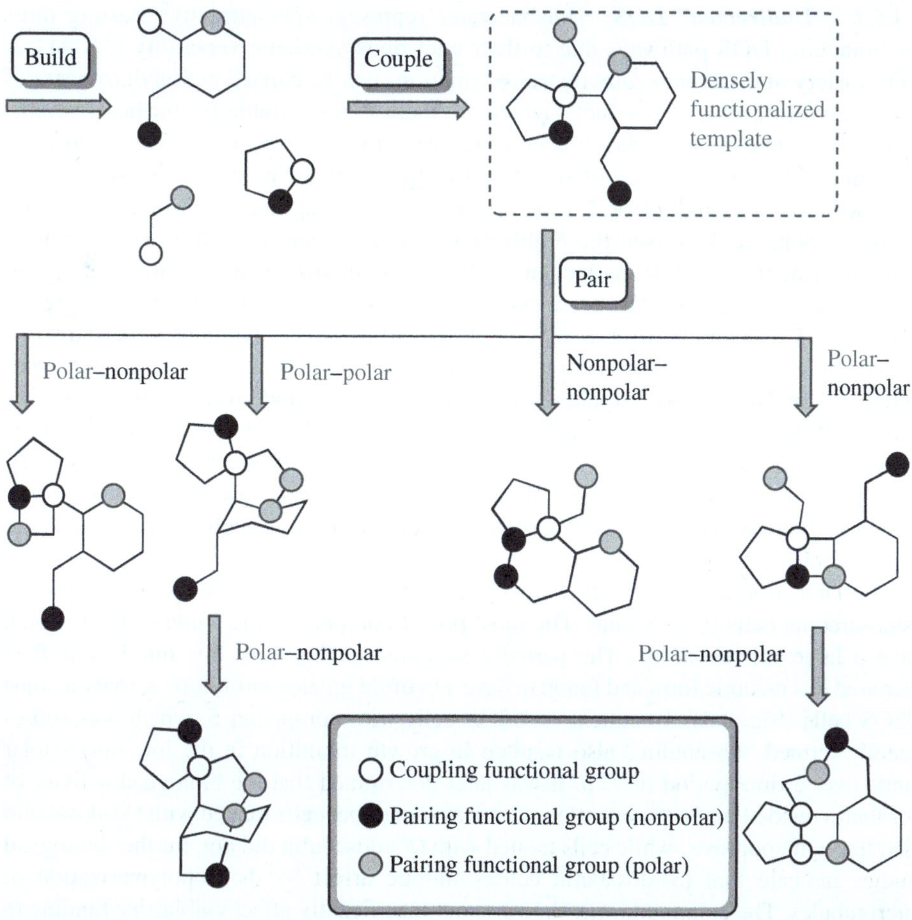


**FIGURE 4.2** The build/couple/pair (B/C/P) synthetic algorithm, a common feature in DOS pathways.

### 4.3.4 Scaffold Diversity and Molecular Type

In a broad sense, most DOS programs can be classified into two categories: (i) "undirected" approaches, which do not target a particular "class" of molecule, or (ii) "directed" approaches, which do target a predefined "class" of molecule (e.g., macrocycles) and thus generate libraries composed of molecules that typically contain certain structural motifs or features (e.g., a ring structure of 12 or more atoms in the case of macrocycles). DOS campaigns that fall into this latter category can arguably be described as being slightly more "focused" in nature, but nevertheless, the generation of high levels of diversity (*including* scaffold) is still sought.

In the remainder of this chapter, recent representative examples of DOS campaigns from each of these categories are discussed, with examples taken from our own research and from the wider community. These examples illustrate the current state of the art in DOS.

### 4.3.5 Examples of DOS Campaigns

***4.3.5.1 "Undirected" DOS*** Diazoacetates represent very attractive starting units for branching DOS pathways due to their enormous synthetic versatility [35, 41]. A wide variety of different synthetic transformations can be carried out on diazoacetates to yield scaffold-diverse products, which are themselves suitable for further diversification [35]. Ibbeson et al. have recently reported two new branching DOS pathways (Schemes 4.1 and 4.2), each of which utilized a different, readily accessible phenyl diazoacetate compound (**1** or **2**, respectively) as a starting material [42]. The second pathway (Scheme 4.2) used the highly functionalized derivative **3** as a key branch-point intermediate, whose molecular scaffold was modified in a regioselective and multidirectional fashion. The products from both pathways were combined to yield a library of 35 structurally diverse and complex small molecules based around 10 distinct molecular scaffolds. The DOS compounds displayed a high $sp^3$ content; indeed, computational analyses indicated that the DOS library had a good level of shape diversity, with scaffolds that are drug-like but are more 3D in character than drugs. This is notable, as the lack of three-dimensionality (low $sp^3$ carbon count) is considered to be one of the major factors why typical small molecule screening collections have performed so poorly when screened against challenging targets such as protein–protein interactions [23, 42].

The DOS library compounds were screened for antimitotic activity against human osteosarcoma cells (U2OS line). The most potent compound was found to be **4**, which gave a large mitotic arrest. The partially saturated analogue of this hit, **5**, was then prepared in a racemic form and found to have a twofold greater antimitotic activity against U2OS cells (Fig. 4.3). Treatment of U2OS cells with compound **5**, which was subsequently termed "dosabulin," also resulted in growth inhibition in the low micromolar range over a time period of 72 h. It was later determined that the biological activity of dosabulin resided in the (*S*)-enantiomer; it was found that cells treated with (*S*)-dosabulin died through apoptosis, while cells treated with (*R*)-dosabulin did not. Further biological studies indicate that (*S*)-dosabulin causes mitotic arrest by the depolymerization of microtubules. The compound was shown to not significantly affect vinblastine binding to tubulin; instead, evidence suggests binding to a site on tubulin vicinal or allosteric to colchicine [42].

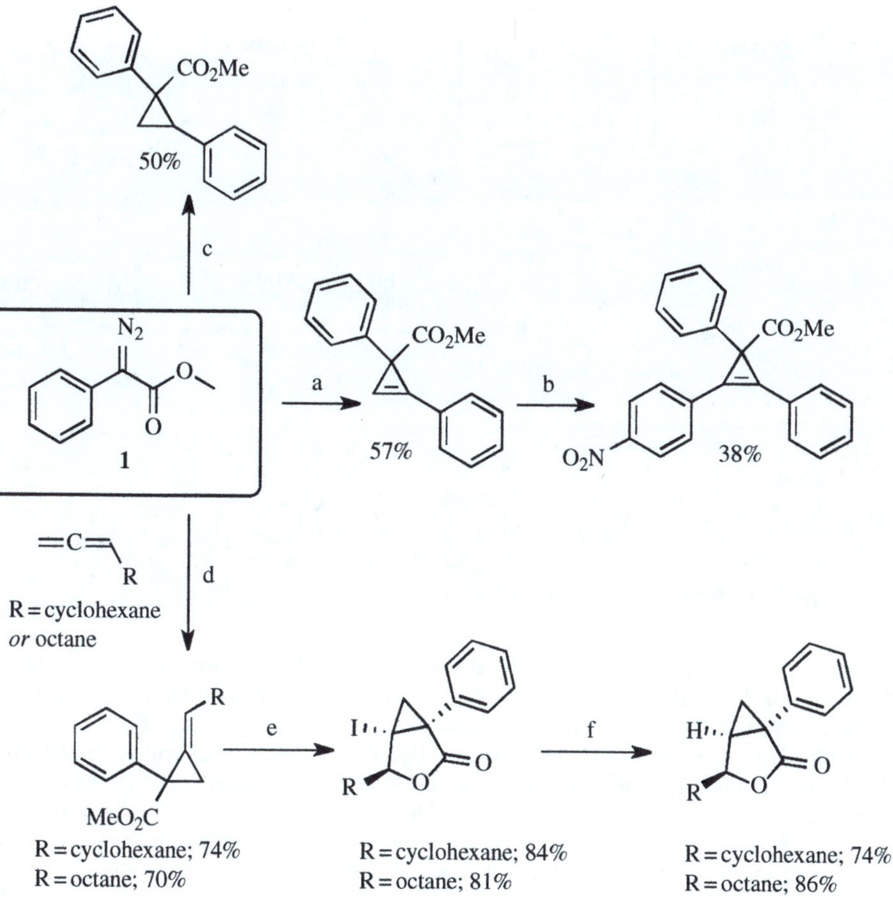


**SCHEME 4.1** DOS library subset generated from compound **1**. Conditions: (a) phenylacetylene, $Rh_2(OAc)_4$ (1 mol%), $CH_2Cl_2$; (b) *p*-nitroiodobenzene, $Pd(OAc)_2$ (10 mol%), $K_2CO_3$, DMF; (c) styrene, $Rh_2(OAc)_4$ (1 mol%), $CH_2Cl_2$; (d) allene, $Rh_2(OAc)_4$ (1 mol%), $CH_2Cl_2$; (e) *N*-iodosuccinimide, MeCN–$H_2O$ (2:1), 50°C; (f) $Bu_3SnH$, AIBN, PhH, 80°C.

Robbins et al. have developed a novel and powerful branching DOS strategy (Scheme 4.3) [43]. Their approach involved the use of symmetrical linear ketone **6** as a starting unit, which could be transformed into 12 distinct scaffolds through an ingenious combination of two-directional synthesis and tandem reactions. The DOS involved the transformation of the central ketone group into various nucleophilic functionalities, which were then able to react with the tethered unsaturated ester groups at the chain termini. In some cases, essentially symmetrical products were produced (e.g., compound **7**). In other cases, tandem reaction sequences were developed in which reaction of the central functionality with one chain end transformed the central functionality such that it was then able to react with the other chain end under a different mode of reactivity, thus creating nonsymmetrical 3D molecules from the simple linear substrate (e.g., compound **8**). The authors suggested that the linear ketone could be considered as a "molecular rope," which was then subjected to 12 separate tandem reactions to "tie the rope in knots" and so

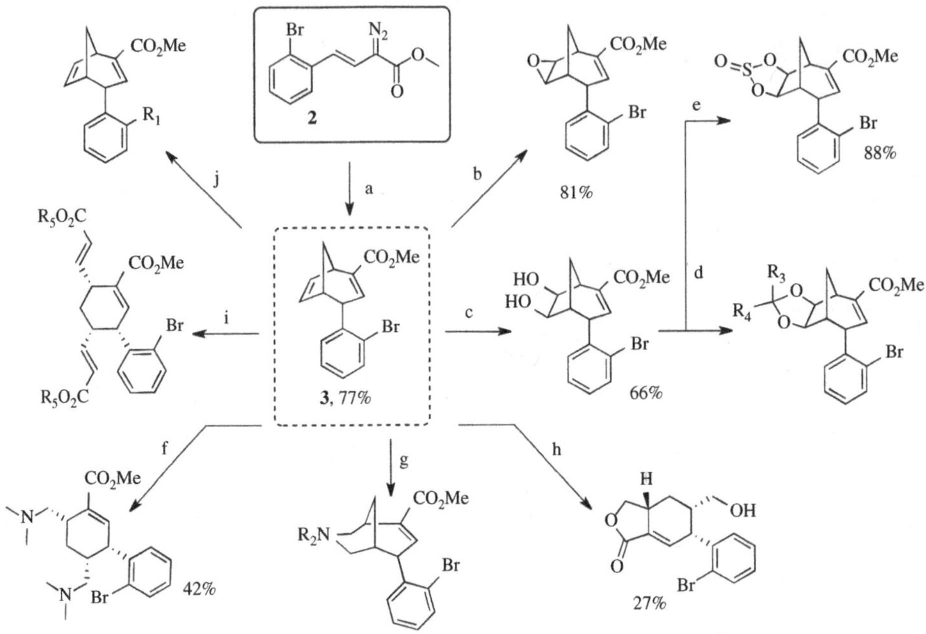

**SCHEME 4.2** DOS library subset generated from compound **2**. Conditions: (a) cyclopentadiene, $Rh_2(OAc)_4$ (1 mol%), $CH_2Cl_2$; (b) mCPBA, $CH_2Cl_2$; (c) $OsO_4$ (2.5 mol%), NMO, acetone/$H_2O$ (9:1); (d) aldehyde/ketone, CSA (10 mol%), 3Å molecular sieves, $CH_2Cl_2$; (e) $SOCl_2$, $CH_2Cl_2$; (f) 2,6-lutidine, NMO, $OsO_4$ (2.5 mol%), $PhI(OAc)_2$, acetone/$H_2O$ (10:1), then dimethylamine, $NaBH(OAc)_3$, $CH_2Cl_2$; (g) 2,6-lutidine, NMO, $OsO_4$ (2.5 mol%), $PhI(OAc)_2$, acetone/$H_2O$ (10:1), then primary amine, $NaBH(OAc)_3$, $CH_2Cl_2$; (h) 2,6-lutidine, NMO, $OsO_4$ (2.5 mol%), $PhI(OAc)_2$, acetone/$H_2O$ (10:1), then $NaBH_4$, MeOH; (i) alkene, Hoveyda–Grubbs (II) catalyst (10 mol%), ethylene, PhMe, 100°C; (j) $Pd(OAc)_2$ (10 mol%), boronic acid, $PPh_3$ (15 mol%), 2N $K_2CO_3$, PhMe, 90°C.

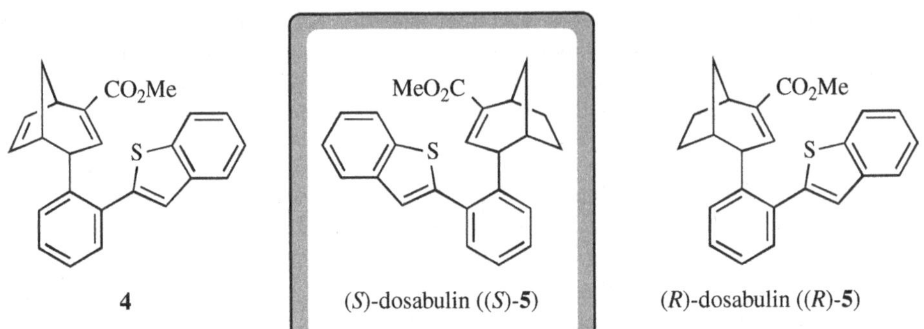


**FIGURE 4.3** Compounds with potent antimitotic activity identified through the DOS campaign of Ibbeson et al.

produce a range of molecular scaffolds. This 12-fold branching DOS strategy enabled the very direct and efficient formation of 3D molecular architectures with natural product-like features (polycyclic ring systems, multiple stereocenters, and a diverse range of functional groups) from a single substrate in an impressive 1.25 steps per scaffold on average.

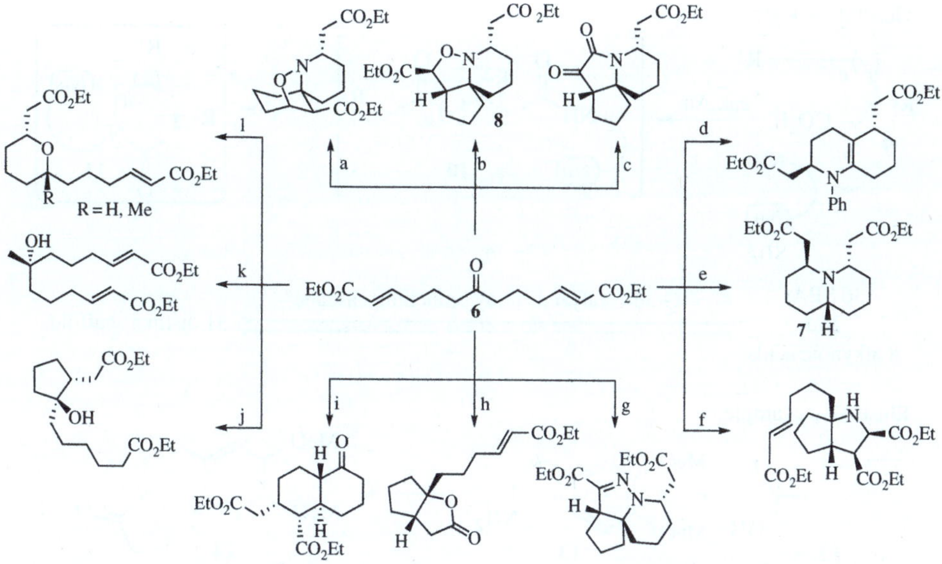

**SCHEME 4.3** The 12-fold branching DOS strategy developed by Robbins et al. Conditions: (a) $NH_2OH.HCl$, NaOAc, MeCN, then PhMe, 140°C, microwave; (b) $NH_2OH.HCl$, NaOAc, MeCN, 60°C; (c) $NH_2OH.HCl$, NaOEt, EtOH; (d) $PhNH_2$, $TiCl_4$, $CH_2Cl_2$; (e) $NaBH_4$, $NH_3$, EtOH, $Ti(OEt)_4$, then AcOH; (f) DIPEA, $H_2NCH_2CO_2Et$; (g) $NH_2NHTs$, toluene, reflux; (h) $SmI_2$ (2 equivalents), THF, MeOH, −78°C; (i) NaH, THF; (j) $SmI_2$ (5 equivalents), THF, MeOH, −78°C; (k) MeMgBr, THF; (l) superhydride, THF.

Patil et al. have recently reported a DOS strategy to access diverse molecular scaffolds based around a so-called relay catalytic branching cascade (RCBC) sequence [44]. The general synthetic concept is outlined in Scheme 4.4. Alkynoic acids (of the general form **9**, a common type of substrate) are reacted with several amine-functionalized scaffold building blocks (SBAs; variable) in the presence of a gold catalyst to form keto amides (of the general form **10**), which can then undergo gold-catalyzed cyclization cascades to produce various chiral multifunctional polyheterocyclic scaffolds (of the general form **11**). A specific example is shown in Scheme 4.4, where acid **12** reacts with amine **13** to generate heterocycle **14**. One of the principal advantages of this strategy is that a large number of SBAs are readily available, thus enabling efficient and rapid access to large scaffold diversity. Overall, the authors reported the DOS of 31 distinct scaffolds using 36 different SBAs and eight different alkynoic acids. All final library compounds were rich in functionality, and many contained structural features reminiscent of those found in natural products. This is undoubtedly a very elegant and powerful method for scaffold diversity generation. The authors suggest that it should be possible to expand the scope of this relay catalytic branching DOS strategy through the use of other reactions rather than just gold-catalyzed processes. For example, the authors highlight the fact that there are a multitude of reactions that can be catalyzed by metal-based catalysts and organocatalysts at the same time. Coupled with the ready availability of the substrates required, this approach clearly offers tremendous potential for the DOS of larger libraries of extremely high levels of structural diversity.

In 2009, the research group of Nelson described an elegant "folding" DOS pathway (Scheme 4.5) [40]. This approach involved the attachments of pairs of unsaturated

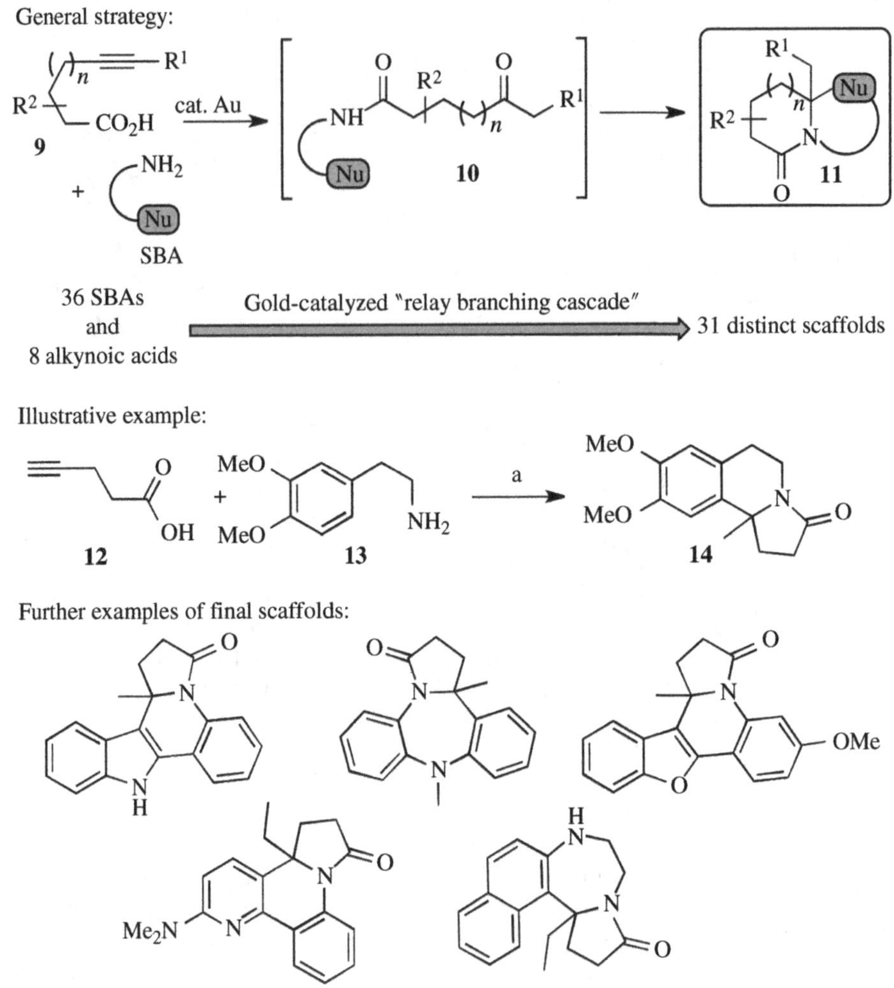


**SCHEME 4.4** DOS strategy to access diverse molecular scaffolds based around a so-called relay catalytic branching cascade (RCBC) sequence. Conditions: (a) $Ph_3PAuOTf$ (5 mol%), $CH_2Cl_2$, 100°C.

functionalized building blocks (called a "propagating" building block in the center of the molecule and a "capping" building block on the end) to a fluorous-tagged linker (a specific allyl alcohol or allyl amine equivalent) to furnish a wide variety of linear intermediates with a dense array of structural features (appendage, functional group, and stereochemical diversity). Treatment with a suitable metathesis catalyst led to intramolecular cyclization reactions that paired unsaturated functional groups together, thus converting a collection of simple and similar substrates into a dense matrix of skeletally complex and diverse products [2, 45]. The fluorous-tagged linker was also designed to take part in the metathesis cascade, with the final cyclization intended to cleave the linker [45]. Consequently, only cyclized products were released from the fluorous tag during the metathesis process, allowing easy generic purification from uncyclized material by fluorous solid-phase extraction [2, 45]. This DOS strategy can be analyzed in terms of the

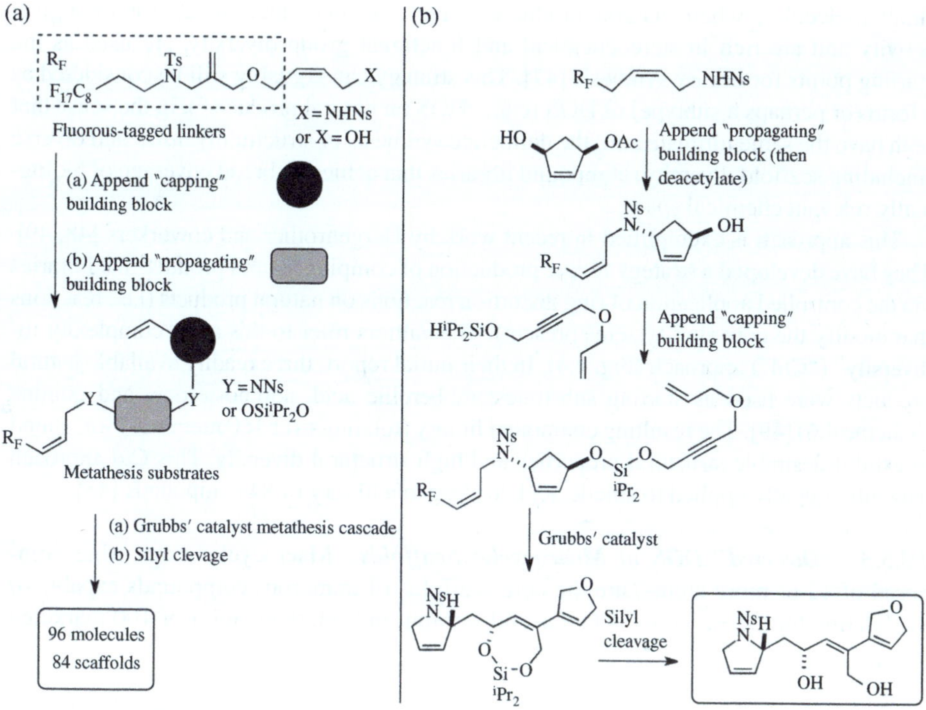


**SCHEME 4.5** Folding DOS pathway developed by Nelson and coworkers. (a) General synthetic scheme. (b) Synthesis of one final library compound.

B/C/P algorithm: the synthesis of the "capping" and "propagating" building blocks can be considered as the "build" phase, formation of the metathesis substrates constitutes the "couple" phase, and the "pair" phase involves intramolecular cyclization reactions. Overall, the DOS of a library of 96 molecules based on a total of 84 distinct molecular scaffolds was reported. The library also contained high levels of functional group and stereochemical diversity, with a diverse range of different 3D features and functionalization motifs present among the compounds [2]. This is currently the largest number of scaffolds present in a deliberately synthesized DOS library. Recently, Maurya et al. have described an interesting extension to this strategy [46]. They demonstrate that it is feasible to exploit triplets of unsaturated building blocks (i.e., various different "initiating," "propagating," and "terminal" building blocks), thus extending the diversity of molecular scaffolds that can be prepared. This modified strategy thus constitutes a more powerful approach to constructing scaffold diversity.

***4.3.5.2 DOS from the "Top Down": Natural Products as Starting Points for the Synthesis of Complex and Diverse Compounds*** DOS has traditionally been viewed as a guiding paradigm for the generation of complexity and diversity in a de novo fashion; that is, "typical" DOS campaigns aim to generate nonnatural (i.e., novel) molecules that are natural product-like in terms of their complexity and functionalization from structurally simpler starting materials (vide supra) [2]. Recent years have witnessed considerable interest in an alternative "top-down" strategy toward the synthesis of complex and diverse small molecules, where natural products, which are inherently biased for biological activity and are rich in stereochemical and functional group diversity, are used as the starting points for library synthesis [47]. This strategy can arguably still be considered as a form (or perhaps a subtype) of DOS (e.g., "DOS on natural products"), in the sense that both have the same ultimate aim: the deliberate synthesis of structurally *novel* and diverse (including scaffold diversity) compound libraries that achieve a broad coverage of biologically relevant chemical space.

This approach is exemplified in recent work by Hergenrother and coworkers [48, 49]. They have developed a strategy for the production of complex natural product-like libraries via the controlled application of ring distortion reactions on natural products (i.e., reactions that modify the core ring systems present). The authors refer to this as a "complexity-to-diversity" ("Ctd") approach (Fig. 4.4). In their initial report, three readily available natural products were used as starting substrates: gibberellic acid, adrenosterone, and quinine (Scheme 4.6) [49]. The resulting compound library (totaling over 160 members) was found to exhibit desirable structural properties and high structural diversity. This Ctd approach was subsequently applied to abietic acid, leading to a library of 84 compounds [48].

***4.3.5.3 "Directed" DOS of Macrocyclic Scaffolds*** Macrocyclic rings (ring composed of 12 or more atoms) are the core scaffolds of numerous compounds capable of modulating biological systems in a useful fashion. Indeed, there are over 100 marketed

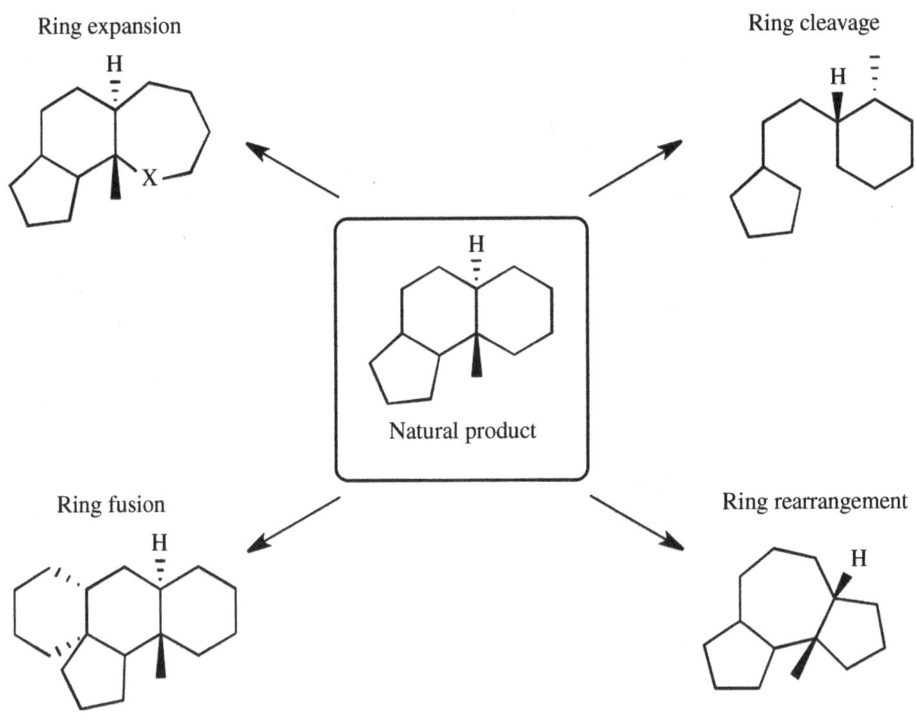


**FIGURE 4.4** The complexity-to-diversity (Ctd) approach for small molecule library construction. The controlled application of ring distortion reactions (ring expansion, ring cleavage, ring fusion, or ring rearrangement) on a natural product starting material yields complex natural product-like compounds.

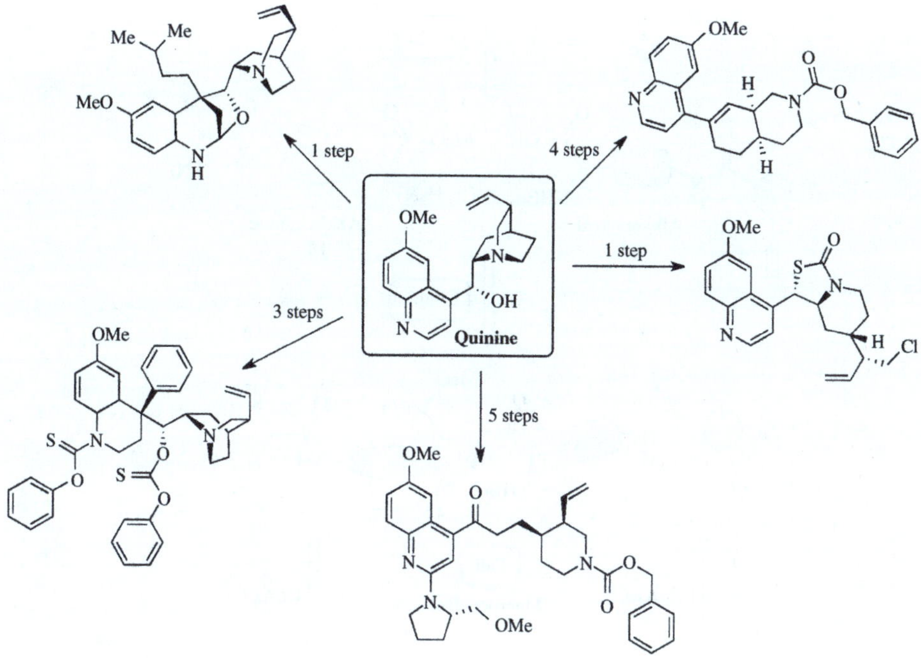


**SCHEME 4.6** Application of the complexity-to-diversity (Ctd) approach to the natural product quinine. Reaction sequences of between one and five steps were used to convert quinine into the five structurally complex and diverse structures shown.

macrocyclic drugs derived from natural products [6, 50, 51]. However, there is widely thought to be a lack of *synthetic* macrocycles in drug discovery. This is in part due to concerns about synthetic intractability [6, 52, 53]. Unsurprisingly, therefore, recent years have witnessed intense interest in the development of new synthetic strategies to efficiently access nonnatural macrocycles, and several DOS programs directed toward this structural type have been reported [6, 54–56].

For example, Isidro-Llobet et al. have reported a DOS pathway for the generation of peptidic macrocyclic compounds (i.e., macrocyclic peptidomimetics; Scheme 4.7) [57]. This DOS strategy was based around the B/C/P algorithm [6]. In the "build" stage, two types of chiral building block were prepared using standard methods: (i) "alkyne acids" of the general form **15**, which contain an alkyne and a carboxylic acid, and (ii) "azido amines" of the general form **16**, which contain a free amine and an azide. In the "couple" stage, these building blocks were combined to produce a range of tripeptides **17**. The subsequent "pair" stage, which provided the basis for the generation of scaffold diversity, was comprised of two cyclization steps. The first step involved two variations of the "click"-type 1,3-dipolar azide–alkyne cycloaddition process, which selectively paired the azide and alkyne functionalities of the linear substrates to generate the desired macrocyclic architectures. The copper-catalyzed azide–alkyne cycloaddition (CuACC) provided the 1,4-isomers **18**, and the ruthenium-catalyzed variant (RuAAC) gave the 1,5-isomers **19**. Further diversity was then introduced into the compound set in the second step of the "pair" stage, which involved an intramolecular reaction between the ester and amine functionalities to generate a diketopiperazine (DKP) motif embedded within the macrocyclic frameworks (molecules of the general form **20** and **21**).

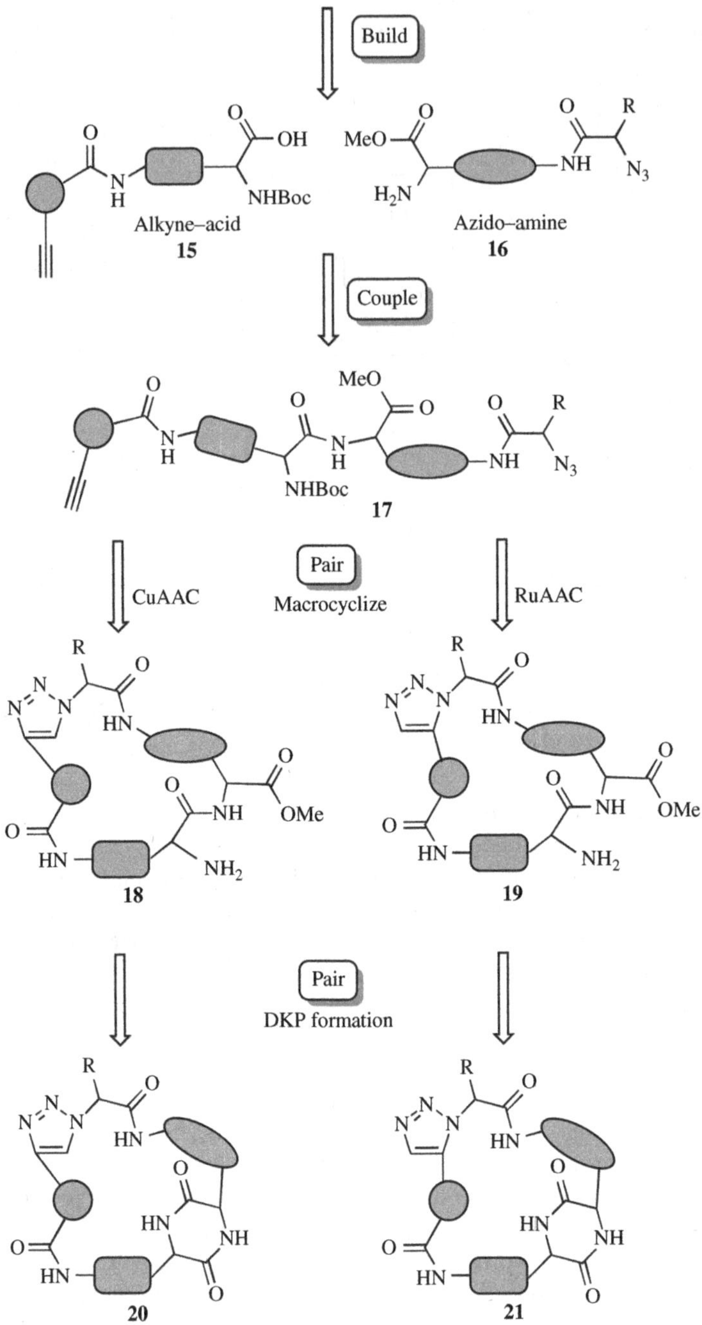


**SCHEME 4.7** Overview of the strategy developed by Isidro-Llobet et al. for the DOS of macrocyclic peptidomimetics. CuAAC, copper-catalyzed azide–alkyne cycloaddition; RuAAC, ruthenium-catalyzed azide–alkyne cycloaddition. The shaded shapes represent major scaffold-defining elements.

In total, a small proof-of-concept library of 14 compounds based around four different macrocyclic scaffolds was generated (viz., *cis*-DKPs, trans-DKPs, 1,4- and 1,5-triazoles) [6].

Beckmann et al. have recently described an innovative and powerful approach for the DOS of macrocyclic scaffolds based around the use of an "advanced" B/C/P strategy [58]. The classical B/C/P strategy involves the synthesis of building blocks in the "build" phase. These are then combined in the "couple" phase using the same reaction type to forge a conserved linking motif between the building blocks (vide supra). The resulting linear precursors are subsequently subjected to scaffold-defining reactions in the "pair" phase. Beckmann and coworkers reasoned that the structural diversity of these linear precursors could be significantly increased through the utilization of a diverse set of branching reactions in the "couple" phase starting from a pluripotent functional group rather than a sole coupling reaction; this approach, which the authors termed "multidimensional coupling," would lead to the introduction of a diverse set of linking motifs between the building blocks (Fig. 4.5). Thus, the structural diversity of the linear precursors would now be defined not only by the nature of the building blocks used but also by the respective linker motif installed. In principle, multidimensional coupling should lead to a greater structural diversity of linear precursors (from a given set of building blocks) than that achieved using the "classical" B/C/P strategy, which in turn should also lead to a significant increase in the overall structural diversity of the resulting compound library.

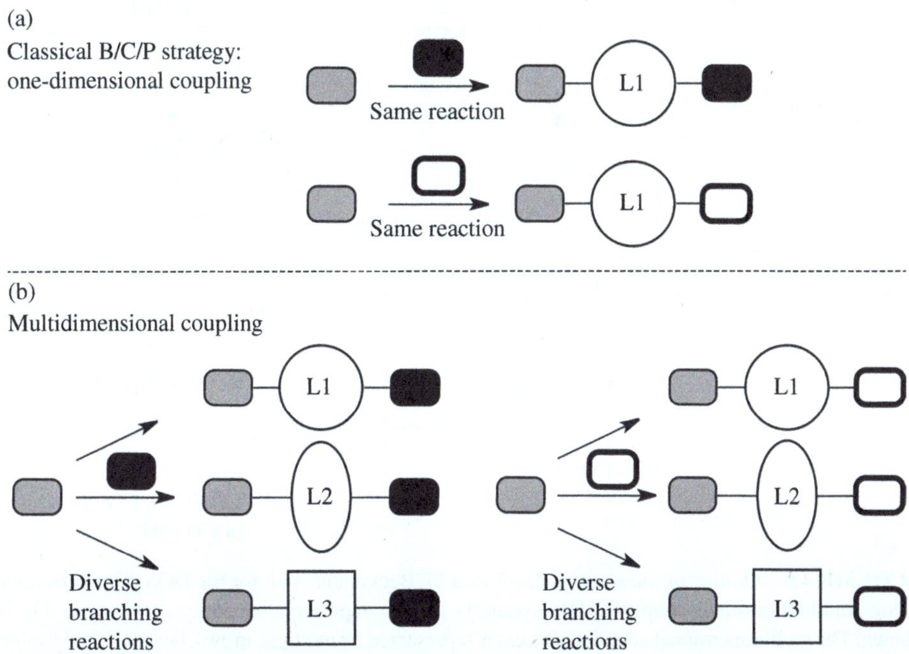


**FIGURE 4.5** Schematic comparison of the "couple" phase in a classical B/C/P strategy with the "multidimensional coupling" phase. In this example, one building block (represented by a gray-filled rectangle) can be coupled with two other different building blocks (represented by black-and-white-filled rectangles) via the formation of three possible different linking motifs (L1–L3). (a) In the classical B/C/P strategy, the building blocks are coupled together using only one linking motif (L1). (b) When multidimensional coupling is used, the three different building blocks are coupled using three different linking motifs (L1–L3).

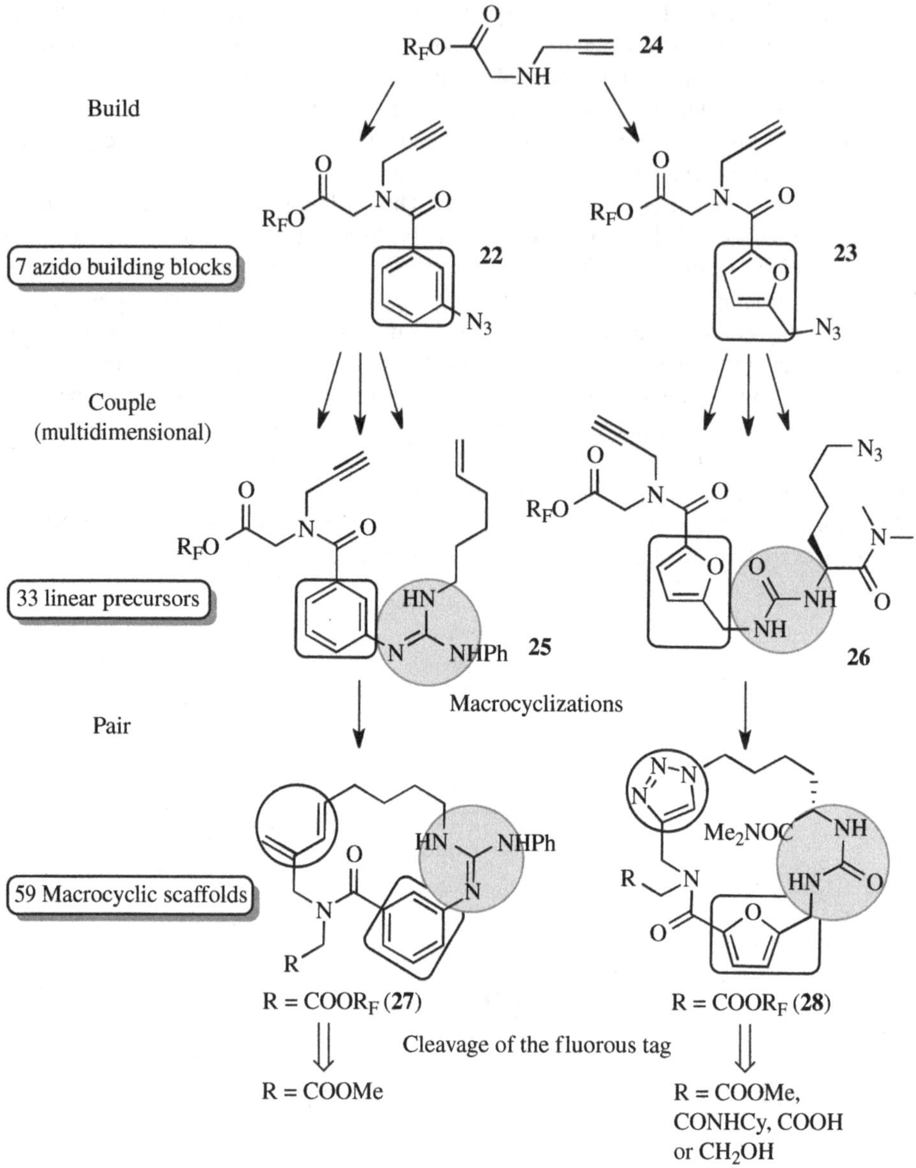


**SCHEME 4.8** Outline of the strategy developed by Beckmann et al. for the DOS of macrocycles using multidimensional coupling. The synthesis of two representative macrocyclic scaffolds is shown. The multidimensional coupling process is represented by multiple arrows, but only one reaction product shown. Major scaffold-defining parameters are the variation of the building block (highlighted with a rectangle), the generation of different linking motifs using multidimensional coupling (highlighted with a gray-filled circle), and divergent macrocyclization (highlighted with a circle).

The authors have reported the application of this multidimensional B/C/P strategy to the DOS of a library of nonpeptidic macrocycles with unprecedented scaffold diversity (Scheme 4.8) [58]. The DOS pathway centered around exploitation of the pluripotent reactivity of azide-derived aza-ylides. In the "build" phase of the DOS, a series of diverse azido building blocks (such as **22** and **23**) were prepared starting from the common fluorous-tagged precursor **24**. In the "multidimensional couple" phase, the azide group was transformed *in situ* into an aza-ylide, which was reacted with a diverse range of electrophiles in aza-Wittig reactions to generate various structural motifs (e.g., guanidine **25** and urea **26**) along with the concomitant installation of either a terminal alkene (as in compound **25**) or a new azide group (as in compound **26**). The azide and alkene groups, together with the initially attached terminal alkyne, served as synthetic handles for macrocyclization in the subsequent "pair" phase. Depending on the functionality present, macrocyclizations were performed either by (i) enyne metathesis (e.g., leading to **27**) or (ii) "click"-type azide–alkyne cycloaddition with copper catalysis (e.g., leading to **28**) or ruthenium catalysis. The presence of a fluorous tag in common precursor **24** allowed standard solution-phase parallel synthesis methods to be coupled with the benefits of fluorous-based purification protocols, allowing the generic purification of products along the DOS pathway by fluorous solid-phase extraction. In the final step of the DOS, the fluorous tag was removed using a diverse range of different methods (transesterification, transamidation, ester hydrolysis, and ester reduction) as a means to incorporate additional appendage and functional group diversity into the library. Overall, through the use of a limited number of structurally diverse building blocks, a DOS of 73 macrocycles based on 59 distinct scaffolds was achieved in no more than 4–5 steps from **24**. This represents an unprecedented level of scaffold diversity in a synthetically derived library of macrocyclic compounds. Computational analysis indicated that the library has a high level of shape diversity and thus overall structural diversity.

## 4.4. CONCLUDING REMARKS

Small molecule screening collections with high levels of structural, and thus functional, diversity are of undoubted value in the discovery of novel biologically active compounds. DOS has firmly established itself as a guiding paradigm for the efficient generation of such compound collections. The field of DOS has evolved rapidly since its "formal" conception some 14 years ago. As evidenced by the cutting-edge examples highlighted in this chapter, DOS planning algorithms have become considerably more sophisticated and powerful, allowing access to an unprecedented range of molecular structures, including many structural types that are underrepresented in standard small molecule screening collections. Numerous biologically active small molecules have been identified through the screening of DOS libraries, including modulators of classically "undruggable" targets and processes, thus validating the usefulness of DOS [2, 6]. However, there remains significant scope for further developments in the field. From a synthetic perspective, an ongoing challenge is to increase the *efficiency* of diversity (and complexity) construction, that is, to reduce the steps (and thus cost) associated with the generation of a library of structurally diverse and complex small molecules [2]. This can only be addressed by the development of new synthetic methodologies of broad utility [2]. A related and arguably much more fundamental issue is the absolute level of structural diversity that is desirable in new small molecule collections [2]. The goal of a DOS should be to efficiently and specifically access biologically *relevant* chemical space, rather than chemical space devoid of biologically useful molecules [2]. As evidenced by the disappointing performance of current low-diversity small molecule collections in drug discovery screens, structural diversity is undoubtedly an important feature of small molecule libraries in this regard. However, if a specific biomolecule or family of biomolecules is targeted, some level of library structural bias will be valuable. In addition, a common criticism of DOS is that many of the physiochemical properties of compounds in DOS libraries are not "drug-like" [20]. This may be attributed to the fact that DOS typically places an emphasis upon efficiently generating structural diversity rather than tailoring the properties of the library members. Thus, even if such compounds appear as hits in a screen, they may be unsuitable for development into lead compounds for further testing [20]. Therefore, it could be argued that future DOS campaigns should aim to generate compounds that fall within preselected structural limits. However, it is unclear how this level of structural bias should be defined, and this remains an area of some debate [2].

## REFERENCES

[1] O'Connor, C. J., Laraia, L., Spring, D. R. (2011) Chemical genetics. *Chemical Society Reviews*, **40**, 4332–4345.

[2] Galloway, W. R. J. D., Isidro-Llobet, A., Spring, D. R. (2010) Diversity-oriented synthesis as a tool for the discovery of novel biologically active small molecules. *Nature Communications*, **1**, 80.

[3] Walsh, D. P., Chang, Y. T. (2006) Chemical genetics. *Chemical Reviews*, **106**, 2476–2530.

[4] O'Connell, K. M. G., Galloway, W. R. J. D., Spring, D. R. The Basics of Diversity-Oriented Synthesis. *Diversity-Oriented Synthesis: Basics and Applications in Organic Synthesis, Drug Discovery, and Chemical Biology*, ed Trabocchi, A., John Wiley & Sons, Inc., Hoboken, NJ, 2013, pp. 253–287.

[5] Tan, D. S. (2005) Diversity-oriented synthesis: exploring the intersections between chemistry and biology. *Nature Chemical Biology*, **1**, 74–84.

[6] Galloway, W. R. J. D., Spring, D. R. (2013) Towards drugging the undruggable: enhancing the scaffold diversity of synthetic small molecule screening collections using diversity-oriented synthesis. *Diversity Oriented Synthesis*, **1**, 21–28.

[7] Walters, W. P., Namchuk, M. (2003) Designing screens: how to make your hits a hit. *Nature Reviews. Drug Discovery*, **2**, 259–266.

[8] O'Connell, K. M. G., Galloway, W. R. J. D., Ibbeson, B. M., Isidro-Llobet, A., O'Connor, C. J., Spring, D. R. Diversity-Oriented Synthesis. *Solid-Phase Organic Synthesis: Concepts, Strategies and Applications*, eds Toy, P. H. and Lam, Y., John Wiley & Sons, Inc., Hoboken, NJ, 2012, pp. 131–150.

[9] Li, J. W., Vederas, J. C. (2009) Drug discovery and natural products: end of an era or an endless frontier? *Science*, **325**, 161–165.

[10] Harvey, A. L. (2008) Natural products in drug discovery. *Drug Discovery Today*, **13**, 894–901.

[11] Cragg, G. M., Newman, D. J. (2013) Natural products: a continuing source of novel drug leads. *Biochimica et Biophysica Acta-General Subjects*, **1830**, 3670–3695.

[12] Butler, M. S. (2005) Natural products to drugs: natural product derived compounds in clinical trials. *Natural Product Reports*, **22**, 162–195.

[13] Galloway, W. R. J. D., Spring, D. R. (2009) Is synthesis the main hurdle for the generation of diversity in compound libraries for screening? *Expert Opinion on Drug Discovery*, **4**, 467–472.

[14] Borman, S. (2004) Rescuing Combichem. *Chemical and Engineering News: Science and Technology*, **82**, 32–40.

[15] Burke, M. D., Berger, E. M., Schreiber, S. L. (2003) Generating diverse skeletons of small molecules combinatorially. *Science*, **302**, 613–618.

[16] Burke, M. D., Schreiber, S. L. (2004) A planning strategy for diversity-oriented synthesis. *Angewandte Chemie, International Edition*, **43**, 46–58.

[17] Phillip Kennedy, J., Williams, L., Bridges, T. M., Nathan Daniels, R., Weaver, D., Lindsley, C. W. (2008) Application of combinatorial chemistry science on modern drug discovery. *Journal of Combinatorial Chemistry*, **10**, 345–354.

[18] Sauer, W. H., Schwarz, M. K. (2003) Molecular shape diversity of combinatorial libraries: a prerequisite for broad bioactivity. *Journal of Chemical Information and Modeling*, **43**, 987–1003.

[19] O'Connor, C. J., Beckmann, H. S. G., Spring, D. R. (2012) Diversity-oriented synthesis: producing chemical tools for dissecting biology. *Chemical Society Reviews*, **41**, 4444–4456.

[20] Hajduk, P. J., Galloway, W. R. J. D., Spring, D. R. (2011) A question of library design. *Nature*, **470**, 42–43.

[21] Haggarty, S. J. (2005) The principle of complementarity: chemical versus biological space. *Current Opinion in Chemical Biology*, **9**, 296–303.

[22] Lipinski, C., Hopkins, A. (2004) Navigating chemical space for biology and medicine. *Nature*, **432**, 855–861.

[23] Lovering, F., Bikker, J., Humblet, C. (2009) Escape from flatland: increasing saturation as an approach to improving clinical success. *Journal of Medicinal Chemistry*, **52**, 6752–6756.

[24] Langdon, S. R., Brown, N., Blagg, J. (2011) Scaffold diversity of exemplified medicinal chemistry space. *Journal of Chemical Information and Modeling*, **51**, 2174–2185.

[25] Hert, J., Irwin, J. J., Laggner, C., Keiser, M. J., Shoichet, B. K. (2009) Quantifying biogenic bias in screening libraries. *Nature Chemical Biology*, **5**, 479–483.

[26] Lipkus, A. H., Yuan, Q., Lucas, K. A., Funk, S. A., Bartelt, W. F., III, Schenck, R. J., Trippe, A. J. (2008) Structural diversity of organic chemistry. A scaffold analysis of the CAS Registry. *Journal of Organic Chemistry*, **73**, 4443–4451.

[27] Rask-Andersen, M., Almen, M. S., Schioth, H. B. (2011) Trends in the exploitation of novel drug targets. *Nature Reviews. Drug Discovery*, **10**, 579–590.

[28] Lipinski, C. A., Lombardo, F., Dominy, B. W., Feeney, P. J. (1997) Experimental and computational approaches to estimate solubility and permeability in drug discovery and development settings. *Advanced Drug Delivery Reviews*, **23**, 3–25.

[29] Schreiber, S. L. (2009) Organic chemistry: molecular diversity by design. *Nature*, **457**, 153–154.

[30] Clemons, P. A., Bodycombe, N. E., Carrinski, H. A., Wilson, J. A., Shamji, A. F., Wagner, B. K., Koehler, A. N., Schreiber, S. L. (2010) Small molecules of different origins have distinct distributions of structural complexity that correlate with protein-binding profiles. *Proceedings of the National Academy of Sciences of the United States of America*, **107**, 18787–18792.

[31] Nielsen, T. E., Schreiber, S. L. (2008) Diversity-oriented synthesis—towards the optimal screening collection: a synthesis strategy. *Angewandte Chemie, International Edition*, **47**, 48–56.

[32] Drews, J. (2006) What's in a number? *Nature Reviews. Drug Discovery*, **5**, 975.

[33] Altmann, K. H., Buchner, J., Kessler, H., Diederich, F., Krautler, B., Lippard, S., Liskamp, R., Müller, K., Nolan, E. M., Samorì, B., Schneider, G., Schreiber, S. L., Schwalbe, H., Toniolo, C., van Boeckel, C. A. A., Waldmann, H., Walsh, C. T. (2009) The state of the art of chemical biology. *ChemBioChem*, **10**, 16–29.

[34] Hung, A. W., Ramek, A., Wang, Y., Kaya, T., Anthony Wilson, J., Clemons, P. A., Yung, D. W. (2011) Route to three-dimensional fragments using diversity-oriented synthesis. *Proceedings of the National Academy of Sciences of the United States of America*, **108**, 6799–6804.

[35] Galloway, W. R. J. D., Bender, A., Welch, M., Spring, D. R. (2009) The discovery of antibacterial agents using diversity-oriented synthesis. *Chemical Communications*, **18**, 2446–2462.

[36] Lee, D., Sello, J. K., Schreiber, S. L. (2000) Pairwise use of complexity-generating reactions in diversity-oriented organic synthesis. *Organic Letters*, **2**, 709–712.

[37] Schreiber, S. L. (2000) Target-oriented and diversity-oriented organic synthesis in drug discovery. *Science*, **287**, 1964–1969.

[38] Spaller, M. R., Burger, M. T., Fardis, M., Bartlett, P. A. (1997) Synthetic strategies in combinatorial chemistry. *Current Opinion in Chemical Biology*, **1**, 47–53.

[39] Spandl, R. J., Rudyk, H., Spring, D. R. (2008) Exploiting domino enyne metathesis mechanisms for skeletal diversity generation. *Chemical Communications*, **26**, 3001–3003.

[40] Morton, D., Leach, S., Cordier, C., Warriner, S., Nelson, A. (2009) Synthesis of natural-product-like molecules with over eighty distinct scaffolds. *Angewandte Chemie, International Edition*, **48**, 104–109.

[41] Wyatt, E. E., Fergus, S., Galloway, W. R. J. D., Bender, A., Fox, D. J., Plowright, A. T., Jessiman, A. S., Welch, M., Spring, D. R. (2006) Skeletal diversity construction via a branching synthetic strategy. *Chemical Communications*, **31**, 3296–3298.

[42] Ibbeson, B. M., Laraia, L., Alza, E., O'Connor, C. J., Tan, Y. S., Davies, H. M. L., McKenzie, G., Venkitaraman, A. R., Spring, D. R. (2014) Diversity-oriented synthesis as a tool for identifying new modulators of mitosis. *Nature Communications*, **5**, Article number 3155.

[43] Robbins, D., Newton, A. F., Gignoux, C., Legeay, J.-C., Sinclair, A., Rejzek, M., Laxon, C. A., Yalamanchili, S. K., Lewis, W., O'Connell, M. A., Stockman, R. A. (2011) Synthesis of natural-product-like scaffolds in unprecedented efficiency via a 12-fold branching pathway. *Chemical Science*, **2**, 2232–2235.

[44] Patil, N. T., Shinde, V. S., Sridhar, B. (2013) Relay catalytic branching cascade: a technique to access diverse molecular scaffolds. *Angewandte Chemie, International Edition*, **52**, 2251–2255.

[45] Galloway, W. R. J. D., Diaz-Gavilan, M., Isidro-Llobet, A., Spring, D. R. (2009) Synthesis of unprecedented scaffold diversity. *Angewandte Chemie, International Edition*, **48**, 1194–1196.

[46] Maurya, S. K., Dow, M., Warriner, S., Nelson, A. (2013) Synthesis of skeletally diverse alkaloid-like molecules: exploitation of metathesis substrates assembled from triplets of building blocks. *Beilstein Journal of Organic Chemistry*, **9**, 775–785.

[47] Morrison, K. C., Hergenrother, P. J. (2014) Natural products as starting points for the synthesis of complex and diverse compounds. *Natural Product Reports*, **31**, 6–14.

[48] Rafferty, R. J., Hicklin, R. W., Maloof, K. A., Hergenrother, P. J. (2014) Synthesis of complex and diverse compounds through ring distortion of abietic acid. *Angewandte Chemie, International Edition*, **53**, 220–224.

[49] Huigens, R. W., III, Morrison, K. C., Hicklin, R. W., Flood, T. A., Jr., Richter, M. F., Hergenrother, P. J. (2013) A ring-distortion strategy to construct stereochemically complex and structurally diverse compounds from natural products. *Nature Chemistry*, **5**, 195–202.

[50] O'Connell, K. M. G., Beckmann, H. S. G., Laraia, L., Horsley, H. T., Bender, A., Venkitaraman, A. R., Spring, D. R. (2012) A two-directional strategy for the diversity-oriented synthesis of macrocyclic scaffolds. *Organic & Biomolecular Chemistry*, **10**, 7545–7551.

[51] Driggers, E. M., Hale, S. P., Lee, J., Terrett, N. K. (2008) The exploration of macrocycles for drug discovery-an underexploited structural class. *Nature Reviews. Drug Discovery*, **7**, 608–624.

[52] Madsen, C. M., Clausen, M. H. (2011) Biologically active macrocyclic compounds—from natural products to diversity-oriented synthesis. *European Journal of Organic Chemistry*, **17**, 3107–3115.

[53] Wessjohann, L. A., Ruijter, E., Garcia-Rivera, D., Brandt, W. (2005) What can a chemist learn from nature's macrocycles? A brief, conceptual view. *Molecular Diversity*, **9**, 171–186.

[54] Wessjohann, L. A., Voigt, B., Rivera, D. G. (2005) Diversity oriented one-pot synthesis of complex macrocycles: very large steroid-peptoid hybrids from multiple multicomponent reactions including bifunctional building blocks. *Angewandte Chemie, International Edition*, **44**, 4785–4790.

[55] Marcaurelle, L. A., Comer, E., Dandapani, S., Duvall, J. R., Gerard, B., Kesavan, S., Lee, M. D., IV, Liu, H., Lowe, J. T., Marie, J.-C., Mulrooney, C. A., Pandya, B. A., Rowley, A., Ryba, T. D., Suh, B.-C., Wei, J., Young, D. W., Akella, L. B., Ross, N. T., Zhang, Y.-L., Fass, D. M., Reis, S. A., Zhao, W.-N., Haggarty, S. J., Palmer, M., Foley, M. A. (2010) An aldol-based build/couple/pair strategy for the synthesis of medium- and large-sized rings: discovery of macrocyclic histone deacetylase inhibitors. *Journal of the American Chemical Society*, **132**, 16962–16976.

[56] Kopp, F., Stratton, C. F., Akella, L. B., Tan, D. S. (2012) A diversity-oriented synthesis approach to macrocycles via oxidative ring expansion. *Nature Chemical Biology*, **8**, 358–365.

[57] Isidro-Llobet, A., Murillo, T., Bello, P., Cilibrizzi, A., Hodgkinson, J. T., Galloway, W. R. J. D., Bender, A., Welch, M., Spring, D. R. (2011) Diversity-oriented synthesis of macrocyclic peptidomimetics. *Proceedings of the National Academy of Sciences of the United States of America*, **108**, 6793–6798.

[58] Beckmann, H. S. G., Nie, F., Hagerman, C. E., Johansson, H., Tan, Y. S., Wilcke, D., Spring, D. R. (2013) A strategy for the diversity-oriented synthesis of macrocyclic scaffolds using multidimensional coupling. *Nature Chemistry*, **5**, 861–867.

# 5

# SOLID-PHASE COMBINATORIAL CHEMISTRY

Marcel Patek, Martin Smrcina, Eric Wegrzyniak, Victor Nikolaev, and Andres Mariscal

*Tucson Research Center, Sanofi US, Tucson, AZ, USA*

## 5.1 INTRODUCTION

In the early 1990s, a series of fundamental changes were adopted into the early drug discovery process that ultimately altered the traditional ways in which drug candidates were identified. Researchers involved in commercial drug research were now setting out to explore the chemical space of vast compound collections through the use of high-throughput screening (HTS) of drug targets that were increasingly derived from the new field of genomics.

The largest provenance of new screening collections was the young and quickly evolving field of combinatorial chemistry. Combinatorial approaches based on the methods of both solid- and solution-based synthesis not only enabled production of a large number of compounds but also their concomitant screening in an automated fashion. It was a time when the prospect of finding new drug candidates within shorter timelines created enormous excitement in the research and business community. As a consequence of this enthusiasm, rapid integration of such combinatorial technologies was pursued by many pharmaceutical and biotechnology organizations.

Unfortunately, over the following years, productivity data did not indicate the predicted improvement in R&D efficiency [1, 2]. Rightly or wrongly, the field of combinatorial technologies has been often perceived as at least partially responsible for this downturn in productivity and therefore a failure [3]. As disappointment grew stronger in the late 1990s, combinatorial chemistry was retooled as parallel synthesis, and most

*Small Molecule Medicinal Chemistry: Strategies and Technologies*, First Edition.
Edited by Werngard Czechtizky and Peter Hamley.

combichem units in industry vanished by 2008. Furthermore, continuation of the negative trend has been recently magnified by industry-wide patent expiration causing major disruption in the pharma industry.

Is a failure of combinatorial chemistry a contributory factor to the unfulfilled productivity expectations of the industry? One may reasonably ask if building large collections of compounds and devising methods for their interrogation has any merit. On the other hand, did the pharma industry abandon combinatorial technologies prematurely and without good reason? Is combichem worth being revived? This will be the subject of the current chapter.

## 5.2 CHAPTER OUTLINE

Rather than attempt a comprehensive review of all solid-phase combinatorial chemistry approaches, we will present, in addition to a general overview, our perspective and experience that goes back to split–mix one-bead–one compound technology and involves two decades dedicated to discovery of small drug-like molecules in various industrial settings, from small start-up to big pharma [4]. To provide context, we first outline the key milestones in the history of combinatorial chemistry and then discuss assumptions and guiding principles in respect to design and synthesis of compound collections. We then turn to the main focus, a description of the essential components necessary for a sustainable growing screening collection based on combinatorial solid-phase libraries. We highlight notable outcomes within our integrated lead generation unit and also point out some issues that may have hampered broader adoption of solid-phase combinatorial chemistry within drug discovery. We conclude with an outlook on the field and its anticipated progression.

## 5.3 COMBINATORIAL CHEMISTRY IN RETROSPECT

Combinatorial chemistry has its earliest origins in the solid-phase synthesis of peptides pioneered by R. Bruce Merrifield in the early 1960s [5, 6]. In the following 20 years, researchers began to apply solid-phase methods to the synthesis of other types of molecules, including small organic molecules [7], carbohydrates [8], and oligonucleotides [9, 10] (Fig. 5.1). The first report mentioning combinatorial chemistry principles was the work of Mario Geysen in 1984. In its original implementation, peptide arrays were synthesized in parallel on 8 × 12 matrices of polypropylene pins, each pin carrying one peptide sequence [12]. In 1985, Richard Houghten introduced the “tea bag” method for facile synthesis of multiple peptides [13].

These and other advances in multiple peptide synthesis [14] led to one noteworthy synthetic approach, the split–mix one-bead–one compound method. In the 1980s and early 1990s, Furka and Lam, respectively, formulated an efficient technique that enabled the chemical synthesis of millions of peptides attached to a single polymer bead [15, 16]. However, as the method inherently provided random mixtures of peptides, an efficient deconvolution method that would identify biologically active sequences was the critical prerequisite. In the case of linear peptides, standard peptide sequencing was the method of choice. In other examples (unnatural amino acids, capping groups, release of peptide from the solid support), library design called for either specialized coding techniques [17, 18] or specific iterative algorithms. Examples of the latter included positional

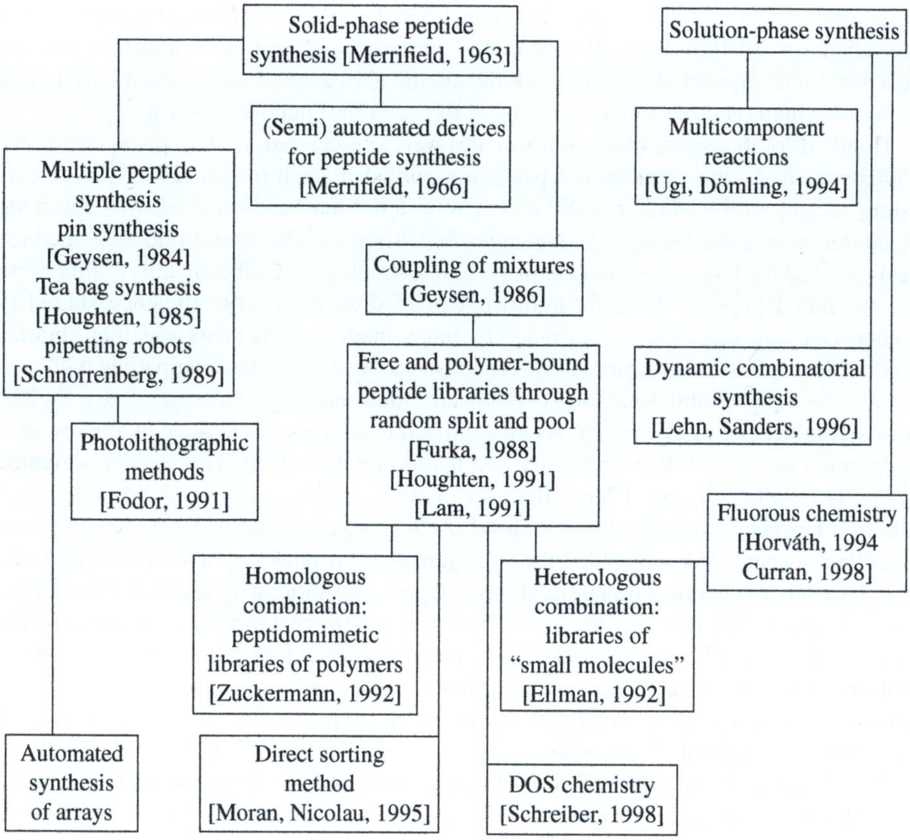


**FIGURE 5.1** Historical overview of early combinatorial chemistry methods. Adapted from Ref. [11].

scanning, orthogonal deconvolution of mixtures, or recursive deconvolution methods [19]. In addition, to allow for a greater flexibility in biological screening, new types of multiple cleavable linkers were developed [20, 21].

Up until 1992, the field of combinatorial chemistry was solely the domain of peptide- and oligonucleotide-based chemistry. Clearly, the challenge of applying similarly powerful techniques to synthesis of drug-like molecules stood in the way of discovery of new drugs. The first report describing solid-phase synthesis of small molecules appeared in 1992 in a study by Bunin and Ellman [22]. Since Ellman's seminal work, combinatorial solid-phase chemistry has quickly evolved into a multitude of technologies focused on the generation of large libraries of oligomers (peptides, "peptoids," oligonucleotides) and small molecules.

For small molecules, even more so than in the case of peptide libraries, tracking each compound and its connectivity to solid support beads presented a huge challenge. This was particularly true as the size of the library increased. In the following years, several tagging methods were devised, and some of these were implemented in industrial settings [23–25]. Before long, several start-up companies were founded around the new combichem platforms that were based on bead coding (Pharmacopeia) or on a new method of directed sorting (Irori). The latter approach used radio frequency identification (RFID) chips, 2D bar-coded containers, or 2D/3D spatially addressed arrays of solid support objects to track the chemical history of each solid-phase bound molecule. Such methods cleverly removed the need for chemical coding.

Hundreds of thousands of small molecules were synthesized on solid phase during the 1990s with many new structural types being published each month [26–39]. An interesting account of the whole era and its key milestones has been provided by Lebl in the inaugural issue of the *Journal of Combinatorial Chemistry* [40], the introduction of which also reflected the level of excitement in the new technologies. Unfortunately, as time went on, the initial hope of dramatic improvement of drug discovery efficiency started to wither. One dawning observation made by many medicinal chemists was that libraries produced by solid-phase combinatorial chemistry lacked properties compatible with progression as drug candidates. Compounds were often too large, too structurally similar, poorly soluble, and of low purity. As an approach to address this challenge, the research community has turned its attention back to natural products [41]. The diversity-oriented synthesis (DOS) concept of large libraries of molecules having complexity of natural products has been formulated and realized by Schreiber's group in 1998 [42, 43]. This and other concepts [44] expanded the strategies of solid-phase combinatorial synthesis. The repertoire of chemical transformations and protocols ultimately spanned from simple linear skeletons, heterocycles, and peptidomimetics to $sp^3$-rich topologies of oxygen-rich and natural product-like compounds. Other notable methods for generating diverse sets of molecules have been developed along with solid-phase methods. Those include multicomponent reactions [45, 46], fluorous chemistry [47], and the ever-growing field of dynamic combinatorial libraries [48].

Nonetheless, the "golden age" of solid-phase combinatorial chemistry and large combichem libraries was about to change. Disappointment with the lack of drug discovery acceleration originated by combinatorial chemistry led to a shift from large discovery libraries to small arrays of compounds. By the late 1990s, synthesis of target-focused combinatorial libraries became the prevailing trend in combinatorial chemistry. Integration of automation, informatics, and structure-guided design paved the way to new focused "smart" combinatorial libraries. Iterative optimization processes were integrated to enable fast multiparametric profiling (e.g., toxicology, ADME, PK) along with optimization of compound potency. Overall, the focus of combichem has slowly moved from making vast discovery libraries to optimization of hits that have already been identified. During this transition, the original sort-and-combine techniques were replaced by methods of parallel synthesis.

As the trend toward smaller synthetic arrays continued, the original field of combinatorial chemistry has slowly fallen into the shadows, particularly at pharma companies. By 2010, combinatorial chemistry can be seen as a lost art, and many practitioners of combichem have long ago moved to other fields. Only a handful of industrial labs still utilize solid-phase combinatorial synthesis, although there are many more practitioners in academia. One prominent combichem journal changed its name, and so did the topics and names of many conferences [3]. A notable exception and example of true commitment to the field is the free online journal *Combinatorial Chemistry*, which provides monthly literature reviews from Elsevier-printed journals [49]. A graphical representation of scientific literature related to "combinatorial chemistry" is depicted in Figure 5.2. The peak in combichem publications appears around 2002–2003 and is followed by a steady decline. Further analysis reveals that combinatorial techniques

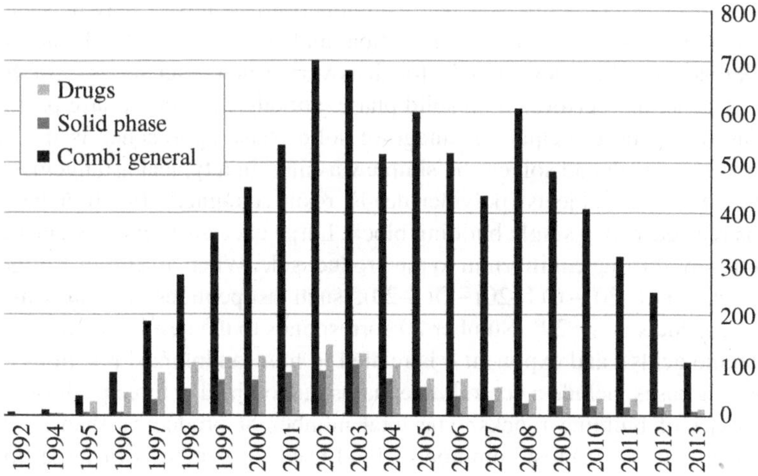


**FIGURE 5.2** Annual number of publications related to combinatorial chemistry. Keywords used in CAS search were "combinatorial chemistry" (excluding materials), "solid phase AND combinatorial chemistry," and "combinatorial chemistry AND drugs."

spread into other areas, such as drug absorption, process development, formulation, dynamic libraries, and computational approaches.

Nonetheless, the field of combinatorial technologies has provided many powerful tools, technological breakthroughs, and platforms for rapid creation of large combichem libraries [11, 50–52]. At the same time, robust automation systems have emerged to support HTS, analytical assessment of compound collections, and physical storage and handling of library compounds.

Where does that leave combinatorial chemistry and solid-phase synthesis approaches? Combinatorial chemistry has gone through changes in expectations, implementations, and perceptions of its utility. Ever-increasing pressure on the pharmaceutical industry to reduce research costs has also played an important role in the retreat from the technique. This retreat is in our opinion premature. Clearly, compound collections are still being maintained, grown, and augmented by compounds of diverse origin. Such continuous endeavors provide a strong indication that in the foreseeable future, small molecule drug discovery will rely heavily on compound libraries and HTS. In the following sections, we will argue that solid-phase combinatorial chemistry is still one of the best approaches to generation of such compound libraries.

## 5.4 FOUNDATIONS OF SOLID-PHASE SYNTHESIS OF COMBINATORIAL CHEMISTRY

Combinatorial chemistry comprises methods of chemical synthesis that enable the generation of complete or partial permutations of building blocks and core scaffolds. The ultimate goal is to find molecules with optimal properties. By performing multiple reactions in one reaction vessel, sets of individual compounds or mixtures can be prepared in a very efficient manner.

From the drug discovery perspective, combinatorial chemistry may be seen as a set of strategies, synthesis techniques, automation, and computational tools, all optimized to provide chemical libraries suitable for discovering new lead series. Combinatorial synthesis is generally performed on solid phase, although solution-phase protocols can be used as well. The principal advantage of solid-phase approaches is the ability to remove excess reagent and solvent by simple washing. In a typical arrangement, a mixture of functionalized objects (polymer beads, resin containers, functionalized plastic objects) is reacted with a single building block. Large excess of reagents ensures favorable kinetics by shifting equilibrium to the product side. When assembling oligomers of type A(1–20)–B(1–20)–C(1–20)–D(1–20), such as peptides, the maximum total number of products is $N=20^4$. Number 20 corresponds to the number of varied building blocks (amino acids), and exponent 4 is related to four randomized positions (A, B, C, D). Prevailing cases include a central core that is randomized at three positions by a different number of building blocks. The total number of product is then A×B×C. A simple schematic of the most common method used for synthesis of large libraries is shown in Figure 5.3. To illustrate the concept, synthesis of a small combinatorial library with two points of randomization and three building blocks in each randomization is depicted. The resulting number of compounds is 3×3=9. Synthesis starts with three batches of resin placed in three containers (step A). In this specific example, we will use Irori (now Brooks Automation, Inc.) NanoKan™ and MiniKan™ systems. 2D bar-coded lids on each container allow complete tracking of each object throughout the synthesis. Resin in each container (Kan) is reacted with the first set of building blocks (■,▲,◆) charged into separate reaction vessels (steps B and C). After thorough washing of objects in each vessel, Kans are combined and sorted (step D) into three different flasks (step F). In step E, resins in each vessel are reacted with the second set of building blocks (▲,■,◆) to provide the full permutation of nine compounds. After washing, the compound from each container is cleaved into one cleavage vessel (test tube, step H). Overall, six reactions (3C + 3F) are performed to provide nine compounds in nine unique vessels. If one were to employ parallel synthesis strategy, synthesis of the same nine

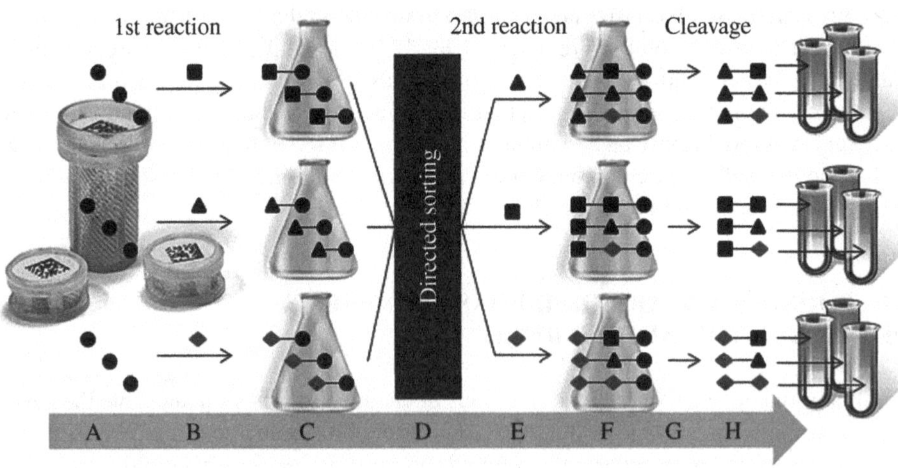


**FIGURE 5.3** Directed sorting approach to synthesis of a combinatorial library of nine compounds.

compounds would require nine reaction steps. Both directed sorting and parallel synthesis approaches became especially popular with the introduction of advanced automation. Those methods provide large amounts of synthesized compounds (8–100 μmol) of known chemical history and hence known composition.

Combichem has the ability to generate a large number of different molecules in just a few reaction steps. With the total number of compounds growing exponentially along the synthesis pathway, this approach offers rapid access to large collections of molecules. The ultimate objective is to identify compounds with the most desirable properties for a given purpose (e.g., molecular recognition, catalytic activity, or optical properties). For the purpose of this chapter, discussed combichem methods are directed toward drug discovery applications such as generation of compound libraries and HTS.

When designing a library, several theoretical concepts come into play, such as diversity and drug-like character. Numerous studies and reviews have been dedicated to those topics with no clear consensus formulated as to the choice of the most adequate approach to compound selection. On one hand, there are valid grounds to believe that larger libraries have a greater chance of finding new starting points. This view may be sensible for scenarios when there is no information about the target to be screened. The reasoning typically refers to one of several estimates of drug-like molecules that vary from $10^{23}$ to $10^{60}$ molecular entities. The latest estimate of the size of Lipinski's chemical space is around $10^{33}$ compounds [53]. With the current synthetic capabilities and the enumeration capability of powerful computers ($10^{14}$ compounds/year), those numbers are still beyond the realm of possibility. A more practical perspective is to consider the number of compounds available on the shelves of screening facilities worldwide, estimated to be in the order of 60,000,000–120,000,000 samples [54]. On the other hand, when the biological target is well understood, design can be narrowed to yield a "targeted library," which is based on hypotheses formed by cheminformatics and *in silico* models. Targeted libraries typically comprise 50–100 examples of purified compounds generated by parallel synthesis. Chemical structures are carefully chosen to answer structural hypotheses and employ essential elements (e.g., scaffolds) that were shown to convey biological activity. Regardless of the initial library bias, the important issue is still the selection of molecules that will be biologically active and optimizable.

To make synthesis, manipulation, and testing of large compound sets manageable, numerous algorithms have been devised for optimal sample pooling and testing. While peptide/oligomer libraries share well-established synthetic protocols, solid-phase synthesis of small molecules often requires a labor-intensive adaptation of classical solution-phase protocols. Methods optimized to solid phase must be reliable, span a wide variety of building blocks, and provide high yields of material. Hence, chemistry imposes practical limitations on diversity and analytical quality of libraries.

### 5.4.1 Ingredients of Solid-Phase Chemistry

Solid-phase methodology used in combinatorial synthesis has three primary components. First is the solid support that typically consists of cross-linked and insoluble polymer. It carries a functional group for covalent attachment of linker, scaffold, or other multifunctional building block. Second is the linker, which connects the solid support and the target molecule. The third essential element is the scaffold, which serves as a central molecular core for further substitution. It is the scaffold composition that defines the topological properties of final chemotypes. An inherent advantage of solid-phase chemistry is the simple purification possible between each reaction step. Reagents, unattached side products, and solvents are efficiently removed by employing multiple washing steps at the end of reaction. Since reaction intermediates do not need to be isolated, the whole synthetic process is vastly accelerated. However, when compared to solution-phase synthesis, reactions on solid-phase support are much more dependent on molecular diffusion. Sound choice of the resin and solvent is thus critical to establish optimal swelling and efficient diffusion. Furthermore, from the kinetic point of view, access of reagents in higher concentration is desirable for pushing reaction equilibrium to the product side.

***5.4.1.1 Polymer Supports*** Over the years, a wide variety of resin types have been described [55, 56], and many have become commercially available. There are solid supports with distinct hydrophobic [5] or hydrophilic character [57], PEG-polystyrene-grafted polymers [58], functionalized silica [59], and polysaccharides [60, 61], to name just a few. After many years of usage and diverse applications, two types of resins have stood the test of time: the hydrophobic cross-linked polystyrene (I) and polyethylene glycol attached to the polystyrene core (II) (Fig. 5.4).

Typical loading (degree of functionalization) of these resins is in the range of 0.5–1.5 mmol/g with bead size ranging from 40 to 150 μm. Another type of solid support worth mentioning is a polypropylene object grafted with a thin layer of polystyrene [25, 62, 63]. Those have been commercialized by Mimotopes Ltd. as SynPhase lanterns (III) and are shown in Figure 5.4. While there are many specialty solid supports and other excellent resins available, they have found only a limited use in the synthesis of combinatorial libraries.

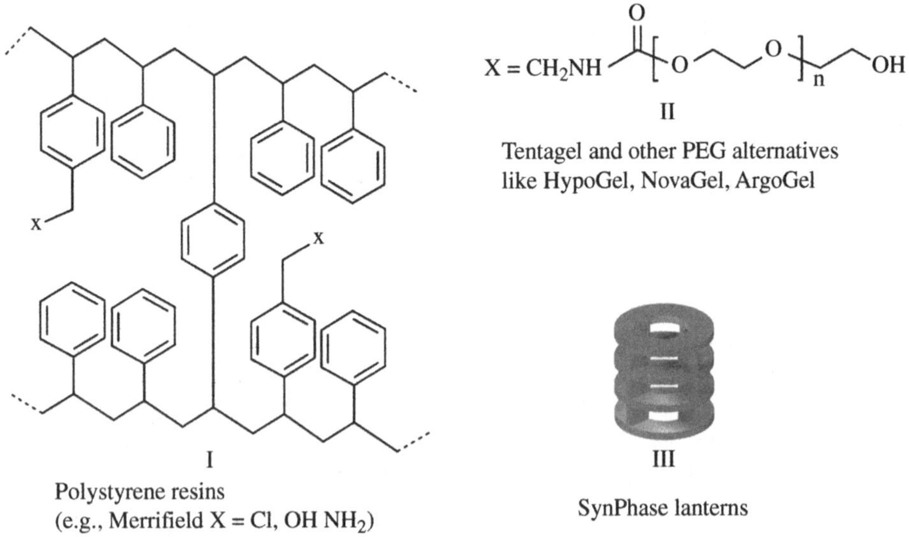


**FIGURE 5.4** General structure of popular polymer supports.

***5.4.1.2 Linkers*** Covalent attachment of the synthesized molecule to polymeric support is the key prerequisite for efficient solid-phase synthesis. A special type of chemical linkage is used to ensure the firm connection throughout the synthesis. Such linkers can be regarded as bifunctional spacers that on one end incorporate features of a protecting group and on the other allow coupling to functionalized supports. The most important and general features of linkers are the efficiency of attachment of organic molecules to the solid support, their stability toward a multitude of reaction conditions, and their ease of cleavage under conditions that guarantee structural integrity of the final product.

As the nature of the compounds synthesized on solid supports progressed from peptides to more complex structures, there has been a significant increase in demands on linkers [64]. There is now a large variety of specialty linkers that include photo-cleavable units and safety catch, traceless, and enzyme-cleavable linkers with many others featuring attachment via sulfur, boron, silicon, tin, bismuth, and some transition metal elements [65]. To affect the cleavage from solid support, numerous cleavage conditions including saponification, transesterification, and aminolysis have been introduced and optimized [55, pp. 33–131]. The most robust and widely used protocols are based on proton-assisted cleavage. Kinetics of the cleavage can be fine-tuned by employing acids of varying strength. Stability of the attachment point is directly linked to the stability of the created cation. Accordingly, most of these linkers are based on trityl, benzhydryl, and benzyl groups as depicted in Figure 5.5. Stability of the corresponding cations can be further fine-tuned by electron-donating alkoxy groups or halogen atoms.

Since the topic of linkers has been discussed in many comprehensive reviews [64, 66, 67] and monographs [55, 65, 68], we only wish to comment on highly dependable and frequently used representatives. The most robust linking strategies encompass connections to solid support via esters, amides, and alcohols.

Depending on the type of linker and cleavage conditions, cleaved products can either feature terminal acids, amides, and alcohols, or they can afford molecular constructs containing lactam or lactone bonds. The latter case is referred to as a "cyclative cleavage" and is present in many popular solid-phase transformations. Most traditional ester-based anchors include linkers such as Wang ester linkage, HMPB, PAM, Rink acid, and

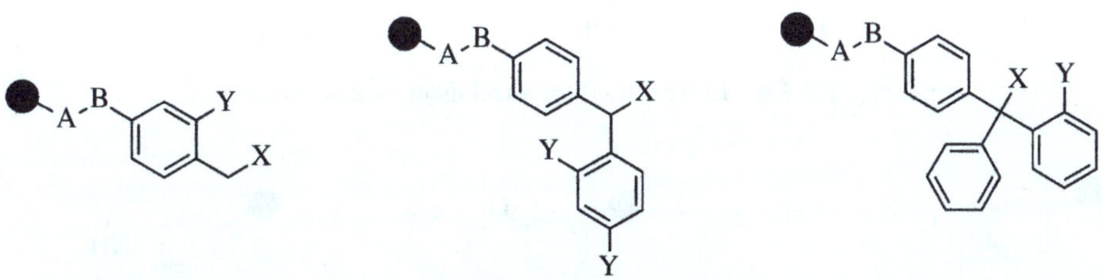


Benzyl-based linkers | Benzhydryl-based linkers | Trityl-based linkers

X : Functionality for attachment of the synthesized molecule
Y : Functionality to fine-tune the electronic nature
A+B : Attachment points for anchoring to support

● Resin

**FIGURE 5.5** General structures of most common linkers.

chlorotrityl linker. Cleavage conditions vary from 0.1% TFA for Rink acid linker up to liquid HF for the PAM linker (Fig. 5.6).

Rink amide (RAM) linker, Sieber 9-xanthenyl linker, and trityl linker allow a wide variety of chemical transformations supported by the stable amide attachment (Fig. 5.7).

Cleavage conditions vary from 2% TFA for Sieber up to 95% TFA for the RAM linker.

The attachment of alcohols and phenols to solid supports is predominantly accomplished by reaction with trityl chloride, Rink chloride, or dihydropyranyl (DHP) linkers (Fig. 5.8). Cleavage conditions for resin-bound ethers are usually very mild and range from 2% TFA for the trityl linker to 95%TFA/water for the DHP linker.

Another important linking strategy involves primary or secondary amines. Ameba, BAL, HMPB (Fig. 5.6), and chlorotrityl linkers are usually used for the amine attachment (Fig. 5.9). Cleavage conditions are generally stronger than those for acids and alcohols. In this case, acid conditions vary from 50% TFA in DCM for the trityl-based linkers to 100% TFA with the ameba linker.

To enable access to compounds of lower molecular weight having no residual functionality, the concept of traceless linkers was introduced in the mid-1990s [69, 70]. First linkers were based on silicon directed ipso-substitution. The concept was subsequently

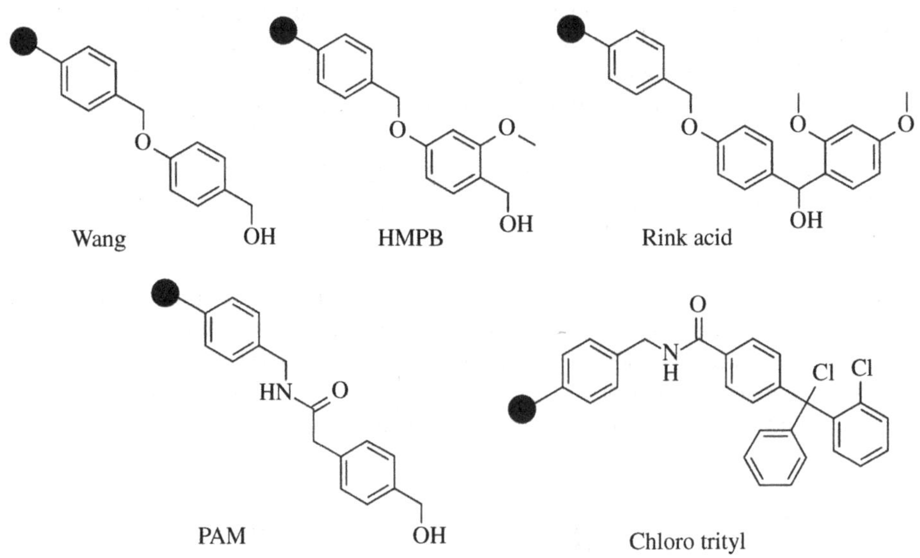


**FIGURE 5.6** Linkers used for attachment of acids as esters.

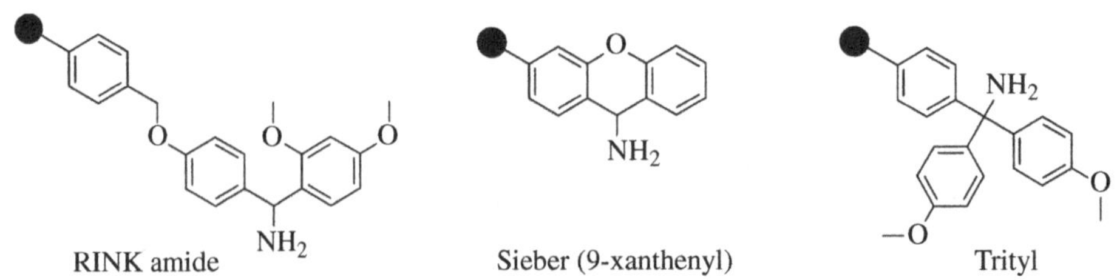


**FIGURE 5.7** Linkers used for attachment of acids as amides.

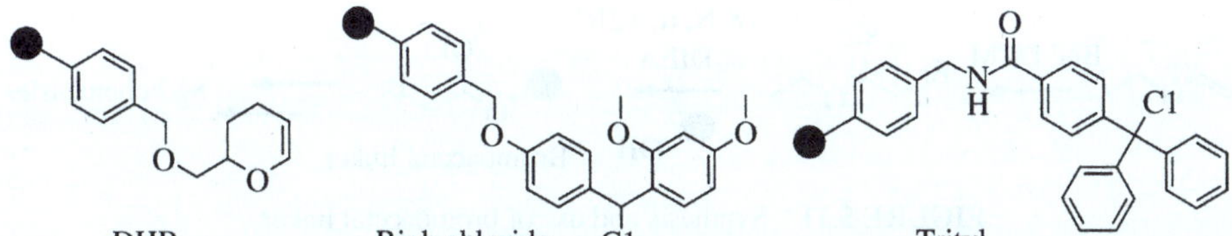

DHP Rink chloride Trityl

**FIGURE 5.8** Linkers used for attachment of alcohols.

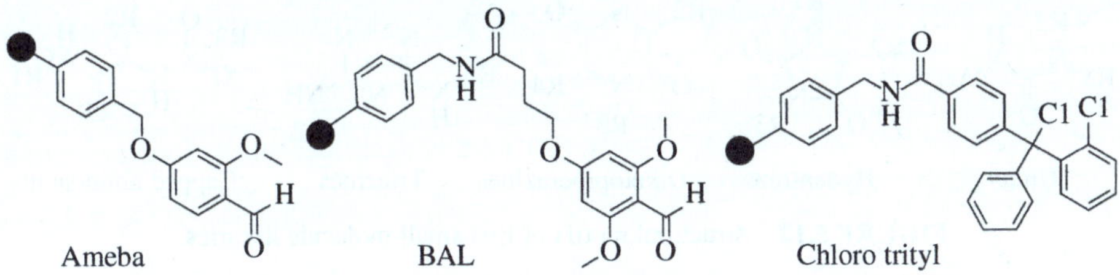

Ameba BAL Chloro trityl

**FIGURE 5.9** Linkers used for attachment of amines.

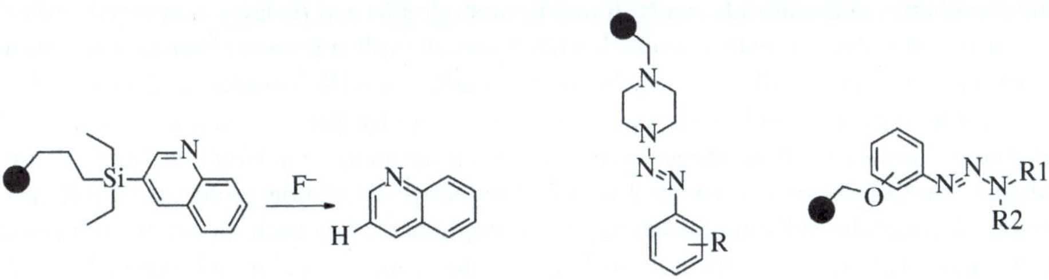

Piperazinyl-type T1 resin T2-triazene resin

**FIGURE 5.10** Examples of traceless linkers.

extended to include phosphonium [71], germanium [72], and selenide linkages [73, 74]. Probably the most versatile traceless linker has been the triazene unit introduced by Moore [75] and Tours [76]. The linker concept has been extensively elaborated and adapted to a numerous solid-phase protocols and types by Bräse [77, 78] (Fig. 5.10).

For many years, we [79, 80] and others [81–83] have been exploring the use of bromoacetal linkers to build a diverse set of libraries via the cyclative cleavage strategy (Fig. 5.11).

The bromoacetal linker is available both on hydroxymethyl polystyrene and TentaGel-type of resins. Cleavage conditions require a medium strength acid (e.g., formic acid) to accomplish a single- or multicyclization steps via *N*-acyliminium ion intermediates [84–92]. Several examples of *N*-acyliminium ion chemistry are shown in the next section (Fig. 5.18).

*5.4.1.3* ***Transformations*** There are many comprehensive reviews [8, 93–97] and entire books [19, 50–52, 55, 68, 98, 99] dedicated to the topic of solid-phase chemistries. We wish to highlight chemistries and chemical transformations that have found a broad utility

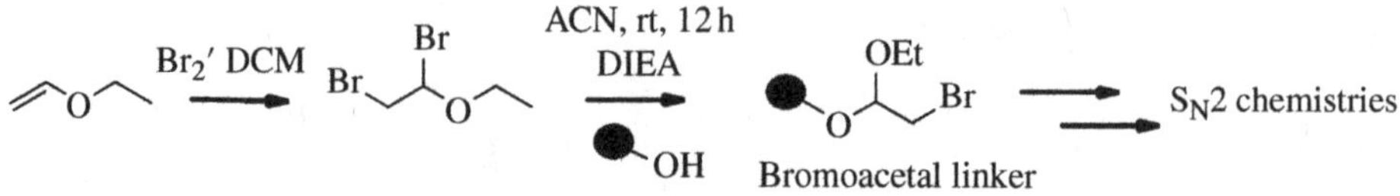


**FIGURE 5.11** Synthesis and use of bromoacetal linker.

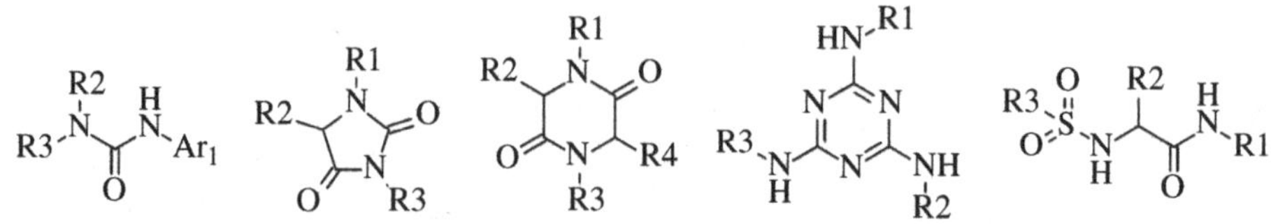


**FIGURE 5.12** Structural motifs of first small molecule libraries.

in the solid-phase synthesis of larger libraries. Unless specifically indicated, references to the chemistries and scaffolds can be found in monographs and reviews mentioned earlier.

Early solid-phase transformations leading to small molecules were based on acylation reactions affording amides, ureas, thioureas, urethanes, sulfonamides, and guanidines. The complexity of solid-phase chemistries has rapidly progressed to another level enabling synthesis of aromatic cores based on aromatic nucleophilic substitution, Mitsunobu alkylation, and a wide variety of reductive alkylation protocols. While aromatic electrophilic substitution has not been implemented to such an extent, transition metal-mediated aromatic substitutions have become widely popular and well elaborated. Examples include C—C bond-forming reactions such as Heck, Stille, and Suzuki couplings, as well as the C—N coupling protocols of Buchwald and Hartwig. In the late 1990s, chemistry procedures expanded their scope to master Knoevenagel, Pictet–Spengler, and aldol condensations as well as many other reaction types, such as cationic cyclizations, olefin metathesis, Horner–Emmons reaction, Pauson–Khand cyclization, 3+2 cycloadditions, Diels–Alder reaction, Sonogashira coupling, and Baylis–Hillman reaction.

In the early days of combichem, design and synthesis of libraries were mostly driven by available and high-yielding chemistries, often providing highly decorated structures with correspondingly suboptimal properties. Even though the scaffolds were present in many known drugs, the brute-force decoration of the core resulted in highly hydrophobic and poorly soluble compounds of high molecular weights. Typical structures of early libraries included ureas, guanidines, hydantoins, thiazolidines, diketopiperazines, and simple N-capped amino acids (Fig. 5.12).

To increase both the number of biologically active compounds and the efficiency of screening campaigns, the field moved toward smaller libraries using the so-called privileged motifs [22, 100]. Many libraries were synthesized around heterocyclic scaffolds such as purines, benzimidazoles, benzodiazepinones, benzodiazepinediones, and also peptides with nonnatural amino acids including beta-turn mimetics. As expected, the strategy of privileged motifs led to a significant increase in "actives" and to new lead series [26–39]. However, despite its clear success, the strategy was frequently facing challenges of structural similarity to known molecules and consequent restricted intellectual property. Representative examples of privileged scaffolds are shown in Figure 5.13.

To further increase structural diversity and library complexity, the concept of "randomized scaffold" was introduced in the mid-1990s [101]. One example shown in Figure 5.14 describes the library of hydroxyaryamides. In this case, a set of 30 bifunctional scaffolds (hydroxyaromatic acids) can be used to build libraries of three randomizations of 12,000 drug-like compounds [102].

Another approach to greater diversity is the "library-from-library" concept introduced by Houghten in 1996 [103]. By synthetic postmodification of an existing library, new "global" randomization is introduced on all library members, giving rise to a new set of compounds. One example that illustrates this concept is a library of thiazolidines, which was converted to library of the corresponding sulfoxides (Fig. 5.15) [104].

Continuous improvements in molecular topology led to small molecules of globular and more rigid structures. The notion of more polar, chiral, and $sp^3$-rich molecules has been emphasized multiple times over the last two decades of solid-phase combinatorial chemistry [105]. Such chemotypes (Fig. 5.16) can be exemplified by structures of tropanes [106], seglins [107], spiroethers [108], and macrocycles [109].

The expanding field of chemical biology required additional sources of chemical probes and ligands. Therefore, from the late 1990s, both large and small libraries were

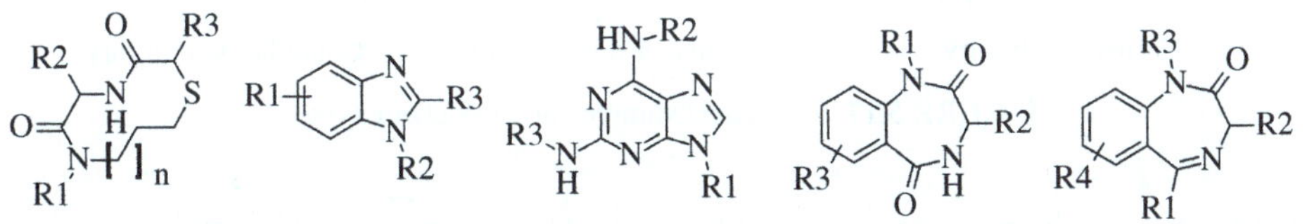

Beta-turn mimetics Benzimidazoles Purines Benzodiazepindiones Benzodiazepinones

**FIGURE 5.13** Structures of privileged scaffolds and the corresponding libraries.

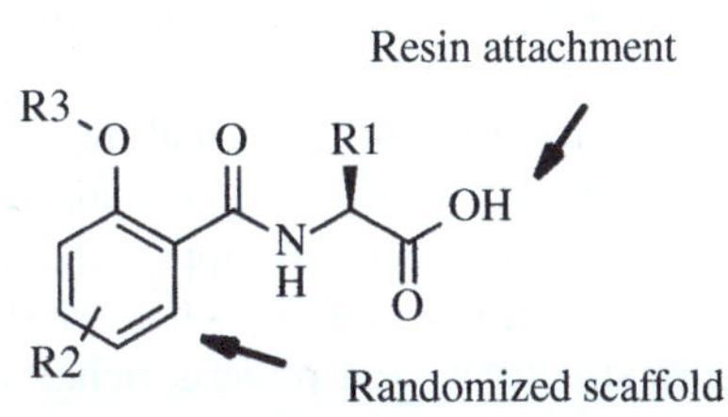


**FIGURE 5.14** The concept of randomized scaffold illustrated on scaffold cores.

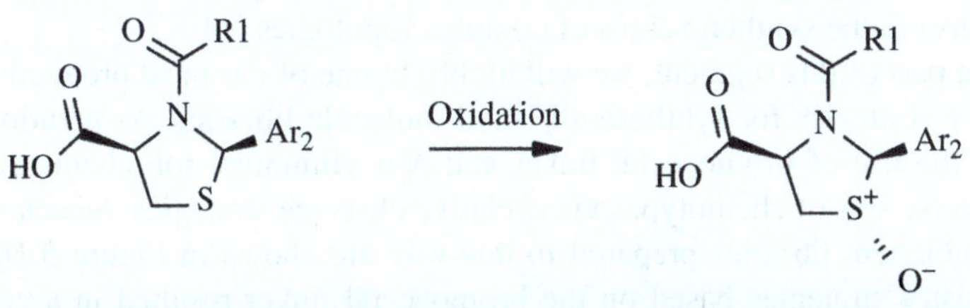


**FIGURE 5.15** Library-from-library concept.

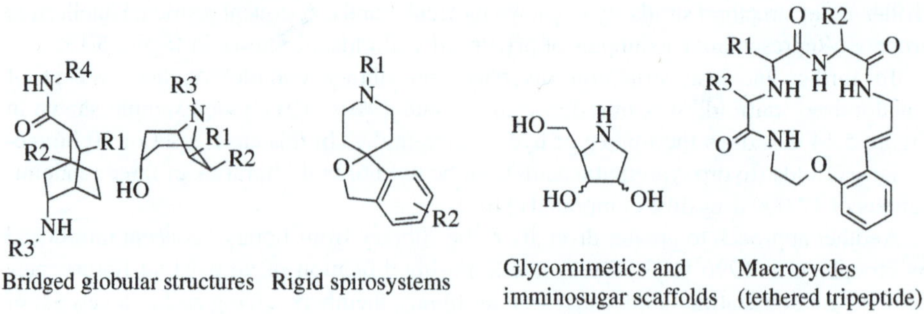


**FIGURE 5.16** Examples of globular and $sp^3$-rich chemotypes.

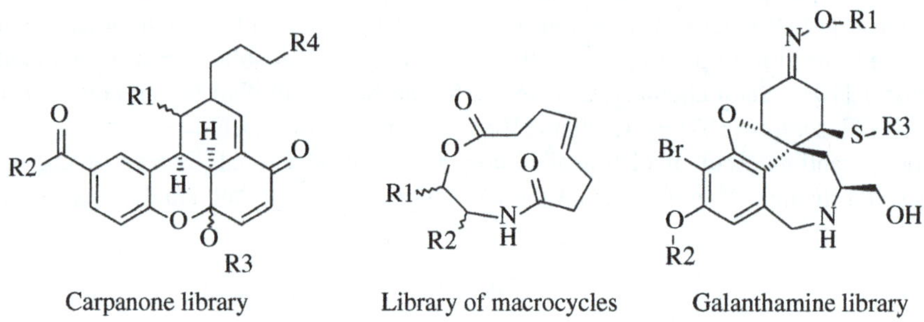


**FIGURE 5.17** Selected examples of DOS chemotypes.

synthesized by pharma companies and academia to discover new active molecules that bind to a variety of target classes. By either standard target-oriented parallel synthesis or by expanding around known binding motifs, many useful or novel lead series were discovered for kinases, GPCR, ion channels, and nuclear hormone receptors. In the fast response to this new paradigm, a broad variety of target-specific collections have also become available from commercial sources.

In order to address challenging and the so-called "undruggable" targets via phenotypic screens, a new approach to design of libraries and potential ligands/probes was introduced by Schreiber in the late 1990s [42, 44]. This approach, known as "DOS," involves the preparation of small libraries of organic molecules that are structurally more complex, have a greater variety of core structures, and possess richer stereochemical variations (Fig. 5.17). It essentially merges the concepts of privileged motifs, randomized scaffold, and library from library into one consistent approach. DOS methodology tries to improve upon the shortcomings of classical combinatorial chemistry by producing large variations (high diversity) in the synthetic cores of complex topologies [110].

In the last part of this segment, we will highlight one of our most productive and successful linker strategies for synthesis of small molecule libraries. As mentioned earlier (Fig. 5.11), the use of bromoacetal linker and *N*-acyliminium ion chemistries enables access to diverse sets of chemotypes via cyclative cleavage strategies. Selected examples of large combichem libraries prepared in this way are shown in Figure 5.18. Over the years, chemistry strategies based on the bromoacetal linker resulted in a collection of 144,000 diverse compounds from 15 different libraries.

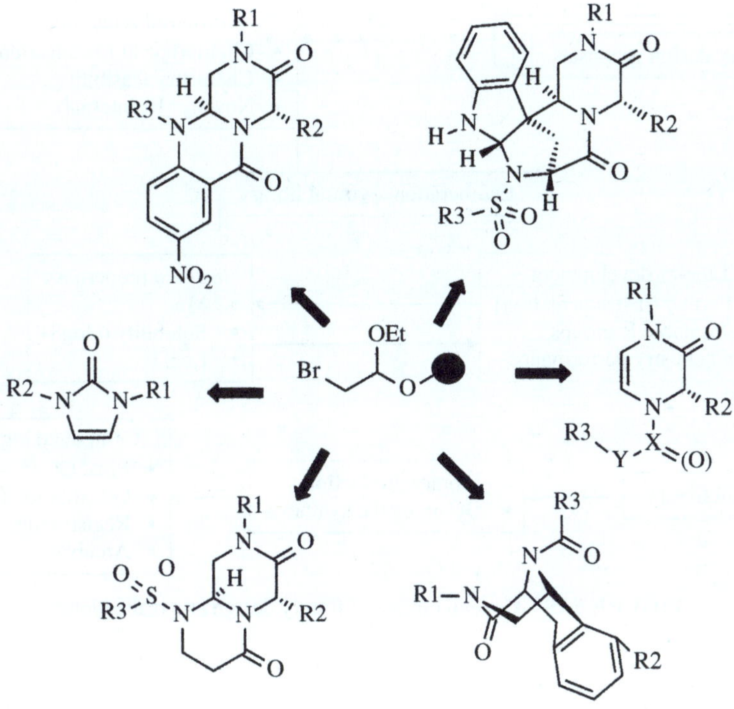

**FIGURE 5.18** Examples of chemistries utilizing bromoacetal linker.

Overall, the majority of synthetic transformations have been successfully adapted to solid-phase synthesis, becoming an invaluable part of the chemist's toolbox for combinatorial library synthesis. Over the 20 years of its existence, solid-phase combinatorial chemistry and related technologies have reach a mature state and become a valuable technique for finding new starting points in many drug discovery campaigns.

### 5.4.2 Library Development and Production

When planning a library, the principal factors to be considered are chemistry protocols, available synthons, solid-phase linkers, and screening strategy. Figure 5.19 outlines several interlocking steps used during the realization of a library. The process typically starts by generating and selecting suitable proposals. In the next step, chemical structures (chemotypes) are assessed with respect to chemical feasibility, novelty, and topological redundancy. The aforementioned design steps precede electronic enumeration of the full combinatorial set (VL). At this stage, it is quite common to use one of the many software packages (e.g., ACD/Labs, QikProp) to predict physicochemical properties of enumerated structures (e.g., lipophilicity, solubility, polar surface area). Only a subset of compounds that passes various property filters is taken into the next step of library development. Selection of a format, availability of synthesis inputs, and actual chemical optimization sit at the heart of library realization. Since it is the performance of chemistry and availability of reagents that

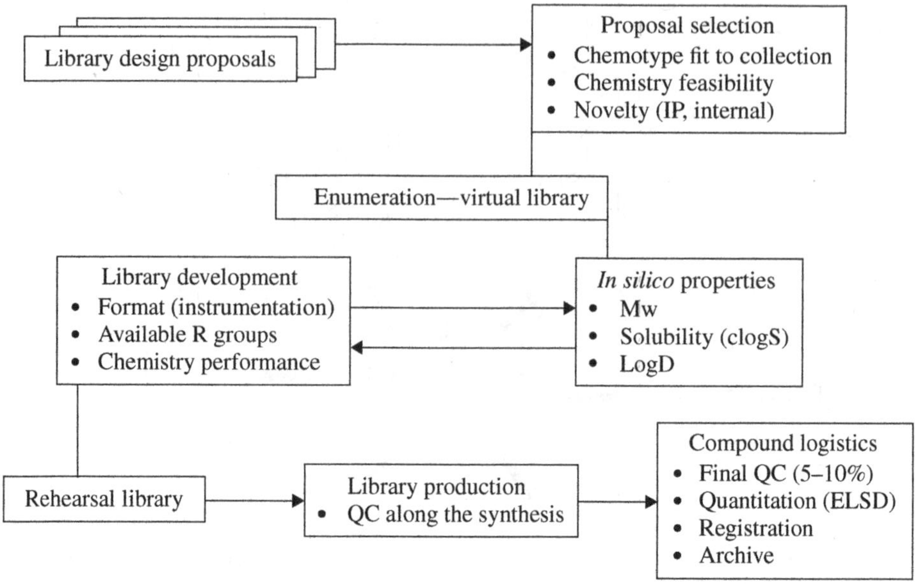


**FIGURE 5.19** General outline of library design and realization.

limit the scope of intended chemical set, it is not unusual to reenumerate the whole library and run selection of final compounds several times.

Most libraries today are produced in one of the two formats: spatially addressable and encoded. The former method is based on a spatial (physical) separation of synthesis objects (solid support beads, functionalized plastic objects) and allows for a straightforward determination of reaction products. Parallel synthesis in 96-well plates or plastic syringes illustrates this approach well. The encoded method relies on using various nonchemical tags (RFID chips, 2D bar-coded resin containers) to drive chemical synthesis and to facilitate assignment of structure container upon synthesis completion. To avoid costly failures at the stage of library production, optimized chemistry protocols require a rehearsal library (RL) where experimental conditions and performance of building blocks used at each randomization are validated. The library synthesis itself is relatively short. Even then, analysis of intermediates along the synthetic pathway is an important part of the library production. Once final compounds are cleaved from the solid support, either random (5–10%) or complete analysis of library samples is performed. Chromatographic and mass spectrometric methods (LC/MS) are frequently used for sample characterization. The last step involves electronic registration of library structures and their archival in the storage system.

To exemplify typical steps of library design and production, we will discuss an enumerated library of aryloxyalkylpyrazoles as illustrated in Figure 5.20. This particular library has four randomization sites including randomized scaffold R2 (arylhydroxy acids) and randomized tether R3. Furthermore, R4 randomization can be accomplished by either [2 + 3] cycloaddition of azomethine ylides to (homo)propargylethers or by direct *O*-alkylation with protected alkylhalopyrazines. Assuming all permutations of building blocks and scaffolds are represented, the theoretical number of compounds is 102,856.

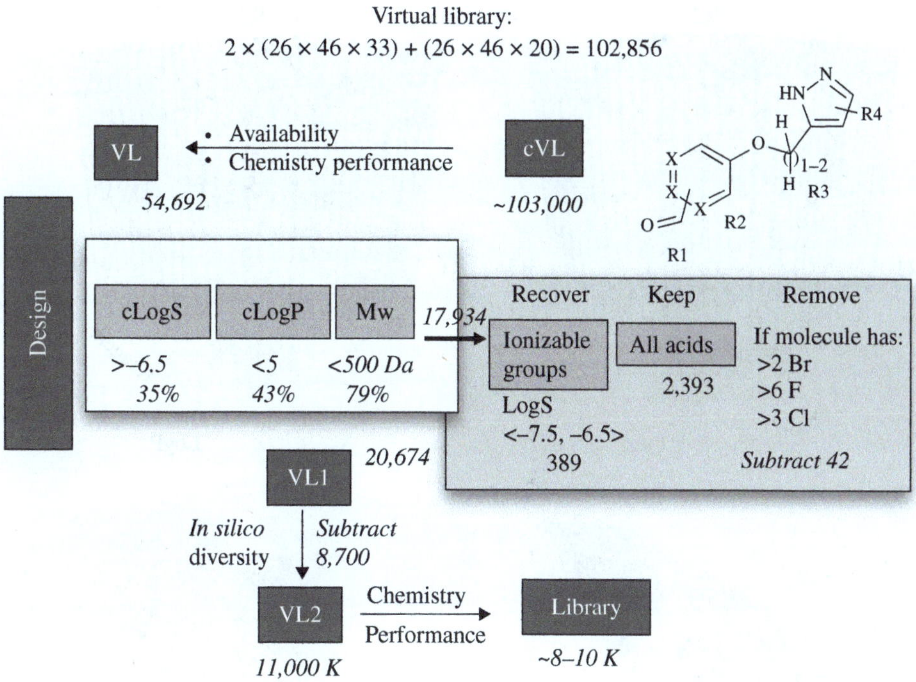


**FIGURE 5.20** Example of library design and selection of the library set.

For R1 = 26, R2 = 46, R4 (azomethine ylides) = 33, and R3 = 2 (methylene, ethyl), the first sublibrary comprises 26 × 46 × 33 × 2 = 78,936 possible compounds. For R4 (alkylhalopyrazines) = 20 and the same number of R1 and R2 building block, we add another 26 × 46 × 20 = 23,920 possible compounds.

Each R1 set includes the "zero" building block, which in this case is a terminal amide bond (CONH2). After a preliminary survey of available building blocks and compatibility of each set of reagents, the calculated virtual library (cVL) is trimmed by about 50%. Virtual library (VL) now represents a feasible set of molecules. The next steps include removal of compounds that do not meet preset criteria of drug-like molecules. As indicated in Figure 5.20, about 65% of compounds have predicted solubility cLogS lower than −6.5. Solubility, molecular weight, and lipophilicity filters lead to removal of 67% of compounds. A detailed inspection of the removed molecules reveals that many of them feature ionizable groups (amines, acids) that are important pharmacophores and modulators of ADME properties. Therefore, about 2800 compounds are added back to the pool of desirable compounds. In addition, polyhalogenated molecules are removed before arriving to the virtual set of feasible and drug-like molecules (VL1). Analysis of structural diversity is performed in the following step to remove similar molecules. Building blocks and scaffolds of the resulting virtual set VL2 are then systematically tested for chemical performance. This is the most labor-intensive part of the library development. It involves iterative chemical optimization of each set of building blocks in a format that is as close as possible to ultimate production conditions. After 4–6 months of chemical adaptation (per chemist), optimized protocols are ready for RL to be followed by the production of

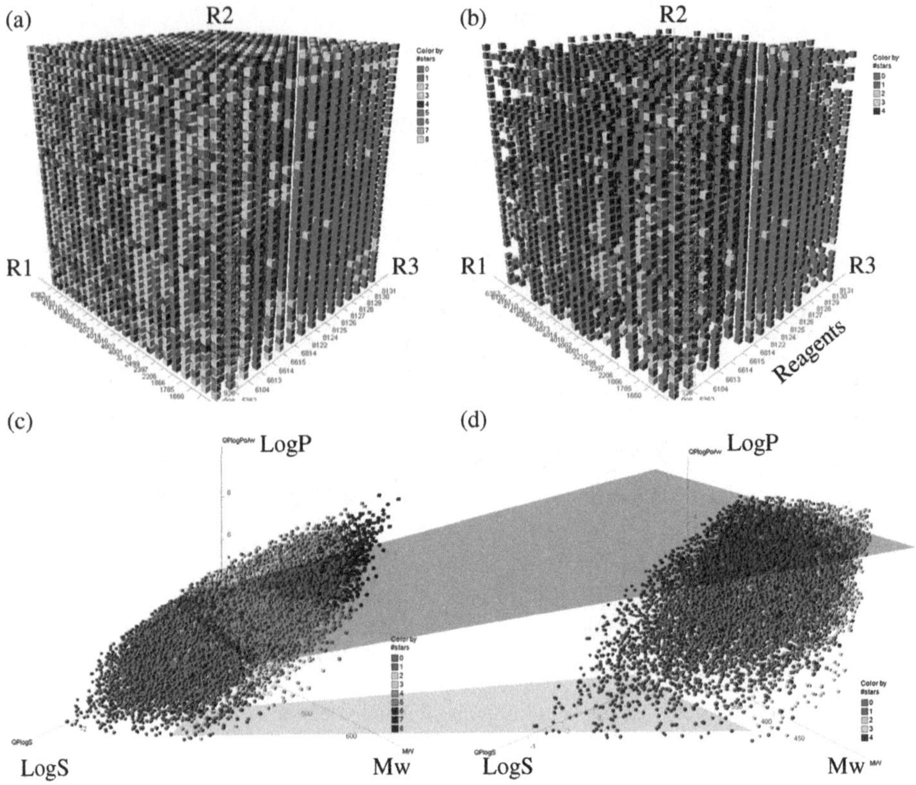


**FIGURE 5.21** Physicochemical properties mapped to virtual libraries. (a and c) Visual comparison of the original 3D scatter plots and (b and d) the final discovery library indicates.

8,000–10,000 compounds. A graphical representation of properties mapped to virtual libraries VL and VL1 is shown in Figure 5.21. Color coding reflects a violation of drug-like properties as calculated by the QikProp package from Schrödinger (number of stars). The greater the number, the least desirable properties are associated with a compound. Undesirable properties in panels (a) and (b) are indicated by points of very dark gray color (apparent along the R1 axis). Areas of undesirable compound properties in panels (c) and (d) are located at the tip (upper portion) of the point distribution (very dark and light gray shades).

***5.4.2.1 Library Production Techniques and Formats*** There are four principal methods for synthesis of solid-phase combinatorial libraries: split and mix techniques [15, 16], spatially addressable synthesis [12], directed sorting methods [24, 111, 112], and spatially addressable sorting methods [25]. These methods are detailed in the following.

*Split and Mix Approaches* Split and mix methodology is an ideal method for synthesis of large random peptide libraries. In the typical arrangement, solid support is divided into *N* equal portions (*N* is the number of amino acids at each step) and coupled with a given Fmoc amino acid. After removal of the Fmoc group, the resins are combined and washed. The procedure is repeated for each synthetic step until the last reaction step, where each portion of the resin is cleaved independently, yielding multiple batches of compound mixtures. To facilitate structure elucidation of the active compounds on a bead or to avoid labor-intensive deconvolution process, various detection methods, such as bead tagging [23, 113] or encoding methodologies [112, 114], have been developed.

While the split and mix method constitutes a very powerful way to produce millions of compounds, requirements for additional structure deconvolution limit its use. To alleviate limitations of structure determination, directed sorting and parallel synthesis technologies were developed in the early 1990s. These have remained the most prevalent methods for library production both in industry and academia.

*Spatially Addressable Parallel Synthesis* In this approach, compounds are synthesized in an array format, most conveniently in reaction blocks, such as deep-well plates or custom reaction blocks. The chemical history of each compound is simply determined by its location in the array.

Smaller combinatorial libraries (<500 compounds) can be prepared without any automation devices using standard deep-well plates (24/48/96 wells) or closed vessel reactions blocks. Also, there are several semimanual tools available for manual solid-phase synthesis [115, 116]. Reagents and resins are added manually in patterns of rows and columns using a manual pipettor. Solvent and reagents can be removed either by filtration or by aspiration from the top of the well using single of multiple-channel cannulas. Bottom filtration for solvent removal or compound collection is usually preferred over the aspiration for smaller runs that are frequently attended by the operator. Efficient filtration requires the utilization of membrane-based reaction blocks [117] to avoid leakage of reagents during the reaction (Fig. 5.22).

Production of larger combinatorial libraries (2000–5000 compounds) is a much more labor-intensive process and requires specific automation. The key steps (a–e) are detailed in the following.

**FIGURE 5.22** Custom 48-well reaction block for bottom filtration.

(a) Bulk Resin Preparation
There are several possibilities for efficient treatment of resin batches with the corresponding reagents. The following example describes our own adaptation of Torviq's domino blocks [118]. From the operational point of view, preparation of large batches of modified resin (first randomization step) is preferred to transfer and aliquoting of unsubstituted resin. In a typical washing cycle, a negative pressure is applied to plastic reactors allowing solvents to be delivered from a solvent reservoir kept at atmospheric pressure. Reactors are shaken for a predefined mixing time followed by the evacuation of all solvents and optional drying of the resin under vacuum (Fig. 5.23). All operations are semiautomated with throughput of up to four 200 g resin batches per run.

(b) Resin Transfer into Reaction Blocks
Resin transfer to deep-well plates can be done manually in the case of relatively small permutational libraries. This operation is rather labor intensive and error prone for large but incompletely randomized combichem libraries, the so-called sparse matrix libraries. For such libraries, solutions for automated resin delivery into deep-well plates have been developed. Resins can be delivered as slurry with a peristaltic pump connected to a xyz robotic platform or as a dry powder. In our example, the xyz robotic platform fitted with resin dispensing bowl feeder is used to deliver precise amounts of dry resin according to a predefined pattern. Such systems can be used to deliver resin into various reaction blocks (deep-well plates) or synthesis objects (Irori NanoKans, X-Kans).

For large exploratory libraries, 15–20 mg of resin is typically dispensed into each well.

**FIGURE 5.23** Custom resin washer.

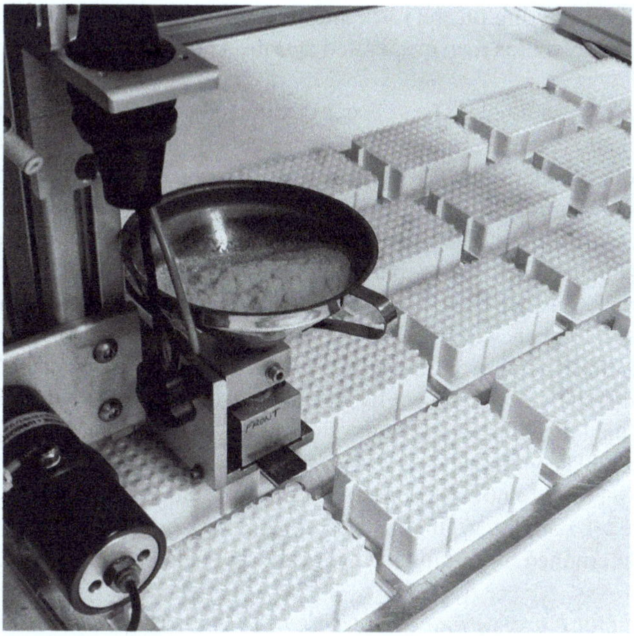

**FIGURE 5.24** Custom dry resin dispenser.

Amounts can be increased for hit resynthesis or for smaller arrays intended for high-throughput purification (Fig. 5.24).

(c) Reagent Addition
In its simplest adaptation, the addition of reagents is performed manually using multichannel pipettes, while liquid handling robots are preferred in the case of large sparse matrix libraries. In the latter case, custom software tools are developed to allow editing of liquid transfer command files and recording of the chemical history of each well in the array.

(d) Plate Washing
For a large number of plates and long unattended runs, solvent removal by top aspiration is a preferred solution that minimizes risk of filter clogging. Solvent is conveniently added by 8- or 96-channel moving cannulas, while 96-channel aspirating head is used to remove the solvent from the top of the wells. There are several implementations of dispenser/aspiration devices, most of them being adapted by labs for their specific needs. A simple set of plate washing and solvent delivery tools is available at the Torviq website [115].

(e) Cleavage and Extraction
The final stage of library production is compound cleavage. After the previous step of washing, individual library compounds are still attached to the solid support via a linker. Cleavage step not only accomplishes release of the compound from the solid support but also the deprotection of functional groups that had to be protected throughout the synthesis. Linker cleavage is usually followed by compound extraction. A cleavage reagent, for example, trifluoroacetic acid (TFA), is added to the reaction well followed by incubation and gentle shaking. Compounds are

obtained by filtering off the resin and by extraction into 96-well daughter plates. Extraction solvent is then evaporated and the sample is dried under vacuum.

*Directed Sorting Technology* The concept of directed sorting is based on tracking and deterministic planning of synthesis steps that selectively associate reagents with specific solid support objects. This enables full control over reagent admixture and allows the designer to easily eliminate building block combinations that, if utilized together, would yield products with undesirable properties. In this way, a fully permuted design can be selectively depopulated, yielding what is termed a "sparse matrix" library. Resin containers (e.g., Irori Kans, tea bags) [24] or synthesis objects such as SynLantern [119] are fitted with a radio frequency (RF) tag or 2D bar-coded lid. Upon reading the object identifier (ID), the object is directed toward a specific reactor. Analogously, after reaction completion and resin washing, object ID is read again and the object is directed toward a new predetermined reaction vessel. At the end of the synthesis, objects are placed into an array format for final compound cleavage. Location within the array is again driven by the object ID. The method combines the power of the split and mix technique with the deterministic nature of parallel synthesis. Reaction history (and identity) of synthesized compound is determined by reading the tag of the corresponding container.

*General Method for Library Production Using Irori Kan System* The list of compounds to be synthesized is first loaded into the system database. One container (NanoKan, X-Kan) is assigned to each individual compound. To allow for on-process quality control, extra Kans are added to permit cleavage and sample analysis. The availability of numerous Kan types supports multiple library strategies and library sizes. With a capacity of about 8 mg of resin per Kan, the NanoKan technology is best suited for large exploratory libraries. Mini X-Kans and Micro X-Kans can accommodate up to 35 and 70 mg of resin, respectively. At such a scale, X-Kans are suitable for smaller exploratory libraries as well as for resynthesis of putative hits.

Micro-Kans (300 mg of resin) will be preferred when larger amounts of crude material are needed. Flexible scalability allowing seamless transition from one type of Kan to another is a major advantage of the Irori technology. Importantly, chemistry performance is about the same between different Kan types. Figure 5.25 outlines the typical cycle of operations during synthesis of a library using the Irori synthesis platform.

Library synthesis begins by filling the NanoKans with solid support. A ceramic lid with 2D bar code is placed over the top of the NanoKan and pressed down to seal the container (A). This process is automated and performed in the dedicated resin filler. NanoKans are then sorted into bins (B) based on the common reagent. Synthesis is carried out outside the system in regular reaction glassware. Upon completion of the reaction step, NanoKans are pooled and all excess reagents are eliminated by multiple washings in NanoKan washer (C). The full cycle (B $\leftrightarrow$ C) is repeated for each reaction step. An extra sorting cycle can be added to allow for a specific reaction within a subset of NanoKans. For example, milder acidic conditions can be used for the final cleavage from resin to minimize unintended debenzylation of benzylethers when strong acid media would be used for the whole library. In the last reaction cycle, NanoKans are arranged into a 96- or 48-well format in the plate loader (D). In the next step, plates are transported into the cleavage and evaporation unit (E) for final cleavage and extraction. Cleavage reagent and solvents are evaporated, and compound is obtained in receiving blocks (F).

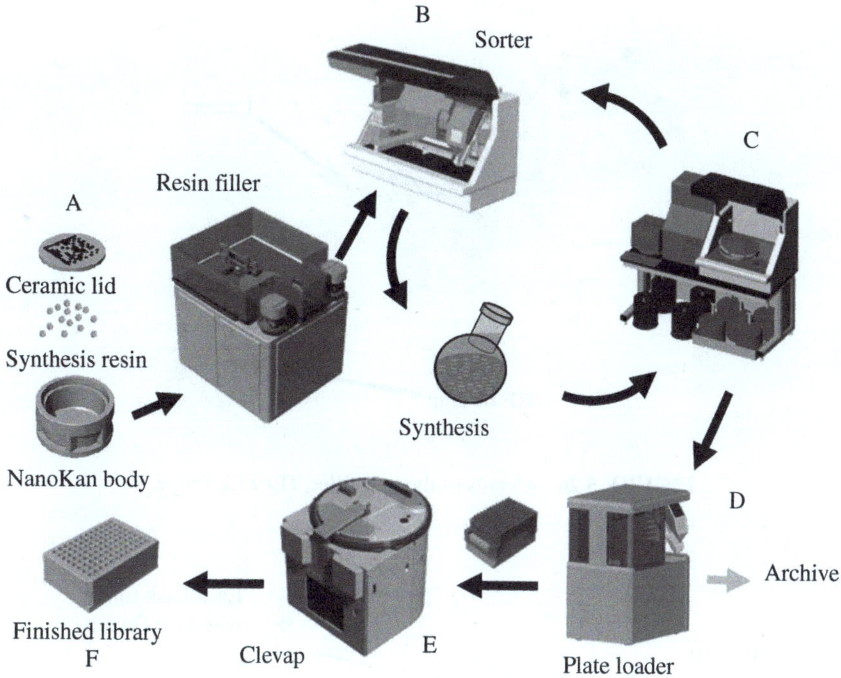


**FIGURE 5.25** The NanoKan directed sorting technology as adapted in Irori synthesis platform.

Since the synthesis progress of each NanoKan followed a unique path, each well of the receiving block contains discreet compound. Moreover, chemical composition of the compound is obtained by merging the structure file with the compound identity (alphanumeric string) and location file. Final quality control by LC/MS verifies integrity of the directed sorting algorithm and chemical composition of compounds.

*Spatially Addressed Sorting Technology* This particular approach, developed at the former Selectide Corporation, is based on manipulation of spatially arranged plastic frames charged with SynPhase lanterns as a macroscopic solid support [25]. Lanterns are friction fitted into an orderly two by eight array (Fig. 5.26). Additional features of frames include two V-shaped dents for accurate positioning in reactor cassettes and a 45° chamfer on the top to make spatial orientation unambiguous. To enable secondary (backup) tracking of frames, a groove is included into the frame design to accommodate an RF microchip. Depending on their type, substitution of SynPhase lanterns varies from 15 to 75 µmol. Those amounts translate into 7–37 mg of compound at an average molecular weight of 500 Da. Lantern series L (15 µmol) are suited for the frame adaptation, while series D and A (35 µmol and 75 µmol, respectively) are used for synthesis of larger amounts of a compound.

The use of 2D frames within a 3D spatial arrangement is illustrated by the example of a hypothetical library of $3 \times 3 \times 3$ inputs = 27 compounds (Fig. 5.27). Twenty-seven lanterns, nine in each flask, are first reacted with reagents A1–C1 and then placed into nine frames, such that three different lanterns are in each frame. In the next step, stacks of three frames are placed into reactors, each having a different reagent A2–C2. After

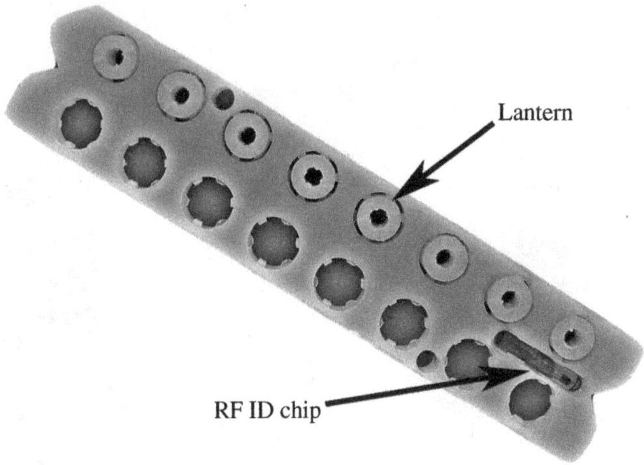


**FIGURE 5.26** Details of the half-filled TEFZEL frame.

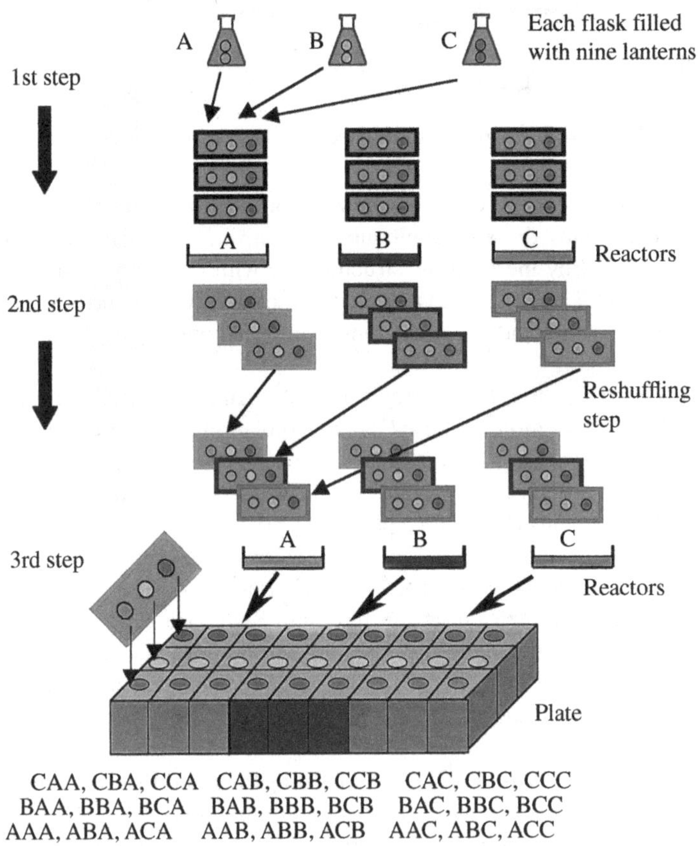


**FIGURE 5.27** Synthesis of a hypothetical 3×3×3 component library.

the stacks are removed from the reactors, one frame is taken from each stack, forming a new stack group of three frames. Each group of frames is then immersed into three reactors, each reactor having a different reagent A3–C3. After the final synthesis step and washing, lanterns are pushed out from the frames and placed into a labeled cleavage plate. Spatial address is thus mapped onto a hypothetical 27-well plate. Cleavage from lanterns would provide 27 unique compounds, where identity of each one can be deduced from the recorded path of the particular frame. It should be noted that for a library of three randomization inputs, only one redistribution process needs to take place.

The "frame" concept is similar to directed sorting techniques, in such a way that at each randomization step, the frames are arranged in stacks that are transported to unique reactors for chemical transformation. A major difference is that spatial address is maintained during the washing steps. The chemical history of each lantern/well is determined by the initial position of the frame within the stack and its location in the cleavage plate.

In summary, four key approaches to the synthesis of combinatorial libraries have been described. These methods vary by concept, level of automation, cost, and library scale. While the use of split and mix methodologies is usually reserved for special situations, automated and semiautomated parallel synthesis is often a method of choice for academic labs and smaller research enterprises. Fully automated parallel synthesis or directed sorting technologies are quite prevalent in larger industrial settings. In the context of our combichem strategy, it was the use of automated synthesis in deep-well plates and directed sorting technologies that were essential for building our screening collection of 900,000 small molecule compounds over the last 13 years.

***5.4.2.2 Library Realization*** In this section, we will present our own perspective on the process of library realization. Our experience includes production of about 400 libraries of small molecules. Methods of library production vary from manual to semiautomated and also include many production campaigns run in a fully automated mode.

*Library Format* Decision about the library format is done early, typically during the library development. Format is dictated by considerations such as:

(a) Purpose of the library (hit finding, structurally biased for a target class, hit optimization)
(b) Size of the library (also known as complexity)
(c) Fully permutational set versus sparse matrix (certain permutations are removed)
(d) Required amounts of library compounds
(e) Expected purity of library compounds
(f) Estimated cost of reagents, scaffolds, and materials

In general, large exploratory libraries (3000–6000 compounds) are produced as nonpurified samples in amounts up to 10 μmol/compound. Due to the number of library members, one of the directed sorting approaches is used. For amounts in the range of 3–5 mg, NanoKans (III) or lanterns in frames (IV) are the method of choice. For the fully permutational sets and simpler chemistries, plate synthesis (II) is the preferred production format. If library design calls for larger quantities, X-Kans, Macro-Kans, or parallel synthesis in reaction blocks will be considered as the first choice for library realization. Compatibility of chemistry with the production format is also an important consideration; however, it is usually secondary to the aforementioned considerations.

*Steps of Library Realization* General steps of library design and realization have already been outlined. Here, we provide practical details pertinent to chemistry optimization and library production. The initial step of library development involves assessment of chemical feasibility for the proposed synthetic route. Sometimes, alternative routes have to be explored to identify an approach that is compatible with methods of library production. During this stage, various sets of inputs (building blocks) are tested for each randomized site of the scaffold and reaction conditions evaluated for the optimal yield and purity. Any major side products are characterized and recorded to facilitate assignment of active components later on. A set of standard compounds is generally prepared and fully characterized during this initial step. Chemical properties, such as kinetic solubility and LogD, are acquired for representative molecules to assess predictiveness of *in silico* tools. The chemical stability of the scaffold is also evaluated to insure integrity during long-term storage in DMSO. Based on the information collected during the initial development steps, *in silico* design of the library (VL1) is completed (see Fig. 5.20).

In the next stage, building blocks for each randomized position are refined, and the list is submitted to chemical procurement tools for pricing and availability. Based on the building block performance and its availability, new lists of reagents are used for *in silico* enumeration to generate an "optimized" VL (VL2). Depending on the similarity of selected building blocks, *in silico* reenumeration may include additional diversity selection. Visualization and selection tools are utilized to complete the composition of the VL (Fig. 5.21).

To maximize the success of a library production, the chemist continues the optimization of reaction conditions to assure consistent yields and high purity of cleaved compounds. Chemical performance of building blocks can be evaluated either by testing all inputs by positional scanning or by a careful clustering and testing of representative building blocks within each cluster. Such evaluation of building blocks allows elimination of poorly performing inputs that were not identified during the initial chemistry evaluation.

Finally, a RL is prepared to test all "validated" inputs in a production environment. To be fully predictive, the RL must consist of final compounds where each randomized position is present in a statistically significant redundancy [120]. The number of compounds to be synthesized is obtained by the product of the largest input set by the requested redundancy. For example, for a three randomization library where R1 = 30, R2 = 50, and R3 = 20, a set of 250 (50 × 5) compounds is synthesized with each component present at least five times. After cleavage and analysis of all compounds, the design of the production library is finalized (VL2′). Providing that no unexpected changes in conditions and procedures were introduced during the library production, the purity profile of the final library will be nearly identical to the profile of the RL (Fig. 5.28). By our internal measures, combinatorial solid-phase library passes purity criteria if 75% of the library has purity (UV 220 nm) greater than 50%.

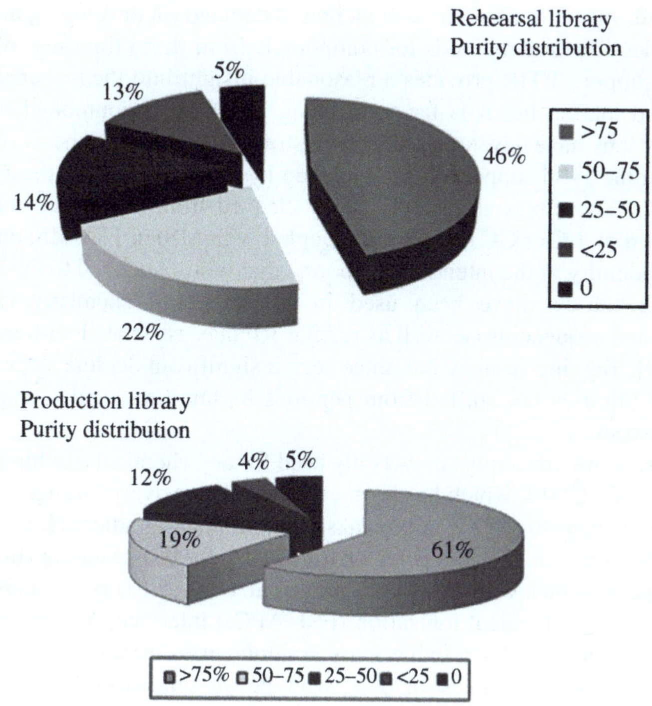


**FIGURE 5.28** Correlation of purity profiles between rehearsal and production libraries.

In summary, the realization of a solid-phase combinatorial library is a multistep process involving the initial evaluation of solid-phase chemistry, library development, selection of library format, virtual design based on physicochemical properties of library members, synthesis of an RL, and the ultimate library production. Optionally, the results of on-process or final quality control may lead to removal of poor quality samples before library registration and use in screening.

### 5.4.3 Analytical Chemistry and Solid-Phase Synthesis of Libraries

There are a number of reviews concerning analytical aspects of chemical combinatorial libraries [118, 121–124]. They are dedicated to listing and briefly describing all existing analytical methods that have been applied to the characterization of libraries. Some of those analytical methods include MALDI–MS, FTIR, and magic-angle spinning NMR for on-bead compounds. For cleaved compounds, high-throughput flow injection analysis (FIA)/MS, high-throughput NMR, LC/MS, capillary electrophoresis/MS (CE/MS), and other techniques have been employed. Some attention has also been paid to which data should be collected in order to provide a sufficient level of confidence about purity, quantity, and content of huge collections of chemicals.

Other reviews are more focused on requirements for analytical techniques as well as their suitability for different types of chemical combinatorial libraries. Since some libraries are synthesized on solid support only (without the possibility of cleaving the final compounds in solution), there is an increased demand for developing more sensitive and reliable identification methods for compounds from those libraries. With the right polymer as a support, FTIR provides a reasonable insight into the presence or absence of the intended matter, but it is far from being a definitive compound identification method. Many labs have developed their own strategies where synthesis of the desired compounds on the solid support is accompanied by simultaneous synthesis of tracking tags [125, 126]. These tags are usually easily cleaved from the carrier and are conveniently analyzed by LC or GC (sometimes coupled with MS) or FIA/MS, thus providing unambiguous identity of the intended compound that is still attached to the solid support bead. Oligonucleotides have been used in combinatorial chemistry (followed by amplification and sequencing) as well as regular RF tags associated with resin or object containers [24]. Tagging strategy has since seen a significant decline since the focus of combinatorial libraries has shifted from peptides to small molecules employing very diverse chemistries.

There is one technique almost universally used for any chemical combinatorial library analyses—that of LC/MS, which has been employed extensively throughout the history of combinatorial chemistry [127]. A typical setup is a binary gradient HPLC system with coupled UV–Vis or a diode-array detector followed by a fast-scanning (usually single-quad) mass spectrometer with an electrospray ionization (ESI) or a combination with atmospheric pressure chemical ionization (ESI–APCI) interface. A scan rate of at least 10 kDa/s is recommended for reliable mass assignment of closely eluting compounds. For concentration (or quantity) estimation, two types of detectors are used in most laboratories: evaporative light-scattering detector (ELSD) and chemiluminescence nitrogen detector (CLND). Other detectors have been proposed to quantify chemical libraries, including UV and charged aerosol detector (CAD) but their use is not widespread. In our lab, we have deployed both ELSDs and CLNDs, but the necessity of having a nitrogen atom in a molecule, ambiguity of signal output [128] for single and double nitrogen–nitrogen bonds (as in many nitrogen-containing heterocycles), and high maintenance requirements have led to exclusive use of ELS detectors.

Further development of the field has led to the development of fully automated analytical and preparative LC/MS systems for analysis and mass-triggered purification of combinatorial libraries [129, 130]. Parallel analysis and purification are achieved by using multiprobe autosamplers, multichannel UV detectors, and multiplexed mass-spectrometer interfaces (MUX). There are a few other useful classical HPLC detection techniques that can be applied for combinatorial library analysis. Those include fluorescence, electrochemical, refraction index, and polarimetric detectors. Special attention should be paid to keeping all flow cell volumes to a minimum and carefully eliminating all dead volumes from the flow paths. This results in high LC resolution and reduction of the analysis time for each sample. In our experience, the Agilent 1200 HPLC system coupled with Agilent 6140 or 6150 single-quadrupole mass spectrometer provides an adequate solution for library characterization without the need to employ ultrahigh-pressure systems. Rapid resolution columns and/or cartridges from ZORBAX (such as Eclipse XDB-C18 RRHT; 1.8 μm; 2.1 × 30 mm) produce exceptional chromatography for very different classes of compounds, usually resulting in peak widths of about 2–3 s. When an alternative separation is necessary, one can use the whole spectrum of other phases available in small columns from different vendors. For our purposes, we find the most effective to be supplemental fluorinated phase columns (like MacMod's ACE

Excel C18-PFP; 2.0 or 3.0 μm) and, for very hydrophilic compounds, the very unique Thermo Hypercarb columns. High-throughput HILIC separation with MS detection has also been performed in our lab for characterization of several chemical combinatorial libraries.

A few different strategies exist for sample selection for a combinatorial library QC. With earlier libraries reaching tens or even hundreds of thousands of compounds, it was obviously not feasible to have 100% of all library components characterized even if opted for FIA. Random sample selection covering 5–10% of the whole library is practically the only method for split–mix libraries where the spatial position of each component is unknown. Libraries obtained via parallel synthesis where each expected component's exact position is known can be sampled "diagonally" ("hockey stick") [131] or in a "cross-sectional" pattern [118]. We settled on using quasirandom sample selection, wherein the samples are selected automatically using a customized MS Excel VBA macro so that every building block at every single randomization point of the library is tested at least a few times. Redundancy is specified by the analyst for the randomization point that has the most building blocks. It is based on the number of those building blocks and the total size of the final library so that the number of samples that need to be analyzed is kept reasonable (usually no >500). The results of LC/MS runs are then processed and are compiled into Excel worksheets. Purity by UV at 220 nm; concentration by ELSD; observed masses, including by-products; and retention times are then further processed and averaged to estimate them against each building block used in the synthesis. That allows "cleaning" of the final library by removing compounds containing poorly performing blocks or prediction of possible by-products that could be expected in library wells found to be active in biological screening.

Fast data acquisition stipulates fast data processing and result reporting. The software packages that came standard with LC/MS systems in the 1990s to early 2000s were mainly focused on operating the equipment and processing acquired data one sample at a time. Occasionally, those software tools allowed for batch processing of the total ion current (TIC) signal for concentration estimates (for metabolic stability studies). None of them satisfied the need for high-throughput data processing to estimate the purity of multiple samples based on a UV signal as well as concentrations/quantity based on ELSD or CLND signal in a single run. In the beginning, it was quite a challenging task, one that we solved with multiple programs running in batch mode to perform all the calculations and reporting. Acquired LC/MS data from Power PC was processed in batch mode using Apple Script controlling the manufacturer's data processing program and Microsoft Excel where the results were collected and combined into a single spreadsheet. Those spreadsheets contained important data (such as retention time [RT], percentage purity by selected UV signal and concentration estimate based on an ELSD signal output, and molecular masses observed in every UV peak of interest) for multiple samples and also included bitmaps of chromatograms for each sample. After switching to the PC platform, we had to actively search for an affordable data processing package that would be able to read raw data from multiple OEM instruments, handle hundreds of analyses simultaneously, and report combined results for batches and not just single compounds. Virscidian's Analytical Studio [132] has proved to be such a software solution, and it is being extensively used in our lab for almost all processing and reporting QC results straight to the Laboratory Information Management System (LIMS).

For drug discovery applications, it becomes very important to estimate the properties of newly synthesized organic matter as soon as possible [133]. This task is usually achieved by performing early ADME studies on all promising new compounds, evaluating new compound solubility, pKa, cell permeability, lipophilicity, Cyp inhibition, metabolic/tissue/plasma stability, and metabolic profile. At least some of these studies can also be easily performed with the same tools that are used for routine analysis of combinatorial libraries. That led to development of high-throughput ADME methods employing already available LC/MS systems [134, 135].

## 5.5 THE OUTCOME OF TUCSON COMBINATORIAL CHEMISTRY AT SANOFI

This section summarizes selected outcomes and observations that pertain to our specific implementation of solid-phase combinatorial chemistry at the Sanofi Research Center in Tucson, a drug discovery unit that had its origins in start-up company Selectide Corporation before being incorporated into Hoechst, now part of Sanofi [4].

There are many variables at play that influence drug discovery success rates and compound attrition, which ultimately render measures of productivity difficult to interpret. Such variables include business decisions around clinical indications, the company's strategic focus, target-related issues (*in vivo* efficacy and selectivity), and the failure of further lead optimization (typically in respect to toxicity). Nevertheless, there are some specific lessons and high-level conclusions that we can derive from our experience of combinatorial chemistry in a drug discovery setting.

### 5.5.1 Overall Strategy

Our implementation of solid-phase combinatorial chemistry has the following attributes. The number of employees at our site has historically ranged between 60 and 70 full-time associates, including 20–25 chemists and about 20 biologists. Our lead discovery unit comprises four key functions—chemistry, biology, analytics, and informatics with local ADME support. All functions are fully integrated and constantly optimized for the generation of leads and their subsequent transfer to global units within the company for further optimization.

Since 2000, solid-phase combinatorial chemistry has provided about 900,000 single compounds that represent the bulk of our combichem screening collection (currently 1.2 million samples in total). In addition, about 70,000 HPLC-purified samples were produced that sample the diversity of the larger screening collection. The main tactical approaches supporting our strategy are summarized in Figure 5.29 in the three areas of combinatorial chemistry, screening, and medicinal chemistry.

A cartoon of the typical process with screening as input and lead compounds as output is shown in Figure 5.30. As compounds progress along the stages of Exploration and Optimization, their optimization becomes gradually constrained by selected ADME and physicochemical properties. In this standard multiparametric optimization approach, methods of medicinal chemistry, parallel synthesis, and combichem are employed to increase the value of the lead compounds.

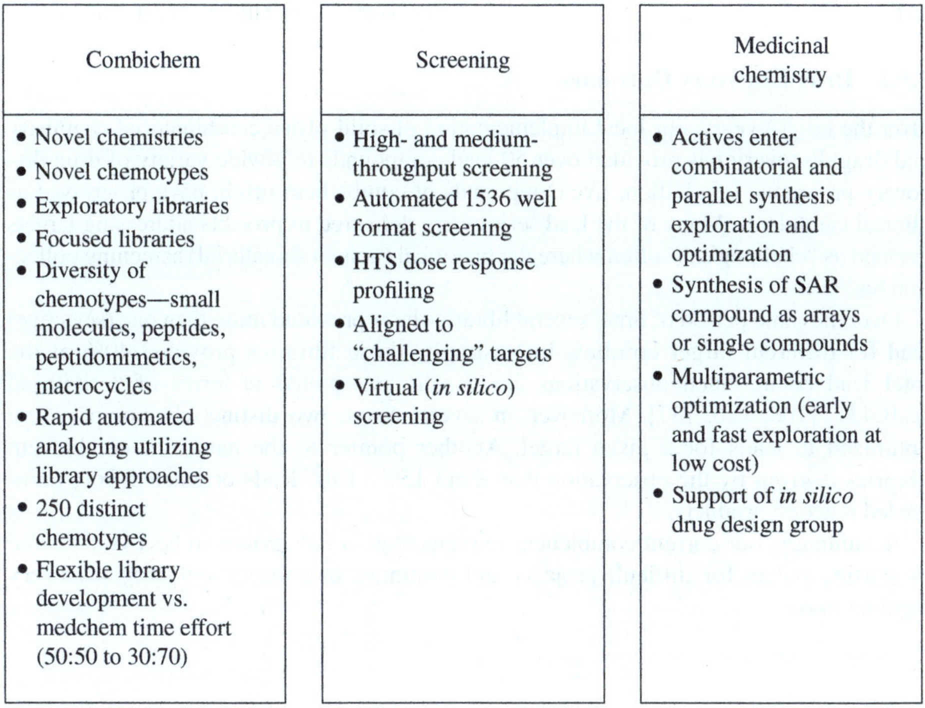


**FIGURE 5.29** Parallel approaches to lead discovery.

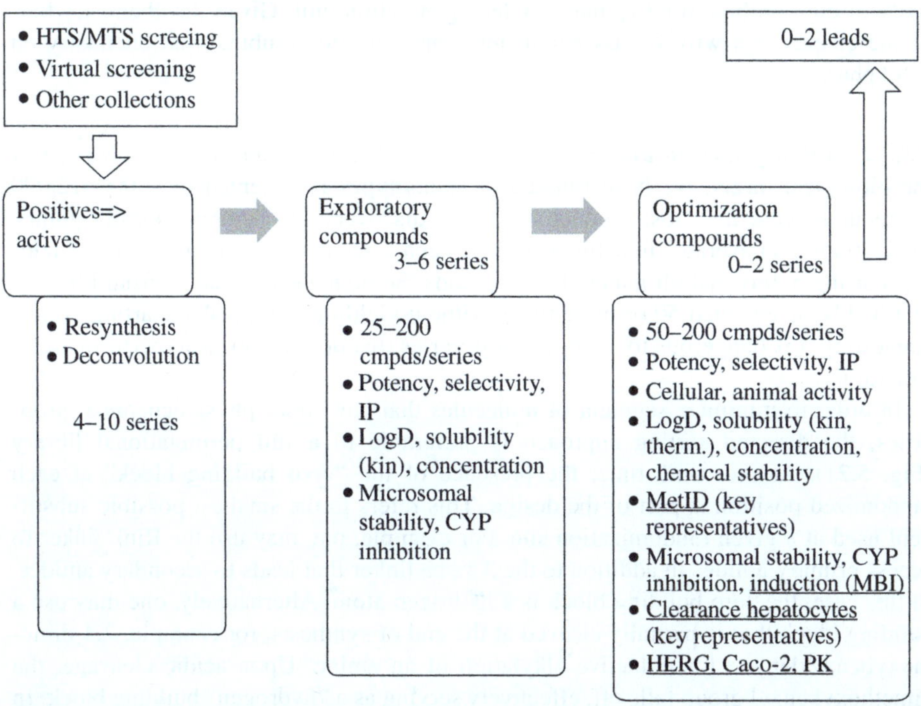


**FIGURE 5.30** Compound flow and main stages of lead discovery process. Underlined ADME tests are provided only for key representatives of each series.

### 5.5.2 Drug Discovery Outcomes

Over the last 12 years, our local implementation of solid-phase combinatorial chemistry and drug discovery has provided over 50 lead compounds to a wide variety of drug discovery programs. In addition, five compounds of combichem origin have progressed as clinical candidates. Many of the lead series were delivered to projects addressing targets deemed as "challenging," often where the historical (noncombinatorial) screening collection had failed.

Over the same period of time, several libraries have provided more than one (selective) lead for different target families. Interestingly, those libraries provided 40% of the total lead count. Such observations are usually interpreted in terms of "privileged scaffolds" [100, 136, 137]. Moreover, in several cases, two distinct chemotypes were optimized to leads for a given target. Another pointer to the nature of combichem libraries is given by the observation that about 15% of the leads originated from unintended reaction products.

In summary, our current combichem implementation has proven to be a vital source of starting points for difficult projects and continues to enhance our drug discovery pipeline today.

### 5.5.3 Key Parameters of Combichem Productivity

For many years, we have been looking for the key principles that correlate with productivity of our combinatorial technology lead generation unit. Given the chemistry bias of the chapter, we will discuss parameters connected to combinatorial chemistry on solid phase.

***5.5.3.1 Library Design and Synthesis*** In the absence of any prior knowledge of the biological target, we have found that chemotypes represented by 3,000–12,000 compounds were the most productive for the discovery of chemical starting points. Interestingly, on average, new libraries have been screened on 32 targets before yielding starting points that ultimately became leads. Some leads have come from libraries that had been screened 50 or more times without yielding optimizable starting points, suggesting it is dangerous to assume that libraries that do not appear to perform are of low value.

In order to minimize segment of molecules that have poor physicochemical properties, the directed sorting approach is preferable to a full permutational library (Fig. 5.21). At the same time, the presence of the "zero building block" at each randomized position is part of the design. This refers to the smallest possible substituent used at a given randomization site. For example, one may use the Rink linker to access primary amides in addition to the Ameba linker that leads to secondary amides. In this case, the zero building block is a hydrogen atom. Alternatively, one may use a building block that is partially cleaved at the end of synthesis, for example, 2,4-dimethoxybenzaldehyde for reductive alkylation of an amine. Upon acidic cleavage, the dimethoxybenzyl group falls off, effectively serving as a "hydrogen" building block. In these cases, larger 3-randomization libraries include three smaller 2-randomization libraries as an inherent part of the synthesis. Even though it is more labor intensive, the concept of scaffold randomization is often favored, preferably at the stage of library proposal selection (Fig. 5.14).

The intrinsic advantage of densely populated libraries is the fast and efficient analoging around identified hits. Since a robust chemistry protocol is already developed, its adaptation to similar chemistries is faster. Also, as a result of dense elaboration of the scaffold, structure–activity relationship (SAR) patterns frequently emerge directly from the library actives.

***5.5.3.2 Purity of Combichem Libraries*** There is a common perception that the main drawback of solid-phase combinatorial libraries is their impure character. In our experience, the major problem of nonpurified libraries is rather the fluctuation of compound concentration. Determination of dose–response values (e.g., IC50) for compounds of uneven concentration, which is further influenced by the noisy character of HTS assays, brings even more uncertainty to the final values. To compensate for the lower confidence in activity data, testing larger libraries of structurally related compounds is a sensible approach. In general, impurities do not create a significant problem especially when HPLC deconvolution of sample wells is available. Indeed, ironically, about 25% of our active series originated from unintended side product present in the screening well. More importantly, as the consequence of these variables, screening results cannot be used to establish "negative" SAR. This is not to say that considerations of library purity can be completely disregarded. Indeed, our internal criteria for the level of purity are set for 75% of the library members to have purity better than 50% by UV at 220 nm. Overall, a delicate balance of compound quality, quantity, and HTS methods is best struck through the use of pragmatic approaches. Finding the next example in a well-populated series is not as important as not missing a new series. Methods have been optimized to discover even weak actives by retesting deeply into the tail of the normal activity distribution.

## 5.6 CONCLUSIONS AND OUTLOOK

In this chapter, we have provided the reader with examples of solid-phase combinatorial chemistry as practiced in early drug discovery. Although solid-phase chemistry has made great advances, its tactical use for generation of unbiased screening collections has not been fully realized. Nonetheless, while the field has been nearly abandoned by larger pharmaceutical organizations, there are many examples that justify its continued use and development.

The fundamentals of combinatorial chemistry share the same principles that underlie biological systems: diversity of form and function. By adhering to a few key practices, outcomes of drug discovery can be significantly improved. Those practices include process of library design, novelty and richness of scaffold topology, synthetic chemistry, purity and quantity criteria, and the matching of technology to specific tasks. Approaches that employ randomized scaffold libraries, the selection of robust solid-phase reactions, and an emphasis on structural novelty are critical for unbiased screens and development of useful and productive libraries.

The discovery portfolios of many pharma and biotech companies contain numerous challenging targets. Indeed, to successfully "drug" targets such as class B GPCR receptors, intracellular signaling complexes, misfolded proteins, and transcription factors will require the use of every research asset and approach available. The composition of diverse screening collections is thus of strategic importance to enabling the discovery of new drug candidates. Compound libraries will continue to be enriched and enlarged by new compounds of diverse origin by using methods such as combichem platforms along with parallel synthesis, acquisition of commercial collections, collection exchanges, and use of large virtual sets.

The pendulum is swinging back to building diverse collections of small molecules and with it comes the challenge of finding the most efficient ways of producing them. For many years, we stayed confident that solid-phase combinatorial chemistry was one of them, and we believe that today this is more the case than ever before.

## REFERENCES

[1] Scannell, J. W., Blanckley, A., Boldon, H., Warrington, B. (2012). Diagnosing the decline in pharmaceutical R&D efficiency. *Nat. Rev. Drug Discov.* **11** (3), 191–200.

[2] FDA/CDER (2011). Comparison of NMEs Approved in 2010 to Previous Years, http://www.fda.gov/downloads/Drugs/DevelopmentApprovalProcess/HowDrugsareDevelopedandApproved/DrugandBiologicApprovalReports/UCM242695.pdf, accessed on: September 25, 2013.

[3] Kodadek, T. (2011). The rise, fall and reinvention of combinatorial chemistry. *Chem. Commun.* **47** (35), 9757–9763.

[4] Koppal, T. (2003). Combichem's desert hideaway. *Drug Discov. Devel.* **6** (7), 30.

[5] Merrifield, R. B. (1963). Solid phase peptide synthesis. I. The synthesis of a tetrapeptide. *J. Am. Chem. Soc.* **85** (14), 2149–2154.

[6] Merrifield, R. B., Stewart, J. M., Jernberg, N. (1966). Instrument for automated synthesis of peptides. *Anal. Chem.* **38** (13), 1905–1914.

[7] Leznoff, C. C. (1978). The use of insoluble polymer supports in general organic synthesis. *Acc. Chem. Res.* **11**, 327–333.

[8] Seeberger, P. H., Haase, W. C. (2000). Solid—phase oligosaccharide synthesis and combinatorial carbohydrate libraries. *Chem. Rev.* **100** (12), 4349–4393.

[9] Fodor, S. P., Read, J. L., Pirrung, M. C., Stryer, L., Lu, A. T., Solas, D. (1991). Light-directed, spatially addressable parallel chemical synthesis. *Science* **251** (4995), 767–773.

[10] Frank, R., Heikens, W., Heisterberg-Moutsis, G., Blocker, H. (1983). A new general approach for the simultaneous chemical synthesis of large numbers of oligonucleotides: segmental solid supports. *Nucleic Acids Res.* **11** (13), 4365–4377.

[11] Beck-Sickinger, A., Weber, P., Soderman, M., Wier, A. (2002). *Combinatorial Strategies in Biology and Chemistry*, John Wiley & Sons, Ltd., West Sussex.

[12] Geysen, H. M., Meloen, R. H., Barteling, S. J. (1984). Use of peptide synthesis to probe viral antigens for epitopes to a resolution of a single amino acid. *Proc. Natl. Acad. Sci. U. S. A.* **81** (13), 3998–4002.

[13] Houghten, R. A. (1985). General method for the rapid solid-phase synthesis of large numbers of peptides: specificity of antigen-antibody interaction at the level of individual amino acids. *Proc. Natl. Acad. Sci. U. S. A.* **82** (15), 5131–5135.

[14] Houghten, R. A., Pinilla, C., Blondelle, S. E., Appel, J. R., Dooley, C. T., Cuervo, J. H. (1991). Generation and use of synthetic peptide combinatorial libraries for basic research and drug discovery. *Nature* **354**, 84–86.

[15] Furka, A., Sebestyen, F., Asgedom, M., Dibo, G. (1991). General method for rapid synthesis of multicomponent peptide mixtures. *Int. J. Pept. Protein Res.* **37** (6), 487–493.

[16] Lam, K. S., Salmon, S. E., Hersh, E. M., Hruby, V. J., Kazmierski, W. M., Knapp, R. J. (1991). A new type of synthetic peptide library for identifying ligand-binding activity. *Nature* **354**, 82–84.

[17] Czarnik, A. W. (1997). Encoding strategies in combinatorial chemistry. *Proc. Natl. Acad. Sci. U. S. A.* **94** (24), 12738–12739.

[18] Barnes, C., Balasubramanian, S. (2000). Recent developments in the encoding and deconvolution of combinatorial libraries. *Curr. Opin. Chem. Biol.* **4**, 346–350.

[19] Gordon, E. M., Kerwin, J. F. (1998). *Combinatorial Chemistry and Molecular Diversity in Drug Discovery*, Wiley-Liss, New York.

[20] Kocis, P., Krchnak, V., Lebl, M. (1993). Symmetrical structure allowing the selective multiple release of a defined quantity of peptide from a single bead of polymeric support. *Tetrahedron Lett.* **34**, 7251–7252.

[21] Lebl, M., Patek, M., Kocis, P., Krchnak, V., Hruby, V. J., Salmon, S. E., Lam, K. S. (1993). Multiple release of equimolar amounts of peptides from a polymeric carrier using orthogonal linkage-cleavage chemistry. *Int. J. Pept. Protein Res.* **41** (2), 201–203.

[22] Bunin, B. A., Ellman, J. A. (1992). A general and expedient method for the solid phase synthesis of 1,4-benzodiazepine derivatives. *J. Am. Chem. Soc.* **114**, 10997–10998.

[23] Ohlmeyer, M. H. J., Swanson, R. N., Dillard, L. W., Reader, J. C., Asouline, G., Kobayashi, R., Wigler, M., Still, W. C. (1993). Complex synthetic chemical libraries indexed with molecular tags. *Proc. Natl. Acad. Sci. U. S. A.* **90**, 10922–10926.

[24] Nicolaou, K. C., Xiao, X. Y., Parandoosh, Z., Senyei, A., Nova, M. P. (1995). Radiofrequency encoded combinatorial chemistry. *Angew. Chem. Int. Ed. Engl.* **34** (20), 2289–2291.

[25] Patek, M., Safar, P., Smrcina, M., Wegrzyniak, E., Bjergarde, K., Weichsel, A., Strop, P. (2004). 2D and 3D spatially addressed arrays for high-throughput automated synthesis of combinatorial libraries. *J. Comb. Chem.* **6** (1), 43–49.

[26] Dolle, R. E. (1998). Comprehensive survey of combinatorial libraries with undisclosed biological activity: 1992–1997. *Mol. Divers.* **4** (4), 233–256.

[27] Dolle, R. E. (1997). Comprehensive survey of chemical libraries yielding enzyme inhibitors, receptor agonists and antagonists, and other biologically active agents: 1992 through 1997. *Mol. Divers.* **3** (4), 199–233.

[28] Dolle, R. E., Nelson, K. H. (1999). Comprehensive survey of combinatorial library synthesis: 1998 [review]. *J. Comb. Chem.* **1** (4), 235–282.

[29] Dolle, R. E. (2000). Comprehensive survey of combinatorial library synthesis: 1999. *J. Comb. Chem.* **2** (5), 383–433.

[30] Dolle, R. E. (2001). Comprehensive survey of combinatorial library synthesis: 2000. *J. Comb. Chem.* **3** (6), 1–41.

[31] Dolle, R. E. (2002). Comprehensive survey of combinatorial library synthesis: 2001. *J. Comb. Chem.* **4** (5), 369–418.

[32] Dolle, R. E. (2003). Comprehensive survey of combinatorial library synthesis: 2002. *J. Comb. Chem.* **5** (6), 693–753.

[33] Dolle, R. E. (2004). Comprehensive survey of combinatorial library synthesis: 2003. *J. Comb. Chem.* **6** (5), 623–679.

[34] Dolle, R. E. (2005). Comprehensive survey of combinatorial library synthesis: 2004. *J. Comb. Chem.* **7** (6), 739–798.

[35] Dolle, R. E., Le, B. B., Morales, G. A., Moriarty, K. J., Salvino, J. M. (2006). Comprehensive survey of combinatorial library synthesis: 2005. *J. Comb. Chem.* **8** (5), 597–635.

[36] Dolle, R. E., Le, B. B., Goodman, A. J., Morales, G. A., Salvino, J. M., Zhang, W. (2007). Comprehensive survey of chemical libraries for drug discovery and chemical biology: 2006. *J. Comb. Chem.* **9** (6), 855–902.

[37] Dolle, R. E., Le, B. B., Goodman, A. J., Morales, G. A., Thomas, C. J., Zhang, W. (2008). Comprehensive survey of chemical libraries for drug discovery and chemical biology: 2007. *J. Comb. Chem.* **10** (6), 753–802.

[38] Dolle, R. E., Le, B. B., Goodman, A. J., Morales, G. A., Thomas, C. J., Zhang, W. (2009). Comprehensive survey of chemical libraries for drug discovery and chemical biology: 2008. *J. Comb. Chem.* **11** (5), 739–790.

[39] Dolle, R. E., Bourdonnec, B. L., Worm, K., Morales, G. A., Thomas, C. J., Zhang, W. (2010). Comprehensive survey of chemical libraries for drug discovery and chemical biology: 2009. *J. Comb. Chem.* **12** (6), 765–806.

[40] Lebl, M. (1999). Parallel personal comments on "classical" papers in combinatorial chemistry. *J. Comb. Chem.* **1** (1), 3–24.

[41] Verdine, G. L. (1996). The combinatorial chemistry of nature. *Nature* **384** (6604 Suppl), 11–13.

[42] Tan, D. S., Foley, M. A., Shair, M. D., Schreiber, S. L. (1998). Stereoselective synthesis of over two million compounds having structural features both reminiscent of natural products and compatible with miniaturized cell-based assays. *J. Am. Chem. Soc.* **120** (33), 8565–8566.

[43] Spring, D. R. (2003). Diversity-oriented synthesis: a challenge for synthetic chemists. *Org. Biomol. Chem.* **1** (22), 3867–3870.

[44] Burke, M. D., Berger, E. M., Schreiber, S. L. (2003). Generating diverse skeletons of small molecules combinatorially. *Science* **302** (5645), 613–618.

[45] Domling, A. (2000). The discovery of new isocyanide-based multi-component reactions. *Curr. Opin. Chem. Biol.* **4** (3), 318–323.

[46] Keating, T. A., Armstrong, R. W. (1995). Molecular diversity via a convertible isocyanide in the Ugi four-component condensation. *J. Am. Chem. Soc.* **117**, 7842–7843.

[47] Curran, D. P. (1998). Strategy-level separations in organic synthesis—from planning to practice. *Angew. Chem. Int. Ed. Engl.* **37** (9), 1175–1196.

[48] Joost, N. H. R., Sijbren, O. (2010). *Dynamic Combinatorial Chemistry*, Wiley-VCH, Weinheim.

[49] Terrett, N. K. (2009). Combinatorial Chemistry—an Online Journal, http://www.sciencedirect.com/science/journal/14643383/, accessed on: September 1, 2013.

[50] Wilson, S. R., Czarnik, A. W. (1997). *Combinatorial Chemistry: Synthesis and Application*, Wiley, New York.

[51] Fassina, G., Stanislaus, M. (eds) (1999). *Combinatorial Chemistry and Technology: Principles, Methods and Applications*, Marcel Dekker, Inc., New York/Basel.

[52] Stockwell, R. B. (2011). *The Quest for the Cure: The Science and Stories Behind the Next Generation of Medicines*, 1st edition, Columbia University Press, New York.

[53] Polishchuk, P. G., Madzhidov, T. I., Varnek, A. (2013). Estimation of the size of drug-like chemical space based on GDB-17 data. *J. Comput. Aided Mol. Des* **27** (8), 675–679.

[54] Hann, M. M., Oprea, T. I. (2004). Pursuing the leadlikeness concept in pharmaceutical research. *Curr. Opin. Chem. Biol.* **8** (3), 255–263.

[55] Dörwald, F. Z. (ed.) (2000). *Organic Synthesis on Solid Phase: Supports, Linkers, Reactions*, Wiley-VCH, Weinheim.

[56] Winter, M., Warrass, R. (2000). Resins and anchors for solid phase organic synthesis, in *Combinatorial Chemistry: A Practical Approach*, **6** (ed. Fenniri, H.), Oxford University Press, Oxford/New York, pp. 117–138.

[57] Li, W., Yan, B. (1998). Effects of polymer supports on kinetics of solid-phase organic reactions: a comparison of polystyrene and TentaGel based resins. *J. Org. Chem.* **63**, 4092–4097.

[58] Adams, J. H., Cook, R. M., Hudson, D., Jammalamadaka, V., Lyttle, M. H., Songster, M. F. (1998). A reinvestigation of the preparation, properties, and applications of aminomethyl and 4-methylbenzhydrylamine polystyrene resins. *J. Org. Chem.* **63** (11), 3706–3716.

[59] Singh, J., Allen, M. P., Ator, M. A., Gainor, J. A., Whipple, D. A., Solowiej, J. E., Treasurywala, A. M., Morgan, B. A., Gordon, T. D., Upson, D. A. (1995). Validation of screening immobilized peptide libraries for discovery of protease substrates. *J. Med. Chem.* **38** (2), 217–219.

[60] Merrifield, R. B. (1995). Solid-phase synthesis, in *Peptides: Synthesis, Structures, and Applications*, **3** (ed. Gutte, B.), Academic Press, San Diego, pp. 159–194.

[61] Eichler, J., Bienert, M., Stierandova, A., Lebl, M. (1991). Evaluation of cotton as a carrier for solid-phase peptide synthesis. *Pept. Res.* **4** (5), 296–307.

[62] Rasoul, F., Ercole, F., Pham, Y., Bui, C. T., Wu, Z., James, S. N., Trainor, R. W., Wickham, G., Maeji, N. J. (2000). Grafted supports in solid-phase synthesis. *Biopolymers* **55** (3), 207–216.

[63] Gerritz, S. W., Norman, M. H., Barger, L. A., Berman, J., Bigham, E. C., Bishop, M. J., Drewry, D. H., Garrison, D. T., Heyer, D., Hodson, S. J., Kakel, J. A., Linn, J. A., Marron, B. E., Nanthakumar, S. S., Navas, F. J., III (2003). High-throughput manual parallel synthesis using SynPhase crowns and lanterns. *J. Comb. Chem.* **5** (2), 110–117.

[64] Guillier, F., Orain, D., Bradley, M. (2000). Linkers and cleavage strategies in solid-phase organic synthesis and combinatorial chemistry. *Chem. Rev.* **100** (6), 2091–2157.

[65] Lebreton, S., Patek, M. (2009). Safety-catch linker units, in *Linker Strategies in Solid-Phase Organic Synthesis* (ed. Scott, P. J. H.), John Wiley & Sons, Ltd, Chichester, pp. 195–220.

[66] James, I. W. (1999). Linkers for solid phase organic synthesis [review]. *Tetrahedron* **55** (16), 4855–4946.

[67] Patek, M., Lebl, M. (1998). Safety-catch and multiply cleavable linkers in solid-phase synthesis. *Biopolymers* **47**, 353–363.

[68] Fenniri, H. (ed.) (2000). *Combinatorial Chemistry: A Practical Approach*, Oxford University Press, Oxford/New York.

[69] Plunkett, M. J., Ellman, J. A. (1995). A silicon-based linker for traceless solid-phase synthesis. *J. Org. Chem.* **60**, 6006–6007.

[70] Hu, Y. H., Porco, J. A., Labadie, J. W., Gooding, O. W., Trost, B. M. (1998). Novel polymer-supported trialkylsilanes and their use in solid-phase organic synthesis. *J. Org. Chem.* **63** (13), 4518–4521.

[71] Hughes, I. (1996). Application of polymer-bound phosphonium salts as traceless supports for solid phase synthesis. *Tetrahedron Lett.* **37**, 7595–7598.

[72] Plunkett, M. J., Ellman, J. A. (1997). Germanium and silicon linking strategies for traceless solid-phase synthesis. *J. Org. Chem.* **62** (9), 2885–2893.

[73] Ruhland, T., Andersen, K., Pedersen, H. (1998). Selenium-linking strategy for traceless solid-phase synthesis: direct loading, aliphatic C–H bond formation upon cleavage and reaction monitoring by gradient MAS NMR spectroscopy. *J. Org. Chem.* **63**, 9204–9211.

[74] Nicolaou, K. C., Winssinger, N., Hughes, R., Smethurst, C., Cho, S. Y. (2000). New selenium-based safety-catch linkers: solid-phase semisynthesis of vancomycin. *Angew. Chem. Int. Ed. Engl.* **39** (6), 1084–1088.

[75] Young, J. K., Nelson, J. C., Moore, J. S. (1994). Synthesis of sequence specific phenylacetylene oligomers on an insoluble support. *J. Am. Chem. Soc.* **116**, 10841–10842.

[76] Jones, L., Schumm, J. S., Tour, J. M. (1997). Rapid solution and solid phase syntheses of oligo(1,4-phenylene ethynylene)s with thioester termini—molecular scale wires with alligator clips—derivation of iterative reaction efficiencies on a polymer support. *J. Org. Chem.* **62** (5), 1388–1410.

[77] Brase, S., Kobberling, J., Enders, D., Lazny, R., Wang, M. F., Brandtner, S. (1999). Triazenes as robust and simple linkers for amines in solid-phase organic synthesis. *Tetrahedron Lett.* **40** (11), 2105–2108.

[78] Brase, S., Dahmen, S., Pfefferkorn, M. (2000). Solid-phase synthesis of urea and amide libraries using the T2 triazene linker. *J. Comb. Chem.* **2** (6), 710–715.

[79] Vojkovsky, T., Weichsel, A., Patek, M. (1998). Solid-phase synthesis of heterocycles containing a 1-acyl-3-oxopiperazine skeleton. *J. Org. Chem.* **63** (10), 3162–3163.

[80] Rosse, G., Strickler, J., Patek, M. (2004). Efficient solid-phase synthesis of disubstituted 1,3-dihydro-imidazol-2-ones. *Synlett* (12), 2167–2168.

[81] Todd, M. H., Ndubaku, C., Bartlett, P. A. (2002). Amino acid derived heterocycles: lewis acid catalyzed and radical cyclizations from peptide acetals. *J. Org. Chem.* **67** (12), 3985–3988.

[82] Veerman, J. J. N., Rutjes, F. P. J. T., van Maarseveen, J. H., Hiemstra, H. (1999). A novel acid stable/base labile carbamate linker for *N*-acyliminium ion reactions on solid support. *Tetrahedron Lett.* **40** (33), 6079–6082.

[83] Speckamp, W. N., Moolenaar, M. J. (2000). New developments in the chemistry of *N*-acyliminium ions and related intermediates. *Tetrahedron* **56** (24), 3817–3856.

[84] La, V., Lemrova, B., Krchnak, V. (2013). Regioselective incorporation of backbone constraints compatible with traditional solid-phase peptide synthesis. *ACS Comb. Sci.* **15** (1), 59–72.

[85] La, V., Dolensky, B., Krchnak, V. (2013). Polymer-supported stereoselective synthesis of tetrahydro-2H-oxazolo[3,2-a]pyrazin-5(3H)-ones from N-(2-Oxo-ethyl)-derivatized dipeptides via eastbound iminiums. *ACS Comb. Sci.* **15** (3), 162–167.

[86] Eguchi, M., Lee, M. S., Nakanishi, H., Stasiak, M., Lovell, S., Kahn, M. (1999). Solid-phase synthesis and structural analysis of bicyclic β-turn mimetics incorporating functionality at the i to i+3 positions. *J. Am. Chem. Soc.* **121** (51), 12204–12205.

[87] Kim, H. O., Nakanishi, H., Lee, M. S., Kahn, M. (2000). Design and synthesis of novel conformationally restricted peptide secondary structure mimetics. *Org. Lett.* **2** (3), 301–302.

[88] Groth, T., Meldal, M. (2001). N-Terminal peptide aldehydes as electrophiles in combinatorial solid phase synthesis of novel peptide isosteres. *J. Comb. Chem.* **3** (1), 45–63.

[89] Veerman, J. J. N., Klein, J., Aben, R. W. M., Scheeren, H. W., Kruse, C. G., van Maarseveen, J. H., Rutjes, F. P. J. T., Hiemstra, H. (2002). Solid-phase synthesis of piperidines by N-acyliminium ion chemistry. *Eur. J. Org. Chem.* (18), 3133–3139.

[90] Sun, H., Martin, C., Kesselring, D., Keller, R., Moeller, K. D. (2006). Building functionalized peptidomimetics: use of electroauxiliaries for introducing *N*-acyliminium ions into peptides. *J. Am. Chem. Soc.* **128** (42), 13761–13771.

[91] Lee, S. C., Park, S. B. (2007). Practical solid-phase parallel synthesis of 5-2-oxopiperazines via *N*-acyliminium ion cyclization. *J. Comb. Chem.* **9** (5), 828–835.

[92] Le, Q., Petersen, R., Meldal, M., Nielsen, T. E. (2010). *N*-acyliminium intermediates in solid-phase synthesis. *Biopolymers* **94** (2), 242–256.

[93] Krchnak, V., Holladay, M. W. (2002). Solid phase heterocyclic chemistry. *Chem. Rev.* **102** (1), 61–91.

[94] Franzen, R. G. (2000). Utilization of Grignard reagents in solid-phase synthesis: a review of the literature [review]. *Tetrahedron* **56** (5), 685–691.

[95] Wendeborn, S., De Mesmaeker, A., Brill, W. K. D., Berteina, S. (2000). Synthesis of diverse and complex molecules on the solid phase [review]. *Acc. Chem. Res.* **33** (4), 215–224.

[96] Hall, S. E. (1999). Recent advances in solid phase synthesis [review]. *Mol. Divers.* **4** (2), 131–142.

[97] Booth, S., Hermkens, P. H. H., Ottenheijm, H. C. J., Rees, D. C. (1998). Solid-phase organic reactions III: a review of the literature Nov 96 Dec 97 [review]. *Tetrahedron* **54** (51), 15385–15443.

[98] Scott, P. J. H. (ed.) (2009). *Linker Strategies in Solid-Phase Organic Synthesis*, John Wiley & Sons, Ltd, Chichester.

[99] Miertus, S., Fassina, G. (eds) (1999). *Combinatorial Chemistry and Technology: Principles, Methods and Applications*, Marcel Dekker, New York.

[100] Horton, D. A., Bourne, G. T., Smythe, M. L. (2003). The combinatorial synthesis of bicyclic privileged structures or privileged substructures. *Chem. Rev.* **103** (3), 893–930.

[101] Lebl, M., Krchnak, V., Patek, M., Lam, K. S. (1996). Synthetic combinatorial libraries of scaffolded and unordered structures. *Methods Mol. Cell. Biol.* **6**, 41–55.

[102] Krchnak, V., Flegelova, Z., Weichsel, A. S., Lebl, M. (1995). Polymer-supported Mitsunobu ether formation and its use in combinatorial chemistry. *Tetrahedron Lett.* **36**, 6193–6196.

[103] Houghten, R. A., Blondelle, S. E., Dooley, C. T., Dorner, B., Eichler, J., Ostresh, J. M. (1996). Libraries from libraries: generation and comparison of screening profiles. *Mol. Divers.* **2** (1–2), 41–45.

[104] Patek, M., Drake, B., Lebl, M. (1995). Solid-phase synthesis of "small" organic molecules based on thiazolidine scaffold. *Tetrahedron Lett.* **36** (13), 2227–2230.

[105] Lovering, F., Bikker, J., Humblet, C. (2009). Escape from flatland: increasing saturation as an approach to improving clinical success. *J. Med. Chem.* **52** (21), 6752–6756.

[106] Koh, J. S., Ellman, J. A. (1996). Palladium mediated three-component coupling strategy for the solid phase synthesis of tropane derivatives. *J. Org. Chem.* **61**, 4494–4495.

[107] Horne, G., Wilson, F. X., Tinsley, J., Williams, D. H., Storer, R. (2011). Iminosugars past, present and future: medicines for tomorrow. *Drug Discov. Today* **16** (3–4), 107–118.

[108] Goto, Y., Arai-Otsuki, S., Tachibana, Y., Ichikawa, D., Ozaki, S., Takahashi, H., Iwasawa, Y., Okamoto, O., Okuda, S., Ohta, H., Sagara, T. (2006). Identification of a novel spiropiperidine opioid receptor-like 1 antagonist class by a focused library approach featuring 3D-pharmacophore similarity. *J. Med. Chem.* **49** (3), 847–849.

[109] Mallinson, J., Collins, I. (2012). Macrocycles in new drug discovery. *Future Med. Chem.* **4** (11), 1409–1438.

[110] Tan, D. S. (2007). Diversity-oriented synthesis, in *Chemical Biology*, **2** (ed. Schreiber, S. L., Kapoor, T. M., Wess, G.), Wiley-VCH, Weinheim, pp. 483–518.

[111] Czarnik, A. W., Nova, M. P. (1997). No static at all: electronic encoding in combinatorial organic synthesis. *Chem. Brit.* **33**, 39.

[112] Moran, E. J., Sarshar, S., Cargill, J. F., Shahbaz, M. M., Lio, A., Mjalli, A. M. M., Armstrong, R. W. (1995). Radio frequency tag encoded combinatorial library method for the discovery of tripeptide-substituted cinnamic acid inhibitors of the protein tyrosine phosphatase PTP1B. *J. Am. Chem. Soc.* **117**, 10787–10788.

[113] Brenner, S., Lerner, R. A. (1992). Encoded combinatorial chemistry. *Proc. Natl. Acad. Sci. U. S. A.* **89**, 5381–5383.

[114] Geysen, H. M., Wagner, C. D., Bodnar, W. M., Markworth, C. J., Parke, G. J., Schoenen, F. J., Wagner, D. S., Kinder, D. S. (1996). Isotope or mass encoding of combinatorial libraries. *Chem. Biol.* **3**, 679–688.

[115] TORVIQ (2008). TORVIQ: Personal Chemistry Tools, http://www.torviq.com/index.html, accessed on: December 28, 2013.

[116] Krchnak, V., Burritt, A. (2002). Simple tools for manual parallel solid phase synthesis, in *Combinatorial Library: Methods and Protocols* (ed. English, L. B.), Humana Press, Totowa, NJ, pp. 41–60.

[117] Harness, J. R. (1996). Automation of high-throughput synthesis: automated laboratory workstations designed to perform and support combinatorial chemistry, in *Molecular Diversity and Combinatorial Chemistry: Libraries and Drug Discovery* (ed. Chaiken, I. M., Janda, K. D., Cambridge Healthtech Institute), American Chemical Society, Washington, DC, pp. 188–198.

[118] Sepetov, N., Issakova, O. (1999). Analytical characterization of synthetic organic libraries, in *Combinatorial Chemistry and Technology: Principles, Methods and Applications*, **8** (ed. Miertus, S., Fassina, G.), Marcel Dekker, New York, pp. 169–203.

[119] SynPhase™ Publications (2008). SynPhaseTM Publications from Mimotopes Pty Ltd, Collaborators and Independent Laboratories, http://www.mimotopes.com/files/editor_upload/File/CombinatorialChemistry/SynPhase_Publications.pdf, accessed on: December 18, 2013.

[120] Ge, N., Fogel, P., Young, S. S., Austin, R. E., Wegrzyniak, E., Connelly, J. A. (2006). Analysis and prediction of combinatorial chemistry synthesis and screening data. *Comb. Chem. High Throughput. Screen.* **9** (6), 465–472.

[121] Yan, B., Chu, Y. H., Shapiro, M., Richmond, R., Chin, J., Liu, L., Yu, Z. (2000). Analytical methods in combinatorial chemistry, in *Combinatorial Chemistry: A Practical Approach*, **8** (ed. Fenniri, H.), Oxford University Press, Oxford/New York, pp. 263–285.

[122] Snyder, L. R., Kirkland, J. J, Glajch, J. L. (eds) (1997). *Practical HPLC Method Development*, Wiley-Interscience, New York.

[123] Fang, L., Dreyer, M., Chen, Y., Plunkett, M., Hocker, M. (1997). Analysis of small organic combinatorial libraries by LC-MS. *Proceedings of the 45th ASMS Conference on Mass Spectrometry and Allied Topics*, June 1–5, 1997, Palm Springs, California, American Society for Mass Spectrometry, Santa Fe, NM, p. 401.

[124] Issakova, O., Nikolaev, V., Ma, N., Wade, S., Sepetov, N. (1997). Analytical characterization of one-bead-one-compound combinatorial libraries by LC/MS and LC/MS/MS. *Proceedings of the 45th ASMS Conference on Mass Spectrometry and Allied Topics*, June 1–5, 1997, Palm Springs, California, American Society for Mass Spectrometry, Santa Fe, NM, p. 403.

[125] Janda, K. D. (1994). Tagged versus untagged libraries: methods for the generation and screening of combinatorial chemical libraries. *Proc. Natl. Acad. Sci. U. S. A.* **91** (23), 10779–10785.

[126] Nefzi, A., Ostresh, J. M., Houghten, R. A. (1997). The current status of heterocyclic combinatorial libraries. *Chem. Rev.* **97** (2), 449–472.

[127] Kassel, D. B. (2001). Combinatorial chemistry and mass spectrometry in the 21st century drug discovery laboratory. *Chem. Rev.* **101** (2), 255–267.

[128] Yan, B., Zhao, J., Leopold, K., Zhang, B., Jiang, G. (2007). Structure-dependent response of a chemiluminescence nitrogen detector for organic compounds with adjacent nitrogen atoms connected by a single bond. *Anal. Chem.* **79** (2), 718–726.

[129] Zeng, L., Wang, X., Wang, T., Kassel, D. B. (1998). New developments in automated PrepLCMS extends the robustness and utility of the method for compound library analysis and purification. *Comb. Chem. High Throughput. Screen.* **1** (2), 101–111.

[130] Coffey, P., Ramieri, J., Garr, C., Schultz, L. (1999). An open, multivendor specification for high-throughput organic chemistry (HTOC): a flexible production process for pure, characterized, quantified organic compounds. *Am. Lab.* **31**, 57–71.

[131] Kyranos, J. N., Hogan, J. C. (1998). High-throughput characterization of combinatorial libraries generated by parallel synthesis. *Anal. Chem.* **70** (11), 389A–395A.

[132] Virscidian Inc. (2013). Analytical Software Solutions Customized to Fit Your Natural Laboratory, http://www.virscidian.com/, accessed on: December 16, 2013.

[133] Kerns, E., Di, L. (2008). *Drug-Like Properties: Concepts, Structure Design and Methods: From ADME to Toxicity Optimization*, Academic Press, Amsterdam/Boston.

[134] Kassel, D. B. (2004). Applications of high-throughput ADME in drug discovery. *Curr. Opin. Chem. Biol.* **8** (3), 339–345.

[135] Chiang, P. C., Hu, Y. (2009). Simultaneous determination of LogD, LogP, and pK(a) of drugs by using a reverse phase HPLC coupled with a 96-well plate auto injector. *Comb. Chem. High Throughput. Screen.* **12** (3), 250–257.

[136] Evans, B. E., Rittle, K. E., Bock, M. G., DiPardo, R. M., Freidinger, R. M., Whitter, W. L., Lundell, G. F., Veber, D. F., Anderson, P. S., Chang, R. S. (1988). Methods for drug discovery: development of potent, selective, orally effective cholecystokinin antagonists. *J. Med. Chem.* **31** (12), 2235–2246.

[137] Welsch, M. E., Snyder, S. A., Stockwell, B. R. (2010). Privileged scaffolds for library design and drug discovery. *Curr. Opin. Chem. Biol.* **14** (3), 347–361.

# 6

# RECENT ADVANCES IN MULTICOMPONENT REACTION CHEMISTRY: APPLICATIONS IN SMALL MOLECULE DRUG DISCOVERY

Christopher Hulme, Muhammad Ayaz, Guillermo Martinez-Ariza, Federico Medda, and Arthur Shaw

*University of Arizona, Tucson, AZ, USA*

## 6.1 INTRODUCTION

The efficiency of translational small molecule medicinal chemistry campaigns is directly related to the "iterative speed" or "iterative efficiency" of the skilled drug hunter, who in a quest for improved potency, selectivity, optimal physicochemical properties, and many other parameters navigates the "hypothesis–synthesis–screening" loop gradually moving molecules along the value chain. As such, modern-day file enhancement strategies aim to incorporate compounds possessing fragment-, lead-, or drug-like properties measured by a variety of metrics that have been constantly evolving. Indeed, these corporate collections in conjunction with high-throughput screening (HTS) are a primary driver for the generation of new biologically active small molecules to aid initiation of drug discovery campaigns in conjunction with *de novo* design and smart iterative screening exercises.

This chapter specifically discusses the pros and cons of employing multicomponent reactions (MCRs) to both enhance the molecular diversity of in-house screening collections and expedite forward progression of lead molecules toward key milestones associated with "drugging" macromolecular targets of interest for therapeutic benefit. Sections 6.2 through 6.9 of this chapter focus heavily on both well-established MCRs in the

*Small Molecule Medicinal Chemistry: Strategies and Technologies*, First Edition.
Edited by Werngard Czechtizky and Peter Hamley.

pharmaceutical arena and recently discovered enticing reactions, including asymmetric versions, yet to realize their full potential. The chemistry underpinning these transformations will be discussed with reference to their biological relevance. As such, it must be noted that the last 20 years has seen a huge upswing in MCR research, involving both the development of new MCRs and the use of existing MCRs through subsequent product manipulation to generate further intriguing pharmacologically relevant diversity. Nevertheless, many published MCRs fail to produce what one would consider, through a "rearview mirror" mind-set, molecules that would be viewed as currently pharmacologically relevant and these have thus been omitted. However, this statement in no way is meant to diminish the value of more baroque structures when one considers the tremendous need to develop new unexplored molecular diversity to address the plethora of emerging new targets that currently fall into the "undruggable class" where attempts to find appropriate small molecule partners have fallen short. The reader is thus directed to more comprehensive reviews by several groups [1]. Sections 6.10.1 through 6.10.6 pay particular attention to MCRs utilized to discover clinical candidates or marketed drugs, separated into well-known target family categories that clearly demonstrate the value of employing MCR technologies in drug discovery.

So, what exactly is an MCR? The most common definition is that MCRs are chemical transformations where three or more reactants combine in one pot to afford a single product, with each of the reactants contributing matter to the final product [1]. Relative to classical multistep syntheses, they are generally high yielding, operationally friendly, time- and cost-effective transformations, often being highly atom-economical [2a] reactions meeting many of the "green chemistry criteria" [2b]. Furthermore, MCRs often represent mechanistically nonobvious one-step transformations to the "MCR-nonspecialist" medicinal chemistry practitioner and are ideal molecular complexity generation tools, delivering products with multiple points of diversity from readily available reagents functionalized with the necessary pharmacophoric features required to probe small molecule–macromolecular interactions. Together, all these features make MCRs ideal for diversity-oriented synthesis (DOS) [2c], function-oriented synthesis (FOS) [2d], target family-oriented synthesis (TOS) [2e], serendipity-oriented synthesis (SOS) [2f]—the latter may also be coined "combinatorial reaction finding"—and parallel/sequential synthesis encompassed within the small molecule drug discovery domain. In short, MCRs produce molecules with tremendous "iterative efficiency potential" [2g] that have the ability to dramatically reduce the number of iterations required for value chain progression in drug discovery [3, 4]. This plethora of pros is counteracted with a nominal number of cons, personally encountered over time spent in both the private and public sectors: (i) There is often a reluctance to take up isocyanide-based MCR (IMCR) technologies because of the unpleasant odor of relatively low molecular weight reagents. When handled correctly, with a beaker of methanolic HCl on standby, one can operate on scale or in parallel with few issues. (ii) As a private sector medicinal chemist taking up the technology, one must realize that your productivity will dramatically increase and you are very likely to discover a multitude of publishable protocols in a short time span. One must be cognizant of colleagues/hierarchy not skilled in the field assuming resources are being overly focused on MCRs to the detriment of critical path drug discovery and therefore manage expectations appropriately. (iii) There remains a great need for methodology development of asymmetric variants of MCRs described herein.

## 6.2 CLASSICAL MULTI-COMPONENT REACTIONS (MCRs)

The most common sequential steps in MCRs are a condensation of an amine with a carbonyl input followed by a nucleophilic addition, and this very simple sequence is of direct industrial relevance for production of α-aminonitriles (precursors to α-amino acids) via the first ever documented MCR, the Strecker reaction [5]. From a synthetic perspective, tremendous developments covering catalytic and asymmetric versions of the Strecker reaction have been seen in recent years [6]. Table 6.1 (Schemes 6.1 through 6.11) highlights some of the more prevalent and heavily mined MCRs [5–18] in a chronological order.

Likewise, the Hantzsch [7], Biginelli [8], Mannich [9], Asinger [11, 19], and Gewald [13] reactions have directly led to well-established therapeutics and have been widely utilized in modern-day file enhancement. However, significant efforts over the last 15 years have focused on post-MCR modifications, that is, where MCR products are subjected to subsequent transformations, predominantly involving intramolecular cyclization, that have led to biologically appealing new scaffolds. Highlighting these advances, the following sections focus on the venerable Italian Passerini reaction (1921), which subsequently spawned the classical "suite" of Ugi reactions (1958). More recent developments over 10–15 years ago discuss exploitation of the Van Leusen, Petasis, and Groebke–Bienaymé–Blackburn reactions, followed by recently discovered MCRs (Sections 6.8.1 through 6.8.4). The reader is directed to key compilations of accessible structures in Figures 6.1, 6.2, and 6.4.

## 6.3 THE PASSERINI REACTION (MARIO PASSERINI, 1921)

The Passerini reaction was the first reported IMCR and comprises reaction of an aldehyde or ketone, a carboxylic acid, and an isocyanide to afford α-acyloxy carboxamides (Scheme 6.5, Table 6.1) [10]. Postcondensation modifications of the Passerini reaction have delivered heterocycles such as butenolides (via Passerini/Horner–Emmons–Wadsworth sequences) [10c], oxazoles (via a Passerini/Davidson cyclization), and PADAM methodology (*P*asserini *D*eprotection *A*cyl *M*igration *D*eprotection), the latter affording the ability to rapidly target proteases (Section 6.10.2) [10d]. However, the monumental contribution of the Passerini reaction is that it set the stage for the discovery of the versatile Ugi reaction. Giving due credit to Passerini, Ivar Ugi [12] writes, "Had Passerini been conversant with present day views on reaction mechanisms while studying the reaction which now bears his name, he would probably have added ammonia or the primary amines, to his three starting materials and thereby discovered the α-amino alkylation of isonitriles and acids."

## 6.4 UGI REACTION

The Ugi reaction [12] represents one of the most powerful C—C and C—N bond-forming reactions in the MCR toolbox. A typical reaction involves the condensation of an amine, a carbonyl input, a carboxylic acid, and an isocyanide leading to the formation of α-amido-amides (Scheme 6.7, Table 6.1). The isocyanide behaves with a divalent nature, being both a nucleophile and then an electrophile in the form of a nitrilium ion. Mechanistically, the reaction is driven forward by the final step, a Mumm rearrangement, which is irreversible [20]. Moreover, the Ugi reaction is actually a "suite of reactions" in that the classical

**TABLE 6.1 Classical Multicomponent Reactions**

| Name | Reaction |
|---|---|
| Strecker [5] | 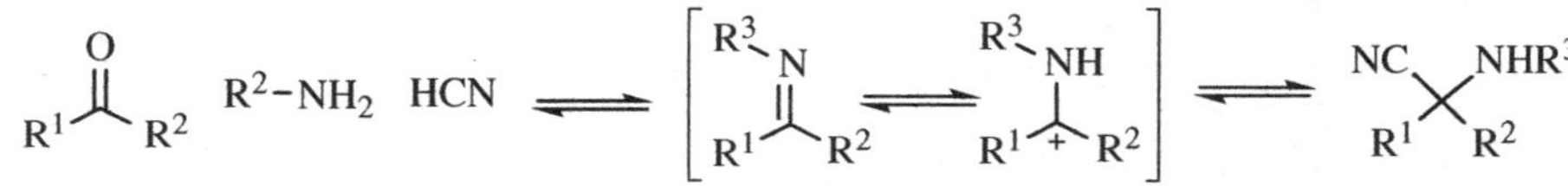 **SCHEME 6.1** The Strecker reaction. |
| Hantzsch [7] | 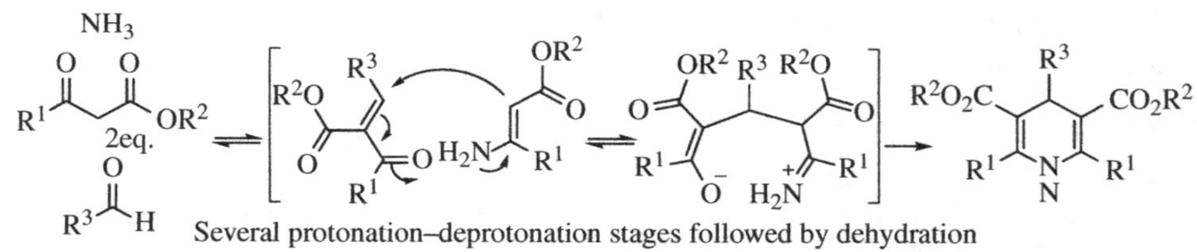  **SCHEME 6.2** The Hantzsch reaction. |
| Biginelli [8] | 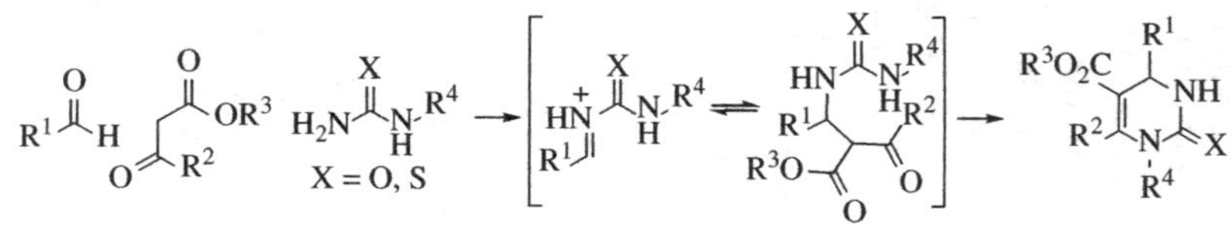  **SCHEME 6.3** The Biginelli reaction. |
| Mannich [9] | 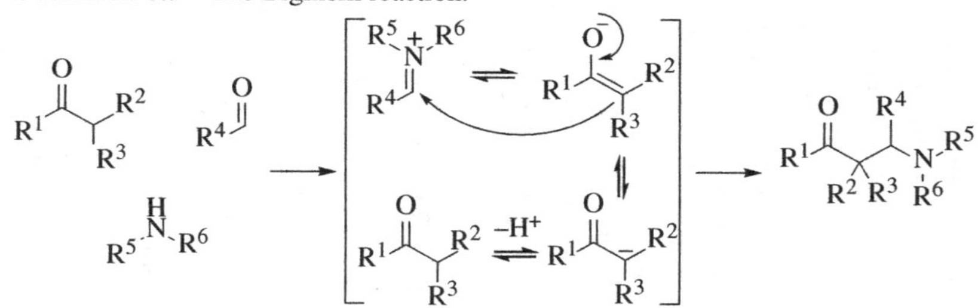 **SCHEME 6.4** The Mannich reaction. |

| | |
|---|---|
| Passerini [10a] | 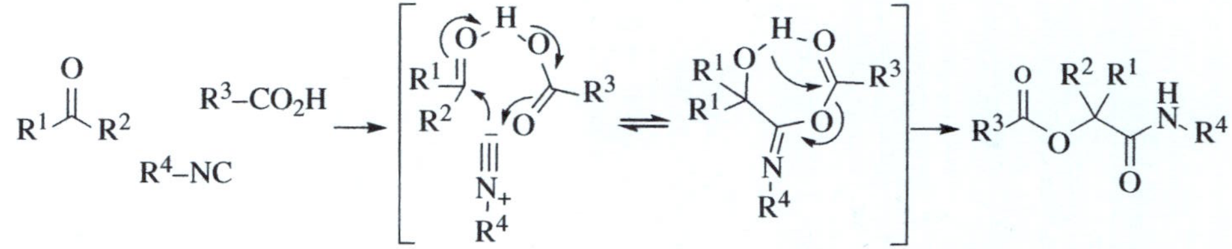 **SCHEME 6.5** The Passerini reaction. |
| Asinger [11] | 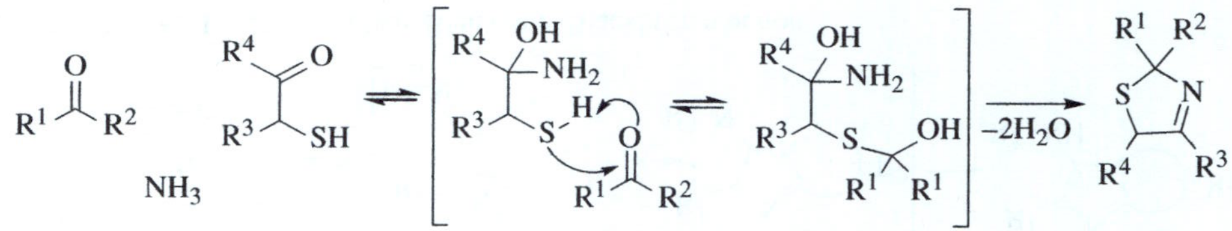 **SCHEME 6.6** The Asinger reaction. |
| Ugi [12] | 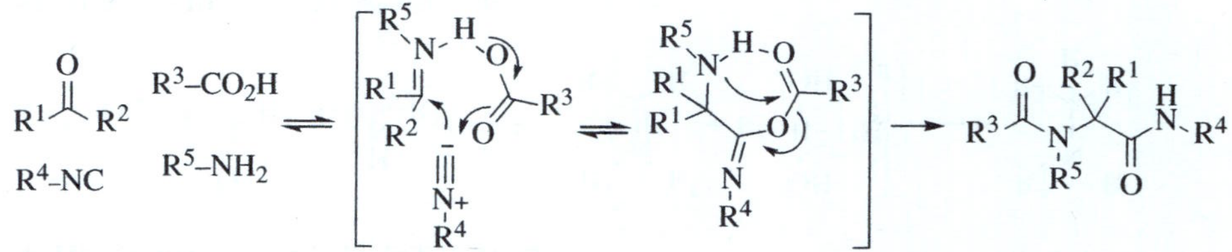 **SCHEME 6.7** The Ugi reaction. |
| Gewald [13] | 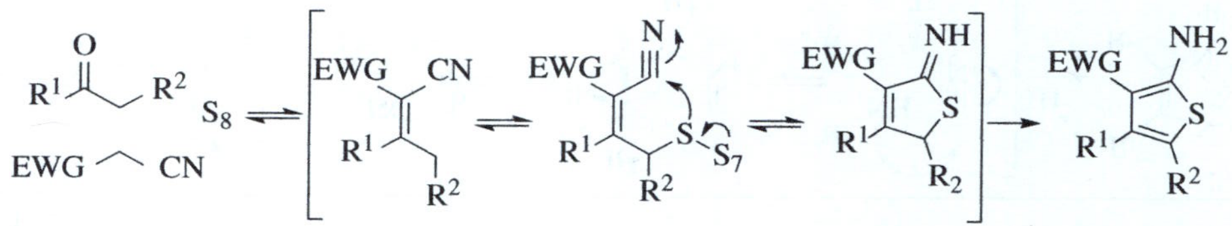 **SCHEME 6.8** The Gewald reaction. |

(*Continued*)

**TABLE 6.1** (*Continued*)

| Name | Reaction |
|---|---|
| Van Leusen [14] | 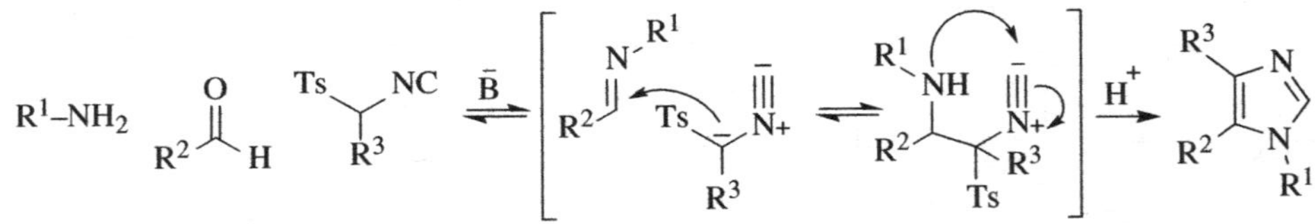<br>**SCHEME 6.9** The Van-Leusen reaction. |
| Petasis [15] | 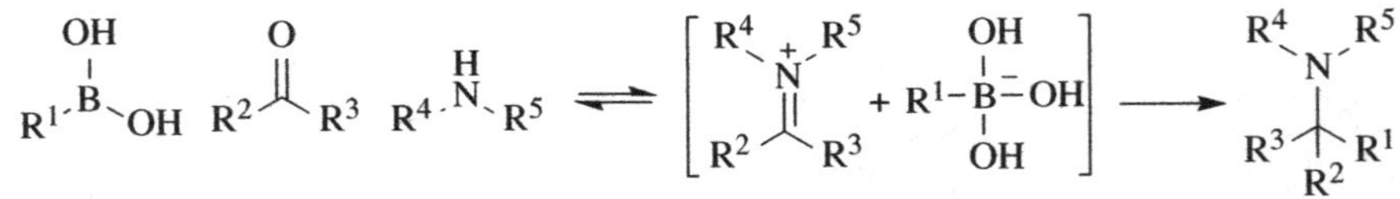<br>**SCHEME 6.10** The Petasis reaction. |
| Groebke–Bienaymé–Blackburn [16] | 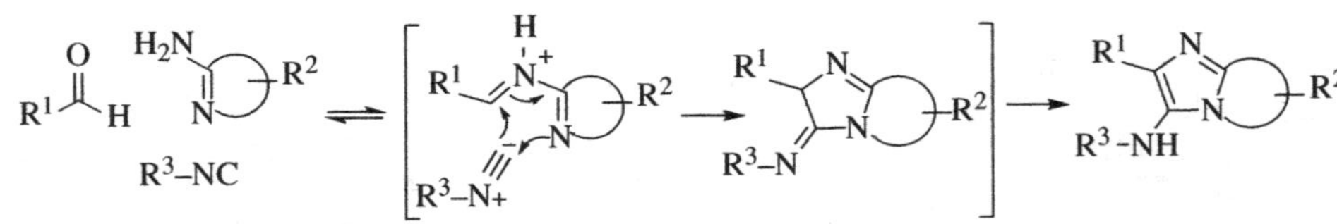<br>**SCHEME 6.11** The Groebke–Bienayme–Blackburn reaction. |

**FIGURE 6.1** Scaffold diversity generated by UDC strategies using 1–3 internal nucleophiles.

carboxylic acid can be replaced with alternate nucleophiles that trap out the intermediate nitrilium ion: $HN_3$, HNCO, HNCS, HNCSe, $H_2S_2O_3$, $H_2Se$, and phenol or water represent merely a selection. The amine input may also be substituted by hydrazines, hydrazides, ureas, hydrazones, or sulfonamides. Impressively, the majority of the previously modified reactions were outlined by Ugi in his initial report [12]. The prime example of therapeutics from the one-step Ugi reactions is the class of local anaesthetics represented by Xylocaine **179** and produced in one step, where water acts as a carboxylic acid replacement in a 100% atom-economical reaction. However, the bolus of methodology development with the Ugi reaction over the last 15 years has focused on using secondary reactions or postcondensation modifications (predominantly intramolecular) to constrain the Ugi adduct into a variety of heterocycles of medicinal worth [21]. In this regard, methodology coined Ugi–deprotect–cyclize (UDC) (Section 6.4.1) [21] has delivered a plethora of two-step routes to both known and novel chemotypes often with chemistry compatible to 96-well plate production in solution, supported by partial purification with scavenging resins. Utilizing tethered Ugi reagents, named the bifunctional approach (BIFA) (Section 6.4.2) [2f, 17, 22] in conjunction with UDC, generates even greater access to new chemical space in a remarkably facile manner.

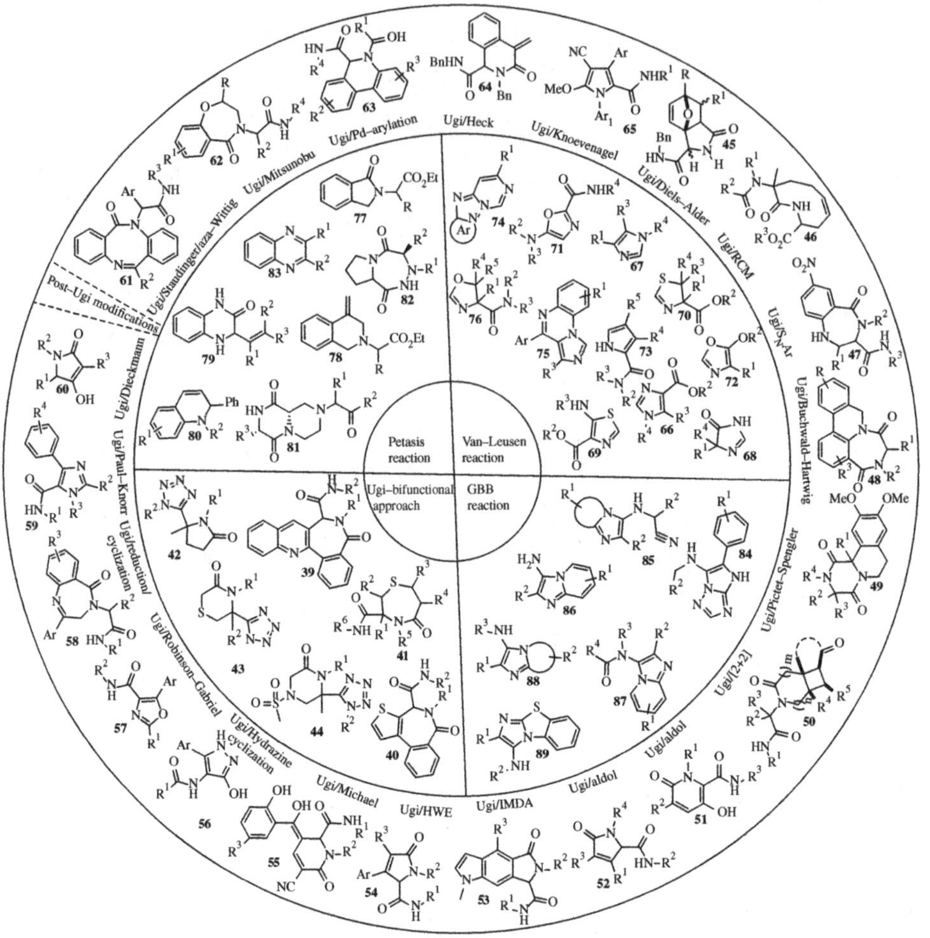

**FIGURE 6.2** Example chemotypes generated via (i) post-Ugi modifications, (ii) the Ugi-bifunctional approach, (iii) the Petasis reaction, (iv) the Van Leusen reaction, and (v) the GBB reaction.

### 6.4.1 The Ugi-Deprotect-Cyclize (UDC) Strategy

The UDC strategy utilizes internal masked amino nucleophiles, typically Boc protected, and linked to the amine, isocyanide, aldehyde, or carboxylic acid inputs in the Ugi 4-component reaction (Ugi 4CR). After the completion of the Ugi reaction, solvent is removed, followed by acidic treatment (TFA/DCE) of the crude product, which triggers deprotection and cyclization steps furnishing a multitude of heterocyclic scaffolds. The UDC concept initially utilized the convertible isonitrile **2** (Scheme 6.12) [23], yet with appropriate positioning of amine nucleophiles, both carbonyls in the Ugi skeleton were found to be available for ring closure [24, 25]. Similarly, such electrophilic sites could be preorganized in the side chains of the Ugi inputs enabling access to further arrays of heterocyclic systems.

The UDC strategy can be classified by the number of internal amino nucleophiles present on one or more of the starting materials; the most representative examples are shown in

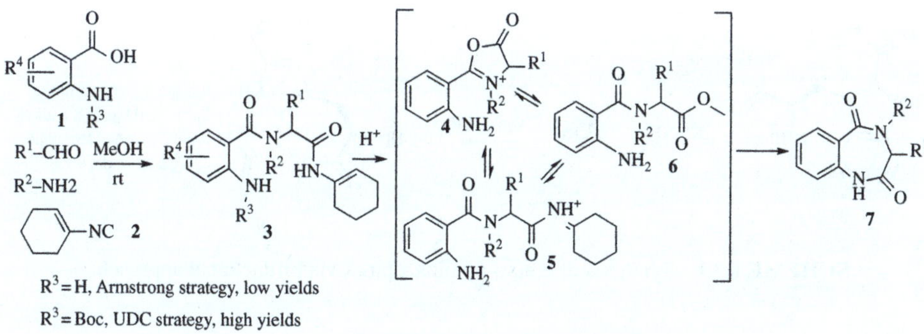

**SCHEME 6.12** 1,4-Benzodiazepines derived from the UDC strategy.

Figure 6.1. Using one internal amino nucleophile and cyclohexenylisocyanide generated heterocyclic templates that include diketopiperazines (**8-a,b**) [25b], 1,4-benzodiazepine-2,5-diones (**9-a,d**) [25a], γ-lactams (**10-a,e**) [26], bicyclic lactams (**11-a,f,g**) [26], ketopiperazines (**12-a,h**) [27], and dihydroquinoxalinones (**13-a,i**) [28]. Using ethyl glyoxylate (**b**) as a source of a side-chain electrophilic carbonyl, the carboxamide congeners of the aforementioned heterocycles were also generated [29]. The same strategy also affords concise routes to imidazolines (**14-c**), benzimidazoles (**15-j,16-i**) [24, 30], quinoxalinones (**17-i,k**) [31], and dihydroquinazolines (**18-l,19-m**) [32]. The use of $TMSN_3$ as an acid equivalent in conjunction with the same strategy furnished bicyclic azepine-tetrazoles (**20-n**) [33] and tetrazolo-ketopiperazines (**21-n**) [34]. Utilizing two internal nucleophiles allows the synthesis of more sophisticated chemotypes with at least two fused rings or hybrid heterocycles, exemplified by triazadibenzoazulenones (**22-d,i**) [35], dihydroquinazolines (**23-d,l,24-d,m**) [32], benzimidazole-quinoxalines (**25,26-b,i,j,k**) [36], *bis*-benzimidazoles (**27-i,j**) [36, 37], *bis*-benzimidazole-dihydroquinoxalines (**28-i,j,k**) [36], *bis*-benzodiazepines (**29-a,b,d,l**) [38], and quinoxaline-benzodiazepines (**30-a,b,d,m**) [38]. Remarkably, employing different tethered keto acids or aldehyde acids (known as bifunctional reagents) [2f, 22] under UDC conditions rendered benzimidazole-carboxamides (**31-i,o**) and benzimidazole-quinazolines (**32-j,l,o**) [39]. In addition, tricyclic scaffolds **33-b,p,l** and **34-b,p,i** were generated using a similar strategy [39]. Very recently, the most complex and ordered examples of UDC make use of three internal nucleophiles rendering complex polycyclic molecular frameworks (**35-b,p,j,l** and **36-b,p,g,j**) [40]. More detailed information about UDC methodology can be found in seminal reviews [21, 41].

### 6.4.2 Bi-Functional Approach (BIFA)

A common approach to constrained MCR products is the use of bifunctional reagents, which are embedded with two functional groups that participate in the Ugi reaction. In simple terms, constrained cores are often preferred over flexible scaffolds in file enhancement, as rigidity diminishes entropic barriers associated with receptor–ligand binding [42]. A recent example is the one-pot diastereoselective synthesis of amino-indoloazepinones (**38**), using 2-formyl-*L*-tryptophan (**37**) as a bifunctional building block, via an Ugi three-component reaction (Ugi 3CR) (Scheme 6.13) [43]. Furthermore, novel heterocyclic dihydro-benzoazepines (**39, 40**) were generated by using bifunctional

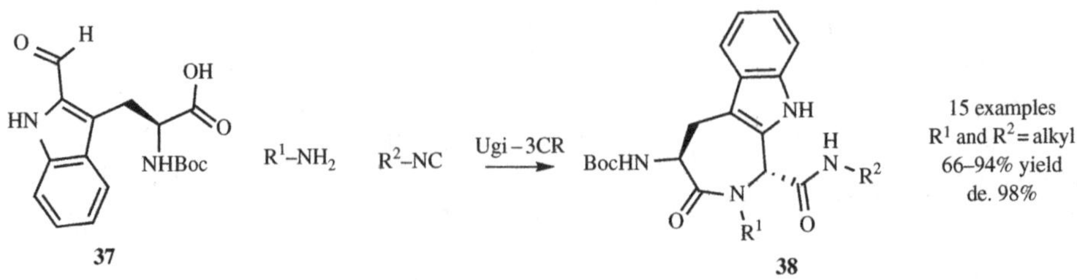


**SCHEME 6.13** Synthesis of amino-indoloazepines via bifunctional approach.

inputs in the Ugi 3CR (Figure 6.2) [44]. An excellent example of the intramolecular Ugi reaction employing bifunctional inputs has been reported by Marcaccini producing the 1,4-thiazepinone scaffold (**41**) [45]. Finally, this laboratory has published applications of the bifunctional Ugi-azide MCR coupled with the postcondensation modifications to generate *bis*-heterocyclic lactam tetrazoles (**42–44**) (Figure 6.2) [46].

### 6.4.3 Miscellaneous Post-Ugi Condensations

In addition to UDC and bifunctional approaches, the complete arsenal of organic methodology is available to both design starting materials and subsequently constrain the Ugi adduct that has delivered an overwhelming number of structurally enticing chemotypes. A major reason for the success of this approach is the generality of the pivotal Ugi reaction that proceeds under mild conditions tolerating extensive decoration of reaction inputs. Most noteworthy post-Ugi condensation modifications include Ugi/Diels–Alder [47–51], Ugi/RCM [52–58], Ugi/$S_NAr$ [59–63], Ugi/Buchwald–Hartwig [64–68], Ugi/Pictet–Spengler [69–73], Ugi/[2+2] [74], Ugi/Aldol [75], Ugi/IMDA [76], Ugi/HWE [77], Ugi/Michael [78], Ugi/hydrazine-mediated cyclization [79], Ugi/Robinson–Gabriel [80], Ugi/nitro-reduction/cyclization [81], Ugi/Paul–Knorr [82], Ugi/Dieckmann [83], Ugi/ Staudinger/aza-Wittig [84–88], Ugi/Mitsunobu [89–94], Ugi/Pd-arylation [95], Ugi/Heck [96–99], and Ugi/Knoevenagel sequences [100–102], delivering chemotypes **45** through **65** (Fig. 6.2).

## 6.5 VAN LEUSEN REACTION

Discovered in 1977, the Van Leusen reaction exploits only the electrophilicity of the bivalent isocyanide input [14, 103]. A typical reaction involves the condensation of aldehyde and amine input with a class of isocyanides that bear an acidic proton and a tosyl leaving group at the α-position (Table 6.1, Scheme 6.9). The base-catalyzed reaction leads to the formation of medicinally relevant imidazoles with three points of diversity. Given the tolerance of the transformation for modified reaction inputs, the Van Leusen reaction has been utilized to generate a range of small heterocycles such as imidazoles [104], imidazolones [105], thiazoles [106], thiazolidines [107], oxazoles [108], pyrroles [109], azolopyrimidines [110], and, more recently, imidazoquinoxalines (Figure 6.2) (structures **66–76**) [111]. Most notably was its utilization to generate some of the early p38 inhibitors exemplified by **154**.

## 6.6 PETASIS REACTION

The Petasis reaction (Table 6.1, Scheme 6.10) is a relatively modern transformation that replaces the typical carbonyl-based nucleophile of the Mannich reaction with a boronic acid and, as such, is sometimes denoted the Borono–Mannich reaction (BMR) [15, 112]. The reaction employs an aldehyde, an amine, and a boronic acid where the latter usually requires a suitably located additional anchoring heteroatom. Post-Petasis modifications (Fig. 6.2) include palladium-catalyzed carbonylation–amination or allenylation–amination sequences generating amino acid derivatives **77** and **78**, respectively [113]. Making use of protected amino acids, β-turn mimetics with generic structures **81** and **82** have also been prepared using post-Petasis transformations [114, 115]. Similarly, post-Petasis transformations have enabled access to quinolines, **80**, quinoxalinones **79** and quinoxalines **83** in good yields [116–118].

## 6.7 GROEBKE–BLACKBURN–BIENAYMÉ (GBB) REACTION

Reported by three groups in 1998, the Groebke–Blackburn–Bienaymé (GBB) reaction offers facile assembly of medicinally important heterocycles, namely, imidazoazines in a single operation [16]. A typical transformation undergoes a [4+1] cycloaddition of an isonitrile to a Schiff base derived from 2-aminopyridines and aldehydes (Table 6.1, Scheme 6.11). The fundamental strength of this reaction lies in the fact that it is compatible with a plethora of α-amino *N*-heterocycles and hence subsequent libraries [119a] are extremely high in skeletal diversity. Furthermore, the fused imidazole ring embedded in all these products typically imparts products with highly desirable physicochemical properties. Noteworthy is an extension of the amine pool by Krasavin who employed TMSCl-mediated catalysis to promote reaction progression [119b] and replacement of the traditional isocyanide with trimethylsilyl cyanide (TMSCN) to obtain primary amines [120]. Several scaffolds [120–124] (**84–89**) obtained from the GBB reaction or modification thereafter are depicted (Fig. 6.2).

## 6.8 RECENTLY DISCOVERED NOVEL MCRs

Ongoing discovery of new MCRs has continued unabated over the last ten years with a well-established worldwide community of scientists participating in these efforts. As such, each new MCR delivers potentially new biologically active products directly or indirectly through subsequent post-MCR modifications [125–127]. A collection of recently developed MCRs producing biologically appealing cores is described in this section, categorized on the basis of common starting material.

### 6.8.1 Cyclic Anhydride-Based MCRs

Reaction of cyclic anhydrides with imines has recently been reviewed by Shaw [128, 129]. First reported in 2003 in an ionic liquid-mediated version (**91**) [130], it proceeds with an aldehyde, amine, and homophthalic anhydride (**90**) to give isoquinoline carboxylic acids (**94**) in a highly *syn*-diastereoselective fashion. The mechanism (Scheme 6.14) is analogous to many anhydride annulations yielding (**94**) in good yield. Other catalysts reported for this 3CR include $InCl_3$ [130] and $Yb(OTf)_3$ [131].

A more complex 4CR, discovered by Shaw and coworkers [132], involves the highly diastereoselective reaction (Scheme 6.15) between aldehydes, amines, thiols (**95**), and maleic anhydrides (**96**) to yield γ-lactams (**97**). Remarkably, this high-yielding process allows the control of three stereogenic centers and has broad reagent scope and stereochemical diversity. Mechanistically, it involves an analogous annulation to the Yadav MCR, mediated by an iminium ion intermediate. Indeed, this exquisite 4CR has already been used to produce biologically significant molecules (Section 6.10.4).

### 6.8.2 1-Azadiene-Based MCRs

1-Azadienes **100** are versatile building blocks for the synthesis of nitrogen-containing heterocycles [133] and offer several modes of reactivity being able to react as nucleophiles and electrophiles in 1,2- and 1,4-additions and as heterodienes in cycloadditions (Fig. 6.3).

Initially prepared from phosphonates **98**, nitriles **99**, and aldehydes via an MCR reported by Kiselyov [134], 1-azadienes **100** have recently been used in novel MCRs to afford new scaffolds via *S*ingle *R*eactant *R*eplacement (SRR) strategies [135] (Scheme 6.16). They react with α-arylacetonitriles **101** to deliver 2-aminopyridines **103** and with enolates **102** to produce trisubstituted pyridines **104** [134]. Similarly, employing amidines with 1-azadienes affords highly substituted pyrimidines **106** [136]. Orru [137, 138] has used isocyanates **107**

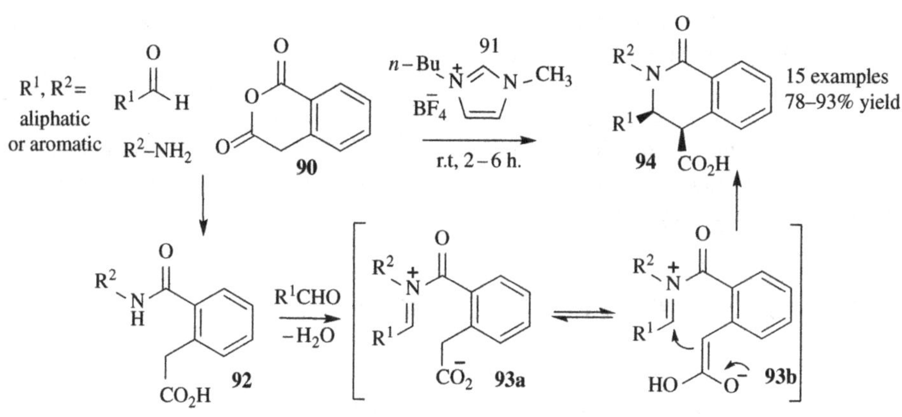


**SCHEME 6.14** Yadav 3CR to access isoquinoline carboxylic acids.

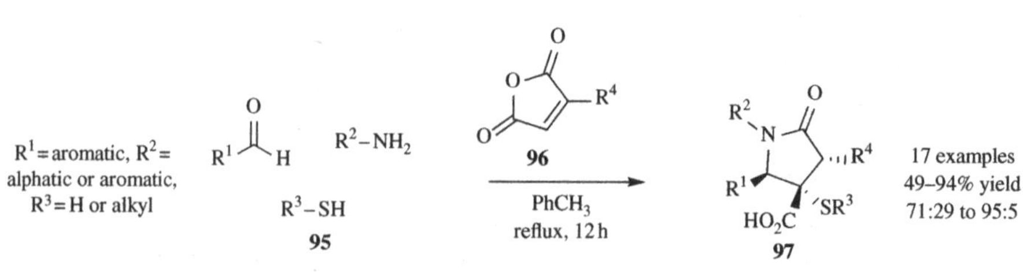


**SCHEME 6.15** Synthesis of γ-lactams using Shaw's 4CR.

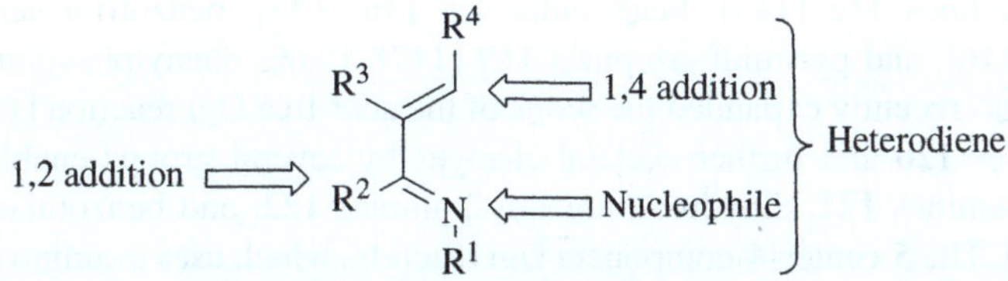


**FIGURE 6.3** Reactivity of 1-azadienes.

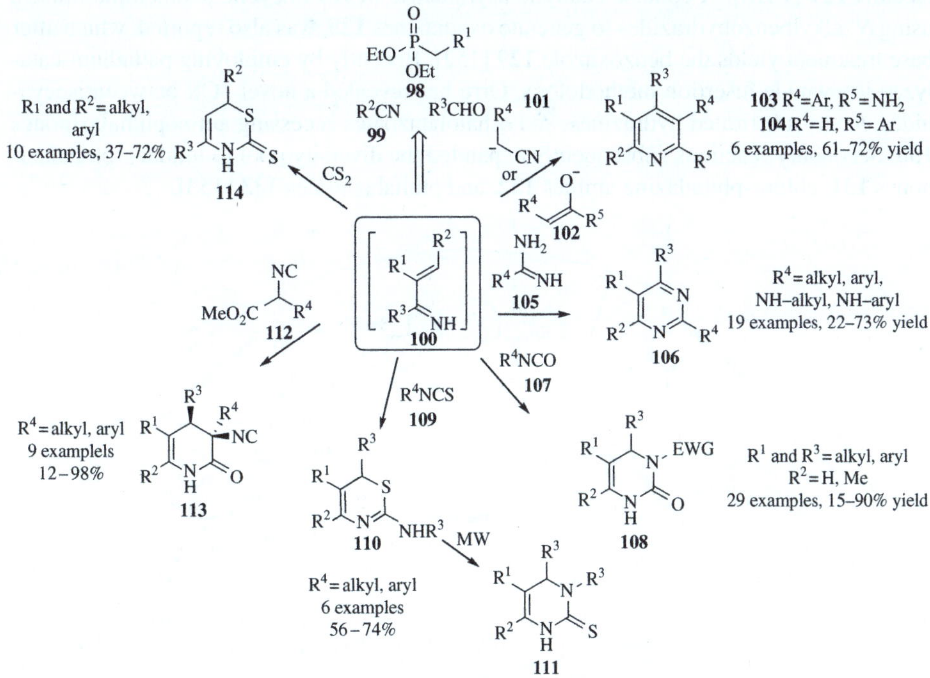

**SCHEME 6.16** 1-Azadienes in MCRs.

in conjunction with 1-azadienes to afford 3,4-dihydropyrimidin-2(1*H*)-ones **108** in good yields and with broad scope. Analogously, isothiocyanates **109** react with 1-azadienes, albeit via a different mechanism, resulting in the formation of iminothiazines **110** that may be further converted to dihydropyrimidinethiones **111** [139]. Another noteworthy example from the Orru group is the 4CR of 1-azadienes with methyl isocyanoacetate **112**, which affords dihydropyridinones **113** with intact isonitrile functionality [140]. Reaction of the 1-azadienes with $CS_2$ also produces thiazinones **114** [141].

### 6.8.3 Recent IMCRs and Secondary Reactions

Several new variants of Ugi- and Passerini-type reactions [142] are shown in Figure 6.4. One truly exquisite example is the Ugi–Smiles 4CR reported by El Kaim [143]. This MCR employs phenols or thiophenols in place of the carboxylic acid used in the Ugi reaction, and several postcondensation modifications have been reported delivering dihydroquinoxalines **115** [144], benzimidazoles **116** [145], benzotriazoles **117** [145], indoles **118** [146], and pyrimido-azepines **119** [147]. Using phenylphosphinic acid as a catalyst, List has recently expanded the scope of the acid-free Ugi reaction [148] to access α-amino amides **120** and further skeletal changes by several groups enabled access to benzoxazin-2-amines **121**, dihydroquinoxalin-2-amines **122**, and benzothiazin-2-amines **123** [149, 150]. The 5-center-4-component Ugi reaction, which uses α-amino acids instead of classical carboxylic acids to generate iminodicarboxamides **124** [151a], has also been exploited by Dömling giving pyrrolidinediones **125**, oxoisoindolines **126**, and tetracyclic lactams **127** [151b]. A concise catalytic asymmetric 3CR of acyclic azomethine imines, using *N′*-alkylbenzohydrazides to generate oxadiazines **128**, was also reported, which after base treatment yields the benzoxazole **129** [152]. Recently by employing palladium-catalyzed isocyanide insertion methodology, Orru has revealed a novel 3CR between isocyanides, *mono*-substituted hydrazines, and *o*-halobenzoates accessing aminophthalazinones **130**. Secondary reactions subsequently expanded the diversity pool to imidazo-phthalazinones **131**, chloro-phthalazine amines **132**, and phthalazinones **133** [153].

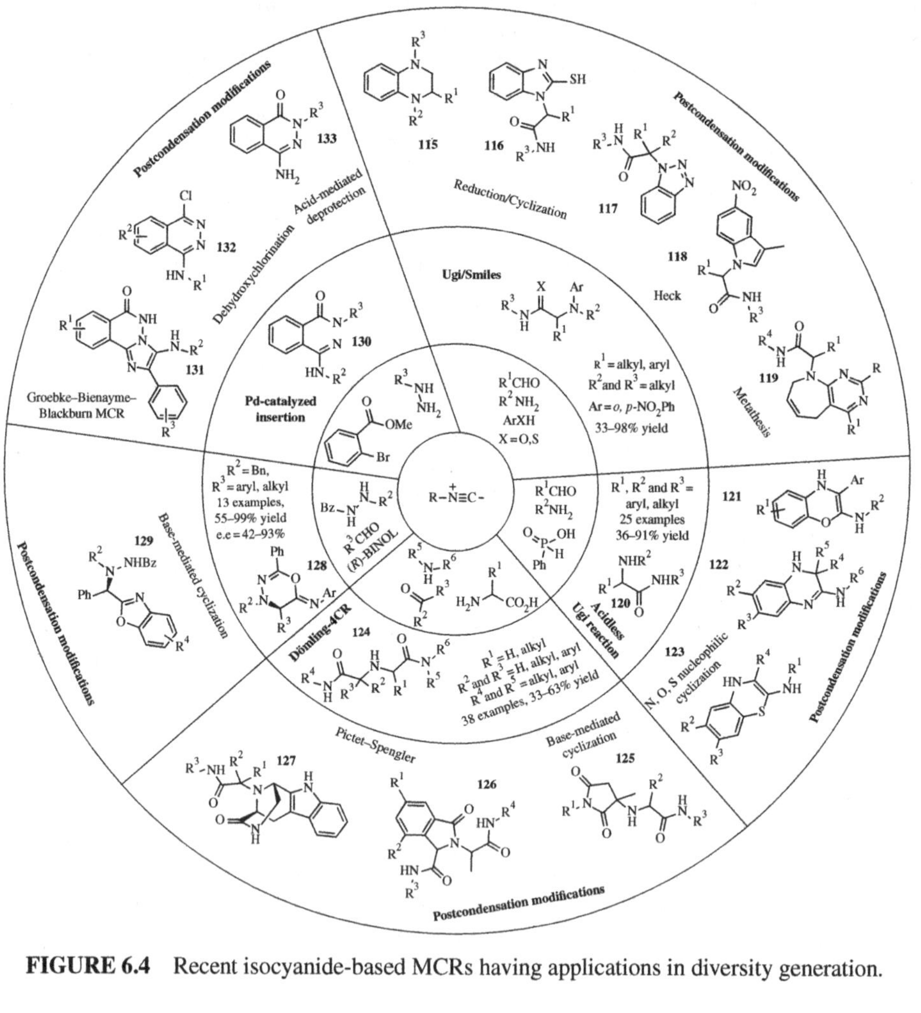


**FIGURE 6.4** Recent isocyanide-based MCRs having applications in diversity generation.

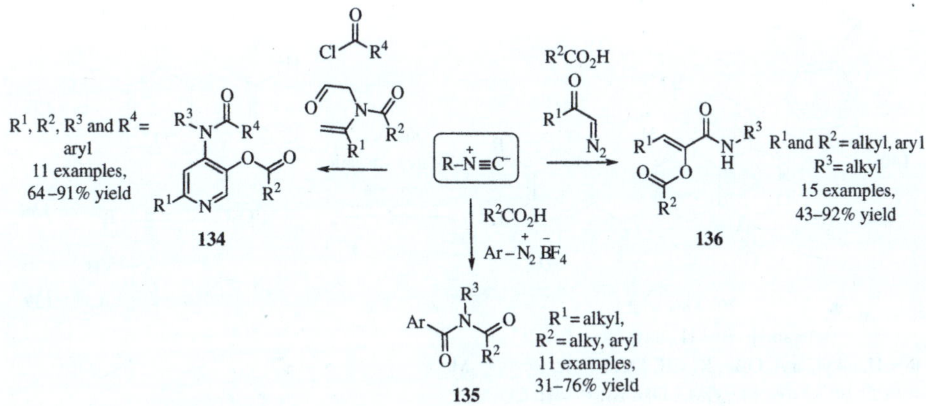

**SCHEME 6.17** Recent isocyanide-based three-component reactions.

In summary, the aforementioned MCRs and chemotypes derived from concomitant secondary reactions are represented in Figure 6.4. Moreover, a bolus of new MCRs are waiting to be exploited for complexity generation through secondary reactions (Scheme 6.17). Exemplifying these MCRs, Zhu and Wang describe an unprecedented α-addition of aldehydes and enamides to isonitriles that form pyridines **134** through a [5+1] cycloaddition [154]. A 3CR metal-free arylation of isonitriles to generate imides **135** using carboxylic acids and aryldiazonium salts has also been recently reported (Scheme 6.17) [155], and in 2012, Basso and Banfi revealed a unique 3CR between diazoketones, carboxylic acids, and isocyanides to generate olefins **136** (Scheme 6.17) [156].

### 6.8.4 Miscellaneous MCRs

An intriguing recent example of an MCR is the sequential one-pot 3CR between 3-formyl chromones (**137**), alkynes (**138**), and tryptamine (**139**) to furnish indoloquinolizines (**140**). This one-pot cascade reaction involves 12 chemical transformations, 2 modes of catalysis, and 6 points of diversity. Furthermore, the generated scaffold was found to be a modulator of centrosome integrity and mitosis (Scheme 6.18) [157].

The year 2013 witnessed a report of a novel 3CR from the Hulme group [158a] involving a unique indole *N*-1-alkylation process of α-iminoketones. The reaction of arylglyoxaldehydes **141** with indoles **142** at elevated temperature promoted *N*-alkylation of α-iminoketones furnishing indole derivatives **143**. A subsequent acid-promoted deprotection of an internal amino nucleophile paved the way for intramolecular cyclization delivering *N*-1 quinoxaline-indoles (Scheme 6.19). The protocol is ideally suited for parallel synthesis, and the generic core has been reported in the kinase inhibitor literature [158b].

## 6.9 ASYMMETRIC MCRs

Optimized methods to produce stereochemically pure/enriched compounds are of high demand in the small molecule pharmaceutical sector [159a–d], and although significant efforts have been made to produce optically pure products from MCR chemistry, the

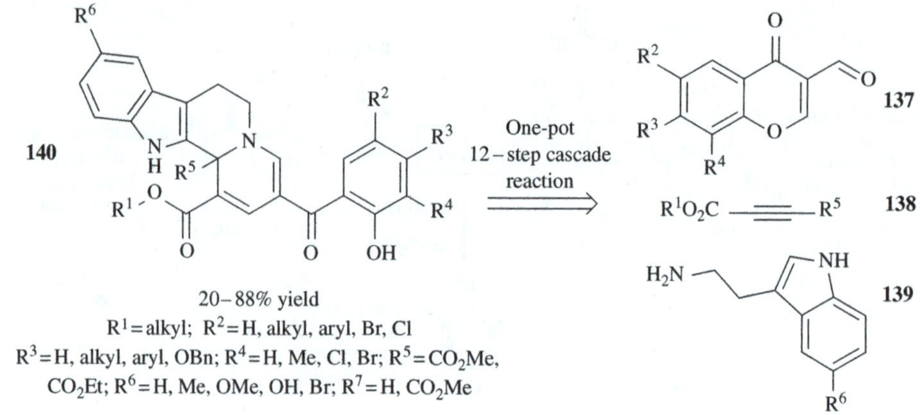


**SCHEME 6.18** 3CR involving a cascade of 12 steps.

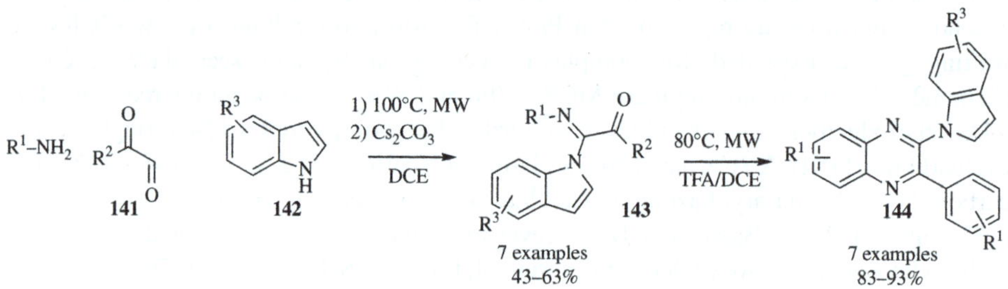


**SCHEME 6.19** Novel 3CR involving *N*-1 alkylation of indoles with α-iminoketones.

area remains a challenging one [160]. Despite this, catalytic enantioselective MCRs have progressed in recent times with successful deployment of biocatalysts [161a], metallo-organic catalysts [161b], and more recently organocatalysts [161c] that induce stereochemical enrichment of products. Indeed, a number of catalytic enantioselective variants of MCRs [162–166] are available, yet methods for asymmetric MCRs are scarce [167, 168].

As such, two successful examples of catalytic approaches to MCRs are highlighted. An asymmetric P-3CR from Schreiber [169a] was enabled with Cu(II)pybox **145a** affording products **148** in excellent yield and high enantioselectivities (Scheme 6.20). Interestingly, the same asymmetric induction with comparable efficacy is also possible using an aluminum–salen catalyst system **145b** reported by Zhu [169b].

Using a binol-derived chiral phosphoric acid **152** as catalyst and isocyanide **151**, Zhu and colleagues [170] were able to perform a highly enantioselective variant of the Ugi reaction in good yields (Scheme 6.21).

## 6.10 APPLICATIONS OF MCRs IN MEDICINAL CHEMISTRY

The goal of this section is to survey many of the most therapeutically advanced small molecules developed via MCR-based methodologies. Due to space limitations, particular attention will be devoted to marketed drugs or molecules currently undergoing clinical

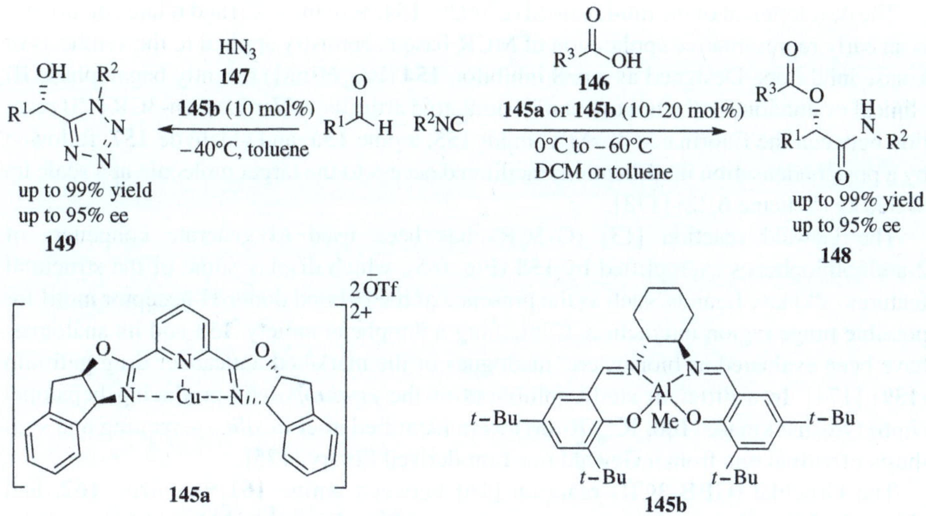

**SCHEME 6.20** Catalytic enantioselective Passerini reactions.

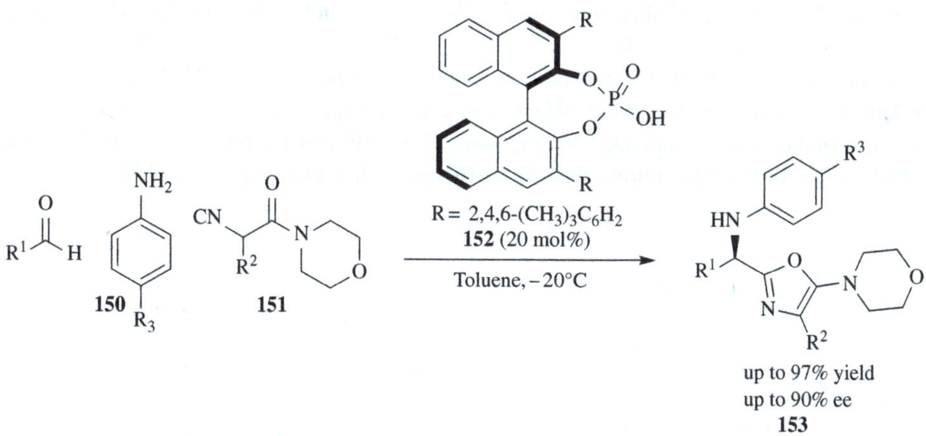

**SCHEME 6.21** Catalytic enantioselective variant of the Ugi reaction.

development. For a more detailed and comprehensive analysis of biologically active small molecules generated via MCR-based chemistry, the reader is referred to several excellent reviews in the public domain [42, 171].

### 6.10.1 Kinase Inhibitors

With 20% of all research programs in the private sector based on kinase-dependent pathologies, rapid generation of diverse arrays of small molecule modulators of this broad class of enzymes has become an essential part of file enhancement activities. Moreover, considering the highly conserved nature of the ATP pocket, access to novel kinase inhibitor space is at a premium for orthosteric type I inhibition [172].

The development of the imidazole SB220025 (**154**, Scheme 6.22) and related analogues is an early representative application of MCR-based chemistry applied to the synthesis of kinase inhibitors. Designed as a p38 inhibitor, **154** ($IC_{50}$ 60 nM) recently began phase III clinical evaluation for the treatment of rheumatoid arthritis. A Van Leusen-3CR [13] reaction between the fluorinated TosMIC input **155**, amine **156**, and aldehyde **157**, followed by a postcondensation modification, facilitated access to the target molecule at a scale up to 500 kg (Scheme 6.22) [173].

The Gewald reaction [13] (G-3CR) has been used to generate congeners of 2-aminothiophenes exemplified by **158** (Fig. 6.5), which display some of the structural features of kinase ligands, such as the presence of the H-bond donor/H-acceptor motif for possible hinge region interaction. Containing a thiophene moiety, **158** and its analogues have been evaluated as bioisosteric analogues of the marketed anticancer drug gefitinib (**159**) [174]. In a different study, inhibitors of the *erythropoietin*-producing hepatoma (Eph) tyrosine kinase (**160**, $IC_{50}$ 76 μM) were identified after *in silico* screening and synthesis of virtual hits from a Gewald reaction-derived library [175].

The Groebke (GBB-3CR) reaction [16] between amine **161**, isonitrile **162**, and aldehyde **163** offers rapid entry into arrays of bicyclic imidazo[1,2-*x*]-heterocyclic congeners (**164a–d**; Scheme 6.23). Again containing a classical H-bond donor/acceptor moiety analogous to ATP, **164a–d**, derived from a secondary reaction on GBB MCR products, displayed potent *in vitro* activity and different selectivity against a broad panel of kinases [176].

Collections of potent inhibitors of the cell cycle serine/threonine kinase Plk-1 were generated at Amgen in one step via the $TMSN_3$-Ugi reaction (Fig. 6.6) [177]. Interestingly, employing the spiropiperidine moiety embedded in spiperone [178] afforded micromolar inhibitors **165a**, which were subsequently optimized to afford **165b–c** with improved potency [177].

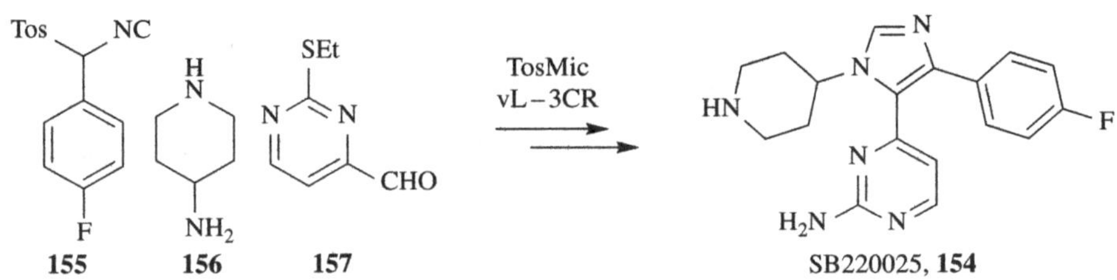


**SCHEME 6.22** Structure and synthesis of SB220025.

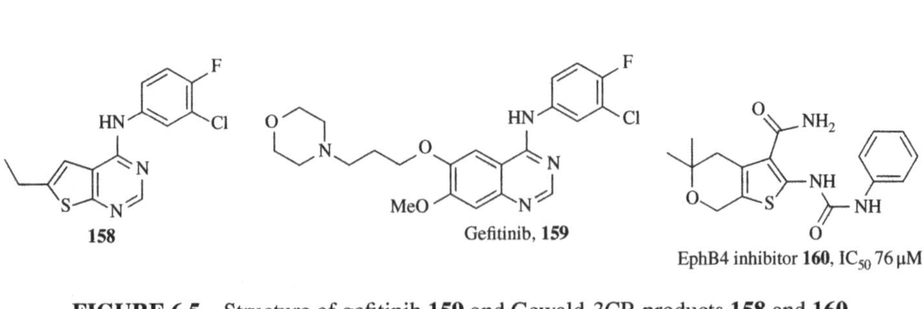


**FIGURE 6.5** Structure of gefitinib **159** and Gewald-3CR products **158** and **160**.

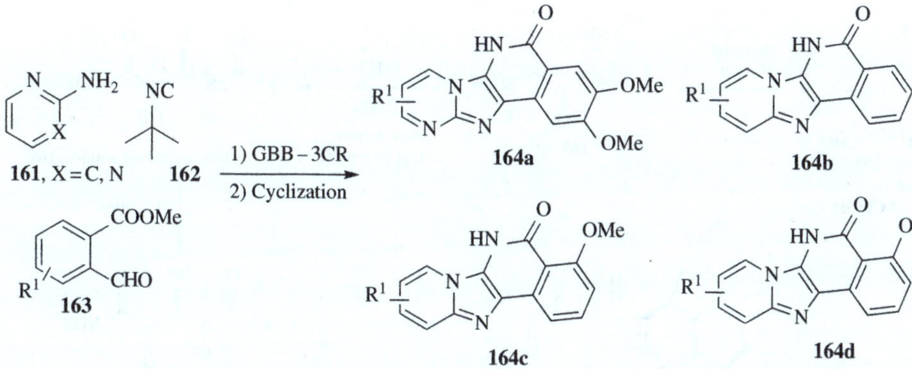


**SCHEME 6.23** Novel kinase inhibitors **164a–d** prepared via the Groebke GBB-3CR.

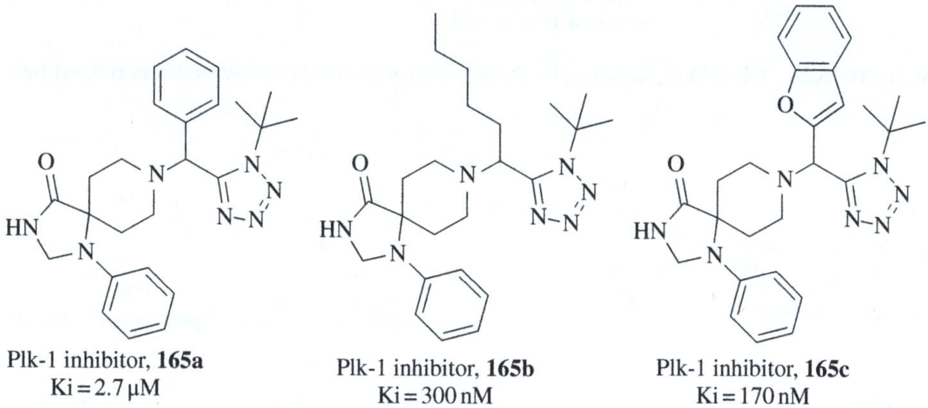


**FIGURE 6.6** Structures of ATP-competitive Plk-1 inhibitors **165a–c** generated via the $TMSN_3$-Ugi reaction.

### 6.10.2 Protease Inhibitors

Currently, the effects of modulating more than 10% of the greater than 500 known human proteases are under investigation in the pharmaceutical industry [179]. The most common mechanism of small molecule inhibition of this class of enzymes is based on the "transition-state mimic," where a "noncleavable" ligand mimics the shape of the substrate usually cleaved by the enzyme. A common structural feature of known protease inhibitors is the secondary hydroxyl "statine" moiety, which is able to engage in hydrogen bonds with key amino acids involved in the enzymatic catalytic process, such as serine/threonine and aspartate residues.

Large arrays (>9000 members) of novel protease inhibitors have been prepared using "PADAM" methodology (*P*asserini *D*eprotection *A*cyl *M*igration *D*eprotection), where a Passerini product **167**, bearing a masked amine nucleophile derived from aminoaldehyde **166**, is treated with acid (Scheme 6.24) [180]. After removal of the Boc group, acyl migration is promoted under basic conditions, leading to final hydroxymethyl-amide products (**168**), which can be further oxidized to keto amides **169**. A representative example of a structurally complex keto amide protease inhibitor

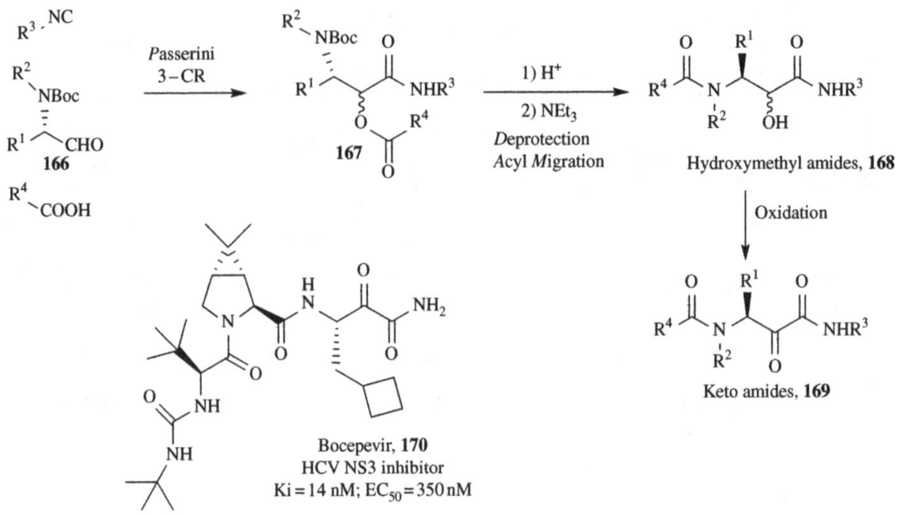


**SCHEME 6.24** PADAM methodology for the assembly of potential protease inhibitors **168** and **169**.

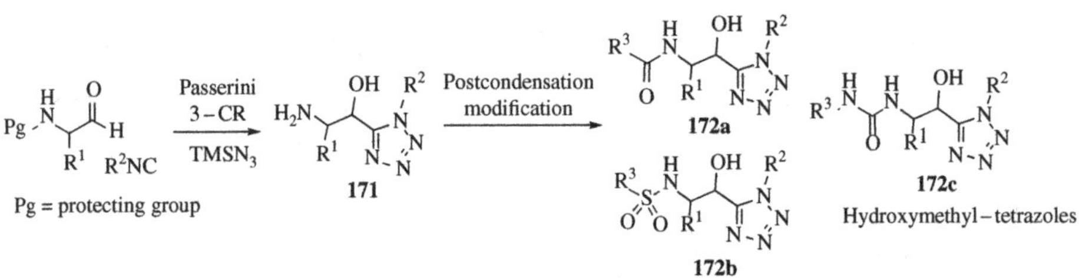


**SCHEME 6.25** Hydroxymethyl-tetrazole protease inhibitors **172a–c** assembled via Passerini $TMSN_3$ reaction.

assembled via PADAM methodology is the approved HCV NS3 inhibitor boceprevir (**170**; Scheme 6.24), which demonstrated positive results in clinical trials for the treatment of hepatitis C [181].

A variant of the Passerini reaction has also been used to assemble libraries of protease inhibitors based on the hydroxymethyl-tetrazole scaffold (**172a–c**; Scheme 6.25). Deprotection of the Passerini product affords **171**, which was then functionalized to introduce a third diversity point (**172a–c**) [182].

Vildagliptin (**173**; Scheme 6.26), approved for the treatment of type II diabetes, is an inhibitor of the serine protease dipeptidyl peptidase IV (DPP-IV). The α-aminonitrile moiety of vildagliptin, common to many other DPP-IV inhibitors, can be accessed using variations of the Ugi 4CR (**174**; Scheme 6.26) [183].

The piperazine nucleus of Crixivan™ (**176**, indinavir; Scheme 6.27), a marketed aspartyl protease inhibitor for the treatment of HIV, was prepared via an Ugi 4CR to afford **177** followed by an enantioselective hydrogenation to yield the desired chiral piperazine (Scheme 6.27). With a prior synthesis of **176** relying on approximately 20 steps, this is a remarkable example of MCR-based methodology dramatically shortening the number of steps to marketed drug and associated "cost of goods" [184].

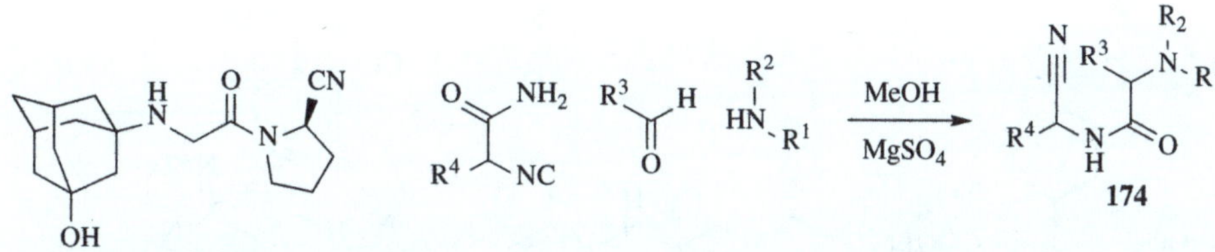


Vildagliptin, 173
DPP-IV inhibitor, $IC_{50}=2\,nM$

**SCHEME 6.26** Assembly of the α-aminonitrile nucleus **174** of vildagliptin **173**.

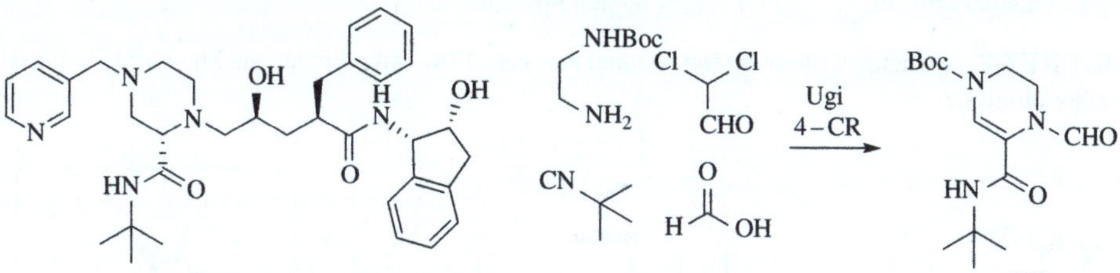


Crixivan™ (Indinavir, **176**)
HIV-1, $IC_{50}=0.52\,nM$
HIV-2, $IC_{50}=3.3\,nM$

**SCHEME 6.27** Assembly of the precursor (**177**) to the piperazine core of Crixivan (**176**) via Ugi 4CR reaction.

### 6.10.3 Ion Channel Inhibitors

Ion channels are transmembrane proteins that mediate transport of ions across the cell membrane, thus translating electrical signals into biochemical events. Disruption of these signals, in particular those associated with $Ca^{2+}$, $Na^{+}$, and $K^{+}$ channels, has been implicated in several disease states [185]. The impact of MCRs on this target family has been impressively exemplified by the Hantzsch reaction [7], which enabled the generation of the marketed antihypertensive L-type $Ca^{2+}$ channel blocker nifedipine (**178**; Fig. 6.7) [186]. A second example of an MCR-derived ion channel inhibitor is the local anesthetic Xylocaine (**179**; Fig. 6.7) that antagonizes voltage-gated $Na^{+}$ channels [187]. Furthermore, the dihydropyrazolopyrimidine **180**, a bioavailable and potent Kv1.5 channel blocker, is currently undergoing clinical development for atrial fibrillation (AF) and can be rapidly prepared via a Biginelli 3CR [188].

### 6.10.4 Protein–Protein Interaction Inhibitors

Protein–protein interactions occur at large interfaces between proteins and only in recent years has it been recognized that a subset of these contacts can be disrupted by small molecules targeting deep, often hydrophobic pockets in one of the binding partners [189]. Interestingly, MCR products or derivatives from subsequent reactions have proven quite fruitful in targeting PPIs [171, 189].

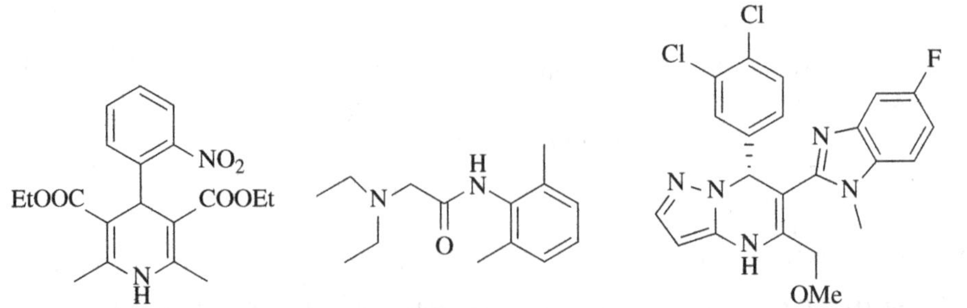


Nifedipine, **178**
L–type $Ca^{++}$
channel blocker

Xylocaine, **179**
fast voltage-gated
$Na^+$ channel blocker

Dihydropyrazolopyrimidine, **180**
$K_V$ 1.5, $K^+$ channel blocker

**FIGURE 6.7** Structures of potent ion channel blockers **178–180** prepared via known MCR-based methodologies.

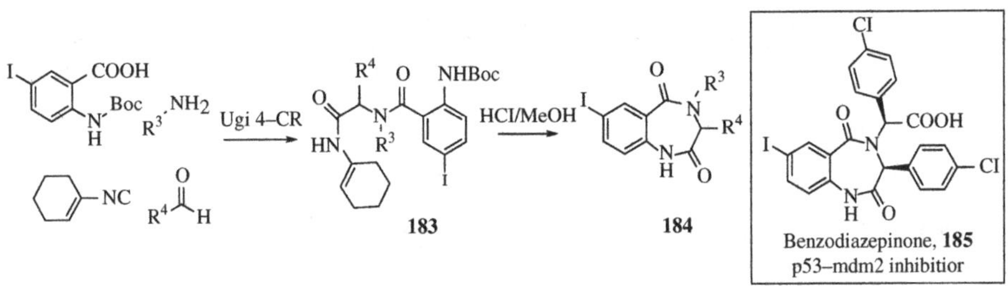


**SCHEME 6.28** UDC methodology toward benzodiazepines **184** and **185** via an Ugi 4CR (Scheme 6.29).

Interactions between the so-called "guardian of the genome" p53 and mdm2 are implicated in the chemo- and radiation resistance of cancers, and thus, disruption of the interaction is of high value in oncology [190]. Benzodiazepinones of generic structure **184** were identified by HTS of a 16,000-member MCR-derived library at Johnson & Johnson and are among the most potent p53–mdm2 inhibitors reported (Scheme 6.28). UDC methodology, employing Boc-protected anthranilic acids in conjunction with cyclohexenylisocyanide, afforded the intermediary Ugi product **183**, which upon treatment with acid yielded the post-MCR benzodiazepine **184** [191]. Further optimization garnered the highly active ($IC_{50}$ 0.4 μM) analogue **185**.

Approaching the same therapeutic target, the MCR-focused company Priaxon more recently reported the discovery of **184** (Scheme 6.29) derived from Shaw's 5CR [132] involving benzylamine **185**, dione **186**, thiol **187**, aldehyde **188**, and secondary amine **189**, followed by post-MCR amide bond formation [192].

### 6.10.5 Tubulin Polymerization Inhibitors

MCRs have been extensively used to prepare complex biologically active natural product analogues, such as inhibitors of tubulin polymerization for cancer treatment that elicit cell cycle arrest [193]. Tubugis **190** are analogues of tubulysin tetrapeptides, potent tubulin

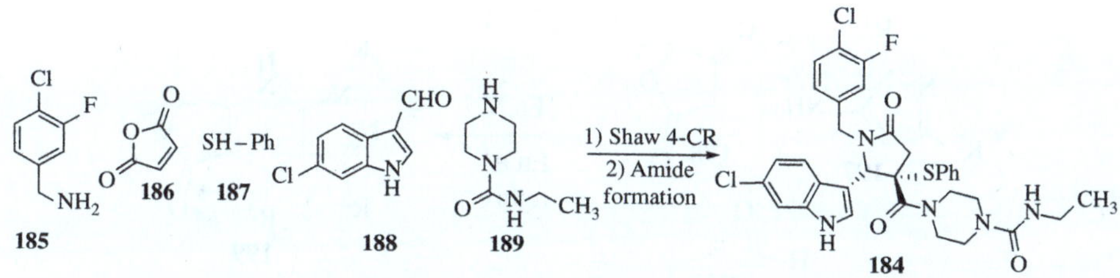

**SCHEME 6.29** Compound **184**, assembled by Priaxon via a Shaw 4CR.

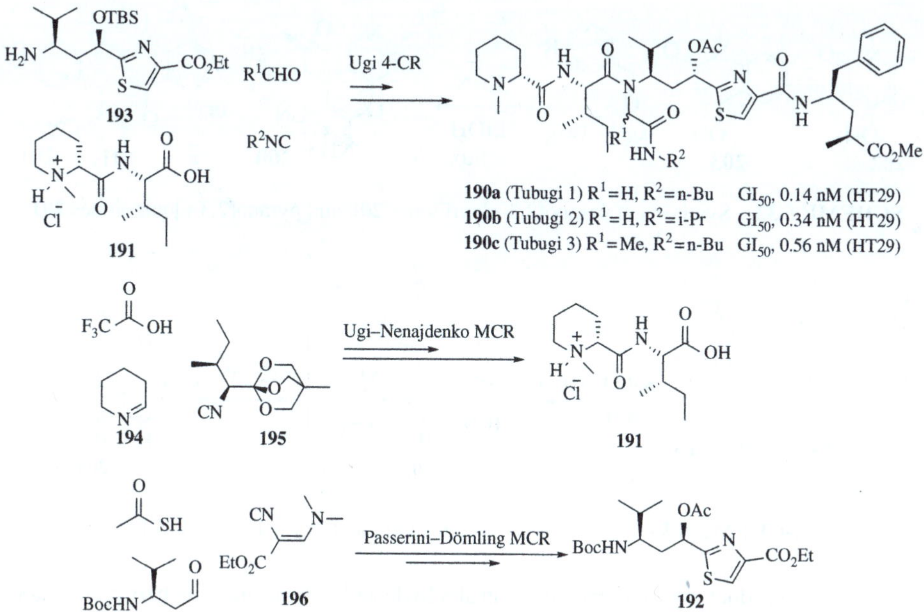

**SCHEME 6.30** Synthesis of tubugi analogues **190a–c**.

polymerization inhibitors with cytotoxic activity against drug-resistant cancer lines (Scheme 6.30) [194]. Wessjohann employed three known MCRs to synthesize tubugis **190a–c** (Scheme 6.30) [195]. The key step is a late stage Ugi 4CR with diversity elements derived from supporting aldehydes and isonitriles, followed by amide bond formation. The other two components of the Ugi 4CR are generated from an Ugi–Nenajdenko 3CR followed by hydrolysis and reductive amination to afford carboxylic acid **191** and a Passerini–Dömling 3CR to afford the thiazole **192**, which upon Boc deprotection yields the amine input **193**.

The natural product podophyllotoxin is another tubulin polymerization inhibitor [196]. Routes to analogues **199** have been reported by Kornienko et al. who used a 3CR between aminopyrazoles **197**, tetramic acid **198**, and an aldehyde (Scheme 6.31). The reaction proceeds via Knoevenagel condensation of **198** with aldehyde, followed by a Michael addition/ring closure process to give congeners of structure **199** [197].

The same research group also employed a 3CR reaction to prepare antitubulin agents, such as pyrano[3,2-*c*3]pyridones **200** and pyrano[3,2-*c*]quinolones **201**. A Knoevenagel

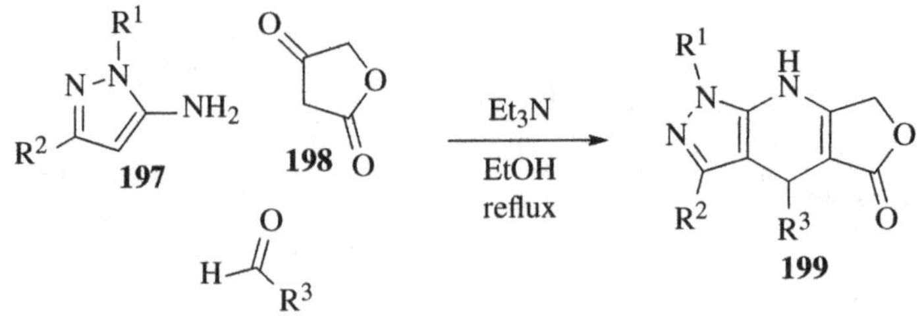

**SCHEME 6.31** Synthesis of podophyllotoxin analogues **199**.

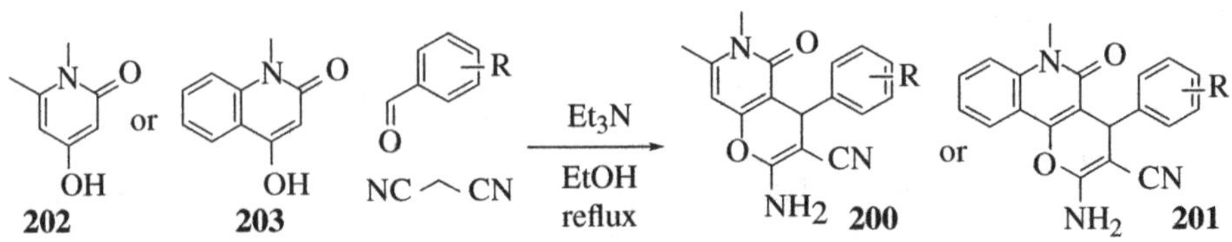

**SCHEME 6.32** Synthesis of pyrano[2,3-*c*]pyridones **200** and pyrano[2,3-*c*]quinolones **201**.

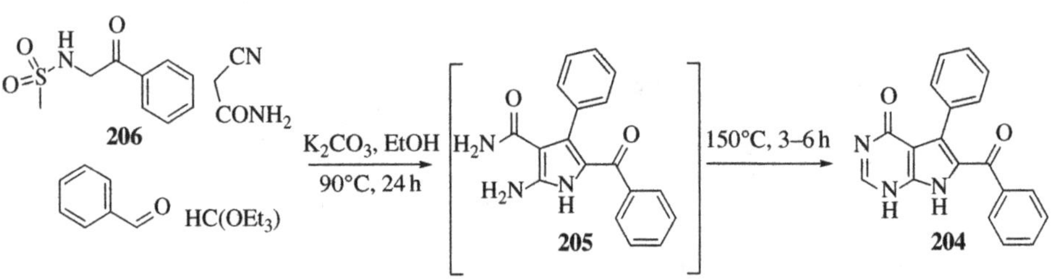

**SCHEME 6.33** Synthesis of 7-deazaxanthine **204** via a one-pot 4CR.

condensation product was generated from an aldehyde and malonitrile, which then undergoes Michael addition of pyridone **202** or **203**, followed by ring closure (Scheme 6.32) [198].

More recently, Kornienko et al. have developed novel tubulin-targeting agents inspired by the "rigidin" family of marine alkaloids [199, 200]. The 7-deazahypoxanthine scaffold **204** was synthesized via a "four-component one-pot two-step" process (Scheme 6.33).

Early studies by Pinney et al. identified benzo[*b*]thiophenes as weak inhibitors of tubulin polymerization [201]. Subsequently, Flynn et al. used a palladium-catalyzed MCR to prepare a series of benzo[*b*]furans **207** that proved to be potent tubulin polymerization inhibitors (Scheme 6.34) [202]. Initial reaction between *o*-iodophenol **208**, alkyne **209**, and methylmagnesium chloride generated **210**. Aryl iodide **211** was utilized in the presence of carbon monoxide (CO) to affect a final carbonylative heteroannulative coupling giving **207**.

### 6.10.6 G-Protein-Coupled Receptors

G-protein-coupled receptors (GPCRs) are popular targets, and their importance is highlighted by the fact that approximately 25% of all drugs in the marketplace elicit their pharmacological effects through them [203, 204]. Interestingly, there is a plethora of activity in the field initiated by molecules derived from MCRs.

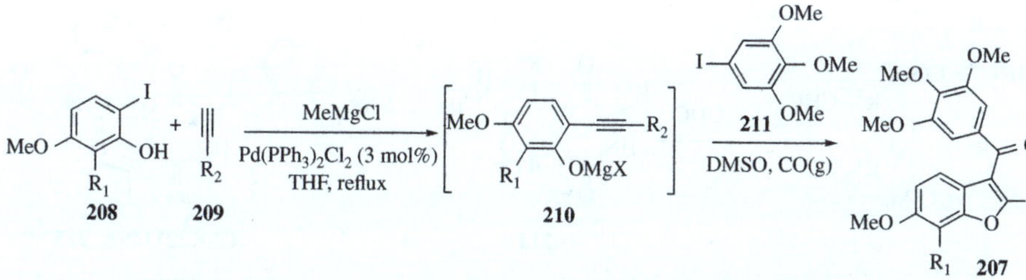


**SCHEME 6.34** Palladium-catalyzed synthesis of benzo[*b*]furans **207**.

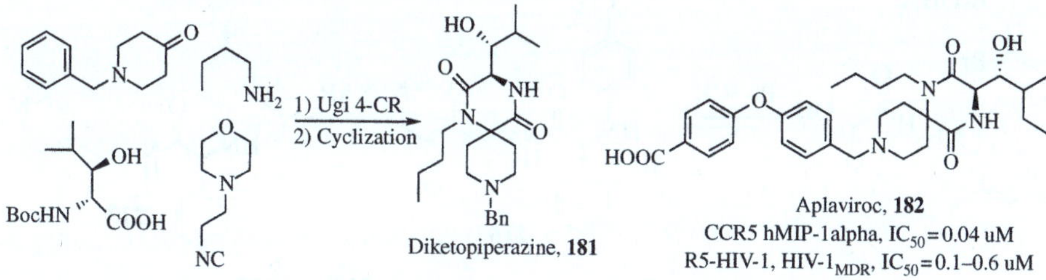


**SCHEME 6.35** Synthesis of the diketopiperazine (**181**) core of Aplaviroc (**182**) via Ugi 4CR.

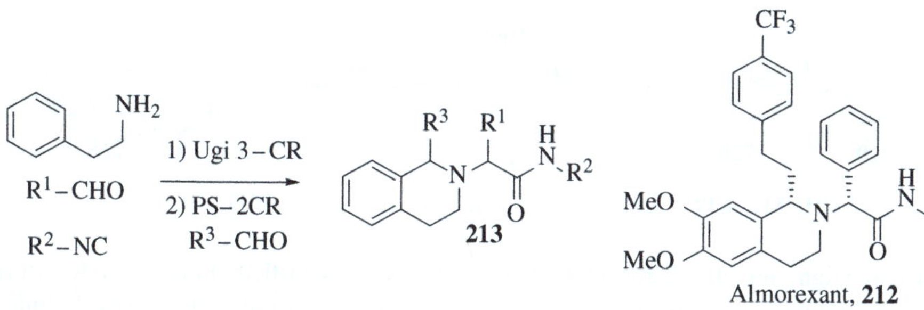


**SCHEME 6.36** 1,2,3,4-Tetrahydroisoquinolines **212–213** as orexin receptor antagonists.

CCR5 antagonists have been shown to be involved in limiting HIV infection. Compounds based on the spirocyclic diketopiperazine core **181** (Scheme 6.35) were synthesized using UDC methodology delivering promising early stage hits for the chemokine GPCR. Subsequent iterations using the same chemistry improved antagonistic activity and ultimately resulted in the discovery of Aplaviroc™, which progressed to phase II clinical trials before withdrawal due to hepatotoxicity (**182**; Scheme 6.35) [205].

Orexins are neuropeptide hormones responsible for basic physiological states, including the regulation of wakefulness and food intake [206, 207]. Almorexant **212** is a first-in-class orexin receptor antagonist, currently undergoing phase III clinical trial for the treatment of insomnia [208]. Access to the 1,2,3,4-tetrahydroisoquinoline core **213** was achieved via a tandem Ugi/Pictet–Spengler reaction sequence (Scheme 6.36) [209].

Preterm labor is the major cause of neonatal morbidity and has been directly linked to the action of oxytocin (OT) on the OT receptor, which stimulates contractions and is widely used for the induction of labor [210]. The UDC strategy was applied to the construction of arrays of diketopiperazines **214** (Scheme 6.37) at GSK. Following an

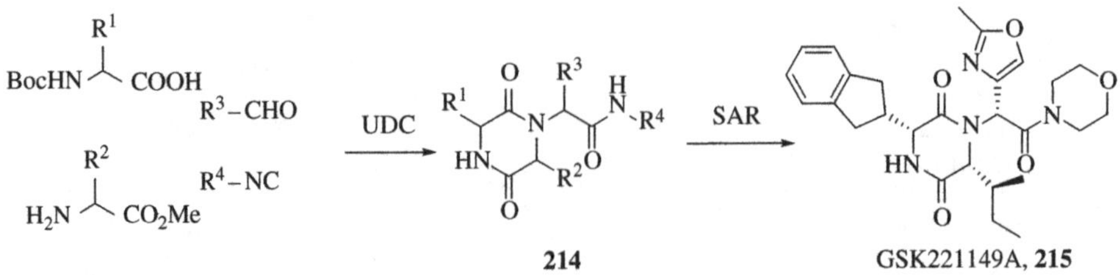

**SCHEME 6.37** UDC methodology delivering the OT receptor antagonist GSK221149A.

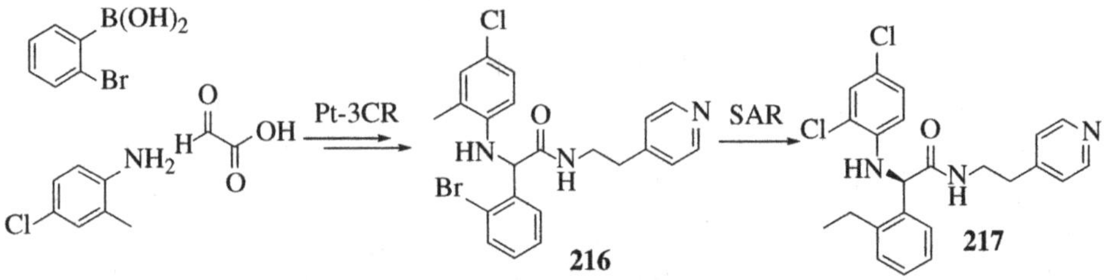

**SCHEME 6.38** CRF receptor antagonists **216** and **217**.

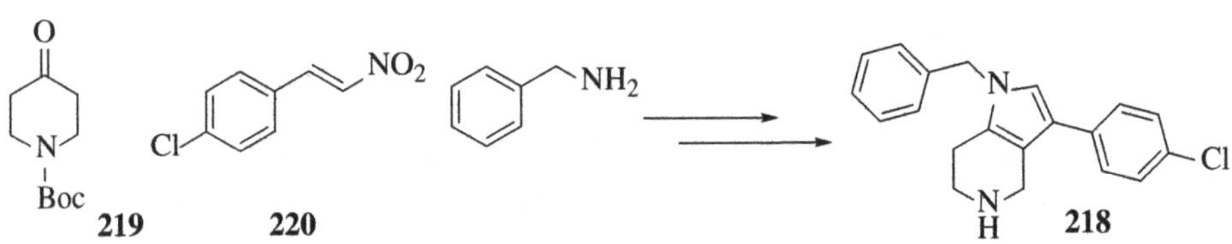

**SCHEME 6.39** Tetrahydropyrido[3,2-*c*]pyrrole **218** as 5-HT$_7$ antagonist.

HTS campaign, novel OT receptor antagonists were identified. Further SAR efforts afforded the first small molecule antagonist of the OT receptor entering phase I clinical trials. GSK221149A (**215**, Ki = 650 pM; Scheme 6.37) is currently undergoing phase II studies for the treatment of preterm labor [211].

Corticotropin-releasing factor (CRF) is a 41-amino-acid hormone involved in stress response [212]. Small molecule antagonists are currently under investigation as potential treatments for generalized anxiety disorders and alcoholism. The impact of MCRs on this target was realized through the discovery of **216**, synthesized in a two-step process involving Petasis 3CR, followed by amidation (Scheme 6.38). Subsequently, **217** was discovered and shown to be a potent CRF receptor antagonist (Ki = 154 nM) [213].

The 5-HT$_7$ receptor is the most studied member of the serotonin receptor family, being involved in thermo- and endocrine regulation [214]. Rudolph et al. used a 3CR to synthesize a series of tetrahydropyrido[3,2-*c*]pyrroles, which afforded the 5-HT$_7$ antagonist **218** [215]. Thus, after Schiff base formation between ketone **219** and benzylamine, a Michael reaction with the nitrostyrene **220** was followed by Boc deprotection to give the tetrahydropyrido[3,2-*c*]pyrrole scaffold **218** (Ki = 14 nM) (Scheme 6.39).

Melanin-concentrating hormone (MCH), a cyclic peptide expressed in the hypothalamus, is involved in several physiological processes, such as energy homeostasis and food intake regulation. Thus, the development of MCH receptor antagonists represents a

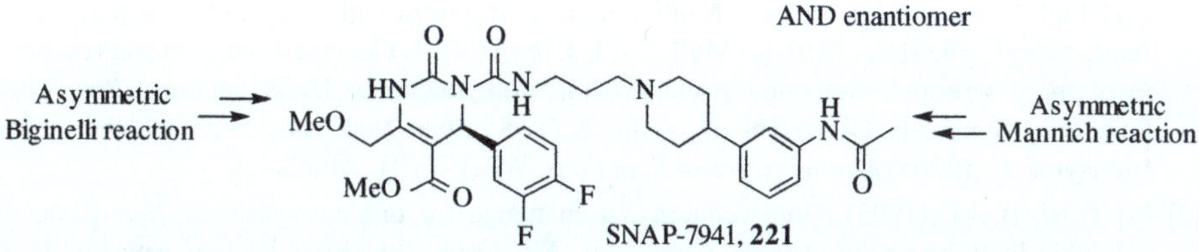


**FIGURE 6.8** Synthesis of SNAP-7941 (**221**) via Biginelli and Mannich reactions.

potential treatment of obesity, depression, and anxiety [216]. Dihydropyrimidone MCH receptor antagonist SNAP-7941 (**221**) is currently undergoing preclinical evaluation (Fig. 6.8), being derived from MCR methodology utilized by Schaus and Goss who developed two organocatalytic enantioselective syntheses of **221** using a Biginelli- and a Mannich-based MCR, respectively [217].

## 6.11 SUMMARY

After a 40-year time span during which MCR research was considered a niche field, the last 15 years have seen a resurgence of interest in developing new MCRs to access unique molecular diversity. Frequently, these efforts have produced small molecule libraries with desirable properties, serving as screening sets to spur hit to lead efforts. In this chapter, we have extensively detailed key advances in scaffold generation and demonstrated impact in medicinal chemistry campaigns spanning several target families. Indeed, MCRs afford compounds with built-in "iterative efficiency potential," a factor increasingly being recognized with associated downstream benefits in medicinal chemistry campaigns. Such campaigns continue to drive ongoing studies of MCR reactions in both academic and industrial sectors, and a plethora of novel scaffolds derived from straightforward 1–2-step protocols remain to be developed. Avoiding such strategic pursuits by simply buying nonexclusive compound sets without assessment of chemical tractability or more importantly "iterative efficiency potential" [2f] may in the long run prove detrimental to the discovery of small molecule partners for drug targets, either because new "activity space" is not accessed or compounds are not novel (do not have freedom to operate). Moreover, solely following a "rearview mirror" approach in designing targeted family libraries, without the balance of a proprietary, chemocentric, diversity-oriented strategy, is a philosophy in need of reevaluation in light of the exciting developments in the MCR arena. The employment of MCR methodologies is a cost-effective and still relatively unexploited strategy to augment small molecule compound collections with new chemical diversity and therefore ideally complements targeted approaches.

## REFERENCES

[1] (a) Touré, B. B. and Hall, D. G. (2009) Natural product synthesis using multicomponent reaction strategies. *Chem. Rev.*, **109**, 4439–4486. (b) Ugi, I. and Heck, S. (2001) The multicomponent reactions and their libraries for natural and preparative chemistry. *Comb. Chem. High. Throughput Screen*, **4**, 1–34. (c) Knapp, J. M., Kurth, M. J., Shaw, J. T. and Younai, A. (2013) *Diversity-Oriented Synthesis*, John Wiley & Sons, Inc., Hoboken, NJ, 27–57.

(d) Dömling, A. and Ugi, I. (2000) Multicomponent reactions with isocyanides. *Angew. Chem. Int. Ed. Engl.*, **39**, 3168–3210. (e) Müller, T. J. J. (ed.) (2014) General discussion and reactions involving a carbonyl compound as electrophilic component. In *Multicomponent Reactions*; Science of Synthesis, Georg Thieme Verlag K.G., Stuttgart/New York, 5–27. (f) Zhu, J. and Bienaymé, H. (2005) *Multicomponent Reactions*, Wiley-VCH, Weinheim.

[2] (a) Trost, B. M. (1995) Atom economy: a challenge for organic synthesis: homogeneous catalysis leads the way. *Angew. Chem. Int. Ed. Engl.*, **34**, 259–281. (b) Anastas, P. T. and Warner, J. C. (1998) *Green Chemistry: Theory and Practice*, Oxford University Press, New York. (c) Heiner, E. (2012) Diversity oriented syntheses of conventional heterocycles by smart multi component reactions (MCRs) of the last decade. *Molecules*, **17**, 1074–1102. (d) Wender, P. A., Verma, V. A., Paxton, T. J. and Pillow, T. H. (2007) Function-oriented synthesis, step economy, and drug design. *Acc. Chem. Res.*, **41**, 40–49. (e) Schreiber, S. L. (2000) Target-oriented and diversity-oriented organic synthesis in drug discovery. *Science*, **287**, 1964–1969. (f) Weber, L., Illgen, K. and Almstetter, M. (1999) Discovery of new multi component reactions with combinatorial methods. *Synlett*, **1999**, 366–374. (g) Hulme, C. and Dietrich, J., (2009) Emerging molecular diversity from the intra-molecular Ugi reaction: iterative efficiency in medicinal chemistry. *Mol. Divers.*, **13**, 195–207.

[3] Bienaymé, H., Hulme, C., Oddon, G. and Schmitt, P. (2000) Maximizing synthetic efficiency: multi-component transformations lead the way. *Chem. Eur. J.*, **6**, 3321–3329.

[4] Tuch, A. and Wallé, S. (2005) *Handbook of Combinatorial Chemistry*, Wiley-VCH, Weinheim, 685–705.

[5] Strecker, A. (1850) Ueber die künstliche Bildung der Milchsäure und einen neuen, dem Glycocoll homologen Körper. *Ann. Chem.*, **75**, 27–45.

[6] (a) Chuang, T., Yang, C., Chang, C. and Fang, J. (1990) Base-catalyzed autoxidation of α-aminonitriles: an efficient method for conversion of aldehydes to amides and 2-amino-2-sulfenylacetonitrile to carbamates. *Synlett*, **12**, 733–734. (b) Baeza, A., Najera, C. and Sansano, J. M. (2007) Solvent-free synthesis of racemic α-aminonitriles. *Synthesis*, **8**, 1230–1234. (c) Shen, Z., Ji, S. and Loh, T. (2008) Indium (III) iodide-mediated Strecker reaction in water: an efficient and environmentally friendly approach for the synthesis of α-aminonitrile via a three-component condensation. *Tetrahedron*, **64**, 8159–8163. (d) Surendra, K., Krishnaveni, N. S., Mahesh, A. and Rao, K. R. (2006) Supramolecular catalysis of Strecker reaction in water under neutral conditions in the presence of β-cyclodextrin. *J. Org. Chem.*, **71**, 2532–2534. (e) Cruz-Acosta, F., Santos-Exposito, A., De Armas, P. and Garcia-Tellado, F. (2009) Lewis base-catalyzed three-component strecker reaction on water. An efficient manifold for the direct small α-cyanoamination of ketones and aldehydes. *Chem. Commun.*, **44**, 6839–6841. (f) Li, Z., Ma, Y., Xu, J., Shi, J. and Cai, H. (2010) One-pot three-component synthesis of α-aminonitriles using potassium hexacyanoferrate(II) as an eco-friendly cyanide source. *Tetrahedron Lett.*, **51**, 3922–3926.

[7] (a) Hantzsch, A. (1882) Synthesis of pyridine derivatives from acetoacetic ester and aldehydeammoniak. *Liebigs Ann. Chem.*, **215**, 1–82. (b) Saini, A., Kumar, S. and Sandhu, S. J. (2008) Hantzsch reaction: recent advances in Hantzsch 1,4-dihydropyridines. *J. Sci. Ind. Res.*, **67**, 95–111.

[8] (a) Biginelli, P. (1893) The urea-aldehyde derivatives of acetoacetic esters. *Gazz. Chim. Ital.*, **23**, 360–416. (b) Suresh, S. and Sandhu, J. S. (2012) Past, present and future of the Biginelli reaction: a critical perspective. *Arkivoc*, **1**, 66–133.

[9] (a) Mannich, C. (1917) Synthesis of β-ketonic bases. *Arch. Pharm.*, **255**, 261–276. (b) Robinson, R. (1917) A synthesis of tropinone. *J. Chem. Soc. London*, **111**, 762–768.

[10] (a) Passerini, M. (1921) Isonitriles II. Compounds with aldehydes or with ketones monobasic organic acids. *Gazz. Chim. Ital.*, **51**, 181–189. (b) Marcaccini, S., Miguel, D., Torroba, T.

and Garcia-Valverde, M. (2003) 1,4-thiazepines, 1,4-benzothiazepin-5-ones, and 1,4-benzothioxepin orthoamides via multicomponent reactions of isocyanides. *J. Org. Chem.*, **68**, 3315–3318. (c) Beck, B., Magnin-Lachaux, M., Herdtweck, E. and Dömling, A. (2001) A novel three-component butenolide synthesis. *Org. Lett.*, **3**, 2875–2878. (d) Bossio, R., Marcaccini, S. and Pepino, R. (1991) Studies on isocyanides and related compounds. Synthesis of oxazole derivatives via the Passerini reaction. *Liebigs Ann. Chem.*, **10**, 1107–1108.

[11] (a) Asinger, F. and Thiel, M. (1958) Einfache Synthesen und chemisches Verhalten neuer heterocyclischer Ringsysteme. *Angew. Chem.*, **70**, 667–683 (b) Ugi, I. and Wischhöfer, E. (1962) Isonitrile, XI. Synthese einfacher Penicillansäure-Derivate. *Chem. Ber.*, **95**, 136–140.

[12] Ugi, I., Meyr, R., Fetzer, U. and Steinbruckner, C. (1959) Studies on isonitriles. *Angew. Chem.*, **71**, 386–389.

[13] Gewald, K., Schinke, E. and Bottcher, H. (1966) Heterocycles from CH-acidic nitriles VII 2-aminothiophene, carbonyl compounds, and sulfur. *Chem. Ber. Recl.*, **99**, 94–100.

[14] Van Leusen, A. M., Wildeman, J. and Oldenziel, O. H. (1977) Chemistry of sulfonylmethyl isocyanides. 12. Base-induced cycloaddition of sulfonylmethyl isocyanides to carbon, nitrogen double bonds. Synthesis of 1,5-disubstituted and 1,4,5-trisubstituted imidazoles from aldimines and imidoyl chlorides. *J. Org. Chem.*, **42**, 1153–1159.

[15] Petasis, N. A. and Zavialov, I. A. (1997) A new and practical synthesis of -amino acids from alkenyl boronic acids. *J. Am. Chem. Soc.*, **119**, 445–446.

[16] (a) Groebke, K., Weber, L. and Mehlin, F. (1998) Synthesis of imidazo[1,2-*a*] annulated pyridines, pyrazines and pyrimidines by a novel three-component condensation. *Synlett*, **6**, 661–663. (b) Bienaymé, H. and Bouzid, K. (1998) A new heterocyclic multicomponent reaction for the combinatorial synthesis of fused 3-aminoimidazoles. *Angew. Chem. Int. Ed. Engl.*, **37**, 2234–2237. (c) Blackburn, C. (1998) A three-component solid-phase synthesis of 3-aminoimidazo[1,2-*a*]azines. *Tetrahedron Lett.*, **39**, 5469–5472.

[17] Ugi, I., Dömling, A. and Hörl, W. (1994) Multicomponent reactions in organic chemistry. *Endeavour*, **18**, 115–122.

[18] Brauch, S., Van Berkel, S. S. and Westermann, B. (2013) Higher-order multicomponent reactions: beyond four reactants. *Chem. Soc. Rev.*, **42**, 4948–4962.

[19] Dömling, A. and Ugi, I. (1993) The seven-component reaction. *Angew. Chem. Int. Ed. Engl.*, **32**, 563–564.

[20] Marcaccini, S. and Torroba, T. (2007) The use of the Ugi four-component condensation. *Nat. Protoc.*, **2**, 632–639.

[21] Ayaz, M., De Moliner, F., Dietrich, J. and Hulme, C. (2012) *Isocyanide Chemistry*, Wiley-VCH, Weinheim, 335–384.

[22] Basso, A., Banfi, L., Riva, R. and Guanti, G. (2005) A novel highly selective chiral auxiliary for the asymmetric synthesis of L- and D α-amino acid derivatives via a multicomponent Ugi reaction. *J. Org. Chem.*, **70**, 575–579.

[23] Keating, T. A. and Armstrong, R. W. (1996) A remarkable two-step synthesis of diverse 1,4-benzodiazepine-2,5-diones using the Ugi four-component condensate ion products: 1-Isocyanocyclohexene as a convertible isocyanide. Mechanism of conversion, synthesis of diverse structures and demonstration of resin capture. *J. Am. Chem. Soc.*, **118**, 2574–2583.

[24] Hulme, C., Ma, L., Romano, J. and Morrissette, M. M. (1999) Remarkable three step-one pot solution phase preparation of novel imidazolines utilizing a UDC (Ugi/de-Boc/cyclize) strategy. *Tetrahedron Lett.*, **40**, 7925–7928.

[25] (a) Hulme, C., Peng, J., Tang, S., Burns, C. J., Morize, I. and Labaudiniere, R. (1998) Improved procedure for the solution phase preparation of 1,4-benzodiazepine-2,5-dione libraries via Armstrong's convertible isonitrile and the Ugi reaction. *J. Org. Chem.*, **83**, 8021–8023.

(b) Hulme, C., Morrissette, M. M., Volz, F. A. and Burns, C. J. (1998) The solution phase synthesis of diketopiperazine libraries via the Ugi reaction: novel application of Armstrong's convertible isonitrile. *Tetrahedron Lett.*, **39**, 1113–1116.

[26] Hulme, C., Ma, L., Cherrier, M. P., Romano, J., Morton, G., Duquenne, C., Salvino, J. and Labaudiniere, R. (2000) Novel applications of convertible isonitriles for the synthesis of mono- and bi-cyclic γ-lactam via a UDC strategy. *Tetrahedron Lett.*, **41**, 1883–1887.

[27] Hulme, C., Peng, J., Louridas, B., Menard, P., Krolikowski, P. and Kumar, N. V. (1998) Applications of *N*-BOC-diamines for the solution phase synthesis of ketopiperazine libraries utilizing a Ugi/De-BOC/Cyclization (UDC) strategy. *Tetrahedron Lett.*, **39**, 8047–8050.

[28] Hulme, C., Peng, J., Morton, G., Salvino, J. M., Herpin, T. and Labaudiniere, R. (1998) Novel safety-catch linker and its application with a Ugi/De-Boc/Cyclization (UDC) strategy to access carboxylic acids, 1,4-benzodiazepines, diketopiperazines, ketopiperazines and dihydroquinoxalinones. *Tetrahedron Lett.*, **39**, 7227–7230.

[29] Hulme, C. and Cherrier, M. P. (1999) Novel applications of ethyl glyoxylate with the Ugi MCR. *Tetrahedron Lett.*, **40**, 5295–5299.

[30] Tempest, P., Ma, V., Thomas, S., Hua, Z., Kelly, M. G. and Hulme, C. (2001) Two-step solution-phase synthesis of novel benzimidazoles utilizing a UDC (Ugi/De-Boc/Cyclize) strategy. *Tetrahedron Lett.*, **42**, 4959–4962.

[31] Nixey, T., Tempest, P. and Hulme, C. (2002) Two-step solution-phase synthesis of novel quinoxalinones utilizing a UDC (Ugi/De-Boc/Cyclize) strategy. *Tetrahedron Lett.*, **43**, 1637–1639.

[32] Dietrich, J., Kaiser, C., Meurice, N. and Hulme, C. (2010) Concise two-step solution phase syntheses of four novel dihydroquinazoline scaffolds. *Tetrahedron Lett.*, **51**, 3951–3955.

[33] Nixey, T., Kelly, M., Semin, D. and Hulme, C. (2002) Short solution phase preparation of fused azepine-tetrazoles via UDC (Ugi/De-Boc/Cyclize) strategy. *Tetrahedron Lett.*, **43**, 3681–3684.

[34] Nixey, T., Kelly, M. and Hulme, C. (2000) The one-pot solution-phase preparation of fused tetrazole-ketopiperazines. *Tetrahedron Lett.*, **41**, 8729–8733.

[35] Hulme, C., Chappeta, S., Griffith, C., Lee, Y. and Dietrich, J. (2009) An efficient solution phase synthesis of triazadibenzoazulenones: "designer isonitrile free" methodology enabled by microwaves. *Tetrahedron Lett.*, **50**, 1939–1942.

[36] Xu, Z., Shaw, A., Dietrich, J., Cappelli, A., Nichol, G. and Hulme, C. (2012) Facile, novel two-step syntheses of benzimidazoles, *bis*-benzimidazoles, and *bis*-benzimidazole-dihydroquinoxalines. *Mol. Divers.*, **16**, 73–79.

[37] Xu, Z., Shaw, A., Nichol, G., Cappelli, A. and Hulme, C. (2012) Applications of *ortho*-phenylisonitrile and *ortho*-*N*-Boc aniline for the two-step preparation of novel *bis*-heterocyclic chemotypes. *Mol. Divers.*, **16**, 607–612.

[38] Xu, Z., Dietrich, J., Shaw, A. Y. and Hulme, C. (2010) Two-step syntheses of fused quinoxaline-benzodiazepines and *bis*-benzodiazepines. *Tetrahedron Lett.*, **51**, 4566–4569.

[39] Xu, Z., Ayaz, M., Cappelli, A. P. and Hulme, C. (2012) General one-pot, two-step protocol accessing a range of novel polycyclic heterocycles with high skeletal diversity. *ACS Comb. Sci.*, **14**, 460–464.

[40] Xu, Z., De Moliner, F., Cappelli, A. P., Ayaz, M. and Hulme, C. (2014) Expeditious routes to polycyclic molecular frameworks via one-pot, two-step Ugi ring-closing sequences. *Synlett*, **25**, 225–228.

[41] (a) Dömling, A. (2006) Recent developments in isocyanide based multicomponent reactions in applied chemistry. *Chem. Rev.*, **106**, 17–89. (b) Hulme, C. and Gore, V. (2003) "Multi-component reactions: emerging chemistry in drug discovery" "from Xylocain to Crixivan". *Curr. Med. Chem.*, **10**, 51–80.

[42] Ruben, A. J., Kiso, Y. and Freire, E. (2006) Overcoming roadblocks in lead optimization: a thermodynamic perspective. *Chem. Biol. Drug. Des.*, **67**, 2–4.

[43] Jida, M., Betti, C. Urbanczyk-Lipkowska, Z., Tourwé, D. and Ballet, S. (2013) Highly diastereoselective synthesis of 1-carbamoyl-4-aminoindoloazepinone derivatives via the Ugi reaction. *Org. Lett.*, **15**, 5866–5869.

[44] Shkirando, M., Kysil, V. M., Potapov, V. V. and Ivachtchenko, A. V. (2009) One-step assembly of novel carbamoyl-substituted heterocycle-fused 1-oxo-2,3dihydro-1*H*-benzo[c]azepines. In Mironov, M. A. (ed.) *Proceedings of the 4th International Conference on Multi-Component Reactions and Related Chemistry, Ekaterinburg, Russia*, Springer, New York, 104–105.

[45] Marcaccini, S., Miguel, D., Torroba, T. and Garcia-Valverde, M. (2003) 1,4-thiazepines, 1,4-benzothiazepin-5-ones and 1,4-benzothioxepinorthoamides via multicomponent reactions of isocyanides. *J. Org. Chem.*, **68**, 3315–3318.

[46] Gunawan, S. and Hulme, C. (2013) Bifunctional building blocks in the Ugi-azide condensation reaction: a general strategy toward exploration of new molecular diversity. *Org. Biomol. Chem.*, **11**, 6036–6046.

[47] Paulvannan, K. (1999) Preparation of tricyclic nitrogen heterocycles via tandem four-component condensation/intramolecular Diels-Alder reaction. *Tetrahedron Lett.*, **40**, 1851–1854.

[48] Paulvannan, K. (2004) An atom-economical approach to conformationally constrained tricyclic nitrogen heterocycles via sequential and tandem Ugi/intramolecular Diels–Alder reaction of pyrrole. *J. Org. Chem.*, **69**, 1207–1214.

[49] Wright, D. L., Robotham, C. V. and Aboud, K. (2002) Studies on the sequential multi-component coupling/Diels–Alder cycloaddition reaction. *Tetrahedron Lett.*, **43**, 943–946.

[50] Ilyin, A., Kysil, V., Krasavin, M., Kurashvili, I. and Ivachtchenko, A. V. (2006) Complexity-enhancing acid-promoted rearrangement of tricyclic products of tandem Ugi 4CC/intramolecular Diels–Alder reaction. *J. Org. Chem.*, **71**, 9544–9547.

[51] Huang, X. and Xu, J. (2009) One-pot facile synthesis of substituted isoindolinones via an Ugi four-component condensation/Diels–Alder cycloaddition/deselenization-aromatization sequence. *J. Org. Chem.*, **74**, 8859–8861.

[52] Piscopio, A. D., Miller, J. F. and Koch, K. (1999) Ring closing metathesis in organic synthesis: evolution of a high speed, solid phase method for the preparation of β-turn mimetics. *Tetrahedron*, **55**, 8189–8198.

[53] Vo-Thanh, G., Boucard, V., Sauriat-Dorizon, H. and Guibé, F. (2001) Ring closing metathesis of unsaturated amides as a route to short and medium-sized unsaturated lactams and to ethylenic pseudopeptides. *Synlett*, 37–40.

[54] Banfi, L., Basso, A., Guanti, G. and Riva, R. (2003) Application of tandem Ugi reaction/ring-closing metathesis in multicomponent synthesis of unsaturated nine-membered lactams. *Tetrahedron Lett.*, **44**, 7655–7658.

[55] Dietrich, S. A., Banfi, L., Basso, A., Damonte, G., Guanti, G. and Riva, R. (2005) Application of tandem Ugi multi-component reaction/ring closing metathesis to the synthesis of a conformationally restricted cyclic pentapeptide. *Org. Biomol. Chem.*, **3**, 97–106.

[56] Krelaus, R. and Westermann, B. (2004) Preparation of peptide-like bicyclic lactams via a sequential Ugi reaction—olefin metathesis approach. *Tetrahedron Lett.*, **45**, 5987–5990.

[57] Basso, A., Banfi, L., Riva, R. and Guanti, G. (2006) Preparation of optically pure fused polycyclic scaffolds by Ugi reaction followed by olefin and enyne metathesis. *Tetrahedron*, **62**, 8830–8837.

[58] Ribelin, T. P., Judd, A. S., Akritopoulou-Zanze, I., Henry, R. F., Cross, J. L., Whittern, D. N. and Djuric, S. W. (2007) Concise construction of novel bridged bicyclic lactams by sequenced Ugi/RCM/Heck reactions. *Org. Lett.*, **9**, 5119–5122.

[59] Tempest, P., Ma, V., Kelly, M. G., Jones, W. and Hulme, C. (2001) MCC/SNAr methodology. Part 1: novel access to a range of heterocyclic cores. *Tetrahedron Lett.*, **42**, 4963–4968.

[60] Tempest, P., Pettus, L., Gore, V. and Hulme, C. (2003) MCC/SNAr methodology. Part 2: novel three-step solution phase access to libraries of benzodiazepines. *Tetrahedron Lett.*, **44**, 1947–1950.

[61] Cristau, P., Vors, J. and Zhu, J. (2001) A rapid access to biaryl ether containing macrocycles by pairwise use of Ugi-4CR and intramolecular SNAr-based cycloetherification. *Org. Lett.*, **3**, 4079–4082.

[62] Cristau, P., Vors, J. and Zhu, J. (2003) Rapid and diverse route to natural product-like biaryl ether containing macrocycles. *Tetrahedron*, **59**, 7859–7870.

[63] Cristau, P., Vors, J. and Zhu, J. (2003) Solid-phase synthesis of natural product-like macrocycles by a sequence of Ugi-4CR and SNAr-based cycloetherification. *Tetrahedron Lett.*, **44**, 5575–5578.

[64] Cuny, G., Bois-Choussy, M. and Zhu, J. (2003) One-pot synthesis of polyheterocycles by a Palladium-catalyzed intramolecular *N*-arylation/C-H activation/aryl-aryl bond-forming domino process. *Angew. Chem. Int. Ed. Engl.*, **42**, 4774–4777.

[65] Cuny, G., Bois-Choussy, M. and Zhu, J. (2004) Palladium- and copper-catalyzed synthesis of medium- and large-sized ring-fused dihydroazaphenanthrenes and 1,4-benzodiazepine-2,5-diones. Control of reaction pathway by metal-switching. *J. Am. Chem. Soc.*, **126**, 14475–14484.

[66] Salcedo, A., Neuville, L., Rondot, C., Retailleau, P. and Zhu, J. (2008) Palladium-catalyzed domino intramolecular *N*-arylation/ intermolecular C—C bond formation for the synthesis of functionalized benzodiazepinediones. *Org. Lett.*, **10**, 857–860.

[67] Kalinski, C., Umkehrer, M., Ross, G., Kolb, J., Burdack, C. and Hiller, W. (2006) Highly substituted indol-2-ones, quinoxalin-2-ones and benzodiazepin-2,5-diones via a new Ugi(4CR)-Pd assisted N-aryl amidation strategy. *Tetrahedron Lett.*, **47**, 3423–3426.

[68] Bonnaterre, F., Bois-Choussy, M. and Zhu, J. (2006) Rapid access to oxindoles by the combined use of an Ugi four-component reaction and a microwave-assisted intramolecular Buchwald–Hartwig amidation reaction. *Org. Lett.*, **8**, 4351–4354.

[69] Erb, W., Neuville, L. and Zhu, J. (2009) Ugi-post functionalization, from a single set of Ugi-adducts to two distinct Heterocycles by microwave-assisted Palladium-catalyzed cyclizations: Tuning the reaction pathways by ligand switch. *J. Org. Chem.*, **74**, 3109–3115.

[70] El Kaim, L., Gageat, M., Gaultier, L. and Grimaud, L. (2007) New Ugi/Pictet-Spengler multicomponent formation of polycyclic diketopiperazines from isocyanides and α-keto acids. *Synlett*, **2007**, 500–502.

[71] Wang, W., Ollio, S., Herdtweck, E. and Dömling, A. (2011) Polycyclic compounds by Ugi–Pictet–Spengler sequence. *J. Org. Chem.*, **76**, 637–644.

[72] Wang, W., Herdtweck, E. and Dömling, A. (2010) Polycyclic indole alkaloid-type compounds by MCR. *Chem. Commun.*, **46**, 770–772.

[73] Znabet, A., Zonneveld, J., Janssen, E., De Kanter, F. J. J., Helliwell, M., Turner, N. J., Ruijter, E. and Orru, R. V. A. (2010) Asymmetric synthesis of synthetic alkaloids by a tandem biocatalysis/Ugi/Pictet–Spengler-type cyclization sequence. *Chem. Commun.*, **46**, 7706–7708.

[74] Akritopoulou-Zanze, I., Whitehead, A., Waters, J. E., Henry, R. F. and Djuric, S. W. (2007) Synthesis of novel and uniquely shaped 3-azabicyclo[4.2.0]octan-4-one derivatives by sequential Ugi/[2+2] ene–enone photocycloadditions. *Org. Lett.*, **9**, 1299–1302.

[75] (a) Xu, Z., De Moliner, F., Cappelli, A. P. and Hulme, C. (2012) Ugi/aldol sequence: expeditious entry to several families of densely substituted nitrogen heterocycles. *Angew. Chem. Int. Ed. Engl.*, **51**, 8037–8040. (b) Xu, Z., De Moliner, F., Cappelli, A. P. and Hulme, C. (2013) Aldol reactions in multicomponent reaction based domino pathways: a multipurpose enabling tool in heterocyclic chemistry. *Org. Lett.*, **15**, 2738–2741.

[76] Lu, K., Luo, T. P., Xiang, Z., You, Z. J., Fathi, R., Chen, J. H. and Yang, Z. (2005) A concise and diversity oriented strategy for the synthesis of benzofurans and indoles via Ugi and Diels-Alder reactions. *J. Comb. Chem.*, **7**, 958–967

[77] Beck, B., Picard, A., Herdtweck, E. and Dömling, A. (2004) Highly substituted pyrrolidinones and pyridones by 4-CR/2-CR sequence. *Org. Lett.*, **6**, 39–42

[78] Neo, A. G., Carrillo, R. M., Barriga, S., Moman, E., Marcaccini, S. and Marcos, C. F. (2007) Multicomponent synthesis of highly substituted 2-pyridones. *Synlett*, **2**, 327–329.

[79] Shaw, A. Y., McLaren, J. A., Nichol G. S. and Hulme, C. (2012) Hydrazine-mediated cyclization of ugi products to synthesize novel 3-hydroxypyrazoles. *Tetrahedron Lett.*, **53**, 2592–2594.

[80] Shaw, A. Y., Xu, Z. and Hulme, C. (2012) Ugi/Robinson–Gabriel reactions directed toward the synthesis of 2, 4, 5-trisubstituted oxazoles. *Tetrahedron Lett.*, **53**, 1998–2000.

[81] Marcaccini, S., Miliciani, M. and Pepino, R. (2005) A facile synthesis of 1, 4-benzodiazepine derivatives via Ugi four-component condensation. *Tetrahedron Lett.*, **46**, 711–713.

[82] (a) Zhang, C., Moran, E. J., Woiwode, T. F., Short, K. M. and Mjalli, A. M. M. (1996) Synthesis of tetrasubstituted imidazoles via α-(*N*-acyl-*N*-alkylamino)-β-ketoamides on Wang resin. *Tetrahedron Lett.*, **37**, 751–754. (b) Sung, K. S., Wu, S. H. and Chen, P. I. (2002) Facile two-pot syntheses of novel alternating benzene/imidazole systems. *Tetrahedron*, **58**, 5599–5602.

[83] Spatz, J. H., Welsch, S. J., Duhaut, D. E., Jager, N., Boursier, T., Fredrich, M., Allmendinger, L., Ross, G., Kolb, J. and Burdack, C. (2009) Tetramic acid derivatives via Ugi-Dieckmann-reaction. *Tetrahedron Lett.*, **50**, 1705–1707.

[84] Corres, N., Delgado, J. J., García-Valverde, M., Marcaccini, S., Rodríguez, T., Rojo, J. and Torroba, T. (2008) Synthesis of pseudopeptidic [(5*H*)-6-oxodibenzo[*b*,*f*][1,5]diazocine-5-yl]aryl-glycinamides by an Ugi 4CC/Staudinger/aza-Wittig sequence. *Tetrahedron*, **64**, 2225–2232.

[85] Sañudo, M., Marcaccini, S., Basurto, S. and Torroba, T. (2006) Synthesis of 3-hydroxy-6-oxo[1,2,4]triazin-1-yl alaninamides, a new class of cyclic dipeptidyl ureas. *J. Org. Chem.*, **71**, 4578–4584.

[86] Lecinska, P., Corres, N., Moreno, D., García-Valverde, M., Marcaccini, S. and Torroba, T. (2010) Synthesis of pseudopeptidic (*S*)-6-amino-5-oxo-1,4-diazepines and (*S*)-3-benzyl-2-oxo-1,4-benzodiazepines by an Ugi 4CC/Staudinger/aza-Wittig sequence. *Tetrahedron*, **66**, 6783–6788.

[87] Zhong, Y., Wang, L. and Ding, M. (2011) New efficient synthesis of 2,3,4-trisubstituted 3,4-dihydroquinazolines by a Ugi 4CC/Staudinger/aza-Wittig sequence. *Tetrahedron*, **67**, 3714–3723.

[88] He, P., Nie, Y., Wu, J. and Ding, M. (2011) Unexpected synthesis of indolo[1,2-*c*]quinazolines by a sequential Ugi 4CC-Staudinger-aza-Wittig-nucleophilic addition reaction. *Org. Biomol. Chem.*, **9**, 1429–1436.

[89] Banfi, L., Basso, A., Guanti, G., Lecinska, P. and Riva, R. (2006) Multicomponent synthesis of dihydrobenzoxazepinones by coupling Ugi and Mitsunobu reactions. *Org. Biomol. Chem.*, **4**, 4236–4240.

[90] Banfi, L., Basso, A., Cerulli, V., Guanti, G., Lecinska, P., Monfardini, I. and Riva, R. (2010) Multicomponent synthesis of dihydrobenzoxazepinones, bearing four diversity points, as potential α-helix mimics. *Mol. Divers.*, **14**, 425–442.

[91] Banfi, L., Basso, A., Guanti, G., Kielland, N., Repetto, C. and Riva, R. (2007) Ugi multicomponent reaction followed by an intramolecular nucleophilic substitution: convergent multicomponent synthesis of 1-sulfonyl 1,4-diazepan-5-ones and of their benzo-fused derivatives. *J. Org. Chem.* **72**, 2151–2160.

[92] Banfi, L., Basso, A., Guanti, G., Lecinska, P. and Riva, R. (2008) Multicomponent synthesis of benzoxazinones via tandem Ugi/Mitsunobu reactions: an unexpected *cine*-substitution. *Mol. Divers.*, **12**, 187–190.

[93] Banfi, L., Basso, A., Giardini, L., Riva, R., Rocca, V. and Guanti, G. (2011) Tandem Ugi MCR/Mitsunobu cyclization as a short, protecting-group-free route to benzoxazinones with four diversity points. *Eur. J. Org. Chem.*, **2011**, 100–109.

[94] Xing, X., Wu, J., Feng, G. and Dai, W. (2006) Microwave-assisted one-pot U-4CR and intramolecular *O*-alkylation toward heterocyclic scaffolds. *Tetrahedron*, **62**, 6774–6781.

[95] Ma, Z., Xiang, Z., Luo, T., Lu, K., Xu, Z., Chen, J. and Yang, Z. (2006) Synthesis of functionalized quinolines via Ugi and Pd-catalyzed intramolecular arylation reactions. *J. Comb. Chem.*, **8**, 696–704.

[96] Xiang, Z., Luo, T., Lu, K., Cui, J., Shi, X., Fathi, R., Chen, J. and Yang, Z. (2004) Concise synthesis of isoquinoline via the Ugi and Heck reactions. *Org. Lett.*, **6**, 3155–3158.

[97] Ugi, I., Demharter, A., Hörl, W. and Schmid, T. (1996) Ugi reactions with trifunctional α-amino acids, aldehydes, isocyanides and alcohols. *Tetrahedron*, **52**, 11657–11664.

[98] Umkehrer, M., Kalinski, C., Kolb, J. and Burdack, C. (2006) A new and versatile one-pot synthesis of indol-2-ones by a novel Ugi-four-component-Heck reaction. *Tetrahedron Lett.*, **47**, 2391–2393.

[99] Kalinski, C., Umkehrer, M., Schmidt, J., Ross, G., Kolb, J., Burdack, C., Hiller, W. and Hoffmann, S. D. (2006) A novel one-pot synthesis of highly diverse indole scaffolds by the Ugi/Heck reaction. *Tetrahedron Lett.*, **47**, 4683–4686.

[100] Bossio, R., Marcaccini, S. and Pepino, R. (1994) Studies on isocyanides and related compounds: a novel synthesis of pyrroles via Ugi reaction. *Synthesis*, **1994**, 765–766.

[101] Marcosa, C. F., Marcaccini, S., Pepino, R., Poloa, C. and Torroba, T. (2003) Studies on isocyanides and related compounds: a facile synthesis of functionalized 3(2H)-pyridazinones via Ugi four-component condensation. *Synthesis*, 691–694.

[102] Marcaccini, S., Pepino, R., Pozo, M. C., Basurto, S., García-Valverde, M. and Torroba, T. (2004) One-pot synthesis of quinolin-2-(1*H*)-ones via tandem Ugi–Knoevenagel condensations. *Tetrahedron Lett.*, **45**, 3999–4001.

[103] Sisko, J. (1998) A one-pot synthesis of 1-(2,2,6,6-tetramethyl-4-piperidinyl)-4-(4-fluorophenyl)-5-(2-amino-4-pyrimidinyl)- imidazole: A potent inhibitor of p38 MAP kinase. *J Org. Chem.*, **63**, 4529–4531.

[104] (a) Possel, O. and Van Leusen, A. M. (1977) Synthesis of oxazoles, imidazoles and pyrroles with the use of mono-substituted tosylmethyl isocyanides. *Heterocycles*, **7**, 77–80. (b) Van Leusen, A. M., Siderius, H., Hoogenboom, B. E. and Van Leusen, D. (1972) A new and simple synthesis of the pyrrole ring system from Michael acceptors and tosylmethylisocyanides. *Tetrahedron Lett.*, **13**, 5337–5340. (c) Have, R., Leusink, F. R. and Van Leusen, A. M. (1996) An efficient synthesis of substituted 3(4)-nitropyrroles from nitroalkenes and tosylmethyl isocyanides. *Synthesis*, 871–876.

[105] (a) Schollkopf, U., Hausberg, H. H., Hoppe, I., Segal, M. and Reiter, U. (1978) Asymmetrische synthese von α-alkyl-α-aminocarbonsäuren durch alkylierung von 1-*chiral*-substituierten 2-imidazolin-5-one. *Angew. Chem.*, **90**, 136–138. (b) Schollkopf, U., Hausberg, H. H., Segal, M., Reiter, U., Hoppe, I., Saenger, W. and Lindner, K. (1981) Asymmetrische Synthesen über heterocyclische Zwischenstufen, IV Asymmetrische Synthese von α-Methylphenylalanin und seinen Analoga durch Alkylieren von 1-chiral substituiertem 4-Methyl-2-imidazolin-5-one. *Liebigs Ann. Chem.*, **3**, 439–458.

[106] (a) Suzuki, M., Moriya, T., Matsumoto, K. and Miyoshi, M. (1982) Synthesis of 7,10-disubstituted benzo[b]phenazine-6,11-quinones. *Synthesis*, 874–875. (b) Solomon, D. M., Rizvi, R. K. and Kaminski, J. J. (1987) Observations on the reactions of isocyanoacetane esters with isothiocyanates and isocyanates. *Heterocycles*, **26**, 651–674.

[107] Meyer, R., Schollkopf, U., Madawinata, K. and Stafforst, D. (1978) Total synthese von 2,2-Diethyl-3-methyl-und-2,2-Diethyl-3,6-dimethyl-6-(phenoxyacetamido)penam-3-carbonsaure. *Liebigs Ann. Chem.*, **8**, 1982–1989.

[108] Chupp, J. P. and Leschinsky, K. L. (1980) Heterocycles from substituted amides. VIII Oxazole derivatives from reaction of isocyanates with 2-isocyanoacetamides. *J. Heterocycl. Chem.*, **17**, 711–715.

[109] (a) Hoppe, I. and Schollkopf, U. (1976) Cycloaddition von "Isocyanketen" an Benzophenonanilund Thioimidsäureester zu β-Lactamen. *Chem. Ber.*, **109**, 482–487. (b) Ito, Y., Sawamura, M., Hamashima, H., Emura, T. and Hayashi, T. (1989) Asymmetric aldol reaction of α-ketoesters with isocyanoacetate and isocyanoacetamide catalyzed by a chiral ferrocenylphosphine-gold(I) complex. *Tetrahedron Lett.*, **30**, 4681–4684.

[110] Beaza, A., Mendiola, J., Burgos, C., Alvarez-Builla, J. and Vaquero, J. J. (2005) Heterocyclizations with tosylmethyl isocyanide derivatives. A new approach to substituted azolopyrimidines. *J. Org. Chem.*, **70**, 4879–4882

[111] De Moliner, F. and Hulme, C. (2012) Straightforward assembly of phenylimidazoquinoxalines via a one-pot two-step MCR process. *Org. Lett.*, **14**, 1354–1357.

[112] Shi, X., Kiesman, W. F., Levina, A. and Xin, Z. (2013) Catalytic asymmetric Petasis reactions of vinylboronates. *J. Org. Chem.*, **78**, 9415–9423.

[113] Grigg, R., Sridharan, V. and Thayaparan, A. (2003) Synthesis of novel cyclic α-amino acid derivatives via a one-pot sequential Petasis reaction/palladium catalysed process. *Tetrahedron Lett.*, **44**, 9017.

[114] Mieczkowski, A., Kozminski, W. and Jurczak, J. (2010) A traceless, solid-supported synthesis of β-turn mimetics based on the hexahydropyrazino[1,2-*a*]pyrazine-1,2-dione scaffold. *Synthesis*, **2**, 221–232.

[115] Neogi, S., Roy, A. and Naskar, D. (2010) One-pot synthesis of new fused 4,5-bridged 1,2,5-triazepine-3,6-diones, 1,2,5-triazepine-3,7-diones heterocycles by Petasis reaction. *J. Comb. Chem.*, **12**, 75–83.

[116] Petasis, N. A. and Patel, Z. D. (2000) Synthesis of piperazinones and benzopiperazinones from 1,2-diamines and organoboronic acids *Tetrahedron Lett.*, **41**, 9607–9611.

[117] Petasis, N. A. and Butkevich, A. N. (2009) Synthesis of *2H*-chromenes and 1,2-dihydroquinolines from aryl aldehydes, amines, and alkenylboron compounds. *J. Organomet. Chem.*, **694**, 1747–1753.

[118] Ayaz, M. and Hulme, C. (2011) A novel route to synthesize libraries of quinoxalines via Petasis methodology in two synthetic operations. *Tetrahedron. Lett.*, **52**, 4821–4823.

[119] (a) Hulme, C. and Lee, Y. (2008) Emerging approaches for the syntheses of bicyclic imidazo[1,2-*x*]-heterocycles. *Mol. Divers.*, **12**, 1–15. (b) Tsirulnikov, S., Kysil, V., Ivachtchenko, A. and Krasavin, M. (2009) Trimethylsilyl chloride as a promoter for the Groebke–Blackburn reaction: preparation of imidazo[2,1-*b*][1,3]benzothiazoles. *Synth. Commun.*, **40**, 111–119.

[120] Schwerloske, J., Masquelin, T., Perun, T. and Hulme, C. (2005) New multicomponent reaction accessing 3-aminoimidazo[1,2-*a*]pyridines. *Tetrahedron Lett.*, **46**, 8355–8357.

[121] Umkehrer, M., Ross, G., Jäger, N., Burdack, C., Kolb, J., Hu, H., Alvim-Gaston, M. and Hulme, C. (2007) Expeditious synthesis of imidazo[1,2-*c*]pyrimidines via a [4+1]-cycloaddition. *Tetrahedron Lett.*, **48**, 2213–2216.

[122] Ghose, A. K., Herbertz, T., Salvino, J. M. and Mallamo, J. P. (2006) Knowledge based chemoinformatics approaches to drug discovery. *Drug Discov. Today*, **11**, 110–118.

[123] Parchinsky, V. Z., Koleda, V. V., Shuvalova, O., Kravchenko, D. V. and Krasavin, M. (2006) Air-oxidized products of multi-component reactions between 3-amino-1,2,4-triazole, aromatic aldehydes and isonitriles. *Tetrahedron Lett.*, **47**, 6891–6894.

[124] Sun, C., Shun-Jun, J. and Liu, Y. (2008) A novel, simple and efficient synthesis of 3-amino-benzo[*d*]imidazo[2,1-*b*]thiazole derivatives via a multicomponent procedure. *J. Chin. Chem. Soc.*, **55**, 292–296.

[125] Biggs-Houck, J. E., Younai A. and Shaw, J. T. (2010) Recent advances in multicomponent reactions for diversity-oriented synthesis. *Curr. Opin. Chem. Biol.*, **14**, 371–382.

[126] Ruijter, E., Scheffelaar, R. and Orru, R. V. A. (2011) Multicomponent reaction design in the quest for molecular complexity and diversity. *Angew. Chem. Int. Ed. Engl.*, **50**, 6234–6246.

[127] Van, D. H., Ruijter, E. and Orru, R. V. A. (2013) Efficiency, diversity, and complexity with multicomponent reactions. *Synlett*, **24**, 666–685.

[128] Gonzalez-Lopez, M. and Shaw, J. T. (2009) Cyclic anhydrides in formal cycloadditions and multicomponent reactions. *Chem. Rev.*, **109**, 164–189.

[129] Castagnoli, N. (1969) Condensation of succinic anhydride with N-benzylidene-N-methylamine. Stereoselective synthesis of *trans*- and *cis*-1-methyl-4-carboxy-5-phenyl-2-pyrrolidinone. *J. Org. Chem.*, **34**, 3187–3189.

[130] Yadav, J. S., Reddy, B. V. S., Saritha-Raj, K. and Prasad, A. R. (2003) Room temperature ionic liquids promoted three-component coupling reactions: a facile synthesis of *cis*-isoquinolonic acids. *Tetrahedron*, **59**, 1805–1809.

[131] Wang, L., Liu, J., Tian, H., Qian, C. and Sun, J. (2005) One-pot synthesis of *cis*-isoquinolonic acid derivatives via three-component reaction of homophthalic anhydride with aldehydes and amines using ytterbium(III) triflate as catalyst. *Adv. Synth. Catal.*, **347**, 689–694.

[132] Wei, J. and Shaw, J. T. (2007) Diastereoselective synthesis of γ-lactams by a one-pot, four-component reaction. *Org. Lett.*, **9**, 4077–4080.

[133] Groenendaal, B., Ruijter, E. and Orru, R. V. A. (2008) 1-Azadienes in cycloaddition and multicomponent reactions towards *N*-heterocycles. *Chem. Commun.*, **43**, 5474–5489.

[134] Kiselyov, A. S. (1995) Reaction of *in-situ* generated α,β-unsaturated imines with CH-nucleophiles: a novel and efficient synthesis of pyridines. *Tetrahedron Lett.*, **36**, 9297–9300.

[135] Ganem, B. (2009) Strategies for innovation in multicomponent reaction design. *Acc. Chem. Res.*, **42**, 463–472.

[136] Kiselyov, A. S. (2005) One-pot synthesis of polysubstituted pyrimidines. *Tetrahedron Lett.*, **46**, 1663–1665.

[137] Vugts, D. J., Jansen, H., Schmitz, R. F., de Kanter, F. J. J. and Orru, R. V. A. (2003) A novel four-component reaction for the synthesis of functionalized dihydropyrimidines. *Chem. Commun.*, **20**, 2594–2595.

[138] Vugts, D. J., Koningstein, M. M., Schimitz, R. F., de Kanter, F. J. J., Groen, M. B. and Orru, R. V. A. (2006) Multicomponent synthesis of dihydropyrimidines and thiazines. *Chem. Eur. J.*, **12**, 7178–7189.

[139] Glasnov, T., Vugts, D. J., Koningstein, M. M., Desai, B., Fabian, W. M. F., Orru, R. V. A and Kappe, C. (2006) Microwave-assisted dimroth rearrangement of thiazines to dihydropyrimidinethiones: synthetic and mechanistic aspects. *QSAR Comb. Sci.*, **25**, 509–518.

[140] Paravidino, M. Bon, R. S. Scheffelaar, R., Vugts, D. J., Znabet, A., De Kanter, F. J. J., Lutz, M., Spek, A. L., Groen, M. B and Orru, R. V. A. (2006) Diastereoselective multicomponent synthesis of dihydropyridones with an isocyanide functionality. *Org. Lett.*, **8**, 5369–5372.

[141] Kruithof, A., Ploeger, M. L., Janssen, E., Helliwell, M., Kanter, F. J. J., Ruijter, E. and Orru, R. V. A. (2012) Multicomponent Synthesis of 3,6-Dihydro-2H-1,3-thiazine-2-thiones. *Molecules*, **17**, 1675–1685.

[142] El Kaim, L. and Grimaud, L. (2009) Beyond the Ugi reaction: less conventional interactions between isocyanides and iminium species. *Tetrahedron*, **65**, 2153–217.

[143] (a) El Kaim, L., Grimaud, L. and Oble, J. (2005) Phenol ugi-smiles systems: strategies for the multicomponent *N*-arylation of primary amines with isocyanides, aldehydes, and phenols.

*Angew. Chem. Int. Ed. Engl.*, **44**, 7961–7964. (b) El Kaim, L. and Grimaud, L. (2010) Ugi-Smiles couplings: new entries to *N*-aryl carboxamide derivatives. *Mol. Divers.*, **14**, 855–867.

[144] Oble, J., El Kaim, L., Gizzi, M. and Grimaud, L. (2007) Ugi-Smiles access to quinoxalines derivatives. *Heterocycles*, **73**, 503–517.

[145] El Kaim, L., Grimaud, L. and Coffinier, D. (2009) New benzotriazole and benzimidazole scaffolds from Ugi-Smiles couplings of isocyanides. *Org. Lett.*, **11**, 995–997.

[146] El Kaim, L., Gizzi, M. and Grimaud, L. (2008) New MCR-Heck-isomerization cascade towards indole. *Org. Lett.*, **10**, 3417–3419.

[147] El Kaim, L., Grimaud, L. and Oble, J. (2007) New Ugi-Smiles-metathesis strategy towards the synthesis of pyrimido-azepines. *J. Org. Chem.*, **72**, 5835–5838.

[148] Pan, S. C. and List, B. (2008) Catalytic three-component Ugi reaction. *Angew. Chem. Int. Ed. Engl.*, **47**, 3622–3625.

[149] Heravi, M., Baghernejad, B. and Oskooie, H. (2009) A novel three-component reaction for the synthesis of 3-aryl-4H-benzo [1,4] oxazin-2-amine. *Mol. Divers.*, **13**, 395–398.

[150] Mofakham, H., Hezarkhani, Z. and Shaabani, A. (2012) Cellulose-$SO_3H$ as a biodegradable solid acid catalyzed one-pot three-component Ugi reaction: synthesis of α-amino amide, 3,4-dihydroquinoxalin-2-amine, 4H-benzo[*b*][1,4]thiazin-2-amine and 1,6-dihydropyrazine-2,3-dicarbonitrile derivatives. *J. Mol. Cat. A: Chem.*, **360**, 26–34.

[151] (a) Khoury, K., Sinha, M. K., Nagashima, T., Herdtweck, E. and Dömling, A. (2012) Efficient assembly of iminodicarboxamides by a "truly" four-component reaction. *Angew. Chem. Int. Ed. Engl.*, **51**, 10280–10283. (b) Sinha, M. K., Khoury, K., Herdtweck, E. and Dömling, A. (2013) Various cyclization scaffolds by a truly Ugi 4-CR. *Org. Biomol. Chem.*, **11**, 4792–4796.

[152] Hashimoto, T., Kimura H., Kawamata, Y. and Maruoka, K. (2012) A catalytic asymmetric Ugi-type reaction with acyclic azomethine imines. *Angew. Chem. Int. Ed. Engl.*, **51**, 7279–7281.

[153] Vlaar, T., Mampuys, P., Helliwell, M., Maes, B. U. W., Orru, R. V. A. and Ruijter, E. (2013) Multicomponent synthesis of 4-aminophthalazin-1(2*H*)-ones by palladium-catalyzed isocyanide insertion. *J. Org. Chem.*, **78**, 6735–6745.

[154] Lei, C., Wang, D., Zhao, L., Zhu, J. and Wang, M. (2013) Synthesis of substituted pyridines from cascade [1+5] cycloaddition of isonitriles to *N*-formylmethyl-substituted enamides, aerobic oxidative aromatization, and acyl transfer reaction. *J. Am. Chem. Soc.*, **135**, 4708–4711.

[155] Basavanag, U. M. V., Dos Santos A., El Kaim, L., Gámez-Montaño, R. and Grimaud, L. (2013) Three-component metal-free arylation of isocyanides. *Angew. Chem. Int. Ed. Engl.*, **52**, 7194–7197.

[156] Basso, A., Banfi, L., Garbarino, S. and Riva, R. (2013) Ketene three-component reaction: a metal-free multicomponent approach to stereodefined captodative olefins. *Angew. Chem. Int. Ed. Engl.*, **52**, 2096–2099.

[157] Dückert, H., Pries, V., Khedkar, V., Menninger, S., Bruss, H., Bird, A. W., Maliga, Z., Brockmeyer, A., Janning, P., Hyman, A., Grimme, S., Schürmann, M., Preut, H., Hübel, K., Ziegler, S., Kumar, K. and Waldmann, H. (2012) Natural product-inspired cascade synthesis yields modulators of centrosome integrity. *Nat. Chem. Biol.*, **8**, 179–184.

[158] (a) Martinez-Ariza, G., Ayaz, M. and Hulme, C. (2013) A simple one-pot 2-step *N*-1-alkylation of indoles with α-iminoketones toward the expeditious 3-step synthesis of *N*-1-quinoxaline-indoles. *Tetrahedron Lett.*, **54**, 6719–6721. (b) McCall, J. M., John, R., Donnal, L. and Clair, M. (2011) Heterocyclic compounds for the inhibition of PASK. US Patent 2012165318.

[159] (a) Mori, K. (2000) Organic synthesis and chemical ecology. *Acc. Chem. Res.*, **33**, 102–110. (b) Brenna, E., Fuganti, C. and Serra, S. (2003) Enantioselective perception of chiral

odorants. *Tetrahedron Asymmetry*, **14**, 1–42. (c) Lovering, F., Bikker, J. and Humblet, C. (2009) Escape from flatland: increasing saturation as an approach to improving clinical success. *J. Med. Chem.*, **52**, 6752–6756. (d) Lovering, F. (2013) Escape from flatland 2: complexity and promiscuity. *Med. Chem. Commun.*, **4**, 515–519.

[160] (a) De Graaff, C., Ruijter, E. and Orru, R. V. A. (2012) Recent developments in asymmetric multicomponent reactions. *Chem. Soc. Rev.*, **41**, 3969–4009. (b) Ramón, D. J. and Yus, M. (2005) Asymmetric multicomponent reactions (AMCRs): the new frontier. *Angew. Chem. Int. Ed. Engl.*, **44**, 1602–1634.

[161] (a) Znabet, A. (2012) Biocatalysis and multicomponent reactions: the ideal synergy. Asymmetric synthesis of substituted proline derivatives. Ph.D Thesis, University of Amsterdam, De Boelelaan, Amsterdam. (b) Clavier, H. and Pellissier, H. (2012) Recent developments in enantioselective metal-catalyzed Domino reactions. *Adv. Synth. Catal.*, **354**, 3347–3403. (c) Guillena, G., Ramón, D. J. and Yus, M. (2007) Organocatalytic enantioselective multicomponent reactions (OEMCRs). *Tetrahedron Asymmetry*, **18**, 693–700.

[162] Wang, J., Liu, X., and Feng, X. (2011) Asymmetric Strecker reactions. *Chem. Rev.*, **111**, 6947–6983.

[163] (a) Xiao-Hua, C. and Xie, B. (2013) Recent advances on organocatalysed asymmetric Mannich reactions, *Arkivoc* (i), 264–293. (b) Xiao-Hua, C., Hui, G. and Bing, X. (2012) Recent progress in the asymmetric Mannich reaction. *Eur. J. Org. Chem.*, **3**, 258–266.

[164] Heravi, M. M., Asadi, S. and Lashkariani, B. M. (2013) Recent progress in asymmetric Biginelli reaction. *Mol. Divers.*, **17**, 389–407.

[165] Candeias, N. R., Montalbano, F., Cal, P. M. S. D. and Gois, P. M. P. (2010) Boronic acids and esters in the Petasis-borono mannich multicomponent reaction. *Chem. Rev.*, **110**, 6169–6193.

[166] Jie-Ping, W. and Yunyun, L. (2012) Recent advances in new multicomponent synthesis of structurally diversified 1,4-dihydropyridines. *RSC Adv.*, **2**, 9763–9777.

[167] (a) Bertelsen, S. and Jørgensen, K. A. (2009) Organocatalysis—after the gold rush. *Chem. Soc. Rev.*, **38**, 2178–2189. (b) Wende, R. C. and Schreiner, P. R. (2012) Evolution of asymmetric organocatalysis: multi- and retrocatalysis. *Green Chem.*, **14**, 1821–1849.

[168] (a) Lelais, G. and MacMillan, D. W. C. (2006) Modern strategies in organic catalysis: the advent and development of iminium activation. *Aldrichimca Acta*, **39**, 79–87. (b) Miyabe, H. and Takemoto, Y. (2008) Discovery and application of asymmetric reaction by multifunctional thioureas. *Bull. Chem. Soc. Jpn.*, **81**, 785–795. (c) Zhang, Z. and Schreiner, P. R. (2009) (Thio)urea organocatalysis-what can be learnt from anion recognition? *Chem. Soc. Rev.*, **38**, 1187–1198. (d) Peng, F. and Shao, Z. (2011) Asymmetric organocatalytic cascade reactions catalyzed by chiral bronsted acids. *Curr. Org. Chem.*, **15**, 4144–4160. (e) Enders, D., Niemeier, O. and Henseler, A. (2007) Organocatalysis by *N*-heterocyclic carbenes. *Chem. Rev.*, **107**, 5606–5655. (f) Beeson, T. D., Mastracchio, A., Hong, J., Ashton, K. and MacMillan, D. W. C. (2007) Enantioselective organocatalysis using SOMO activation. *Science*, **316**, 582–585. (g) Bertelsen, S., Johansen, R. L. and Jørgensen, K. A. (2008) Controlling the formation of 1 out of 64 stereoisomers using organocatalysis. *Chem. Commun.*, **26**, 3016–3018.

[169] (a) Andreana, P. R., Liu, C. C. and Schreiber, S. L. (2004) Stereochemical control of the Passerini reaction. *Org. Lett.*, **6**, 4231–4233. (b) Wang, S., Wang, M., Wang, D. and Zhu, J. (2008) Catalytic enantioselective Passerini three-component reaction. *Angew. Chem. Int. Ed. Engl.*, **47**, 388–391.

[170] Yue, T., Wang, M., Wang, D., Masson, G. and Zhu, J. (2009) Brønsted acid catalyzed enantioselective three-component reaction involving the α-addition of isocyanides to imines. *Angew. Chem. Int. Ed. Engl.*, **48**, 6717–6721.

[171] (a) Dömling, A., Wang, W. and Wang, K. (2013) Chemistry and biology of multicomponent reactions. *Chem. Rev.*, **112**, 3083–3135. (b) Slobbe, P., Ruijter, E. and Orru, R. V. A. (2013)

Recent applications of multicomponent reactions in medicinal chemistry. *Med. Chem. Comm.*, **3**, 1189–1218. (c) Akritopoulou-Zanze, I. (2008) Isocyanide-based multicomponent reactions in drug discovery. *Curr. Opin. Chem. Biol.*, **12**, 324–331.

[172] (a) Lapenna, S. and Giordano, A. (2009) Cell cycle kinases as therapeutic targets for cancer. *Nat. Rev. Drug Discov.*, **8**, 547–566. (b) Adams, J. A. (2001) Kinetic and catalytic mechanism of protein kinases. *Chem. Rev.*, **101**, 2271–2290. (c) Vieth, M., Higgs, R. E., Robertson, D. H., Shapiro, M., Gragg, E. A. and Hemmerle, H. (2004) Kinomics—structural biology and chemogenomics of kinase inhibitors and targets. *Biochim. Biophys. Acta, Proteins Proteomics*, **1697**, 243–257.

[173] (a) Sisko, J. and Mellinger, M. (2002) Development of a general process for the synthesis of highly substituted imidazoles. *Pure Appl. Chem.*, **74**, 1349–1358. (b) Wang, Z., Canagarajah, B. J., Boehm, J. C., Kassisà, S., Cobb, M. H., Young, P. R., Abdel-Meguid, S., Adams, J. L. and Goldsmith, E. J. (1998) Structural basis of inhibitor selectivity in MAP kinases. *Structure*, **6**, 1117–1128.

[174] (a) Barnes, D. M., Haight, A. R., Hameury, T., McLaughlin, M. A., Mei, J., Tedrow, J. S. and Riva Toma, J. D. (2006) New conditions for the synthesis of thiophenes via the Knoevenagel-Gewald reaction sequence. Application to the synthesis of a multitargeted kinase inhibitors. *Tetrahedron*, **62**, 11311–11319. (b) Dai, Y., Guo, Y., Frey, R. R., Ji, Z., Curtin, M. L., Ahmed, A. A., Albert, D. H., Arnold, L., Arries, S. S., Barlozzari, T., Bauch, J. L., Bouska, J. J., Bousquet, P. F., Cunha, G. A., Glaser, K. B., Guo, J., Li, J., Marcotte, P. A., Marsh, K. C., Moskey, M. D., Pease, L. J., Stewart, K. D., Stoll, V. S., Tapang, P., Wishart, N., Davidsen, S. K. and Michaelides, M. R. (2005) Thienopyrimidines ureas as novel and potent multitargeted receptor tyrosine kinase inhibitors. *J. Med. Chem.*, **48**, 6066–6083.

[175] (a) Pasquale, E. B. (2010) Eph receptors and ephrins in cancer: bidirectional signaling and beyond. *Nat. Rev. Cancer*, **10**, 165–180. (b) Kolb, P., Kipouros, C. B., Huang, D. and Caflisch, A. (2008) Structure-based tailoring of compound libraries for high-throughput screening: discovery of novel EphB4 kinase inhibitors. *Proteins: Struct., Funct., Bioinf.*, **73**, 11–18.

[176] Che, C., Xiang, J., Wang, G.-X., Fathi, R., Quan, J.-M. and Yang, Z. (2007) One-pot synthesis of quinoline-based tetracycles by a tandem three-component reaction. *J. Comb. Chem.*, **9**, 982–989.

[177] Nixey, T., Boylan, J., Hulme, C., Powers, D., Smith, A. and Wong, A. (2006) Inhibitors of the polo-like protein kinase Plk-1. 231st ACS National Meeting, Atlanta, GA, MEDI 277.

[178] (a) Ismaiel, A. M., Dukat, M., Nelson, D. L., Lucaites, V. L. and Glennon, R. A. (1996) Binding of $N_2$-substituted pyrido[4,3-*b*]indole analogues of spiperone at human 5-$HT_{2A}$, 5-$HT_{2B}$, and 5-$HT_{2C}$ serotonin receptors. *Med. Chem. Res.*, **6**, 197–211. (b) Metwally, K. A., Dukat, M., Egan, C. T., Smith, C., DuPre, A., Gauthier, C. B., Herrick-Davis, K., Teitler, M. and Glennon, R. A. (1998) Spiperone: influence of spiro-ring substituents on 5-HT2A serotonin receptor binding. *J. Med. Chem.*, **41**, 5084–5093.

[179] (a) Southan, C. (2001) A genome perspective on human proteases as drug targets. *Drug Discov. Today*, **6**, 681–688. (b) Turk, B. (2006) Targeting proteases: success, failure and future prospects. *Nat. Rev. Drug Discov.*, **5**, 785–799.

[180] (a) Banfi, L., Guanti, G. and Riva, R. (2000) Passerini multicomponent reaction of protected α-aminoaldehydes as a tool for combinatorial synthesis of enzyme inhibitors. *Chem. Commun.*, **2000**, 985–986. (b) Semple, J. E., Owens, T. D., Nguyen, K. and Levy, O. E. (2000) New synthetic technology for efficient construction of α-hydroxy-β-amino amides via the Passerini reaction. *Org. Lett.*, **2**, 2769–2772. (c) Banfi, L., Guanti, G., Riva, R., Basso, A. and Calcagno, E. (2002) Short synthesis of protease inhibitors *via* modified Passerini condensation of *N*-Boc-α-aminoaldehydes. *Tetrahedron Lett.*, **43**, 4067–4069.

[181] (a) Velázquez, F., Venkatraman, S., Blackman, M., Pinto, P., Bogen, S. P., Sannigrahi, M., Chen, K., Pichardo, J., Hart, A., Tong, X., Girijavallabhan, V. and Njoroge, F. G. (2009)

Design, synthesis and evaluation of oxygen-containing macrocyclic peptidomimetics as inhibitors of HCV NS3 protease. *J. Med. Chem.*, **52**, 700–708. (b) Venkatraman, S., Velazquez, F., Wu, W., Blackman, M., Chen, K. X., Bogen, S., Nair, L., Tong, X., Chase, R., Hart, A., Agrawal, S., Pichardo, J., Prongay, A., Cheng, K.-C., Girijavallabhan, V., Piwinski, J., Shih, N.-Y. and Njoroge, F. G. (2008) Discovery and structure-activity relationships of P1-P3 ketoamide derived macrocyclic inhibitors of hepatitis C virus NS3 protease. *J. Med. Chem.*, **52**, 336–346. (c) Velazquez, F., Venkatraman, S., Wu, W., Blackman, M., Prongay, A., Girijavallabhan, V., Shih, N.-Y. and Njoroge, F. G. (2007) Application of ring-closing metathesis for the synthesis of macrocyclic peptidomimetics as inhibitors of NCV NS3 proteases. *Org. Lett.*, **9**, 3061–3064. (d) Prongay, A. J., Guo, Z., Yao, N., Pichardo, J., Fischmann, T., Strickland, C., Myers, J., Weber, P. C., Beyer, B. M., Ingram, R., Hong, Z., Prosise, W. W., Ramanathan, L., Taremi, S. S., Yarosh-Tomaine, T., Zhang, R., Senior, M., Yang, R.-S., Malcolm, B., Arasappan, A., Bennett, F., Bogen, S. L., Chen, K., Jao, E., Liu, Y.-T., Lovey, R. G., Saksena, A. K., Venkatraman, S., Girijavallabhan, V., Njoroge, F. G. and Madison, V. (2007) Discovery of the HCV NS3/4A protease inhibitor (1*R*,5*S*)-*N*- [3-Amino-1-(cyclobutylmethyl)-2,3-dioxopropyl]-3[2(*S*)[[[(1,1dimethylethyl)amino]carbonyl]amino]-3,3-dimethyl-1-oxobutyl]-6,6-dimethyl-3-azabicyclo[3.1.0]hexan-2(*S*)-carboxamide(Sch 503034) II. Key steps in structure-based optimization. *J. Med. Chem.*, **50**, 2310–2318.

[182] Nixey, T. and Hulme, C. (2002) Rapid generation of *cis*-constrained norstatine analogs using TMSN3-modified Passerini MCC-N capping strategy. *Tetrahedron Lett.*, **43**, 6833–6835.

[183] (a) Oefner, C., D'arcy, A., Mac Sweeney, A., Pierau, S., Gardiner, R. and Dale, G. E. (2003) High-resolution structure of human apo dipeptidyl peptidase IV/CD26 and its complex with 1-[({2-[(5-iodopyridin-2-yl)amino]-ethyl}amino)-acetyl]-2-cyano-(*S*)-pyrrolidine). *Acta Crystallogr. D Biol. Crystallogr.*, **59**, 1206–1212. (b) Behnke, D., Taube, R., Illgen, K., Nerdinger, S. and Herdtweck, E. (2004) Substituted 2-(cyanomethyl-amino)-amides via a novel three component reaction. *Synlett*, **2004**, 688–692. (c) Elders, N., Ruijter, E., De Kanter, F. J. J., Janssen, E., Lutz, M., Spek, A. L. and Orru, R. V. A. (2009) A multicomponent reaction towards *N*-(Cyanomethyl)amides. *Chem. Eur. J.*, **15**, 6096–6099.

[184] Rossen, K., Pye, P. J., Dimichele, L. M., Volante, R. P. and Reider, P. J. (1998) An efficient asymmetric hydrogenation approach to the synthesis of Crixivan piperazine intermediate. *Tetrahedron Lett.*, **39**, 6823–6826.

[185] (a) Raymond, V. and Sattelle, D. B. (2002) Novel animal-health drug targets from ligand-gated chloride channels. *Nat. Rev. Drug Discov.*, **1**, 427–436. (b) Kawate, T., Michel, J. C., Birdsong, W. T. and Gouaux, E. (2009) Crystal structure of the ATP gated P2X4 ion channel in the close state. *Nature*, **460**, 592–598.

[186] Goldmann, S. and Stoltefuss, J. (1991) 1,4-dihydropyridines: effects of chirality and conformation on the calcium antagonist and calcium agonist activities. *Angew. Chem. Int. Ed. Engl.*, **30**, 1559–1578.

[187] Ugi, I. and Steinbrueckner, C. (1961) Aktiebolaget Astra Apotekarnes Kemiska Fabriker. Ger. Patent DE1103337.

[188] (a) Vaccaro, W., Huynh, T., Lloyd, J., Atwal, K., Finlay, H. J., Levesque, P., Conder, M. L., Jenkins-West, T., Shi, H. and Sun, L. (2008) Dihydropyrazolopyrimidines inhibitors of Kv 1.5 (Ikur). *Bioorg. Med. Chem. Lett.*, **18**, 6381–6385. (b) Lloyd, J., Finlay, H. J., Atwal, K., Kover, A., Prol, J., Yan, L., Bhandaru, R., Vaccaro, W., Huynh, T., Huang, C. S., Conder, M., Jenkins-West, T., Sun, H., Li, D. and Levesque, P. (2009) Dihydropyrazolopyrimidines containing benzimidazoles as Kv 1.5 potassium channel antagonists. *Bioorg. Med. Chem. Lett.*, **19**, 5469–5463.

[189] (a) Arkin, M. R. and Wells, J. A. (2004) Small-molecule inhibitors of protein–protein interactions: progressing towards the dream. *Nat. Rev. Drug Discov.*, **3**, 301–317.

(b) Jacoby, E., Boettcher, A., Mayr, L. M., Brown, N., Jenkins, J. L., Kallen, J., Engeloch, C., Schopfer, U., Furet, P., Masuya, K. and Lisztwan, J. (2009) *Chemogenomics: Methods and Applications*, Springer, Heidelberg. (c) Reynès, C., Host, H., Camproux, A.-C., Laconde, G., Leroux, F., Mazars, A., Deprez, B., Fahraeus, R., Villoutreix, B. O. and Sperandio, O. (2010) Designing focused chemical libraries enriched in protein-protein interaction inhibitors using machine-learning methods. *PLoS Comput. Biol.*, **6**, e1000695.

[190] Wade, M. and Wahl, G. M. (2009) Targeting mdm2 and mdmx in cancer therapy: better living through medicinal chemistry? *Mol. Cancer Res.*, **7**, 1–11.

[191] (a) Grasberger, B. L., Lu, T., Schubert, C., Parks, D. J., Carver, T. E., Koblish, H. K., Cummings, M. D., Lafrance, L. V., Milkiewicz, K. L., Calvo, R. R., Maguire, D., Lattanze, J., Franks, C. F., Zhao, S., Ramachandren, K., Bylebyl, G. R., Zhang, M., Manthey, C. L., Petrella, E. C., Pantoliano, M. W., Deckman, I. C., Spurlino, J. C., Maroney, A. C., Tomczuk, B. E., Molloy, C. J. and Bone, R. F. (2005) Discovery and cocrystal structure of benzodiazepinedione HMD2 antagonists that activate p53 in cells. *J. Med. Chem.*, **48**, 909–912. (b) Parks, D. J., Lafrance, L. V., Calvo, R. R., Milkiewicz, K. L., Gupta, V., Lattanze, J., Ramachandren, K., Carver, T. E., Petrella, E. C., Cummings, M. D., Maguire, D., Grasberger, B. L. and Lu, T. (2005) 1,4-Benzodiazepine-2,5-diones as small molecule antagonists of the HDM2–p53 interaction: discovery and SAR. *Bioorg. Med. Chem. Lett.*, **15**, 765–770. (c) Marugan, J. J., Leonard, K., Raboisson, P., Gushue, J. M., Calvo, R., Koblish, H. K., Lattanze, J., Zhao, S., Cummings, M. D., Player, M. R., Schubert, C., Maroney, A. C. and Lu, T. (2006) Enantiomerically pure 1,4-benzodiazepine-2,5-diones as HDM2 antagonists. *Bioorg. Med. Chem. Lett.*, **16**, 3115–3120. (d) Parks, D. J., Lafrance, L. V., Calvo, R. R., Milkiewicz, K. L., José Marugán, J., Raboisson, P., Schubert, C., Koblish, H. K., Zhao, S., Franks, C. F., Lattanze, J., Carver, T. E., Cummings, M. D., Maguire, D., Grasberger, B. L., Maroney, A. C. and Lu, T. (2006) Enhanced pharmacokinetic properties of 1,4-benzodiazepine-2,5-dione antagonists of the HDM2-p53 protein–protein interaction through structure-based drug design. *Bioorg. Med. Chem. Lett.*, **16**, 3310–3314.

[192] Burdack, C., Kalinsky, C., Ross, G., Weber, L. and Khazak, V. (2010) Pyrrolidin-2-ones as hdm2 ligands. Int. Patent WO2010028862.

[193] Dumontet, C. and Jordan, M. A. (2010) Microtubule-binding agents: a dynamic field of cancer therapeutics. *Nat. Rev. Drug Discov.*, **9**, 790–803.

[194] Sasse, F., Steinmetz, H., Heil, J., Höfle, G. and Reinbach, H. (2000) Tubulysins, new cytostatic peptides from myxobacteria acting on microtubuli. Production, isolation, physico-chemical and biological properties. *J. Antibiot.*, **53**, 879–885.

[195] Pando, O., Stark, S., Denkert, A., Porzel, A., Preusentanz, R. and Wessjohann, L. A. (2011) The multiple multicomponent approach to natural product mimics: tubugis, *N*-substituted anticancer peptides with picomolar activity. *J. Am. Chem. Soc.*, **133**, 7692–7695.

[196] Xu, H., Lv, M. and Tian, X. (2009) A review on hemisynthesis, biosynthesis, biological activities, mode of action, and structure-activity relationship of podophyllotoxins: 2003–2007. *Curr. Med. Chem.*, **16**, 27–349.

[197] Magedov, I. V., Manpadi, M., Van Slambrouck, S., Steelant, W. F. A., Rozhkova, E., Przheval'skii, N. M., Rogelj, S. and Kornienko, A. (2007) Discovery and investigation of antiproliferative and apoptosis-Inducing properties of new heterocyclic Podophyllotoxin analogues accessible by a one-step multicomponent synthesis. *J. Med. Chem.*, **50**, 5183–5192.

[198] (a) Magedov, I. V., Frolova, L., Manpadi, M., Bhoga, U., Tang, H., Evdokimov, N. M., George, O., Hadje Georgiou, K., Renner, S., Getlic, M., Kinnibrugh, T. L., Fernandes, M. A., Van Slambrouck, S., Steelant, W. F. A., Shuster, C. B., Rogelj, S., van Otterlo, W. A. L. and Kornienko, A. (2011) Anticancer properties of an important drug lead Podophyllotoxin can be efficiently

mimicked by diverse heterocyclic scaffolds accessible via one-step synthesis. *J. Med. Chem.*, **54**, 4234–4246. (b) Magedov, I. V., Manpadi, M., Evdokimov, N. M., Elias, E. M., Rozhkova, E., Ogasawara, M. A., Bettale, J. D., Przheval'skii, N. M., Rogelj, S. and Kornienko, A. (2007) Antiproliferative and apoptosis inducing properties of pyrano[3,2-*c*]pyridones accessible by a one-step multicomponent synthesis. *Bioorg. Med. Chem. Lett.*, **17**, 3872–3876.

[199] Magedov, I. V., Manpadi, M., Ogasawara, M. A., Dhawan, A. S., Rogelj, S., Van Slambrouck, S., Steelant, W. F. A., Evdokimov, N. M., Uglinskii, P. Y., Elias, E. M., Knee, E. J., Tongwa, P., Antipin, M. Y. and Kornienko, A. (2008) Structural simplification of bioactive natural products with multicomponent synthesis. 2. Antiproliferative and antitubulin activities of pyrano[3,2-*c*]pyridones and pyrano[3,2-*c*]quinolones. *J. Med. Chem.*, **51**, 2561–2570.

[200] Evdokimov, N. M., Van Slambrouck, S., Heffeter, P., Tu, L., Le Calve, B., Lamoral-Theys, D., Hooten, C. J., Uglinskii, P. Y., Rogelj, S., Kiss, R., Steelant, W. F. A., Berger, W., Yang, J. J., Bologa, C. G., Kornienko, A. and Magedov, I. V. (2011) Structural simplification of bioactive natural products with multicomponent synthesis. 3. Fused uracil-containing heterocycles as novel topoisomerase-targeting agents. *J. Med. Chem.*, **54**, 2012–2021.

[201] Pinney, K. G., Bounds, A. D., Dingeman, K. M., Mocharla, V. P., Pettit, G. R., Bai, R. and Hamel, E. (1999) A new anti-tubulin agent containing the benzo[*b*]thiophene ring system. *Bioorg. Med. Chem. Lett.*, **9**, 1081–1086.

[202] Flynn, B. L., Gill, G. S., Grobelny, D. W., Chaplin, J. H., Paul, D., Leske, A. F., Lavranos, T. C., Chalmers, D. K., Charman, S. A., Kostewicz, E., Shackleford, D. M., Morizzi, J., Hamel, E., Jung, M. K. and Kremmidiotis, G. (2011) Discovery of 7-hydroxy-6-methoxy-2-methyl-3-(3,4,5-trimethoxybenzoyl)benzo[*b*]furan (BNC105), a tubulin polymerization inhibitor with potent antiproliferative and tumor vascular disrupting properties. *J. Med. Chem.*, **54**, 6014–6027.

[203] Tyndall, J. D. A., Pfeiffer, B., Abbenante, G. and Fairlie, D. P. (2005) Over one hundred peptide-activated G protein-coupled receptors recognize ligands with turn structure. *Chem. Rev.*, **105**, 793–826.

[204] Jacoby, E., Bouhelal, R., Gerspacher, M. and Seuwen, K. (2006) The 7-TM G-protein-coupled receptor target family. *ChemMedChem.*, **1**, 760–782.

[205] (a) Habashita, H., Kokubo, M., Hamano, S.-I., Hamanaka, N., Toda, M., Shibayama, S., Tada, H., Sagawa, K., Fukushima, D., Maeda, K. and Mitsuya, H. (2006) Design, synthesis, and biological evaluation of the combinatorial library with a new spirodiketopiperazine scaffold. Discovery of novel potent and selective low-molecular-weight CCR5 antagonists. *J. Med. Chem.*, **49**, 4140–4152. (b) Nishizawa, R., Nishiyama, T., Hisaichi, K., Matsunaga, N., Minamoto, C., Habashita, H., Takaoka, Y., Toda, M., Shibayama, S., Tada, H., Sagawa, K., Fukushima, D., Maeda, K. and Mitsuya, H. (2007) Spirodiketopiperazine-based CCR5 antagonists: lead optimization from biologically active metabolite. *Bioorg. Med. Chem. Lett.*, **17**, 727–731.

[206] Boss, C., Brisbare-Roch, C. and Jenck, F. (2009) Biomedical application of orexin/hypocretin receptor ligands in neuroscience. *J. Med. Chem.*, **52**, 891–903.

[207] Gatfield, J., Brisbare-Roch, C., Jenck, F. and Boss, C. (2010) Orexin receptor antagonists: a new concept in CNS disorders? *ChemMedChem*, **5**, 1197–1214.

[208] Brisbare-Roch, C., Dingemanse, J., Koberstein, R., Hoever, P., Aissaoui, H., Flores, S., Mueller, C., Nayler, O., Van Gerven, J., De Haas, S. L., Hess, P., Qiu, C., Buchmann, S., Scherz, M., Weller, T., Fischli, W., Clozel, M. and Jenck, F. (2007) Promotion of sleep by targeting the orexin system in rats, dogs and humans. *Nat. Med.*, **13**, 150–155.

[209] Aissaoui, H., Cappi, M., Clozel, M., Fischli, W. and Koberstein, R. (2001) Preparation of 1,2,3,4-tetrahydroisoquinolines as orexin receptor antagonists. Int. Patent WO2001068609.

[210] Wathes, D. C., Borwick, S. C., Timmons, P. M., Leung, S. T. and Thornton, S. (1999) Oxytocin receptor expression in human term and preterm gestational tissues prior to and following the onset of labor. *J. Endocrinol.*, **161**, 143–151.

[211] Liddle, J., Allen, M. J., Borthwick, A. D., Brooks, D. P., Davies, D. E., Edwards, R. M., Exall, A. M., Hamlett, C., Irving, W. R., Mason, A. M., Mccafferty, G. P., Nerozzi, F., Peace, S., Philp, J., Pollard, D., Pullen, M. A., Shabbir, S. S., Sollis, S. L., Westfall, T. D., Woollard, P. M., Wu, C. and Hickey, D. M. B. (2008) The discovery of GSK221149A: a potent and selective oxytocin antagonist. *Bioorg. Med. Chem. Lett.*, **18**, 90–94.

[212] Bale, T. L. and Vale, W. W. (2004) CRF and CRF receptors: role in stress responsivity and other behaviors. *Annu. Rev. Pharmacol. Toxicol.*, **44**, 525–557.

[213] Molteni, V., Penzotti, J., Wilson, D. M., Termin, A. P., Mao, L., Crane, C. M., Hassman, F., Wang, T., Wong, H., Miller, K. J., Grossman, S. and Grootenhuis, P. D. J. (2004) *N*-phenylphenylglycines as novel corticotropin releasing factor receptor antagonists. *J. Med. Chem.*, **47**, 2426–2429.

[214] Vanhoenacker, P., Haegeman, G. and Leysen, J. E. (2000) 5-HT7 receptors: current knowledge and future prospects. *Trends Pharmacol. Sci.*, **21**, 70–77.

[215] Rudolph, D. A., Dvorak C. A., Dvorak, L., Nepomuceno D., Bonaventure, P., Lovenberg, T. W. and Carruthers, N. I. (2011) Novel tetrahydropyrido[3,2-*c*]pyrroles as 5-HT7 antagonists. *Bioorg. Med. Chem. Lett.*, **21**, 42–44.

[216] Chung, S., Parks, G. S., Lee, C. and Civelli, O. (2011) Recent updates on the melanin-concentrating hormone (MCH) and its receptor system: lessons from MCH1R antagonists. *J. Mol. Neurosci.*, **43**, 115–121.

[217] Goss, J. M. and Schaus, S. E. (2008) Enantioselective synthesis of SNAP-7941: chiral dihydropyrimidone inhibitor of MCH1-R. *J. Org. Chem.*, **73**, 7651–7656.

# PART III

# SCREENING STRATEGIES

# 7

# COMPUTATIONAL TECHNIQUES TO SUPPORT HIT TRIAGE

DOUGLAS B. KITCHEN and HÉLÈNE Y. DECORNEZ
*Albany Molecular Research Inc. (AMRI), Albany, NY, USA*

## 7.1 LEAD FINDING PROCESS: OVERVIEW AND CHALLENGES

### 7.1.1 The Need for Triage

High-throughput screening (HTS) is still the cornerstone of lead generation for small molecules. The triage process is a critical step in HTS campaigns to ensure that high-quality hits are identified and proceed to more expensive later stage experiments. The task involves the assessment and diagnosis of potentially enormous amounts of chemical and biological data, wherein the prognosis for the ultimate success or failure of a compound or series of compounds is estimated. To manage this, computational chemistry and cheminformatics are routinely used to help identify, prioritize, and make decisions on structures worthy of further consideration. Methods are based on chemical structure, calculated properties, rule-based filters, and the concept of binding efficiency. These different computational tools and cheminformatic approaches are now routinely combined to form efficient triage computer systems in many organizations. This chapter will describe the various strategies and methods currently used.

### 7.1.2 The Lead Generation Process

Hand in hand with the genomics revolution, the drug discovery process has evolved over the last few decades to become an extremely complex, multiyear, multidiscipline approach [1]. To improve efficiency and innovation, new technologies such as HTS and virtual screening (VS) were developed to search broad regions of chemical space and identify new starting points. The hit identification process is illustrated in Figure 7.1: a protein is

*Small Molecule Medicinal Chemistry: Strategies and Technologies*, First Edition.
Edited by Werngard Czechtizky and Peter Hamley.

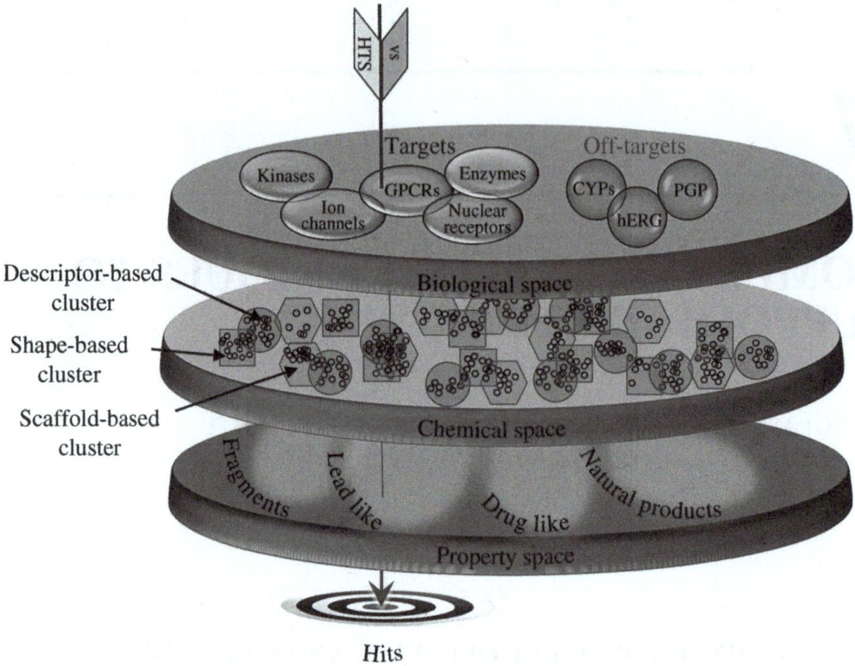


**FIGURE 7.1** The various classification schemes used during hit identification to triage compounds and identify the most promising hits.

chosen as a potential therapeutic target, and chemical space is searched to identify chemical structures (hits) corresponding to a region where a structure–activity relationship (SAR) exists; in turn, this region must correspond to good drug-like properties and must avoid undesirable biological off targets.

Figure 7.2 illustrates the process in more detail, starting with chemical library design, followed by screening where large numbers of actives are identified and verified as true hits (active to hit (A2H)). The process of hit analysis or hit triage selects series of compounds (hit series) for additional modification. The hits that survive triage enter hit to lead (H2L) that involves many rounds of synthetic exploration around each hit series to provide one or ideally several robust lead series with predictable SAR and promising drug-like properties that can progress into the lead optimization (LO) phase. In triage, the criteria are "soft" and intended to maximize the number of viable hits and are distinguished from the "hard" criteria used during LO stages. Examples of both types of criteria and decisions and how they differ in principle are also given in Figure 7.2.

To mitigate risk, the drug discovery process has been adapted to make critical decisions early: the so-called "fail early, fail cheap" approach. Usually, some preliminary measurements of ADME, toxicity, and/or other *in vivo* experiments provide evidence that *in vitro* activity can translate into *in vivo* activity suggestive of target engagement and desirable drug properties. This lead series can then undergo optimization with the goal of identifying a compound suitable for clinical development.

Screening technologies have kept pace with the growing collections of compounds and allow miniaturization and reduction of the cost of screening such that millions of wells can now be efficiently measured. Primary screens nowadays include not just single-target protein

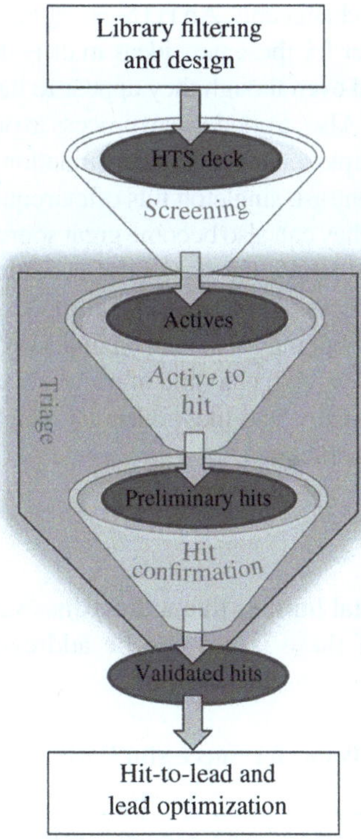


| | Triage (Soft decisions) | Lead-op (Hard decisions) |
|---|---|---|
| Potency cutoff | Weak (up to 10 mM) | <100 nM |
| Properties | Broad ranges; Failures acceptable | Drug like; Limit failures |
| Solubility | Moderate | High |
| ADMET | Liabilities acceptable optimization expected | No liabilities |
| Off-targets | Acceptable; Generally optimizable | Minimize |
| SAR | Not necessary | Necessary not flat |
| Reactive groups | Not acceptable | |

**FIGURE 7.2** Schematic representation of early drug discovery using large sources of compound collections either from experimental high-throughput screens or through virtual screening. Due to the large amount of data generated, results invariably undergo a triage process to categorize, prioritize, and classify compounds first as potential actives and then as hits. The contrasting soft approach to decision-making during hit triage is compared to the hard decisions necessary in subsequent H2L and LO campaigns. (*See insert for color representation of the figure.*)

assays but increasingly cell-based assays involving various forms of high-content screening [2, 3]. Such complex screens can make the analysis and selection of hits more complicated.

In tandem with experimental developments, computational techniques have developed and emerged as powerful aids to early phase drug discovery. Techniques such as VS reviewed in this book have themselves contributed to the identification of new chemical starting points. Computational evaluation has also been heavily used to redefine optimum HTS collections of compounds to be screened [4–6]. Concurrently, computational tools have been developed to distinguish structural features associated with poor HTS outcomes (e.g., frequent hitters, interfering compounds) and to permanently remove these compounds from consideration or inclusion. Just as importantly, the analysis of HTS data at various stages with computational techniques allows the analysis not just by statistical outcomes but by considering the relationships between the chemical content of the samples and their biological readout. In summary, computational chemists are now integral partners in early phase drug discovery [7].

### 7.1.3 Hit Triage: From Actives to Hits to Hit Series

Compounds with a positive readout in an HTS are often labeled "actives," and the process to validate these actives is often termed *hit identification* or *A2H* [8]. Unfortunately, the terminology surrounding hits and actives is often used interchangeably. An HTS can generate large numbers of hits, and the triage process is the first filter for these new ideas in drug discovery. This process needs to allow hits to proceed forward even though they appear to have a low probability of success purely based on their activities. Also, a good triage process avoids needless expenditure of effort on a series composed of multiple related hits since validation of a few hits is sufficient to validate the entire hit series. In contrast, singleton hits often require more effort to validate the activity and to develop SAR, but they can also become great sources of new chemical matter [9]. During H2L, new compounds can be the product of evolution, expansion, and merging of the original hit series. In the process of H2L and LO, compounds typically grow in size in order to improve binding to a target but may also take on undesirable physical property values and characteristics. Therefore, a desirable outcome of triage should be to select hit series with structural room to maneuver (typically lead-like molecular weight) in order to solve various drug property issues as they are encountered.

### 7.1.4 Challenges to Successful Lead Finding

Table 7.1 outlines many of the decisions and experimental hurdles that a drug discovery program faces to identify useful compounds. Some of these issues can be addressed experimentally, others computationally [10].

**TABLE 7.1 Organizational Considerations and Decision Points for Successful Early Drug Discovery**

| Organizational Resources | Computational/Cheminformatics |
|---|---|
| *What is required to determine if the hits are valid candidate series for hit to lead?* | *Are hits amenable to hit to lead?* |
| Internal resources | |
| • Computational/cheminformatics | • Lead-like? |
| • Biochemical/physical | • Ligand efficient? |
| • Chemical sample | • Is it a singleton? (Higher cost but potentially higher benefit: better IP, novel, but weak activities) |
| External resources | *Are there common challenges?* |
| • Purchasable compound database (rapid SAR exploration, analog purchase) | • Reactive groups<br>• Frequent-hitter functionality<br>• High lipophilicity (logP/logD, MW)<br>• Poor solubility |
| Biochemical/Physical | Chemical Analysis/Sample Validation |
| *Are all/any compounds valid or artifacts?* | |
| • Hit validation | • Are the compound structures correct? |
| • Assay confirmation | • Series identified and amenable to SAR development? |
| • Orthogonal biochemical measurements | |
| • False positive identification | • Does it contain a privileged substructure? |
| • Mechanism of action: inhibitor, competitive, orthosteric, allosteric | • Is IP crowded?<br>• Resynthesis: common synthetic routes? |
| • Structural proof of binding | • Does resynthesis validate? |
| • Biophysical methods: X-ray, NMR | |
| • Off-target activity, drug–drug interactions (DDIs) | |

The first challenge is HTS data quality. While HTS technology can be robust and allows rapid measurements, data is often of low precision and accuracy, with many potential complications, including single concentration runs, small numbers of replicate measurements, and fairly small amounts and inaccurate weights of test samples. The effect of experimental uncertainties has been quantified with regard to concentration and initial sample weight errors [11, 12], measurement error, and reproducibility [13] and the subsequent effect on confirmation rates [14]. It has been shown that confirmation experiments significantly reorder activities yielding different conclusions about hit potencies. Therefore, simply sorting compounds by primary screening activities and taking the most active percentage ("top-X") will likely result in many false conclusions about actives and hits. Triage processes need to allow tolerance for these unavoidable uncertainties while simultaneously attempting to make the most of expensive HTS data. Definitive conclusions about specific compounds need to wait for the confirmation of preliminary results. Subsequent standard confirmation steps involve a retest of the active wells, retest from freshly dissolved stock samples, and ultimately resynthesis and testing of the active structures. Computational techniques that group compounds by similar chemical features help confirm the activity of certain groups of compounds despite potentially low activity and experimental uncertainties. It has been found that preprocessing based on chemical structure is superior to the "top-X" approach, which wastes precious resources on uninteresting, irreproducible, and/or unprogressable hit series [15].

### 7.1.5 Frequent Hitters

There are many examples of compounds that are identified as active in multiple different and unrelated biological screens. These compounds are often termed "frequent hitters" or promiscuous compounds. A concern is that such a compound's activity is simply due to common artifacts: fluorescent interference, aggregation, and/or simple covalent reactivity specific to the assay conditions. Many types of artifactual behaviors especially aggregation have been published—characteristics that can be significant warning signs include steep dose–response curves [16] and synergy among mixtures of compounds [17]. Additionally, high-throughput assays have been developed for the sole purpose of indicating aggregation [18]. Artifactual hits, such as nonstoichiometric enzyme inhibitors, can be identified using orthogonal assays or various adaptations to the primary assay with the addition of detergent, light-scattering measurements to test aggregation, and testing with higher enzyme concentrations. Several experimental approaches exist to identify promiscuous compounds and distinguish them from aggregators, and these experiments can be performed in a fairly efficient manner (see, e.g., [19–21]).

### 7.1.6 Implications of Human Decision-Making

Recently, decision-making processes in drug development were reviewed and compared to the size of the R&D effort [22]. While many of the conclusions are more relevant to later stage discovery decisions, several important observations were made with regard to poor decision-making. First, there is a tendency toward "progression-seeking behavior" because progression of compounds to the next level is often rewarded, whereas killing projects early is not. Secondly, anecdotal successes tend to be used as examples that allow a project or lead compound to be "save[d] from oblivion" [22]. Attempting such salvage sometimes occurs when an impurity is the cause of activity in a well, and then inordinate amounts of resources are devoted to identifying various impurities and testing each one. Often, the active component is never discovered, or if it is identified, it is of no further interest. Others have highlighted similar human factors in decision-making—Maggiora et al. studied the tendency of medicinal chemists to disagree on the quality of hits [23], and other authors draw similar conclusions [24]. Computational approaches derived from analysis of past drug development experience can alleviate some of the shortcomings and biases inherent to human selection of compounds by automating procedures based on agreed criteria.

## 7.2 CHEMICAL STRUCTURE ANALYSIS OF HIT LISTS

The similarity principle states that similar chemical structures are likely to have similar biological effects [25, 26]. The principle needs to be applied with some care because the choice of chemical descriptors used to calculate similarity can have significant effects on the results within a data set [27]. Typical SAR representations are an application of the similarity principle because a series of compounds share a common core and share similar activities. The organization of hits by similar chemical structure is a powerful approach to triage and visualization of chemical space coverage. Making decisions on long lists of individual compounds is inefficient, tedious, and error prone. Decision-making is most efficient when made based on chemical substructures shared by many compounds. For example, a stoplight analysis can be applied in a series to its leading compounds resulting in the advancement or removal of an entire series based on the identification of problematic features or properties. Additionally, analysis of a hit list by substructure class facilitates intellectual property (IP) analysis, and common synthetic chemistry is more easily identified to accelerate hit exploration.

Perhaps the most overlooked advantage of clustering hits is that false positives and false negatives are more easily identified based on the similarity principle. Among a group of related compounds, some appear just slightly above, and others just below an activity cutoff in that HTS. When compounds are grouped by chemical similarity, these relatively small variations can be observed, and a true hit recovered that would have been missed by simply applying the "top-X" approach to hit selection (e.g., reconfirm only the top 1% of hits). In this way, false negatives can be rescued because of their chemical similarity to true positives. Similarly, if a large number of related compounds are typically found to be in the experimental noise and only one compound is above the cutoff, then it may be a false positive. The value of such an approach to hit confirmation has been exemplified in different labs and can greatly improve confirmation rates [28, 29]. Finally, grouping hits by structural similarity allows latent hit series that are composed of moderately weak responses to rise to the top of the priority list rather than being removed from further investigation.

Short lists of compounds are easily arranged manually by common substructures and functional groups, whereas the long lists of compounds are often unmanageable. When hit lists are somewhat randomly arranged, it is quite difficult to see useful patterns. To assist this effort, there are many cheminformatic methods to cluster or group compounds for later analysis. In substructure-based methods, common substructures, also called scaffolds, are identified among a group of chemical structures, and the results are chemically intuitive to a group of chemists. Other common methods for grouping similar compounds are clustering and partitioning. In these approaches, compounds are grouped together by numerical similarity calculations among chemical structures, by rules based on properties, or by partitioning of descriptor space. In similarity-based clustering, heterogeneous clusters arise where compounds are grouped together based on mathematical similarity but possess little substructure similarity. Frequently, some compounds are singletons, that is, compounds with no very similar structures or compounds with no common substructures in the data set. Large numbers of singletons in HTS data analysis is not unusual because one of the design criteria for HTS decks is chemical diversity.

### 7.2.1 Similarity-Based Clustering

Clustering of chemical compounds has long been pursued, and there are two types of clustering methods: hierarchical and nonhierarchical. Hierarchical methods develop a tree structure that defines ever-narrowing membership and family relationships (illustrated in Fig. 7.3 as nodes labeled $H_1$; $H_{1,1}$; etc.).

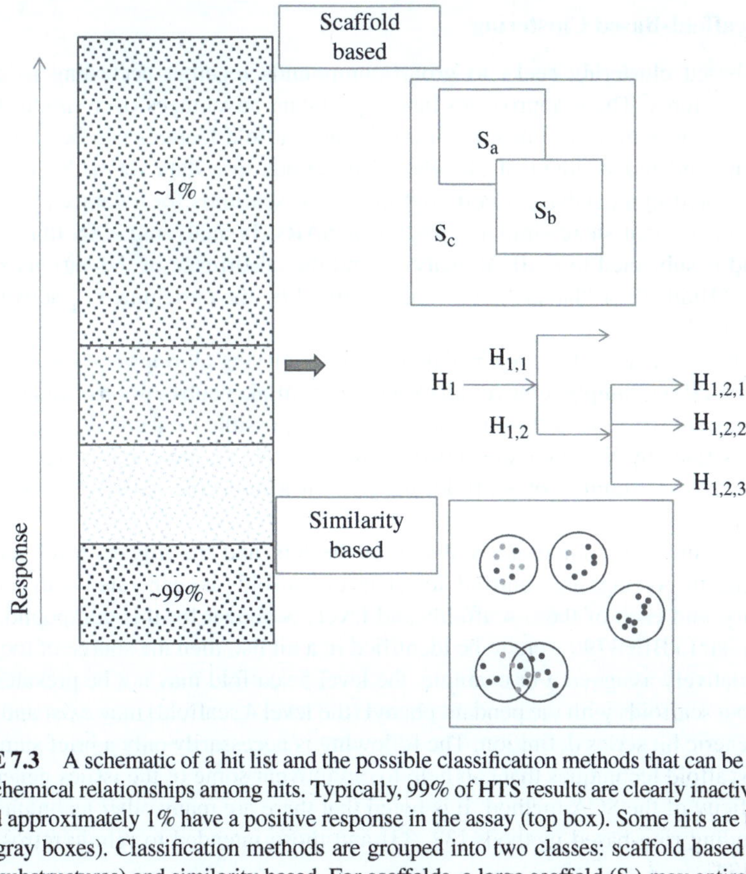


**FIGURE 7.3** A schematic of a hit list and the possible classification methods that can be applied to identify chemical relationships among hits. Typically, 99% of HTS results are clearly inactive (bottom box), and approximately 1% have a positive response in the assay (top box). Some hits are borderline (middle gray boxes). Classification methods are grouped into two classes: scaffold based (common ring and substructures) and similarity based. For scaffolds, a large scaffold ($S_c$) may entirely encompass other scaffolds ($S_a$ and $S_b$). Similarity-based clustering methods are hierarchical (the dendrogram in this figure) or agglomerative (represented by the circles). Actives (+) and inactives (–) can fall into any of the clusters. In both clustering methods compounds may belong to overlapping clusters.

Nonhierarchical methods such as self-organizing maps, neural nets, and others group compounds together into distinct and separate clusters possibly with no necessary relationship between clusters as illustrated by the circles in Figure 7.3 each containing numbers of active and inactive compounds. Several reviews of similarity calculations and the available choices among descriptor and fingerprint sets have been published [27, 30]. Depending on the similarity method used, clusters may be quite nonintuitive to many chemists because the numerical similarities group compounds together with no obvious structural similarity. Therefore, numerical clustering methods while very helpful in preorganizing long lists of compounds with fairly simple algorithms are often not easy to define visually. At some point, the chemical structures that occupy that cluster need to be exemplified in such a way that the actual hit series is described. An additional difficulty with similarity-based methods is that different similarity scoring methods and descriptors/fingerprints will yield different clusters. A third and very practical difficulty is that clustering depends on calculations among all compounds in the hit list and to the screening deck requiring pairwise calculations of similarities to potentially millions of compounds.

### 7.2.2 Scaffold-Based Clustering

Scaffold-based clustering seeks to group compounds together according to common structural features. These approaches have significant advantages over similarity-based clustering because they are linear algorithms, and scaffold features can be identified for each compound independent of all other compounds in a data set with the resulting scaffolds stored in a database. Useful definitions of scaffolds for a data set group compounds together that share similar activities or SARs. In analyzing HTS libraries, each compound is subjected to scaffold analysis, and the occurrence of the various scaffolds tabulated. "Hidden" or "latent" SARs are identified by the enrichment of actives within various scaffolds.

Scaffolding methods define scaffolds by first identifying all rings and the connections among them. If one simply removes all exocyclic, nonlinker substitutions from molecules, the resulting substructure is then the largest scaffold possible for a compound. This is the approach defined by Xu, and these substructures describe a compound's natural chemical scaffold [31]. An example of scaffold analysis on a molecule, BIRB796, is shown in Figure 7.4.

In this figure, the compound is shown at the top, and the highest-level scaffold is shown beneath. As rings are removed, lower-level scaffolds are generated with decreasing complexity, and each of these scaffolds and levels is shown for this compound. If a hit series similar to BIRB796 were to be identified in a hit list, then the source of the activity can be putatively assigned. For example, the level 5 scaffold may not be prevalent in the data set, but scaffolds with the pendant phenyl (the level 4 scaffold) may exist and suggest a more generic hit series definition. The following is necessarily only a brief summary of selected scaffold techniques that can help to circumvent some of the issues generated by the specificity of the SCA method. It is noted that there are many other techniques available including ring-based methods [32, 33] and those intended to mimic retrosynthetic analysis [34].

In order to maximize the chances of identifying common scaffolds among active compounds, the scaffold tree [35] and HierS methods [36] define multiple scaffolds for each compound. In the scaffold tree method, each compound has as many scaffolds as rings,

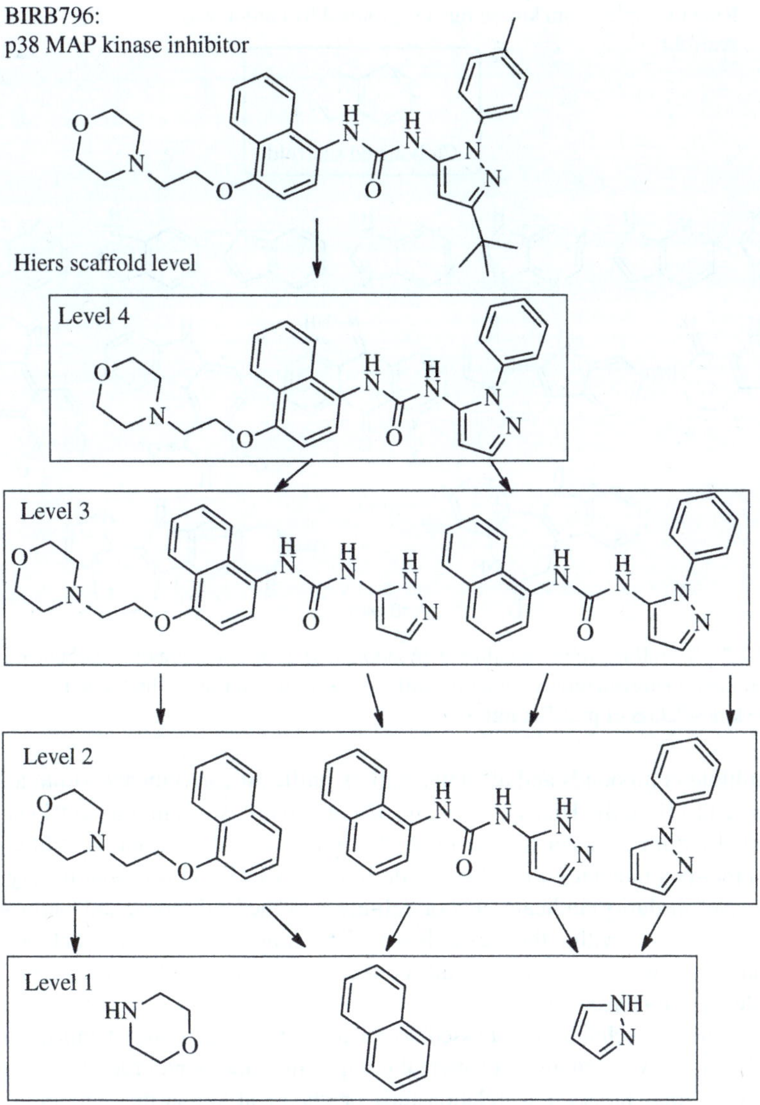


**FIGURE 7.4** Examples of the scaffolding process.

and there is a natural relationship between scaffolds and their descendants. At each scaffold level, only one scaffold is chosen among several options, and therefore, the number of scaffolds equals the number of ring systems. A more complete definition of all possible scaffolds for a set of compounds is found in the HierS method where the full hierarchy of scaffolds is retained. In this method, the number of scaffold levels per compound approximates the number of ring systems. The scaffold tree and HierS methods allow some flexibility at identifying scaffolds, but in each, the compounds need to be grouped from the most complex scaffolds to the simplest in order to optimize the number of compounds per scaffold. The scaffold network method is an adaptation of the scaffold tree method where the scaffolds that survive at each level are chosen based on biological activity

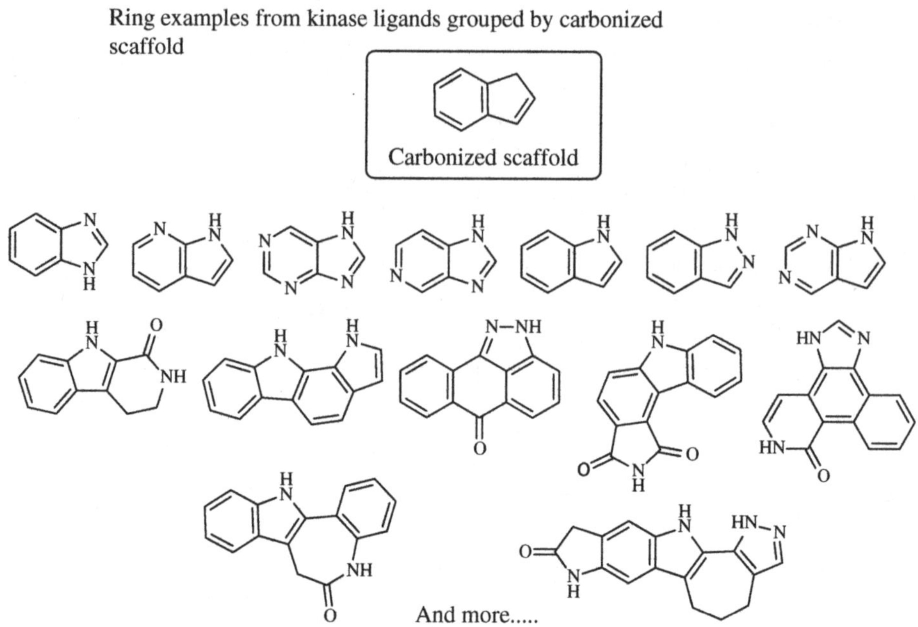


**FIGURE 7.5** Simplified representation of an indole ring as its corresponding carbonized scaffold and related kinase substructures interacting with a similar orientation to the hinge region of members of the kinase class of protein targets.

across multiple compounds and all statistically significant enrichments within a scaffold are preserved [37]. In the HierS method, the average pairwise Tanimoto (APT) similarity is calculated within a cluster to determine the tightness of the cluster. A tight cluster of active compounds indicates a good series definition. A broad cluster that is very diverse (low pairwise similarity) indicates a poor definition of the lead series, and therefore, little enrichment of actives within the cluster is seen. This latter case indicates that the scaffold is too small in relation to the compounds containing the scaffold, and therefore, a larger scaffold definition is necessary.

The scaffolding techniques discussed so far leave the underlying chemical structures completely intact by maintaining elemental composition and bond orders. These types of scaffolds may miss bioisosteric relationships or chemical similarities among scaffolds. Generalizing the scaffold of a compound into its carbon framework can overcome some of the difficulties with previously described scaffold techniques. Bemis and Murcko published an early approach that defines the carbon framework as the highest-level scaffold with all heteroatoms changed to carbon atoms combining related ring systems into one family [38, 39]. These frameworks provide an intuitive method of grouping similar scaffolds together and grouping similar compounds with similar activities. The selection of rings in Figure 7.5 illustrates how intuitively related ring systems such as indole and benzimidazole can be identified as separate scaffolds when activity patterns may suggest otherwise.

Closely related to the framework technique is a method that defines each portion of a scaffold in generic terms such as reduced graphs [40]. In this scenario, the graph of the molecule is annotated with features such as donor and acceptor, and these features are encoded in each scaffold. The combination of frameworks, reduced graphs, and similar generalizations of a scaffold allow active compounds to be compiled in one group, whereas more exacting scaffold definitions separate these active compounds. Finally, it should be noted that clusters can also be formed by combining fingerprint similarities and maximum common substructure methods [41]. After a computational scaffold approach has been completed, a project team then groups or ungroups scaffolds and substructures by manual inspection in order to decide the fate of each hit series.

### 7.2.3 Application of Clustering Classification Methods

The simplest application of clustering is to identify groups of molecules that have a higher enrichment for activity. These models can be tested for statistical significance very easily by examining the entire HTS data set, confirmation data, or preliminary $IC_{50}$ data. If the frequency of hits within a scaffold is greater than the hit rate for the assay, then the model may be significant and suggests that an early SAR has been identified. These calculations can then be rigorously tested based on their statistical significance such as monitoring the local hit rate [42]. This rigor can only be achieved with large clusters containing a high number of actives. However, if one out of five compounds in a cluster is "active" (i.e., a 20% hit rate), that may suggest that the cluster is identifying key structural descriptors for activity but cannot be supported by any rigorous statistics.

In any method, the size of a cluster can be misleading and should not be the only measure of interest in a cluster. If a large number of chemotypes are desired, then small clusters and singletons need to be given higher priority than large clusters. Scaffold-oriented methods are typically better at grouping compounds by biological activity; however, larger numbers of clusters each with very low occupancy are a possible outcome. Chemical similarity clustering can produce smaller numbers of clusters, and the number of clusters is often defined as part of the algorithm [43]. However, these types of clusters may have lower commonality among biological activities [44]. Using the actual data values for compounds rather than active–inactive designations can improve the statistics of various scaffold enrichments [35]. Both clustering and scaffold methods improve the rate of hit conformation when used to build lists of compounds to undergo confirmation experiments [42, 45, 46].

## 7.3 RULES AND FILTERS

Computational techniques that can rapidly identify the good compounds in a large data set are a common goal to HTS and VS. Simple rules that predict the likely fate of hit series in H2L can be a powerful tool for rapid evaluation of screening decks, which often number in the millions. When properly applied these rules allow increases in MW and clogP in a lead series by medicinal chemistry teams during H2L and LO in order to achieve greater potency or solve ADMET problems. Rules are often intended to be simple observations of a specific set of compounds, such as Lipinski's rule of 5 (Ro5) derived from studies of the orally available set of marketed drugs [47] or the lead-like rules applicable to early drug discovery, a more stringent set of criteria [48]. It is certainly wise to give higher priority to lead-like hits; however, when the data suggests that valuable compounds exist outside lead-like and even drug-like property space, these should not be ignored. A traffic light classification with only two colors (green and yellow) can be used, and the outcome is rarely the removal of a chemical class for reasons other than obvious reactivity or other such promiscuous behavior. The reasons for this are obvious: even the Ro5 criteria were the results of calculations of the properties of known drugs where each individual rule was satisfied by only 90% of the compounds in the study set (oral drugs only)—the usefulness of the 10% outside the acceptable range defined by the rules is undeniable as they were also marketed drugs. At the triage stage, it is clear that we would not wish to exclude those same drugs in the 10% set. Two well-known drugs, atorvastatin and montelukast, are as classic examples of compounds that fail more than two of Lipinski's Ro5 classifications [49]. Compounds that fall outside drug-like classification but are still viewed as valuable can be assembled for use in more challenging targets such as protein–protein interaction where *any* hit is valuable [50].

### 7.3.1 Computational Descriptors for Property Assessment

The field of assessing drug properties has evolved extensively, and many additional chemical descriptors are now considered suitable markers for assessing the fragment-, lead-, and drug-likeness of compounds. Some of these descriptors are intended to be estimates of physical properties, such as logP. Other descriptors are not physical observables, such as polar surface and rotatable bonds [51], and are better described as constitutive descriptors, deriving solely from 2D or 3D structures of molecules. The list of possible calculated properties that can be used is extremely long (see, e.g., Testa et al. [52]), and the proliferation of descriptors and filters was criticized for the disconnect between application and success in the clinic [53]. A recent paper provides additional warnings regarding the "inflation of correlation" observed when trying to use descriptors to formulate new rules [54]. Additionally, the accuracy and precision of these calculations are often suspect. In the case of the parameter, calculated logP (clogP), we routinely observe a minimum error of 0.5 and significant variation among logP calculators. Hajduk et al. presented an excellent summary of various calculation methods for property assessment, and this assists in the selection of the best computational methods to balance speed, accuracy, and physical meaning [55]. Because a triage process can eliminate many useful compounds incorrectly based on poorly chosen property criteria, it is important to use properties that are most closely related to success at each stage and of the drug discovery process. The various rules currently finding favor for compound evaluation are given in Table 7.2. More recent correlations with *in vitro* and *in vivo* drug disposition and toxicity experiments [64] continue to reaffirm the usefulness of various descriptors to keep compounds within desirable property space.

Descriptors are aggregated into different rules for different purposes: drug-like, fragment-like, permeability, solubility, general ADMET, CYP, etc. They can also take into account the target of interest and its location. It has long been recognized that CNS drugs and antibacterial and oncology agents have different property distributions. Multiple groups continue to use the large amount of information in their screening collections to investigate the impact of various descriptors and properties on screening outcomes, and recommendations have emerged based on these experimental results [65]. These recent analyses are appealing because the relationship between calculated properties and HTS experiments is likely to be more significant than comparisons to descriptor values for compounds in clinic or in the marketplace. In many cases, however, firm assignment of

**TABLE 7.2 Rules and Filters for Compounds Assessment in HTS Triage**

| | Simple Rules | Formulation | Description |
|---|---|---|---|
| Soft Filters | Drug-like (rule of 5) [56] | • H-bond acceptors (N+O) ≤10<br>• H-bond donors ≤5<br>• (MW) ≤500<br>• ClogP ≤5 | Properties of compounds likely to be *orally* available in humans |
| | Lead-like (rule of 3) [57] | • H-bond acceptors ≤3<br>• H-bond donors ≤3<br>• MW ≤300<br>• ClogP ≤ 3<br>• Rotatable bonds ≤3 | Based on analysis of difference between 96 lead-like and drug pairs of compounds |
| | Fragment-like [58] | • ClogP ≤ 3<br>• MW ≤300<br>• Hydrogen bond donors ≤3<br>• Hydrogen bond acceptors ≤3 | Analysis of a set of 40 fragments from high-throughput X-ray (additional filters: rotatable bonds ≤3; PSA ≤ 60 Å$^2$) |
| | Rule of 4/400 [59] | • ClogP ≤ 4<br>• MW ≤ 400 | Higher clogP and MW correlated with poor ADMET profiles |
| | PSA/rotatable bonds [51]<br>or<br>HBA + HBD/ rotatable bonds [51] | • PSA ≤ 140 Å$^2$<br>• Rotatable bonds ≤10<br>or<br>• Hydrogen bond donors + acceptors ≤12<br>• Rotatable bonds ≤10 | Rat oral bioavailability >20–40% |
| | Golden triangle [60] | Within triangle defined by:<br>• Baseline at MW = 200 and logD from −2.0 to 5<br>• Apex at MW 450 and logD between 1.0 and 2.0 | *In vitro* permeability and low clearance |
| | Rule of 3/75 [61] | • ClogP ≤ 3<br>• PSA ≥ 75 Å$^2$ | Combined low PSA and clogP indicate high lipophilicity Correlation with toxicity |
| | Lipophilicity [62, 63] | • 0.8 ≤ clogP ≥ 3.2 | Required for membrane permeability and renal clearance |
| | More Complex Rules | | |
| | CNS MPO | See Section 7.3.4 | Single parameter for CNS penetration |
| | QED | See Section 7.3.4 | Single parameter estimate of drug-likeness |
| Hard Filters | Structural Filters | | |
| | PAINs | See Section 7.3.5 | Assay interference liability |
| | Reactive groups | See Section 7.3.6 | General binder liability |

acceptable descriptor ranges is problematic due to the small number of observations with extreme descriptor values because cutoffs may have been applied previously to the data set. Descriptors of particular interest for triage fall into three broad classes and their recommended ranges are given in Table 7.3.

**TABLE 7.3 Typical Simple Calculated Descriptors and Recommended Ranges in HTS Triage**

| | | Triage Range | Outside Range Liability |
|---|---|---|---|
| Soft Decisions | Structural Descriptors | | |
| | Molecular weight (MW) | 200 to 500 | Low MW: high concentrations needed for binding. High MW: frequently high lipophilicity, SAR follow-up more challenging, decreased permeability, target promiscuity, and more |
| | Heavy atoms | 20 to 50 | |
| | N+O | 2 to 12 | Higher number gives polar compounds. Higher clearance through renal elimination |
| | Total charge | −2 to 2 | Difficult formulation. Ionization state of active compound challenging to determine |
| | Structural Shape Descriptors | | |
| | Fsp3 | >0.2 | Low fsp3 "flat" hits acceptable with clear SAR to improve value during hit to lead |
| | Rotatable bonds | 0 to 9 | High number of rotatable bonds indicates high entropy of binding. Active conformation hard to determine, challenging SAR development |
| | Aromatic rings | 0 to 5 | Higher lipophilicity, high number of aromatic rings a CYP, and solubility liability, among others |
| | Physicochemical Parameters | | |
| | LogP/LogD | −2 to 5/−2 to 3 | LogP >4 high lipophilicity: lower solubility and increased safety risk. Optimum ranges: logP 0–3; logD at pH 7.4 1–3 (MW dependent) |
| | LogS | > −4 | Less soluble compounds more lipophilic. |
| | Polar surface area (PSA) | 20 to120 | LogP<3 and PSA<75 Å$^2$ a safety risk. Optimum range vary and must balance absorption, CNS penetration, etc. |

### 7.3.2 Lipophilicity and Other Physicochemical Descriptors

One of the current challenges for a medicinal chemist is the tendency to increase potency through strategies that increase MW and clogP. SAR exploration efforts often result in increases in parameters such as MW, heavy atom, or bond counts. These descriptors in turn are correlated with other important physicochemical parameters such as logP. Indeed, gains in binding potency often come from entropic contributions, which have been correlated with a host of ills such as promiscuity and low solubility [66].

While satisfying all recommended parameters and rules is important at the LO stages [67], in triage compound, profiling is primarily used for classification rather than for inclusion or exclusion from further consideration. However, clogP, which is the calculated values for octanol–water partitioning, warrants more attention (clogP for neutral compounds and clogD at pH 7.4 for charged species). The early analysis of Lipinski emphasized the effect of high lipophilicity (logP) on compound attrition because of low oral bioavailability, permeability, and absorption for compounds with high (>5) clogP values [56]. Since then, multiple studies have correlated clogP values with various outcomes (ADMET, promiscuity, solubility, etc.), and invariably, the range of recommended clogP values has narrowed. The multitude of implications of high (and also in some cases low) lipophilicity with respect to a host of measurable drug properties have been reviewed by Waring [62, 63]. LogP combined with a high number of aromatic rings (NAR) (>3) has shown to have particularly low solubility outcomes, which adds to ADMET concerns of the expected high attrition associated with lipophilic compounds [68, 69]. Therefore, high clogP compounds should be flagged during triage because it can be challenging to optimize during later stages and because high logP often indicates low solubility and thus increased uncertainty in measurements.

### 7.3.3 Structural and Shape Descriptors

Structural descriptors, sometimes termed constitutive descriptors, indicate the number of heavy atoms, rings, atom types, and other simple quantities to describe a molecule. These descriptors can be sufficient to reflect changes in properties from fragments to lead-like to drug-like, which parallel the expected changes to a series that might occur in the course of SAR exploration efforts. To the extent that changes in simple descriptors can be examined within a clustered set of compounds and their biological activity, one can obtain an indication of the requirements for activity. Monitoring hydrogen bond donors or acceptors might give an indication of specific interactions with a target, whereas aromatic content might indicate the shape of the pocket occupied in the protein. These observations are especially useful when confined to a common substructure within a compound cluster.

There are several simple 2D estimates of the 3D character of compounds. These descriptors have been used to describe "flatness" of molecules, something that is considered detrimental to compound progression in the late stages of drug discovery [70] because of the relationship of 3D structures to aqueous solubility. Fsp3 is the ratio of sp3-hybridized carbons to the total number of carbons [70]. It can however significantly over- or underestimate 3D content. Structural descriptors such as low rotatable bond counts and high aromatic atom counts correlate strongly with low fsp3. Ar-fsp3 is an additional measure of the complementary nature between fsp3 content and aromatic content of compounds and is defined as the difference in the number of aromatic carbons to sp3-hybridized carbons [71]. Beyond aromatic ring counts, the type of aromatic ring was shown to be important for downstream success and indicates that shape descriptors are best considered with other constitutive descriptors for compound evaluation [72].

The ability of a compound to fit into a protein binding site normally requires significant 3D shape so that it can efficiently interact with the common types of functionality of enzyme active sites or receptors. Hann has developed some measures to determine the drugability of a binding site and the corresponding requirements of complexity in the 3D structures of molecules [73]. Direct measures of 3D content are preferable to 2D approximations, but conformational dependence adds significant variability to the calculation of any 3D descriptors. One convenient 3D description of molecules is the normalized components of the principal moments of inertia (PMI) with ratios of the longest and shortest axes of a molecule. The values of the PMI components have been used to classify molecules as rodlike, disclike, and sphere-like [74]. Because PMIs are conformation-dependent parameters, individual compounds with rotatable bonds can have multiple PMIs but are described by a single fsp3 value.

PMI analysis, fsp3, and Ar-fsp3 can only provide approximate guides to problem compounds such as overly rigid or flat molecules. The use of these shape descriptors is challenging in the triage setting. The cutoffs for these descriptors were largely determined from the rate of progress of compounds through the clinic, and such criteria are likely too presumptive to be used to eliminate hits from a screen. For example, kinase inhibitors generally have lower fsp3 content because of the narrow binding site occupied by the flat adenine ring of the natural substrate ATP. By contrast, steroid drugs that target hormone nuclear receptors have high fsp3 content. Incorporating compounds that cover the range of shape is desirable in a general purpose collection. In contrast, a focused screening collection can use these same descriptors within a narrow range usually associated with the target class.

### 7.3.4 Multiparameter Calculations: MPO and QED

The large number of parameters that are simultaneously optimized during drug discovery has provided an impetus to develop composite calculations that provide a single value to indicate the quality of a compound. This process, multiparameter optimization (MPO), has led to the development of several mathematical equations that combine various calculated properties into a single numeric quantity.

Figure 7.6 illustrates an MPO function derived by Wager et al. for the propensity of a compound to enter the CNS [75]. Six individual parameters are calculated and combined by linear functions such that if a property is within acceptable limits, that term is given a full weight (in this case 1.0). If the property is outside allowed limits, then the value for the property is 0.0. A gradient is applied to each property value if the value is a slight violation (the middle boxed region on the graphs of Fig. 7.6). The sum of each of these properties then has a maximum of 6.0 in this case and a minimum of 0. The authors suggest that a value of 4 is sufficiently comparable to known CNS agents to indicate that the compound has a strong chance to enter the CNS. This type of function permits uncertainty in the calculated properties and also on the filter values. A compound can be on the border of all the properties and still achieve the minimum score. Other mathematical formulations have been envisaged such as a probability weighting in the method termed quantitative estimation of drug-likeness (QED) [49]. By transforming multiple filters and values into a single column of numbers, MPOs provide a significant advantage in sorting large numbers of compounds avoiding multiple column sorting procedures. Also, MPOs

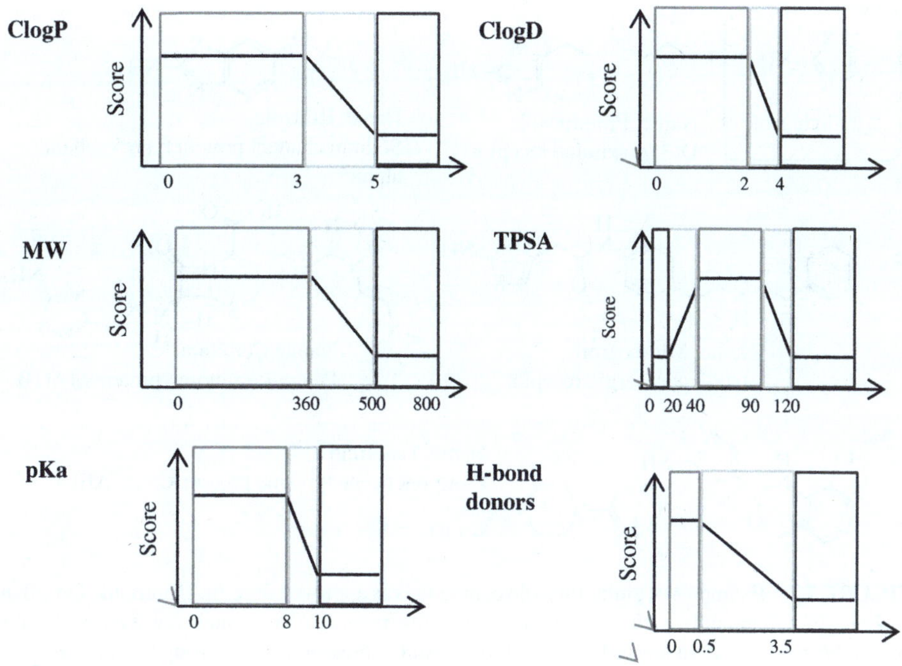


CNS MPO
Left-most boxes: fully within acceptable limits
Center-most boxes: partially inside limits
Right-most boxes: beyond acceptable limits

**MPO score** $= \sum_{\{X\}} \frac{fx}{fx_{max}}$

$$f_x = w \quad x < y_0$$

$$f_x = \frac{w(x - y_o)}{(y_{max} - y_0)}$$

$$f_x = 0 \quad x > y_{max}$$

**FIGURE 7.6** Visual representation of an MPO as designed by Wager et al. for CNS drugs. Red boxes indicate property values in unacceptable ranges, green boxes are completely acceptable, and yellow boxes indicate a favorability with a maximum contribution of the weight on the that property (typically 1.0) and linearly diminishing to 0.0. (*See insert for color representation of the figure.*)

are flexible ways to select important properties for each project and assign a single assessment specific to the project so that the importance of each property is balanced. Once sorted, a team can investigate in more detail as to whether the numeric value for a compound is truly justified by looking at individual components of the score.

### 7.3.5 Frequent-Hitter Analysis

It is efficient to eliminate as many likely interfering compounds early in the hit analysis stage with computer searches when causative substructures can be clearly identified. The first collections of substructure-based filters were assembled in the 1990s. Some of the earliest mentions of reactive compounds and frequent hitters in HTS processes were

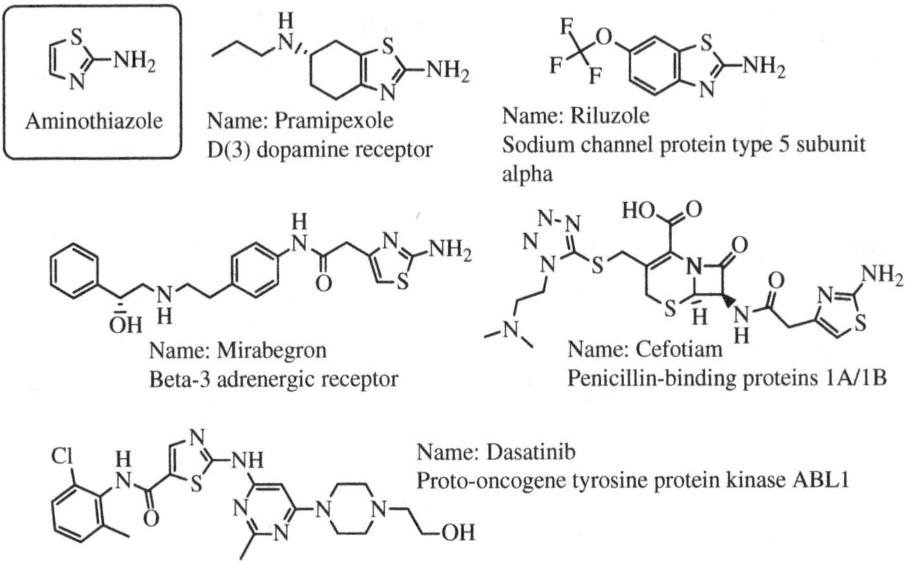


**FIGURE 7.7** Examples of aminothiazole compounds as approved drugs in DrugBank. DrugBank contains 63 compounds with an aminothiazole substructure, 20 of which are approved drugs. Aminothiazoles are often labeled frequent hitters, but the presence of this group in a hit is not necessarily indicative of a hit's likelihood of success in later stages.

the early works of Rishton [76] and of the Vertex group (rapid elimination of swill, REOS) [77, 78]. Over the course of many years and refinement by many different organizations, the specifics of these filters have undergone additional debate and refinement. These lists have often started from suspected or known reactive groups [79, 80]. Based on activity measurements across a small set of assays against varying targets, a set of substructure filters termed Pan Assay Interference compounds (PAINS) were developed [81]. A later analysis compared 15 enzyme assays and identified relatively few frequent hitters, and in some cases, compounds flagged by previously identified PAINS filters were not actual frequent hitters [82].

Some level of complexity is required to properly deal with the risks associated with overfiltering or underfiltering. For example, the Lilly filters, which were based on a large number of screens, illustrate difficulties due to chemical contamination. "Color" rules were difficult to define because the interference from a sample in an optical assay was often due to colored impurities or origin of the sample and was not caused by the predominant chemical species attributed to the testing well. In many other cases, the cause of the promiscuity was determined to be chemical contamination [83]. Other groups have developed warning systems for PAINS, and frequent hitters that like the Lilly software go well beyond simple labels of reactivity [81, 84]. Some examples of substructures that are likely PAINS filters are shown in Figure 7.7.

Another concern is that some types of frequent hitters can be useful and therefore difficult to separate from useless, promiscuous compounds. Some compounds exhibit "polypharmacology" or are "dirty drugs" because they bind potently to multiple targets. Often these compounds contain a substructure (a so-called privileged structure) from which their polypharmacology derives. The aminothiazole group is found in many

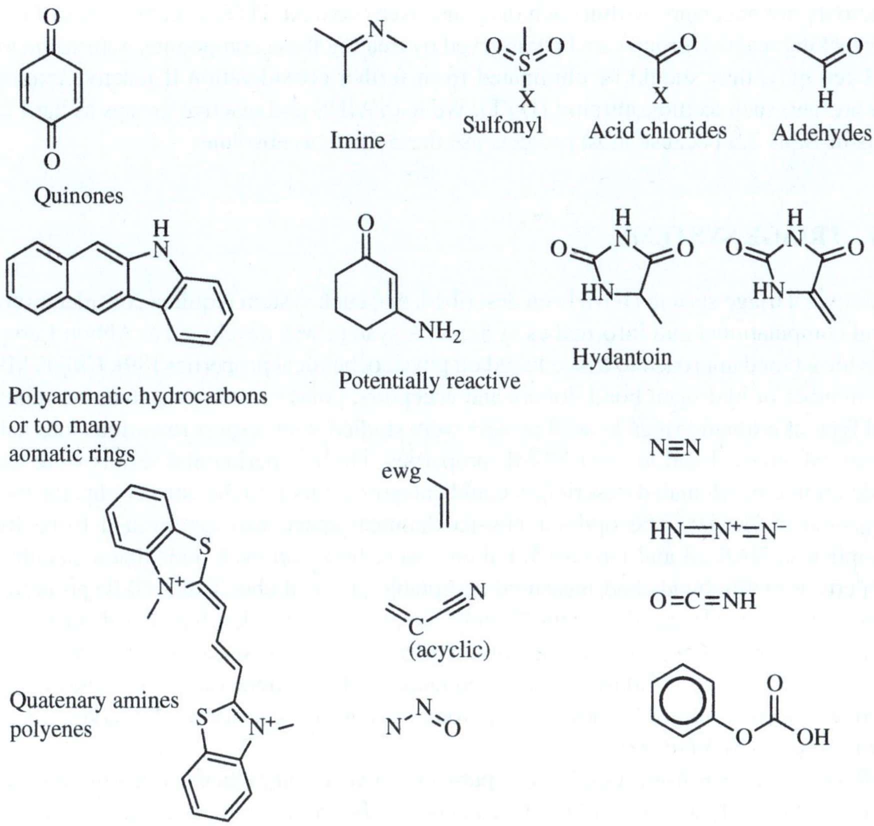


**FIGURE 7.8** Examples of typically reactive or interfering groups.

screening libraries and is sometimes identified as a "frequent-hitter" substructure, but this group has been found in compounds that bind 5-HT1a, adenosine A1, muscarinic receptors, dopamine uptake, sodium channel, and antibacterial targets [85].

Figure 7.8 shows a sampling of aminothiazoles from 20 approved drugs with the aminothiazole substructure reported by DrugBank [86]. This type of polypharmacology can be considered for its detrimental or useful characteristics on a case-by-case basis.

### 7.3.6 Reactive Group Analysis

Reactive compounds were found to be a problem early in the history of HTS [77]. While a term such as reactivity would seem to have unequivocal definition, there are now some disputes regarding the reactivity of many chemical groups. Some groups are partially reactive and crystallized with their proteins [87], and other groups react or irreversibly bind with their targets in the active site [88]. Indeed, while reaction of compounds with protein surface-exposed cysteine residues is a liability, reactivity of cysteines or other reactive side chains are binding site dependent. The broad or selective promiscuity of reactive groups remains difficult to discern and makes decisions on these compounds challenging in a triage setting. Ultimately, target knowledge and assays to determine reactivity are necessary within each drug discovery context. HTS screens with sufficient hits lacking reactive groups are better served by treating these compounds with suspicion, and certainly, they should be eliminated from further consideration if reactive reagents are present, such as dithiothreitol (DTT). We list PAINS and reactive groups as hard criteria in Table 7.3 because most projects use these filters as absolutes.

## 7.4 TRIAGE SYSTEMS

Automated triage systems have been described, and each system requires consideration of local computational and informatics systems. A system was developed at Abbott Labs to provide a tiered approach to triage based on physicochemical properties [89]. ClogP, MW, the number of hydrogen bond donors and acceptors, polar surface area, and the number and type of aromatic rings as well as fsp3 were studied with respect to various internally measured physical and *in vitro* ADME properties. These experimental results were used to determine if calculated descriptors could categorize hits into the correct bins for these measured properties. The optimal physicochemical space was determined to be Ro4 compliance, NAR$\leq$3 and fsp3$>$0.5. Failures were based on even odds that a calculated property for compounds had measured acceptable physical chemical/ADME properties. Rather than considering these cutoffs only, seven tiers were developed with increasing descriptor failure. For example, the optimal physicochemical space was assigned tier 1, while tier 2 allowed a failure in fsp3. Sequentially, lower tiers varied the types of rule failures that were allowed. These authors found that nearly one-fourth of marketed drugs fell into tier 7 (Ro5 failure).

Recently, a team from AstraZeneca published a very comprehensive list of criteria to divide hits into 11 categories [24]. Ten categories label compounds with varying amounts of problems, including “bland” (e.g., only one heteroatom), PAINS (e.g., dyes), and difficult to develop classes (e.g., steroids and natural products). Compounds that are not in the 10 negative categories and pass a broader range of drug-like filters are considered core compounds and remain for further consideration. The Lilly group has also provided an extensive list of undesirable groups and software that allows further flexibility through a demerit system [83]. For example, 51 rules specify acylating agents and 29 rules specify different aldehydes.

## 7.5 LIGAND EFFICIENCY INDICES

Ligand efficiency has become an important concept in triage and is described in detail elsewhere in this book. In summary, it has been observed that chemical series tend toward an asymptotic maximal affinity with regard to the number of atoms [90] and that preceding this is a linear range where it is possible to obtain a linear addition of free energy of binding for every heavy atom added to a molecule. Andrews [91] observed that common functional groups could be used as simple descriptors to fit large numbers of binding affinities. Observations such as these have suggested ways to normalize binding potency for variations in MW and number of heavy atoms [92]. There are many good thermodynamic reasons for nonmonotonic behavior of such curves, but the simplest explanation is that on most occasions the addition of another functional group to a molecule is nonoptimal or even detrimental to binding to a target. Therefore, when all other comparisons are equal, the smaller (more ligand-efficient) starting points are better than larger and slightly more potent compounds.

The principles of ligand efficiency are very important to identify fragment-like molecules that are unlikely to be noticed when sorted by activity alone. Additionally, summary values for a potential lead series are helpful, such as median for a cluster of compounds, minimum/maximums, and other simple statistical calculations to appreciate the optimization opportunity within possible lead series [93]. Starting points for H2L are often driven by entropic and presumably hydrophobic interactions [66]. A better starting point would likely have higher enthalpic contributions. Isothermal titration calorimetry (ITC) is often used to determine experimentally enthalpic and entropic contributions to binding. At the triage stage, an early estimate of the possibility of enthalpic driving forces can be found in the number of hydrogen bond donors and acceptors and the polar surface area.

## 7.6 HIT SERIES ANALYSIS

### 7.6.1 Latent Hit Series and Singletons

The aspects and challenges of converting a large number of hit molecules to several viable hits and hit series are exemplified in the terms latent hits and latent hit series. In latent hit series [94], a group of related compounds all have very weak activity but are measurably above the signal-to-noise ratio. No individual compound possesses potent activity, and a superficial analysis would indicate that no further work should be performed on the series. There is evidence that these series can still be good starting points. The outcome of modifications to latent hits and latent hit series is difficult to foresee because there are many other factors to be optimized besides the primary potency. Therefore, during the remainder of the discovery process, weak hits produce latent hits [95] and a collection of chemically related weak hits then forms the basis of a latent hit series [94].

Identifying singletons is also critical to the success of lead finding. A lone compound with significant activity can be easily overlooked as the starting point for a lead series, and this series' success may depend on relatively few measurements that are often uncertain. Therefore, hit exploration for these compounds by computational similarity and substructure searching can be very effective at identifying similar hits without extensive and expensive resynthesis. More challenging to identify are individual latent hits. A key structure can emerge during the hit analysis and enrichment process where the original hit is quite weak. After various chemical modifications, a new, significantly more potent compound emerges that had previously been hidden (latent) from view [95].

### 7.6.2 Rapid Hit Exploration and Compound Set Enrichment

Hit exploration is often necessary because screening collections are built to maximize diversity, and therefore, the number of analogs around a given hit is likely to be limited. Expanding the chemical surroundings by VS, similarity search, substructure queries, and docking or pharmacophore searching are inexpensive computational tools used to improve or confirm activity within a series or fill in SAR holes. Such techniques are sometimes referred to as "SAR by catalog," where chemically related compounds can be identified computationally and purchased more inexpensively than one-by-one synthesis. This can be an efficient process to validate that structure is related to activity (a true SAR) and/or that the primary activity can be improved. The value of related compounds and their impact on SAR can be validated by methods used during triage such as monitoring the local hit rate for a compound cluster [42]. Additionally, techniques such as recursive partitioning and cluster SAR can be applied to perform sequential screening to produce iterative improvements to the number and quality of hits [96]. Obtaining too many experimental measurements for early hits likely wastes too much time and money, so computational methods can give a probability-based argument for retention of possible lead series until the chemical search space is narrow enough to warrant experimental triage.

### 7.6.3 SAR Analysis

SAR analysis techniques have evolved as another set of computational tools for further hit analysis. A tractable SAR is an important criterion for progression to H2L status because it suggests that a series will lead to potent compounds through predictable chemical modifications. It has been found that the binding geometry is often conserved as a molecule is grown or evolved in H2L and LO processes [97]. A good example is the PKA–PKB kinase system as illustrated in Figure 7.9 [98].

The phenyl pyrazole structure 1 was the fragment lead for this program. As the structure was examined, additional groups were added to the para position of the benzene ring and activity was improved. Importantly, the position of the phenyl pyrazole was found to be unchanged in the kinase binding site. Therefore, biophysical structural methods can assist in elevating the status of a potential series.

A criterion that states that the SAR is not "flat" is often too general and open to individual bias. SAR landscape analysis attempts to monitor changes in structure and their effects on biological activity. Several cheminformatic methods are now available to quantify SAR landscapes: the structure–activity landscape index (SALI) [99], SAR maps [100], and SARI [101]. Such methods identify connections among compounds within a chemical series and also among different series. Regions are indicated where changes in structure do not change biological activity (flat SARs) and where small changes in structure greatly improve (or diminish) activity (activity cliffs, [99]). One of the complications during hit analysis is that there are few very highly active compounds and many weak hits. Landscape analysis need not be quantified in this case, and simply identifying areas in a molecule that diminish activity will make early H2L more efficient. Additional methods have been developed to group compounds by substructure and then to define SARs for rather coarse HTS hit sets. Reduced graph queries are based on clustering and substructure searching [40, 102] with each ring system generalized. As an example, consider a generalized indole ring and examples of related ring systems in known kinase inhibitors that were shown previously in Figure 7.5. Matched molecular pairs (MMPs) are a fairly new approach to analyzing SARs. In this method, compounds with only one position that varies are compared [103]. From a computational perspective, the maximum common substructure of a group of compounds is identified, and only positions that change are tracked. Recently, MMPs have been studied extensively and attempts made to identify the prevalence of good, neutral, or bad changes on activity and pharmaceutical properties [104].

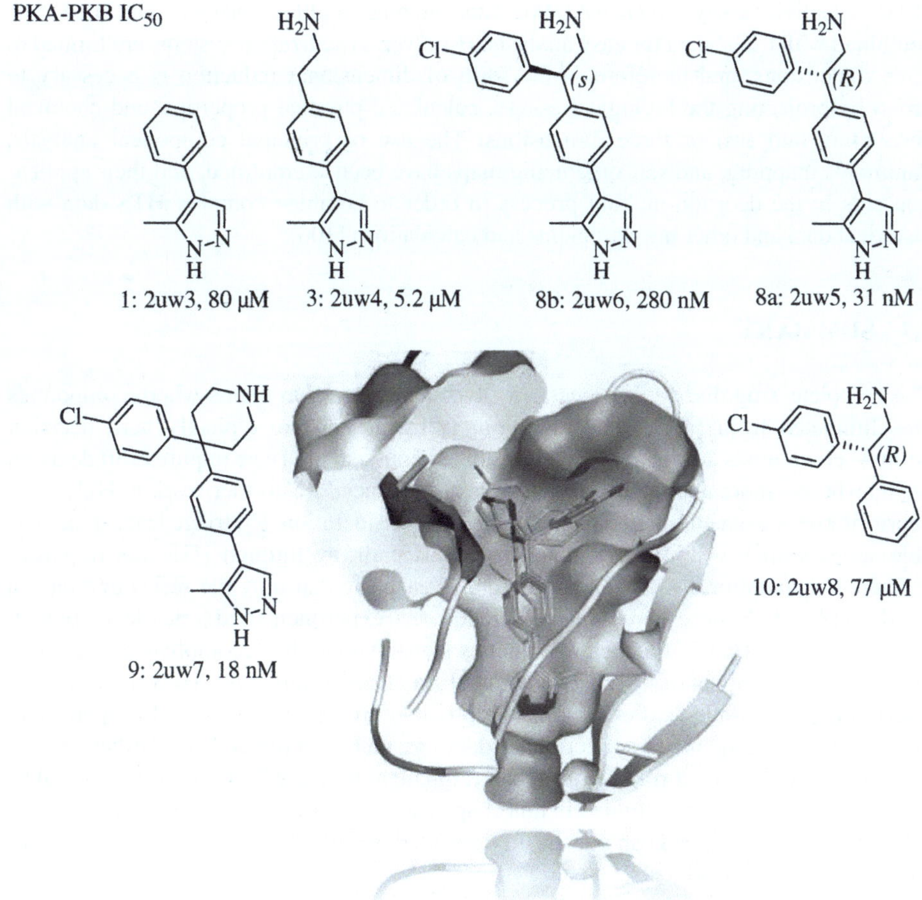


**FIGURE 7.9** Example taken of a protein kinase B inhibitor series and binding to the active site. The crystal structures of these inhibitors illustrate that the binding of a group of analogs has its SAR dictated by key binding interactions that are preserved in each analog. In this case, the pyrazole ring makes hydrogen bonds to the hinge of the kinase (backbone hydrogen bonding groups that are used frequently in kinase inhibitors). (*See insert for color representation of the figure.*)

### 7.6.4 Data Volume, Integration, Retrieval, and Visualization

The information and decisions made in the early stages of a project need to be accessible to downstream project teams. Decisions need to be derived from this data providing room for SAR improvements in over a multiyear and multicompound effort. While single output biological data is most commonly used as the comparator for clustering, there is an increasing use of multiplexed data such as phenotypic assays, pathway analysis, and similar high-content screening assays. The process of clustering compounds is still performed with similar techniques, but identifying signals across multiple assay outputs can be more complex. Several examples have been published that illustrate methods to provide simpler database visualizations of multidimensional HTS data and chemical clusters in order to select fruitful directions and to ensure that the mechanisms of action are appropriate

[105]. Selectivity assays, *in vitro* ADME data, and other multidimensional data need to be combined with triage data for easy analysis. However, visualization systems are limited to three dimensions, and therefore, some form of dimensional reduction is necessary to group by projecting the biological assays, calculated physical properties, and chemical descriptors into two or three dimensions. The use of principal component analysis, Sammon's mapping, and self-organizing maps have been exemplified, and their application aids in the decision-making process in order to combine complex HTS data with historical data and other measurements and calculations [106].

## 7.7 SUMMARY

The complete drug discovery process involves many decision points where compounds are eliminated from further consideration. H2L and LO are typically hard decision points—compounds are "in" or they are "out." In contrast, hit triage requires soft decision points where a more abstractly defined series is advanced for further work in H2L, or is deprioritized or even eliminated from any further consideration. Hit triage leads to actionable series with key limitations noted so that a pathway through H2L can be easily defined. Computational methods are critical in ensuring that only the series that have a good likelihood for success are subjected to additional experiments and series development. The key computational and cheminformatics steps that can be accomplished after HTS are organizing compounds by similar chemical structures, filtering for likely nuisance and interfering compounds, and evaluating the possible hit series for physical property and descriptors. This organized collection of data can then be provided for further experimental analysis and confirmation and then refinement by early SAR analysis. The rules, filters, and procedures highlighted in this chapter were the fruit of the experience of many different project teams and can now be implemented in fairly simple processes throughout early drug discovery projects.

## REFERENCES


[1] Schnecke, V., Bostrom, J. (2006) Computational chemistry-driven decision making in lead generation. *Drug Discov. Today* **11**, 43–50.

[2] Younis, Y., Street, L. J., Waterson, D., Witty, M. J., Chibale, K. (2013) Cell-based medicinal chemistry optimization of high throughput screening its for orally active antimalarials. Part 2: hits from SoftFocus kinase and other libraries. *J. Med. Chem.* **56**(20), 7750–7754.

[3] Chatterjee, A. K. (2013) Cell-based medicinal chemistry optimization of high-throughput screening (HTS) hits for orally active antimalarials. Part 1: challenges in potency and absorption, distribution, metabolism, excretion/pharmacokinetics (ADME/PK). *J. Med. Chem.* **56**(20), 7741–7749.

[4] Decornez, H. Y., Duffy, B. C., Kitchen, D. B. (2011) Design and analysis of HTS libraries. SP2, 38–39.

[5] Bakken, G. A., Bell, A. S., Boehm, M., Everett, J. R., Gonzales, R., Hepworth, D., Klug-McLeod, J. L., Lanfear, J., Loesel, J., Mathias, J., Wood, T. P. (2012) Shaping a screening file for maximal lead discovery efficiency and effectiveness: elimination of molecular redundancy. *J. Chem. Inf. Model.* **52**, 2937–2949.

[6] Brown, N., Zehender, H., Azzaoui, K., Schuffenhauer, A., Mayr, L. M., Jacoby, E. (2006) A chemoinformatics analysis of hit lists obtained from high-throughput affinity-selection screening. *J. Biomol. Screen.* **11**, 123–130.

[7] Duffy, B. C., Zhu, L., Decornez, H., Kitchen, D. B. (2012) Early phase drug discovery: cheminformatics and computational techniques in identifying lead series. *Bioorg. Med. Chem.* **20**, 5324–5342.

[8] Davis, A. M., Keeling, D. J., Steele, J., Tomkinson, N. P., Tinker, A. C. (2005) Components of successful lead generation. *Curr. Top. Med. Chem.* **5**, 421–439.

[9] Keseru, G. M., Makara, G. M. (2006) Hit discovery and hit-to-lead approaches. *Drug Discov. Today* **11**, 741–748.

[10] Freeman-Cook, K. D., Kung, D. W. (2009) Hit triage: medicinal chemistry strategies. *Top. Med. Chem.* **1**, 141–174.

[11] Popa-Burke, I., Novick, S., Mantilla, M., McGrath, K. (2013) Analysis of compound weighing precision in drug discovery. *J. Biomol. Screen.* **18**, 331–340.

[12] Popa-Burke, I. G., Issakova, O., Arroway, J. D., Bernasconi, P., Chen, M., Coudurier, L., Galasinski, S., Jadhav, A. P., Janzen, W. P., Lagasca, D., Liu, D., Lewis, R. S., Mohney, R. P., Sepetov, N., Sparkman, D. A., Hodge, C. N. (2004) Streamlined system for purifying and quantifying a diverse library of compounds and the effect of compound concentration measurements on the accurate interpretation of biological assay results. *Anal. Chem.* **76**, 7278–7287.

[13] Wu, Z., Liu, D., Sui, Y. (2008) Quantitative assessment of hit detection and confirmation in single and duplicate high-throughput screenings. *J. Biomol. Screen.* **13**, 159–167.

[14] Gubler, H., Schopfer, U., Jacoby, E. (2013) Theoretical and experimental relationships between percent inhibition and IC50 data observed in high-throughput screening. *J. Biomol. Screen.* **18**, 1–13.

[15] McFayden, I., Walker, G., Alvarez, J. (2004) Enhancing hit quality and diversity within assay throughput constraints. In Oprea, T. I., ed. *Cheminformatics in Drug Discovery*. Wiley-VCH: Weinheim, 143–173.

[16] Shoichet, B. K. (2006) Interpreting steep dose-response curves in early inhibitor discovery. *J. Med. Chem.* **49**, 7274–7277.

[17] Feng, B. Y., Shoichet, B. K. (2006) Synergy and antagonism of promiscuous inhibition in multiple-compound mixtures. *J. Med. Chem.* **49**, 2151–2154.

[18] Feng, B. Y., Shelat, A., Doman, T. N., Guy, R. K., Shoichet, B. K. (2005) High-throughput assays for promiscuous inhibitors. *Nat. Chem. Biol.* **1**, 146–148.

[19] Habig, M., Blechschmidt, A., Dressler, S., Hess, B., Patel, V., Billich, A., Ostermeier, C., Beer, D., Klumpp, M. (2009) Efficient elimination of nonstoichiometric enzyme inhibitors from HTS hit lists. *J. Biomol. Screen.* **14**(6), 679–689.

[20] Ryan, A. J., Gray, N. M., Lowe, P. N., Chung, C. W. (2003) Effect of detergent on "promiscuous" inhibitors. *J. Med. Chem.* **46**, 3448–3451.

[21] Che, J., King, F. J., Zhou, B., Zhou, Y. (2012) Chemical and biological properties of frequent screening hits. *J. Chem. Inf. Model.* **52**, 913–926.

[22] Ringel, M., Tollman, P., Hersch, G., Schulze, U. (2013) Does size matter in R&D productivity? If not, what does? *Nat. Rev. Drug Discov.* **12**, 901–902.

[23] Lajiness, M. S., Maggiora, G. M., Shanmugasundaram, V. (2004) Assessment of the consistency of medicinal chemists in reviewing sets of compounds. *J. Med. Chem.* **47**, 4891–4896.

[24] Cumming, J. G., Davis, A. M., Muresan, S., Haeberlein, M., Chen, H. (2013) Chemical predictive modelling to improve compound quality. *Nat. Rev. Drug Discov.* **12**, 948–962.

[25] Johnson, M. A., Maggiora, G. M. (1990) *Concepts and Applications of Molecular Similarity*, 1st edn. John Wiley & Sons: New York.

[26] Martin, Y. C., Kofron, J. L., Traphagen, L. M. (2002) Do structurally similar molecules have similar biological activity? *J. Med. Chem.* **45**, 4350–4358.

[27] Maggiora, G., Vogt, M., Stumpfe, D., Bajorath, J. (2014) Molecular similarity in medicinal chemistry. *J. Med. Chem.* **57**(8), 3186–3204.

[28] Yan, S. F., Asatryan, H., Li, J., Zhou, Y. (2005) Novel statistical approach for primary high-throughput screening hit selection. *J. Chem. Inf. Model.* **45**, 1784–1790.

[29] Engels, M. F., Wouters, L., Verbeeck, R., Vanhoof, G. (2002) Outlier mining in high throughput screening experiments. *J. Biomol. Screen.* **7**, 341–351.

[30] Willett, P., Barnard, J. M., Downs, G. M. (1998) Chemical similarity searching. *J. Chem. Inf. Comput. Sci.* **38**(6), 983–996.

[31] Xu, J. (2002) A new approach to finding natural chemical structure classes. *J. Med. Chem.* **45**, 5311–5320.

[32] Lewell, X. Q., Jones, A. C., Bruce, C. L., Harper, G., Jones, M. M., McLay, I. M., Bradshaw, J. (2003) Drug rings database with web interface. A tool for identifying alternative chemical rings in lead discovery programs. *J. Med. Chem.* **46**, 3257–3274.

[33] Nilakantan, R., Nunn, D. S., Greenblatt, L., Walker, G., Haraki, K., Mobilio, D. (2006) A family of ring system-based structural fragments for use in structure-activity studies: database mining and recursive partitioning. *J. Chem. Inf. Model.* **46**, 1069–1077.

[34] Lewell, X. Q., Judd, D. B., Watson, S. P., Hann, M. M. (1998) RECAP—retrosynthetic combinatorial analysis procedure: a powerful new technique for identifying privileged molecular fragments with useful applications in combinatorial chemistry. *J. Chem. Inf. Comput. Sci.* **38**, 511–522.

[35] Varin, T., Gubler, H., Parker, C. N., Zhang, J. H., Raman, P., Ertl, P., Schuffenhauer, A. (2010) Compound set enrichment: a novel approach to analysis of primary HTS data. *J. Chem. Inf. Model.* **50**, 2067–2078.

[36] Wilkens, S. J., Janes, J., Su, A. I. (2005) HierS: hierarchical scaffold clustering using topological chemical graphs. *J. Med. Chem.* **48**, 3182–3193.

[37] Varin, T., Schuffenhauer, A., Ertl, P., Renner, S. (2011) Mining for bioactive scaffolds with scaffold networks: improved compound set enrichment from primary screening data. *J. Chem. Inf. Model.* **51**, 1528–1538.

[38] Bemis, G. W., Murcko, M. A. (1996) The properties of known drugs. 1. Molecular frameworks. *J. Med. Chem.* **39**, 2887–2893.

[39] Bemis, G. W., Murcko, M. A. (1999) Properties of known drugs. 2. Side chains. *J. Med. Chem.* **42**, 5095–5099.

[40] Birchall, K., Gillet, V. J., Harper, G., Pickett, S. D. (2008) Evolving interpretable structure-activity relationship models. 2. Using multiobjective optimization to derive multiple models. *J. Chem. Inf. Model.* **48**, 1558–1570.

[41] Stahl, M., Mauser, H. (2005) Database clustering with a combination of fingerprint and maximum common substructure methods. *J. Chem. Inf. Comput. Sci.* **45**(3), 542–548.

[42] Posner, B. A., Xi, H., Mills, J. E. (2009) Enhanced HTS hit selection via a local hit rate analysis. *J. Chem. Inf. Model.* **49**, 2202–2210.

[43] Schuffenhauer, A., Brown, N., Ertl, P., Jenkins, J. L., Selzer, P., Hamon, J. (2007) Clustering and rule-based classifications of chemical structures evaluated in the biological activity space. *J. Chem. Inf. Model.* **47**, 325–336.

[44] Schuffenhauer, A., Ertl, P., Roggo, S., Wetzel, S., Koch, M. A., Waldmann, H. (2007) The scaffold tree—visualization of the scaffold universe by hierarchical scaffold classification. *J. Chem. Inf. Model.* **47**(1), 47–58.

[45] Pu, M., Hayashi, T., Cottam, H., Mulvaney, J., Arkin, M., Corr, M., Carson, D., Messer, K. (2012) Analysis of high-throughput screening assays using cluster enrichment. *Stat. Med.* **31**, 4175–4189.

[46] Glick, M., Jenkins, J. L., Nettles, J. H., Hitchings, H., Davies, J. W. (2006) Enrichment of high-throughput screening data with increasing levels of noise using support vector machines, recursive partitioning, and laplacian-modified naive bayesian classifiers. *J. Chem. Inf. Model.* **46**, 193–200.

[47] Lipinski, C. A., Lombardo, F., Dominy, B. W., Feeney, P. J. (1997) Experimental and computational approaches to estimate solubility and permeability in drug discovery and development settings. *Adv. Drug Deliv. Rev.* **23**, 3–25.

[48] Oprea, T. (2002) Current trends in lead discovery: Are we looking for the appropriate properties? *J. Comput. Aided Mol. Des.* **16**, 325–334.

[49] Bickerton, G. R., Paolini, G. V., Besnard, J., Muresan, S., Hopkins, A. L. (2012) Quantifying the chemical beauty of drugs. *Nat. Chem.* **4**, 90–98.

[50] Cox, P. B., Vasudevan, A. (2012) Generating a high-quality compound collection. In Wigglesworth, M., Wood, T., eds. *Management of Chemical and Biological Samples for Screening Applications*, 1st edn. Wiley-VCH: Weinheim, 9–22.

[51] Veber, D. F., Johnson, S. R., Cheng, H. Y., Smith, B. R., Ward, K. W., Kopple, K. D. (2002) Molecular properties that influence the oral bioavailability of drug candidates. *J. Med. Chem.* **45**, 2615–2623.

[52] Vistoli, G., Pedretti, A., Testa, B. (2008) Assessing drug-likeness-what are we missing? *Drug Discov. Today* **13**, 285–294.

[53] Shultz, M. D. (2013) Setting expectations in molecular optimizations: strengths and limitations of commonly used composite parameters. *Bioorg. Med. Chem. Lett.* **23**(21), 5980–5991.

[54] Kenny, P. W., Montanari, C. A. (2013) Inflation of correlation in the pursuit of drug-likeness. *J. Comput. Aided Mol. Des.* **27**, 1–13.

[55] Muchmore, S. W., Edmunds, J. J., Stewart, K. D., Hajduk, P. J. (2010) Cheminformatic tools for medicinal chemists. *J. Med. Chem.* **53**, 4830–4841.

[56] Lipinski, C. A. (2000) Drug-like properties and the causes of poor solubility and poor permeability. *J. Pharmacol. Toxicol. Methods* **44**, 235–249.

[57] Oprea, T. I., Davis, A. M., Teague, S. J., Leeson, P. D. (2001) Is there a difference between leads and drugs? A historical perspective. *J. Chem. Inf. Comput. Sci.* **41**, 1308–1315.

[58] Congreve, M., Carr, R., Murray, C., Jhoti, H. (2003) A "rule of three" for fragment-based lead discovery? *Drug Discov. Today* **8**, 876–877.

[59] Gleeson, M. P. (2008) Generation of a set of simple, interpretable ADMET rules of thumb. *J. Med. Chem.* **51**, 817–834.

[60] Johnson, T. W., Dress, K. R., Edwards, M. (2009) Using the Golden Triangle to optimize clearance and oral absorption. *Bioorg. Med. Chem. Lett.* **19**, 5560–5564.

[61] Hughes, J. D., Blagg, J., Price, D. A., Bailey, S., Decrescenzo, G. A., Devraj, R. V., Ellsworth, E., Fobian, Y. M., Gibbs, M. E., Gilles, R. W., Greene, N., Huang, E., Krieger-Burke, T., Loesel, J., Wager, T., Whiteley, L., Zhang, Y. (2008) Physicochemical drug properties associated with in vivo toxicological outcomes. *Bioorg. Med. Chem. Lett.* **18**, 4872–4875.

[62] Waring, M. J. (2010) Lipophilicity in drug discovery. *Expert Opin. Drug Discov.* **5**, 235–248.

[63] Waring, M. J. (2009) Defining optimum lipophilicity and molecular weight ranges for drug candidates-molecular weight dependent lower logD limits based on permeability. *Bioorg. Med. Chem. Lett.* **19**, 2844–2851.

[64] Sutherland, J. J., Raymond, J. W., Stevens, J. L., Baker, T. K., Watson, D. E. (2012) Relating molecular properties and in vitro assay results to in vivo drug disposition and toxicity outcomes. *J. Med. Chem.* **55**, 6455–6466.

[65] Pope, A. (2012) Screening Heursitcs and Chemical Propery Bias: New Directions for Lead Identification and Optimization. Available at: http://www.slideshare.net/andypopeuk/screening-heuristics-popefinal (accessed May 14, 2015).

[66] Hann, M. M., Keseru, G. M. (2012) Finding the sweet spot: the role of nature and nurture in medicinal chemistry. *Nat. Rev. Drug Discov.* **11**, 355–365.

[67] Meanwell, N. A. (2011) Improving drug candidates by design: a focus on physicochemical properties as a means of improving compound disposition and safety. *Chem. Res. Toxicol.* **24**, 1420–1456.

[68] Hill, A. P., Young, R. J. (2010) Getting physical in drug discovery: a contemporary perspective on solubility and hydrophobicity. *Drug Discov. Today* **15**(15/16), 648–655.

[69] Young, R. J., Green, D. V., Luscombe, C. N., Hill, A. P. (2011) Getting physical in drug discovery II: the impact of chromatographic hydrophobicity measurements and aromaticity. *Drug Discov. Today* **16**(17–18), 822–830.

[70] Lovering, F., Bikker, J., Humblet, C. (2009) Escape from flatland: increasing saturation as an approach to improving clinical success. *J. Med. Chem.* **52**, 6752–6756.

[71] Leeson, P. D., St-Gallay, S. A., Wenlock, M. C. (2011) Impact of ion class and time on oral drug molecular properties. *Med. Chem. Commun.* **2**, 91–105.

[72] Ritchie, T. J., Macdonald, S. J., Young, R. J., Pickett, S. D. (2011) The impact of aromatic ring count on compound developability: further insights by examining carbo- and hetero-aromatic and -aliphatic ring types. *Drug Discov. Today* **16**(3–4), 164–171.

[73] Hann, M. M., Leach, A. R., Harper, G. (2001) Molecular complexity and its impact on the probability of finding leads for drug discovery. *J. Chem. Inf. Comput. Sci.* **41**, 856–864.

[74] Sauer, W. H., Schwarz, M. K. (2003) Molecular shape diversity of combinatorial libraries: a prerequisite for broad bioactivity. *J. Chem. Inf. Comput. Sci.* **43**, 987–1003.

[75] Wager, T. T., Hou, X., Verhoest, P. R., Villalobos, A. (2010) Moving beyond rules: the development of a central nervous system multiparameter optimization (CNS MPO) approach to enable alignment of druglike properties. *ACS Chem. Neurosci.* **1**, 435–449.

[76] Rishton, G. M. (1997) Reactive compounds and in vitro false positives in HTS. *Drug Discov. Today* **2**, 382–384.

[77] Charifson, P. S., Walters, W. P. (2002) Filtering databases and chemical libraries. *J. Comput. Aided Mol. Des.* **16**, 311–323.

[78] Walters, W. P., Murcko, M. A. (2002) Prediction of "drug-likeness". *Adv. Drug Deliv. Rev.* **54**, 255–271.

[79] Brenk, R., Schipani, A., James, D., Krasowski, A., Gilbert, I. H., Frearson, J., Wyatt, P. G. (2008) Lessons learnt from assembling screening libraries for drug discovery for neglected diseases. *ChemMedChem* **3**, 435–444.

[80] Metz, J. T., Huth, J. R., Hajduk, P. J. (2007) Enhancement of chemical rules for predicting compound reactivity towards protein thiol groups. *J. Comput. Aided Mol. Des.* **21**, 139–144.

[81] Baell, J. B., Holloway, G. A. (2010) New substructure filters for removal of pan assay interference compounds (PAINS) from screening libraries and for their exclusion in bioassays. *J. Med. Chem.* **53**, 2719–2740.

[82] Mok, N. Y., Maxe, S., Brenk, R. (2013) Locating sweet spots for screening hits and evaluating pan-assay interference filters from the performance analysis of two lead-like libraries. *J. Chem. Inf. Model.* **53**(3), 534–544.

[83] Bruns, R. F., Watson, I. A. (2012) Rules for identifying potentially reactive or promiscuous compounds. *J. Med. Chem.* **55**, 9763–9772.

[84] Pearce, B. C., Sofia, M. J., Good, A. C., Drexler, D. M., Stock, D. A. (2006) An empirical process for the design of high-throughput screening deck filters. *J. Chem. Inf. Model.* **46**, 1060–1068.

[85] Annadurai, S., Martinez, R., Canney, D. J., Eidem, T., Dunman, P. M., Abou-Gharbia, M. (2012) Design and synthesis of 2-aminothiazole based antimicrobials targeting MRSA. *Bioorg. Med. Chem. Lett.* **22**(24), 7719–7725.

[86] Law, V., Knox, C., Djoumbou, Y., Jewison, T., Guo, A. C., Liu, Y., Maciejewski, A., Arndt, D., Wilson, M., Neveu, V., Tang, A., Gabriel, G., Ly, C., Adamjee, S., Dame, Z. T., Han, B., Zhou, Y., Wishart, D. S. (2014) DrugBank 4.0: shedding new light on drug metabolism. *Nucleic Acids Res.* **42**(D1), D1091–D1097.

[87] Patch, R. J., Searle, L. L., Kim, A. J., De, D., Zhu, X., Askari, H. B., O'Neill, J. C., Abad, M. C., Rentzeperis, D., Liu, J., Kemmerer, M., Lin, L., Kasturi, J., Geisler, J. G., Lenhard, J. M., Player, M. R., Gaul, M. D. (2011) Identification of diaryl ether-based ligands for estrogen-related receptor alpha as potential antidiabetic agents. *J. Med. Chem.* **54**, 788–808.

[88] Zhang, C., Srinivasan, Y., Arlow, D. H., Fung, J. J., Palmer, D., Zheng, Y., Green, H. F., Pandey, A., Dror, R. O., Shaw, D. E., Weis, W. I., Coughlin, S. R., Kobilka, B. K. (2012) High-resolution crystal structure of human protease-activated receptor 1. *Nature* **492**, 387–392.

[89] Cox, P. B., Gregg, R. J., Vasudevan, A. (2012) Abbott Physicochemical Tiering (APT)—a unified approach to HTS triage. *Bioorg. Med. Chem.* **20**, 4564–4573.

[90] Kuntz, I. D., Chen, K., Sharp, K. A., Kollman, P. A. (1999) The maximal affinity of ligands. *Proc. Natl. Acad. Sci. U. S. A.* **96**, 9997–10002.

[91] Andrews, P. R., Craik, D. J., Martin, J. L. (1984) Functional group contributions to drug-receptor interactions. *J. Med. Chem.* **27**, 1648–1657.

[92] Hopkins, A. L., Groom, C. R., Alex, A. (2004) Ligand efficiency: a useful metric for lead selection. *Drug Discov. Today* **9**, 430–431.

[93] Shultz, M. D. (2013) The thermodynamic basis for the use of lipophilic efficiency (LipE) in enthalpic optimizations. *Bioorg. Med. Chem. Lett.* **23**(21), 5992–6000.

[94] Varin, T., Didiot, M. C., Parker, C. N., Schuffenhauer, A. (2012) Latent hit series hidden in high-throughput screening data. *J. Med. Chem.* **55**(3), 1161–1170.

[95] Mestres, J., Veeneman, G. H. (2003) Identification of "latent hits" in compound screening collections. *J. Med. Chem.* **46**(16), 3441–3444.

[96] Wang, X. S., Salloum, G. A., Chipman, H. A., Welch, W. J., Young, S. S. (2007) Exploration of cluster structure-activity relationship analysis in efficient high-throughput screening. *J. Chem. Inf. Model.* **47**, 1206–1214.

[97] Murray, C. W., Verdonk, M. L., Rees, D. C. (2012) Experiences in fragment-based drug discovery. *Trends Pharmacol. Sci.* **33**, 224–232.

[98] Saxty, G., Woodhead, S. J., Berdini, V., Davies, T. G., Verdonk, M. L., Wyatt, P. G., Boyle, R. G., Barford, D., Downham, R., Garrett, M. D., Carr, R. A. (2007) Identification of inhibitors of protein kinase B using fragment-based lead discovery. *J. Med. Chem.* **50**, 2293–2296.

[99] Guha, R., Van Drie, J. H. (2008) Structure–activity landscape index: identifying and quantifying activity cliffs. *J. Chem. Inf. Model.* **48**, 646–658.

[100] Kolpak, J., Connolly, P. J., Lobanov, V. S., Agrafiotis, D. K. (2009) Enhanced SAR maps: expanding the data rendering capabilities of a popular medicinal chemistry tool. *J. Chem. Inf. Model.* **49**(10), 2221–2230.

[101] Peltason, L., Bajorath, J. (2007) SAR index: quantifying the nature of structure-activity relationships. *J. Med. Chem.* **50**, 5571–5578.

[102] Birchall, K., Gillet, V. J., Harper, G., Pickett, S. D. (2008) Evolving interpretable structure-activity relationships. 1. Reduced graph queries. *J. Chem. Inf. Model.* **48**, 1543–1557.

[103] Dossetter, A. G., Griffen, E. J., Leach, A. G. (2013) Matched molecular pair analysis in drug discovery. *Drug Discov. Today* **18**(15–16), 724–731.

[104] Leach, A. G., Jones, H. D., Cosgrove, D. A., Kenny, P. W., Ruston, L., MacFaul, P., Wood, J. M., Colclough, N., Law, B. (2006) Matched molecular pairs as a guide in the optimization of pharmaceutical properties; a study of aqueous solubility, plasma protein binding and oral exposure. *J. Med. Chem.* **49**(23), 6672–6682.

[105] Heyse, S. (2002) Comprehensive analysis of high-throughput screening data. *Proc. SPIE* **4626**, 535–547.

[106] Maniyar, D. M., Nabney, I. T., Williams, B. S., Sewing, A. (2006) Data visualization during the early stages of drug discovery. *J. Chem. Inf. Model.* **46**, 1806–1818.

# 8

# FRAGMENT-BASED DRUG DISCOVERY

JEAN-PAUL RENAUD, THOMAS NEUMANN, and LUC VAN HIJFTE
*NovAliX, BioParc, Illkirch Cedex, France*

## 8.1 INTRODUCTION

In the last decades, output of new marketed drugs in the pharmaceutical industry has been declining, while costs have increased enormously. The industry is since seeking new approaches in order to increase the probability of success within the drug discovery process [1]. The 1990s were the era of combinatorial chemistry, focusing on numbers, assuming that if one needs 10,000 novel substances for one marketed drug, one only needed to multiply the "numbers" at the start to increase the output. As we have realized today, this has been a completely false assumption, as focus went toward numbers, with less creative chemistry, while the quality (purity, information content) was significantly reduced. Medicinal chemists had stepped away from the rational design–make–test paradigm, where in a by far more rational, structure-based approach, compounds are optimized in a multifactorial way, to introduce not only the right activity on the target but also ensure that the compounds have the right properties in order to have a good ADME and toxicological behavior.

Realizing that these properties are often hurdles in the drug discovery process, people have engaged in the exploration of novel strategies to improve the drug-likeness of molecules. In the late 1990s, these efforts led to the emergence of fragment-based drug discovery, in which the starting points for the drug discovery projects are chemical entities with molecular masses ranging from 150 to 250/300 Da or alternatively up to 18 heavy atoms [2–12]. Although a strategy was already clearly established in which the chemists were looking for the "minimum pharmacophore" within HTS hits [13], this approach was initially hampered by the lack of sensitive screening methods. The initial hits typically display activities in the micromolar to millimolar range, so sensitive detection methods

*Small Molecule Medicinal Chemistry: Strategies and Technologies*, First Edition.
Edited by Werngard Czechtizky and Peter Hamley.

are required to identify these relatively weak interactions. The first breakthrough was reported by Shuker et al., with a pioneering work on SAR by NMR [14]. Since then, a rapid development occurred in the biophysical arena, with novel "molecular" screening paradigms based on X-ray crystallography, nuclear magnetic resonance (NMR), noncovalent mass spectrometry, surface plasmon resonance (SPR), isothermal calorimetry (ITC), and differential scanning fluorimetry (DSF) (to name the more established ones).

Since its inception, fragment-based drug discovery has clearly demonstrated its value as an efficient drug discovery process, as numerous compounds that were developed with this strategy have proceeded to clinical development (See Table 8.1) [12, 15, see http://practicalfragments.blogspot.fr/2013/01/fragments-in-clinic-2013-edition.html].

A number of the presented targets were considered as intractable, such as BACE-1 (β-secretase 1) and BCL-2/BCL-XL, where all attempts using classical approaches had given disappointing results. Applying the rational FBDD approach, scientists were able to identify low-molecular-mass and low-affinity but still efficient binders that were developed into druggable lead compounds with the required properties to ensure a clinical development. The nicest example that can be quoted today is the discovery and development of vemurafenib (Plexxikon), today a marketed drug. The whole process was accomplished in a record time of 6 years, from initiation of the project to market approval of the compound in 2011 for metastatic melanoma.

What makes fragment-based lead discovery attractive and successful? First, one observes that chemical space is better sampled with small fragments than with larger, lead-like molecules. The number of plausible chemical structures with MM < 250 Da has been estimated in the $10^{14}$ range, whereas the number is estimated in the $10^{60}$ range for compounds with MM < 500 Da. It was calculated that the number of structures increases by a factor of 8,3 per heavy atom [16]. A collection of 1,000 fragments therefore covers much more chemical space than 1,000,000 lead-like motifs.

Secondly, fragment binding, though weaker in affinity, often shows a higher ligand efficiency and can be considered as more universal, matching more likely a given hot spot

**TABLE 8.1 Selected Clinical-Stage Beyond, Phase 2 and Beyond, Originating from FBDD**

| | | |
|---|---|---|
| Approved | | |
| Vemurafenib (PLX4032) | Plexxikon | B-Raf(V600E) inhibitor |
| Phase 2/3 | | |
| MK-8931 | Merck | BACE1 inhibitor |
| Phase 2 | | |
| AT13387 | Astex | HSP90 inhibitor |
| AT7519 | Astex | CDK1,2,4,5 inhibitor |
| AT9283 | Astex | Aurora, Janus kinase 2 inhibitor |
| AUY922 | Vernalis/Novartis | HSP90 inhibitor |
| Indeglitazar | Plexxikon | pan-PPAR agonist |
| Linifanib (ABT869) | Abbott | VEGF and PDGFR inhibitor |
| LY2886721 | Lilly | BACE1 inhibitor |
| LY517717 | Lilly/Protherics | FXa inhibitor |
| Navitoclax (ABT 263) | Abbott | Bcl-2/Bcl-xL inhibitor |
| PLX3397 | Plexxikon | FMS, KIT, and FLT-3-ITD inhibitor |

Reproduced with permission from Erlanson and Zartler [12]

in some binding site, whereas mismatches exponentially increase with the complexity of a molecule. Today, it is clear that identifying fragment hits, based on current biophysical assays, is no longer the problem. It is rather the evolution of these fragments toward highly drug-like lead compounds.

Another big advantage of the fragment-based lead generation approach is the opportunity to gradually build in drug-like properties, with a main emphasis on molecular mass ("begin small, stay small") and cLogP. It also provides more freedom to access novel chemical space, thereby creating novel IP. Also, as scientists work toward high ligand efficiency for a given protein-binding site, fragment-derived lead structures have a tendency for higher selectivity. Finally, the whole process is highly structure driven, which often reduces the number of compounds needed to advance the program toward high-quality leads, making the overall process more efficient.

## 8.2 FRAGMENT LIBRARIES

It is obvious to scientists that the quality of a given fragment-derived lead will depend on the quality of the fragment hit. Therefore, many pharmaceutical companies have undertaken considerable efforts to establish high-quality fragment libraries [12, 17, 18]. As fragment libraries are by far smaller (from some hundreds up to some thousands of molecules) than classical screening collections, evolving the library toward more privileged structures is by far easier to accomplish. Astex has reported to work with a sixth-generation library that has been optimized in order to further improve the probability of success of screening campaigns [15]. In analogy with the well-known Lipinski's rule of 5, scientists have come up with guidelines for privileged fragment space, which most people follow with respect to molecular mass and lipophilicity. There is a trend to move to even smaller, more hydrophilic fragments, as justified by a study presented at FBLD 2012 from Astex showing an increase of hit rates. Other groups have started to pay attention to the 3D properties of the molecules, aiming to escape the famous "flat land" partially established in the combinatorial chemistry era. It has been reported though that hit rates tend to decrease with higher three-dimensionality in the fragments, but they would be preferential for PPIs [12, see discussion of October 8th]. Finally, some groups have introduced "natural product-derived" fragments [19] to access what they call a more biologically relevant space. It is still unclear today which shapes will finally provide the best starting points for drug discovery purposes. It is our understanding that a good mixture of all of the previously mentioned fragment types within a reasonably sized library will provide tractable chemical starting points for most targets.

## 8.3 BIOPHYSICAL SCREENING TECHNOLOGIES

The combination of binding fragments to yield a high-affinity ligand was in essence put forward in 1981 by Jencks et al. using the theoretical consideration of additivity of binding energies [20]. In order to measure interactions at the higher micromolar up to millimolar levels, it was necessary to develop biophysical methods able to detect weak binding, as very often the biochemical readout at such low affinities remains impossible. It took some time to put this into practice; the seminal publication of Shuker et al. in 1996 was a landmark in the

FBDD field since it was the first report of high-affinity ligands built from fragments using a "SAR by NMR" approach [14]. In this study, NMR was the method of choice to identify fragment hits, but since then, numerous biophysical techniques have been developed, adapted, or invented to detect fragment binding. A major challenge in the field was the development of screening techniques combining high-throughput, low-protein consumption and low false positive/negative rates. The second challenge faced was the development of high-content techniques yielding detailed information on the target/fragment interaction. In this respect, all techniques have their respective advantages and limitations (Table 8.2).

Based on the inverse correlation between information content and throughput, one has to select the best compromise for a given target in a certain context (screening vs. follow-up). The current trend consists in choosing a robust primary technique adapted to the chosen target and to complement it by an orthogonal, secondary technique allowing to remove false positives as well as bringing information on the binding mode. In our hands, this setup leads to high success rates in the subsequent crystallographic campaigns where only validated hits are soaked/cocrystallized with the target of interest. This final step not only definitely prove the binding of the selected fragments, but also yields the structural basis for the hit-to-lead optimization process.

According to the results of a poll with approximately 100 respondents, the most popular techniques applied by practitioners in 2013 were SPR, ligand-observed NMR, thermal shift, X-ray crystallography, and functional screening [21], four out of five being biophysical techniques. A survey of the recent literature indicates that the most popular primary screening techniques nowadays are ligand-observed NMR and SPR.

In their seminal study, Shuker et al. first used protein-observed NMR as the primary screening technique [14]. Direct crystallographic screening then became popular in spite of the low success rate-to-cost ratio because the outcome for positive hits directly indicates the binding mode. Nowadays, a widespread approach consists in solving the crystal structure of target–fragment complexes only for fragment hits cross validated by at least two orthogonal techniques [22].

The following sections describe a selection of established biophysical techniques including SPR, NMR, X-ray crystallography, DSF, and MS, with an emphasis on chemical microarray SPR and noncovalent mass spectrometry, which are extensively used at NovAliX. Although we do use ITC, its long measurement times and high-protein consumption limit its applicability as a primary screening technique, even with an automated apparatus. In general, ITC is rather used to validate strong fragment hits and for thermodynamic profiling of more advanced compounds.

### 8.3.1 Surface Plasmon Resonance (SPR)

Among the optical biophysical techniques used for drug discovery and fragment-based discovery in particular, SPR technology has gained widespread popularity among scientists [23]. SPR is known to be a versatile technique for studying biomolecular interactions in a label-free manner and has distinct advantages over other optical assays. The surface-sensitive technology detection principle avoids artifacts and interferences known from, for example, fluorescence-based assays and does not require further reporter assays or labels. While, typically, one binding partner is immobilized on a sensor surface, the binding of a soluble analyte induces a change in refractive index in close vicinity to the noble metal sensor surface (Fig. 8.1).

**TABLE 8.2 Comparison between Different Biophysical Screening Technologies**

| Screening Technique | Pros | Cons | Specific Binding Assessment | Use with Membrane Proteins |
|---|---|---|---|---|
| *SPR (channel based) (see also Table* 8.3*)* | Access to $k_{on}$, $k_{off}$, $K_d$; relatively high throughput | Potential artifacts due to protein immobilization | Y | Y but technically delicate |
| *SPR (chemical microarray) (see also Table* 8.3*)* | Ultrahigh throughput (12,000 interactions in triplicate/day/instrument) | No information on the binding site except from competition experiments | Y | Not yet adapted to membrane proteins |
| *X-ray crystallography* | Atomic-scale details on the binding mode; no false positives | Needs reproducible well-diffracting protein crystals; possible false negatives due to crystal packing | Y | Y but technically delicate |
| *NMR (ligand-observed)* | Relatively high throughput; no limitation on protein size or nature | High-affinity ligands not detected | N | Y |
| *NMR (protein-observed)* | Identifies the residues involved in the binding; few false positives and false negatives | Needs relatively high amounts of soluble, labeled protein; protein size practically limited to ~30–40 kDa | Y | N |
| *DSF (TSA)* | Easy setup; high throughput; low-protein consumption | Not applicable to all targets; risks of high false positive and false negative rates | N | Y but technically delicate; needs fluorescent dyes |
| *Noncovalent MS* | Relatively high throughput; low-protein consumption; high information content | Unspecific binding limits the detection of weak binders | Y | Not yet adapted to membrane proteins |

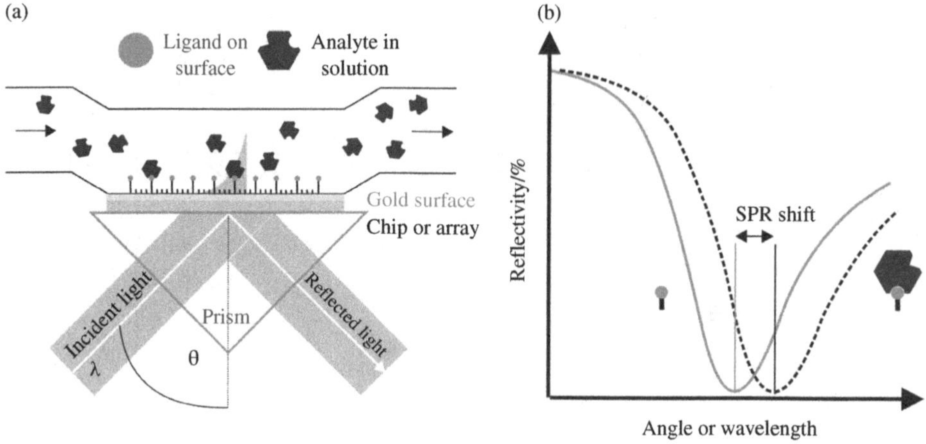


**FIGURE 8.1** (a) Schematic of SPR systems comprising an optical readout system, the coupling interface (grating or prism coupling), and the sensor chip or array, including the surface chemistry, an immobilized binding partner, and flow channel. (b) The angle- or wavelength-dependent position of the SPR minimum shifts due to binding of the analyte to the immobilized ligand. (*See insert for color representation of the figure.*)

Such changes alter the resonance conditions for surface plasmons (SPs) and can be read out optically in a surface-sensitive manner in real time. SPR instruments enable researchers to perform detailed studies of biomolecular interactions, such as protein–protein (and in particular protein–antibody), protein–DNA, and of course protein–small molecule interactions as starting points for drug discovery.

A wealth of published case studies can be found in the literature reflecting the broad range of applications of SPR and its impact on studying biomolecular interactions. More than a thousand publications on studies using commercial biosensor instrumentations are published per year. This flood of reports has been reviewed by David Myszka [24] on a regular basis, clearly indicating SPR technology as a well-established method in the researcher's biophysical toolbox.

While biosensor data seem to be generated quickly and apparently with ease, the recording of meaningful, artifact-free, and reliable SPR data requires experience and well-defined experimental conditions. Nevertheless, the label-free, biophysical analysis of the interaction between immobilized and soluble binding partners delivers a high information content, which can be exploited in all phases throughout the (fragment-based) drug discovery workflow (screening, hit confirmation, hit-to-lead optimization). Depending on the key features of the different formats (regular SPR spectroscopy, imaging, arrays) such as sensitivity, limit of detection, robustness of the assay, as well as the throughput, SPR technology can contribute to all phases mentioned previously.

The high information content provided by SPR tool ranges from:

- Rather simple yes/no binding assays for thousands of fragment library compounds in the primary screening phase
- Secondary screening/hit characterization, determination of affinity ($K_D$), specificity, and stoichiometry for typically hundreds of advanced/confirmed hits
- Kinetic measurements ($k_{on}$ and $k_{off}$ rates) for a limited number of lead compounds in later stages of the development process

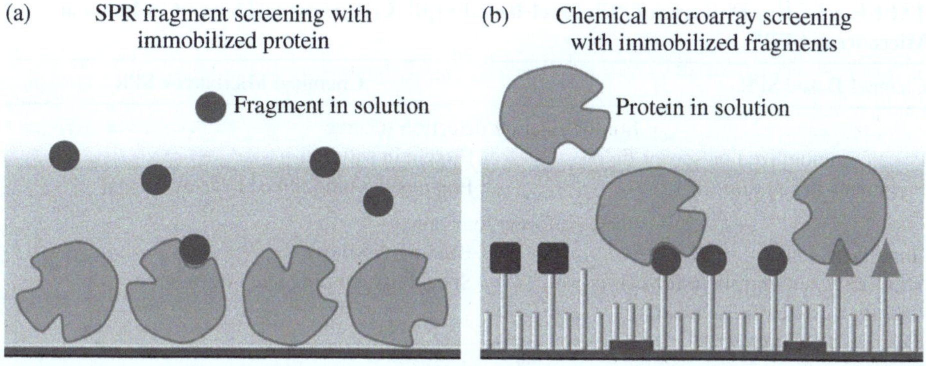


**FIGURE 8.2** Comparison between two major direct assay principles applied to SPR fragment screening: (a) channel-based SPR, immobilization of the protein target to the SPR sensor chip surface and detection of soluble fragments; (b) CM-SPR, screening of proteins against fragment collections immobilized on chemical microarrays. (*See insert for color representation of the figure.*)

Two direct-binding assay schemes have been developed, differing in the choice of the binding partner, namely, channel-based SPR, with the protein immobilized, or chemical microarray SPR (CM-SPR), with the small molecule/fragment immobilized on the sensor (Fig. 8.2).

Both approaches have their strengths and weaknesses, provide different information content, and can be compared in light of using SPR as a tool for primary screening or secondary hit characterization (Table 8.3).

***8.3.1.1 Channel-Based SPR Spectroscopy*** In the case of channel-based SPR systems with a limited number of individual sensor fields, the protein target of interest is attached to a chip surface in order to monitor its binding to low-molecular-mass analytes. This approach can be applied both to the generation of hits in a medium-throughput mode and for the more detailed characterization of a binding phenomenon.

For screening purposes, a number of library compounds (up to a few thousands) are analyzed at a limited number of concentrations or at single concentration only. Report points can be defined after a defined injection time and are helpful for quickly defining hit thresholds and delivering binding (yes/no) answers.

The typical result of a binding study, a sensorgram, is shown in Figure 8.3, plotting the recorded SPR response *versus* time throughout the different phases of the binding experiment: (i) baseline detection, (ii) association phase in which the analyte is injected over the chip at a defined concentration and binds to the immobilized entity until reaching equilibrium, (iii) dissociation upon washing the chip with analyte-free running buffer, and (iv) regeneration of the chip surface, removal of bound analyte, and recovery of the baseline signal.

Identified fragment hits can be confirmed and validated by SPR protocols and orthogonal biophysical methods. For such follow-up experiments during the hit characterization phase, more detailed measurements (affinity $K_D$, kinetics, association and dissociation rates) need to be repeated at multiple analyte concentrations (such as 10 concentrations around the expected $K_D$). The association and dissociation profile of the sensorgrams can then be simultaneously fitted to the rate constants ($k_{on}/k_{off}$), and the $K_D$ of the interaction can be calculated ($=k_{off}/k_{on}$).

**TABLE 8.3 Key Features of "Channel-Based SPR" Compared to Those of "Chemical Microarray SPR"**

| Channel-Based SPR | Chemical Microarray SPR |
|---|---|
| Immobilization/detection scheme | |
| Protein immobilized on sensor fields | Protein in solution |
| Fragments in solution | Fragments immobilized in array format |
| Information content | |
| Binding: yes/no | Binding: yes/no |
| Specificity, binding site information | Specificity, binding site information |
| Affinity $K_D$ and affinity ranking | |
| Kinetics, stoichiometry | |
| Throughput | |
| Sequential measurements cycles | Highly parallel readout of 9,216 sensors per array |
| High: 100–1000 measurements/day depending on instrument and assay | Ultrahigh: 12,000 fragment interactions in triplicates per day/instrument |
| Main use for fragment-based screening | |
| Primary and secondary screening, compound characterization | Fast primary screening of large low-molecular-mass compound collections |
| Sensitivity/SPR data quality | |
| Sensitive enough for fragment detection | Sensitive enough for fragment detection |
| Low-molecular-mass compound as analyte can lead to lower signal/noise ratio | Large protein as analyte generates higher SPR signal |
| Setup of experiments/screening | |
| Setup of assay and immobilization can be tricky | Buffer optimization for protein in solution required |
| Surface chemistries widely available | Optimized sensor surface |
| Modern instruments enable parallel experiment with off-targets/control proteins | Single-protein species per array experiment. Control experiments require separate screening |
| Sensitive to purity of reagents, DMSO concentrations | Robust assay, mainly dependent on target properties |
| Fragment solubility and library | |
| Fragments need to be soluble at high concentrations | Immobilized fragments not limited in their physicochemical profile |
| Commercial fragments libraries can be readily screened | Attachment of fragments generates initial effort |
| False positives; hits need to be confirmed | False positives; hits need to be confirmed |
| | False negatives due to linker |
| Activity of target protein | |
| Inactivation possible due to immobilization conditions | Protein required to be active in solution only |
| Immobilized protein stability to be checked with tool compounds | Fresh protein used for each array |
| Accessibility of binding site? Orientation of protein restricted | All protein-binding sites accessible in solution |
| Not all targets feasible | A broad range of targets were measured |
| Tool compound required to validate assay | Tool compounds not required, but favorable |

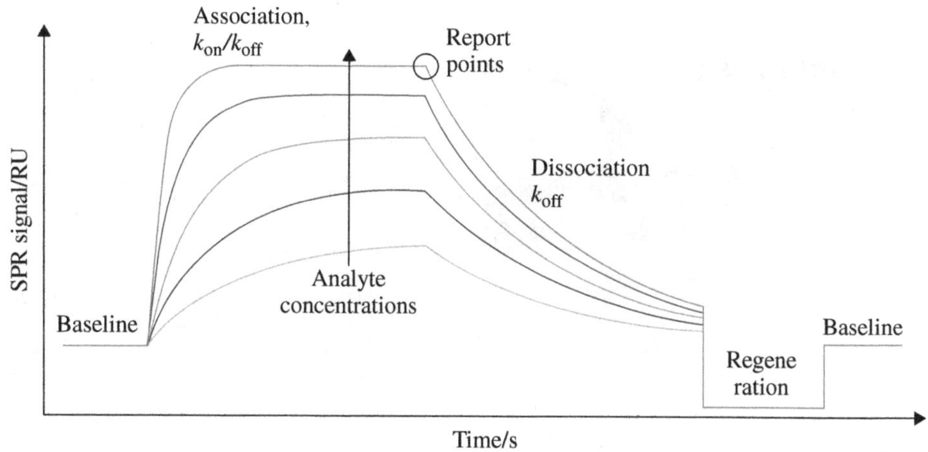


**FIGURE 8.3** Generic channel-based SPR sensorgram as obtained from a typical binding experiment: baseline detection, association, and dissociation phase, followed by regeneration of the chip surface and recovery of the baseline signal.

Alternatively, steady-state $K_D$ values are derived from Langmuir plots of the equilibrated SPR response versus the applied analyte concentrations.

For fragments—which show in general rapid association and dissociation profiles—the determination of rate constants can become difficult or simply nonapplicable. Binding site information can be deduced from competition/inhibitory assays with known tool compounds, if available.

Since the event of the first commercially available SPR-based biosensor system by Biacore (acquired by GE) in 1990, the popularity of SPR as a biophysical tool for studying biomolecular interactions has grown dramatically. More than 20 companies (GE, BioRad, Fuji, SensiQ to name only a few) are marketing SPR systems in different formats, mostly comprising a number of sensor fields addressed by flow channels. Comprehensive reviews on SPR instruments, their historic background, and their applications can be found in the literature [23].

***8.3.1.2 Chemical Microarray SPR*** In the 1980s, studies by Knoll [25] and others demonstrated that laterally structured coatings on metal films could be analyzed by surface plasmon microscopy (SPM) or imaging. The technology utilizes an expanded beam of parallel light to illuminate the entire sensor area covering several square centimeters. The reflected light is then captured by means of a charge-coupled device (CCD) camera. While scanning over a range of wavelengths or incidence angles covering the SPR resonance conditions, reflection images are recorded stepwise. Grayscale analysis of the obtained pictures yields SPR resonance curves (as in Fig. 8.1b) for all spots on the chip surface [26]. Thus, arrays of thousands of spots/sensor fields can be analyzed, and SPM of such array-type sensors has been successfully applied to the characterization and high-throughput analysis of biomolecular interactions [27]. Today, a number of SPR imaging systems are either commercially available from different sources (Reichert, IBIS, GWC, etc.) or are used as internal screening platform (NovAliX, etc.).

Chemical microarray SPR represents a unique format of SPR imaging, exploiting arrays of fragment libraries instead of immobilized protein. The construction, readout, and application of this universal binding platform was described in detail elsewhere

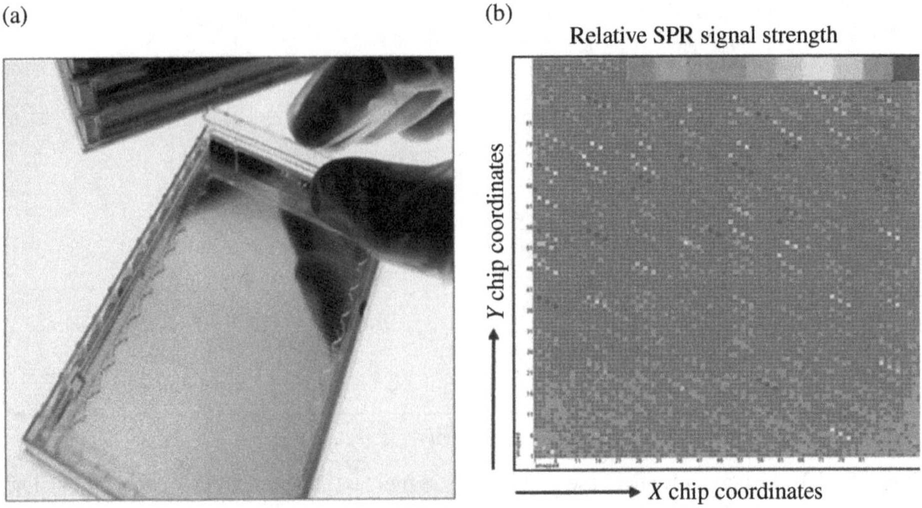


**FIGURE 8.4** (a) 3000 fragments are spotted on gold chips to construct chemical microarrays. (b) Fragment hits identified in one screening experiment are indicated by triplicate pattern on a relative color scale plot. (*See insert for color representation of the figure.*)

[28–33]. Briefly, for the construction of arrayed fragment libraries, the 25,000 members of a tagged fragment library are covalently immobilized to a chemically modified gold chip surface by pin-tool spotting to deposit nanoliter volumes onto defined areas of the chip. The target protein under near-native conditions is then allowed to interact with the ligands constituting the chemical microarrays. With greater than 9000 sensor fields per array (Fig. 8.4), the technology can simultaneously generate affinity data for more than 3000 immobilized fragments spotted in triplicates per array.

The binding of proteins induces large changes in the refractory index, making this system highly sensitive and thus applicable for the screening of fragments. NovAliX's SPR imagers read out up to four arrays per instrument, and fragment hits are rapidly identified as triplicate pattern in relative color scale plots. In addition to more common target classes such as kinases, proteases, metabolic enzymes, etc., difficult target classes including protein–protein interaction (PPI) targets and intrinsically disordered proteins have also been addressed (publication in preparation).

The chemical microarray SPR technology can equally provide information on the binding site, as on-array competition studies with established competitor molecules can be executed. This allows the differentiation between active site and allosteric site binders, as shown in Figure 8.5.

***8.3.1.3 Other Optical Biosensor Platforms*** In addition to SPR-based biosensors, a variety of other optical technologies can be applied to screening and hit confirmation in drug discovery and biology. The detection in those systems is not based on surface plasmons, but on other optical principles. In common is, however, label-free sensing of analytes triggered by an altered refractive index in evanescent fields close to the sensor surface. Examples of popular platforms are the Octet Red by ForteBio, based on biolayer interferometry (BLI); Corning's EPIC system; and SRU BIND employing resonant waveguiding (RW) via grating excitation in a microtiter plate (MTP) format [23].

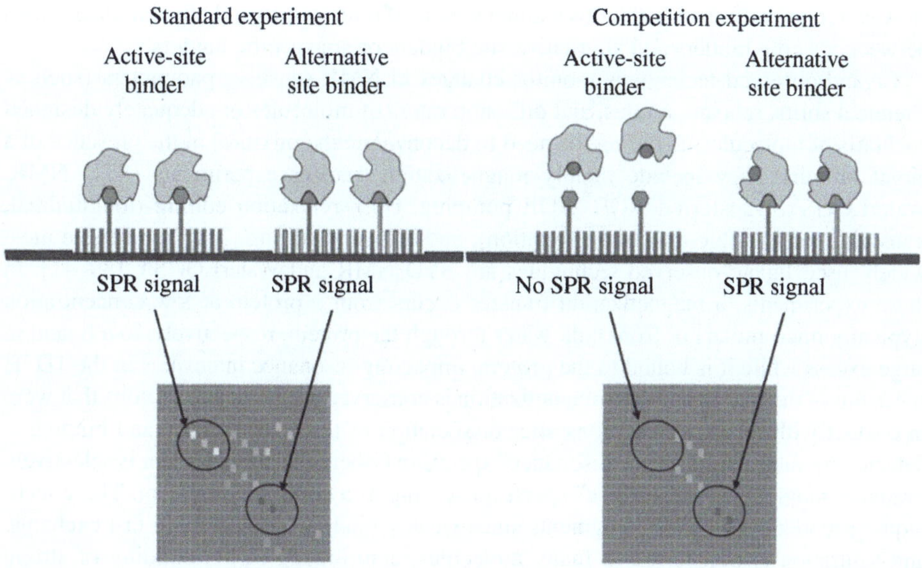


**FIGURE 8.5** Binding site mapping by an on-array competition study: in the standard experiment, all available binding sites give rise to an SPR array signal upon binding. Blocking known sites with tool compounds enables the discrimination between, for example, active and alternative site binders. (*See insert for color representation of the figure.*)

MTP-based systems, providing multiparallel readouts, are advantageous in terms of throughput. The robustness against analyte/compound artifacts seems to be higher compared to SPR assays, and the assay setup is considerably rapid. However, the amount of protein needed to immobilize the protein target on the bottom of the MTP wells is not negligible. Also, a slightly lower overall sensitivity requires sufficient affinity between ligand and analyte if low-molecular-mass compounds are to be detected. Screening of fragment libraries with weak affinities can therefore become troublesome.

### 8.3.2 Nuclear Magnetic Resonance (NMR)

The use of NMR screening in drug discovery (and more particularly in FBDD) has already been widely reviewed [34–37]. The seminal study of Shuker et al. in 1996 was a landmark in the FBDD field since it was the first report of high-affinity ligands built from fragments using a "SAR-by-NMR" approach, with NMR being the method of choice to identify fragment hits [14]. Since then, NMR has probably become the most popular fragment screening technique, primarily because of its easy setup and high sensitivity. These advantages apparently overcome the major drawbacks, which are the relatively high amounts of protein needed and the medium throughput. NMR experiments can grossly be divided into two categories: ligand-observed techniques and protein-observed techniques (Table 8.3). The former offer a higher throughput because data acquisition times are shorter (mostly 1D spectra recorded), much less protein is consumed, and there is no limit in target size. Furthermore, the protein does not require labeling. There are limitations, though; the information content is rather low, since no information is obtained on the binding site.

However, competition experiments using a validated reference ligand allow to distinguish between specific binders and alternative site binders or nonspecific binders.

Ligand-observed techniques monitor changes in NMR-sensitive parameters (such as chemical shifts, relaxation rates, and diffusion rates) of molecules or adequately designed cocktails of molecules (avoiding the need to deconvolute the mixture) in the presence of a target protein. They include mainly magnetization transfer experiments (STD NMR, WaterLOGSY, transferred NOE, NOE pumping, etc.), relaxation editing (longitudinal, transverse, and double-quantum relaxation), and diffusion editing [34, 37, 38]. The most widely used ligand-observed techniques are STD NMR and WaterLOGSY [39–41]. In these experiments, a magnetization transfer occurs from a protein at low concentration (typically micromolar) or from bulk water through the protein, respectively, to a ligand in large excess while it is bound to the protein, impacting resonance intensities in the 1D $^{1}$H spectrum of the free ligand (the magnetization is conserved by the ligand protons that were in contact with the protein protons after dissociation of the complex). Ligand binding is detected by subtracting an "on-resonance" spectrum (where the protein/water is selectively irradiated) and an "off-resonance" spectrum serving as a control experiment. These techniques are ideally suited for fragments since weakly binding ligands are in fast exchange and saturation is transferred to many molecules, contributing to the building of strong signals. Conversely, they are unable to detect strong binders ($K_d < 100$ nM). High-affinity fragments are rare but should not be missed. To detect them, it is possible in this case to perform competition experiments in the presence of a low- to medium-affinity reporter ligand (see, for instance, [42, 43]). Competition experiments also serve to assess the binding specificity of hits by adding a high-affinity reference compound that competitively displaces only fragments binding to the targeted active site and are routinely coupled to screening campaigns. A very powerful development of competition experiments involves a fluorinated reporter ligand. $^{19}$F NMR *per se* offers unique advantages due to the 100% natural abundance of $^{19}$F, its high NMR sensitivity (83% of the proton sensitivity), its wide range of chemical shifts (leading practically to the absence of spectral overlaps), and the absence of background signals from proteins and most buffers, additives and detergents. Ligand-observed $^{19}$F NMR methods have been developed [44], but direct $^{19}$F ligand-observed screening requires the availability of fluorinated compound libraries. Fluorine chemical shift anisotropy and exchange for screening (FAXS) [45] is a fast competition screening method using a fluorinated reporter ligand (the spy molecule), where changes in the transverse relaxation rate $R_2$ of the $^{19}$F resonance of the spy molecule are monitored in the presence of a cocktail of compounds. In the absence of ligand, the spy molecule is in equilibrium between the free and the bound states, resulting in a broad $^{19}$F resonance. If the cocktail contains a ligand that displaces the spy molecule from the target protein, the resonance becomes sharp (Fig. 8.6).

A weak binder is selected as spy molecule to allow the detection of low-affinity hits, and cocktails of more than 100 compounds can be used thanks to the absence of spectral overlap. However, a good compromise has to be found with the subsequent deconvolution process.

Apart from competition experiments, ligand-observed techniques cannot provide information on the binding site. On the other hand, in some cases, it is possible to obtain information on the binding orientation of the ligand: group epitope mapping (GEM), which derives information on the distance between the protons of the compound and the binding site from their variable degrees of saturation, allows the mapping of the ligand's binding epitope [41, 46]; interligand NOE for pharmacophore mapping (INPHARMA) is

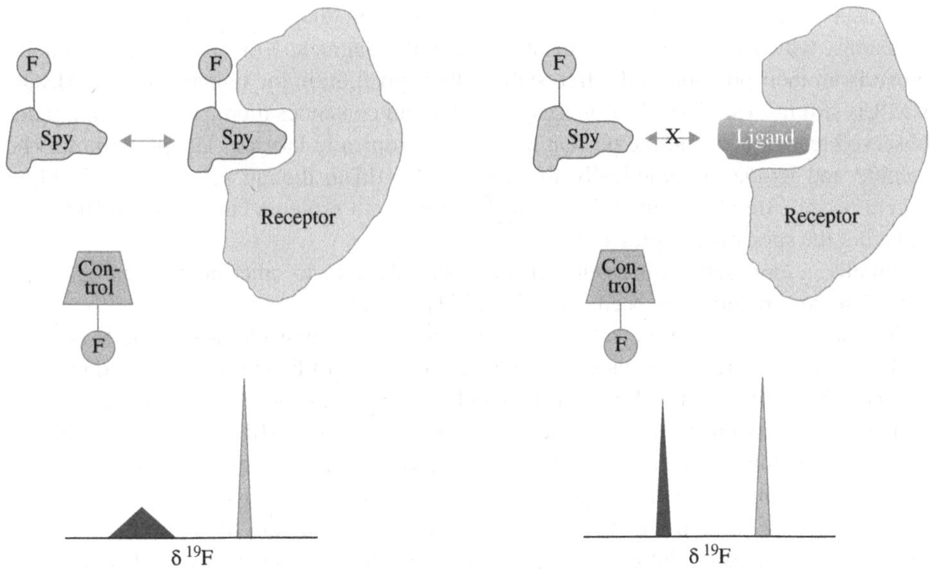


**FIGURE 8.6** Principle of the FAXS competition experiment. Reproduced with permission from Dalvit et al. [44] (*See insert for color representation of the figure.*)

based on the measurement of protein-mediated NOEs between two competing fragments that bind consecutively to the same target protein to determine their relative orientations provided a structural model for the binding site is available [47]; SALMON, a method derived from WaterLOGSY, allows one to determine the orientation of a bound ligand by mapping its solvent accessibility [48]. The newest developments of STD NMR have been reviewed recently [49]. Finally, the measurement of ligand binding affinities using ligand-observed techniques has been discussed [50].

Protein-observed techniques follow changes in NMR signals arising from proteins (chemical shifts and peak widths) upon ligand binding. The most widely used technique is two-dimensional $^1H/^{15}N$ (and also $^1H/^{13}C$) heteronuclear single quantum correlation (HSQC) NMR. 2D HSQC NMR is preferred to 1D NMR because the large size of proteins enhances the problem of resonance overlap. In standard $^1H/^{15}N$ HSQC spectra, essentially resonances of $^{15}N$-labeled amide groups are observed, and their position and shape are very sensitive to their environment, in particular upon ligand binding, meaning that there will be very few false positives. Binding is detected when one or more cross peaks in the 2D spectrum are shifted (chemical shift perturbation) in the presence of a compound or mixture of compounds. This is sufficient to perform a screening, if one only wants to identify binders, because peak assignment is not necessary. However, when the protein 3D structure is known and the resonances could be assigned, the technique also provides information on the binding site (chemical shift mapping), which is the major advantage of protein-observed techniques. Indeed, chemical shift mapping can help to identify the protein active site when it is unknown and serves to distinguish ligands that bind to the active site from ligands that bind to alternative sites such as allosteric sites. Moreover, 2D HSQC NMR gives access to the binding affinity by performing titration experiments with increasing ligand concentrations. In case of slow exchange (compared to the NMR experiment time scale), the signals of the affected residues in the free state will gradually decrease, while the signals in the bound state will gradually increase. In case of fast exchange, which is expected for most low-affinity fragments, the peaks will gradually move from their position in the free state to their position in the bound state. 2D HSQC NMR is still fast (total acquisition time ~10 min) but consumes more protein than ligand-observed techniques, and the protein has to be isotopically labeled. The protein must be soluble and its size is practically limited to 30–40 kDa, though selective $^{13}$C-labeling schemes can extend the suitability of the technique to proteins larger than 100 kDa by reducing the spectral complexity [37].

Finally, a thorough comparison between NMR-based and other biophysical methods for fragment screening has been carried out [51].

To take only one example, a target considered barely druggable, KRas, the K isoform of the human oncogene, was tackled using a combination of ligand-observed and protein-observed NMR techniques. A 3300-fragment library was screened by STD NMR, yielding numerous primary hits further validated by 2D $^{1}$H/$^{15}$N-HSQC NMR. Among the confirmed hits, 25 fragments induced chemical shift perturbations at a common site different from the nucleotide-binding site. This novel binding site, revealed through the crystal structure determination of KRas-fragment complexes, is adjacent to the Ras–SOS interface. Indeed, selected compounds inhibit nucleotide exchange by blocking the interaction between Ras and SOS, preventing Ras activation [52].

### 8.3.3 X-Ray Crystallography

The concept of fragment cocktail crystallography was first put forward in 1990 by Wim Hol, though its application to a real case was disappointing due to technical limitations at the time [53]. Then crystallography was extensively used as a primary screening technique by some companies such as Astex in the early days of industrial FBDD (see, for instance, [54–56]), although the throughput was relatively low [57]. Since then, new technical developments (with respect to hardware and software) have dramatically improved the throughput, making crystallography now very attractive for primary screening [57–59]. X-ray crystallography screening obviously requires that the target is amenable to crystallographic study, which is not always granted. It also requires the availability of a robust gene-to-crystal pipeline since hundreds of crystals are typically needed for a screening campaign: ideally, easily reproducible, well-diffracting crystals of the target in the apo form must be readily available to allow the soaking of hundreds/thousands of ligands, pooled in cocktails of typically four to eight compounds. These have to be sufficiently chemically diverse to avoid ambiguities in the interpretation of the electron density maps that would slow down the process by requiring cocktail deconvolution. When crystals of the apoprotein are not available for soaking, cocrystallization of the target with individual compounds can be performed, but because the process is much more tedious, this should be done only with hits validated by other screening techniques.

The obvious advantage of crystallography is the access to atomic-scale 3D information on the fragment mode of binding as direct output of the screening process, and this information can be immediately used to design optimized ligands. Although there are no false positives, false negatives may arise from crystal packing (for instance, when crystal packing precludes access to a binding site). When no hits are found, it may be advisable to test other crystal forms, if available.

### 8.3.4 Noncovalent Mass Spectrometry

Mass spectrometry allows the determination of the molecular mass of an ionized species from its mass-to-charge ratio. The development of soft ionization techniques such as matrix-assisted laser desorption/ionization (MALDI) and electrospray ionization (ESI) in the early 1990s has opened the way to the analysis of macromolecules by mass spectrometry with femtomolar sensitivity. Furthermore, two pioneering studies [60, 61] also showed that noncovalent complexes can remain intact in the gas phase using ESI, allowing the direct observation of protein–protein and/or protein–ligand interactions as they exist in solution. By measuring the mass of a complex, knowing the species present in solution, the composition of the complex can be deduced since the mass of each component constitutes an intrinsic label. This particular approach is called noncovalent/supramolecular/native/nondenaturing mass spectrometry. At NovAliX, this involves the use of a mass spectrometer with an analyzer suited for high m/z ratios such as time-of-flight (TOF) or quadrupole-time-of-flight (QTOF) analyzers and of a TriVersa NanoMate robot from Advion (http://www.advion.com/products/triversa-nanomate/) coupled to the source of the spectrometer for the nanoelectrospray ionization infusion of the samples. The TriVersa NanoMate offers several benefits with its microchip of single-use nanonozzles for sample infusion: no cross-contamination, no more column clogging, and better signal-to-noise ratios, thanks to a more efficient desolvation upon injection of a smaller volume (nanoelectrospray). To avoid protein denaturation and to maintain noncovalent interactions, the use of organic solvents should be avoided. Instead, the protein has to be conditioned in a volatile buffer requested by electrospray ionization. Thus, prior to mass spectrometry analysis, the protein purification buffer has to be exchanged against a volatile buffer (at concentrations as low as possible), most of the time ammonium acetate (or ammonium carbonate if a higher pH is needed). Finally, two parameters of the mass spectrometer have to be fine-tuned to preserve the noncovalent complexes and to optimize the quality of the spectra: $V_C$, the cone voltage in the interface between the source and the analyzer, and $P_i$, the residual pressure in the interface that controls the rate and energy of collisions with the residual inert gas (usually $N_2$) in the interface [62, 63]. Fine-tuning of these parameters requires an experimented user, probably explaining why noncovalent mass spectrometry has not become more widespread.

Obviously, the basic information obtained is the detection of binding: at first, a protein solution is analyzed alone and its molecular mass ($M_P$) can be measured (from the corresponding m/z peaks observed in the raw mass spectrum). The solution is then analyzed again after incubation with a compound of mass $M_L$. If the compound does not interact with the target protein in the experimental conditions of the solution, only the m/z peaks corresponding to $M_P$ will be detected. If the compound interacts with the target protein, new peaks corresponding to a new mass $M_{PL}$ (the sum of $M_P$ and $M_L$ assuming a simple 1:1 stoichiometry) will be detected. In case of a very high affinity, the compound may saturate the target and only the peak corresponding to $M_{PL}$ will be observed in the deconvoluted spectrum. In most cases involving fragments, the affinity is weak and only partial binding is achieved, so both peaks corresponding to $M_P$ and $M_{PL}$ are observed. Using the peak heights ($h$) or areas ($A$), the $h_{PL}/(h_{PL}+h_P)$ or $A_{PL}/(A_{PL}+A_P)$ ratios indicate the percentage of binding. Theoretically, one could estimate the $K_d$ from a single point but of course the result would be a gross approximation. For instance, using a 10 µM protein concentration and a fivefold excess of ligand (50 µM total ligand concentration) and assuming the simplest binding model P + L ⇌ PL, 30% binding ($h_{PL}/h_{PL} + h_P = 30\%$) would correspond to [PL] ~ 3 μM, [P] ~ 7 μM, and [L] ~ 47 μM, and thus, $K_d$ = [P] × [L]/[PL] ~ 100 μM, a typical affinity for fragments. A better estimate of $K_d$ can be obtained by performing titration experiments using different ligand concentrations. Noncovalent mass spectrometry is well suited for fragment screening, though the practical limit of detection is around 500 μM, not as good as for NMR (a few millimolar), but the big advantage is the very small amount of protein needed (10–20 pmol/assay) as well as a decent medium throughput. In our in-house setup, the whole process (mixing of the protein and the ligands, infusion in the source, spectrum acquisition, spectrum interpretation) was automated, and a 96-well plate can be screened in one day. The protein consumption can be further decreased and the throughput further increased by using 384-well plates, and compound cocktails (practically up to five compounds) provided the mass of the different compounds in the same cocktail are sufficiently different (at least 20 Da) to avoid peak overlaps, which make the interpretation of spectra difficult.

A second inherent information is the stoichiometry of binding, given by the ratio $(M_{PLn} - M_P)/M_L$ in case of multiple binding. Sometimes mono-, bis-, and higher-order adducts with decreasing peak intensities can be observed simultaneously, which is the signature for unspecific binding (even though one ligand molecule may bind specifically). In fact, noncovalent mass spectrometry conveniently tells whether a compound binds to a target, but does not bring structural information on the binding. However, competition experiments with a reference compound (a high-affinity *bona fide* ligand when available) can distinguish between specific binding (at the same site as the reference compound) and unspecific binding or specific binding at an alternative site (for instance, an allosteric site). In the latter case, ternary complexes between the target and two ligands binding simultaneously at different sites can be observed (V. Vivat and J.P. Renaud, unpublished results).

The limiting factors of the technique are the following: (i) the stability of the protein target in the ammonium buffer at low concentration (but a concentration of ammonium buffer up to several hundreds of millimolar is tolerated if the protein is unstable in low salt); (ii) the necessity to strictly avoid detergents (therefore the study of membrane proteins remains a real challenge); (iii) occurrence of unspecific binding making the detection of low-affinity binders (which may still be interesting starting points for fragment evolution) more difficult. However, unspecific binding particularly occurs with rather apolar fragments, while the need for fragment evolution tends to favor higher functionalized fragments, which are thus generally rather polar.

From our experience, there is a good correlation between specific binding assessed by noncovalent mass spectrometry (validated by competition experiments) and the success rate of subsequent complex crystallization (by soaking or cocrystallization) for well-behaved targets. Furthermore, besides screening, noncovalent mass spectrometry allows for the detailed characterization of protein–ligand interactions [stoichiometry, binding specificity, binding affinity, gas-phase stability as an indicator of the enthalpic contribution to the binding (estimated through $V_C50$ values derived from collision-induced experiments), and covalent character of the binding], which has been reviewed elsewhere [63].

However, noncovalent mass spectrometry is by no means the only way in which mass spectrometry has been used in FBDD. Tethering [64, 65] detects binding fragments through the creation of a covalent link: a library of disulfide-containing fragments is allowed to react with a native or an engineered cysteine at the targeted binding site of a protein, and the fragments captured through disulfide exchange are identified by mass spectrometry. Then the crystal structures of the target with the bound fragments (with and without the disulfide linker) guide the elaboration of the fragments into noncovalent high-affinity ligands. In indirect affinity selection-mass spectrometry (AS-MS) techniques, the target protein is incubated with a cocktail of compounds; then the nonbinders are separated from the protein–binder complexes, in general by size exclusion chromatography; and the binders are identified by mass spectrometry, for instance, through complex denaturation and analysis by LC-MS. These techniques have a high throughput and can be applied to solubilized membrane proteins, but they are better suited for the detection of high-affinity ligands because weakly binding fragments are mostly lost during the size exclusion chromatography step due to their high off-rates. Weak affinity chromatography (WAC), where fragments flowing over a protein immobilized in a capillary column are eluted more slowly when they interact with the target (and are then detected by mass spectrometry), appears to be more suited for fragment screening [66].

### 8.3.5 Differential Scanning Fluorimetry (DSF)

DSF, also called thermal shift assay (TSA), is a fluorescence-based technique that monitors the unfolding of a target protein with increasing temperature, allowing to measure the denaturation midpoint ($T_m$) [67, 68]. It involves an environment-sensitive fluorescent dye (in general SYPRO Orange) that in general binds to hydrophobic regions of the unfolded protein. A shift in the denaturation midpoint or thermal shift ($\Delta T_m$) induced by the presence of a compound is indicative of binding. This assay can be performed on a cheap apparatus combining precise temperature control and fluorescence measurement in a miniaturized format. Ligand binding usually increases the $T_m$; however, the $\Delta T_m$ values are not directly related to binding affinities and cannot be used for ranking ligands—the binding affinity at a given temperature is also function of the binding enthalpy [67]. Interestingly, some ligands induce negative $\Delta T_m$, either because they destabilize the protein upon binding (as in the case of formation of a covalent bond [69]) or when the ligand binds more strongly to the unfolded state than to the native state [70]) or because of artifacts from data processing.

Due to its ease of setup and relatively high throughput, DSF is also used as a primary screening method [71], though less popular than NMR or SPR; however, it suffers from some drawbacks and its use in fragment screening remains controversial [72]. It is not applicable to all targets, for instance, to hydrophobic proteins, which give rise to high fluorescence backgrounds, at least in the basic setup. DSF can be adapted to hydrophobic proteins and membrane proteins using alternative dyes [73], but there has been up to now no reported case of fragment screening on a membrane protein using DSF. It may give rise to numerous false negatives (for instance, when ligand binding does not induce any change in the $T_m$ due to some compensatory effect or when weak fragment binding leads to a $\Delta T_m$ below the chosen threshold (typically 1°C) or when binding fragments themselves quench the fluorescence). For weakly binding fragments, high concentrations needed to measure significant $\Delta T_m$ values (as in the case of biochemical assays to detect measurable activities) are limited by compound solubility and may cause aggregation problems. False positives arise when fragments stabilizing the protein bind outside the active site. Furthermore, data interpretation may be difficult when the melting curve does not exhibit a clear sigmoidal shape, for instance, in the case of a multiple-state transition. However, DSF remains useful as a fast and inexpensive primary screening method when highly soluble fragments are used, provided hits are cross-validated by orthogonal techniques.

### 8.3.6 Biophysical Techniques for Fragment Screening against Membrane Proteins

Membrane proteins play essential roles in physiology and disease. GPCRs alone mediate the effects of approximately 40% of all marketed drugs, whereas only approximately 10% of the approximately 500 nonolfactory GPCRs have been targeted so far. However, GPCRs, ion channels, and other membrane proteins—especially integral membrane proteins—are not easily amenable to the standard approaches of structure-based drug design because these were developed for soluble proteins. Apart from solubility problems, the amount of purified protein needed for biophysical assays is another limiting factor since membrane proteins are notoriously more difficult to overproduce, to purify, and to stabilize. Furthermore, it is crucial to include appropriate control experiments to eliminate false positives caused by the presence of detergents. Finally, the lack of crystallographic information for many membrane proteins does not enable the crucial structure-based hit-to-lead progression. Currently, mostly two techniques are practically used to screen fragments on membrane proteins: SPR and NMR.

Concerning SPR, only the protein-immobilized version is suitable up to now. The setup is even more delicate for membrane proteins than for soluble proteins (where artifacts from immobilization are possible), since the detergent-solubilized membrane protein has to be immobilized on a biosensor surface under conditions where it remains functional [74–76]. The control has to be selected carefully, so as to be as close as possible to the target [74]. NovAliX is currently dedicating efforts to adapt the chemical microarray SPR screening technology (see Section 8.3.1) to GPCRs.

Concerning NMR, the proprietary target immobilized NMR screening (TINS) technique appears to be the most efficient fragment screening technique because of the low amount of protein needed (a single sample for the whole library). A mixture of fragments is flowed through parallel sample cells containing the immobilized target (the solubilized target membrane protein) and a reference (a control membrane protein solubilized in the same conditions to account for nonspecific interactions with the detergent), respectively, and binding is detected by comparing the 1D $^1$H spectra. (Note that TINS can also be used with soluble proteins.) The challenges in adapting NMR screening techniques in general to membrane proteins have been discussed elsewhere [77].

The most spectacular advances in the field come from Heptares, where both protein-immobilized SPR and TINS have been used to screen fragments against thermostabilized GPCRs [78–80], resulting in the first crystal structures of a GPCR in complex with ligands identified from fragment screening [80].

### 8.3.7 Biophysical Techniques for Fragment Screening against PPIs

There is an increasing interest in protein–protein interfaces as potential new therapeutic targets even though finding PPI inhibitors remains a challenging task [81–84]. PPI surfaces tend to be large, flat, and featureless compared to classical, concave binding sites of enzymes and receptors, and PPIs were therefore considered rather undruggable for many decades. However, alanine-scanning mutagenesis studies showed that most of the PPI free energy is contributed by groups of a few amino acids called hot spots, which exhibit a high propensity for interaction with diverse partners and may thus constitute privileged sites for binding [85].

In general, a first screening is performed against one partner (especially when one partner is well-structured or more robust) to find binders and, in a second step, the hits are tested for PPI disruption in another assay. It is certainly favorable if the structure of the protein–protein complex is known. In this case, X-ray crystallography or protein-observed NMR is ideal to look for fragments binding to the PPI. However, exposure of hydrophobic patches on a large interaction surface may compromise the solubility and stability of the partner. It may therefore become necessary to engineer stable mutants. Alternatively, a screening can monitor the inhibition of the protein–protein complex formation in the presence of a test compound, but this approach may be more difficult with fragments because their inherent weak affinity may not be sufficient to disrupt a PPI.

The use of different biophysical techniques for fragment screening against PPIs has been reviewed [86]. NMR seems to be particularly well adapted because of its ability to detect weak binders [87]. The NovAliX chemical microarray SPR imaging system is also well suited to screen PPIs (see Section 8.3.1). Recent examples involving DSF [88] and NMR [89] have been published.

## 8.4 FRAGMENT EVOLUTION STRATEGIES

Many researchers in the area rightfully claim that today identifying fragment hits is not any longer an obstacle, but the issues are situated in "what to do with them," in the sense that converting a fragment into a lead is far from evident.

Once a hit has been found, it is optimized into a lead by several different strategies: linking, growing, or merging. Though linking two fragments binding to different hot spots within a given binding site may appear as a very interesting approach, it has a high likelihood of failure. Firstly, the correct orientation of the two independent fragments may be perturbed upon linking, and secondly, the induced fit will render a fragment a highly efficient binder to its specific hot spot, but this interaction mode may be considerably influenced when already another fragment binds to an adjacent hot spot.

The strategy that has been more successful is to grow fragments in a highly rational, structure-guided approach, whereby in a first instance small libraries of functionalized fragments may be designed and prepared, or eventually acquired ("SAR by catalog"), with subsequent testing, either biophysically or biochemically, in order to explore the different adjacent hot spots. Needless to say, it is very important to obtain structural information on the binding mode of the new ligands, in order to understand the behavior and flexibility of the protein induced by the binding and to guide step by step the fragment evolution.

A third approach is called merging, whereby the chemist will use structural information from different fragments in order to come up with novel entities. This appears to be a highly interesting approach to escape existing IP space or to come up with novel ligands with even higher affinities. This approach requires more efforts from a pure synthetic chemistry perspective.

As mentioned before, the aim of the approach is to obtain highly drug-like leads, which of course requires the right physicochemical properties, especially molecular mass and ClogP. During the overall process of fragment evolution, it is crucial to keep an eye on these different physicochemical parameters. It is known that metabolic instability and permeability tend to decrease with increasing MM and/or ClogP. In order to quantify the descriptors linked to the molecular properties, a number of different parameters have been proposed, such as ligand efficiency (LE), fit quality (FQ), ligand-lipophilicity efficiency (LLE), and group efficiency (GE), to quote the more commonly applied ones. A nice overview of these parameters, and their application in the fragment evolution process, has been published by Schultes et al. [90].

## 8.5 FBDD CASE STUDIES

Today, we may state that every issue of any medicinal chemistry journal will report examples of FBDD, which clearly indicates to what extent this new strategy has been integrated in the drug discovery efforts. In addition to a number of examples already mentioned before to demonstrate the use of the different biophysical technologies, a selection was made among a number of case studies, in order to clearly demonstrate the way of working within the FBDD process, namely, keeping a very close eye on the efficiency of binding, the molecular mass, and cLogP, in addition to a strong structure-based component in the optimization of the activity. The first case study presented here belongs to the field of kinase inhibitors, and the second one to the area of PPIs, namely, the bromodomain/histone tail interaction.

### 8.5.1 Aurora Kinase Inhibitors

Scientists at Astex [91] had observed that a series of pyrazole–benzimidazole compounds, such as fragment **1** and the corresponding benzamide derivative 2, derived from an FBDD approach on CDK2, also showed interesting activities on Aurora A (Fig. 8.7). Fragment **1** even did display better LE on Aurora A than on CDK2 and was thus considered an interesting starting point for a FBDD approach on Aurora A. In order to support its optimization, a soakable crystal form of activated Aurora A was developed. The X-ray structures of both complexes (with fragment **1** and with its benzamide derivative **2**) led to the hypothesis that additional functionalities could be introduced at positions 5 or 6 of the benzimidazole moiety, whereas the benzamide group appeared to bind in a cleft with significant structural differences as compared to CDK2. Both directions were fully explored by continuously paying attention to ligand efficiency, ClogP, cellular potency, and drug-likeness of the derivatives. Introduction of the weakly basic morpholine functionality proved to enhance cellular potency, whereas converting the benzamide function into a series of urea derivatives helped decrease the plasma protein binding, without adding additional molecular mass.

The overall process, starting from the pyrazole–benzimidazole fragment with an excellent LE of 0.59, led to the development of compound 3 with an interesting multikinase inhibitory profile, which was selected for further clinical testing. Continuous attention was paid to preserve the important interactions (see Fig. 8.7) and to keep the molecules well within the molecular mass and lipophilicity guidelines for drug-likeness.

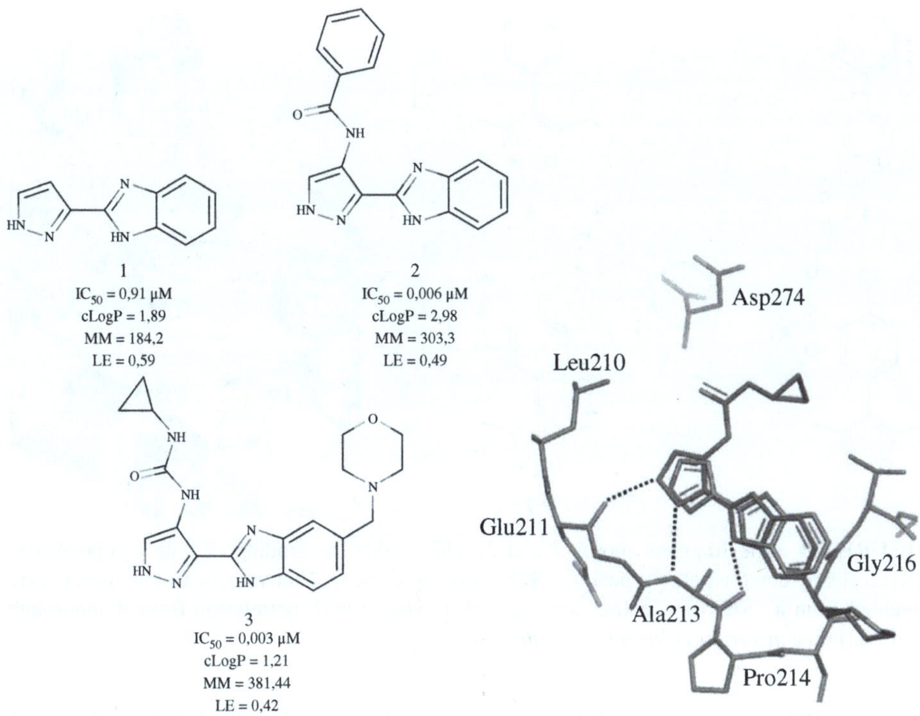


**FIGURE 8.7** Structures of ligands; X-ray crystal structure of 3 (green) complexed with Aurora A (blue) superimposed with the binding mode of fragment 1 (orange). Adapted with permission from Howard et al. [91] (*See insert for color representation of the figure.*)

### 8.5.2 Tackling PPIs: Fragment-Based Discovery of Bromodomain Inhibitor Leads

Bromodomain-containing proteins are key epigenetic regulators of gene transcription [92]. They are part of the famous reader elements that do recognize specific histone signatures via a PPI, in order to initiate or block downstream signaling. Bromodomains specifically recognize acetylated lysine motifs on histone tails. Blocking the interaction between acetylated histone tails and bromodomains will alter the process of chromatin remodeling resulting in an activation or inhibition of gene transcription [92]. Scientists at GSK [93] have generated a chemically diverse, focused fragment set of putative acetyl-lysine mimetics, guided by structural insights and computational methods. This library was then screened with a fluorescence anisotropy (FA) assay against the three tandem BET (BRD2, BRD3, BRD4) constructs. Numerous starting points were obtained. Isoxazole **4** (Fig. 8.8) was one of the weaker hits with around 30% activity at 200 μM against BRD3 and BRD4, but was selected for further optimization as it was small, chemically attractive, novel, and showed a decent LE of about 0.4. In order to guide the optimization, the compound was soaked into apo crystals of the N-terminal bromodomain of BRD2. The binding mode was compared to the position of the acetylated lysine. As the compound is quite symmetrical, even at the less than 2 Å resolution of the X-ray structure, the respective position of the N and O could not be observed. For this reason, the ethyl analog **5** was prepared and crystallized with the BRD2 construct. With the larger ethyl

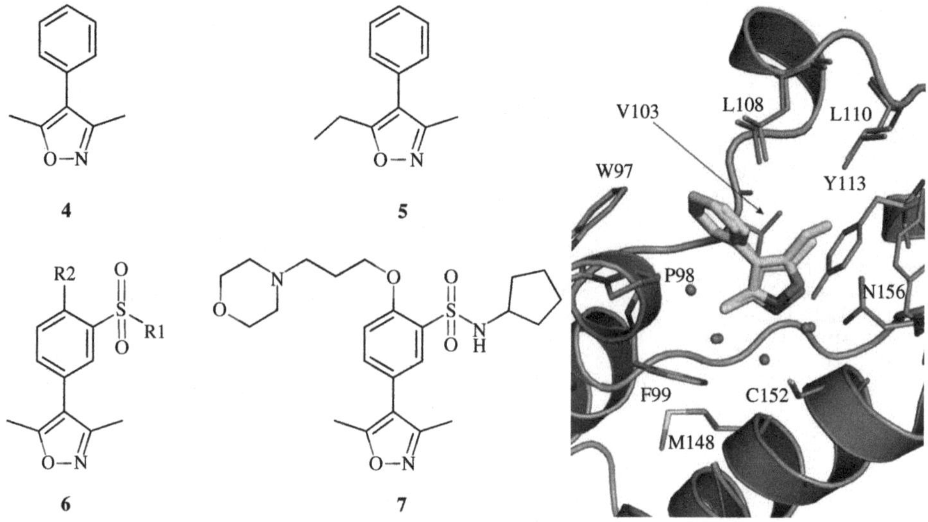


**FIGURE 8.8** The fragment motifs 4 and 5, with (right) the binding of both compounds to BRD2, clearly demonstrate the position of the nitrogen in the oxazole moiety. Compounds 6 were obtained from a "SAR-by-catalog" approach. Reproduced with permission from Bamborough et al. [93] (*See insert for color representation of the figure.*)

residue being clearly visible in the electron density map, the exact binding mode of compound **5**, and thus also of **4**, could be deduced (Fig. 8.8).

Based on structural information and a pharmacophoric model derived from the available structures, a search for commercially available compounds was performed, following the so-called and widely used "SAR by catalog" approach. A cluster of sulfonamide-substituted isoxazoles **6** was ordered and tested. The compounds showed encouraging potency in both FA and TR-FRET assays, with IC50s in the low micromolar range. During the overall process, the LE was preserved, staying around 0,4, a value comparable to the LE of the starting hit **4**, making this series an interesting starting point to pursue.

A number of sulfonamides were then soaked within the apo bromodomain crystals, in order to guide further optimization. Based on the structures, it was deduced that R1 was a promising position to improve potency. Since the chemical series was hampered by a limited solubility, the position R2 was further explored to introduce polar groups. This optimization led to the discovery of compound **7**, with an IC50 of 1–2 μM against the three bromodomains studied.

In order to obtain label-free confirmation and an equilibrium binding constant of the binding to the BET bromodomain, a number of analogs were tested in an SPR-based assay. Compound **7** was found to have a $K_D$ of 2–3 μM, consistent with the TR-FRET IC50 values.

Although a number of bromodomain inhibitors had been already reported, this team was the first to show the applicability of an FBDD approach to discover inhibitors of the bromodomain/acetylated histone tail interaction, based on crystallography, SPR, and biochemical testing, thus opening the way to explore other members of this important target class.

## 8.6 THE FUTURE

After a decade of FBDD, it may already be stated that this novel drug discovery approach has evolved into a mature research area that is widely recognized as a viable method for lead identification. FBDD has already delivered above its expectations, clearly demonstrated by a number of successful developments of clinical candidates on targets that were deemed undruggable by classical approaches. Even more, The "FBDD mindset" has found an entry into the more classical medicinal chemistry community. It is interesting to observe that the use of some of the "FBDD descriptors" in more classical medicinal chemistry projects is gaining interest. A very nice example of such an approach is the work presented by Tanaka et al. [94], in which the selection of lead-like starting points for optimization on FAAH is rather driven by LE consideration than the affinities of the different ligands. Having said that, it should not be neglected that the successful application of FBDD requires a large amount of expertise, the right technologies, the right library, and the right target.

Another evolution along the same lines is an increased reflection on "molecular descaffolding," for example, an approach to disconnect large (and ugly) lead-sized binding motifs into smaller entities, having in mind a search for the "minimum pharmacophoric elements" required for binding and the subsequent optimization of these fragments (or lower-molecular-mass species) to access novel structures. Some work has recently been reported on this type of approach while also questioning the ability of binding of the dissected moieties to the protein, as—rightfully—a larger lead-like motif may induce a considerable perturbation of the protein upon binding [95]. We at NovAliX have also had considerable success in such an approach during the generation of novel selective leads for PIM-1 kinase. A number of fragments derived from lead-sized hits, obtained with the chemical microarray SPR technology on our immobilized ligand library, have proven very successful starting points for a fragment-to-lead campaign (Paola Ciapetti, Daniel Schwitzer et al., unpublished results).

From a number of recent communications (see, for instance, a presentation by N. Skelton at the DDC meeting in San Diego, based on the work published by Gianetti et al. [96]), it is becoming clear that FBDD and the more classical HTS approach should not be in competition. In contrary, tackling a target with both approaches in parallel leads to complementary data, which has proven successful in the identification of highly interesting novel starting points for entry into an LO campaign.

FBDD has found an increasing interest in the area of tool compound development when exploring completely novel, unexplored targets. Indeed, as FBDD heavily relies on biophysical assays, one is capable of identifying compounds that interact at the molecular level with the target (binders to a catalytic site, inhibitors of PPIs). These molecules will then be excellent tool compounds for the pharmacologists to unravel the biochemical activities and understand phenotypic readouts induced by inhibiting or modulating the given targets.

Receptor-informed fragment assembly approaches, such as *in situ* fragment assembly techniques based on click chemistry [97] and dynamic combinatorial libraries [98] are also receiving attention by the scientific community. While this approach is scientifically very interesting, the scope and limitations of such approaches still need to be fully established.

In line with the evolution of FBDD, there is a continued search for different, complementary biophysical tools, be it to increase sensitivity, to be able to study certain more difficult target classes such as membrane bound proteins, or—last but not least—to view protein complexes with multiple partners, since numerous novel targets rely on dynamics in these systems. In this context, electron microscopy seems to receive an increased interest from the scientific community as a novel structure elucidation technique.

During the course of the writing of this manuscript, a very interesting viewpoint on "fragonomics" has been published by Teddy Zartler [99]. We may indeed state that FBDD has already proven its usefulness as a DD approach, but it should not be considered as the miracle solution to solve every single problem. As Rod Hubbard pointed out, "FBDD does not make drug design easier but it does give you more options."

## REFERENCES


[1] (a) Paul SM, Mytelka DS, Dunwiddie CT, Persinger CC, Munos BH, Lindborg SR, Schacht AL (2010). How to improve R&D productivity: the pharmaceutical industry's grand challenge. *Nat. Rev. Drug Discov.* **9**: 203–214. (b)Pammolli F, Magazzini L, Riccaboni M (2011). The productivity crisis in pharmaceutical R&D. *Nat. Rev. Drug Discov.* **10**: 428–438.

[2] Hajduk PJ, Greer J (2007). A decade of fragment-based drug design: strategic advances and lessons learned. *Nat. Rev. Drug Discov.* **6**: 211–219.

[3] Jahnke W, Erlanson DA (Eds) (2006). *Fragment-based Approaches in Drug Discovery* (Weinheim: Wiley-VCH).

[4] Erlanson DA, McDowell RS, O'Brien T (2004). Fragment-based drug discovery. *J. Med. Chem.* **47**: 3463–3482.

[5] Warr W (2009). Fragment-based drug discovery. *J. Comput. Aided Mol. Des.* **23**: 453–458.

[6] Congreve M, Chessari G, Tisi D, Woodhead AJ (2008). Recent developments in fragment-based drug discovery. *J. Med. Chem.* **51**: 3661–3680.

[7] Chessari G, Woodhead AJ (2009). From fragment to clinical candidate—a historical perspective. *Drug Discov. Today* **14**: 668–675.

[8] Coyne AG, Scott DE, Abell C (2010). Drugging challenging targets using fragment based approaches. *Curr. Opin. Chem. Biol.* **14**: 299–307.

[9] Whittaker M, Law RJ, Ichihara O, Hesterkamp T, Hallett D (2010). Fragments: past, present and future. *Drug Discov. Today* **7**: 163–171.

[10] Murray W, Verdonk M, Rees DC (2012). Experiences in fragment-based drug discovery. *Trends Pharmacol. Sci.* **33**(5): 224–232.

[11] Hubbard RE, Murray JB (2011). Experiences in fragment-based lead discovery. *Methods Enzymol.* **493**: 509.

[12] Erlanson DA, Zartler E. Available at: http://practicalfragments.blogspot.fr/2015/01/fragments-in-clinic-2015-edition.html (accessed July 28, 2015).

[13] Michne WF (1996). Hit-to-lead chemistry: a key element in new lead generation. *Pharmaceut. News* **3**: 19–21.

[14] Shuker SB, Hajduk PJ, Meadows RP, Fesik SW (1996). Discovering high-affinity ligands for proteins: SAR by NMR. *Science* **274**: 1531–1534.

[15] Baker M (2013). Fragment-based lead discovery grows up. *Nat. Rev. Drug Discov.* **12**: 5–7.

[16] Fink T, Reymond JL (2007). Virtual exploration of the chemical universe up to 11 atoms of C, N, O, F: assembly of 26.4 million structures (110.9 million stereoisomers) and analysis for new ring systems, stereochemistry, physicochemical properties, compound classes, and drug discovery. *J. Chem. Inf. Model.* **47**: 342–353.

[17] Boyd S, Turnbull A, Walse B (2012). Fragment library design considerations. *WIREs Comput. Mol. Sci.* **2**: 868–885.

[18] Schulz MN, Landström J, Bright K, Hubbard RE (2011). Design of a fragment library that maximally represents available chemical space. *J. Comput. Aided Mol. Des.* **25**: 611–620.

[19] Over B, Wetzel S, Grütter C, Nakai Y, Renner S, Rauh D, Waldmann H (2013). Natural-product-derived fragments for fragment-based ligand discovery. *Nat. Chem.* **5**: 21–28.

[20] Jencks W (1981). On the attribution and additivity of binding energies. *Proc. Natl. Acad. Sci. U. S. A.* **78**: 4046–4050.

[21] Erlanson D (2014). Poll results: affiliation, fragment-finding methods, and library size. Available at: http://practicalfragments.blogspot.fr/2014/01/poll-results-affiliation-fragment.html (accessed May 13, 2015).

[22] Hennig M, Ruf A, Huber W (2011). Combining biophysical screening and X-ray crystallography for fragment-based drug discovery. *Top. Curr. Chem.* **317**: 115–143.

[23] Schasfoort RBM, Tudos AJ (Eds) (2008). *Handbook of Surface Plasmon Resonance* (Cambridge: RSH Publishing).

[24] Rich RL, Myszka DG (2008). Survey of the year 2007 commercial optical biosensor literature. *J. Mol. Recognit.* **21**(6): 355–400.

[25] Rothenhausler B, Knoll W (1988). Surface-plasmon microscopy. *Nature* **332**: 615–617.

[26] Zizlsperger M, Knoll W (1998). Multispot parallel on-line monitoring of interfacial binding reactions by surface plasmon microscopy. *Progr. Colloid Polym. Sci.* **109**: 244–253.

[27] Campell CT, Kim G (2007). SPR microscopy and its applications to high-throughput analyses of biomolecular binding events and their kinetics. *Biomaterials* **28**: 2380–2392.

[28] Neumann T, Junker HD, Schmidt K, Sekul R (2007). SPR-based fragment screening: advantages and applications. *Curr. Top. Med. Chem.* **7**(16): 1630–1642.

[29] Dickopf S, Frank M, Junker HD, Maier S, Metz G, Ottleben H, Rau H, Schellhaas N, Schmidt K, Sekul R, Vanier C, Vetter D, Czech J, Lorenz M, Matter H, Schudok M, Schreuder H, Will DW, Nestler HP (2004). Custom chemical microarray production and affinity fingerprinting for the S1 pocket of factor VIIa. *Anal. Biochem.* **335**(1): 50–57.

[30] Heim-Riether A, Taylor SJ, Liang S, Gao DA, Xiong Z, Michael August E, Collins BK, Farmer BT, Haverty K, Hill-Drzewi M, Junker HD, Mariana Margarit S, Moss N, Neumann T, Proudfoot JR, Keenan LS, Sekul R, Zhang Q, Li J, Farrow NA (2009). Improving potency and selectivity of a new class of non-Zn-chelating MMP-13 inhibitors. *Bioorg. Med. Chem. Lett.* **19**(18): 5321–5324.

[31] Neumann T, Junker HD, Keil O, Burkert K, Ottleben H, Gamer J, Sekul R, Deppe H, Feurer A, Tomandl D, Metz G (2005). Discovery of thrombin inhibitor fragments from chemical microarray screening. *Lett. Drug Des. Discovery* **2**: 563–566.

[32] Neumann T, Sekul R (2011). SPR screening of chemical microarrays for fragment based discovery. In *Label-Free Technologies for Drug Discovery*, Ed. LM Mayr, MA Cooper (Chichester: John Wiley & Sons, Ltd), pp 42–45.

[33] Arnold M, Bittermann H, Kalbfuss-Zimmermann B, Neumann T, Schmidt K, Sekul R, Hilbrig F, Ludolph H, Freitag R (2011). Antibody purification by affinity chromatography based on small molecule affinity ligands identified by SPR-based screening of chemical microarrays. *J. Chromatogr. A* **1218**(29): 4649–4659 (in eng).

[34] Stockman BJ, Dalvit C (2002). NMR screening techniques in drug discovery and drug design. *Prog. Nucl. Magn. Reson. Spectrosc.* **41**: 187–231.

[35] Lepre CA, Moore JM, Peng JW (2004). Practical aspects of NMR-based fragment screening. *Chem. Rev.* **104**: 3641–3676.

[36] Lepre CA (2011). Theory and applications of NMR-based screening in pharmaceutical research. *Methods Enzymol.* **493**: 219–239.

[37] Pellecchia M, Bertini I, Cowburn D, Dalvit D, Giralt E, Jahnke W, James TL, Homans SW, Kessler H, Luchinat C, Meyer B, Oschkinat H, Peng J, Schwalbe H, Siegal S (2008). Perspectives on NMR in drug discovery: a technique comes of age. *Nat. Rev. Drug Discov.* **7**: 738–745.

[38] Hajduk PJ, Olejniczak ET, Fesik SW (1997). One-dimensional relaxation- and diffusion-edited NMR methods for screening compounds that bind to macromolecules. *J. Am. Chem. Soc.* **119**: 12257–12261.

[39] Mayer M, Meyer B (1999). Characterization of ligand binding by saturation transfer difference NMR spectroscopy. *Angew. Chem. Int. Ed.* **38**: 1784–1788.

[40] Dalvit C, Fogliatto G, Stewart A, Veronesi M, Stockman B (2001). WaterLOGSY as a method for primary NMR screening: practical aspects and range of applicability. *J. Biomol. NMR* **21**: 349–359.

[41] Meyer B, Peters T (2003). NMR spectroscopy techniques for screening and identifying ligand binding to protein receptors. *Angew. Chem. Int. Ed.* **42**: 864–890.

[42] Dalvit C, Flocco M, Knapp S, Mostardini M, Perego R, Stockman BJ, Veronesi M, Varasi M (2002). High-throughput NMR-based screening with competition binding experiments. *J. Am. Chem. Soc.* **124**: 7702–7709.

[43] Jahnke W, Floersheim P, Ostermeier C, Zhang X, Hemmig R, Hurth K, Uzunov DP (2002). NMR reporter screening for the detection of high-affinity ligands. *Angew. Chem. Int. Ed.* **41**: 3420–3423.

[44] Dalvit C (2007). Ligand- and substrate-based 19F NMR screening: principles and applications to drug discovery. *Prog. Nucl. Magn. Reson. Spectrosc.* **51**: 243–271.

[45] Dalvit C, Fagerness PE, Hadden DT, Sarver RW, Stockman BJ (2003). Fluorine-NMR experiments for high-throughput screening: theoretical aspects, practical considerations, and range of applicability. *J. Am. Chem. Soc.***125**: 7696–7703.

[46] Mayer M, Meyer B (2001). Group epitope mapping by saturation transfer difference NMR to identify segments of a ligand in direct contact with a protein receptor. *J. Am. Chem. Soc.* **123**: 6108–6117.

[47] Sanchez-Pedregal VM, Reese M, Meiler J, Blommers MJ, Griesinger C, Carlomagno T (2005). The INPHARMA method: protein-mediated interligand NOEs for pharmacophore mapping. *Angew. Chem. Int. Ed.* **44**: 4172–4175.

[48] Ludwig C, Michiels PJ, Wu X, Kavanagh KL, Pilka E, Jansson A, Oppermann U, Günther UL (2008). SALMON: solvent accessibility, ligand binding, and mapping of ligand orientation by NMR spectroscopy. *J. Med. Chem.* **51**: 1–3.

[49] Wagstaff JL, Taylor SL, Howard MJ (2013). Recent developments and applications of saturation transfer difference nuclear magnetic resonance (STD NMR) spectroscopy. *Mol. Biosyst.* **9**: 571–577.

[50] Śledź P, Abell C, Ciulli A (2012). Ligand-observed NMR in fragment-based approaches. In *NMR of Biomolecules: Towards Mechanistic Systems Biology*, Ed. I Bertini, KS McGreevy, G Parigi (Weinheim: Wiley-VCH), pp 264–280.

[51] Dalvit C (2009). NMR methods in fragment screening: theory and a comparison with other biophysical techniques. *Drug Discov. Today* **14**: 1051–1057.

[52] Maurer T, Garrenton LS, Oh A, Pitts K, Anderson DJ, Skelton NJ, Fauber BP, Pan B, Malek S, Stokoe D, Ludlam MJ, Bowman KK, Wu J, Giannetti AM, Starovasnik MA, Mellman I, Jackson PK, Rudolph J, Wang W, Fang G (2012). Small-molecule ligands bind to a distinct pocket in Ras and inhibit SOS-mediated nucleotide exchange activity. *Proc. Natl. Acad. Sci. U. S. A.* **109**: 5299–5304.

[53] Verlinde CL, Fan E, Shibata S, Zhang Z, Sun Z, Deng W, Ross J, Kim J, Xiao L, Arakaki TL, Bosch J, Caruthers JM, Larson ET, Letrong I, Napuli A, Kelly A, Mueller N, Zucker F,

Van Voorhis WC, Buckner FS, Merritt EA, Hol WG (2009). Fragment-based cocktail crystallography by the medical structural genomics of pathogenic protozoa consortium. *Curr. Top. Med. Chem.* **9**: 1678–1687.

[54] Nienaber VL, Richardson PL, Klighofer V, Bouska JJ, Giranda VL, Greer J (2000). Discovering novel ligands for macromolecules using X-ray crystallographic screening. *Nat. Biotechnol.* **18**: 1105–1108.

[55] Blundell TL, Jhoti H, Abell C (2002). High-throughput crystallography for lead discovery in drug design. *Nat. Rev. Drug Discov.* **1**: 45–54.

[56] Lesuisse D, Lange G, Deprez P, Bénard D, Schoot B, Delettre G, Marquette JP, Broto P, Jean-Baptiste V, Bichet P, Sarubbi E, Mandine E (2002). SAR and X-ray. A new approach combining fragment-based screening and rational drug design: application to the discovery of nanomolar inhibitors of Src SH2. *J. Med. Chem.* **45**: 2379–2387.

[57] Davies TG, Tickle IJ (2012). Fragment screening using X-ray crystallography. *Top. Curr. Chem.* **317**: 33–59.

[58] Jhoti H, Cleasby A, Verdonk M, Williams G (2007). Fragment-based screening using X-ray crystallography and NMR spectroscopy. *Curr. Opin. Chem. Biol.* **11**: 485–493.

[59] Wasserman SR, Koss JW, Sojitra ST, Morisco LL, Burley SK (2012). Rapid-access, high-throughput synchrotron crystallography for drug discovery. *Trends Pharmacol. Sci.* **33**: 261–267.

[60] Ganem B, Li YT, Henion JD (1991). Detection of noncovalent receptor-ligand complexes by mass spectrometry. *J. Am. Chem. Soc.* **113**: 6294–6296.

[61] Katta V, Chait BT (1991). Observation of the heme-globin complex in native myoglobin by electrospray-ionization mass spectrometry. *J. Am. Chem. Soc.* **113**: 8534–8535.

[62] Sanglier S, Atmanene C, Chevreux G, Van Dorsselaer A (2008). Nondenaturing mass spectrometry to study noncovalent protein/protein and protein/ligand complexes: technical aspects and application to the determination of binding stoichiometries. In *Functional Proteomics: Methods and Protocols*, Methods in Molecular Biology, vol. **484**, Ed. C Schaeffer-Reiss, JD Thompson, M Ueffing (Totowa, NJ: Humana Press), pp 217–243.

[63] Vivat Hannah V, Atmanene C, Zeyer D, Van Dorsselaer A, Sanglier-Cianférani S (2010). Native MS: an "ESI" way to support structure- and fragment-based drug discovery. *Future Med. Chem.* **2**: 35–50.

[64] Erlanson DA, Braisted AC, Raphael DR, Randal M, Stroud RM, Gordon EM, Wells JA (2000). Site-directed ligand discovery. *Proc. Natl. Acad. Sci. U. S. A.* **97**: 9367–9372.

[65] Erlanson DA, Wells JA, Braisted AC (2004). Tethering: fragment-based drug discovery. *Annu. Rev. Biophys. Biomol. Struct.* **33**: 199–223.

[66] Duong-Thi MD, Meiby E, Bergström M, Fex T, Isaksson R, Ohlson S (2011). Weak affinity chromatography as a new approach for fragment screening in drug discovery. *Anal. Biochem.* **414**: 138–146.

[67] Lo MC, Aulabaugh A, Jin G, Cowling R, Bard J, Malamas M, Ellestad G (2004). Evaluation of fluorescence-based thermal shift assays for hit identification in drug discovery. *Anal. Biochem.* **332**: 153–159.

[68] Niesen FH, Berglund H, Vedadi M (2007). The use of differential scanning fluorimetry to detect ligand interactions that promote protein stability. *Nat. Protoc.* **2**: 2212–2221.

[69] Dai R, Wilson DJ, Geders TW, Aldrich CC, Finzel BC (2014). Inhibition of Mycobacterium tuberculosis transaminase BioA by aryl hydrazines and hydrazides. *Chembiochem* **15**: 575–586.

[70] Cimmperman P, Baranauskiene L, Jachimoviciūte S, Jachno J, Torresan J, Michailoviene V, Matuliene J, Sereikaite J, Bumelis V, Matulis D (2008). A quantitative model of thermal stabilization and destabilization of proteins by ligands. *Biophys. J.* **95**: 3222–3231.

[71] Mashalidis EH, Śledź P, Lang S, Abell C (2013). A three-stage biophysical screening cascade for fragment-based drug discovery. *Nat. Protoc.* **8**: 2309–2324.

[72] Davis BJ, Erlanson DA (2013). Learning from our mistakes: The "unknown knowns" in fragment screening. *Bioorg. Med. Chem. Lett.* **23**: 2844–2852.

[73] Alexandrov AI, Mileni M, Chien EY, Hanson MA, Stevens RC (2008). Microscale fluorescent thermal stability assay for membrane proteins. *Structure* **16**: 351–359.

[74] Navratilova I, Besnard J, Hopkins AL (2011). Screening for GPCR ligands using surface plasmon resonance. *ACS Med. Chem. Lett.* **2**: 549–554.

[75] Seeger C, Christopeit T, Fuchs K, Grote K, Sieghart W, Danielson UH (2012). Histaminergic pharmacology of homo-oligomeric β3 γ-aminobutyric acid type A receptors characterized by surface plasmon resonance biosensor technology. *Biochem. Pharmacol.* **84**: 341–351.

[76] Aristotelous T, Ahn S, Shukla AK, Gawron S, Sassano MF, Kahsai AW, Wingler LM, Zhu X, Tripathi-Shukla P, Huang XP, Riley J, Besnard J, Read KD, Roth BL, Gilbert IH, Hopkins AL, Lefkowitz RJ, Navratilova I (2013). Discovery of β2 adrenergic receptor ligands using biosensor fragment screening of tagged wild-type receptor. *ACS Med. Chem. Lett.* **4**: 1005–1010.

[77] Yanamala N, Dutta A, Beck B, van Vliet B, Hay K, Yazbak A, Ishima R, Doemling A, Klein-Seetharaman J (2010). NMR-based screening of membrane protein ligands. *Chem. Biol. Drug Des.* **75**: 237–256.

[78] Congreve M, Rich RL, Myszka DG, Figaroa F, Siegal G, Marshall FH (2011). Fragment screening of stabilized G-protein-coupled receptors using biophysical methods. *Methods Enzymol.* **493**: 115–136.

[79] Chen D, Errey JC, Heitman LH, Marshall FH, Ijzerman AP, Siegal G (2012). Fragment screening of GPCRs using biophysical methods: identification of ligands of the adenosine A(2A) receptor with novel biological activity. *ACS Chem. Biol.* **7**: 2064–2073.

[80] Christopher JA, Brown J, Doré AS, Errey JC, Koglin M, Marshall FH, Myszka DG, Rich RL, Tate CG, Tehan B, Warne T, Congreve M (2013). Biophysical fragment screening of the β1-adrenergic receptor: identification of high affinity arylpiperazine leads using structure-based drug design. *J. Med. Chem.* **56**: 3446–3455.

[81] Wells JA, McClendon CL (2007). Reaching for high-hanging fruit in drug discovery at protein-protein interfaces. *Nature* **450**: 1001–1009.

[82] Mullard A (2012). Protein-protein interaction inhibitors get into the groove. *Nat. Rev. Drug Discov.* **11**: 173–175.

[83] Thompson AD, Dugan A, Gestwicki JE, Mapp AK (2012). Fine-tuning multiprotein complexes using small molecules. *ACS Chem. Biol.* **7**: 1311–1320.

[84] Valkov E, Sharpe T, Marsh M, Greive S, Hyvönen M (2012). Targeting protein-protein interactions and fragment-based drug discovery. *Top. Curr. Chem.* **317**: 145–179.

[85] Moreira IS, Fernandes PA, Ramos MJ (2007). Hot spots—a review of the protein-protein interface determinant amino-acid residues. *Proteins* **68**: 803–812.

[86] Winter A, Higueruelo AP, Marsh M, Sigurdardottir A, Pitt WR, Blundell TL (2012). Biophysical and computational fragment-based approaches to targeting protein-protein interactions: applications in structure-guided drug discovery. *Q. Rev. Biophys.* **45**: 383–426.

[87] Barile E, Pellecchia M (2014). NMR-based approaches for the identification and optimization of inhibitors of protein-protein interactions. *Chem. Rev.* **114**(9): 4749–4763.

[88] Scott DE, Ehebauer MT, Pukala T, Marsh M, Blundell TL, Venkitaraman AR, Abell C, Hyvönen M (2013). Using a fragment-based approach to target protein-protein interactions. *Chembiochem* **14**: 332–342.

[89] Dias DM, Van Molle I, Baud MGJ, Galdeano C, Geraldes CFGC, Ciulli A (2014). Is NMR fragment screening fine-tuned to assess druggability of protein-protein interactions? *ACS Med. Chem. Lett.* **5**: 23.

[90] Schultes S, de Graaf C, Haaksma E, de Esch I, Leurs R, Kramer O (2010). Ligand efficiency as a guide in fragment hit selection and optimization. *Drug Discov. Today* **7**: 157–162.

[91] Howard S, Berdini V, Boulstridge JA, Carr MG, Cross DM, Curry J, Devine LA, Early TR, Fazal L, Gill AL, Heathcote M, Maman S, Matthews JE, McMenamin RL, Navarro EF, O'Brien MA, O'Reilly M, Rees DC, Reule M, Tisi D, Williams G, Vinkovic′ M, Wyatt P (2009). Fragment-based discovery of the pyrazol-4-yl urea (AT9283), a multitargeted kinase inhibitor with potent aurora kinase activity. *J. Med. Chem.* **52**: 379–388.

[92] Filippakopoulos P, Knapp S (2014). Targeting bromodomains: epigenetic readers of lysine acetylation. *Nat. Rev. Drug Discov.* **13**: 337–356.

[93] Bamborough P, Diallo H, Goodacre JD, Gordon L, Lewis A, Seal JT, Wilson DM, Woodrow MD, Chung C (2012). Fragment-based discovery of bromodomain inhibitors part 2: optimization of phenylisoxazole sulfonamides. *J. Med. Chem.* **55**: 587–596.

[94] Tanaka D, Tsuda Y, Shiyama T, Nishimura T, Chiyo N, Tominaga Y, Sawada N, Mimoto T, Kusunose N (2011). A practical use of ligand efficiency indices out of the fragment-based approach: ligand efficiency-guided lead identification of soluble epoxide hydrolase inhibitors. *J. Med. Chem.* **54**: 851–857.

[95] Barelier S, Pons J, Marcillat O, Lancelin JM, Krimm I (2010). Fragment-based deconstruction of Bcl-xL inhibitors. *J. Med. Chem.* **53**: 2577–2588.

[96] Giannetti AM, Zheng X, Skelton NJ, Wang W, Bravo BJ, Bair KW, Baumeister T, Cheng E, Crocker L, Feng Y, Gunzner-Toste J, Ho Y, Hua R, Liederer BM, Liu Y, Ma X, O'Brien T, Oeh J, Sampath D, Shen Y, Wang C, Wang L, Wu H, Xiao Y, Yuen P, Zak M, Zhao G, Zhao Q, Dragovich PS (2014). Fragment-based identification of amides derived from trans-2-(pyridin-3-yl)cyclopropanecarboxylic acid as potent inhibitors of human nicotinamide phosphoribosyltransferase (NAMPT). *J. Med. Chem.* **57**(3): 770–792.

[97] Hirose T, Maita N, Gouda H, Koseki J, Yamamoto T, Sugawara A, Nakano H, Hirono S, Shiomi K, Watanabe T, Taniguchi H, Sharpless KB, Omura S, Sunazuka T (2013). Observation of the controlled assembly of preclick components in the in situ click chemistry generation of a chitinase inhibitor. *Proc. Natl. Acad. Sci. U. S. A.* **110**(40): 15892–15897.

[98] Lehn JM, Eliseev AV (2001). Dynamic combinatorial chemistry. *Science* **291**(5512): 2331–2332.

[99] Zartler E (2014). Fragonomics: the -omics with real impact. *ACS Med. Chem. Lett.* **5**(9): 952–953.

# 9

# VIRTUAL SCREENING

KARL-HEINZ BARINGHAUS and GERHARD HESSLER
*Sanofi-Aventis Deutschland GmbH, Frankfurt am Main, Germany*

## 9.1 INTRODUCTION

Drug discovery has gone through a significant change in recent decades from a phenotypic approach to a target-based approach. Traditionally, drug discovery used authentic biological systems, such as animals or cells, to determine the biological activity of novel compounds. In the last decades, the focus of drug discovery shifted to hypothesis-driven, target-based approaches in which a specific gene product (associated with a disease) is targeted. The advent of high-throughput technologies like combinatorial chemistry and high-throughput screening allowed synthesis and biological testing of thousands of compounds for their biological activity on the molecular target in question. Biological test systems are established, which allow identification of compounds, modulating the activity of the target protein as desired.

The identification of lead structures is a crucial step in drug discovery, since this selection critically determines the likelihood of success for subsequent lead optimization. A key parameter for lead selection is the biological activity to the target of interest. Nevertheless, many other properties need to be considered to be successful in later optimization steps [1]. The so-called absorption, distribution, metabolism, and excretion (ADME) properties describe what the body does to the drug and how well the drug can reach its site of action. Physicochemical properties of the compounds like polarity and lipophilicity strongly influence properties such as solubility or permeability and thus determine how the compound can be administered later. Additionally, metabolic lability strongly modifies the pharmacokinetic profile of drug candidates. Also, affinity to undesired biological targets has to be considered (e.g., hERG channel) to reduce likelihood of side effects. Therefore, it is necessary to carefully watch several parameters in the lead

*Small Molecule Medicinal Chemistry: Strategies and Technologies*, First Edition.
Edited by Werngard Czechtizky and Peter Hamley.

selection process in order to take up compounds with as few potential liabilities as possible.

In target-based drug discovery, one can distinguish between experimental and *in silico* approaches. High-throughput screening, a well-established lead identification strategy, is based on experimental testing of large compound collections. Depending on the quality of the assays, only few active compounds are identified with hit rates in the range from 0.1 to 2% [2].

*In silico* drug discovery approaches employ information about the molecular target or make use of already known ligands. In structure-based drug design efforts, X-ray crystal structures of the target protein with a ligand are used to optimize ligand affinity, since a detailed picture of protein–ligand interactions becomes available from such structures. On the other hand, information from known ligands of the molecular target can also be used in ligand-based design efforts to derive molecular features of the ligands, which are necessary for good interaction with the target protein.

Virtual screening, which is the *in silico* counterpart of high-throughput screening, has become an established lead generation strategy. Compound collections are virtually (in the computer) screened to predict their putative activity against a target protein. Available information from protein structures or available ligands is applied to select a smaller set of compounds out of larger databases for subsequent biological testing. In contrast to high-throughput screening, virtual screening generates only a small subset of compounds for subsequent determination of biological activity. Thus, virtual screening aims to explore large chemical databases at minimal costs. The used rationale either is the known protein structure of the target (structure-based virtual screening) or employs information from known actives (ligand-based virtual screening).

Virtual screening approaches can be distinguished as shown in Figure 9.1. Three-dimensional (3D) approaches summarize virtual screening technologies, which make use of any sort of 3D structural information either from ligands (ligand-based) or from protein (structure based). In contrast to that, many ligand-based approaches only use topological information such as the chemical substructure of a biologically active ligand or similarity to a biologically active ligand. Such techniques, which do not need a 3D structure, are called 2D approaches. As indicated in Figure 9.1, virtual screening strongly requires target and/or ligand information upfront, which is carefully analyzed to select the virtual screening methods.

Virtual screening is meanwhile a well-accepted method, which is frequently applied in hit and lead identification. This established position is reflected by numerous reviews describing different facets of virtual screening [3–16]. In this chapter, we will focus on requirements and aspects of virtual screening, which we consider as being important for a successful application.

### 9.1.1 Goals of Virtual Screening

The aim of virtual screening is to identify molecules active against a certain protein. A wealth of different techniques and software programs is available for virtual screening. For example, substructure searching uses a defined chemical substructure to identify other molecules with the same chemical fragments. Various similarity searching methods

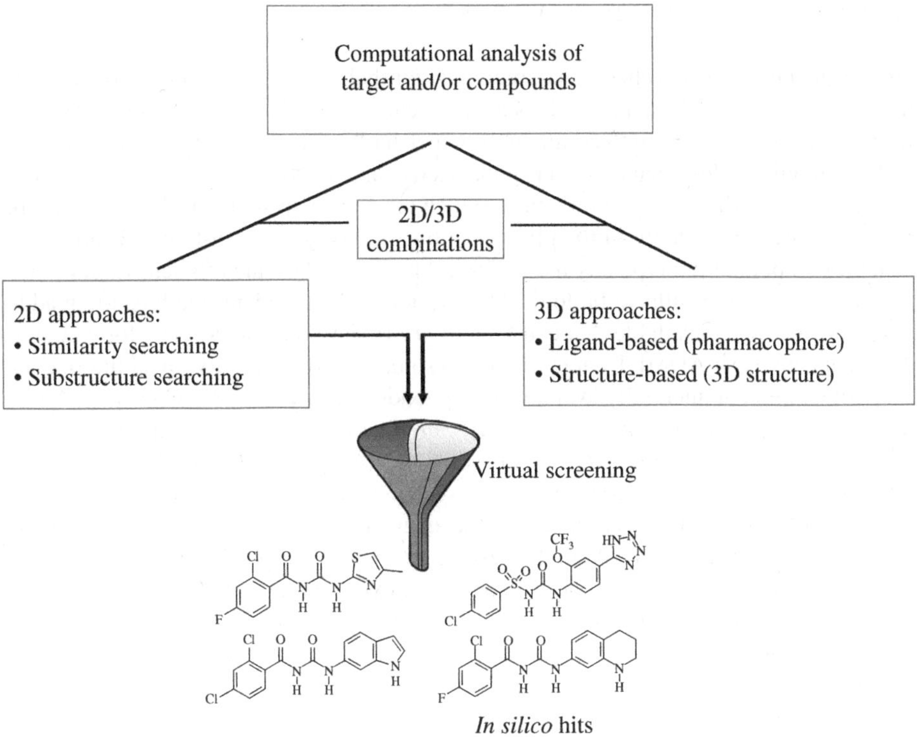


**FIGURE 9.1** Virtual screening is a computational filtering procedure, which uses available knowledge to filter out compounds with a low probability to be biologically active against the target of interest. To this end, three-dimensional (3D) information is used from ligand and protein structures in the so-called 3D approaches, while 2D approaches rely on topological information from ligands, such as the chemical substructure. Often, 2D and 3D approaches are combined.

recognize chemically related molecules as hits of virtual screening searches. Thus, generally close analogues of the query structures are found. Accordingly, these methods are used for the identification of additional analogues of a known biologically active molecule in order to complete the SAR picture from available chemical compounds. On the other extreme, it might be necessary to identify novel chemical structures due to threats of the intellectual property position or due to toxicity issues of the actual chemical series. 3D approaches like molecular docking, which do not rely on a chemical structure of a query molecule, can be used to accomplish such a goal. Success criteria are completely different in these situations. In the first example, many active compounds are beneficial to complete the SAR picture, while in the second case, the hit rate is of less importance, but the number of novel chemical scaffolds identified is much more relevant.

In addition, the available knowledge about the target and its ligands can differ significantly starting from only one active ligand to many different molecules together with a detailed understanding of protein–ligand interactions due to available cocrystal structures. Depending on the actual knowledge and goal of the virtual screening, methods and databases have to be carefully chosen.

## 9.2 DATABASES AND DATABASE PREPARATION

Virtual screening assesses chemical compounds for their potential suitability to modulate the activity of the biological target. This analysis can be done on physically existing or on physically nonexisting (virtual) compounds, which have to be synthesized. Since chemical synthesis requires additional time and synthesis resources, virtual screening is often preferentially done with existing compounds, which are readily available in-house or can be purchased easily. Especially in the pharmaceutical industry, large in-house databases of available compounds are typically used. This allows a very fast and efficient validation of the virtual screening results in biological test systems. For the identified hits, often additional information is available that can help in prioritizing the compounds, for example, additional activity or ADMET data pointing to some potential liability. In addition, databases containing commercially available compounds, for example, ZINC [17] or eMolecules [18] can be used. Such compounds can usually be purchased if identified as interesting virtual screening hits. Databases collecting publically available chemical structures associated with biological activity, such as ChEMBL [19], Reaxys® Medicinal Chemistry [20], GVKBio [21], etc., are also of interest in virtual screening, but these compounds are not readily available for testing and have to be synthesized.

Virtual screening requires access to databases of existing molecules, in which the compounds are stored in a manner necessary for the applied method. Different virtual screening software and techniques use different ways to translate the molecular structure into a computer readable format, which result in different molecular descriptors. For example, the UNITY database (UNITY software, Certara, St Louis, MO, USA) has its own format, which is completely different from feature trees [22] format. The CATS descriptor [23] again needs another representation. These few examples, which are depicted in Figure 9.2, illustrate that different software and different techniques employ completely different database formats. Different descriptors usually result in different hit lists, which only partially overlap. Therefore, from a pragmatic point of view, a decision has to be taken, on which tools one would like to focus. In our experience, it turned out that at least two to three different tools should be applied for virtual screening to get more diversity into the hit lists. For the identification of novel scaffolds that are significantly different from the query (scaffold hopping), we have good experiences in using the CATS descriptor, feature trees, and rapid overlay of chemical structures (ROCS) [24, 25].

Each virtual screening database requires a clear definition of how chemical structures are represented in a meaningful way. Especially for 3D databases, important aspects are treatment of stereoisomers, charges and protonation state, tautomers, and conformers. Although the preparation of the virtual screening database significantly influences the results, to our knowledge, no comprehensive analysis of all of these aspects was performed. Nevertheless, some publications clearly demonstrate the effect of the decisions taken at that stage [26–29].

An aspect that has been thoroughly investigated by several authors is the influence of the conformational sampling on the representation of the bioactive conformation in the data set, that means, is the conformation of the ligand bound to the protein structure represented in the conformational ensemble. A ligand bound to different proteins can exist in several bioactive conformations with different conformational energies [30, 31]. In addition, for a significant fraction of examples studied, the bioactive conformation is only present in the conformational ensemble, if large energy thresholds are chosen [32–39].

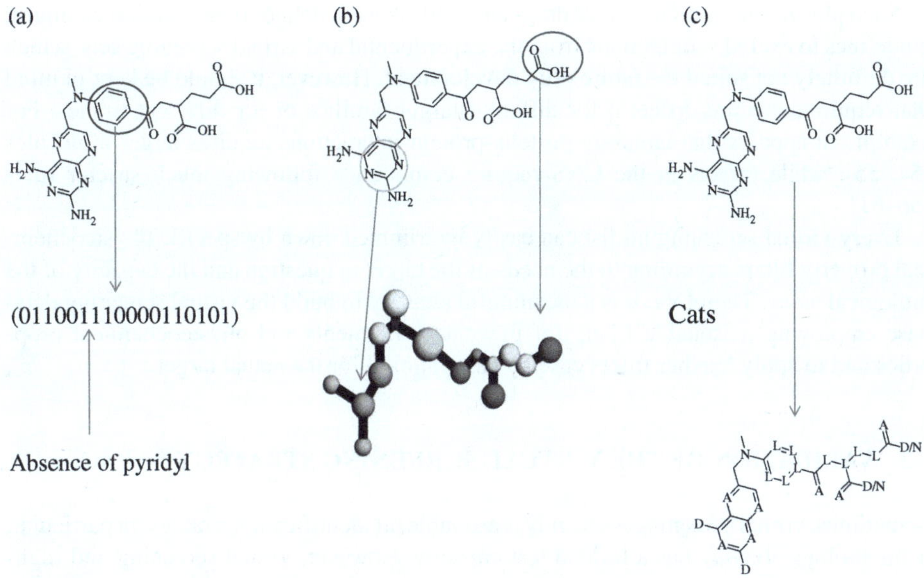


**FIGURE 9.2** Illustration of different descriptors, which are used to calculate chemical similarity. (a) Fingerprints look for absence or presence of chemical fragments in the molecular structure and set a 1 or a 0 in bit string accordingly. (b) Chemical fragments are converted into a node while capturing molecular features like H-bond donors, H-bond acceptors, or size of the corresponding group. The connectivity of the nodes is described in a graph-like structure. (c) The CATS descriptor converts each atom into its corresponding pharmacophoric feature like H-bond donor, H-bond acceptor, lipophilic, etc. The frequency of occurrences of pairs of such features separated by the number of bonds in between is stored in a normalized vector (not shown).

A recent analysis points out that for a large part of protein–ligand complexes, ligand conformational parameters deviate significantly from small molecule X-ray structure reference values. It has also been highlighted that due to errors in ligand geometry of PDB structures, analysis of strain energies should be focused to the protein structures of the highest quality [40]. All studies indicate that conformational ensembles in virtual screening databases should include conformationally diverse ensembles including conformations of higher energy.

As for experimental screening collections, virtual screening databases should not contain undesired compounds, which are most likely not pursued further, if found active. While it is rather easy to find consensus among medicinal chemists on reactive, undesired functional groups, such as acid chlorides, aldehydes, etc. [41–43], less unambiguous agreement is usually observed if physicochemical property filters need to be defined. After the pioneering work of Lipinski [44] defining the rule of 5, several other authors followed this line of thinking. Accordingly, other parameters were introduced, which influence permeability of small molecules, such as the number of rotatable bonds [45] or polar surface area [46]. These rules have been reinvestigated and refined several times [47, 48], and their application to decision-making has been advocated in different embodiments [49–51]. Typically, compounds grow in molecular weight and lipophilicity during an optimization phase [52], which resulted in numerous publications recommending tighter thresholds for physicochemical properties for lead-likeness, for example, the rule of three [53].

Such physicochemical limits for drug- and lead-likeness define some overall meaningful guidelines to exclude compounds from the experimental and virtual screening sets, which are definitely not suited for future drug development. However, it should be kept in mind that requirements are different for different target families or for different tissues. For example, it appears that targeting protein–protein interactions requires larger molecules [54, 55], while targets in the CNS require compounds following much stricter rules [56–61].

Every virtual screening hit list can easily be trimmed down by specific physicochemical property filters according to the needs of the target in question and the capacity of the biological assay. Therefore, it is a meaningful strategy to build the virtual screening database employing reasonable filters for unwanted fragments and physicochemical properties and to apply harsher filters case by case tailored for the actual target.

## 9.3 VALIDATION OF THE VIRTUAL SCREENING STRATEGY

Sometimes, virtual screening is the only reasonable hit identification strategy, in particular, if the biological assay has a limited test capacity. However, virtual screening and high-throughput screening are often done in parallel. Virtual screening can be used to evaluate compounds that are not included in the high-throughput screening set, such as compounds from external databases. For virtual screening, not only the hit rate is important but also its capability to identify novel chemotypes. Additionally, the virtual screening approach also provides the benefit of a rationale for important molecular features of the ligand, for example, the fit to a pharmacophore model or even a binding mode hypothesis in structure-based virtual screening, which can subsequently be used for compound optimization.

Successful virtual screening applications require the definition and the careful validation of the virtual screening strategy. Important aspects of the virtual screening strategy comprise questions like which virtual screening methods are used, whether these methods are applied in parallel or used sequentially in a cascaded virtual screening, and how many compounds can be selected for biological testing. For the validation step, a set of active and inactive compounds is required. Since a virtual screening run generates hits in which the top-ranked compounds are supposed to be biologically active, this list can be compared with the known true biological actives to judge the success of the chosen virtual screening strategy. Biologically active molecules among the top-ranked compounds are true positives (TP), whereas biologically inactive compounds in that list are false positives (FP). True negatives (TN) are compounds that are correctly predicted as biologically inactive by the virtual screening. Biologically active compounds that are predicted to be inactive are called false negatives (FN). Based on this information, the virtual screening can be evaluated and different success criteria can be checked and defined. Important criteria related to the successful identification of biologically active compounds that are frequently used are summarized in Figure 9.3.

The enrichment factor (EF) is calculated from the hit rate found in the virtual screening hit list compared to the hit rate in the complete validation data set. Thus, it specifies how many more actives are identified in the virtual screening in comparison to a random selection of compounds. The accuracy captures how many molecules are correctly

$$\text{Hit rate} = \frac{TP}{n}$$

$$EF = \frac{TP / n}{A / N}$$

$$\text{Accuracy} = \frac{TP + TN}{N}$$

$$\text{Sensitivity} = \frac{\text{selected actives}}{\text{all actives}} = \frac{TP}{TP + FN}$$

$$\text{Specificity} = \frac{\text{discarded inactives}}{\text{all inactives}} = \frac{TN}{TN + FP}$$

*EF*: enrichment factor
*A*: actives
*TP*: true positives and *TN*: true negatives
*FN*: false negatives and *FP*: false positives
*n*: number of compounds in virtual screening hitlist
*N*: number of compounds in validation data set

**FIGURE 9.3** Frequently used success criteria for virtual screening.

classified either as active or inactive, while sensitivity evaluates how well biologically actives were retrieved. Specificity refers to the capability to identify inactives correctly. A predictive virtual screening run can distinguish well between biologically active and inactive molecules, which would mean high specificity and high sensitivity. In a graphical analysis, both sensitivity and specificity are used to generate a so-called receiver operating characteristic (ROC) analysis, which allows a quick analysis of the performance of the virtual screen [62] (Fig. 9.4). The *x*-axis describes the FP rate (1-specificity), while the *y*-axis plots the sensitivity. If the virtual screening run can distinguish well between actives and inactives, a high sensitivity is combined with low FP rate; thus, the plotted curve rises steeply, resulting in a large area under the curve (AUC). For different virtual screening strategies, the area under this curve can be compared to find the optimal virtual screening strategy.

Virtual screening allows testing a smaller set of compounds, while a large database of compounds was evaluated *in silico* for potential biological activity. An efficient virtual screening strategy is characterized by hit lists, which are as small as possible. The ROC analysis does not take into account if the actives are correctly identified in the very top of the virtual screening hit list or at the end of the hit list, which is called the early recognition problem. BEDROC provides an improved version of this type of analysis including early recognition of positives [63]. In a nice example, several pharmacophore models have been developed for 11-βHSD types 3 and 5. Careful validation and subsequent virtual screening yielded several novel inhibitors [64].

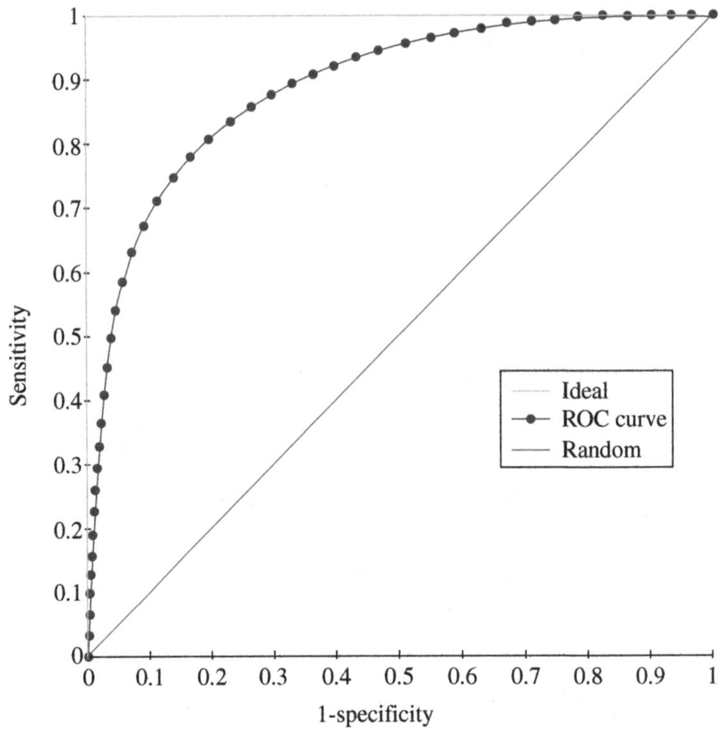


**FIGURE 9.4** In an ROC plot, the true positive rate (sensitivity) is plotted against the false positive rate (1-specifity). In a successful virtual screening run, the curves quickly rise, if many TP are identified among only few false positives. The area under curve (AUC) becomes large. The diagonal corresponds to a random selection of compounds.

Another important criterion is the capability of the virtual screening to identify novel scaffolds. Needless to say, this requires a definition of a scaffold. Often, the concept of the molecular framework is used for that, which was introduced by Bemis and Murcko [65], which comprises ring systems connected by linkers in between. Such frameworks allow to judge how many different chemotypes have been identified within virtual screening. Related definitions have also been used to measure this capability [66], and recently, quantification of scaffold similarity was established [67].

If enough data for a careful virtual screening validation is not at hand, iterative screening [68–70] comes into play, which allows learning cycles for improvement of the virtual screening. In our experience, this increases success rates of virtual screening significantly.

## 9.4 LIGAND-BASED VIRTUAL SCREENING

Ligand-based virtual screening only employs information from available ligands of the target in question, but does not require any knowledge about the 3D protein structure. Methods can be divided into 2D and 3D approaches. While 3D approaches make use of 3D structure of ligands, two-dimensional (2D)-based virtual screenings rely on the use of 2D molecular descriptors, depicted, for example, in Figure 9.2.

### 9.4.1 2D Approaches

The underlying principle of 2D ligand-based virtual screening is the similarity principle stating that similar chemical structures show similar biological activity [71]. In 2D-based approaches, similarity is considered from a ligand point of view, since different chemical structures are compared on the basis of descriptors. Nevertheless, it should be kept in mind that similarity depends on the molecular descriptor. For example, chemical structures can be similar with respect to some physicochemical properties, but dissimilar with respect to other physicochemical properties [72]. The same is frequently observed with respect to activity against protein targets. Similar compounds can be active against some targets but show completely different activity at other targets [73]. Limitations of the similarity concept are also well known by medicinal chemists. If a small change in the chemical structure results in a dramatic change in biological activity, this phenomenon is called an activity cliff [74, 75].

***9.4.1.1 Descriptor Selection and Similarity Searching*** Descriptor types are differently well suited for the task of virtual screening. Searching for analogues is usually well performed with 2D fingerprints, in which absence or presence of chemical fragments are counted and collected into a bit string (Fig. 9.2). Typical examples are UNITY fingerprints (UNITY software, Certara, St Louis, MO, USA), Daylight fingerprints, or MACCS keys (MACCS Structural Keys; Accelrys, San Diego, CA, USA). Different similarity measures can be calculated from these bit strings, for example, the Euclidean distance or the Tanimoto coefficient [76]. The Tanimoto coefficient is the ratio of the number of bits both molecules have in common and the total number of bits. It has higher values if two molecules share many chemical fragments and thus are similar.

2D fingerprints perform well in the identification of close analogues to the query structure. In order to expand the chemical space accessible by fingerprints, the virtual hits of the previous iteration are automatically used as queries for the next iteration [77]. Nevertheless, parameters like similarity radius and number of iterations have to be carefully selected to balance the increased search space against the number of hits.

Other descriptors use a more abstract representation of the chemical structure by applying pharmacophoric patterns as atom representations. A nice example is the CATS descriptor (Fig. 9.2) [23]. The atom types are converted into five generalized atom types (hydrogen-bond donor, hydrogen-bond acceptor, positively charged, negatively charged, lipophilic). The distance between pairs of atoms is determined as the shortest bond path and translated into a histogram. Due to their chemically more abstract atom representation, CATS fingerprints prove to be successful in the identification of novel chemical scaffolds in several applications [78–80].

Feature tree [22] is a different descriptor that uses pharmacophoric features in a graph-like representation of the molecule (Fig. 9.2). Therefore, feature trees can be considered as a 2D pharmacophore of the molecule. Several successful scaffold hoppings were reported for this descriptor [81–83]. Unfortunately, there is no best descriptor available for virtual screening [84–87]. The examples given here are a subjective choice that showed satisfying results in several applications.

In general, 2D ligand-based approaches return a hit list ranked by the similarity to the query structure. A cutoff defines a similarity radius, which separates similar compound from dissimilar compounds. The user has to keep in mind that no general cutoff for different descriptors and different metrics can be applied for compound selection [73]. Different descriptors result in different hit lists, which only partially overlap. Hence, parallel application of several descriptors and several queries is advised. The resulting hit lists can be either merged or treated with different data fusion techniques [88–90].

Very often, the highest similarity found for a compound or the sum of different similarity measures from the various descriptors is used to generate the final ranking. Another strategy is to focus on the compounds identified by more than one method, which is expected to increase the chance to find active compounds. On the other hand, this strategy increases the risk to miss interesting compounds top ranked by only one method.

***9.4.1.2 Machine Learning in Virtual Screening*** Medicinal chemistry experience tells that some parts of the molecule are more important for the interaction to the pharmaceutical target, while other parts of the molecule are not necessary for interaction or even reduce the biological activity. This structure–activity relationship is not employed in similarity-based virtual screenings where every part of the molecule has the same weight for the search, even those parts of the molecule that negatively affect the biological activity. Therefore, different efforts have been applied to introduce some weighing scheme by generating SAR models with machine learning algorithms from data sets of active and inactive molecules. The obtained and validated models are then used for virtual screening.

Such machine learning efforts can be divided into supervised and unsupervised learnings. An unsupervised learning algorithm does not take the target value into account during model development. Within the learning phase, only descriptors are used to group molecules together. The biological activity is then mapped to molecules to reveal groups of biologically active compounds. In the virtual screening effort, novel compounds are projected onto the model, and compounds are predicted to be active if they fall into a group of neighboring active compounds. Examples are principal component analysis (PCA) and self-organizing maps (SOM) [91–94]. An unsupervised learning algorithm does not face the problem of overtraining, but there is no guarantee that compounds group together according to their biological activity. Nevertheless, interesting examples have been presented for the successful use of PCAs [95] and SOMs [96–98] in virtual screening recently.

Supervised learning techniques make use of the available biological data to train a classifier or a regression algorithm. In supervised learning techniques, there is a risk that models are adopted to the training data and thus explain the training data set well, but are not suited anymore to predict novel data different from the training data. This means that the model is overfitted to the training data set. To recognize overfitting, the available data usually must be split into some training and test set. Overfitted models can only poorly predict the test data. If data sets of sufficient size are available, robust models usually can be obtained. Therefore, 2D descriptors in combination with various machine learning algorithms are frequently used to derive global models for ADMET properties like permeability, metabolic liability [99, 100], or hERG inhibition [101–103]. Frequently, decision trees [104], random forest [105], support vector machines (SVM) [106], or Bayesian classifiers [107–109] are trained to obtain SAR models. There are numerous published examples for the successful use of these approaches for virtual screening either alone or in combination with other virtual screening methods [110–118].

### 9.4.2 3D Ligand-Based Approaches

3D ligand-based virtual screening employs the 3D structure of the ligands and thus takes conformational aspects into account. Therefore, the spatial orientation of important molecular features for the interaction with the target, so-called pharmacophoric features, is considered. Only molecules are returned as hits that are capable of achieving the same spatial orientation of pharmacophoric features as required in the query.

Another option is the application of the similarity concept in 3D space. An illustrative example is the ROCS program [24, 25], in which the shape of a molecule is approximated by atom-centered Gaussians. The overlap between Gaussian functions allows a straightforward mathematical approximation of atom overlap. By maximization of the sum of atomistic Gaussian overlaps, ROCS can rapidly superpose two ligands based on their shape and returns a shape similarity score. Additionally, matching of related pharmacophoric features can be considered by a corresponding score. Recently, several successful applications of ROCS in ligand-based virtual screening have been described [119–121]. Numerous other 3D similarity-based methods are available, for example, Flex-S [122] or Surflex-Sim [123], which have been recently compared [124].

The most frequently used technology for 3D ligand-based virtual screening is the pharmacophore model. A recent formal definition comes from C. Wermuth [125]: "A pharmacophore is the ensemble of steric and electronic features that is necessary to ensure the optimal supramolecular interactions with a specific biological target and to trigger or block its biological response." There are numerous well-established commercial programs and algorithms that can be used to derive pharmacophore models, for example, DS Catalyst [126, 127], LigandScout [128], Phase [129], and the pharmacophore tool of MOE [130]. Although the algorithms differ in detail, the same principle is used: In the first step, a conformational analysis of the training data of biologically active ligands is performed. Subsequently, the ligands are aligned to each other. Based on this alignment, common pharmacophoric features are extracted and stored as pharmacophore models (Fig. 9.5). Typically, pharmacophoric features like hydrogen-bond acceptor, hydrogen-bond donor, positively or negatively charged, and lipophilic or aromatic features are recognized as important molecular interaction points, although the detailed definition varies from software to software. In most programs, the default definition of pharmacophoric features can be modified to tackle specific features of the data set. As indicated in Figure 9.5, 3D structures of protein–ligand complexes allow to recognize key pharmacophoric features of protein–ligand interactions. The software LigandScout offers an efficient analysis of protein–ligand complexes in terms of pharmacophoric features.

Important prerequisites to generate meaningful, reliable pharmacophore models exist. First of all, pharmacophore models are derived from an alignment hypothesis of the molecules in the training set. Therefore, the underlying assumption is that all molecules follow the same mode of action and have the same binding site and binding mode. Second, since pharmacophore models use generalized molecular features, a diverse training set is important, in which ideally the same molecular interaction feature is represented by different chemical groups. Finally, the conformational search generates larger sets of different conformers, from which the alignment is generated. Inclusion of less flexible molecules increases the chance to find a reliable pharmacophore model.

In any case, pharmacophore models require careful validation prior to the application in virtual screening. A nice application of pharmacophore models with a thorough validation step before has been described by Wolber et al. [131].

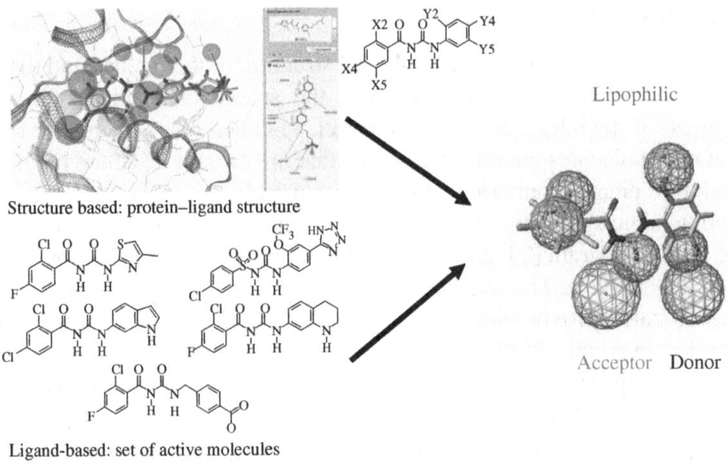


**FIGURE 9.5** Generation of a pharmacophore model for a chemical series of glycogen phosphorylase inhibitors. A 3D structure of a complex between glycogen phosphorylase and an inhibitor is analyzed in the program LigandScout. Hydrogen bonds between ligand and protein are indicated by vectors. Lipophilic ligand features and excluded volumes, which cannot be occupied by the ligand, are represented as spheres. The data set of five different ligands was used to derive several ligand-based pharmacophores. The combination of both types of knowledge allows to select the final pharmacophore. (*See insert for color representation of the figure.*)

One of the big advantages of a pharmacophore model is that it helps in visualizing molecular features, which are important for the interaction to the target. This can stimulate creative ideas for modification of the actual compounds. In addition, pharmacophore models allow deriving alignment rules between different chemotypes, which can be used to transfer SAR from one chemical series to another one.

Classical Pharmacophore software uses defined chemical patterns to assign chemical groups to certain pharmacophoric features. A recent promising development, called FLAPpharm [132], uses molecular fields calculated with the program GRID [133] instead to recognize corresponding molecular features. These interaction fields are used to extract pharmacophoric features after the molecular alignment of the ligands. Extensive validation studies for this approach have been provided [134].

Pharmacophore-based virtual screening requires 3D databases of ligands, which are searched with the pharmacophore model yielding structures in line with the pharmacophore model. Many tools use databases of precomputed conformations, for example, DS CATALYST, LigandScout, or PHASE. On the other hand, UNITY 3D database searches use the direct tweak algorithm [135] to generate conformations of the candidate molecules on the fly. Precomputed databases can be searched more quickly but require a larger disc space for storage.

Usually, the hits of the virtual screening are scored according to their fit to the pharmacophore model. DS CATALYST, for example, calculates fit values based on the ratio between the distance of pharmacophore features and molecular features of the molecule, $D(c)$, and the allowed tolerance $T(c)$ for that feature:

$$\text{Fit} = \sum \text{weight}(f) * \left[ 1 - \left\langle \frac{D(c)}{T(c)} \right\rangle^2 \right]$$

A high score indicates a good fit of the molecule to the pharmacophore model. Thus, the hit lists can be ranked according to their scores. In case of partial matches, not all features of the model have to be met by a compound. This is interesting, if most, but not all active molecules do show a certain feature.

If the 3D structure of a protein–ligand complex is available, it provides a wealth of information, since it directly highlights the interacting features of the ligand. Thus, structure-derived pharmacophore models offer a quick access to reliable pharmacophore models.

## 9.5 STRUCTURE-BASED VIRTUAL SCREENING

A protein structure and in particular the structure of a protein–ligand complex provide a wealth of information about the detailed picture of the protein–ligand interaction for the target in question. The hydrogen-bond network, salt bridges, and hydrophobic interactions are visible and help to derive design rationales. Structure-derived pharmacophore models translate such key interaction features into pharmacophore models, which make this information readily available for 3D-based virtual screening.

Structure-based virtual screening uses the information content provided by an X-ray or NMR structure of the biological target and the premise that shape and molecular interaction pattern of ligand and protein are complementary to each other. To this end, different ligands are placed into the binding cavity of the protein, and the binding pose as well as the quality of the fit into the protein is predicted. This so-called docking procedure searches for a binding pose in two steps, a conformational sampling step and a scoring step. During the sampling step, different conformations of the ligand are investigated for their goodness of fit into the protein structure. Some programs use a database of precomputed conformations, for example, the program DOCK [136] or Fred [137]. The second strategy employed by programs like FlexX [138] and Surflex [139] relies on buildup strategies: The ligand is fragmented, and starting from an anchor fragment, it is rebuilt in the binding site using databases of allowed torsions, for example, the MIMUMBA [140] database. Stochastic searches like Monte Carlo methods are the third possibility used for conformational sampling of the ligands and are applied by programs like GLIDE [141], in which the conformations are generated on the fly by random translational or torsional changes for the ligand. The new conformation is only accepted if the energy improves. Conformations with higher energy are only accepted with a probability proportional to the Boltzmann weight factor $\exp(-E/kT)$. This procedure ensures that the energy of the obtained protein–ligand complexes is distributed according to the Boltzmann distribution.

Genetic algorithms are another widely used method for stochastic sampling of ligand poses and conformations. They are applied by programs like AutoDock [142] and Gold [143]. Other stochastic algorithms such as ant colony optimization in the program PLANTS [144] or stochastic tunneling in the program FlexScreen [145, 146] are used as well.

The second important aspect of docking is the scoring function. Scoring functions have a manifold importance for docking and virtual screening. First of all, scoring functions are used to assess the goodness of the binding pose during the sampling procedure; second, it ranks the ligands according to their putative binding energies, putting the best ligands onto the first ranks while adding worse ligands to the end of the rank list. Finally, scoring functions are used to estimate the binding energy of potential ligands. In the context of virtual screening, scoring up to several hundred thousands of ligands is performed in the docking process. Therefore, the main task of scoring functions is to separate potential binders from nonbinders at an appropriate speed. In principle, three types of scoring functions are used [147, 148]: force field-based scoring functions, knowledge-based scoring functions, and empirical scoring functions.

Force field-based scoring functions use the nonbonding interaction terms of classical force fields to estimate the intermolecular interaction energy between ligand and proteins. The terms comprise electrostatic interactions as well as van der Waals energies. Sometimes, hydrogen-bonding terms or the internal strain of the ligand are also included. Knowledge-based scoring functions are parameterized from a large set of protein–ligand complexes under the assumption that energetically favorable distances between different atom types are more frequently populated than unfavorable distances. Thus, the frequency of occurrence of ligand atom—protein atom pairs is converted into a potential of mean force based on the inverse Boltzmann law. Examples of these scoring functions are DrugScore [149] or PMF [150]. Empirical scoring functions fit various terms important for molecular interaction to predict the experimental binding energies. To this end, a regression equation is trained based on terms describing hydrogen bonds, contact surfaces, etc. to reproduce the experimental affinities for a large diverse set of proteins. Various efforts have been undertaken to develop empirical scoring functions, for example, the SFC score [151], X-Score [152], ChemScore [153], PLP [154], etc.

In numerous studies, the performance of different scoring functions has been assessed and recently been summarized [148]. In short, no scoring function performs equally well for all targets. Pose predictions of experimental binding poses can be quite accurate with up to 80% of the predictions being correct. Unfortunately, the performance is significantly worse with respect to the correct ranking of scoring functions, which is in the range of 60%. The correlation between predicted and experimental affinities is even worse. Different approaches have been applied to improve the situation. Often, tailored scoring functions are used, which can be trained for the particular target of interest. Prerequisite is a large dataset of active and less active compounds. In AFMoC [155], for example, terms from the knowledge-based scoring function DrugScore are calculated for different ligands from the training set and mapped on a grid within the binding pocket. Weights for the different grid points are trained with the known binding affinities of the ligands. Another interesting example was described for a series of fXa inhibitors with a scoring function called Tscore [156], in which individual protein–ligand interaction terms are derived per residue.

Scoring functions typically used for structure-based virtual screening are only approximations to the physical phenomena involved in protein–ligand binding. For example, entropic effects are only captured implicitly in knowledge-based scoring functions, while empirical scoring functions typically have a term for freezing of rotatable bonds of the ligand, but cannot capture entropic changes within the protein. In addition, desolvation is only poorly captured in scoring function. Again, knowledge-based functions tackle desolvation implicitly, while some empirical functions consider the size of lipophilic area covered. Therefore, computationally intensive, more elaborated scoring schemes have been introduced in the scoring of protein–ligand complexes. MM–PBSA [157] or MM–GBSA [158] approaches introduce desolvation effects by the Poisson–Boltzmann [159] or the generalized Born [160] approach. For some examples, good correlation with experimental binding affinities could be obtained, while there are also less convincing results available [148]. Free energy perturbation (FEP) calculations [161] take into account structural ensembles from molecular dynamics simulations and thus are capable of capturing desolvation and entropic effect to some extent. Nevertheless, these calculations are computationally very expensive and thus can only be used for smaller sets of compounds. Hence, the area of application of these algorithms is in compound optimization or in the last step of a cascaded virtual screening approach.

A successful virtual screening requires tailored docking and scoring and has to be performed with care. It is very important to prepare the ligand database correctly with consideration of protonation states of the ligands and potential tautomers and stereoisomers. The same care has to be applied for the setup of the protein structure. Again, the protonation states must be chosen carefully, especially for histidine residues in the active site. Hydrogen-bonding amino acids like serine, threonine, and tyrosine have to be inspected in detail in order to have a meaningful hydrogen-bonding network in the binding cavity. Finally, protein flexibility needs to be tackled. Some docking programs allow specifying rotatable side chains, for example, PLANTS or FlexScreen, which are kept flexible during the docking procedure. In other programs, multiple structures need to be defined for the docking exercise to describe the protein flexibility. In some programs like FlexR or GOLD, ensembles of protein structures can be used. Different protein structures can be available from different protein–ligand complexes, which represent different conformational states. Cross-docking studies [162–166], in which a ligand from one complex structure is docked into the protein structure of a different protein–ligand complex, show that docking results are sensible to the protein structure chosen. Therefore, selection of the protein structure or the ensemble of protein structures for the virtual screening needs careful validation.

Water plays an important role in protein–ligand interactions. It is not only important for solvation effects but is also often found to mediate interactions between the protein and the ligand. Therefore, water molecules critically influence the result of virtual screening runs. Various studies demonstrate that explicit treatment of water molecules in docking studies influences the results [167–169]. Due to the important role of water, the scoring function of GLIDE has been improved to consider water sites in the scoring procedure. In a short molecular dynamics calculation, preferential water positions are identified in WaterMap calculations [170, 171]. This information is used to penalize ligands that show unfavorable interactions at water sites [172].

Due to the diversity of scoring functions, different uncertainties in setting up the protein structure, etc., it is important to perform a thorough validation study in order to find the best settings for the virtual screening. If a diverse set of biologically active ligands is known, the choice of the protein structure, scoring function, etc. can be optimized to give the best enrichment for the training set.

In addition to a careful validation upfront, postdocking filters can improve the success of structure-based virtual screening dramatically by removing potentially FP compounds from the hit list. One very frequently used approach is the combination of different scoring functions into consensus scoring [173], in which predictions of different scoring functions are combined to rank poses. Parts of the improvement that can be achieved by consensus scoring were attributed to the observations that different scoring functions have different systematic errors [174]. Another relevant postprocessing strategy uses constraints defining a feature docking poses have to meet. Hydrogen bonds are typically used as constraints, for example, for docking into the ATP pocket of kinases in which a hydrogen bond to the hinge region is often used as constraint. Structural interaction fingerprints (SIFt) [175] use the complete interaction pattern of reference ligands to define a binary fingerprint capturing protein–ligand interactions. This fingerprint is used to filter out ligands based on similarity to a given reference protein–ligand complex. Thus, ligands that have a similar interaction pattern like the reference ligand are enriched in the final hit list. The strategy has been applied to virtual screening of kinases [176]. TScore follows a related strategy. Based on a training set of active and inactive ligands, interactions to amino acids are weighed by a regression approach [156]. Finally, docking poses often show mismatches between the protein and the ligand surface, for example, when polar substituents are placed into lipophilic environments. However, all scoring procedure and postdocking filters are not reliable enough to avoid human interaction. A very important step in the analysis of docking poses is the visual filtering of the top-ranked hits, considering, for example, ligand conformation and details of protein–ligand interactions.

Although the number of structures in the PDB database is still growing strongly, there is still no structural information for many protein targets. In these situations, homology models offer a good opportunity to apply structure-based drug design. This is especially relevant for membrane proteins where still only a limited number of pharmaceutically relevant proteins are structurally characterized, although crystallography has achieved a significant breakthrough for GPCRs recently [177–179]. In this context, the solution of novel GPCR structures has been used to compare success rates for structure-based virtual screening between a homology model and the actual GPCR structure. This exercise has been performed for the D3 receptor, for which a homology model has been derived from the $\beta_1$ and $\beta_2$ structures. More than three million compounds have been docked into the homology model and into the X-ray structure of the D3 receptor. Hit rates of about 20% and activity ranges in the microMolar range were comparable in both efforts [180]. Interestingly enough, for CXCR4, the same effort resulted in significantly poorer results for the homology model than for the X-ray structure [181]. One reason for the differences has been attributed to the difference in homology between the templates and the target structure, which was 42% for D3 and only 23% for CXCR4.

Virtual screening approaches for GPCRs have nicely been reviewed recently [179], and guidelines for docking have been developed [182]. Overall, these studies confirm the well-established criteria of about 40% sequence homology to be able to derive useful models. Nevertheless, in the field of GPCRs with structural space still sparsely populated, somewhat lower sequence homologies have to be accepted.

## 9.6 OTHER VIRTUAL SCREENING APPLICATIONS

Virtual screening provides a wealth of different technologies to identify compounds that are biologically active against a given target. Various different sources of information can be used in that effort: either known ligands or protein structures of the target of interest. Thus, virtual screening provides means to knowledge-driven lead finding. Nevertheless, the concepts developed for virtual screening also find application in related scientific areas, for example, in chemogenomics or in *in silico* profiling applications.

Chemogenomics is heavily relying on the similarity principle but in a modified perspective: "similar compounds are active against similar targets." On this rationale, even targets can be tackled for which neither active ligands are known nor any protein structural information is available. Compounds and related analogues are selected for testing on homologous or orthologous targets. Such approaches have been extensively reviewed in recent years [183–185]. A typical approach relies on the identification of targets with high homology to the target of interest. In the next step, compounds that are biologically active against the neighboring targets are identified. To broaden the chemical space investigated, analogues of these compounds are typically added by 2D or 3D similarity searches.

G-protein-coupled receptors are a large target family interesting for chemogenomic approaches. There are still about 100 orphan targets in that family for which neither ligand information nor structural knowledge is available. Klabunde et al. have focused the identification of homologous targets to the putative transmembrane binding region by evaluating the similarity between different GPCRs based on the amino acids lining the putative binding site. Based on that similarity, compounds from neighboring targets and their analogues have been selected for chemogenomics-based screening efforts [186]. 3D techniques like pharmacophore models are also used in this effort. An interesting approach applies pharmacophore models for orphan GPCRs based on sequence fingerprints, thus giving rise to first pharmacophore models, which can be used in virtual screening campaigns to identify novel ligands for that receptor [187].

Typically, in late lead finding phases or in lead optimization, compounds are tested against a panel of different targets. This profile of biological activity gives insights into potential side effects of the compounds and helps to identify potential side effects before compounds enter the costly development phase [188, 189]. One aspect is the identification of targets that are clearly associated with side effects, for example, adrenergic receptors and blood pressure or muscarinic receptor 2 and heart rate. In addition to that, efforts are undertaken to correlate the biological activity profile with known side affinities [190–192]. Although the potential benefit of the experimental biological profile is obvious, due to the high costs associated with biological testing in tens or hundreds of assays, experimental profiling can only be done for a few selected compounds. Therefore, *in silico* profiling has gained increasing interest in recent years. The task is the same as in virtual screenings, namely, to determine if a compound is active against a particular target. Instead of testing many compounds for activity against one target, one compound is tested against many targets, but the same technical approaches are used for *in silico* profiling than for virtual screening. Among other methods, similarity searching based on 2D descriptors like PHRAG, FBD, and SHED [193–196] is used within the software CTlink [197], which predicts potential biological activities for a given compound. Based on this methodology novel activities of several drugs can be detected [194]. Next to similarity-based approaches, numerous different machine learning algorithms are also employed to predict *in silico* profiles, for example, Bayesian models [198–200], Kohonen Maps [201, 202], or SVMs [203, 204]. Additionally, 3D methods are also employed for *in silico* predictions. For example, a large panel of receptor-based and ligand-derived pharmacophore models has been derived and made available as PharmacophorDB by Inteligand [205–207]. Although docking against many targets is a quite demanding task, a large project has recently been announced to dock compounds into thousands of protein pockets [208]. A large panel of antitarget has been developed at Sanofi. For every antitarget, for which a significantly large data set is available, regression tree models are generated [209]. In the

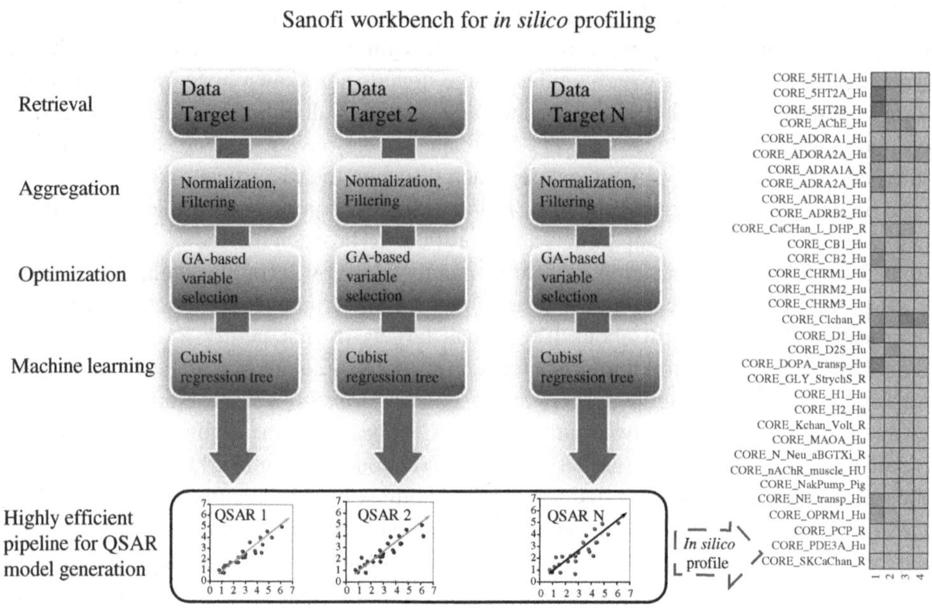


**FIGURE 9.6** Representation of the Sanofi workbench for *in silico* profiling, which is based on large experimental data sets for typical antitargets. (*See insert for color representation of the figure.*)

first step, the data set is carefully filtered, for example, to remove molecules with contradictory data, and normalized. In the second step, a training and test set is generated, which is used to develop the final regression tree model in combination with a genetic algorithm for variable selection [99]. The obtained QSAR models are used to predict the antitarget activity of novel compounds to generate an *in silico* profile for these compounds as shown on the right-hand side of the Figure 9.6.

The capability of *in silico* profiling to propose potential targets against which a compound might be biologically active can also be used for target fishing, for example, for revealing a target against which a hit compound in a phenotypic screen might act. A nice example was presented with the similarity-based method SEA [210, 211] to identify potential targets for a behavioral assay in zebra fish [212]. From the 20 compounds tested, 11 showed nanoMolar to microMolar activity on the proposed targets.

## 9.7 CONCLUSION

Virtual screening comprises a wealth of different approaches covering 2D and 3D methodologies. Although the different technologies have strengths and weaknesses, virtual screening is now a well-established methodology for lead discovery. It offers the particular chance to include prior knowledge into the selection of compounds for biological testing, since it is a data-driven, computational approach. In ligand-based approaches, already known ligands can help to guide the virtual screening into the biologically relevant chemical space for the particular target, while structure-based methods focus on ligands that are complementary to the known binding site. Additionally, the virtual hit list can be further prioritized according to other relevant molecular properties. For example, the use of ADMET models to filter the virtual hit list allows restricting biological testing to the chemical space that has a higher likelihood to yield lead structures.

A recent survey of ligand-based [11] and structure-based [213] virtual screenings clearly showed what can be expected from virtual screening in terms of hit rate, different chemotypes, and activity. Typically, compounds with microMolar activity are identified in virtual screenings. While more structure-based virtual screenings are reported in the literature, there appears to be a tendency that ligand-based virtual screenings result in more active compounds.

Nevertheless, virtual screening has to be carefully applied. Numerous pitfalls are known and described [14]. It is a key aspect that careful validation of the chosen procedure increases likelihood of success significantly. Additionally, visual inspection of hit lists is a crucial final step in compound selection.

Virtual screening will remain an important technique. Increasing computing power will enable the more intensive use of virtual screening techniques for *in silico* profiling approaches. This includes the identification of potential liabilities (side affinities) and the prediction of polypharmacological action.

## REFERENCES


[1] Pearce B.C., Sofia M.J., Good A.C., Drexler D.M., Stock A.D. (2006). An empirical process for the design of high-throughput screening deck filters. *Journal of Chemical Information and Modeling*, **46**, 1060–1068.

[2] Lipinski C.A. Overview of Hit to Lead: The Medicinal Chemist's Role from HTS Retest to Lead Optimization Hand Off. In: Hayward M.M. editor. *Topics in Medicinal Chemistry*, vol. **5**, Springer-Verlag, New York, 2009, pp. 1–24.

[3] Sotriffer C. *Virtual Screening: Principals, Challenges and Practical Guidelines*, Wiley-VCH Verlag GmbH &Co KGaA, Weinheim, 2011.

[4] Clark D.E. (2008). What has virtual screening ever done for drug discovery? *Expert Opinion on Drug Discovery*, **3**, 841–851.

[5] Guido R.V.C., Oliva G., Andricopulo A.D. (2008). Virtual screening and its integration with modern drug design technologies. *Current Medicinal Chemistry*, **15**, 37–46.

[6] Muegge I. (2008). Synergies of virtual screening approaches. *Mini Reviews in Medicinal Chemistry*, **8**, 927–933.

[7] Kontoyianni M., Madhav P., Suchanek E., Seibel W. (2008). Theoretical and practical considerations in virtual screening: A beaten field? *Current Medicinal Chemistry*, **15**, 107–116.

[8] Koeppen H. (2009). Virtual screening: What does it give us? *Current Opinion in Drug Discovery & Development*, **12**, 397–407.

[9] Wildman S.A. (2013). Approaches to virtual screening and screening library selection. *Current Pharmaceutical Design*, **19**, 4787–4796.

[10] Braga R.C., Andrade C.H. (2013). Assessing the performance of 3D pharmacophore models in virtual screening: How good are they? *Current Topics in Medicinal Chemistry*, **13**, 1127–1138.

[11] Stumpfe D., Ripphausen P., Bajorath J. (2012). Virtual compound screening in drug discovery. *Future Medicinal Chemistry*, **4**, 593–602.

[12] Ghemtio L., Pérez-Nueno V.I., Leroux V., Asses Y., Souchet M., Mavridis L., Maigret B., Ritchie D.W. (2012). Recent trends and applications in 3D virtual screening. *Combinatorial Chemistry & High Throughput Screening*, **15**, 749–769.

[13] Cheng T., Li Q., Zhou Z., Wang Y., Bryant S.H. (2012). Structure-based virtual screening for drug discovery: A problem-centric review. *The AAPS Journal*, **14**, 133–141.

[14] Scior T., Bender A., Tresadern G., Medina-Franco J.L., Martinez-Mayorga K., Langer T., Cuanoalo-Contreras K., Agrafiotis D.K. (2012). Recognizing pitfalls in virtual screening: A critical review. *Journal of Chemical Information and Modeling*, **52**, 867–881.

[15] Heikamp K., Bajorath J. (2013). The future of virtual compound screening. *Chemical Biology & Drug Design*, **81**, 33–40.

[16] Kar S., Roy K. (2013). How far can virtual screening take us in drug discovery. *Expert Opinion on Drug Discovery*, **8**, 245–261.

[17] Irwin J.J., Sterling T., Mysinger M.M., Bolstad E.S., Coleman R.G. (2012). ZINC: A free tool to discover chemistry for biology. *Journal of Chemical Information and Modeling*, **52**, 1757–1768.

[18] https://www.emolecules.com/. Accessed May 27, 2015.

[19] http://www.ebi.ac.uk/chembldb/. Accessed May 27, 2015.

[20] http://www.elsevier.com/online-tools/reaxys/about/reaxys_m_c. Accessed May 27, 2015.

[21] http://www.gostardb.com. Accessed May 27, 2015.

[22] Rarey M., Dixon J.S. (1998). Feature trees: A new molecular similarity measure based on tree matching. *Journal of Computer-Aided Molecular Design*, **12**, 471–490.

[23] Schneider G., Neidhart W., Giller T., Schmid G. (1999). "Scaffold-Hopping" by topological pharmacophore search: A contribution to virtual screening. *Angewandte Chemie International Edition*, **38**, 2894–2896.

[24] Grant J.A., Gallardo M.A., Pickup B.T. (1996). A fast method of molecular shape comparison: A simple application of a Gaussian description of molecular shape. *Journal of Computational Chemistry*, **17**, 1653–1666.

[25] Rush T.S., Grant J.A., Mosyak L., Nicholls A. (2005). A shape-based 3-D scaffold hopping method and its application to a bacterial protein-protein interaction. *Journal of Medicinal Chemistry*, **48**, 1489–1495.

[26] Cummings M.D., Arnoult E., Buyck C., Tresadern G., Vos A.M., Wegner J.K. Preparing and Filtering Compound Databases for Virtual and Experimental Screening. In: Sotriffer C. editor. *Virtual Screening: Principals, Challenges and Practical Guidelines*, Wiley VCH Verlag GmbH &Co KGaA, Weinheim, 2011, pp. 35–59.

[27] Cummings M.D., Gibbs A.C., DesJarlais R.L. (2007). Processing of small molecule databases for automated docking. *Medicinal Chemistry*, **3**, 107–113.

[28] Knox A.J., Meegan M.J., Carter G., Lloyd D.G. (2005). Considerations in compound database preparation- "hidden" impact on virtual screening. *Journal of Chemical Information and Modeling*, **45**, 1908–1919.

[29] Bologa C.G., Olah M.M., Oprea T.J. (2006). Chemical database preparation for compound acquisition or virtual screening. *Methods in Molecular Biology*, **316**, 375–388.

[30] Chen I.J., Foloppe N. (2011). Is conformational sampling of drug-like molecules a solved problem? *Drug Development Research*, **72**, 85–94.

[31] Stockwell G.R., Thornton J.M. (2006). Conformational diversity of ligands bound to proteins. *Journal of Molecular Biology*, **356**, 928–944.

[32] Boström J. (2001). Reproducing the conformations of protein-bound ligands: A critical evaluation of several popular conformational searching tools. *Journal of Computer-Aided Molecular Design*, **15**, 1137–1152.

[33] Boström J., Norrby P.O., Liljefors T. (1998). Conformational energy penalties of protein-bound ligands. *Journal of Computer-Aided Molecular Design*, **12**, 383–396.

[34] Perola E., Charifson P.S. (2004). Conformational analysis of drug-like molecules bound to proteins: an extensive study of ligand reorganization upon binding. *Journal of Medicinal Chemistry*, **47**, 2499–2510.

[35] Kirchmair J., Wolber G., Laggner C., Langer T. (2006). Comparative performance assessment of the conformational model generators of omega and catalyst: A large-scale survey on the retrieval of protein-bound ligand conformations. *Journal of Chemical Information and Modeling*, **46**, 1848–1861.

[36] Agrafiotis D.K., Gibbs A.C., Zhu F., Izrailev S., Martin E. (2007). Conformational sampling of bioactive molecules: A comparative study. *Journal of Chemical Information and Modeling*, **47**, 1067–1086.

[37] Chen I.J., Foloppe N. (2008). Conformational sampling of druglike molecules with MOE and catalyst: Implications for pharmacophore modelling and virtual screening. *Journal of Chemical Information and Modeling*, **48**, 1773–1791.

[38] Schwab C. (2010). Conformations and 3D pharmacophore searching. *Drug Discovery Today. Technologies*, **7**, E245–E253.

[39] Sitzmann M., Weidlich I.E., Filippov I., Liao C., Peach M.L., Ihlenfeldt W.D., Karki R.G., Borodina Y.V., Cachau R.E., Nicklaus M.C. (2012). PDB ligand conformational energies calculated quantum-mechanically. *Journal of Chemical Information and Modeling*, **52**, 739–756.

[40] Liebeschuetz J., Hennemann J., Olsson T., Groom C.R. (2012). The good, the bad and the twisted: A survey of ligand geometry in protein crystal structures. *Journal of Computer-Aided Molecular Design*, **26**, 169–183.

[41] Muegge I., Heald S.L., Britelli D. (2001). Simple selection criteria for drug-like chemical matter. *Journal of Medicinal Chemistry*, **44**, 1841–1846.

[42] Martin E.J., Critchlow R.E. (1999). Beyond mere diversity: Tailoring combinatorial libraries for drug discovery. *Journal of Combinatorial Chemistry*, **1**, 32–45.

[43] Rishton G.M. (1997). Reactive compounds and in vitro false positives in HTS. *Drug Discovery Today*, **2**, 382–384.

[44] Lipinski C.A., Lombardo F., Dominy D.W., Feeney P.J. (1997). Experimental and computational approaches to estimate solubility and permeability in drug discovery and development settings. *Advanced Drug Delivery Reviews*, **23**, 3–25.

[45] Veber D.F., Johnson S.R., Cheng H.Y., Smith B.R., Ward K.W., Kopple K.D. (2002). Molecular properties that influence the oral bioavailability of drug candidates. *Journal of Medicinal Chemistry*, **45**, 2615–2623.

[46] Clark D.E., Pickett S.D. (2000). Computational methods for the prediction of "drug-likeness". *Drug Discovery Today*, **5**, 49–58.

[47] Martin C.Y. (2005). A bioavailability score. *Journal of Medicinal Chemistry*, **48**, 3164–3170.

[48] Gleeson M.P. (2008). Generation of a set of simple, interpretable ADMET rules of thumb. *Journal of Medicinal Chemistry*, **51**, 817–834.

[49] Cumming J.G., Davis A.M., Muresan S., Haeberlein M., Chen H. (2013). Chemical predictive modelling to improve compound quality. *Nature Reviews. Drug Discovery*, **12**, 948–962.

[50] Leeson P.D., Springthorpe B. (2007). The influence of drug-like concepts on decision-making in medicinal chemistry. *Nature Reviews Drug Discovery*, **6**, 881–890.

[51] Vistoli G., Pedretti A., Testa B. (2008). Assessing drug-likeness—What are we missing? *Drug Discovery Today*, **13**, 285–294.

[52] Oprea T.I., Davis A.M., Teague S.J., Leeson P.D. (2001). Is there a difference between leads and drugs? A historical perspective. *Journal of Chemical Information and Computer Sciences*, **41**, 1308–1315.

[53] Congreve M., Carr R., Murray C., Jhoti H. (2003). A "rule of three" for fragment-based lead discovery? *Drug Discovery Today*, **8**, 876–877.

[54] Pagliaro L., Felding J., Audouze K., Nielson S.J., Terry R.B., Krog-Jensen C., Butcher S. (2004). Emerging classes of protein-protein interaction inhibitors and new tools for their development. *Current Opinion in Chemical Biology*, **8**, 442–449.

[55] Fischer P.M. (2005). Protein-protein interactions in drug discovery. *Drug Design Reviews*, **2**, 179–207.

[56] Van de Waterbeemed H., Camenisch G., Folkers G., Chretien J.R., Raevsky O.A. (1998). Estimation of blood-brain barrier crossing of drugs using molecular size and shape and H-bonding descriptors. *Journal of Drug Targeting*, **6**, 151–165.

[57] Kelder J., Grootenhuis P.D., Bayada D.M., Delbressine L.P., Ploemen J.P. (1999). Polar molecular surface as a dominating determinant for oral absorption and brain penetration of drugs. *Pharmaceutical Research*, **16**, 1514–1519.

[58] Clark D.E. (2005). Computational prediction of blood-brain barrier permeation. *Annual Reports in Medicinal Chemistry*, **40**, 403–415.

[59] Fagerholm U. (2007). The highly permeable blood-brain barrier: An evaluation of current opinions about brain uptake capacity. *Drug Discovery Today*, **12**, 1076–1082.

[60] Shityakov S., Neuhaus W., Dandekar T., Förster C. (2013). Analysing molecular polar surface descriptors to predict blood-brain barrier permeation. *International Journal of Computational Biology and Drug Design*, **6**, 146–156.

[61] Suenderhauf C., Hammann F., Huwyler J. (2012). Computational prediction of blood-brain barrier permeability using decision tree induction. *Molecules*, **17**, 10429–10445.

[62] Triballeau N., Acher F., Brabet I., Pin J.P., Bertrand H.O. (2005). Virtual screening workflow development guided by the "receiver operating characteristic" approach. Application to high-throughput docking on metabotropic glutamate receptor subtype 4. *Journal of Medicinal Chemistry*, **48**, 2534–2547.

[63] Truchon J.F., Bayly C.I. (2007). Evaluating virtual screening methods: Good and bad metrics for the "early recognition" problem. *Journal of Chemical Information and Modeling*, **47**, 488–508.

[64] Schuster D., Waltenberger B., Kirchmair J., Distinto S., Markt P., Stuppner H., Rollinger J.M., Wolber G. (2010). Predicting cyclooxygenase inhibition by three-dimensional pharmacophore profiling. Part I: Model generation, validation and applicability in ethnopharmacology. *Molecular Informatics*, **29**, 75–86.

[65] Bemis G.W., Murcko M.A. (1996). The properties of known drugs 1. Molecular framework. *Journal of Medicinal Chemistry*, **39**, 2887–2893.

[66] Li R., Stumpfe D., Vogt M., Geppert H., Bajorath J. (2011). Development of a method to consistently quantify the structural distance between scaffolds and to assess scaffold hopping potentials. *Journal of Chemical Information and Modeling*, **51**, 2507–2514.

[67] Renner S., Schneider G. (2005). Scaffold hopping potential of ligand-based similarity concepts. *ChemMedChem*, **1**, 181–185.

[68] Abt M., Lim Y., Sacks J., Xie M., Young S.S. (2001). A sequential approach for identifying lead compound in large chemical databases. *Statistical Science*, **16**, 154–168.

[69] Engels M.F., Venkatarangan P. (2001). Smart screening: Approaches to efficient HTS. *Current Opinion in Drug Discovery & Development*, **4**, 275–283.

[70] Liebschuetz J.W., Stuart S.D., Wiley M.E., Young S.C. Iterative Structure-Based Screening of Virtual Chemical Libraries and Factor Xa: Finding the Orally Available Antithrombotic Candidate LY517717. In: Hubbard R.E. editor. *Structure-Based Drug Discovery: An Overview*, RCS Publishing, Cambridge, 2006, pp. 173–192.

[71] Johnson M.A., Maggiora G.M. *Concepts and Applications of Molecular Similarity*, John Wiley & Sons, Inc., New York, 1990.

[72] Maggiora G., Vogt M., Stumpfe D., Bajorath J. (2014). Molecular similarity in medicinal chemistry. *Journal of Medicinal Chemistry*, **57**, 3186–3204.

[73] Koeppen H., Kriegel J., Lessel U., Tautermann C.S., Wellenzohn B. Ligand-Based Virtual Screening in Virtual Screening. In: Sotriffer C. editor. *Virtual Screening: Principals, Challenges and Practical Guidelines*, Wiley VCH Verlag GmbH &Co KGaA, Weinheim, 2011, pp. 61–85.

[74] Maggiora G.M. (2006). On outliers and activity cliffs—Why QSAR often disappoints. *Journal of Chemical Information and Modeling*, **46**, 1535.

[75] Stumpfe D., Bajorath J. (2012). Exploring activity cliffs in medicinal chemistry. *Journal of Medicinal Chemistry*, **55**, 2932–2942.

[76] Willett P., Barnard J.M., Downs G.M. (1998). Chemical similarity searching. *Journal of Chemical Information and Computer Sciences*, **38**, 983–996.

[77] Hert J., Willett P., Wilton D.J., Acklin P., Azzaoui K., Jacoby E., Schuffenhauer A. (2005). Enhancing the effectiveness of similarity-based virtual screening using nearest-neighbour information. *Journal of Medicinal Chemistry*, **48**, 7049–7054.

[78] Franke L., Schwarz O., Müller-Kuhrt L., Hoernig C., Fischer L., George S., Tanrikulu Y., Schneider P., Werz O., Steinhilber D., Schneider G. (2007). Identification of natural-product-derived inhibitors of 5-lipoxygenase activity by ligand-based virtual screening. *Journal of Medicinal Chemistry*, **50**, 2640–2646.

[79] Reutlinger M., Koch C.P., Reker D., Todoroff N., Schneider P., Rodrigues T., Schneider G. (2013). Chemically advanced template search (CATS) for scaffold-hopping and prospective target prediction for "orphan" molecules. *Molecular Informatics*, **32**, 133–138.

[80] Noeske T., Jirgensons A., Starchenkovs I., Renner S., Jaunzeme L., Trifanova D., Hechenberger M., Bauer T., Kauss V., Parsons C.G., Schneider G., Weil T. (2007). Virtual screening for selective allosteric mGluR1 antagonists and structure–activity relationship investigations for coumarine derivatives. *ChemMedChem*, **2**, 1763–1773.

[81] Kiss R., Sandor M., Gere A., Schmidt E., Balogh G.T., Kiss B., Molnar L., Lemmen C., Keseru G.M. (2012). Discovery of novel histamine H4 and serotonin transporter ligands using the topological feature tree descriptor. *Journal of Chemical Information and Modeling*, **52**, 233–242.

[82] Wellenzohn B., Lessel U., Beller A., Isambert T., Hoenke C., Nosse B. (2012). Identification of new potent GPR119 agonists by combining virtual screening and combinatorial chemistry. *Journal of Medicinal Chemistry*, **55**, 11031–11041.

[83] Berne S., Podobnik B., Zupanec N., Novak M., Kraševec N., Turk S., Korošec B., Lah L., Šuligoj E., Stojan J., Gobec S., Komel R. (2012). Virtual screening yields inhibitors of novel antifungal drug target, benzoate 4-monooxygenase. *Journal of Chemical Information and Modeling*, **52**, 3053–3063.

[84] Hert J., Willett P., Wilton D.J., Acklin P., Azzaoui K., Jacoby E., Schuffenhauer A. (2004). Comparison of fingerprint-based methods for virtual screening using multiple bioactive reference structures. *Journal of Chemical Information and Computer Sciences*, **44**, 1177–1185.

[85] Bender A., Jenkins J.L., Scheiber J., Sukuru S.C.K., Glick M., Davies J.W. (2009). How similar are similarity searching methods? A principal component analysis of molecular descriptor space. *Journal of Chemical Information and Modelling*, **49**, 108–119.

[86] Brown R.D., Martin Y.C. (1997). The information content of 2D and 3D structural descriptors relevant to ligand-receptor binding. *Journal of Chemical Information and Computer Sciences*, **37**, 1–9.

[87] Matter H., Pötter T. (1999). Comparing 3D pharmacophore triplets and 2D fingerprints for selecting diverse compound subsets. *Journal of Chemical Information and Computer Sciences*, **39**, 1211–1225.

[88] Willet P. (2013). Combination of similarity rankings using data fusion. *Journal of Chemical Information and Modeling*, **53**, 1–10.

[89] Whittle M., Gillet V.J., Willet P., Loesel J. (2006). Analysis of data fusion methods in virtual screening: Theoretical model. *Journal of Chemical Information and Modeling*, **46**, 2193–2205.

[90] Whittle M., Gillet V.J., Willet P., Loesel J. (2006). Analysis of data fusion methods in virtual screening: Similarity and group fusion. *Journal of Chemical Information and Modeling*, **46**, 2206–2219.

[91] Schneider G., Baringhaus K.H. *Molecular Design*, Wiley-VCH Verlag GmbH, Weinheim, 2008.

[92] Zupan J., Gasteiger J. *Neuronal Networks in Chemistry and Drug Design*, Wiley-VCH Verlag GmbH, Weinheim, 1999.

[93] Ivanenkov Y.A., Savchuk N.P., Ekins S., Balkin K.V. (2009). Computational mapping tools for drug discovery. *Drug Discovery Today*, **14**, 767–775.

[94] Schneider P., Tanrikulu Y., Schneider G. (2009). Self-organizing maps in drug discovery: Compound library design, scaffold-hopping, repurposing. *Current Medicinal Chemistry*, **16**, 258–266.

[95] Liu S., Fu R., Zhou L.H., Chen S.P. (2012). Application of consensus scoring and principal component analysis for virtual screening against β-secretase (BACE-1), *PLoS One*, **7**, e38086.

[96] Steri R., Achenbach J., Steinhilber D., Schubert-Zsilavecz M., Proschak E. (2012). Investigation of imatinib and other approved drugs as starting points for antidiabetic drug discovery with FXR modulating activity. *Biochemical Pharmacology*, **83**, 1674–1681.

[97] Noeske T., Trifanova D., Kauss V., Renner S., Parsons C.G., Schneider G., Weil T. (2009). Synergism of virtual screening and medicinal chemistry: Identification and optimization of allosteric antagonists of metabotropic glutamate receptor 1. *Bioorganic & Medicinal Chemistry*, **17**, 5708–5715.

[98] Schneider G., Nettekoven M. (2003). Ligand-based combinatorial design of selective purinergic receptor (A2A) antagonists using self-organizing maps. *Journal of Combinatorial Chemistry*, **5**, 233–237.

[99] Baringhaus K.-H., Hessler G., Matter H., Schmidt F. Development and Applications of Global ADMET Models: In Silico prediction of Human Microsomal Lability. In: Bajorath J. editor. *Chemoinformatics for Drug Discovery*, John Wiley & Sons, Inc., Hoboken, 2014, pp. 245–265.

[100] Schwaighofer A., Schroeter T., Mika S., Hansen K., Ter Laak A., Lienau P., Reichel A., Heinrich N., Müller K.R. (2008). A probabilistic approach to classifying metabolic stability. *Journal of Chemical Information and Modeling*, **48**, 785–796.

[101] Shen M., Su B.H., Esposito E.X., Hopfinger A.J., Tseng Y.J. (2011). A comprehensive support vector machine binary hERG classification model based on extensive but biased end point hERG data sets. *Chemical Research in Toxicology*, **24**, 934–949.

[102] Doddareddy R.M., Klaasse E.C., Shagufta E.C., Jzerman A.P., Bender A. (2010). Prospective validation of a comprehensive in silico hERG model and its applications to commercial compound and drug databases. *ChemMedChem*, **5**, 716–729.

[103] Wang M., Yang X.G., Xue Y. (2008). Identifying hERG potassium channel inhibitors by machine learning methods. *QSAR & Combinatorial Science*, **27**, 1028–1035.

[104] Rusinko A., Farmen M.W., Lambert C.G., Brown P.L., Young S.S. (1999). Analysis of a large structure/biological activity data set using recursive partitioning. *Journal of Chemical Information and Computer Sciences*, **39**, 1017–1026.

[105] Svetnik V., Liaw A., Tong C., Culberson J.C., Sheridan R.P., Feuston B.P. (2003). Random forest: A classification and regression tool for compound classification and QSAR modeling. *Journal of Chemical Information and Computer Sciences*, **43**, 1947–1958.

[106] Cristianini N., Shaw-Taylor J. *An Introduction to Support Vector Machines and Other Kernel-Based Learning Methods*, Cambridge University Press, Cambridge, 2000.

[107] Bender A. (2011). Bayesian methods in virtual screening and chemical biology. *Methods in Molecular Biology*, **672**, 175–196.

[108] Chen B., Harrison R.F., Papadatos G., Willett P., Wood D.J., Lewell X.Q., Greenidge P., Stiefl N. (2007). Evaluation of machine-learning methods for ligand-based virtual screening. *Journal of Computer-Aided Molecular Design*, **21**, 53–62.

[109] Klon A.E. (2009). Bayesian modelling in virtual high throughput screening. *Combinatorial Chemistry & High Throughput Screening*, **12**, 469–483.

[110] Pan Y., Chothe P.P., Swaan P.W. (2013). Identification of novel breast cancer resistance protein (BCRP) inhibitors by virtual screening. *Molecular Pharmaceutics*, **10**, 1236–1248.

[111] Fang J., Yang R., Gao L., Zhou D., Yang S., Liu A.L., Du G.H. (2013). Predictions of BuChE inhibitors using support vector machine and naive Bayesian classification techniques in drug discovery. *Journal of Chemical Information and Modeling*, **53**, 3009–3020.

[112] Di-Wu L., Li L.L., Wang W.J., Xie H.Z., Yang J., Zhang C.H., Huang Q., Zhong L., Feng S., Yang S.Y. (2012). Identification of CK2 inhibitors with new scaffolds by a hybrid virtual screening approach based on Bayesian model; pharmacophore hypothesis and molecular docking. *Journal of Molecular Graphics and Modelling*, **36**, 42–47.

[113] Krueger B.A., Weil T., Schneider G. (2009). Comparative virtual screening and novelty detection for NMDA-GlycineB antagonists. *Journal of Computer-Aided Molecular Design*, **23**, 869–881.

[114] Li Y., Tan C., Gao C., Zhang C., Luan X., Chen X., Liu H., Chen Y., Jiang Y. (2011). Discovery of benzimidazole derivatives as novel multi-target EGFR, VEGFR-2 and PDGFR kinase inhibitors. *Bioorganic & Medicinal Chemistry*, **19**, 4529–4535.

[115] Xie Q.Q., Zhong L., Pan Y.L., Wang X.Y., Zhou J.P., Di-Wu L., Huang Q., Wang Y.L., Yang L.L., Xie H.Z., Yang S.Y. (2011). Combined SVM-based and docking-based virtual screening for retrieving novel inhibitors of c-Met. *European Journal of Medicinal Chemistry*, **46**, 3675–3680.

[116] Ren J.X., Li L.L., Zheng R.L., Xie H.Z., Cao Z.X., Feng S., Pan Y.L., Chen X., Wei Y.Q., Yang S.Y. (2011). Discovery of novel Pim-1 kinase inhibitors by a hierarchical multistage virtual screening approach based on SVM model, pharmacophore, and molecular docking. *Journal of Chemical Information and Modeling*, **51**, 1364–1375.

[117] Naeem S., Hylands P., Barlow D. (2012). Construction of an Indonesian herbal constituents database and its use in Random Forest modelling in a search for inhibitors of aldose reductase. *Bioorganic & Medicinal Chemistry*, **20**, 1251–1258.

[118] Paoletta S., Steventon G.B., Wildeboer D., Ehrman T.M., Hylands P.J., Barlow D.J. (2008). Screening of herbal constituents for aromatase inhibitory activity. *Bioorganic & Medicinal Chemistry*, **16**, 8466–8470.

[119] Matter H., Zoller G., Herling A.W., Sanchez-Arias J.A., Philippo C., Namane C., Kohlmann M., Pfenninger A., Voss M.D. (2013). Benzimidazole-carboxamides as potent and bioavailable stearoyl-CoA desaturase (SCD1) inhibitors from ligand-based virtual screening and chemical optimization. *Bioorganic & Medicinal Chemistry Letter*, **23**, 1817–1822.

[120] Distinto S., Esposito F., Kirchmair J., Cardia M.C., Gaspari M., Maccioni E., Alcaro S., Markt P., Wolber G., Zinzula L., Tramontano E. (2012). Identification of HIV-1 reverse transcriptase dual inhibitors by a combined shape-, 2D-fingerprint- and pharmacophore-based virtual screening approach. *European Journal of Medicinal Chemistry*, **50**, 216–229.

[121] Hevener K.E., Mehboob S., Su P.C., Truong K., Boci T., Deng J., Ghassemi M., Cook J.L., Johnson M.E. (2012). Discovery of a novel and potent class of *F. tularensis* enoyl-reductase (FabI) inhibitors by molecular shape and electrostatic matching. *Journal of Medicinal Chemistry*, **55**, 268–279.

[122] Lemmen C., Lengauer T., Klebe G. (1998). FLEXS: A method for fast flexible ligand superposition. *Journal of Medicinal Chemistry*, **41**, 4502–4520.

[123] Jain A.N. (2004). Ligand-based structural hypotheses for virtual screening. *Journal of Medicinal Chemistry*, **47**, 947–961.

[124] Giganti D., Guillemain H., Spadoni J.L., Nilges M., Zagury J.F., Montes M. (2010). Comparative evaluation of 3D virtual ligand screening methods: Impact of the molecular alignment on enrichment. *Journal of Chemical Information and Modeling*, **50**, 992–1004.

[125] Wermuth C.G., Ganellin C.R., Lindberg P., Mitscher L. (1998). Glossary of terms used in medicinal chemistry (IUPAC recommendations 1998). *Pure and Applied Chemistry*, **70**, 1129–1143.

[126] Greene J., Kahn S.D., Savoj H., Sprague P., Teig S.L. (1994). Chemical function queries for 3D database search. *Journal of Chemical Information and Computer Sciences*, **34**, 1297–1308.

[127] Kurogi Y., Güner O.F. (2001). Pharmacophore modeling and three-dimensional database searching for drug design using catalyst. *Current Medicinal Chemistry*, **8**, 1035–1055.

[128] Wolber G., Langer T. (2005). LigandScout: 3D pharmacophores derived from protein-bound ligands and their use as virtual screening filters. *Journal of Chemical Information and Modeling*, **45**, 160–169.

[129] Dixon S.L., Smondyrev A.M., Knoll E.H., Rao S.N., Shaw D.E., Friesner R.A. (2006). PHASE: A new engine for pharmacophore perception, 3D QSAR model development, and 3D database screening: 1. Methodology and preliminary results. *Journal of Computer-Aided Molecular Design*, **20**, 647–671.

[130] http://www.chemcomp.com/. Accessed May 27, 2015.

[131] Schuster D., Markt P., Grienke U., Mihaly-Bison J., Binder M., Noha S.M., Rollinger J.M., Stuppner H., Bochkov V.N., Wolber G. (2011). Pharmacophore-based discovery of FXR agonists. Part I: Model development and experimental validation. *Bioorganic & Medicinal Chemistry*, **19**, 7168–7180.

[132] Cross S., Baroni M., Goracci L., Cruciani G. (2012). GRID-based three-dimensional pharmacophores I: FLAPpharm, a novel approach for pharmacophore elucidation. *Journal of Chemical Information and Modeling*, **52**, 2587–2598.

[133] Goodford P.J. (1985). A computational procedure for determining energetically favorable binding sites on biologically important macromolecules. *Journal of Medicinal Chemistry*, **28**, 849–857.

[134] Cross S., Ortuso F., Baroni M., Costa G., Distinto S., Moraca F., Alcaro S., Cruciani G. (2012). GRID-based three-dimensional pharmacophores II: PharmBench, a benchmark data set for evaluating pharmacophore elucidation methods. *Journal of Chemical Information and Modeling*, **52**, 2599–2608.

[135] Hurst T. (1994). Flexible 3D searching. The directed tweak technique. *Journal of Chemical Information and Computer Sciences*, **34**, 190–196.

[136] Kuntz I.D., Blaney J.M., Oatley S.J., Langridge R., Ferrin T.E. (1982). A geometric approach to macro-molecule ligand interaction. *Journal of Molecular Biology*, **161**, 269–288.

[137] McGann M.R., Almond H.R., Nicholls A., Grant J.A., Brown F.K. (2003). Gaussian docking functions. *Biopolymers*, **68**, 76–90.

[138] Rarey M., Kramer B., Lengauer T., Klebe G. (1996). A fast flexible docking method using an incremental construction algorithm. *Journal of Molecular Biology*, **261**, 470–489.

[139] Jain A.N. (2003). Surflex: Fully automated flexible docking using a molecular similarity-based search engine. *Journal of Medicinal Chemistry*, **46**, 499–511.

[140] Klebe G., Mietzner T. (1994). A fast and efficient method to generate biologically relevant conformations. *Journal of Computer-Aided Molecular Design*, **8**, 583–606.

[141] Friesner R.A., Banks J.L., Murphy R.B., Halgren T.A., Klicic J.J., Mainz D.T., Repasky M.P., Knoll E.H., Shelley M., Perry J.K., Shaw D.E., Francis P., Shenkin P.S. (2004). Glide: A new approach for rapid, accurate docking and scoring. 1. Method and assessment of docking accuracy. *Journal of Medicinal Chemistry*, **47**, 1739–1749.

[142] Morris G.M., Goodsell D.S., Halliday R.S., Huey R., Hart W.E., Belew R.K., Olson A.J. (1998). Automated docking using a Lamarckian genetic algorithm and an empirical binding free energy solution. *Journal of Computational Chemistry*, **19**, 1639–1662.

[143] Jones G., Willet P., Glen R.C., Leach A.R., Taylor R. (1997). Development and validation of a genetic algorithm for flexible docking. *Journal of Molecular Biology*, **267**, 727–748.

[144] Korb O., Stützle T., Exner T.E. (2009). Empirical scoring function for advanced protein-ligand docking with PLANTS. *Journal of Chemical Information and Modeling*, **49**, 84–96.

[145] Merlitz H., Wenzel W. (2002). Comparison of stochastic optimization methods for receptor-ligand docking. *Chemical Physics Letters*, **362**, 271–277.

[146] Merlitz H., Burghardt B., Wenzel W. (2003). Application of the stochastic tunneling method to high throughput database screening. *Chemical Physics Letters*, **370**, 68–73.

[147] Rognan D. Docking Methods for Virtual Screening: Principles and Recent Advances in Virtual Screening. In: Sotriffer C. editor. *Virtual Screening Principles, Challenges and Practical Guidelines*, Wiley VCH Verlag GmbH & Co. KGaA, Weinheim, 2011, pp. 153–176.

[148] Sottrifer C., Matter H. The Challenge of Affinity Prediction: Scoring Functions for Structure-Based Virtual Screening. In: Sotriffer C. editor. *Virtual Screening Principles, Challenges and Practical Guidelines*, Wiley VCH Verlag GmbH & Co. KGaA, Weinheim, 2011, pp. 179–221.

[149] Gohlke H., Hendlich M., Klebe G. (2000). Knowledge-based scoring function to predict protein-ligand interactions. *Journal of Molecular Biology*, **295**, 337–356.

[150] Muegge I. (2006). PMF scoring revisited. *Journal of Medicinal Chemistry*, **49**, 5895–5902.

[151] Sotriffer C.A., Sanschagrin P., Matter H., Klebe G. (2008). SFCscore: Scoring functions for affinity prediction of protein-ligand complexes. *Proteins*, **73**, 395–419.

[152] Wang R., Lai L., Wang S. (2002). Further development and validation of empirical scoring functions for structure-based binding affinity prediction. *Journal of Computer-Aided Molecular Design*, **16**, 11–26.

[153] Eldrige J., Murray C.W., Auton T.R., Paolini G.V., Mee R.P. (1997). Empirical scoring functions: I. The development of a fast empirical scoring function to estimate the binding affinity of ligands in receptor complexes. *Computer-Aided Molecular Design*, **11**, 425–445.

[154] Gelhaar D.K., Verkhivker G.M., Rejto P.A., Sherman C.J., Fogel L.J., Freer S.T. (1995). Molecular recognition of the inhibitor AG-1343 by HIV-1 protease: Conformationally flexible docking by evolutionary programming. *Chemistry & Biology*, **2**, 317–324.

[155] Gohlke H., Klebe G. (2002). DrugScore meets CoMFA: Adaption of fields for molecular comparison (AFMoC) or how to tailor knowledge-based pair-potentials to a particular protein. *Journal of Medicinal Chemistry*, **45**, 4153–4170.

[156] Matter H., Will D.W., Nazare M., Schreuder H., Laux V., Wehner V. (2005). Structural requirements for factor Xa inhibition by 3-oxybenzamids with neutral P1 substituents: Combining X-ray crystallography, 3D-QSAR and tailored scoring functions. *Journal of Medicinal Chemistry*, **48**, 3290–3312.

[157] Kollmann P.A., Massova I., Reyes C., Kuhn B., Huo S., Chong L., Lee M., Lee T., Duan Y., Wang W., Donini O., Cieplak P., Srinivassan J., Case D.A. (2000). Calculating structures and free energies of complex molecules: Combining molecular mechanics and continuum models. *Accounts of Chemical Research*, **33**, 889–897.

[158] Huang N., Kalyanaraman C., Irwin J.J., Jacobson M.P. (2006). Physics-based scoring of protein-ligand complexes: Enrichment of known inhibitors in large-scale virtual screening. *Journal of Chemical Information and Modeling*, **46**, 243–253.

[159] Baker N.A. Biomolecular applications of Poisson–Boltzmann methods. In: Lipkowitz K.B., Larter R., Cundrai T.R. editors. *Reviews in Computational Chemistry*, vol. **21**, John Wiley & Sons, Inc., Hoboken, NJ, 2005, pp. 349–379.

[160] Gosh A., Rapp C.S., Friesner R.A. (1998). Generalized Born model based on a surface integral formulation. *The Journal of Physical Chemistry. B*, **102**, 10983–10990.

[161] Aqvist J., Medina C., Samuelson J.E. (1994). A new method for predicting binding affinity in computer-aided drug design. *Protein Engineering*, **7**, 385–391.

[162] Murray C.W., Baxter C.A., Frenkel A.D. (1999). The sensitivity of the results of molecular docking to induced fit effects: Application to thrombin, thermolysin and neuraminidase. *Journal of Computer-Aided Molecular Design*, **6**, 547–562.

[163] Duca J.S., Madison V.S., Voigt J.H. (2008). Cross-docking of inhibitors into CDK2 structures. 1. *Journal of Chemical Information and Modeling*, **4**, 659–668.

[164] Sutherland J., Nandigam R.K., Erickson J.A., Vieth M. (2007). Lessons in molecular recognition. 2. Assessing and improving cross-docking accuracy. *Journal of Chemical Information and Modeling*, **47**, 2293–2302.

[165] Zhao Y., Sanner M.F. (2009). Protein–ligand docking with multiple flexible side chains. *Journal of Computer-Aided Molecular Design*, **22**, 673–679.

[166] Jain A.N. (2009). Effects of protein conformation in docking: Improved pose prediction through protein pocket adaptation. *Journal of Computer-Aided Molecular Design*, **23**, 355–374.

[167] Roberts B.C., Mancera R.L. (2008). Ligand-protein docking with water molecules. *Journal of Chemical Information and Modeling*, **48**, 397–408.

[168] Kirchmair J., Spitzer G.M., Liedl K.R. Consideration of Water and Solvation Effects in Virtual Screening. In: Sotriffer C. editor. *Virtual Screening Principles, Challenges and Practical Guidelines*, Wiley VCH Verlag GmbH & Co. KGaA, Weinheim, 2011, pp. 263–289.

[169] Lenselink E.B., Beuming T., Sherman W., van Vlijmen H.W.T., IJzerman A.P. (2014). Selecting an optimal number of binding site waters to improve virtual screening enrichments against the adenosine A2A receptor. *Journal of Chemical Information and Modeling*, **54**, 1737–1746.

[170] Young T., Abel R., Kim B., Berne B., Friesner R.A. (2007). Motifs for molecular recognition exploiting hydrophobic enclosure in protein-ligand binding. *Proceedings of the National Academy of Sciences of the United States of America*, **104**, 808–813.

[171] Abel R., Young T., Farid R., Berne B., Friesner R.A. (2008). Role of the active-site solvent in the thermodynamics of factor Xa ligand binding. *Journal of the American Chemical Society*, **130**, 2817–2831.

[172] Repasky M.P., Murphy R.B., Banks J.L., Greenwood J.R., Tubert-Brohman I., Bhat S., Friesner R.A. (2012). Docking performance of the glide program as evaluated on the Astex and DUD datasets: A complete set of glide SP results and selected results for a new scoring function integrating WaterMap and glide. *Journal of Computer-Aided Molecular Design*, **26**, 787–799.

[173] Charifson P.S., Corkery J.J., Murcko M.A., Walters W.P. (1999). Consensus scoring: A method for obtaining improved hit rates from docking databases of three-dimensional structures into proteins. *Journal of Medicinal Chemistry*, **42**, 5100–5109.

[174] Baber J.C., Shirley W.A., Gao Y., Feher M. (2006). The use of consensus scoring in ligand-based virtual screening. *Journal of Chemical Information and Modeling*, **46**, 277–288.

[175] Deng Z., Chuaqui C., Singh J. (2004). Structural interaction fingerprints (SIFt): A novel method for analysing three-dimensional protein-ligand binding interactions. *Journal of Medicinal Chemistry*, **47**, 337–344.

[176] Deng Z., Chuaqui C., Singh J. (2005). Interaction profiles of protein kinase–inhibitor complexes and their application to virtual screening. *Journal of Medicinal Chemistry*, **48**, 121–133.

[177] Kolb P., Klebe G. (2011). The golden age of GPCR structural biology: Any impact on drug design? *Angewandte Chemie International Edition*, **50**, 1153–11575.

[178] Congreve M., Langmead J., Mason J.S., Marshall F.H. (2011). Progress in structure based drug design for G protein coupled receptors. *Journal of Medicinal Chemistry*, **54**, 4283–4311.

[179] Andrews S.P., Brown G.A., Christopher J.A. (2014). Structure-based and fragment-based GPCR drug discovery. *ChemMedChem*, **9**, 256–275.

[180] Carlsson J., Coleman R.G., Setola V., Irwin J.J., Fan H., Schlessinger A., Sali A., Roth B.L., Shoichet B.K. (2011). Ligand discovery from a dopamine D3 receptor homology model and crystal structure. *Nature Chemical Biology*, **7**, 769–778.

[181] Mysinger M.M., Weiss D.R., Ziarek J.J., Gravel S., Doak A.K., Karpiak J., Heveker N., Shoichet B.K., Volkmann B.F. (2012). Structure-based ligand discovery for the protein-protein interface of chemokine receptor CXCR4. *Proceedings of the National Academy of Sciences of the United States of America*, **109**, 5517–5522.

[182] Beuming T., Sherman W. (2012). Current assessment of docking into GPCR crystal structures and homology models: Successes, challenges, and guidelines. *Journal of Chemical Information and Modeling*, **52**, 3262–3277.

[183] Kubinyi H., Müller G. *Chemogenomics in Drug Discovery: A Medicinal Chemistry Perspective*, Wiley-VCH Verlag GmbH & Co. KGaA, Weinheim, 2004.

[184] Gloriam D.E. (2012). Chemogenomics of allosteric binding sites in GPCRs. *Drug Discovery Today. Technologies*, **10**, e307–e313.

[185] Bieler M., Koeppen H. (2012). The role of chemogenomics in the pharmaceutical industry. *Drug Development Research*, **73**, 357–364.

[186] Baringhaus K.H., Evers A., Giegerich C., Jaeger R., Klabunde K. Computer-Assisted Lead Finding of GPCR Modulators in http://www.unicam.it/farmacia/symposium/abstractBook2010.pdf. Accessed May 13, 2015.

[187] Klabunde T., Giegerich C., Evers A. (2009). Sequence-derived three-dimensional pharmacophore models for G-protein-coupled receptors and their application in virtual screening. *Journal of Medicinal Chemistry*, **52**, 2923–2932.

[188] Peters J.U., Hert J., Bissantz C., Hillebrecht A., Gerebtzoff G., Bendels S., Tillier F., Migeon J., Fischer H., Guba W., Kansy M. (2012). Can we discover pharmacological promiscuity early in the drug discovery process? *Drug Discovery Today*, **17**, 325–335.

[189] Peters J.U. (2013). Polypharmacology—Foe or friend? *Journal of Medicinal Chemistry*, **56**, 8955–8971.

[190] Campillos M., Kuhn M., Gavin A.C., Jensen L.J., Bork P. (2008). Drug target identification using side-effect similarity. *Science*, **321**, 263–266.

[191] Lounkine E., Keiser M.J., Whitebread S., Mikhailov D., Hamon J., Jenkins J.L., Lavan P., Weber E., Doak A.K., Cote S., Shoichet B.K., Urban L. (2012). Large-scale prediction and testing of drug activity on side-effect targets. *Nature*, **486**, 361–367.

[192] Garcia-Serna R., Mestres J. (2010). Anticipating drug side effects by comparative pharmacology. *Expert Opinion on Drug Metabolism & Toxicology*, **6**, 1253–1263.

[193] Vidal D., Mestres J. (2010). In silico recepterome. Screening of antipsychotic drugs. *Molecular Informatics*, **29**, 543–551.

[194] Vidal D., Garcia-Serna R., Mestres J. (2011). Ligand-based approaches to in silico pharmacology. *Methods in Molecular Biology*, **672**, 489–502.

[195] Gregori-Puigjane E., Mestres J. (2006). SHED: Shannon entropy descriptors from topological feature distributions. *Journal of Chemical Information and Modeling*, **46**, 1615–1622.

[196] Mestres J., Martin-Couce L., Gregori-Puigjane E., Cases M., Boyer S. (2006). Ligand-based approach to in silico pharmacology: Nuclear receptor profiling. *Journal of Chemical Information and Modeling*, **46**, 2725–2736.

[197] Gregori-Puigjané E., Mestres J. (2008). A ligand-based approach to mining the chemogenomic space of drugs. *Combinatorial Chemistry and High Throughput Screening*, **11**, 669–676.

[198] Glick M., Davies J.W., Jenkins J.L. (2006). Prediction of biological targets for compounds using multiple-category Bayesian models trained on chemogenomics databases. *Journal of Chemical Information and Modeling*, **46**, 1124–1133.

[199] Nigsch F., Bender A., Jenkins J.L., Mitchell J.B.O. (2008). Ligand-target prediction using Winnow and naive Bayesian algorithms and the implication of overall performance statistics. *Journal of Chemical Information and Modeling*, **48**, 2313–2325.

[200] Niijima S., Yabuuchi H., Okuno Y. (2011). Cross-target view to feature selections: Identification of molecular interaction features in ligand-target space. *Journal of Chemical Information and Modeling*, **51**, 15–24.

[201] Schneider P., Stutz K., Kasper L., Haller S., Reutlinger M., Reisen F., Geppert T., Schneider G. (2011). Target profile prediction and practical evaluation of a Biginelli-type dihydropyrimidine compound library. *Pharmaceuticals*, **4**, 1236–1247.

[202] Reker D., Rodrigues T., Schneider P., Schneider G. (2014). Identifying the macromolecular targets of de novo-designed chemical entities through self-organizing map consensus. *Proceedings of the National Academy of Sciences of the United States of America*, **111**, 4067–4072.

[203] Heikamp K., Bajorath J. (2014). Support vector machines for drug discovery. *Expert Opinion on Drug Discovery*, **9**, 93–104.

[204] Wang F., Liu D., Wang H., Luo C., Zheng M., Liu H., Zhu W., Luo X., Zhang J., Jiang H. (2011). Computational screening for active compounds targeting protein sequences: Methodology and experimental validation. *Journal of Chemical Information and Modeling*, **51**, 2821–2828.

[205] Steindl T.M., Schuster D., Laggner C., Langer T. (2006). Parallel screening: A novel concept in pharmacophore modelling and virtual screening. *Journal of Chemical Information and Modeling*, **46**, 2146–2157.

[206] Steindl T.M., Schuster D., Wolber G., Laggner C., Langer T. (2006). High-throughput structure-based pharmacophore modelling as a basis for successful parallel virtual screening. *Journal of Computer-Aided Molecular Design*, **20**, 703–715.

[207] Markt P., Schuster D., Kirchmair J., Laggner C., Langer T. (2007). Pharmacophore modelling and parallel screening for PPAR ligands. *Journal of Computer-Aided Molecular Design*, **21**, 575–590.

[208] http://drugable.com/. Accessed May 27, 2015.

[209] Loh W.-Y. (2011). Classification and regression trees. *WIREs Data Mining and Knowledge Discovery*, **1**, 14–23.

[210] Keiser M.J., Roth B.L., Armbruster B.N., Ernsberger P., Irwin J.J., Shoichet B.K. (2007). Relating protein pharmacology by ligand chemistry. *Nature Biotechnology*, **25**, 197–206.

[211] Hert J., Keiser M.J., Irwin J.J., Oprea T.I., Shoichet B.K. (2008). Quantifying the relationships among drug classes. *Journal of Chemical Information and Modeling*, **48**, 755–765.

[212] Laggner C., Kokel D., Setola V., Tolia A., Lin H., Irwin J.J., Keiser M.J., Cheung C.Y., Minor D.L., Roth B.L., Peterson R.T., Shoichet B.K. (2012). Chemical informatics and target identification in a zebra fish phenotypic screen. *Nature Chemical Biology*, **8**, 144–146.

[213] Ripphausen P., Stumpfe D., Bajorath J. (2012). Analysis of structure-based virtual screening studies and characterization of identified active compounds. *Future Medicinal Chemistry*, **4**, 603–613.

# 10

# PHENOTYPIC SCREENING

Michelle Palmer
*Broad Institute of Harvard and MIT, Cambridge, MA, USA*

## 10.1 INTRODUCTION

Phenotypic screening is not a new approach to drug discovery but rather has undergone a resurgence based on a retrospective analysis of the success of developing new drugs with novel mechanisms, advances in model systems, and target deconvolution tools that enable more rapid progression from hit to lead candidate and ultimately an approved drug [1, 2]. Phenotypic screening had fallen out of favor due to challenges such as relevant assays, scalability, assessment of off-target effects, and target deorphaning. Progress in molecular biology, structural biology, and biochemical assay methods complemented by a rational approach to understanding the relationship between target and drug propelled target-based screening to the forefront. The current state of screening is much more a thoughtful approach to using complementary approaches utilizing both target based and phenotypic screening strategies.

Human genetics is uncovering a wealth of data relating both common and rare variants to human disease, but few have been translated to therapeutic interventions. Most common human diseases are not linked to a single-gene defect [3]. Genome-wide association studies (GWAS) have uncovered a multitude of gene variants that may be contributing to complex diseases, such a schizophrenia, coronary artery disease, and diabetes [4–6]. Many of the implicated genetic loci point to noncoding regions that may be involved in gene regulation or to genes that converge on pathways. Implementation of target-based screening at this stage may not be feasible. Validation of many of these genes involves the use of phenotypic assays to probe effects of genetic manipulation such as mutations, knockdowns, and overexpression. This can be an early starting point for a phenotypic screen using small molecules that can probe the biology and potentially correct the defect.

*Small Molecule Medicinal Chemistry: Strategies and Technologies*, First Edition.
Edited by Werngard Czechtizky and Peter Hamley.

A phenotypic screening approach incorporates the ability to identify a small molecule that can modulate the phenotype and probe multiple mechanisms in an unbiased fashion. This approach requires that a critical path for progression of molecules incorporates the ability to sort through possible mechanisms and translates those back to the human state.

This chapter will discuss the advantages and challenges of using phenotypic screening strategies for discovery of small molecule probes and drug leads. The topics will include selection of model system, critical assay path design, assay methods, and target identification strategies.

## 10.2 HISTORY AND PAST SUCCESSES

More than 25 years ago before target-based drug discovery became an established strategy, many lead discovery efforts were dependent on phenotypic approaches that utilized cells or whole organisms. A phenotype is an observable characteristic that can be measured in response to treatment with different chemicals or genetic perturbagens. Phenotypic screening has its roots in classical pharmacology. Once a small molecule was found to have a desired activity on the phenotype of interest, the focus changed to understanding the clinical potential in terms of efficacy and safety. The use of whole organisms was prevalent, especially in the discovery of anti-infectives, but the number of compounds screened was significantly smaller than today's million plus chemical libraries. This approach did successfully lead to the development of drugs such as ivermectin, FK506, early NSAIDs, the first thiazolidinedione, and the hypertension drug Aldomet [7–10]. The mechanism of action (MOA) of these drugs was discovered long after approval for human use. One challenge to continued use of this approach is that advances in understanding of diseases have demonstrated that many animal models show poor translation to the human disease.

## 10.3 IMPACT OF PHENOTYPIC SCREENING

The development of new technologies for cell culture and cell-based assays has enhanced the use of cell models and elevated the phenotypic assay to meet the needs of modern drug discovery. The ability to integrate an understanding of genetic abnormalities in targets, pathology of disease in relevant cell models, systems biology, and pharmacology makes this tried and true old successful approach new again. In a cell-based assay, the measured phenotype is a consolidation of many targets working in concert, and screening uncovers the susceptible pathway nodes [11]. Although a biochemical assay can capture the native function of a target, it is more challenging to capture the context-specific state that may be relevant for modulation of the disease, for example, the presence of the native substrate, activation state of target, protein complex formation, and posttranslational modification.

For any screening campaign, careful thought must be given to not only the primary assay design and validation but also the downstream screening cascade that will be used to triage and prioritize the most interesting hits based on a grouping by mechanism and overlap with the desired lead candidate profile [12]. Considerations of the cell model that best represents the disease biology include lineage, validated presence of disease targets or pathway, genotype, suitability of cells to assay format, scalability (propagation and miniaturization), and others. During primary assay development, planning for counterscreens and secondary assays should be underway. A screening cascade of assays (example in Fig. 10.1) should be developed with criteria set for compounds advancing to the next tier. As implied by the funnel-shaped figure, this cascade is designed to reduce and refine the compounds to the most interesting set of scaffolds for exploratory chemistry efforts. The assays incorporated in this funnel should be selected based on deep biological knowledge, designed to eliminate unwanted mechanisms, and confirm or further elucidate a desired mechanism. The criteria set for compounds to pass to the next tier of assays incorporates potency and/or selectivity. The interesting chemical scaffolds that emerge from the funnel can then be considered for use in target identification and lead optimization efforts.

Before embarking on full-scale screening, a pilot with a screening validation set (bioactives, subset of full library collection or promiscuous hitters) is helpful to set parameters for hit selection and anticipate possible assay artifacts. It is possible for some phenotypic screens to yield very high hit rates (>1%), most often due to off-target effects such as cytotoxicity. The insertion of a matched cell line that can be used to normalize activity of hits is useful for conducting counterscreening. For example, a screen to find molecules with selective killing of a mutant gene dependency in a cancer cell line requires a well-matched counterscreening cell line to filter nonspecific hits. The criteria set for selectivity is best assessed using a dose–response technique, not single concentration, and

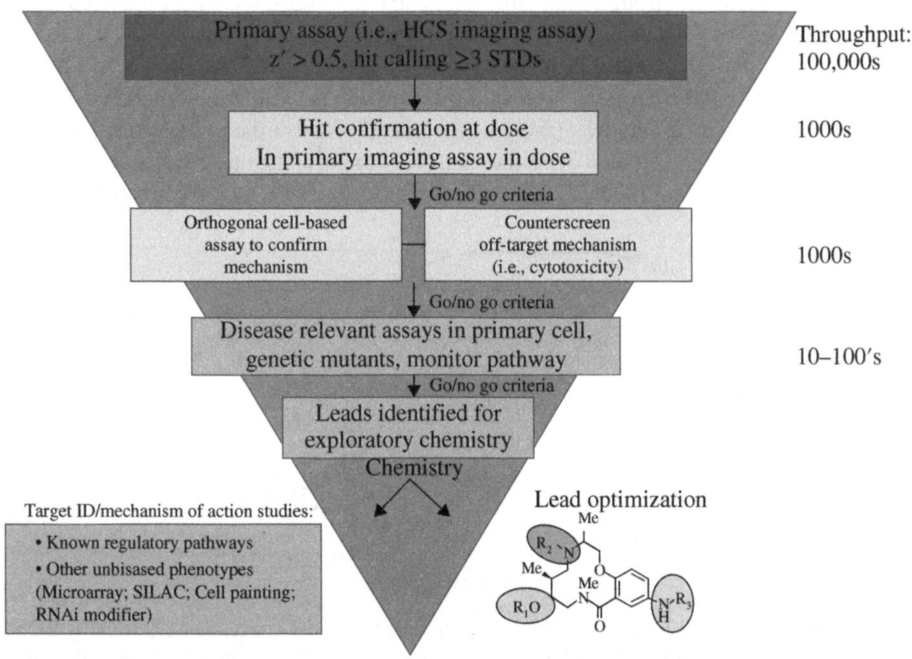


**FIGURE 10.1** An example of a screening cascade. The screening cascade is analogous to a funnel filter with the top tier representing the primary screening assay with capacity to screen large number of compounds. Moving through the funnel, the assays provide the stringency to remove or permit passage of compounds to the next tier below. The assays used in the lower tiers may not have a high-throughput capacity and thus need to be placed appropriately in the cascade to support screening programs with large numbers of compounds.

comparing area under the curve (AUC) rather than IC50s due to differences in curve shapes, slope, and max inhibition (see Fig. 10.2) [13, 14].

Phenotypic screening approaches benefit from a deep understanding of the biology, so that perturbagens that affect phenotype can be rapidly translated to assays to assess efficacy of compounds on relevant biology. The expression of a given phenotype and the effect of a drug on that phenotype in different cellular models may change because of a variety of system-dependent factors (cellular metabolism, species, genetic variability, environmental conditions, etc.), and this leads to a reliance on a control where equal cellular effects can be compared. When assessing medicinal chemistry series over a number of iterations, it is often necessary to include control compounds in every batch to benchmark within a given assay or across assays. Parameters such as differences in cellular density of receptors or efficiencies in receptor coupling can make it difficult to compare potency and specificity between two different cell lines.

Because a given assay may have inherent biases, it is important to include orthogonal assays that can assess the desired phenotype by an alternative assay method. For example, inhibition of cell proliferation can occur by a number of mechanisms so orthogonal assays that can assess alternative mechanisms of cell number reduction are needed. Cell killing, using assays to determine cell number (nuclei counting, BRd incorporation), is complemented by assays that assess effects on cellular ATP levels, cellular senescence, or induction of apoptosis.

There is the risk that through medicinal chemistry iterations, the MOA may change or pick up additional target activities. It is critical to include secondary assays in the cascade that assess toxicity, other pathway endpoints, or nodes that can be spot-checked during the course of medicinal chemistry. A successful approach is to utilize biomarkers for both on and off mechanism. If one was trying to specifically inhibit the expression and secretion of a target protein such as a cytokine, then it is important to assess the effect of the test compound on general mechanisms such as transcript stability, protein stability, and

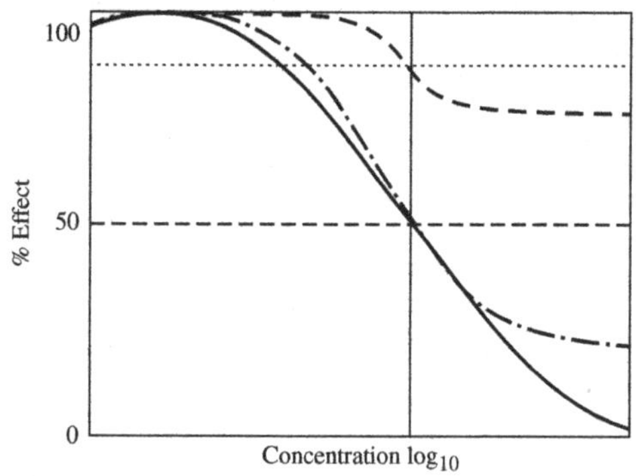


**FIGURE 10.2** Using AUC captures differences in dose–response curves not captured by EC50 statistics [13] The solid and dash/dot line curves achieve 50% reduction in cell effect at the same compound concentration, but the solid curve achieves 100% effect at higher compound concentrations. Both curves would yield the same IC50 value (dotted line at 50%), while the AUC captures the greater sensitivity of the assay to the sample represented by the solid curve. The dashed line curve illustrates a sample with less than 50% maximal reduction in cell effect at high compound concentrations. The EC50 statistic would be the same for the dashed and solid curves (vertical dotted line), while the AUC statistic captures the increased response of the solid curve.

secretion mechanisms (trafficking, PTMs, cleavage, folding). The choice of the counter-screening assay should be matched such that the stability of transcript or protein levels for the counter target has similar stability. For example, short-lived mRNAs (proto-oncogenes, nuclear transcription factors, and cytokines) are significantly enriched among regulatory genes, while genes with long-lived mRNAs are enriched among housekeeping genes [15, 16]. Therefore, the use of a housekeeping gene to compare specificity could be misleading. An assay to measure transferrin secretion is effective at eliminating compounds with nonspecific effects on secretion, but it would also be important to measure the impact on protein levels in related pathways.

## 10.4 MODEL SYSTEMS FOR PHENOTYPIC ASSAYS

The types of model systems that can be used in phenotypic screening are quite diverse and have varied benefits and challenges. The largest class is of course cell lines. However, examples have grown for primary and stem cells, cell coculture systems, 3D cell systems, tissues, organs on a chip, and whole organisms, including zebrafish, *Drosophila*, and *Caenorhabditis elegans* [17–20]. This section will focus on cell lines, primary cells, and 3D cell systems.

### 10.4.1 Cell Lines

The vast majority of phenotypic screening has relied on the use of cell lines, which are immortalized lines [21] that can be continuously passaged. Good quality control dictates that all lines should be mycoplasma tested [22] and DNA fingerprinted to assure reliability and identity [23]. ATCC has developed a comprehensive database of short tandem repeat (STR) DNA profiles for all of their human cell lines to identify quality control [24]. Establishment of passage number performance as a component of assay development and a carefully described cell culture protocol is important for robustness across the entire screening campaign. It is helpful to freeze down many batches of cells at an early passage number for large campaigns to assure each screening run has good reproducibility [25]. Many phenotypic assays have been developed in commonly used lines such as HeLa and HEK293 due to broad availability and citations in the literature. This approach has been successful especially for common pathways or recombinant applications where a reporter or tagged target has been introduced [26]. If one is concerned about cellular context as it relates to relevant substrates or protein interactions, then choosing a cell line that is representative of the tissue lineage is best. There are efforts underway to expand the representation of cell types in collections through immortalization of new human isolates [21]. A toolbox of perturbagens such as pathway or target selective small molecule inhibitors and genetic editors, such as RNAi, are useful to assess relevance of the cell models and assays.

### 10.4.2 Primary and Stem Cells

Primary human cells have become more readily available for certain cell types such as hepatocytes, T cells, PBMCs, endothelial, fibroblasts, preadipocytes, and muscle. Although primary cells may be considered closer to the human state, they are cultured in media and on an artificial matrix. The relevance to human biology must therefore be weighed against the cost and sometimes greater heterogeneity of response compared to cell lines. There are a number of examples where human-derived cells have been obtained in quantities to allow for screening of larger chemical libraries. Donor heterogeneity can lead to large variations in biological response in the assay. To achieve the day-to-day, batch-to-batch consistency required to assure robust screening results, strategies such as mixing populations of donor cells [27, 28] and implementing a batch pretest before screening increase success rates [29]. The results of a screen to identify compounds that could mediate the production of IFN-alpha using PBMCs isolated from about 200 human donors was described in 2009 [27]. Using electroluminescence, a homogeneous detection technology, over 1 million compounds were screened in a 384-well format. A reasonable hit rate and retest rate were observed despite 20% of the PBMC batches failing due to no or low response with the positive control resiquimod. This example of a very large screen was able to solve the problem of availability and cost to obtain primary cells.

### 10.4.3 Cocultures

Primary cells are more relevant to human biology but their use in assays in tissue culture plate is not representative of the native environment. The number of available primary cell types has increased, and the use of cocultures technologies (2D and 3D) has greatly improved the translational potential of these assays. Cocultures provide a more physiologically relevant environment to assess small molecule effects. Factors produced by one cell may influence the function of another cell in the coculture, thus uncovering a mechanism that would be unavailable in a single-cell system. The assay method chosen must be able to distinguish the appropriate biological readout in the specific cell. This can be achieved using imaging features or biomarkers that are specific for the desired cell readout. There are a number of examples where this approach has been successfully applied in the identification of small molecules in a screening approach [30–32]. In an effort to develop more effective therapies for acute leukemia, a phenotypic assay that utilized primary leukemic stem cells (LSCs) in coculture with bone marrow stromal cells was screened against 15 K compounds. The goal of selectively killing LSCs in a protective stromal cell environment required distinguishing killing of one cell type over the other. A microscopy imaging signature was developed that distinguished the formation of a cobblestone phenotype, indicative of burrowing of primitive cells beneath the stromal cells [32, 33]. A counterscreen using biologically similar hematopoietic stem cells was also utilized to select for LSC-specific killing and avoid dose-limiting toxicity of standard drugs such as cytarabine and daunorubicin.

The availability of disease-relevant primary cells is limiting in sufficient quantity and at a cost to allow for large-scale high-throughput screening (HTS). The promise of stem cell systems producing more faithful models of disease has increased interest in the use of phenotypic screening. The possibility of utilizing a patient-derived cell that has been differentiated to a disease cell type shortens the translational path and avoids the ethical issues [34, 35]. Advances in the development of induced pluripotent stem cells (iPSCs) have already produced some genetically relevant cell models that can be scaled in quantity for HTS. Success in differentiating iPSCs to neurons (motor, crest precursors, striatal, dopaminergic, astrocytes), hepatocytes, pancreatic beta cells, hematopoietic, progenitors, cardiomyocytes, and adipocytes is well documented, but standardization in protocols is still lacking [36–38]. Careful characterization of the differentiated cells using biomarkers for maturation, cell type, and functional integrity would provide useful standards for comparison. This is a new technology and it will take time to understand the full potential and limitations. As it stands today, iPSC has many of the benefits of immortalized cell lines such as abundance, transfection efficiency, and hearty to freeze thaw but still suffers from cost, population heterogeneity, and limited characterization [39].

### 10.4.4 3D Cell Models

In drug discovery, the assay chosen strives to recapitulate the biology within practical limitations, but it is well recognized that the standard 2D cell-based model systems used for most phenotypic screening do not always translate to the human clinical situation. There have been many advances in 3D cell culture models that are now being implemented in screening, high-throughput ADME, and stem cell production [40]. The most common formats use adherent cell culture on 3D supports such as sponges, gels, and microcarriers. The support incorporates complex biological gels or coatings for growth of mono or cocultures. Fluorescent readouts and high-content imaging are the detection method of choice. Challenges exist to obtain uniform multicellular tumor spheroids in a scalable manner and develop imaging algorithms to efficiently analyze the images. A colorectal carcinoma soft agar colony formation assay was used to screen a collection of natural products using laser-scanning cytometry. A colon stromal cell and cancer cell 3D coculture was used to confirm results. Compounds were identified that selectively killed the cancer cells in the 3D format that had not been previously found in standard 2D culture conditions [17]. The use of 3D matrices and microfluidics has been used in tissue engineering and generation of relevant models for safety assessment. Early models for the lung, heart, and intestine are also in development [41]. It is early in the implementation of the technology to know how successful these screening approaches will be, but it is believed to more closely mirror the environment experienced by normal cells in the body [42, 43].

## 10.5 ASSAYS

### 10.5.1 Assay Technologies

How one measures the perturbation of the phenotype in a cell can vary greatly and may be as straightforward as measuring a change in the proliferation of the cell or a more complex measurement of the change in the expression of defined biomarkers. Phenotypic assays can be thought of as low content or high content; Table 10.1 groups assay technologies based on this type of classification. The lower content assays are typically most amenable to miniaturization and very large-scale screening of greater than 1 million compounds, whereas the higher-content assays tend to be more expensive to run, may be difficult to miniaturize beyond 384-well format, and are often best suited to focused screens of smaller chemical libraries. The choice of assay is dictated by the phenotype being measured, compatibility with the chosen cell model, and ability to obtain a robust signal window. Screening for a molecule that can selectively kill a cancer cell line with a given

**TABLE 10.1 Phenotypic Assay Technologies**

| Execution Complexity | Category | Assay Type | Assay Technologies | Throughput | Biological function/ Mechanism | Reference |
|---|---|---|---|---|---|---|
| | | Cell viability | Alamar Blue, calcein AM, chemiluminescent | 100,000s | Mitochondrial function, cellular ATP | [44, 45] |
| | | Cell death | Caspase-Glo, annexin binding | 100,000s | Caspase activity | [46, 47] |
| | | Receptor/channel activity | Calcium flux dyes, cAMP detection, membrane potential dyes | 100,000s | Channel or receptor activity, second messengers, intracellular ion flux | [48–50] |
| Moderate | Low Content | | | | | |
| | | Gene or protein expression | Reporter gene (luciferase, beta-lactamase) | 100,000s | Monitor pathways, gene expression, intracellular or secreted | [51] |
| | | Receptor binding | Cell ELISA, radiometric | 10,000s | Detect cell binding of ligand | [52] |
| | | Posttranslational modification | ALPHA-LISA, HTRF, SPA | 100,000s | Phosphorylation and other post translational modifications | [53, 54] |
| | | Trafficking | Fluorescent protein tagging, immunohistochemistry, reporters | 10,000–100,000s | Internalization, intracellular relocalization, nuclear translocation, membrane association | [55, 56] |
| | | Morphology changes | Imaging dyes and antibodies, mass redistribution | 10,000–100,000s | Differentiation, activation, adherence, neurite outgrowth | [57, 58] |

| | | | | | | |
|---|---|---|---|---|---|---|
| | | Gene or protein expression | qPCR, gene signatures (bead arrays, microfluidics) | 1,000–10,000s | Measure mRNA copy number by exponential amplification/ detection | [59] |
| | | Biosensor | Ion sensing, granularity, protein interaction (BRET, split enzyme, two-hybrid) | | Synaptic activation, surface marker expression, kinetic marker expression, exocytosis, cell interaction | [60, 61] |
| Complex | Medium to high content | Biophysical, electrophysiology | Mea, auto patch, impedence | 1,000–10,000s | Channel activation, neuronal network activity, cardiac cell beating | [62] |
| | | Object intensity | Fluorescence, immunohistochemistry | 10,000–100,000s | ROS generation, autophagy, actin dynamics, biomarker, localization, and trafficking | [63, 64] |
| | | Protein expression | Flow cytometry, static cytometry, ALPHA-LISA, HTRF, SPA | 10,000–100,000s | Pathway readout, protein stabilization | [65] |
| | | Migration, invasion | Impedence, wave guide, luminescent | 1,000–10,000s | Cellular motility and adhesion | [66] |

genotype can be carried out using a simple proliferation assay. However, if there was knowledge of a target pathway dependency as the preferred target mechanism, one might choose a pathway-specific gene or protein biomarker as the assay endpoint. Either way, the assay has provided an unbiased context in which to test a chemical library that allows bioactive molecules to find their preferred target and determine if that small molecule target interaction results in modification of the phenotype.

Assay development may be initiated with different levels of understanding of the disease biology and relevant mechanisms. One is typically starting with observations from the underlying biology based on differences unveiled comparing the disease versus normal, wild type versus mutant, and perturbations with knockdown, overexpression, and chemical biology tools. The assay technologies and model system choice may be limited to one or two, but more often, there are numerous options. Beginning with a definition of the attributes of the molecules you are trying to uncover is helpful in deciding the assay format.

For example, the hits desired must selectively inhibit cell proliferation or induce differentiation to a given cell state based on the therapeutic hypothesis. The first phenotype can be assessed using an assay that determines cell number, whereas the second phenotype requires a cellular marker that would indicate differentiation, such as a morphology change or gene expression. Additional knowledge allows one to focus on the measurement of a particular pathway or cellular function such as trafficking, gene expression, protein stability, or membrane potential.

### 10.5.2 Assay Development Considerations

When embarking on assay development, there are many variables to consider. There are many excellent references available that provide valuable information on assay optimization [67], but it is worth emphasizing some key aspects that are required when the assay must also support medicinal chemistry efforts. Robustness and reproducibility is highly dependent on the stability of the relevant biology, which is most often linked to passage number of the cells and cell culture conditions. Defining the cell passage window (both lower and upper limits) and a highly detailed SOP for cell culturing can avoid a disruption in the project due to changing IC50 values in controls. Assay artifacts are also a common problem that can derail projects. Trace metal contaminants such as palladium and copper can affect assays that measure metabolites such as FAD, NADH. It is useful to have a set of reagents to test effects of such artifacts. The most common artifact in phenotypic assays is mycoplasma contamination. A rigorous testing of a cell line at regular intervals is important using a sensitive method such as PCR-based methods [68].

Once a robust assay has been established, then validation for assay window (Z′, dose dependence, curve steepness), reproducibility, and hit selection criteria using an appropriate control is required. A key validation step is to screen a validation set of compounds of a least 2000 molecules composed of both bioactives and representative molecules from the full screening collection to get a sense of assay performance under production conditions. Assay performance, as assessed by Z′, hit criteria (3 STD from the DMSO control), and hit reproducibility can be determined over multiple runs.

The variance in the DMSO control can sometimes be as high as 15% thus using a control that gives 100% inhibition is often needed to get a robust Z′ value of 0.5 or greater

[69]. Incomplete inhibition upon dose testing is often seen with early screening hits due to low potency and poor solubility, but sometimes, it is also the result of the biology. Compensatory biology and sensitivity of the assay may confound this.

Illustrative examples of phenotypic screening in the literature that have yielded advanced leads or even resulted in marketed drugs are limited, but a few examples are given in the following. These examples illustrate the benefit and challenges in execution of these types of projects. They are resource intensive and deep knowledge of the disease biology and potential target classes are essential in the downstream experimentation needed to define mechanisms and support the medicinal chemistry efforts.

### 10.5.3 Example 1: Selective Killing of Breast Cancer Stem Cells

This example illustrates the outcome of the most straightforward phenotypic screening method [70–72], cell viability, as a successful approach to identify small molecules with selective killing of breast cancer stem cells (CSCs).

Breast cancer subpopulations of cells drive tumor growth, and this rare population is resistant to chemotherapeutics. Gupta et al. developed a stemlike cell line by inducing epithelial-to-mesenchymal transdifferentiation in normal mammary epithelial cells resulting in an enriched cell population that demonstrated stemlike properties [70].

A number of promising scaffolds, including salinomycin, were identified in the screen of greater than 350 K compounds. The greatest shortcoming of this strategy is that this screen captures both selective and nonselective mechanisms of killing. The hit rate at a standard cutoff of three sigma was greater than 2%. Selective triaging of hits by chemists and profiling in both the screening line and an isogenic control cell line reduced the number of scaffolds for characterization to a manageable number. The natural product salinomycin was shown to selectively kill the stem cell population. This result was validated through a number of secondary assays including killing in a mammosphere assay. Salinomycin is a commonly used antibiotic in agriculture and is characterized as a potassium ionophore; however, the MOA in CSCs has not been described.

### 10.5.4 Example 2: CFTR Potentiator Drug

A hallmark example of the use of phenotypic screening in the successful development of a drug is the story behind development of VX-770 [73, 74], an orally available cystic fibrosis transmembrane conductance regular (CFTR) chloride channel potentiator. Scientists at Vertex developed an assay using NIH-3T3 cells expressing defective DeltaF508-CFTR to identify modulators that could increase anion channel flux. This mutation impairs protein folding, trafficking, and channel gating in epithelial cells. The hope was to capture both correctors and potentiators in this phenotypic assay. An increase of the mutant channel activity at the cell membrane could be measured using a fluorescent energy transfer between membrane-soluble voltage-sensitive dye and a plasma membrane-localized fluorescent-labeled phospholipid. The assay was designed to run in two different modes: one to monitor correctors and one to monitor potentiators. Ultimately, molecules were found that rescued defective trafficking or gating of the defective channel. One of the biggest challenges faced by this project during lead optimization was defining structure–activity relationship (SAR) using EC50 or potency alone. Because the small molecule corrector was increasing, copy number of the channel at the membrane efficacy was also an important parameter in guiding SAR. Efficacy was normalized to the low temperature control as 100%. Chemistry optimization was able to identify modifications that improved both potency and efficacy. Orthogonal assays to monitor cell surface density and current density by electrophysiology were also employed. SAR assessment for potentiators was further dissected using electrophysiology and modulators of channel gating. The most potent molecule identified in the potentiator screen, VRT-532, was only active in the presence of forskolin, indicating it was not an activator and its mode of action was determined to be increasing the channel open duration of the activated channel.

Examples 1 and 2 illustrate the importance of having a well-defined critical path with gating criteria for selection of hits and progression criteria for compounds. The identification of bioactive compounds during pilot screening provides some hints as to the targets or pathways that may yield active molecules. It also allows one to validate the overall critical path especially if there are targets and pathways one wishes to avoid. Some key questions to consider in assay design and critical path layout are: Will my assay find the type of molecules I am looking for? How will I triage my list of hit scaffolds? What assays will I use to drive medicinal chemistry?

## 10.6 DEORPHANING

Phenotypic screening has the potential to find a disease-modifying molecule with a novel MOA, but it comes with the big challenge of defining the target at the center of the MOA. Using a phenotypic screening approach, one will identify compounds that are cell permeable and potent enough to modulate a target in the cellular context. Target deconvolution or deorphaning is still a very difficult process that requires significant resources. This process is complicated by weak potency and poor biophysical properties of the molecule at this early stage, polypharmacology, and compensatory biology. Although drug development can be successful in the absence of a defined mechanism, having that information in hand supports a rational medicinal chemistry path and reduces risk associated with safety.

Some key challenges in target identification that are not always clearly understood when entering this phase is the role of medicinal chemistry and the importance of implementing an unbiased strategy in the relevant biological context.

Uncovering the target of the small molecule identified in a phenotypic screen requires multiple experimental approaches and technologies to converge on a limited number of testable hypotheses. A systematic approach can be undertaken using established methods that include affinity pulldown, chemical and genetic profiling, functional profiling, computational analysis, and biochemical/biophysical analysis (Fig. 10.3).

### 10.6.1 Affinity-Based Proteomics

The power of phenotypic screening lies in the unbiased nature of the method; however, defining the target mechanism at an early stage of the project can be challenging. Ligand–target affinity-based proteomic methods can be applied here, but depending on the method chosen, a chemical modification of the protein target or the probe molecule will be required to provide a handle for capture.

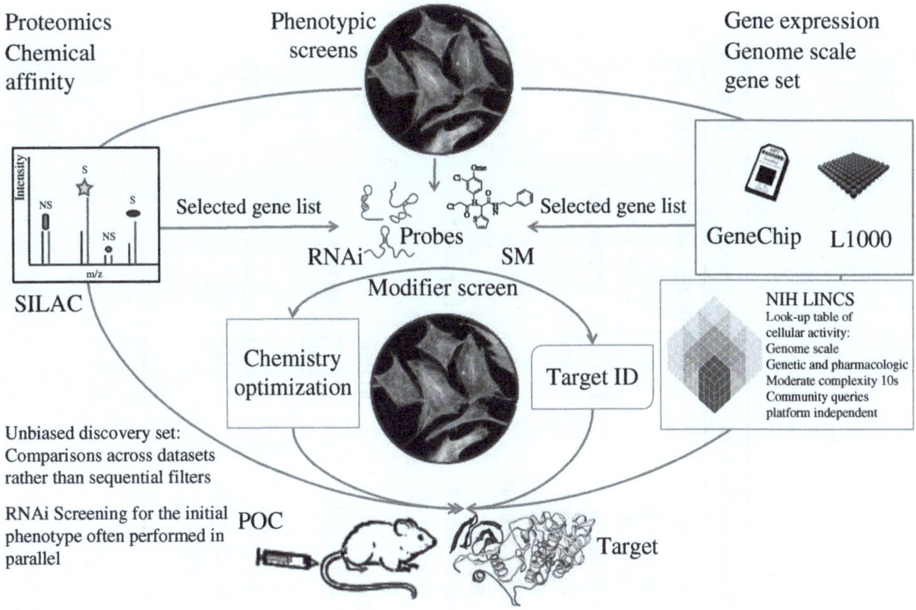


**FIGURE 10.3** Phenotypic screening requires an integrated technology approach. The schematic illustrates the integration of multiple technologies that are required to support phenotypic screening programs and target deconvolution. (*See insert for color representation of the figure.*)

Successful identification of putative targets is limited by weak probe target affinities and low target abundance so a screening hit is unlikely to be a good starting point for target identification. Using an alternative stable isotope-labeling method can mitigate low target abundance [75]. Development of these techniques has led to target class enrichment strategies that can profile the probe ligand affinities in a cellular context. One such method uses weak kinase inhibitors immobilized on beads to enrich for kinases resulting in a kinome profile of probe selectivity in a cellular context [76].

A good starting point for chemical proteomics is after there has been some limited SAR determined through medicinal chemistry efforts to optimize the probe and reach a probe affinity or EC50 in the submicromolar range. Modification of the probe molecule at a non-critical position may be through attachment of a linker for solid support capture or a chemical cross-linking functional group. Where to place the linker and maintain activity in the cell requires knowledge of the SAR and therefore continued input from medicinal chemistry [77]. An important element of any affinity method is to include a negative control, an inactive probe, as a competitor to filter out nonspecific binding targets through subtraction methods. The ideal competitor molecule is an inactive enantiomer or diastereomer of the active molecule. The active and inactive molecules are closely matched in both structural building blocks and biophysical properties simplifying interpretation of nonspecific binders.

A quantitative chemical proteomics approach that provides unbiased proteome-wide target identification is best matched to phenotypic screening. Working in the same cell type as the original screening assay is important and also using lysis buffer conditions that would not disrupt physiologically relevant interactions. There are a number of chemical proteomic strategies that utilize differential labeling conditions (see Table 10.2). The stable

**TABLE 10.2 Proteomic Method and the Small Molecule Probe Type**

| Method | Probe Modification | Protein Modification | Separation Method | Examples | Reference |
|---|---|---|---|---|---|
| Affinity purification | Linker to chromatographic resin | None, cell lysate | Enzymatic activity of fractions | HDAC, thalidomide | [78, 79] |
| FITGE | Chemical or photoaffinity cross-linking group | Cross-linked by probe, tagged with fluorescent dye | Differential fluorescent labeling and 2D gel-based separation | Live cells, HSP60, and tubulin | [80] |
| Silac | Linker to capture beads, soluble competitors | Heavy isotope labeling of cells | Stable isotope labeling of cells | GRB2-SH2 protein | [81, 82] |
| ICAT | Active probe linked to capture beads, inactive compound linked to beads as control | Chemical labeling | Isotope-coded affinity tag | Malate dehydrogenase/ E7070 | [83] |
| iTRAQ | Soluble competitor | Target class enrichment with Kino beads, GTPases, methyltransferases | Isobaric tags labeling after digestion | Yeast proteomics, HDACs | [84, 85] |

isotope labeling by amino acids in cell culture (SILAC) method has proven successful in identifying target proteins [81]. This strategy identifies and quantifies cellular target protein interactions using a modified probe molecule. By combining this method with affinity purification, it is also possible to determine dissociation constants for the probe ligand with the proteins identified in the pulldown and their associated complexes. Addition of affinity measurements provides greater context to prioritize the list of interactors [86–88].

Despite successful outcomes, there are limitations such as bias to higher-affinity interactions, loss of transient protein complexes, incompatibility with membrane-bound protein, incompatibility with use of nonreplicating cells, or treatment of live cells [89]. A potentially more efficient unbiased method, which is compatible with live cell labeling, is fluorescence difference in two-dimensional gel electrophoresis (FITGE) [80]. A photoaffinity tag is added to the chemical probe and to an inactive analogue. Each probe is then used to separately treat live cells. The live cells are then irradiated to cross-link adjacent protein in the cell assuring a relevant context. This alternative proteomic method takes advantage of a modern chemical method; click chemistry, to modify cross-linked active and inactive probe samples differentially with two fluorescent dyes, Cy3 and Cy5. Separation of proteins occurs using difference two-dimensional gel electrophoresis followed by excision of spots for mass spectroscopy (MS) detection. The relevant spots are chosen where differential fluorescent labeling occurred in specific versus nonspecific sample. This technology identified tubulin as the primary target of a novel antiproliferative compound found in a phenotypic screen of Hela cells. The mechanism was confirmed as inhibiting tubulin polymerization and causing cell cycle arrest in Hela cells [80]. This method overcomes a number of the limitations of affinity-based methods.

### 10.6.2 Genetic Profiling

In the past, much of the phenotypic screening focused on infectious disease applications and target identification using genetic methods was often straightforward. Generation of resistant microbial and parasitic strains in the presence of the probe molecule is often successful in identifying the target and MOA using sequencing to identify the resistance causing mutation. Although the cost of sequencing has come down and resistant strain generation in microbial organisms has become routine, more challenging applications with mammalian systems are still rare [90].

A complementary approach to proteomics that can also interrogate chemical probes in an unbiased fashion is genome-wide transcript profiling. Generation of genome-wide expression data has traditionally been generated using gene array chips such as the Affymetrix ones [91]. Like proteomics, this is an expensive undertaking, and therefore, the number of samples tested would be limited. The development of higher-throughput lower-cost methods (L1000) is now allowing this type of analysis to move up earlier in the screening cascade and to test many more compounds for activity on common pathways and cellular functions. One such technology, L1000, measures the expression levels of approximately one thousand landmark genes, and using a computational model trained on more than one hundred thousand microarray experiments can infer the activity of the other genes. This technology reduces the cost to about one-tenth of that of a chip microarray experiment. This technology has been used to expand the Connectivity Map (also known as cmap) [92, 93] data set to now include response to perturbagens in 15 cell types with over 5000 compounds and 3700 genes to generate greater than 1.2 M signatures (http://www.lincsproject.org). Testing of a probe at a given time point and concentration generates a transcription signature of genes that are up- and downregulated. With this signature in hand, one can compare the probe signature to existing signatures of perturbagens (small molecule and genetic) to infer a link to targets of known bioactive molecules or functional pathways/cellular functions [92]. Comparison of your probe-specific gene signature using public tools like gene set enrichment analysis (GSEA) with signatures contained within a large profiling data set will also generate a list of genes/pathways that are highly correlated [94, 95]. Studying the intersection of gene expression and proteomic results can provide further prioritization of targets.

Proteomics profiling data sets for biomarkers and PTMs are beginning to emerge and could be applied in a similar manner to gene expression profiling. Data sets have been generated for acetylation, phosphorylation, ubiquitination, and the secretome under limited conditions [96–99]. Although the data sets are currently limited to a few cell types and perturbagens, there is growing evidence of the value of this type of data in target identification and MOA. The main limitation to greater access to these types of data is a lack of standardized databases and public tools for analysis [100]. There would also be a greater uptake of this type of protein biomarker profiling if like genetic methods the process for sample preparation became more efficient and the availability of user-friendly, dedicated use instruments increased.

### 10.6.3 Target Profiling

Most screening centers have access to large historical data sets where cross assay analysis of hit compounds can provide some insights into possible mechanisms. There are also high-quality public and private screening and bioassay data sets such as Pubchem, Chembl, and Gostar (https://gostardb.com/gostar/) that can be mined [101, 102]. These databases can be put to use in target deconvolution by comparing the relevant profile for the compound of interest with profiles of compounds with known biological activity to infer similar or related MOA. These methods measure similarity between chemical structure clusters based on their biological activity profile. All assays can be treated as equal, but it is more often the case that different types of assays will receive different weighting. Ultimately, this is another hypothesis-generating tool, but the resource required is minimal and is equally useful when looking at a few compounds or larger sets. These methods connect compounds with similar mechanisms of action, enabling prediction of new targets and mechanisms [103].

Large assay panels exist for 100 s of targets such as GPCRs, transporters, RTKs, channels, kinases, phosphatases, proteases, chromatin-modifying enzymes, and many others. Profiling of probe molecules in these panels can be very helpful in classifying or ruling out possible target classes. Many of these panels are limited to competition binding assays, which may be misleading if binding does not translate to functional activity, or allosteric and protein interaction binding sites are missed.

### 10.6.4 Comodifier Profiling

With a limited number of target hypothesis in hand, it is now time to carry out further experimentation. Comodifier profiling is a powerful method that uses coadministration of your chemical probe with selective bioactive compounds or genetic tools such as RNAi to narrow the possible pathway nodes that may be the probe target. Elimination of a target by knockdown or inhibition should phenocopy the probe molecule or a partial inhibition should be additive with your probe molecule. Overexpression of that target would confer resistance to the probe. Assembly of focused chemical probe sets and gene family RNAi sets that can be applied in comodifier screens builds efficiency into the downstream experiments required in many target identification projects.

### 10.6.5 Target Engagement

Although genetic methods are indicative of a target's involvement, it is still wise to validate that the probe molecule binds to the putative target in the context of the cell. Direct binding between the probe and target can be assessed using biophysical methods such as SPR, ITC, DSF, DSC, NMR, and X-ray crystallography if high-quality protein of the target can be obtained. This approach does not account for the cellular context of a phenotypic assay. Cellular thermal shift assay (CETSA) can be used to determine direct binding of a target and the probe in the context of the cell [104]. Small molecule binding to a protein in the presence of ramping temperature will right shift the denaturation temperature of that protein. The change in denaturation temperature is correlated with stabilization of the protein in the presence of the probe and can be assessed using antibody-based detection methods such a gel westerns and ELISAs or by MS of remaining soluble protein after pelleting of precipitated cell components [105]. Although not a high-throughput method, it can be scaled to 10s of compounds or multiple proteins if appropriate antibody reagents are available.

All of the functional methods require a substantial investment of resources to develop the methods and characterize the probe compounds. It is often the case that many of the reagents and assays may already exist as vector sets, probe sets, and genetically modified cell lines within systems biology groups or from external vendors. Although not discussed here in any detail, all of these deorphaning strategies are highly reliant on strong chemo- and bioinformatics resources. Following is a case study describing a comprehensive analysis of mechanism using an integrative approach of proteomics, genomics, and chemical biology to elucidate the mechanism of a small molecule inducer of megakaryocyte differentiation. This example incorporates some of the best features of phenotypic screening; disease-relevant cell models, well-defined high-content phenotype, focused triage path to eliminate and group the most interesting hits by mechanism.

### 10.6.6 Example 3: Elucidating MOA for a Regulator of Polyploidization

A multidisciplinary team undertook a screen to identify small molecule probes that would help to elucidate the mechanisms by which cells overcome mitosis to become polyploidy [106]. Acute megakaryocytic leukemia (AMKL) in adults has a population with a poor prognosis that do not respond to current therapies [107]. AMKL is characterized by expansion of immature megakaryocytes [108]. Discovery of small molecules that could drive immature megakaryocytes toward a mature state and terminal differentiation was proposed as a potential therapeutic strategy.

Mature megakaryocytes display multiple ploidy states and can reach DNA content as high as 128 N. The mechanism underlying this process is not well understood but defects in cell cycle gene expression have been implicated [109]. This polyploidy phenotype can be assessed using nuclear staining dyes to measure nuclei morphology and DNA content. High-content imaging was used as the screening modality. In this project, AMKL patient-derived cells (CMK) were used in the screen, providing high relevance to the disease state [110]. Due to the complexity of the screen, a smaller focused screen was executed against approximately 9000 compounds comprised of bioactives and diversity sets. The assay performed well, meeting HTS standards of Z′ [69]. The assay generated about 200 hits and approximately 50% of these confirmed at dose in the screening assay. Expected hits included microtubule and actin-disrupting compounds, but kinase inhibitors were unexpectedly found. Further experiments focused on understanding if inducers of polyploidization also promoted maturation of megakaryocytes. Many of the confirmed hits inhibited proliferation while inducing polyploid nuclei and apoptosis, but only dimethyl fasudil (diMF) and NP003964 induced expression of megakaryocyte-specific markers CD41 and CD42. This was also confirmed in a panel of megakaryocyte cell lines and primary cells. Characterization was extended to mouse models of AMKL and clinical samples from human DS-AMKL bone marrow specimens. DiMF inhibited proliferation and induced the phenotype in multiple genetic backgrounds found in AMKL predicting a broader applicability of this mechanism.

This project had an advantage over many phenotypic screens in that the bioactive compound, diMF, was annotated as an ATP competitive kinase inhibitor, thus focusing the target deconvolution efforts to determine which kinases were involved. MOA efforts incorporated kinase panels; quantitative proteomics coupled with kinome enrichment and shRNA profiling and modifier screens. All methods converged on a list of 95 distinct genes, which was further prioritized computationally using integrative analysis. Aurora kinase A was at the top of the list for further validation. Additional mechanistic studies in cells and the use of a more selective Aurora kinase A inhibitor, MLN8237, further consolidated a role for this target in megakaryocyte differentiation and maturation.

There are a number of Aurora kinase A inhibitors, including MLN8237, in clinical trials for leukemia, but at the tolerated dose, they do not appear to be effective [111]. However, there has not been a focused trial in AMKL where greater sensitivity to the inhibitors may show benefit.

## 10.7 SUMMARY

Phenotypic screening as a primary screening strategy can be a rational approach to identify a range of mechanisms that can modify the phenotype from a disease state to a normal state. Success is more likely when disease-relevant cell models and the phenotype measured are chosen based on genetic insights. Triaging of screening hits and support of medicinal chemistry efforts requires a suite of downstream assays to quickly build a boundary around possible target and pathway mechanisms. This SAR data package also informs the creation of an effective affinity probe for target deconvolution efforts. The molecules identified by phenotypic methods may lead to drugs, but they also serve as valuable tools at an early stage of a new therapeutic project to validate the disease biology and identify a target for the next iteration in drug discovery.

## REFERENCES


[1] Swinney DC. Phenotypic vs. target-based drug discovery for first-in-class medicines. *Clin Pharmacol Ther.* 2013;**93**(4):299–301.

[2] Swinney DC. The contribution of mechanistic understanding to phenotypic screening for first-in-class medicines. *J Biomol Screen.* 2013;**18**(10):1186–92.

[3] Plenge RM, Scolnick EM, Altshuler D. Validating therapeutic targets through human genetics. *Nat Rev Drug Discov.* 2013;**12**(8):581–94.

[4] Ripke S, Neale BM, Corvin A, Walters JTR, Farh K-H, Holmans PA, et al. Biological insights from 108 schizophrenia-associated genetic loci. *Nature.* 2014;**511**(7510):421–7.

[5] Do R, Willer CJ, Schmidt EM, Sengupta S, Gao C, Peloso GM, et al. Common variants associated with plasma triglycerides and risk for coronary artery disease. *Nat Genet.* 2013;**45**(11):1345–52.

[6] Flannick J, Beer NL, Bick AG, Agarwala V, Molnes J, Gupta N, et al. Assessing the phenotypic effects in the general population of rare variants in genes for a dominant Mendelian form of diabetes. *Nat Genet.* 2013;**45**(11):1380–5.

[7] Campbell WC, Burg RW, Fisher MH, Dybas RD. The discovery of ivermectin and other avermectin. In: Magee PKGMJ, editor. *Pesticide Synthesis Through Rational Approaches.* Washington, DC: ACS; 1984. p. 5–20.

[8] Kino T, Hatanaka H, Hashimoto M, Nishiyama M, Goto T, Okuhara M, et al. FK-506, a novel immunosuppressant isolated from a Streptomyces. I. Fermentation, isolation, and physico-chemical and biological characteristics. *J Antibiot (Tokyo).* 1987;**40**(9):1249–55.

[9] Adams SS. The discovery of Brufen. *Chem Br.* 1987;**23**:1193–5.

[10] Zhou C, Tang C, Chang E, Ge M, Lin S, Cline E, et al. Discovery of 5-aryloxy-2,4-thiazolidinediones as potent GPR40 agonists. *Bioorg Med Chem Lett.* 2010;**20**(3):1298–301.

[11] Lee JA, Chu S, Willard FS, Cox KL, Sells Galvin RJ, Peery RB, et al. Open innovation for phenotypic drug discovery: the PD2 assay panel. *J Biomol Screen.* 2011;**16**(6):588–602.

[12] Peppard JV, Rugg C, Smicker M, Dureuil C, Ronan B, Flamand O, et al. Identifying Small Molecules which Inhibit Autophagy: a Phenotypic Screen Using Image-Based High-Content Cell Analysis. *Curr Chem Genomics Transl Med* 2014;**8**:3–15.

[13] Jang IS, Neto EC, Guinney J, Friend SH, Margolin AA. Systematic assessment of analytical methods for drug sensitivity prediction from cancer cell line data. *Pac Symp Biocomput.* 2014;63–74.

[14] Fallahi-Sichani M, Honarnejad S, Heiser LM, Gray JW, Sorger PK. Metrics other than potency reveal systematic variation in responses to cancer drugs. *Nat Chem Biol.* 2013;**9**(11):708–14.

[15] Vogel C, Marcotte EM. Insights into the regulation of protein abundance from proteomic and transcriptomic analyses. *Nat Rev Genet.* 2012;**13**(4):227–32.

[16] Schwanhäusser B, Busse D, Li N, Dittmar G, Schuchhardt J, Wolf J, et al. Global quantification of mammalian gene expression control. *Nature.* 2011;**473**(7347):337–42.

[17] Horman SR, To J, Orth AP. An HTS-compatible 3D colony formation assay to identify tumor-specific chemotherapeutics. *J Biomol Screen.* 2013;**18**(10):1298–308.

[18] Huang H, Lindgren A, Wu X, Liu N-A, Lin S. High-throughput screening for bioactive molecules using primary cell culture of transgenic zebrafish embryos. *Cell Rep.* 2012;**2**(3): 695–704.

[19] Giacomotto J, Ségalat L. High-throughput screening and small animal models, where are we? *Br J Pharmacol.* 2010;**160**(2):204–16.

[20] Reinhart PH, Kaltenbach LS, Essrich C, Dunn DE, Eudailey JA, DeMarco CT, et al. Identification of anti-inflammatory targets for Huntington's disease using a brain slice-based screening assay. *Neurobiol Dis*. 2011;**43**(1):248–56.

[21] Lipps C, May T, Wirth D. Eternity and functionality—rational access to physiologically relevant cell lines. *Biol Chem*. 2013;**394**(12):1637–48.

[22] Young L, Sung J, Stacey G, Masters JR. Detection of Mycoplasma in cell cultures. *Nat Protoc*. 2010;**5**(5):929–34.

[23] Lorenzi PL, Reinhold WC, Varma S, Hutchinson AA, Pommier Y, Chanock SJ, et al. DNA fingerprinting of the NCI-60 cell line panel. *Mol Cancer Ther*. 2009;**8**(4):713–24.

[24] ATCC. Search the STR Database. Available from: http://www.atcc.org/STR_Database.aspx (accessed May 15, 2015).

[25] Hughes P, Marshall D, Reid Y, Parkes H, Gelber C. The costs of using unauthenticated, over-passaged cell lines: how much more data do we need? *Biotechniques*. 2007;**43**:575–86.

[26] Inglese J, Johnson RL, Simeonov A, Xia M, Zheng W, Austin CP, et al. High-throughput screening assays for the identification of chemical probes. *Nat Chem Biol*. 2007;**3**(8): 466–79.

[27] Dunne A, Jowett M, Rees S. High throughput screening. In: Janzen WP, Bernasconi P, editors. *High Throughput Screening, Methods and Protocols*, Second Edition. Totowa, NJ: Humana Press; 2009. p. 239–57.

[28] Haining WN, Angelosanto J, Brosnahan K, Ross K, Hahn C, Russell K, et al. High-throughput gene expression profiling of memory differentiation in primary human T cells. *BMC Immunol*. 2008;9:44.

[29] Dockendorff C, Aisiku O, Verplank L, Dilks JR, Smith DA, Gunnink SF, et al. Discovery of 1,3-Diaminobenzenes as Selective Inhibitors of Platelet Activation at the PAR1 Receptor. *ACS Med Chem Lett*. 2012;**3**(3):232–7.

[30] Shan J, Schwartz RE, Ross NT, Logan DJ, Thomas D, Duncan SA, et al. Identification of small molecules for human hepatocyte expansion and iPS differentiation. *Nat Chem Biol*. 2013;**9**(8):514–20.

[31] Straussman R, Morikawa T, Shee K, Barzily-Rokni M, Qian ZR, Du J, et al. Tumour micro-environment elicits innate resistance to RAF inhibitors through HGF secretion. *Nature*. 2012;**487**(7408):500–4.

[32] Hartwell KA, Miller PG, Mukherjee S, Kahn AR, Stewart AL, Logan DJ, et al. Niche-based screening identifies small-molecule inhibitors of leukemia stem cells. *Nat Chem Biol*. 2013;**9**(12):840–8.

[33] Breems DA, Blokland EA, Neben S, Ploemacher RE. Frequency analysis of human primitive haematopoietic stem cell subsets using a cobblestone area forming cell assay. *Leukemia*. 1994;**8**(7):1095–104.

[34] Nishikawa S, Goldstein RA, Nierras CR. The promise of human induced pluripotent stem cells for research and therapy. *Nat Rev Mol Cell Biol*. 2008;**9**(9):725–9.

[35] Takahashi K, Yamanaka S. Induction of pluripotent stem cells from mouse embryonic and adult fibroblast cultures by defined factors. *Cell*. 2006;**126**(4):663–76.

[36] Yang YM, Gupta SK, Kim KJ, Powers BE, Cerqueira A, Wainger BJ, et al. A small molecule screen in stem-cell-derived motor neurons identifies a kinase inhibitor as a candidate therapeutic for ALS. *Cell Stem Cell*. 2013;**12**(6):713–26.

[37] Xu X, Zhong Z. Disease modeling and drug screening for neurological diseases using human induced pluripotent stem cells. *Acta Pharmacol Sin*. 2013;**34**(6):755–64.

[38] Rubin LL, Haston KM. Stem cell biology and drug discovery. *BMC Biol*. 2011;**9**(1):42.

[39] Eglen R, Reisine T. Primary cells and stem cells in drug discovery: emerging tools for high-throughput screening. *Assay Drug Dev Technol*. 2011;**9**(2):108–24.

[40] Justice BA, Badr NA, Felder RA. 3D cell culture opens new dimensions in cell-based assays. *Drug Discov Today*. 2009;**14**(1–2):102–7.

[41] Bhatia SN, Ingber DE. Microfluidic organs-on-chips. *Nat Biotechnol*. 2014;**32**(8):760–72.

[42] LaBarbera DV, Reid BG, Yoo BH. The multicellular tumor spheroid model for high-throughput cancer drug discovery. *Expert Opin Drug Discov*. 2012;**7**(9):819–30.

[43] Lama R, Zhang L, Naim JM, Williams J, Zhou A, Su B. Development, validation and pilot screening of an in vitro multi-cellular three-dimensional cancer spheroid assay for anti-cancer drug testing. *Bioorg Med Chem*. 2013;**21**(4):922–31.

[44] Niles AL, Moravec RA, Riss TL. In vitro viability and cytotoxicity testing and same-well multi-parametric combinations for high throughput screening. *Curr Chem Genomics*. 2009;**3**:33–41.

[45] Ramirez CN, Antczak C, Djaballah H. Cell viability assessment: toward content-rich platforms. *Expert Opin Drug Discov*. 2010;**5**(3):223–33.

[46] Tai VW-F, Sperandio D, Shelton EJ, Litvak J, Pararajasingham K, Cebon B, et al. Discovery and structure-activity relationship of 2-phenyl-oxazole-4-carboxamide derivatives as potent apoptosis inducers. *Bioorg Med Chem Lett*. 2006;**16**(17):4554–8.

[47] Kepp O, Galluzzi L, Lipinski M, Yuan J, Kroemer G. Cell death assays for drug discovery. *Nat Rev Drug Discov*. 2011;**10**(3):221–37.

[48] Hodder P, Mull R, Cassaday J, Berry K, Strulovici B. Miniaturization of intracellular calcium functional assays to 1536-well plate format using a fluorometric imaging plate reader. *J Biomol Screen*. 2004;**9**(5):417–26.

[49] Kumar M, Hsiao K, Vidugiriene J, Goueli SA. A bioluminescent-based, HTS-compatible assay to monitor G-protein-coupled receptor modulation of cellular cyclic AMP. *Assay Drug Dev Technol*. 2007;**5**(2):237–45.

[50] Martikkala E, Rozwandowicz-Jansen A, Hänninen P, Petäjä-Repo U, Härmä H. A homogeneous single-label time-resolved fluorescence cAMP assay. *J Biomol Screen*. 2011;**16**(3): 356–62.

[51] Fan F, Wood KV. Bioluminescent assays for high-throughput screening. *Assay Drug Dev Technol*. 2007;**5**(1):127–36.

[52] Rodgers G, Hubert C, McKinzie J, Suter T, Statnick M, Emmerson P, et al. Development of displacement binding and GTPgammaS scintillation proximity assays for the identification of antagonists of the micro-opioid receptor. *Assay Drug Dev Technol*. 2003;**1**(5):627–36.

[53] Khan AP, Schinske KA, Nyati S, Bhojani MS, Ross BD, Rehemtulla A. High-throughput molecular imaging for the identification of FADD kinase inhibitors. *J Biomol Screen*. 2010;**15**(9):1063–70.

[54] Qian J, Mason JL, Holskin BP, Murray KA, Meyer SL, Ator MA, et al. Comparison of two homogeneous cell-based kinase assays for JAK2 V617F: SureFire pSTAT5 and GeneBLAzer fluorescence resonance energy transfer assays. *Assay Drug Dev Technol*. 2012;**10**(2):212–7.

[55] Chan B, Cottrell JR, Li B, Larson KC, Ashford CJ, Levenson JM, et al. Development of a high-throughput AlphaScreen assay for modulators of synapsin I phosphorylation in primary neurons. *J Biomol Screen*. 2014;**19**(2):205–14.

[56] Antczak C, Wee B, Radu C, Bhinder B, Holland EC, Djaballah H. A high-content assay strategy for the identification and profiling of ABCG2 modulators in live cells. *Assay Drug Dev Technol*. 2014;**12**(1):28–42.

[57] Manganelli G, Masullo U, Filosa S. HTS/HCS to screen molecules able to maintain embryonic stem cell self-renewal or to induce differentiation: overview of protocols. *Stem Cell Rev*. 2014;**10**(6):802–19.

[58] Fang Y. Live cell optical sensing for high throughput applications. *Adv Biochem Eng Biotechnol*. 2010;**118**:153–63.

[59] Dykhuizen EC, Carmody LC, Tolliday N, Crabtree GR, Palmer MAJ. Screening for inhibitors of an essential chromatin remodeler in mouse embryonic stem cells by monitoring transcriptional regulation. *J Biomol Screen.* 2012;**17**(9):1221–30.

[60] Couturier C, Deprez B. Setting up a bioluminescence resonance energy transfer high throughput screening assay to search for protein/protein interaction inhibitors in mammalian cells. *Front Endocrinol (Lausanne).* 2012;**3**:100.

[61] Xie W, Pao C, Graham T, Dul E, Lu Q, Sweitzer TD, et al. Development of a cell-based high throughput luciferase enzyme fragment complementation assay to identify nuclear-factor-e2-related transcription factor 2 activators. *Assay Drug Dev Technol.* 2012;**10**(6): 514–24.

[62] Lunn CA. Label-free screening assays: a strategy for finding better drug candidates. *Future Med Chem.* 2010;**2**(11):1703–16.

[63] Tsuganezawa K, Shinohara Y, Ogawa N, Tsuboi S, Okada N, Mori M, et al. Two-colored fluorescence correlation spectroscopy screening for LC3-P62 interaction inhibitors. *J Biomol Screen.* 2013;**18**(9):1103–9.

[64] Zielonka J, Cheng G, Zielonka M, Ganesh T, Sun A, Joseph J, et al. High-throughput assays for superoxide and hydrogen peroxide: design of a screening workflow to identify inhibitors of NADPH oxidases. *J Biol Chem.* 2014;**289**(23):16176–89.

[65] Tegos GP, Evangelisti AM, Strouse JJ, Ursu O, Bologa C, Sklar LA. A high throughput flow cytometric assay platform targeting transporter inhibition. *Drug Discov Today Technol.* 2014;**12**:e95–103.

[66] Scrace S, O'Neill E, Hammond EM, Pires IM. Use of the xCELLigence system for real-time analysis of changes in cellular motility and adhesion in physiological conditions. *Methods Mol Biol.* 2013;**1046**:295–306.

[67] Sittampalam GS, Gal-Edd N, Arkin M, Auld D, Austin C, Bejcek B, et al. *Assay Guidance Manual.* Bethesda, MD: Eli Lilly & Company and the National Center for Advancing Translational Sciences; 2004.

[68] Uphoff CC, Drexler HG. Comparative PCR analysis for detection of mycoplasma infections in continuous cell lines. *In Vitro Cell Dev Biol Anim.* 2002;**38**(2):79–85.

[69] Zhang J-H, Chung TD, Oldenburg KR. A simple statistical parameter for use in evaluation and validation of high throughput screening assays. *J Biomol Screen.* 1999;**4**(2):67–73.

[70] Gupta PB, Onder TT, Jiang G, Tao K, Kuperwasser C, Weinberg RA, et al. Identification of selective inhibitors of cancer stem cells by high-throughput screening. *Cell.* 2009;**138**(4): 645–59.

[71] Carmody LC, Germain AR, VerPlank L, Nag PP, Muñoz B, Perez JR, et al. Phenotypic high-throughput screening elucidates target pathway in breast cancer stem cell-like cells. *J Biomol Screen.* 2012;**17**(9):1204–10.

[72] Germain AR, Carmody LC, Nag PP, Morgan B, Verplank L, Fernandez C, et al. Cinnamides as selective small-molecule inhibitors of a cellular model of breast cancer stem cells. *Bioorg Med Chem Lett.* 2013;**23**(6):1834–8.

[73] Hadida S, Van Goor F, Grootenhuis PDJ. Chapter 10 – CFTR Modulators for the Treatment of Cystic Fibrosis. *Annu Rep Med Chem.* 2010;**45**:157–73.

[74] Van Goor F, Straley KS, Cao D, González J, Hadida S, Hazlewood A, et al. Rescue of DeltaF508-CFTR trafficking and gating in human cystic fibrosis airway primary cultures by small molecules. *Am J Physiol Lung Cell Mol Physiol.* 2006;**290**(6):L1117–30.

[75] Geiger T, Cox J, Ostasiewicz P, Wisniewski JR, Mann M. Super-SILAC mix for quantitative proteomics of human tumor tissue. *Nat Methods.* 2010;**7**(5):383–5.

[76] Xiao Y, Guo L, Wang Y. A targeted quantitative proteomics strategy for global kinome profiling of cancer cells and tissues. *Mol Cell Proteomics.* 2014;**13**(4):1065–75.

[77] Kuai L, Ong S-E, Madison JM, Wang X, Duvall JR, Lewis TA, et al. AAK1 identified as an inhibitor of neuregulin-1/ErbB4-dependent neurotrophic factor signaling using integrative chemical genomics and proteomics. *Chem Biol.* 2011;**18**(7):891–906.

[78] Taunton J, Hassig CA, Schreiber SL. A mammalian histone deacetylase related to the yeast transcriptional regulator Rpd3p. *Science.* 1996;**272**(5260):408–11.

[79] Ito T, Ando H, Suzuki T, Ogura T, Hotta K, Imamura Y, et al. Identification of a primary target of thalidomide teratogenicity. *Science.* 2010;**327**(5971):1345–50.

[80] Park J, Oh S, Park SB. Discovery and target identification of an antiproliferative agent in live cells using fluorescence difference in two-dimensional gel electrophoresis. *Angew Chem Int Ed Engl.* 2012;**51**(22):5447–51.

[81] Ong S-E, Blagoev B, Kratchmarova I, Kristensen DB, Steen H, Pandey A, et al. Stable isotope labeling by amino acids in cell culture, SILAC, as a simple and accurate approach to expression proteomics. *Mol Cell Proteomics.* 2002;**1**(5):376–86.

[82] Lau H-T, Suh HW, Golkowski M, Ong S-E. Comparing SILAC- and stable isotope dimethyl-labeling approaches for quantitative proteomics. *J Proteome Res.* 2014;**13**(9):4164–74.

[83] Oda Y, Owa T, Sato T, Boucher B, Daniels S, Yamanaka H, et al. Quantitative chemical proteomics for identifying candidate drug targets. *Anal Chem.* 2003;**75**(9):2159–65.

[84] Ross PL, Huang YN, Marchese JN, Williamson B, Parker K, Hattan S, et al. Multiplexed protein quantitation in Saccharomyces cerevisiae using amine-reactive isobaric tagging reagents. *Mol Cell Proteomics.* 2004;**3**(12):1154–69.

[85] Bantscheff M, Hopf C, Savitski MM, Dittmann A, Grandi P, Michon A-M, et al. Chemoproteomics profiling of HDAC inhibitors reveals selective targeting of HDAC complexes. *Nat Biotechnol.* 2011;**29**(3):255–65.

[86] Weber C, Schreiber TB, Daub H. Dual phosphoproteomics and chemical proteomics analysis of erlotinib and gefitinib interference in acute myeloid leukemia cells. *J Proteomics.* 2012;**75**(4):1343–56.

[87] Ong S-E, Mann M. Mass spectrometry-based proteomics turns quantitative. *Nat Chem Biol.* 2005;**1**(5):252–62.

[88] Sharma K, Weber C, Bairlein M, Greff Z, Kéri G, Cox J, et al. Proteomics strategy for quantitative protein interaction profiling in cell extracts. *Nat Methods.* 2009;**6**(10):741–4.

[89] Rix U, Superti-Furga G. Target profiling of small molecules by chemical proteomics. *Nat Chem Biol.* 2009;**5**(9):616–24.

[90] Breen L, Keenan J, Clynes M. Generation of lung cancer cell line variants by drug selection or cloning. *Methods Mol Biol.* 2011;**731**:125–33.

[91] Lockhart DJ, Dong H, Byrne MC, Follettie MT, Gallo MV, Chee MS, et al. Expression monitoring by hybridization to high-density oligonucleotide arrays. *Nat Biotechnol.* 1996;**14**(13):1675–80.

[92] Lamb J, Crawford ED, Peck D, Modell JW, Blat IC, Wrobel MJ, et al. The Connectivity Map: using gene-expression signatures to connect small molecules, genes, and disease. *Science.* 2006;**313**(5795):1929–35.

[93] Subramanian A, Golub T. Connectivity Map (CMAP). Cancer Program Resource Gateway. 2014. Available from: http://www.broadinstitute.org/software/cprg/?q=node/12 (accessed May 15, 2015).

[94] Subramanian A, Tamayo P, Mootha VK, Mukherjee S, Ebert BL, Gillette MA, et al. Gene set enrichment analysis: a knowledge-based approach for interpreting genome-wide expression profiles. *Proc Natl Acad Sci U S A.* 2005;**102**(43):15545–50.

[95] Mootha VK, Lindgren CM, Eriksson K-F, Subramanian A, Sihag S, Lehar J, et al. PGC-1alpha-responsive genes involved in oxidative phosphorylation are coordinately downregulated in human diabetes. *Nat Genet.* 2003;**34**(3):267–73.

[96] Udeshi ND, Mertins P, Svinkina T, Carr SA. Large-scale identification of ubiquitination sites by mass spectrometry. *Nat Protoc*. 2013;**8**(10):1950–60.

[97] Molina H, Horn DM, Tang N, Mathivanan S, Pandey A. Global proteomic profiling of phosphopeptides using electron transfer dissociation tandem mass spectrometry. *Proc Natl Acad Sci U S A*. 2007;**104**(7):2199–204.

[98] Rardin MJ, Newman JC, Held JM, Cusack MP, Sorensen DJ, Li B, et al. Label-free quantitative proteomics of the lysine acetylome in mitochondria identifies substrates of SIRT3 in metabolic pathways. *Proc Natl Acad Sci U S A*. 2013;**110**(16):6601–6.

[99] Choudhary A, Hu He K, Mertins P, Udeshi ND, Dančík V, Fomina-Yadlin D, et al. Quantitative-proteomic comparison of alpha and Beta cells to uncover novel targets for lineage reprogramming. *PLoS One*. 2014;**9**(4):e95194.

[100] Jaffe JD, Wang Y, Chan HM, Zhang J, Huether R, Kryukov GV, et al. Global chromatin profiling reveals NSD2 mutations in pediatric acute lymphoblastic leukemia. *Nat Genet*. 2013;**45**(11):1386–91.

[101] Wang Y, Xiao J, Suzek TO, Zhang J, Wang J, Zhou Z, et al. PubChem's BioAssay Database. *Nucleic Acids Res*. 2012;**40**:D400–12.

[102] Bento AP, Gaulton A, Hersey A, Bellis LJ, Chambers J, Davies M, et al. The ChEMBL bioactivity database: an update. *Nucleic Acids Res*. 2014;**42**:D1083–90.

[103] Dančík V, Carrel H, Bodycombe NE, Seiler KP, Fomina-Yadlin D, Kubicek ST, et al. Connecting small molecules with similar assay performance profiles leads to new biological hypotheses. *J Biomol Screen*. 2014;**19**(5):771–81.

[104] Martinez Molina D, Jafari R, Ignatushchenko M, Seki T, Larsson EA, Dan C, et al. Monitoring drug target engagement in cells and tissues using the cellular thermal shift assay. *Science*. 2013;**341**(6141):84–7.

[105] Munteanu B, Meyer B, von Reitzenstein C, Burgermeister E, Bog S, Pahl A, et al. Label-free in situ monitoring of histone deacetylase drug target engagement by matrix-assisted laser desorption ionization-mass spectrometry biotyping and imaging. *Anal Chem*. 2014;**86**(10):4642–7.

[106] Wen Q, Goldenson B, Silver SJ, Schenone M, Dancik V, Huang Z, et al. Identification of regulators of polyploidization presents therapeutic targets for treatment of AMKL. *Cell*. 2012;**150**(3):575–89.

[107] Tallman MS, Neuberg D, Bennett JM, Francois CJ, Paietta E, Wiernik PH, et al. Acute megakaryocytic leukemia: the Eastern Cooperative Oncology Group experience. *Blood*. 2000;**96**(7):2405–11.

[108] Malinge S, Izraeli S, Crispino JD. Insights into the manifestations, outcomes, and mechanisms of leukemogenesis in Down syndrome. *Blood*. 2009;**113**(12):2619–28.

[109] Wen Q, Goldenson B, Crispino JD. Normal and malignant megakaryopoiesis. *Expert Rev Mol Med*. 2011;**13**:e32.

[110] Sato T, Fuse A, Eguchi M, Hayashi Y, Ryo R, Adachi M, et al. Establishment of a human leukaemic cell line (CMK) with megakaryocytic characteristics from a Down's syndrome patient with acute megakaryoblastic leukaemia. *Br J Haematol*. 1989;**72**(2):184–90.

[111] Goldenson B, Crispino JD. The aurora kinases in cell cycle and leukemia. *Oncogene*. 2014;**34**(5):537–45.

# PART IV

# TECHNOLOGIES FOR MEDICINAL CHEMISTRY OPTIMIZATION

# 11

# ADVANCES IN THE UNDERSTANDING OF DRUG PROPERTIES IN MEDICINAL CHEMISTRY

PETER HAMLEY[1] and PATRICK JIMONET[2]

[1] *Sanofi-Aventis Deutschland GmbH, Frankfurt am Main, Germany*
[2] *Sanofi-Aventis R&D, Vitry-sur-Seine, France*

## 11.1 INTRODUCTION

Design of compounds is fundamental to the science of medicinal chemistry, whether at the start of the process to identify active starting points or later, when an active starting point needs to be optimized. A key goal of compound design is always to achieve activity at the target or cellular screen. This is based on inputs at either the compound level (a pharmacophore model based on activity observations in a chemical series or between series) or at the protein level (knowledge of the structure of the binding site) (which of course is not always available). Modification of activities at other targets is almost always required as well: this can be either to achieve a positive effect (to add an additional activity or synergistic effect) or more often to remove unwanted activities (to increase selectivity). However, there are other factors beyond activity that should inform and limit the possible molecules chosen to synthesize, and the appreciation of the importance of these additional factors has increased in recent years. The areas of toxophores, privileged fragments, bioisosteres, covalent binders, etc. are important topics, but the most fundamental area is that of physical properties of drug molecules. This is particularly relevant for oral drugs that have a higher hurdle of intestinal absorption to overcome. The area is likely to have a huge impact on productivity in terms of the numbers of successful new medicines. There are several unexpected and key findings made over recent years that the modern medicinal chemist must be aware of (including an appreciation of the limitations of the available data) and then to apply to new compound design.

*Small Molecule Medicinal Chemistry: Strategies and Technologies*, First Edition.
Edited by Werngard Czechtizky and Peter Hamley.

## 11.2 PROPERTIES AND ORIGINS OF MARKETED DRUGS

### 11.2.1 The Consistent Properties of Oral Drugs

The property range of oral drugs introduced to the market has remained remarkably constant and narrow over a long period of measurement. A study of marketed oral drugs by year of market entry from 1983 to 2002 demonstrated that there was a small increase of 46 Da in the mean value of molecular weight (MW) in the period post-1983 compared with pre-1983 [1]. There were also small increases in hydrogen-bond acceptor (HBA) ability, rotatable bonds (ROTBs), and ring numbers. However, no significant differences were observed in lipophilicity (logP), hydrogen-bond donor ability, or polar surface area (PSA). This analysis suggests a fundamental truth about compound properties: that the balance between polar and nonpolar properties is a characteristic of oral drugs that does not change and is therefore likely to be critical to successful compound design. Similar studies have corroborated these findings while highlighting the unchanging lipophilicity parameter with an apparent steady increase in mean and median MW [2].

This trend of regular MW increase has been confirmed by us when looking at drugs approved for oral administration by the FDA until 2013 (N. Boussac, unpublished results). Statistical analysis of this set of drugs showed an incremental rise of 3 Da/year, resulting in mean MW rising from low 300s in the 1980s as shown by Vieth to around 400 in the 2010s (Fig. 11.1).

This trend is significant despite a quite high dispersion of MWs every year. It is worth noting that the regular increase of the 95% confidence intervals underlines a higher probability to observe high MW compounds in the 2010s versus the 1980s.

The aforementioned considerations were focusing on the oral route of administration, which is the most commonly desired route. In an analysis of 1729 marketed drugs, there were significant differences observed between oral drugs and drugs intended for other forms of administration, particularly injectables [2]. In general, oral drugs were of lower MW and had fewer H-bond donors and acceptors and ROTBs.

### 11.2.2 The Changing Origins of Oral Drugs

This narrow range of properties is all the more remarkable when considering the changing origins of oral drugs. The original source of effective medicines was the natural world—natural products derived from plants have been at the core of the human pharmacopeia for millennia [3]. Aspirin, a fully synthetic but natural compound derived from the bark of the willow, was marketed by Bayer in 1899 [4]. Aspirin has a very low complexity reflected in an MW of 180 Da. As the science of chemistry matured, it became possible to produce the simpler natural products by total synthesis, and fully synthetic compounds, often based on natural templates, became accessible. The chemist was no longer reliant on the designs of nature, but simply limited by the range of available chemical transformations, a range that was rapidly expanding throughout the twentieth century fueled by the discovery of new reactions and the introduction of practical advances such as NMR and chromatography.

The low-throughput nature of biological assays (*in vivo* or simple *in vitro* assays before major advances in molecular biology), coupled with the limited structural knowledge of binding sites, kept the complexity of these molecules relatively low. Exceptions to this

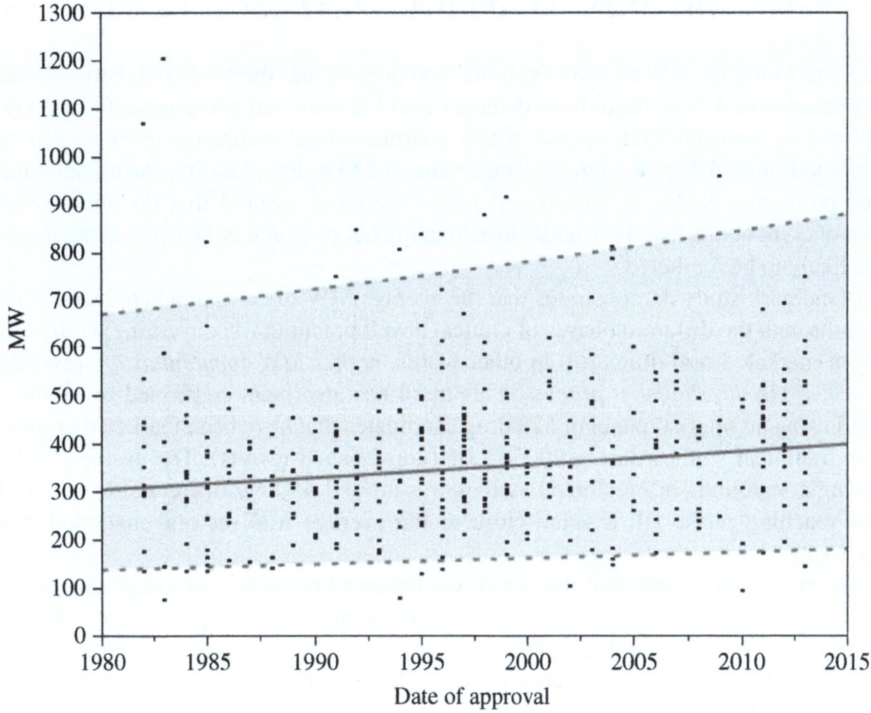


**FIGURE 11.1** MW of FDA-approved drugs from 1983 to 2013. Regression on the MW log is shown as a green line. Ninety-five percent confidence intervals are shown as dotted green lines. (*See insert for color representation of the figure.*)

were antibiotics based on modifications of complex natural products or peptides based on natural hormones that rarely showed bioavailability.

In the 1990s, significant advances in automation, cloning, and miniaturization led to the introduction of high-throughput screening (HTS) as a staple technique for identification of lead compounds. Similar advances in automated high-performance liquid chromatography (HPLC) were being made in the chemistry laboratory. At the same time, chemists were experimenting with combinatorial chemistry, initially using solid-phase techniques to produce large libraries. Solution-phase parallel synthesis followed, ideally suited to the new HPLC techniques. Suddenly, a flow of compounds was available that matched the throughput of HTS, with high expectations of success due simply to the higher number of compounds screened per target.

Diversity was considered a key design parameter in these early libraries, in order to cover as large an amount of chemical space as possible. However, there was little attention paid to the physical properties of compounds. MW was simply dictated by the number of reactions in a sequence and the available building blocks. In addition, structural knowledge of binding sites was becoming more widespread, leading to structure-based design exercises searching for potential new binding interactions without regard to the effect on physical properties.

The question that by the late 1990s had still not been posed was: were the compounds that the industry was studying matching the necessary properties required to achieve a successful market launch?

## 11.3 DRUG PROPERTIES AND ATTRITION IN CLINICAL DEVELOPMENT

The trend to higher MW in investigational compounds and the mismatch between these and marketed oral drugs have been demonstrated [5]. Patented compounds from 2000 to 2010 were compared from several major pharmaceutical companies to marketed oral drugs and found to display higher average values of MW, lipophilicity, and all other measured properties. Studying companies' patent literature ensured that the analysis consisted of compounds that were under investigation but of which only a very small number were likely to be marketed.

A landmark study demonstrates that the average MW of compounds reduces as they move through the different phases of clinical development until converging on the mean MW of marketed oral drugs [6]. In other words, *higher MW compounds are less likely to survive the development process*. This trend has also been confirmed by studies at Sanofi using an internal panel of 323 drug candidates that have been marketed or moved through clinical studies during 2005–2009 (unpublished results). The average MW of ongoing compounds in preclinical studies was around 450 but decreased to about 380 when reaching phase III, a value close to the average MW of oral marketed drugs (Fig. 11.2).

It is interesting to note that this trend was observed regardless of every scientifically driven cause of termination: ADME, efficacy/potency, safety, and toxicity (data not shown). This suggests that high MW compound has a detrimental effect on parameters such as ADME and toxicity profile (see Section 11.6 for discussion). A further study from Sanofi highlights the additional importance of aqueous solubility by comparing internal stopped candidates with a representative set of marketed drugs (N. Boussac, D. Massé, unpublished results). The internal compounds were terminated for reasons related only to ADME/PK (bioavailability/exposure, drug–drug interaction/CYP inhibition or induction) and toxicity (all types). Thermodynamic solubility measurements were taken for all compounds under the same conditions. There was a clear trend: lower solubility for stopped compounds and higher solubility for marketed drugs (Table 11.1).

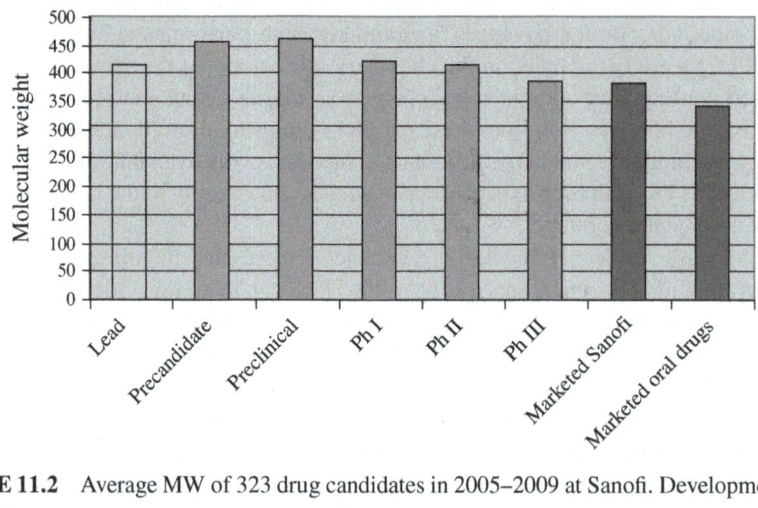


**FIGURE 11.2** Average MW of 323 drug candidates in 2005–2009 at Sanofi. Development phase repartition.

**TABLE 11.1 From 323 Drug Candidates, 41 are Selected that Have Been Stopped in Development for Reasons Related Only to ADME/PK (Bioavailability/Exposure, Drug–Drug Interaction/CYP Inhibition or Induction) and Toxicity (All Types)**

| Table of Observed Frequency: Class by Attrition | | | |
|---|---|---|---|
| Observed Frequency | Attrition | | |
| Solubility (mg/ml) | Stopped | Marketed | Total |
| <0.1 | 25 | 59 | 84 |
| >0.1 and <1 | 12 | 31 | 43 |
| >1 | 4 | 63 | 67 |
| Total | 41 | 153 | 194 |

| Table of Expected Frequency: Class by Attrition | | | |
|---|---|---|---|
| Expected | Attrition | | |
| Solubility (mg/ml) | Stopped | Marketed | Total |
| <0.1 | 18 | 66 | 84 |
| >0.1 and <1 | 9 | 34 | 43 |
| >1 | 14 | 53 | 67 |
| Total | 41 | 153 | 194 |

This set was compared to a selection of 153 marketed oral drugs coming from a pool of 739 compounds that have been clustered into 150 clusters using 14 variables: MW, clogP, clogD, pKa1, pKa2, TPSA, HB donors, HB acceptors, number of rotatable bonds, and combination-derived parameters. Highly correlated variables were discarded. Furthermore, caution was taken to select similar distribution of compounds possessing 0, 1, or 2 pKas in the two groups of compounds. Thermodynamic solubility measurements were performed at three pHs (1, 4.5, and 7.4). Measured solubilities were clustered into three groups of low (<0.1 mg/ml), medium (between 0.1 and 1 mg/ml), and high (>1 mg/ml) solubility (ref to Sanofi colleagues in Analytics and Biostatistics). Trend is significant at pH 7.4 ($p=0.0008$) and is also observed at the other two pHs.

High solubility alone cannot be predictive of success; indeed, compounds with poor solubility in the above panel have still some chance to reach the market. Models are able to predict compounds with low solubility (<0.1 mg/ml) or compounds with high solubility (>1 mg/ml) and are somewhat correlated with logD values. However, poor solubility can lead to attrition in development due to failure to provide an oral formulation as well as contributing to other causes, so it may be an important decision-making parameter. It is of course even more critical for developing compounds for intravenous formulations. A predictive model was built from the Sanofi set looking at attrition simply as a function of MW (Fig. 11.3).

The average probability to be on the market is 63% for a molecule of MW = 350 but only 20% for a molecule of MW = 550. These figures show that internal programs within pharmaceutical companies focused too much on high MW and high hydrophobicity–low solubility molecules. A plausible explanation could be dual: a request for hydrophobic interactions within protein active sites and an "easy" aryl coupling chemistry largely stimulated by the pressure to shorten optimization timelines [7]. This tendency to focus on high MW/high hydrophobicity molecules has clearly been one of the main and largely underconsidered causes of high attrition in development. Aqueous solubility (with the limitation of crystalline/amorphous material knowledge and availability), and therefore logD, should be considered as an essential element of the risk assessment particularly when exceeding the Lipinski limit of 500 Da (see Section 11.4).

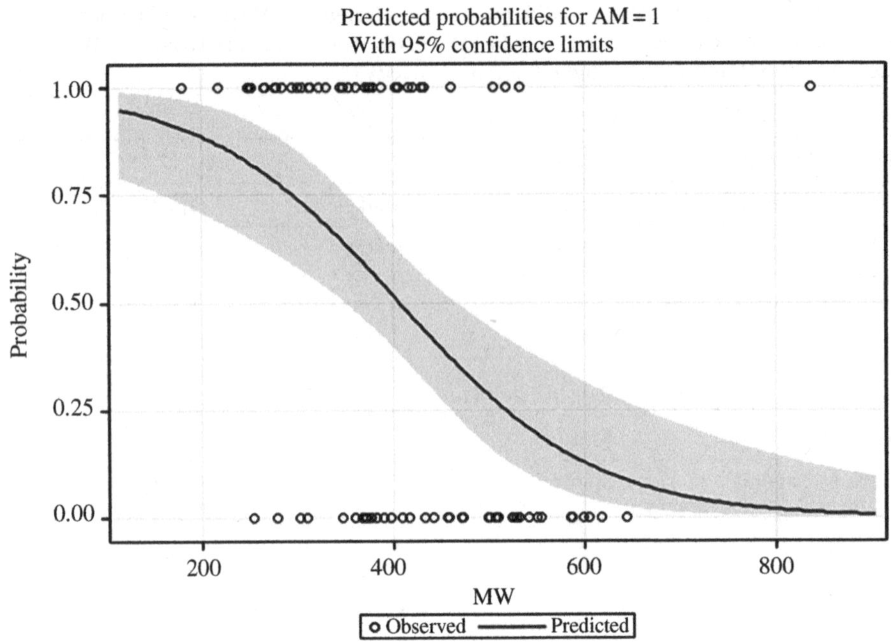


**FIGURE 11.3** 1000 random selections of 41 compounds within the 153 marketed drugs are carried out, followed by 1000 logical analyses with a balanced set of 41 stopped and 41 marketed compounds: 76% of the stopped compounds and 67% of the marketed drugs were predicted well (99.9/100 significant and MW is 100/100 significant).

In the same spirit of trying to improve success in development, two simple and interpretable measures of the complexity of molecules have been proposed [8]. The first is carbon-bond saturation as defined by fraction $sp^3$ ($Fsp^3$), and the second is counting chiral carbons in the molecule. It is further suggested that failures in development are due to an overemphasis on unsaturated chemicals with high melting points responsible for low solubility and poor bioavailability.

## 11.4 THE RULE OF FIVE

### 11.4.1 The Concept

In 1997, observing that the trend in compound design was heading for ever more complex and heavy molecules due to the advent of HTS and combinatorial chemistry, Lipinski formulated his "rule of five" (Ro5), one of the most cited papers in the field of drug discovery [9]. Essentially, the rule predicts that poor absorption or permeation is more likely when there are more than 5 H-bond donors and 10 H-bond acceptors, the MW is greater than 500, and the logP is greater than 5. The Ro5 is so named because the key parameters are divisible by 5, thus making recall easy.

Because the data set used to derive the Ro5 was in fact that of drug candidates that had reached phase II, it is likely that some of these compounds would have failed to reach the market by virtue of their physical properties; therefore, it has been argued that the Ro5 underestimates the real situation [1]. Nevertheless, for some time, the medicinal chemistry community was generally unwilling to adopt the Ro5 in compound design [10]. It took many other studies over the course of the following decade before the rule was regularly applied as a critical factor to assess compound quality. Even then, there have been some controversies about the overzealous use of the rule (*vide infra*).

The authors' own experience shows that most companies have put a lot of effort into biasing screening sets and compound-related milestone decisions toward adhering to Lipinski's rules. This thinking has taken time to take root but is now an accepted part of drug discovery decision making.

The new thinking on compound design has also had a direct effect on new compounds synthesized to populate compound collections [11]. Since corporate compound collections usually now have stringent drug-like or even lead-like property filters, it makes no sense for compound providers to produce compounds failing the commonly used filters. This has arguably had a positive effect on chemical creativity—it is certainly more difficult to use novel chemistry to make a new compound when the final MW must be restricted.

The stringent physicochemical requirements for new compound synthesis, particularly for corporate compound collections, require careful consideration before synthesis begins. Many processes have been described that facilitate good compound design—for example, creating sets of reagents combining high diversity with controlled physical properties [12].

### 11.4.2 Druggability

The Ro5 paradigm has been useful beyond its direct role in compound design by helping to clarify thinking around what is druggable and what is not [13]. As an important example, the Ro5 has been used to enumerate Ro5-compliant small-molecule binding sites among proteins encoded in the human genome [14]. Based on this, consideration of protein families already known to contain such sites, and indications of how many protein sites have been shown so far to be relevant to the treatment of disease, an estimate of the number of druggable and clinical useful human targets was made—in the low hundreds. The authors hypothesize that a much wider range of targets would be accessible if there was more focus on other types of chemical matter (modalities), such as protein drugs, DNA-based therapies, and the advent of nonoral drug delivery systems, in order to target protein sites not responsive to modulation by orally available small molecules.

## 11.5 THE CONCEPT OF LEAD-LIKENESS

In another highly cited drug discovery paper, Teague and Oprea coined the term lead-likeness in 1999 [15, 16]. In their definition, this refers to a compound that has properties suitable for a lead molecule, that is, it is less complex than the final intended drug molecule. This is a desirable quality for a lead due to the observation that MW tends to *increase* during the course of a lead optimization program, particularly for typical screening sets used in HTS (compound series based on natural ligands often have high potency). They postulated that a less complex molecule has a better chance of binding to a receptor [17]. Once this simpler molecule has been identified, introduction of lipophilic groups will increase potency and improve pharmacokinetic (PK) properties by building in new interactions or adding lipophilicity to existing interactions to increase binding energy. An optimal weight range of 100–350 Da was suggested, with logP of 1–3.0. The observation is illustrated by several examples of optimization and has been reproduced by other groups and organizations subsequently [18].

The key implication of the lead-likeness concept is that the starting points for drug discovery should be even lower molecule weight than the Lipinski Ro5 cutoffs. The arguments presented have had a profound effect on the choice of hits and leads in optimization campaigns and even the makeup of screening sets and furthermore have contributed to the popularity of fragment-based screening methodology [19]. An intrinsic advantage of starting from a weakly binding fragment is the opportunity it gives to rapidly increase molecule weight and binding affinity simultaneously, and there are many successful examples of this [20 and cited references]. The so-called rule of 3, which is very close to the lead-like ranges, has been promoted for ideal design of fragment screening sets [21, 22].

A recent analysis of over 30 fragment screening campaigns against a variety of targets underlined a higher hit rate of less complex molecules [23]. The authors attribute this result to a more efficient sampling of chemical space in addition to a broad coverage of the protein classes and therefore protein interaction modalities.

### 11.5.1 The Consequences on Screening and Collections

Since it takes well over a decade to bring a new drug to the market, the impact of the Ro5 and lead-like theories in terms of compound-related attrition is difficult to measure, not only because of the length of development but also because changes in design "culture" via scientific publications take many years to become widespread. This was nicely illustrated by a 2007 study of published and patented compounds from several major drug discovery organizations, where average compound properties, particularly MW and lipophilicity, were found to be significantly higher than those of marketed oral drugs [10]. The exception to this observation was Pfizer, Lipinski's organization, which had significantly lower values and fewer Ro5 violations in the published compounds. This suggests that organizational acceptance of the Ro5 was faster in the originator company. A more detailed follow-up study still revealed considerable differences in MW and lipophilicity between a wide range of pharmaceutical companies (comparing compounds disclosed against the same targets and on different targets [5]).

Again, Pfizer was at the lower end of property values, as were companies where fragment-based screening strategies were prominent in discovery approaches (not surprising considering that such strategies start from very low MW starting points). Whether this has contributed to a measurable lower overall rate of attrition at Pfizer is a so far unanswered question.

## 11.6 INFLUENCE OF DRUG PROPERTIES ON ABSORPTION, DISTRIBUTION, METABOLISM, EXCRETION, AND TOXICITY

As discussed previously, physicochemical properties of drugs, particularly MW, logP/logD, and aqueous solubility, have been shown to impact attrition in development. But what are the biological reasons behind this dramatic end result? Is there any direct influence of these properties on absorption, distribution, metabolism, excretion, and toxicity (ADMET) parameters?

A major analysis using a very large corporate data set has yielded a detailed view on the relationships between ADMET properties of molecules and their molecular properties. Looking at 15 ADMET parameters (solubility, permeability, bioavailability, volume of distribution, plasma protein binding, CNS penetration, brain tissue binding, P-glycoprotein (P-gp) efflux, hERG inhibition, and cytochrome P450 1A2/2C9/2C19/2D6/3A4 inhibition), a detrimental effect of increasing MW and clogP, alone or in combination for almost all the ADMET parameters, was demonstrated; the ionization state of the molecules has a beneficial or detrimental effect depending on the parameter [24]. This observation was also confirmed using the ChEMBL database. Consideration of these two parameters, combined through a score measuring the deviation versus oral drugs' mean values, has been proposed as the best way to prioritize hits with poor ADMET properties resulting from screening [25]. Complementary to this analysis, a comparison of the effects of common substituents on a range of ADMET parameters provided some guidelines on the influence of a given substituent introduction on a core structure [26]. It was concluded that there is no perfect substituent showing a beneficial effect on all parameters as the best balance between properties has to be found during optimization. However, clogD was identified as the most critical substituent physicochemical parameter for most of the ADMET properties: the higher the clogD, the more detrimental the impact.

Unfortunately, the popular target-focused approaches used in the last 15 years have overemphasized the potency at targets as the main priority of medicinal chemistry optimization. Certainly, due to many protein hydrophobic binding sites, the relative ease in improving potency by introduction of lipophilic rings and substituents, as well as efficient high-throughput coupling chemistries, compound "obesity" has increased, leading to high MW and high logP/logD candidates showing poor or somewhat borderline ADMET properties. The need to minimize the negative influence of such parameters has spurred on the development of valuable innovations in pharmaceutical sciences. "Rescue" of some of the parameters, for example, solubility, permeability, and absorption, to provide apparently acceptable PK profiles has been shown on several occasions [27]. But this could be thought of as shortsighted since off-target effects leading to toxicity are negatively influenced by unfavorable physicochemical properties and are often difficult to predict using *in vitro* assays [28].

The relationship between molecular properties and *in vitro* surrogate assays versus *in vivo* properties has been investigated [29]. A database containing 3,773 compounds belonging to 173 series and rodent PK and toxicology data was used. The authors examined the predictive use of simple parameters such as clogP for thermodynamic solubility and MW and H-bond donors for LOAEL and argued against a rigid implementation of the Ro5 that does not consider other surrogates, particularly *in vitro* assays. *In vivo* ADMET was shown to have higher predictive utility.

In another study using a diverse set of 245 preclinical compounds, the *in vivo* tolerance in rats or dogs was analyzed. A correlation was found between physicochemical parameters such as MW, clogP, and TPSA versus toxicity [30]. TPSA was found to give the strongest and most consistent correlation among the descriptors, with low TPSA leading to the highest adverse effects. Furthermore, compounds with high clogP and low TPSA were shown to be about 2.5 times more likely to be toxic than clean. The reverse was also true, highlighting the lower risk of toxicity for compounds with low lipophilicity (clogP$<$3) and high polar surface (TPSA$>$75). The hypothesis of binding promiscuity (off-target activity) has been suggested for the high clogP/low TPSA compounds based on the evaluation of a subset of compounds against a panel of 48 targets.

Strategies to improve *in vivo* toxicology outcomes have been suggested [31]. Using a set of oral compounds possessing basic charges that had been through development or stopped due to preclinical toxicity at AstraZeneca, they could not confirm the observations made previously, but they analyzed a larger range of descriptors and modeling techniques covering size/shape, charge, lipophilicity, polarity, hydrogen bonding, and quantum mechanics. Initial results revealed aromaticity, size, and shape (C-$sp^3$ count, flatness) to be individually important but not sufficient alone to differentiate successful and failed compounds. A multivariate statistical model was therefore developed as a better predictive model, demonstrating that toxicity attrition was decreased by increasing C-$sp^3$ count, whereas increased flatness or aromatic/aliphatic ratio promoted attrition. Several other descriptors, for example, charge related, showed some influence, thus leading to a quite complex picture. Confirmation of this model using a larger set of compounds stopped due to negative preclinical toxicological studies in animals has still to be done. However, there is little doubt that the strategy to decrease the risk of toxicity failure by increasing C-$sp^3$ and/or aliphatic components as well as by reducing non-polar surface area is a valid and valuable one.

## 11.7 BUILDING ON THE Ro5: NEW GUIDELINES FOR COMPOUND DESIGN

There is naturally a strong tendency to regard potency at the primary target as the first and most important factor in judging compounds. If a holistic view is not taken, balancing potency with ADME and physicochemical properties, the result will be potent compounds with high MW and lipophilicity, from which as we have seen it is harder to break away from. To counteract this problem in as simple a manner as possible, there have been several attempts to popularize other methods of comparing compounds, which take potency into account but also simultaneously reflect other physicochemical properties.

### 11.7.1 Ligand Efficiency

Arguably, the simplest and therefore most successful metric for practicing medicinal chemists has been the concept of "ligand efficiency" ($\Delta g$) [32]. This is a simple measure of the binding energy per atom of a compound. It is defined as

$$\Delta g = \Delta G / N$$

where $\Delta G = -\text{RT} \ln \text{Kd}$ (the free energy of ligand binding) and $N$ = number of nonhydrogen atoms.

Since Kd values approximate to the IC50, one can use this more readily available value, particularly if used simply to compare compounds measured in the same assay, multiplied by a constant:

$$\Delta g = \text{approx} - 0.6 \times \ln(\text{IC50}) / (\text{no. of nonhydrogen atoms}) \text{ in kcal/mol}$$

Molecular weight ⟶

IC50/μM ↓

| | 200 | 300 | 400 | 500 |
|---|---|---|---|---|
| 100 | 0,091776 | 0,061184 | 0,045888 | 0,036711 |
| 10 | 0,183553 | 0,122369 | 0,091776 | 0,073421 |
| 1 | 0,275329 | 0,183553 | 0,137665 | 0,110132 |
| 0,1 | 0,367106 | 0,244737 | 0,183553 | 0,146842 |
| 0,01 | 0,458882 | 0,305921 | 0,229441 | 0,183553 |

**FIGURE 11.4** A MW 200 fragment binding at 10 μM has equivalent ligand efficiency to a MW 500 compound binding at 10 nM, but has more room for optimization by increasing MW. Values for $\Delta g$ based on average values for no. of non hydrogen atoms in Pfizer's compound collection.

In fact, a $\Delta G$ change of −1.4 kcal/mol results in a 10-fold increase in potency. Furthermore, compounds with a lower number of "heavy" atoms have a higher binding efficiency for the same measured IC50 (Fig. 11.4).

Ranking compounds in this way emphasizes the value of a lower MW compound with efficient binding to the target. In theory, such a compound is more likely to be successfully optimized to a more potent compound since there is more scope to add MW to increase potency without overstepping the boundaries of lipophilicity.

The first attempt to understand the contributions of a molecule's functional group to its binding was made by Andrews using known values of binding energy for common groups [33]. Ligand efficiency is a simpler and more easily calculated version of this. In many organizations in recent times, when comparing active or hit lists from large screens, compounds are ranked based on this parameter.

### 11.7.2 Ligand Lipophilicity Efficiency and Other Indices

Although approximate ligand efficiency values are relatively easy to calculate, further simplifications have been proposed, for example, potency efficiency index (PEI), which is defined as percentage inhibition divided by MW [34]. This is another way to visualize primary assay percentage inhibition data and is very practical for analyzing the results of a large HTS. Other simple parameters using PSA have also been proposed.

Another well-accepted metric is ligand lipophilicity efficiency (LLE) [35], which is defined as

$$\text{LLE} = \text{pIC50}(\text{or pKi}) - \text{cLogP}(\text{or LogD})$$

LogP can be measured but is readily calculated by various algorithms. A typical candidate drug with a pIC50 of 8 and a LogP of 2 would have a LLE of 6, which is considered an optimum value. LLE is a measurement of potency combined with lipophilicity. The goal should be to have low increases of LLE in a series indicating increased potency with relatively small lipophilicity increases.

### 11.7.3 Chemical Beauty

In this chapter, we have described metrics that are easily calculated as alternatives to judging compounds purely in terms of MW. Clearly, it is possible to obtain more sophisticated metrics based on *in silico* calculations to estimate drug-likeness of compounds. Although it is not a goal of this chapter to describe such algorithms in detail, one method deserves some consideration because of its interesting approach, and that is the concept of chemical beauty, calculated as quantitative estimate of drug-likeness (QED) [36]. The basis for this calculation is the use of eight common, easily calculated, and attrition-relevant molecular properties: MW, logP, HBAs, molecular PSA, number of ROTBs, number of aromatic rings (AROMs), and number of structural alerts (ALERTS). The QED parameter is a function that averages the scores of these properties relative to a data set of successful oral drugs, which is shown to well capture drug-likeness in a set of known drugs.

Such implementation of drug-like algorithms should in the future prove useful for ranking compound series to make better decisions in terms of compound progression, in particular from complex screens that deliver large amounts of activity data. This approach certainly provides a more data-rich, sophisticated method of drug-like ranking than application of the Ro5. However, it should be noted that it does not take account of the concept of lead-likeness, and we have seen that, due to the inflation of MW during an optimization program, building in room to optimize at an early stage is important.

Correlation between QED high- and low-scoring drugs and pharmaceutical and PK profiles in humans has been analyzed recently [37]. Analysis of a set of 199 drugs underlined for the highest-scoring drugs a higher absorption and bioavailability, lower doses, fewer drug–drug interaction warnings, P-gp interactions, and absorption issues due to a food effect. However, no difference was observed between high- and low-scoring compounds for several other ADME and PK parameters such as metabolism.

Practical use of QED in hit selection and optimization still needs to be reported.

## 11.8 ALTERNATIVES, CRITICISMS, AND EXCEPTIONS

There is no doubt that Lipinski's Ro5 has hugely influenced the practice of medicinal chemistry since it was first introduced in 1997. However, there have been a number of doubts and criticisms about the relevance and effect that the new way of thinking has had on the science of drug discovery. These gain credence when combined with the observation that the industry has not become significantly more productive in terms of new chemical entities since the rule was introduced, indeed the reverse. For example, it has been pointed out that the rules, which were intended as guidance and rule of thumb only, have been used as "commandments" in many quarters. Using multiple property filters will eventually limit the number of compounds, in the limiting case to zero. Is there a need for a collective reflection on the wisdom of slavishly following the Ro5?

In defense of drug property observations, it is important to remember that using such property filters as hard cutoffs was not the original intention. Lipinski has pointed out that the driver for the Ro5 was to increase the probability of success, not to distinguish drugs from nondrugs [38]. In addition, since the average time to develop a drug is over

10 years and the new thinking has taken time to show an influence in company collections [1], it is unreasonable to expect to see a tangible increase in approvals at the time of writing.

Nonetheless, it is clear that a growing number of oral marketed drugs lie outside the property space defined by the Ro5, notably some of the CoA reductase inhibitors, for example, Lipitor, the largest selling drug between 1996 and 2012. Several newly introduced kinase inhibitors for oncology indications also lie outside Lipinski space. Many outliers, as pointed out by Lipinski in the original paper, are natural products in origin [9]. Is the reliance on filtered sets of Ro5-compliant compounds contributing to industries' scaling back of natural product research, and is this misguided? Even natural product collections intended for HTS have been in some cases filtered based on Ro5-like parameters [39]. Such strategies will inevitably reduce the opportunities for novel and unexpected discoveries using the power of natural selection present in natural product collections.

In addition, filtering physical property space may have a deleterious effect on an emerging and increasingly biologically attractive target class, that of protein–protein interaction inhibition. It has been shown that the average MW required for activity versus these often large and diffuse binding sites has a tendency to be higher than other target classes [40]. Limiting the opportunities to Ro5- or lead-like-compliant molecules may further increase the difficulty of generating leads in this area. The difficulty of obtaining orally active PPI inhibitors is not only a question of blocking the diffuse binding site but also of obtaining cell permeability and probably also bioavailability and getting the compound successfully through development.

It has been asserted that the number of drugs that cross lipid bilayers by the process of passive diffusion has been hugely overestimated and its existence even questioned [41]. Increasing evidence for active transport, even for fully synthetic molecules, tends to suggest that the deleterious role of higher MW could also have been exaggerated. Therefore, designing molecules that take advantage of active transport could produce compounds outside the normal property range that do not suffer from lack of bioavailability. A recent publication reported very interesting studies about structural requirements that convey active transport-driven permeability to cyclic peptides. Remarkably, the identified bioactive sequence and resulting peptidic conformation were strikingly observed in bioavailable cyclosporine A and somatostatin analogs [42]. Whether this strategy is consistently achievable remains to be demonstrated.

A more subtle effect of limiting chemical space by property filters has been pointed out by Leeson and coworkers [43]. By reducing the MW range available to chemists, the challenge of making truly novel compounds increases. In order to stay within a low MW range, it is not enough to simply add reaction steps. Rather, novel "rescaffolding" chemistry must be designed, which is relatively resource intensive. The simpler alternative is to fill company collections with known scaffolds and "privileged" drug-like structures.

This will inevitably lead to intellectual property issues and hence the (also resource intensive) need to modify structures later in a project lifetime. To avoid this, the emphasis needs to switch to producing and screening novelty early, and there is evidence that this strategy has been put into action by a number of companies in recent years in their efforts to produce novel and differentiated, and therefore competitive, compound collections.

## 11.9 CONCLUSIONS

The observations made since Lipinski's original groundbreaking paper on the influence of molecular properties have become of central importance to modern medicinal chemistry thinking. If oral activity of synthetic molecules is desired, the possibility of success clearly increases if the MW of a compound series is controlled from the start, since MW tends to increase during a medicinal chemistry program. Other parameters such as lipophilicity, number of ROTBs, hydrogen-bonding ability, and subsequent solubility play vital roles in success rates as well.

In applying these guidelines, it must always be born in mind that these are just that—guidelines in order to increase the probability of success. They should not be taken as hard cutoffs and should not rule out the study of chemical space outside their scope. The successful small-molecule outliers, many natural products, the increasing focus on macrocyclic structures, and the growing realization that active transport mechanisms are often important attest to this. In addition, new chemical space outside Ro5 may offer the solution to many hitherto "intractable" targets in the future, and to limit available chemical space will only limit our ability to successfully tackle these, despite the likelihood that a starting point outside this space will have a lower chance of progression to market. Increased research efforts in this area will be critical to successfully achieve the quest for new high-value drugs.

Nonetheless, the guidelines discussed in this chapter are derived from empirical observations on large data sets, and their application is expected to increase the possibility of successfully achieving marketed oral drugs. Therefore, the value of their—sensible—use cannot be overstated.

## REFERENCES

[1] Leeson, P.D., Davis, A.M. (2004). Time-Related Differences in the Physical Property Profiles of Oral Drugs. *Journal of Medicinal Chemistry*, **47**, 6338–6348.

[2] Vieth, M., Siegel, M.G., Higgs, R.E., Watson, I.A., Robertson, D.H., Savin, K.A., Durst, G.L., Hipskind, P.A. (2004). Characteristic Physical Properties and Structural Fragments of Marketed Oral Drugs. *Journal of Medicinal Chemistry*, **47**, 224–232.

[3] Cragg, G.M., Newman, D.J. (2013). Natural Products: A Continuing Source of Novel Drug Leads. *Biochimica et Biophysica Acta*, **1830**, 3670–3695.

[4] Goldberg, D.R. (2009). Aspirin: Turn-of-the-Century Miracle Drug. *Chemical Heritage*, **27**, 26–30.

[5] Leeson, P.D., St-Gallay, S.A. (2011). The Influence of The "Organizational Factor" on Compound Quality in Drug Discovery. *Nature Reviews Drug Discovery*, **10**, 749–765.

[6] Wenlock, M.C., Austin, R.P., Barton, P., Davis, A.M., Leeson, P.D. (2003). A Comparison of Physiochemical Property Profiles of Development and Marketed Oral Drugs. *Journal of Medicinal Chemistry*, **46**, 1250–1256.

[7] Lindsley, C.W., Weaver, D., Bridges, T., Kennedy, J.P. (2012). Lead Optimization in Drug Discovery. *Wiley Encyclopedia of Chemical Biology*, 1–9.

[8] Lovering, F., Bikker J., Humblet, C. (2009). Escape from Flatland: Increasing Saturation as an Approach to Improving Clinical Success. *Journal of Medicinal Chemistry*, **52**, 6752–6756.

[9] Lipinski, C.A., Lombardo, F., Dominy, B.W., Feeney, P.J. (1997). Experimental and Computational Approaches to Estimate Solubility and Permeability in Drug Discovery and Development Settings. *Advanced Drug Delivery Reviews*, **23**, 3–25.

[10] Leeson, P.D., Springthorpe, B. (2007). The Influence of Drug-Like Concepts on Decision-Making in Medicinal Chemistry. *Nature Reviews Drug Discovery*, **6**, 881–890.

[11] Chuprina, A., Lukin, O., Demoiseaux, R. Buzko, A., Shivanyuk, A. (2010). Lead-Likeness, Target Class, and Molecular Diversity Analysis of 7.9 Million Commercially Available Organic Compounds Provided by 29 Suppliers. *Journal of Chemical Information and Modeling*, **50**, 470–479.

[12] Stocks, M.J., Wilden, G.R.H., Pairaudeau, G., Perry, M.W.D., Steele, S., Stonehouse, J.P. (2009). A Practical Method for Targeted Library Design Balancing Lead-Like Properties with Diversity. *ChemMedChem*, **4**, 800–808.

[13] Keller, T.H., Pichota, A., Yin, Z. (2006). A Practical View of "Druggability". *Current Opinion in Chemical Biology*, **10**, 357–361.

[14] Hopkins, A.L., Groom, C.G. (2002). The Druggable Genome. *Nature Reviews Drug Discovery*, **1**, 727–730.

[15] Teague, S.J., Davis, A.M., Leeson, P.D., Oprea, T. (1999). The Design of Leadlike Combinatorial Libraries. *Angewandte Chemie International Edition*, **38**, 3743–3747.

[16] Oprea, T.I., Davis, A.M., Teague, S.J., Leeson, P.D. (2001). Is There a Difference between Leads and Drugs? A Historical Perspective. *Journal of Chemical Information and Computer Sciences*, **41**, 1308–1315.

[17] Hann, M.M., Leach, A.R., Harper, G. (2001). Molecular Complexity and Its Impact on the Probability of Finding Leads for Drug Discovery. *Journal of Chemical Information and Computer Sciences*, **41**, 856–864.

[18] Perola, E. (2010). An Analysis of the Binding Efficiencies of Drugs and Their Leads in Successful Drug Discovery Programs. *Journal of Medicinal Chemistry*, **53**, 2986–2997.

[19] Baker, M. (2013). Fragment-Based Lead Discovery Grows Up. *Nature Reviews Drug Discovery*, **12**, 5–7.

[20] Murray, C.W., Verdonk, M.L., Rees, D.C. (2012). Experiences in Fragment-Based Drug Discovery. *Trends in Pharmacological Science*, **33**, 224–232.

[21] Congreve, M., Carr, R., Murray, C., Jhoti, H. (2003). A Rule of Three for Fragment-Based Lead Discovery? *Drug Discovery Today*, **8**, 876–877.

[22] Mok, N.Y., Brenk, R., Brown, N. (2014). Increasing the Coverage of Medicinal Chemistry-Relevant Space in Commercial Fragments Screening. *Journal of Chemical Information and Modeling*, **54**, 79–85.

[23] Jhoti, H., Williams, G., Rees, D.C., Murray, C.W. (2013). The "Rule of Three" for Fragment-Based Drug Discovery: Where are We Now? *Nature Reviews Drug Discovery*, **12**, 644–645.

[24] Gleeson, M.P. (2008). Generation of a Set of Simple, Interpretable ADMET Rules of Thumb. *Journal of Medicinal Chemistry*, **51**, 817–834.

[25] Gleeson, M.P., Hersey, A., Montanari, D., Overington, J. (2011). Probing the Links Between *in vitro* Potency, ADMET and Physicochemical Parameters. *Nature Reviews Drug Discovery*, **10**, 197–208.

[26] Gleeson, P., Bravi, G., Modi, S., Lowe, D. (2009). ADMET Rules of Thumb II: A Comparison of the Effects of Common Substituents on a Range of ADMET Parameters. *Bioorganic & Medicinal Chemistry*, **17**, 5906–5919.

[27] Byrn, S.R., Henck, J.-O. (2012). Optimizing The Physical form—Opportunities and Limitations. *Drug Discovery Today: Technologies*, **9**, e73–e78.

[28] Tarcsay, A., Keserű, G.M. (2013). Contributions of Molecular Properties to Drug Promiscuity. *Journal of Medicinal Chemistry*, **56**, 1789–1795.

[29] Sutherland, J.J., Raymond, J.W., Stevens, J.L., Baker, T.K., Watson, D.E. (2012). Relating Molecular Properties and in Vitro Assay Results to in Vivo Drug Disposition and Toxicity Outcomes. *Journal of Medicinal Chemistry*, **55**, 6455–6466.

[30] Hughes, J.D., Blagg, J., Price, D.A., Bailey, S., DeCrescenzo, G.A., Devraj, R.V., Ellsworth, E., Fobian, Y.M., Gibbs, M.E., Gilles, R.W., Greene, N., Huang, E., Krieger-Burke, T., Loesel, J., Wager, T., Whiteley, L., Zhang, Y. (2008). Physicochemical Drug Properties Associated with in Vivo Toxicological Outcomes. *Bioorganic & Medicinal Chemistry Letters*, **18**, 4872–4875.

[31] Luker, T., Alcaraz, L., Chohan, K.K., Blomberg, N., Brown, D.S., Butlin, R.J., Elebring, T., Griffin, A.M., Guile, S., St-Gallay, S., Swahn, B.-M., Swallow, S., Waring, M.J., Wenlock, M.C., Leeson, P.D. (2011). Strategies to Improve in Vivo Toxicology Outcomes for Basic Candidate Drug Molecules. *Bioorganic & Medicinal Chemistry Letters*, **21**, 5673–5679.

[32] Hopkins, A.L., Groom, C.R., Alex, A. (2004). Ligand Efficiency: A Useful Metric for Lead Selection. *Drug Discovery Today*, **9**, 430.

[33] Andrews, P.R., Craik, D.J., Martin, J.L. (1984). Functional Group Contributions to Drug-Receptor Interactions. *Journal of Medicinal Chemistry*, **27**, 1648–1657.

[34] Abad-Zapatero, C., Metz, J.T. (2005). Ligand Efficiency Indices as Guideposts for Drug Discovery. *Drug Discovery Today*, **10**, 464–469.

[35] Murray, C.W., Erlanson, D.A., Hopkins, A.L., Keserü, G.M., Leeson, P.D., Rees, D.C., Reynolds, C.H., Richmond, N.J. (2014). Validity of Ligand Efficiency Metrics. *ACS Medicinal Chemistry Letters*, **5**, 616–618 and cited references.

[36] Bickerton, G.R., Paolini, G.V., Besnard, J., Muresan, S., Hopkins, A.L. (2012). Quantifying The Chemical Beauty of Drugs. *Nature Chemistry*, **4**, 90–98.

[37] Ritchie, T.J., Macdonald, S.J.F. (2014). How Drug-Like are "Ugly" Drugs: Do Drug-Likeness Metrics Predict ADME Behavior in Humans? *Drug Discovery Today*, **19**(4), 489–495.

[38] Lipinski, C. (2012). News and Analysis. *Nature Reviews Drug Discovery*, **11**, 900–901.

[39] Quinn, R.J., Carroll, A.R., Pham, N.B., Baron, P., Palframan, M.E., Suraweera, L. (2008). Developing a Drug-Like Natural Product Library. *Journal of Natural Products*, **71**, 464–468.

[40] Guo, W., Wisniewski, J.A., Ji, H. (2014). Hot Spot-Based Design of Small-Molecule Inhibitors for Protein–Protein Interactions. *Bioorganic & Medicinal Chemistry Letters*, **24**, 2546–2554.

[41] Kell, D.B. (2013). Finding Novel Pharmaceuticals in the Systems Biology Era Using Multiple Effective Drug Targets, Phenotypic Screening and Knowledge of Transporters: Where Drug Discovery went Wrong and How to Fix it. *FEBS Journal*, **28**, 5957–5980.

[42] Beck, J.G., Chatterjee, J., Laufer, B., Kiran, M.U., Frank, A.O., Neubauer, S., Ovadia, O., Greenberg, S., Gilon, C., Hoffman, A., Kessler, H. (2012). Intestinal Permeability of Cyclic Peptides: Common Key Backbone Motifs Identified. *Journal of the American Chemical Society*, **134**, 12125–12133.

[43] Leeson, P.D., Davis, A.M., Steele, J. (2004). Drug-Like Properties: Guiding Principles for Design—or Chemical Prejudice? *Drug Discovery Today: Technologies*, **3**, 189–195.

# 12

# RECENT DEVELOPMENTS IN AUTOMATED SOLUTION PHASE LIBRARY PRODUCTION

THOMAS C. MAIER and WERNGARD CZECHTIZKY
*Sanofi-Aventis Deutschland GmbH, Frankfurt am Main, Germany*

## 12.1 INTRODUCTION

### 12.1.1 Introduction and Definitions

In modern drug discovery, automated parallel synthesis is used as a fast and efficient way to generate the molecules needed to progress medicinal chemistry projects. Initially, the development of combinatorial chemistry enabled the rapid generation of hundreds of thousands of structurally relatively similar compounds at low cost [1]. These combinatorial libraries were usually realized on small scale via solid-phase synthesis (see respective chapter). However, the compromises required to produce such large libraries rendered them often unsuitable for high hit rates in screening campaigns of various diverse protein targets; nor were the compounds generated particularly valuable with respect to diversity, beneficial ADME (absorption, distribution, metabolism, excretion [2, 3]) properties or intellectual property [1, 4].

As a consequence, the more recent trend in the industry has tended toward smaller libraries on larger scales (tens of milligrams per compound) often involving a more sophisticated manual synthesis of building blocks [4, 5] with functional groups to be modified by parallel use of standard reaction types such as amide couplings, alkylations, nucleophilic substitutions, reductive aminations, Mitsunobu-type reactions [6], transition metal catalyzed coupling reactions such as Suzuki [7] or Buchwald [8] reactions, etc.—or a sequence thereof. Here, structural diversity is mostly generated in the decoration of the

*Small Molecule Medicinal Chemistry: Strategies and Technologies*, First Edition.
Edited by Werngard Czechtizky and Peter Hamley.

underlying chemical scaffold leading to so-called appendage diversity [9]: variation in structural moieties around a common skeleton (central scaffold) and functional group diversity on the decoration help to vary compound properties or to address additional interaction points of the protein target. The term "library" as it is used in this chapter refers to such compound sets (ca. 20–1000 compounds) covering structurally closely related chemical compounds also referred to as analogues. Using parallel techniques, multiple analogues can be produced with a considerable time saving when compared to serial, one at a time, construction of each final compound.

These libraries can be prepared by solution or solid-phase synthesis. This chapter is focused on library production in solution phase—with the major advantage of maximum synthetic flexibility and a nearly 1:1 adaptation of reaction conditions derived from manual syntheses performed in classical medicinal chemistry labs. There is no real limitation to solution-phase library production except for a few technical restrictions. It can, for example, become impracticable to produce solution-phase libraries efficiently if a certain reaction scheme is prone to very variable reaction times and yields or requiring complicated, nonsystematic workup procedures. Performing multistep syntheses with varying reactants also poses a technical and logistical challenge and is more favorably done on solid phase. In addition, gas-phase reactions (e.g., hydrogenations, carbonylations) or reactions requiring inert conditions (e.g., some transition metal catalyzed reactions, Grignard reactions) have proven inefficient or overly complicated—although suitable equipment is nowadays commercially available and novel techniques facilitate some of these reaction types (see later text, Sections 12.3.2 on microwave reactors and Section 12.4 on flow chemistry).

Individually synthesized libraries, independent of their size and origin, should not be confused with "screening collections," often referred to as "compound collections" or "compound libraries." Such collections are large sets of diverse compounds used to identify chemical matter for medicinal chemistry projects via screening approaches. They consist of single compounds and libraries of structurally related chemotypes [10–12] and are continuously expanded by compound synthesis within corporate medicinal chemistry programs, by combinatorial synthesis programs for collection enhancement, by compound acquisition, and by accessing existing (commercial) collections via various business models. Additionally, isolation of new chemical matter, for example, from natural sources, is used to supplement these collections [13].

### 12.1.2 Library Types

Libraries can serve multiple purposes in drug discovery. A first type is the lead- or drug-like small molecule library for screening collection enhancement [10, 14, 15]. Such libraries can be completely diverse—only limited by synthetic accessibility—or designed as so-called target-oriented libraries with a protein target or protein family in mind [10]. If the design of such libraries fits with the target, the likelihood of detecting actives and receiving immediate information on the corresponding structure–activity relationship (SAR) is high, and leads can be generated within shorter timelines than if the actives are represented only by single compounds. This makes such libraries very valuable for lead generation programs within screening approaches. Their design can be based on target structural data (e.g., crystallographic or NMR data), available sequence and/or mutagenesis data, chemogenomic approaches, or known ligand structures of the same or related proteins [10, 16]. The size of such libraries (often ca. 100–500 compounds around a common central scaffold) is usually a compromise between a sufficient sampling of a chemotype to find hits, the anticipated value of the SAR one can generate within one round of library synthesis, and the cost of goods, purification, analytics, and biological testing [11].

More recently, diversity-oriented synthesis (DOS, [9, 17, 18]) has been promoted as an alternative to one scaffold variations: it allows for the application of a diverse set of reagents and structural transformations (see corresponding chapter). These result in diverging synthetic pathways and create different chemical scaffolds/chemotypes using a common scheme [9]. Libraries with variation in the molecular scaffolds are potentially more valuable than large single-scaffold libraries in terms of biological diversity; it is expected that the overall structural diversity of a library is directly correlated with its overall functional diversity [9].

Compound collection enhancement libraries are usually designed according to Ro5 (rule of five, [14]) requirements combined with further elimination of undesirable molecular features such as known toxicophores, reactive moieties, or motifs known to lead to frequent hitters [10, 11]. However, in recent years, the quest for libraries hitting targets involving protein–protein interactions with larger binding sites has led to a softening of the Ro5 criteria allowing for compounds with higher $sp^3$ character and molecular weights beyond 450–500 [19, 20]. Natural products and corresponding natural product-derived libraries fit well in this area, as these tend to have an inherently higher $sp^3$ character and—sometimes—a higher molecular weight combined with structural novelty [9, 20–23].

A second major class of small molecule libraries are so-called hit-to-lead or lead optimization libraries used in the course of optimizing active starting points. These are fundamental to classical medicinal chemistry programs and differ in size (often ca. 50–250 individual compounds) and in the degree of knowledge about the targeted chemical series. Early in a medicinal chemistry project, for example, during hit-to-lead campaigns, one typically needs a broad elaboration of SARs within a short time to improve compound potency against the target, to increase novelty, and to generate information on physicochemical and ADME properties [2, 3]. Such libraries can also be used to perform "rescaffolding" or "scaffold hopping" in which the core part of the molecule (the scaffold) is replaced by a structurally related, different chemical motif—either for intellectual property or overall compound property reasons. By rapidly synthesizing analogues in a library format, the properties of the new scaffold can be efficiently investigated.

Lead optimization libraries are designed to improve specific weaknesses of a selected chemical series, for example, metabolic lability [24], CYP inhibition [25], low intestinal absorption [26, 27], low solubility [26, 27], unwanted side affinities toward antitargets [28], or typical safety issues such as hERG affinity [29, 30]. A growing knowledge of the SAR as well as physicochemical and ADME characteristics of the chemical series at this stage enables an increasingly sophisticated library design reducing the necessary size of the libraries to be synthesized to often under 50 individual compounds per synthesis cycle.

During the last decade, fragment-based drug design (FBDD, [30–33]) has become a fruitful alternative lead generation approach to high-throughput screening (HTS). Libraries of low molecular weight fragment molecules have been assembled serving as starting points for drug discovery (for further details refer to chapter 9). The elaboration of fragments into hits and leads requires systematic structural expansion of these fragments into hit-sized compounds [30–33], a different challenge to making analogues. This can involve library synthesis with a library design based on crystallographic or NMR data of the fragment's binding mode in the protein. In this context, libraries allow a rapid access to new chemical matter that is not *a priori* present in the classical screening collection of a given pharmaceutical company or academic institution.

Additional libraries for special applications include the derivatization of isolated pure natural products [21–23] or peptides [34] and the incorporation of irreversible binding motifs such as Michael acceptors [35].

### 12.1.3 Chemotypes

Several factors influence the choice of chemical space for which libraries are produced. Synthetic feasibility, avoidance of chemically reactive groups or chemical structures prone to interfere with biological assays, and target-oriented design criteria (particularly within existing medicinal chemistry programs) form the overall basis for a library design. Particular structural features can have a significant influence on the overall properties of designed molecules. The number of aromatic rings, for example, has been identified as a risk factor, while a larger fraction of saturated and chiral centers in a molecule is considered beneficial. For compounds incorporating more than three aromatic rings the attrition risk in development is significant, while the "escape from flatland" increases the likelihood of success of a molecule in the clinic [10].

The previous coverage of a certain chemotype is a further criterion for compound value: due to the need for intellectual property in the pharmaceutical industry, the generation of novel, patentable compounds as starting points for lead generation is an important goal.

Concerning attractiveness of chemotypes, one observes a reemerging interest in peptides, macrocycles, and natural products as lead generation starting points besides the traditional "one scaffold" lead generation libraries. Most current screening collections in the pharmaceutical industry are large assemblies of "one synthesis, one scaffold" sublibraries [9], targeting the traditional target classes such as GPCRs, kinases, ion channels, and proteases. They were generated within the Ro5 [14] paradigm taking molecular size and lipophilicity into consideration as key players for oral bioavailability, in addition to potency, promiscuity versus other targets, ADME profile, and likelihood of toxic side effects [10]. Due to the quest for new approaches, the biological target space has become bigger and more complex; hence, alternative chemotypes covering different chemical space are becoming more and more attractive. Compounds with a certain similarity to natural metabolites have recently been stated to be beneficial to identify higher-quality hits in screening, and the use of conserved chemical ring systems within natural products as starting points for library design has been recommended within the so-called BIOS concept [22]. It has been argued that BIOS libraries around natural product templates should have a higher degree of biological relevance due to the fact that evolutionary pressure has "prevalidated" such molecules to modulate biological processes [9, 22, 23]. According to another investigation, compounds active against Gram-negative bacteria show physicochemical properties that are significantly distinct from those typically associated with drug-likeness, most clearly manifested by a substantially increased polarity [36]. Furthermore—and this is largely due to renewed interest in injectables [37] and recent developments to generate metabolically stable peptides and improved formulation systems [38]—there is a growing need for peptide and macrocycle libraries [39].

## 12.2 LIBRARY PRODUCTION

### 12.2.1 The Library Production Process

To accelerate the drug discovery process in its early stages by efficiently providing compounds in a library format, automation and parallelization of large parts of the synthesis of medicinally relevant molecules is a well-accepted approach [40–43]. There are many models which realize this—from library synthesis in classical medicinal chemistry labs with limited automation to specialized core teams with integrated automation solutions. Most of them are exemplified in pharmaceutical companies [44]. Production of solution-phase libraries (as defined in the introduction) can be summarized in a process consisting of six major steps (see Fig. 12.1):

1. Library concept and design
2. Library logistics (including documentation) and reagent procurement
3. Automated synthesis and workup
4. Purification and postprocessing
5. Analytics
6. Registration and final compound distribution

The subsequent actions, namely, biological and physicochemical characterization (screening against the intended target and antitargets, determination of ADME and physical chemistry properties), can be integrated in a continuous workflow but are distinct from the preparation of the compounds.

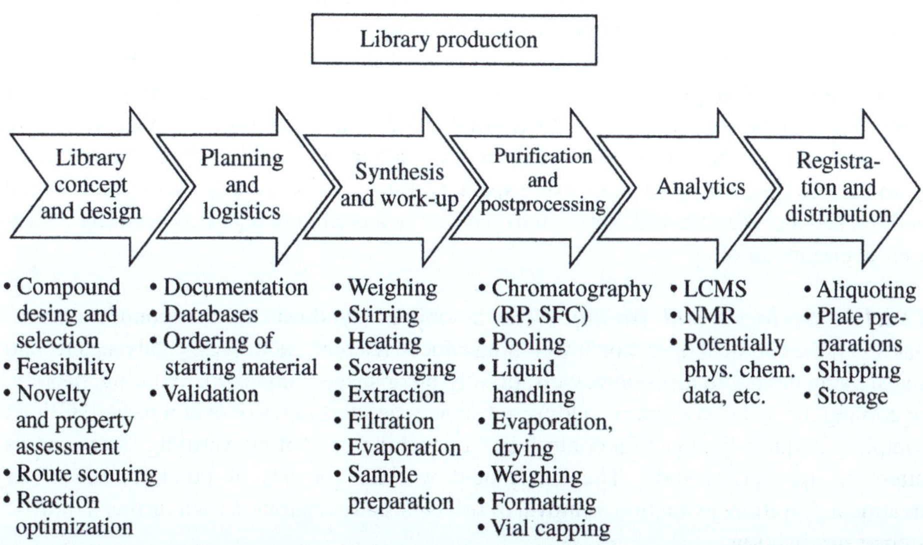


**FIGURE 12.1** Typical processes in library production.

***12.2.1.1 Library Concept and Design*** Library concept and design includes the translation of ideas and concepts into compounds and ways to synthesize them. While for prospective libraries aimed at a whole target class or at enhancing compound collections stringent criteria of lead-likeness such as molecular mass, clogP, calculated solubility, polar surface area (PSA), number of hydrogen bond acceptors or donors (HBA/HBD), etc. are commonly employed, target molecules for dedicated project libraries are usually carefully selected taking into account known SAR, docking and similarity scores, target structure information, and further available information [5, 45]. At this initial stage, synthetic pathways toward the targeted compounds are investigated and established by conducting test reactions and adapting and optimizing conditions. Depending on the objective as well as on the allocated resources and the available timeframe, new chemistries, reaction types, or cascades can be developed (cf. DOS). In a classical approach, although most reactions can principally be conducted in a library format, reliable transformations and reactions for which many reagents are commercially available (such as carboxylic acids, aryl halides, amines, alcohols, etc.) are most commonly encountered. Clearly, the design and planning phase requires sufficient time and multiple skills; the various tasks are usually split between expert teams (e.g., dedicated drug designers, project medicinal chemists, library specialists, etc.). Extensive investigations are thus often decoupled from actual library production.

***12.2.1.2 Library Logistics and Reagent Ordering*** Library logistics and reagent ordering involves the enumeration and registration of the compounds to be made (the "library") in an electronic lab notebook, the creation of a synthesis database, the procurement of reagents (usually from multiple suppliers), their sorting and storage, as well as the planning of the actual synthesis. The "virtual" parts of this stage are supported by various—often tailor-made—IT tools that allow documentation and tracking of libraries and individual library compounds through the whole process and facilitate the registration of the eventually physically obtained substances [46]. Depending on the level of automation used and the compatibility of the individual workstations, the process can be governed by handling and transferring input and output files or by an integrated software protocol [4].

The provision of starting materials/reagents and their preparation in a usable format require a major effort and can take a significant amount of time, particularly if flexibility with regard to the reagent selection is desired and thus delivery times of noncommodity substances have to be considered (see discussion in Section 12.3.1). Whether purchasing from various different vendors or using special arrangements, an order has to be initiated and tracked and the received compounds have to be stored until usage, preventing confusion or commutation.

***12.2.1.3 Synthesis and Workup*** The automated synthesis of compounds encompasses all the classical operations: weighing educts, reagents, additives, catalysts, etc. and transferring them into reaction vessels; dissolving or suspending them; agitating, heating, or cooling the reaction mixture; adding substances at later stages or over a period of time; sampling aliquots for reaction control; and quenching or cooling/warming the reactions after the appropriate time. The subsequent workup consists of potential scavenger treatment, filtration, extraction, removal of solvents, and preparation for a further synthetic step or purification.

Actually conducting the reaction comprises some of the most complex operations in the process of library production and thus—depending on the desired scope, scale, and flexibility of the transformations to be achieved—variable degrees of automation have been realized [42, 43, 47]. Instead of a fully integrated automated "assembly line process," a modular workflow addressing the various tasks separately—with manual interventions where necessary—is arguably more reliable and efficient. While, for example, dispensing of solvents and stock solutions can easily be done by liquid handlers, other operations are most efficiently done by a lab technician. In particular, weighing and dosing of solid or viscous substances are still very difficult to automate if no restrictions are to be made on reagents and reaction types. Alternatively, starting materials can be obtained in the desired quantity and format from vendors offering custom weighing and packaging.

After addition of all necessary reactants, the reaction mixtures are agitated on parallel shakers or stirrers with temperature control (cooling or heating). The use of shakers is often preferred since it avoids addition and removal of stir bars and is compatible with polymer-supported reactants due to smaller shear forces. Reaction progress is usually controlled by LCMS of all library reactions or selected reference vessels. Depending on the setup, aliquots for reaction control are taken manually or automatically employing a suitable liquid handler. It is worth noting that even if high-throughput LCMS measurements with short gradients of less than 2 min are used, sampling every reaction of a medium-sized library takes the better part of a working day. The following quench and/or work up is beneficially done in a parallel format using robots that can directly access the reaction blocks or are loaded with racks retaining the matrix of the reaction blocks [48]. In our laboratories at Sanofi, we have implemented a platform with multiple parallel liquid handling channels and stirrers that is capable of autonomously doing liquid–liquid extractions, filtrations over various cartridges, or "inverse" filtrations using syringe filters—with multiple repetitions of each cycle, if necessary. In a final step, the samples are concentrated and formatted for chromatographic purification.

***12.2.1.4 Purification and Postprocessing*** To avoid the drawbacks of screening mixtures or impure compounds (deconvolution, resynthesis, etc.) and to enhance speed and the quality of data, it is necessary to provide pure compound samples—especially when libraries are used to supplement single compound synthesis during hit and lead optimization. Thus, the purity of samples and the number of compounds that can be purified in a given time became important assets.

During the previous decade, most major pharmaceutical companies set up a specialized and often centralized purification unit or team to cope with the ever increasing demand of purified samples for biological profiling, HTS, etc. [4, 49–52]. In most cases, automated reversed-phase HPLC was the method of choice; however, normal-phase chromatography was partially included in the concept [49]. Recently—triggered by the availability of respective equipment—supercritical fluid chromatography (SFC) has emerged as an additional technique for the high-throughput purification of library compounds (see Section 12.3.3.2). Depending on the degree of automation, a properly equipped laboratory with two technicians is able to purify between 20,000 and 100,000 compounds per year [4].

The desired fractions of each sample are collected using a trigger of various detectors (e.g., MS, UV, ELSD) and are pooled by a robot if necessary. After the removal of solvent, the obtained material is weighed, redissolved, and formatted as required.

***12.2.1.5 Analytics*** Analytical samples taken after purification are commonly characterized by LCMS. To ensure the structural identity and maintain high quality, all library compounds synthesized in our laboratories at Sanofi are additionally analyzed by $^{1}$H-NMR.

***12.2.1.6 Registration and Distribution*** All final compounds with sufficient purity are registered in company databases, added to the repository, and distributed to respective requesters. Most companies have a workflow in place in which selected compound sets are immediately profiled for physicochemical and eADME properties.

### 12.2.2 Process Optimization

Whether library production is best integrated into the classical medicinal chemistry workflow, run in specialized high-tech laboratories, or outsourced to a third party is not only defined by intentions and timelines but is also a question of strategy and depends to a large part on the size of a research operation. In large pharmaceutical research departments, when dealing with libraries for collection enhancement or for very early projects, library production is often outsourced to one or more CROs working in the field [46], while libraries in later stage projects are often handled in-house. To balance time and cost, and to objectively benchmark these two, is difficult since a relevant measure would have to reflect the turnaround time and cost of the overall process from conception/design of compounds to biological results in primary assays. All parameters involved are coupled to throughput and resources for the respective task (often performed in different groups) and are guided by the corresponding project profile; the decision if and when libraries are appropriately involved in the compound production therefore remains with the individual project team.

***12.2.2.1 Timelines and Costs*** Time ranges for library production in literature vastly differ from days to months [53, 54]. Therefore, the parameters of a library as discussed in this section are more precisely specified as follows: we refer to sets of 20–1000 compounds that can be synthesized from similar scaffolds using the same reaction conditions in less than three diversification steps. The most common setting is one diversification step potentially followed by a deprotection step or a smaller diversification followed by a larger orthogonal diversification, sometimes including one or two deprotection steps in the sequence. The typical scale for the final reaction step is about 0.1–0.2 mmol in order to obtain greater than 10 mg of final product—sufficient for several years of screening and first characterizations in case activity is found. We only consider the production of a library, that is, start of synthesis (diversification step(s)) to formatting in storage vials and registration of those, potentially including analytical characterization. This can include some limited test reactions, but selection of target compounds, route scouting ("library concept and design" as defined earlier), and reaction optimization are out of scope. Thus, library production begins with the building of databases and physical selection of starting materials and ends with the final purified compounds bottled in traceable (e. g., barcoded) containers for storage.

*Timelines* In our experience, library production is most efficient in dedicated laboratories with a steady state of incoming library requests and specialized personnel working in close collaboration with traditional medicinal chemists on several projects. Competitive durations for synthesis of such project embedded libraries are anywhere between 2 and 5 weeks—the variability mainly being due to the number of steps (synthesis) and the number of compounds (procurement and purification). While the time needed for procuring starting materials is highly variable (see Section 12.3.1), synthesis and workup typically require 2–5 days per step for ca. 100 compounds. Purification and postprocessing of these batches (weighing, formatting, transfer to storage vials, etc.) take an average of 3–8 days—depending on the purification technique used and the purity desired. Analytics including NMR for all samples then usually requires another 1–3 days. These are, however, ideal time periods that can differ for various reasons such as:

- Insufficient conversion during synthesis requiring elongation of reaction times, need to add further reagent(s)/catalyst(s) and/or increased temperature
- Additional reaction control (LCMS sampling of 100 reactions easily takes greater than 5 h)
- Removal of excess reagents or reactive side products via scavenger treatment
- Precipitation in samples waiting for LC (requiring additional filtration)
- Purification in between library steps, for example, due to low conversion, side products
- Need for orthogonal purification methods to achieve sufficient purity
- Insolubility or volatility of final compounds impeding sample transfer
- Failing reactions or purifications and breakdowns of machinery

Concerning library production at an external partner's labs, turnaround times strongly depend on the type of collaboration. A suitable model for hit finding up to lead identification within a particular project involves the library design by the requestor, while execution is in the external partner's hand [46]. This can be particularly attractive since libraries can be tailored to areas of interest without having to disclose relevant information like intended targets, structural data, constructed pharmacophores, structure-activity relationships (SAR), etc. Such an approach usually leads to timelines that can be roughly compared with the in-house production of libraries. In contrast, early-phase libraries for collection enhancement rather compare to the purchasing of screening compounds. In our experience, depending on the chemistry employed, synthesis, characterization, and shipping of 300–500 compounds usually take 2–3 months for a two-step library or 3–6 months if the synthesis of required scaffolds is necessary (from order to delivery). However, if a collaboration is starting from scratch, much longer periods need to be expected, since negotiations, preparation of contracts and offers, clarification of logistics, etc. have to be taken into account. More efficient timelines can only be realized if the whole process of early discovery is outsourced to a suitable provider in which case the durations mentioned for in-house library production can be applied to the CRO.

*Costs* The picture is even more blurred than for timelines: it is nearly impossible to dissect and price-tag the various contributions of the library production process, especially if library synthesis is an integral part of medicinal chemistry and not only used for collection enhancement, but to support individual projects. For example, how much time is spent thinking about a library design, how much work is invested in building block generation, which part of the existing infrastructure across departments is used for synthesis and processing of library compounds, etc. Overall costs are not even easily calculated when outsourcing the whole process or parts of it to CROs: idea generation, negotiations, alignment of standards, delivery of compounds, registration and transfer to screening sites or laboratories, planning, and supervision of external coworkers—to name but a few tasks—have to be considered and included in any price estimation.

Nevertheless, at the time of writing pricing in the previously mentioned scenario of external synthesis of internally designed libraries is often very competitive—especially in Eastern countries: depending on chemistry and required characterization, $100–150 per compound can be realized (set of 300–500 compounds in custom vials, 10–20 mg, purification via HPLC, LC-MS analytics, including shipping). This compares favorably to the purchase of off-the-shelf screening compounds where prices are in the same range or by a factor of 2–3 cheaper, depending on amount, complexity, and availability of the desired molecules. However, the latter compounds are obviously not exclusive to the buyer.

***12.2.2.2 How to Optimize Timelines*** In our laboratories and others, over the last few years, every part of the library production process was reviewed to increase efficiency and to reduce cycle times of this complex process to a minimum [43, 44]. This resulted in several enhancements of general process organization and logistics and technical changes. At Sanofi, the overall optimization led to the average reduction of library production timelines from sometimes 2 months in the early days to weeks or even days in 2013.

As indicated in Figure 12.2, we have identified the following as the most significant factors for cost and timeline reduction:

**Centralization of Automated Parallel Synthesis and Purification; Production of Libraries in a Dedicated Group with Specialized Equipment and Staff.** Although small libraries (tens of compounds) can be readily realized in classical medicinal chemistry labs using limited automation—for example, parallel heating

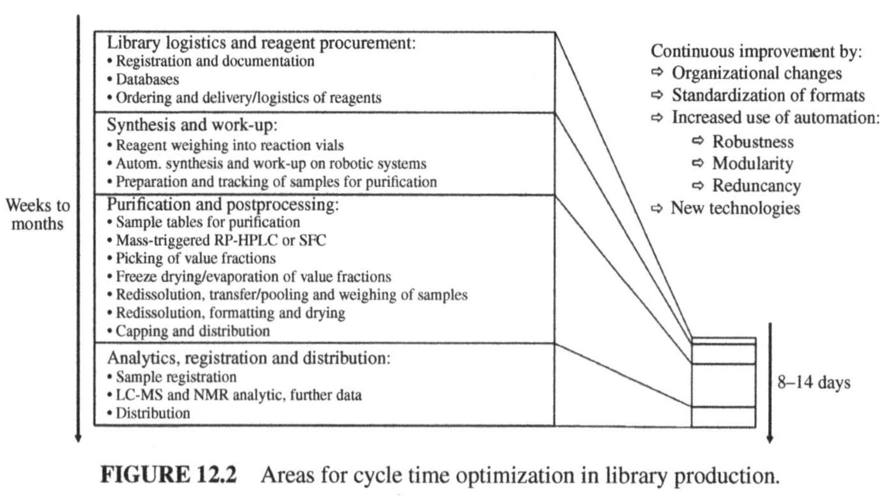


**FIGURE 12.2** Areas for cycle time optimization in library production.

blocks, microwave systems, and parallel extraction modules—larger compound numbers are much more efficiently handled by an "assembly line process." Using an assembly line example, library production is split into logical parts (e.g., logistics, synthesis, purification, etc.). Each task is executed by a specifically trained person with the required specialist equipment. While walk-on machinery can be an alternative for smaller departments or occasional library synthesis, a genuinely high throughput requires a steady state operation with continuously run workstations and smooth handovers between the steps. Each specialist is responsible for a defined part of the process and is intimately familiar with the equipment needed. Sanofi has laboratories focusing on logistics and documentation (including procurement of starting materials, planning, creation of databases, etc.), synthesis and workup, and purification and postprocessing (including chromatography, removal of solvents, sample handling, formatting, etc.). Analytical characterization and distribution to screenings, storage, etc. is done by dedicated personnel in the respective departments.

**Procurement and storage of starting materials** is pivotal to avoid delays. Efficient reagent ordering systems have to be devised, and the chemicals for every library have to be allocated and stored until use. Brokers for chemicals provide cost optimization, reliable catalogues, guaranteed delivery times, and custom packaging—thus simplifying and accelerating the purchasing process and maintaining most of the flexibility of individual orders. However, this comes at a cost, and the timelines involved are often still long. Using state-of-the-art storage technologies such as automatic reagent dispensaries takes the procurement of starting materials out of the critical path and potentially allows for the most significant acceleration of library production.

**Use of Standardized Formats.** Standardizing the dimensions of arrays of plates or vials allows for facile transitions between subprocesses and departments. Library compounds eventually have to end up in a format ready for storage or use by the distribution/screening facility. Standardized formatting avoids reprocessing or duplication of tasks.

**High Degree of Maintenance-Free Automation.** This ensures continuous operation and is particularly valuable for the most time-consuming activities like library purification. Integrated preparative chromatography systems are able to run 24/7 (see Section 12.3.3) with technicians only loading sample racks and removing collection racks and attending to the infrastructure. Sample sets can be directly transferred from one workstation to the next, while appropriate software protocols govern file transfers and individual processes [49, 50]. Automated extraction allows for the unattended processing of complete libraries on a robot including dispensation of solvent to the reaction mixture, mixing, phase separation, removal, and drying/filtration of the desired layer.

**Independent and Simple Workstations for Individual Processes.** As opposed to automated synthesis during the combinatorial chemistry era [42], most laboratories nowadays rely on a modular setup. Individual workstations are used for each processing step (e.g., filtration, extraction, selection of value fractions, etc.). They are easier to use and more reliable than integrated automation solutions [50, 55]: other subprocesses can still run if there is any downtime and the replacement of broken machinery is less expensive. However, the situation might change again in the future with further progress in laboratory automation technology making more sophisticated and affordable solutions available [47].

**Redundancy of All Process-Relevant Instrumentation.** This ensures a continuous service in case of breakdowns [56]. Since the workflow used at Sanofi is designed not only to support lead finding activities but also lead optimization chemistry up to the candidate stage, emphasis has been put on reliability and success rates to ensure the maximum value of sophisticated library precursors/starting scaffolds.

**Use of Accelerating Technologies**. Many solutions that significantly speed up individual steps within the library production process are available. The use of automated reagent dispensaries allows access to decoration reagents in minutes. Reaction times in synthesis can be reduced employing microwave heating technology. The duration of extractive workup can be reduced by using robots with several parallel liquid handling channels and stirrers. Removal of solvents can be accelerated by the use of heated centrifugal dryers or omitting water as an eluent in chromatography. A selection of these aspects is described in more detail in the next section.

More ambitious approaches combining two or more of the steps detailed in Section 12.2.1 have been realized. Most commonly, purification, registration, and distribution are handled together in one laboratory workflow [51, 52], sometimes including analytics and characterization of the compounds. In an example combining synthesis, workup, and purification by integrating a segmented-flow synthesizer and a preparative HPLC-MS system, turnaround times for library production of less than 5 days were realized for a selected set of reactions [53].

There are also efforts to comprise all steps in one continuous albeit sequential process that achieves unrivaled cycle times. However, this approach requires optimization of all processes and is—as of yet—only exemplified in particular case studies [57].

## 12.3 NEW TECHNOLOGIES IN AUTOMATED LIQUID-PHASE LIBRARY SYNTHESIS

### 12.3.1 Provision of Starting Materials: Automated Reagent Dispensaries

Procurement of the starting materials is surely the most time-consuming step in library synthesis unless a special solution is in place. Purchasing the required reagents from any of the numerous existing vendors when the library has been designed and the desired compounds selected requires (i) a significant logistic effort to identify the most effective source based on cost and delivery time, (ii) space and an appropriate system to store the incoming reagents until the time of use, and (iii) interlaced and very flexible planning of several library syntheses including cutoff criteria since—depending on the vendor and/or the selected reagent—delivery times can vary from days to weeks or even months. Pharma companies have come up with several approaches to deal with these issues, all of which have their advantages and disadvantages in terms of speed, cost, and practicability.

The most flexible solution likely is to use one of many chemicals brokers who can procure the desired reagents in a cost-effective way and can also arrange for delivery of all reagents required for one library at the same time—even in specified quantities and/or containers. However, each additional service has its cost, and delivery times can still be prohibitively long.

Another effective approach is building up a collection of preferred reagents—selected according to certain criteria (see later text)—and store them either in-house or at a special provider. Building blocks for library compounds are then selected only or preferably from this limited set of reagents. Since on-site storage of large reagent collections is costly, external partners are often used. Amgen, for example, has a proprietary reagent collection inventoried and maintained by Frontier Scientific Services, Inc. from which a selected set of reagents can be delivered in custom weighed amount and the required vial within 2–3 days [55]. Others have set up similar models in the past [44].

In addition, larger sites are often able to capitalize on historical reagents stocked in centralized stockrooms. Our in-house group at Sanofi has access to roughly 40,000 different reagents or building blocks from previous projects that can be delivered to the bench in less than a day.

On-site storage of reagents with immediate access is by far the most effective way to make reagents available. The library production group at Sanofi is utilizing an automated storage with ca. 10,000 reagents that are immediately accessible (retrieval of 96 compounds in less than half an hour), while other companies have similar solutions in place [43]. In our current setup, only powders and solids are stored at room temperature, but the principle could of course be extended to liquids and lower temperatures if handling of more and thermally less stable reagents is desired. Between 0.1 and 10 g of the reagents are kept in uniform barcoded 20 ml vials arranged in shelves of 181 compounds, which are retrieved by a robotic arm and presented on a tray in the exact matrix also used for synthesis in our laboratory. Weighing and transfer to the reaction vessel is then performed manually to be able to cope with differing consistencies of reagents and to be flexible in terms of scale. After use, the tray is put back in the collection slot, and during the reshelving process, every vial is weighed to track the amount still available. An alert is issued when the amount of a given reagent is below a certain threshold so it can be reordered. Thus, the time-consuming ordering process is decoupled from the critical path of the synthesis.

The automated dispensary is stocked with reagents bearing the most commonly used functional groups in library synthesis such as amines, carboxylic acids, boronic acids, alcohols, sulfonyl chlorides, aryl halogenides, etc. organized in virtual groups. An electronic interface allows for the retrieval of the required reagent matrix. Due to their limited number and to ensure maximal usefulness, selected reagent sets are carefully chosen with respect to their properties and attractiveness—as judged by inspection by experienced medicinal chemists. Customized or commercial software tools are used to filter potential starting materials and reagents according to required functional groups, unwanted moieties (to avoid competing reactivity or predicted pharmacokinetic liabilities), availability/accessibility, and price.

### 12.3.2 Microwave

Although the benefit of microwave heating for organic synthesis is undoubted and a plethora of monographs [58–60] and review articles [61] have been published dealing with its advantages and disadvantages, its use in automated parallel synthesis requires some consideration. Using microwave irradiation as a direct heat source with the capability of superheating solvents in sealed vessels has been shown to reduce reaction times quite significantly, and often, many reaction parameters like temperature, solvents, additives, catalysts, etc. can be evaluated in a few hours making the technique ideally suited for reaction optimization. The most commonly used instrument in synthetic laboratories today is a single-mode microwave [62] containing a cavity of limited size that confines the microwave field to just one modal pattern (standing wave). This ideally results in a predictable field pattern with a known albeit not constant energy distribution within the cavity. To standardize the energy input, the sample to be irradiated is put at a defined location meaning that a set of samples has to be processed sequentially instead of simultaneously. Automated robotic systems with high-capacity mechanic sample changers are available, but using single-mode instruments for library synthesis nevertheless often means losing the advantage of speed [63]. As an example, processing a library of 100 compounds with a reaction time of 20 min in the microwave still takes about 2 days (33 h); for larger libraries or longer reaction times, even the use of several single-mode devices in parallel can result in processing times longer than those for conventional heating where the number of samples is usually not an issue (parallel heating blocks with 48 or more places are common).

A possible alternative is the use of multimode microwaves that can actually be viewed as a step backward since the first instruments investigated for heating chemical reactions were multimode devices—and the regular household microwave oven still is. The cavity of a multimode instrument can be quite large and is able to accommodate all kinds of reaction vessels including microtiter plates, but the microwave field inside it forms an unpredictable pattern due to the existence of many modes and resonance effects. This results in significant and variable discrepancies in energy input and thus temperature ("hot and cold spots"). However, the use of vessels made of strongly microwave absorbing material with high thermal conductivity (e.g., silicon carbide, carbon-doped Teflon®) alleviates the situation and minimizes temperature gradients across the reaction. Employing silicon carbide racks or plates in suitable rotors allows conduction of truly parallel (simultaneous) reactions in the microwave [63, 64]. Obviously, in this case, the heat is transferred "classically" by conduction and convection, but if adequate agitation (stirring) of the reaction medium and temperature control is ensured—and one accepts the nonexistence of nonthermal microwave effects [65]—all microwave advantages translate to the described setup. Further, since the vessel material and not the reaction mixture itself absorbs the microwave energy, the use of media with unfavorable dielectric properties (e.g., unpolar solvents) is possible. Commercial systems using sealable microtiter plates or racks for vials up to 3 ml allow for up to 192 or 96 reactions in parallel at a maximum of 200°C/20 bar [66] resulting in a potential to truly accelerate library production.

### 12.3.3 Library Purification: Automated RP-HPLC and SFC as Orthogonal Methods

***12.3.3.1 Reversed-Phase HPLC*** As mentioned previously, reversed-phase HPLC (RP-HPLC) is the most widely used technique for routine purification of library compounds or single compounds synthesized in the regular medicinal chemistry lab because it is uniquely suited for the compounds usually encountered in pharmaceutical settings (drug-like molecules). Although different organizational models exist (from purification by and in the synthesizing laboratory, via walk on equipment often maintained by dedicated experts, to full-service chromatography units), the most efficient setup is arguably a specialized laboratory that centrally handles the purification of compounds [51, 52]. These purification units usually have a high degree of automation, and the machinery is able to run 24/7. The eluting solvent (acetonitrile or methanol/water/acidic or basic additive) is provided from large reservoirs—in the case of the aqueous part by automatic premixing of acid/base with deionized water (Fig. 12.3)—thus enabling purification sequences to run nonstop for days [49]. Samples are injected by liquid handlers able to cope with racks containing tens or even hundreds of samples.

While the basic instrument parameters such as flow and detection mode are usually kept constant, columns can be switched by suitable valves, and different gradients can be chosen for each sample controlled by the integrated software. In the usual setup, preparative separations of a few milligrams up to ca. 300 mg can be handled in one injection using flow rates of 100–150 ml/min and run times of 5–15 min on 5 μm columns (silica gel C18, 30 × 100 mm). For larger separations up to a few grams, multiple injections on the same system make sense before one has to adapt the dimensions. This covers the amounts usually encountered in library synthesis.

Due to the robustness of the technique, sample preparation often consists of a simple filtration of the reaction mixture taken up in DMF or another appropriate solvent. Using

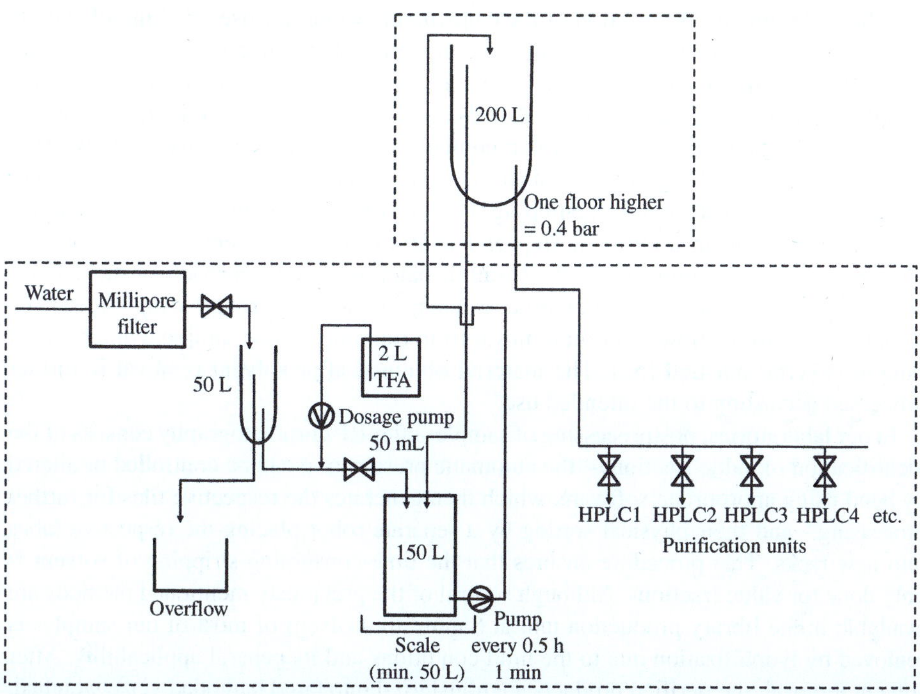


**FIGURE 12.3** Fully automated, maintenance-free water supply and TFA dosage for RP-HPLC.

a precolumn that is regularly replaced, the wear on the main column is low so that up to greater than 10,000 samples can be purified over one column without deterioration of separation.

Collection of the fractions on high-capacity collectors with often greater than 500 fraction tubes is nowadays usually triggered by a mixed MS and UV signal (as opposed to historical UV-only triggers), which enables a precise selection of only the peak containing the desired product. This allows for a low sample to fraction ratio. In most cases, the length of the gradient can be optimized such that just one tube is required per sample. Eluent of a sample that is not collected is either directly transferred into a waste reservoir or is optionally collected in larger containers (0.5–21 depending on flow rate and length of a run) in a waste collector ensuring that no material is lost in case of a malfunction or if the waste has to be reinvestigated. Waste collection containers can be automatically emptied into a centralized sewage system emptying into tanks for removal [49].

Apart from the large amounts of aqueous solvent waste generated by RP-HPLC, solvent removal from the value fractions—especially if the eluent contains significant amounts of water—is actually one of the most time-consuming steps and a long-standing issue in high-throughput purification. Several techniques and devices to accomplish this task have been developed; however, there is still room for improvement. Commonly, evaporation of solvents in high-throughput procedures is accomplished with vacuum centrifugal dryers or blowdown devices using a gas stream directed into the warmed sample tubes [48, 67]. While the former—especially for large parallel applications—can be quite costly due to sophisticated equipment, the latter suffers from the risk of cross-contamination and the potential release of solvent fumes into the environment. A variation of the centrifugal evaporator uses shaking of sample racks in an evacuated cavity and IR irradiation to provide the necessary heat. For small amounts of highly volatile solvents (with a large surface to volume ratio), simple vacuum drying ovens can be used. A very mild high-capacity method, particularly suited for samples with a high water content, is freeze-drying. However, with an increasing fraction of organic solvent in the sample, processing times become longer and the risk of thawing and thus bumping rises. Various combinations of the mentioned techniques require an additional step and additional time, potentially with manual intervention. Overcoming a strictly parallel, batch-oriented approach, some laboratories have realized a sequential process using a modified rotary evaporator [67]. Robotic integration allows solvent removal from fractions of one sample while the next sample is being purified [52]. The material obtained after solvent removal is further processed according to the intended use.

In our laboratories, postprocessing of samples after RP chromatography consists of the identification of value fractions—the automatic preselection can be controlled or altered by hand using appropriate software, which then generates the respective files for further processing—and their physical sorting by a separate robot placing the respective tubes into new racks. This procedure ensures that the time-consuming stripping of solvent is only done for value fractions. Although several of the previously mentioned methods are available in the library production unit at Sanofi, the solvent of most of our samples is removed by lyophilization due to the mild conditions and its general applicability. After solvent removal, the purified products are transferred into tared barcoded vials, weighed, aliquoted (analytics, characterization), and formatted in the appropriate way for further processing. All postprocessing steps are done by dedicated devices following the modular concept and are—with the exception of freeze-drying—identical and independent from the technique used for purification.

*12.3.3.2 SFC* A purification technique complementary to RP-HPLC is SFC, in principle a hybrid of liquid and gas chromatography using supercritical liquids (or at least one component that is fed into the system in its supercritical or close to its critical state) as mobile phase, while the same columns (stationary phases) as for HPLC can be employed. SFC has been investigated since the 1960s [68], but despite great promise, it has only recently been considered for wider laboratory application on preparative scale due to several technical issues and the fact that the physicochemical basics of SFC are still poorly understood [69]. The retention of solutes in SFC is predominantly determined by the compressibility of the mobile phase: the molecular interaction of solute, mobile, and stationary phase depends on the solvent strength of the eluent, which in turn is controlled by the pressure—in general, the higher the pressure, the higher the solvent strength. In reality, however, the relationship is more complicated since the primary intensive variable controlling the properties of a supercritical fluid is its density [70]. Thus, the variation of the mobile phase's density along the column can have a significant influence on retention depending on the experimental conditions. This is one of the major differences of SFC as compared to liquid or gas chromatography.

Although several compressed gases including $NH_3$, $SF_6$, and variously halogenated methane have been employed, the most commonly used mobile phase is supercritical $CO_2$ [70]. The reason for this choice are the physical properties of $CO_2$ with easily accessible critical parameters (critical temperature $T_c$=31.1°C, critical pressure $P_c$=73 bar) in a range where most pharmaceutically relevant molecules are (thermally) stable, a low viscosity of 0.01–0.2 cP (temperature range of 30–75°C), and large diffusion coefficients for relevant solutes as compared to organic solvents. This allows for high flow rates with moderate pressure drops and high efficiencies for separations. Further, $CO_2$ is nonflammable, only toxic in high concentrations, and relatively cheap. To give a rough impression of the eluent cost: at the time of writing, the bulk price for 1 t of $CO_2$ was approximately €150 (3.5 quality, i.e., 99.95% purity) and approximately €450 (4.5 quality, i.e., 99.995% purity), respectively, compared to approximately €3900 for 1000L of acetonitrile, a very common eluent component in RP-HPLC.

Despite the previously mentioned advantages, $CO_2$ is quite unpolar and a poor solvent for most commonly encountered molecules in the pharmaceutical context. To overcome this, polar modifiers are added to fine-tune the properties of the eluent. Usually, methanol or other lower alcohols are employed, but various other modifiers (acetonitrile, THF, DMSO, other organic solvents) have been used depending on the desired polarity [71]. In preparative separations, gradients with up to 80% modifier are encountered, meaning that under regular SFC conditions (temperature 20–40°C, pressure 100–300 bar), the system can clearly be subcritical. This, however, does not really matter since efficient separations can still be achieved [72].

Preparative SFC has now become standard for chiral separations in many pharmaceutical companies [73, 74], mostly due to the significant advantages in speed, ease of recovery of products from the eluent, solvent cost, and ecological superiority as compared to HPLC—and despite the somewhat higher complexity of preparative SFC instruments. However, its use as a standard method in achiral chromatography—especially in a high-throughput setting—has been hampered mainly because of the difficulty to collect the value fractions via an MS trigger in an open bed. A general schematic of a preparative SFC for achiral high-throughput purification is depicted in Figure 12.4 comparing the key features to RP-HPLC.

Mass-directed collection allows for a low sample-to-fraction ratio. It has been introduced in recent years [75] with commercial systems now available. Simple open bed collection at competitive flow rates in standard equipment—as in HPLC—is still a problem [56]. Due to the pressure drop after the back pressure regulator $CO_2$ evaporates. What exits the system is not a liquid but an aerosol, which cannot be simply collected in an open test tube. Some form of a gas–liquid separator (GLS) is needed to coagulate the droplets of the modifier and separate them from the gas stream. Various principles have been used to address this issue, and a range of GLSs have been constructed. Table 12.1 lists the most common GLSs with their properties, but no fully satisfying solution for collection in high-throughput mode is commercially available to date. At Sanofi, we have developed a GLS (see Fig. 12.5) based on a mixture of impact and inertia separation [76]. It enables the almost complete removal of $CO_2$ from the modifier at low pressure for a large range of flow rates and modifier concentrations. This allows fraction collection in standard HPLC equipment without peak broadening and with an excellent recovery. (Using a z-motor positioning the needle below the rim of the test tube is however still recommended to prevent cross-contamination due to residual remains of dissolved $CO_2$.)

Another important parameter influencing separation capacity and quality is the flow rate. While in HPLC—depending on the experimental setup—a compromise between pressure, speed, and quality leads to optimal flow rates in the range of 100–150 ml/min, the low viscosity of the eluent in SFC in principle allows for a very high flow rate. But due to the more common analytical applications and the issues in pumping highly compressible supercritical $CO_2$, as well as the limitations of gas–liquid separation mentioned earlier, current commercial systems operate mostly in a range of 30–100 ml/min. However, using suitable pumps, modifier stream injection, and the GLS described previously, our

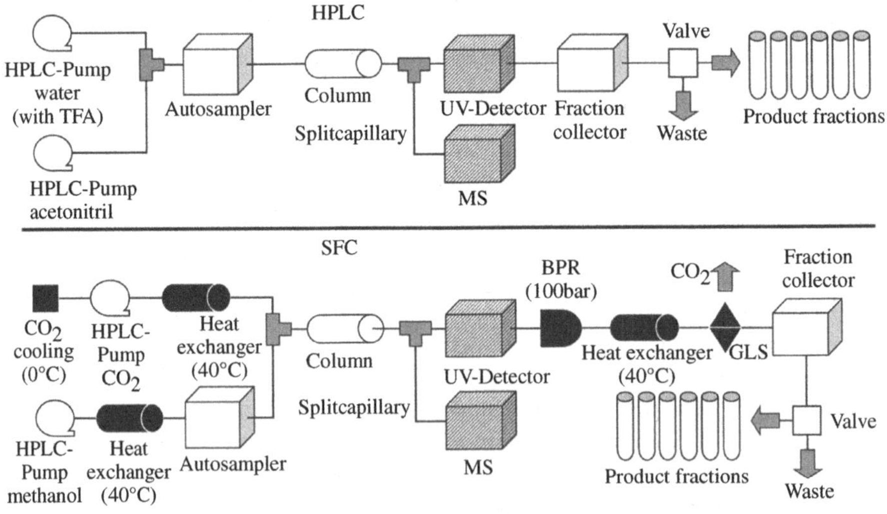


**FIGURE 12.4** Typical flow schemes of HPLC and SFC.

**TABLE 12.1 Different Types of Gas–Liquid Separators (GLS) and Their Properties**

| GLS Type | Advantages | Disadvantages |
|---|---|---|
| Cyclone (centrifugal force separator) | High recovery under optimized conditions (>92%)<br>Suitable for large-scale separations and chiral separations with multiple injections of the same composition | Low recovery under unoptimized conditions (60%)<br>Only for batch separations due to cross-contamination<br>Large surface<br>Weak self-cleansing effect<br>High pressure (20–40 bar)<br>No visual control<br>No dynamic behavior with respect to flow, eluent composition |
| Filter separator/ coalescence separator | High recovery (90–95%)<br>Visual control<br>Simple technology | Only for batch separations due to cross-contamination<br>Large surface<br>Weak self-cleansing effect<br>Danger of clogging |
| Impact separator | High recovery (~95%)<br>Visual control<br>Low pressure (<0.3 bar) | Only for batch separations due to cross-contamination<br>Large surface<br>Weak self-cleansing effect |
| Pressurized impact separator | Low cross-contamination | Mediocre recovery (83–92%)<br>Fraction directly in collection vessel<br>Only limited number of fractions per injection |
| Pressurized impact separator with drain | "Open bed collection" | Mediocre recovery (79–86%)<br>Weak self-cleansing effect (reciprocal addition of MeOH) |
| Impact/inertia separator | Very high recovery (>98%)<br>"Open bed collection"<br>Good self-cleansing effect (starting at 5% MeOH)<br>Dynamic behavior (3–90% MeOH, flow range 100–350 ml/min)<br>Low pressure (<1.5 bar) | Not commercially available |

system is able to operate at flow rates of 200–300 ml/min leading to preparative gradients with high resolution of approximately 5 min duration.

This setup allows collecting tens or even hundreds of fractions. This is mandatory to efficiently purify large sets of compounds in an automated fashion—with every injection being a different sample and capitalizing fully on the advantages of SFC in that area. SFC has therefore become a truly orthogonal alternative to RP-HPLC. Together, they provide a powerful purification toolbox for library compounds; Table 12.2 compiles the major differences between SFC and HPLC.

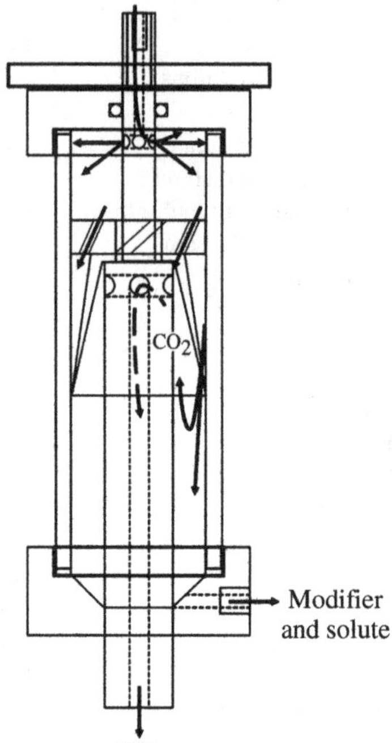


**FIGURE 12.5** Impact/inertia gas–liquid separator (GLS) developed at Sanofi.

**TABLE 12.2 Advantages and Disadvantages of SFC in High-Throughput Purification of Library Compounds Compared to HPLC**

| Advantages of SFC vs. HPLC | Disadvantages of SFC vs. HPLC |
|---|---|
| + Potentially higher loading of columns [77] | – More elaborate sample preparation/ workup necessary (solubility in modifier required, no salts) |
| + Time needed to strip eluent from product | – Instrumentation more complex and expensive |
| + No additive (like TFA) necessary, thus no salt formation of purified product | – Effect of operational parameters on elution more complex, physicochemical knowledge of governing principles less understood |
| + Cheaper, greener, less hazardous eluent | |

## 12.4 FLOW CHEMISTRY AND GAS-PHASE REACTIONS

As opposed to conducting reactions in batch mode, that is, charging a vessel with certain amounts of starting materials, quenching the reaction after a given reaction time, and working up the whole batch at once, flow chemistry allows reactions to be run in a continuous mode. Reagents and catalysts are either fed into the substrate stream as solutions or are immobilized on appropriate columns/cartridges, thus defining the stoichiometry of a process by reagent concentration and the ratio of flow rates as opposed to the volumetric ratio in batch processes.

Although flow reactors have found multiple applications in the production setting for over a century, synthetic organic chemists especially in pharmaceutical research/medicinal chemistry have only recently started to embrace the technology [78–81]. Flow reactors in laboratory use today range from simple systems consisting of inert tubing, connectors, and a pump to sophisticated microreactors with suitable channels cut into glass or similar materials and fully integrated devices.

Conducting reactions in flow has several advantages over classical batch processes concerning reproducibility, process reliability, accessible conditions, and safety. The high surface to volume ratio in a flow reactor and especially in microfluidic devices causes intense liquid-/solid-phase interactions and thus allows for controlled mixing as well as controlled and efficient heat transfer. This means that—while reactivity is an inherent property of the used reagents—reaction conditions (heat and mass transport) can be adapted and often precisely controlled in a flow setting enabling rapid optimization of a given reaction system and leading to more robust results. It should be mentioned that facile in-line monitoring (via UV detectors, LCMS, and/or react-IR) can help to minimize optimization time and maximize reaction performance and allows for fully automated optimizations [80, 82]. Further, flow reactors can be set up as a closed system using a back pressure regulator, which permits the superheating of solvents to broaden the accessible reaction perimeter (see also discussion of microwave technology in Section 12.3.2). Due to the controlled environment, the facile dissipation of heat, and the limited reaction volume inside the reactor at a given time, highly reactive and/or toxic materials can generally be handled safely. This is demonstrated, for example, by the generation of organic azides from either $NaN_3$ and alkyl halides or DPPA and carboxylic acids and their immediate reaction toward 1,2,3-triazoles [83] or protected amines (Curtius rearrangement) [80], respectively, as well as fluorinations [80] in a flow reactor.

Obviously, reactions in a flow setting are suitable for in-line workup or purification by directing the exiting stream of a continuous reactor through scavenger cartridges or chromatography columns. Connecting several reactors and suitable workup modules in-line enables multistep syntheses in one go with minimal manual intervention after appropriate process optimization ("telescoping") [84, 85]. The sequential addition of educt streams at defined stages allows for the channeling of a reaction along a desired path [86].

Another advantage of flow systems is the increased capacity and the simple scale-up or scale-out. Larger product amounts can be obtained by simply running the process longer or using multiple reactors at the same time. Parameters determined in a miniaturized system can often be transferred to larger devices without the need for reoptimization [79].

One major issue of flow reactors is the handling of heterogeneous reaction mixtures containing solid material. The precipitation of intermediates or (side) products can easily lead to clogging of the device. This is obviously more pronounced the smaller the diameter of the reaction channels. Depending on the design of the equipment, removal of a blockage or cleaning the channels is cumbersome to impossible. Another drawback of continuous processing as compared to the very flexible batch approach is the fact that a thorough adaption and optimization of a flow process toward one reaction system is needed before it can be run efficiently. This process is nicely described in a publication dealing with the automated semiparallel flow synthesis of secondary sulfonamides [87] and is particularly impedimental when the rapid and flexible adaptation of different reactions is required to produce a wide variety of compounds or compound arrays in competitive timeframes. For further discussions of advantages and disadvantages, details, and applications of flow chemistry, see the extensive review literature and the citations within [78–81, 88–92].

Parallelization of flow processes can only be realized by the use of multiple reactors. Obviously, a sequential processing is possible using available automation like liquid handlers and collectors [93], but this approach usually requires extensive parameter optimization and in principle suffers from the same drawbacks as microwave irradiation in single-mode devices (see discussion in Section 12.3.2).

For a limited set of compounds and with suitable reactions in an adapted setting, a semi-parallel process can be realized, as was exemplified during one of the very first syntheses of a natural product in flow [94]. En route toward grossamide various amides were synthesized by immobilizing a common scaffold on one of several cartridges as an activated ester and—after switching cartridges—eluting the product with a solution of a suitable amine reaction partner from the first cartridge, while the next was being loaded with the activated intermediate again. Limited parallelization has also been achieved in special cases, for example, by employing a microfluidic reactor with several channels and electroosmotic pumping [95], or in a 4-channel reactor using microwave-assisted, continuous-flow organic synthesis (MACOS) to produce a 1,2,5-thiadiazepane-1,1-dioxide library [96].

A truly parallel solution would be the stacking of individual microfluidic reactor chips or the further miniaturization leading to digital microfluidics and combinatorial chemistry on nanoliter scale where reactants in tiny droplets react with one another on a single chip [97]. However, the former poses a real challenge in terms of logistics and infrastructure since multiple channels/pumps are needed to provide starting material and reagent solutions from respective reservoirs. The latter cannot honestly be referred to as a flow technology.

In our experience, current flow solutions are most efficiently used in the synthesis of building blocks for further derivatization or in the production of small arrays of compounds [98].

### 12.4.1 Reactive Gases in Flow

Due to the mentioned characteristics of flow reactors, they are particularly suited for reactions employing reactive or hazardous gases. An application well established in many laboratories is hydrogenation in flow using the H-cube reactor [99] as a compact and affordable solution. A continuous flow of substrate is combined with hydrogen, generated *in situ* from the electrolysis of water, and the reaction takes place in an exchangeable catalyst cartridge with collection of the product stream, for example, on a liquid handler. A multitude of cartridges containing different catalysts like Pd/C, $PtO_2$, Raney nickel, Rh/C, Ru, etc. is commercially available.

Recently, a membrane-based [100] tube-in-tube design was developed to introduce gaseous reactants into a flow system [101]. An inner tubing made of gas-permeable Teflon AF-2400 (tetrafluoroethylene copolymerized with perfluorodimethyldioxolane) is surrounded by regular tubing with a slightly larger diameter, and the gas stream can be introduced via either of them [102]. This setup allows for an independent flow rate for each phase and the continuous supply of gas throughout the reactor. Various examples for transformations with reactive gases using this design have been reported including ozonolysis of alkenes [101], homogeneous and heterogeneous hydrogenations [103], oxidative couplings [104], carboxylations [105], carbonylations [106], and condensations with ammonia [107].

Employing a similar tube-in-tube reactor containing a gas-permeable Teflon membrane, the generation and reaction of anhydrous diazomethane was safely conducted on laboratory scale [108]. In this case, there are liquid phases in both tubes, and only the generated gas is able to diffuse through the membrane. α-Halo ketones thus obtained from *N*-protected amino acids are essential building blocks for antiretroviral compounds [109].

At the Automated Synthesis and Library Production group at Sanofi, a flow system able to handle gaseous reactants—including potentially hazardous gases like CO and $NH_3$—is set up in a way that it can run continuously without supervision. The flow reactor itself consists of a tube-in-tube reactor as mentioned earlier. Gases are supplied to the reactor from switchable gas cylinders stored in a vented cabinet via inert tubes and an appropriate valve system. Multiple sensors for the used gases are installed at critical locations that trigger an alarm and automatically shut down the system in case of a leakage. The hood containing the system is further equipped with an additional independent exhaust that switches to battery power in case of a power failure. This setup enables the synthesis of building blocks or intermediates that would otherwise be hard to obtain on a variable scale.

## 12.5 CONCLUSION

Libraries are used in various stages and for multiple purposes in modern drug discovery, arguably with the biggest impact in lead finding and optimization. Especially the timely synthesis of high value library compounds in early lead optimization has the potential to significantly accelerate medicinal chemistry projects. Automation and parallelization streamline the synthesis of libraries. During the last two decades, a strong emphasis has been placed on identifying ever more optimized processes with respect to time and cost. Different outsourcing models and lean principles have been introduced in particularly this part of drug discovery R&D. Clearly, organizational changes and continuous process optimization, as well as the innovative use of emerging technologies, have made automated parallel synthesis a very valuable tool for library production. Among the most relevant factors for acceleration of library synthesis are immediate (automated) access to required chemicals and innovations in chromatographic purification (like high-throughput SFC) together with novel ways to conduct synthetic transformations (like the use of microwave heating and flow chemistry). Additionally, reliable automation and an "in-line" profiling of the obtained compounds lead to reduced cycle times and enable much faster progress in drug discovery projects.

## REFERENCES

[1] Lipkin MJ, Stevens AP, Livingstone DJ, Harris CJ (2008). How large does a compound screening collection need to be? *Comb. Chem. High Throughput Screen.* **11**:482–493.

[2] Gleeson MP (2008). Generation of a set of simple, interpretable ADMET rules of thumb. *J. Med. Chem.* **51**:817–834.

[3] Gleeson MP, Hersey A, Montanari D, Overington J (2011). Probing the links between in vitro potency, ADMET and physicochemical parameters. *Nat. Rev. Drug Discov.* **10**:197–208.

[4] Everett J, Gardner M, Pullen F, Smith GF, Snarey M, Terrett N (2001). The application of non-combinatorial chemistry to lead discovery. *Drug Discov. Today* **6**(15):779–785.

[5] Gerencser J, Tymoshenko DO (2011). Parallel synthetic chemistry: Integrating emerging technologies. SP2 Mar/Apr:16–19.

[6] Pelletier JC, Kincaid S (2000). Mitsunobu reaction modifications allowing product isolation without chromatography: Application to a small parallel library. *Tetrahedron Lett.* **41**: 797–800.

[7] Suzuki A (1999). Recent advances in the cross-coupling reactions of organoboron derivatives with organic electrophiles, 1995–1998. *J. Organomet. Chem.* **576**:147–168.

[8] Hartwig JF (1997). Palladium-catalyzed amination of aryl halides: Mechanism and rational catalyst design. *Synlett* **4**:329–340.

[9] Galloway WRJD, Spring DR (2009). Is synthesis the main hurdle for generation of diversity in compound libraries for screening? *Expert Opin. Drug Discov.* **4**(5):567–472.

[10] Renner S, Popov M, Schuffenhauer A, Roth HJ, Breitenstein W, Marzinzik A, Lewis I, Krastel P, Nigsch F, Jenkins J, Jacoby E (2011). Recent trends and observations in the design of high-quality screening collections. *Future Med. Chem.* **3**(6):751–766.

[11] Bakken GA, Bell AS, Boehm M, Everett JR, Gonzales R, Hepworth D, Klug-McLeod JL, Lanfear J, Loesel J, Mathias J, Wood TP (2012). Shaping a screening file for maximal lead discovery efficiency and effectiveness: Elimination of molecular redundancy. *J. Chem. Inf. Model.* **52**:2937–2949.

[12] Spencer RW (1998). High-throughput screening of historic collections: Observations on file size, biological targets, and file diversity. *Biotechnol. Bioeng.* **61**(1):61–67.

[13] Newman DJ, Cragg GM (2012). Natural products as sources of new drugs over the 30 years from 1981 to 2010. *J. Nat. Prod.* **75**:311–335.

[14] Lipinski C (2001). Experimental and computational approaches to estimate solubility and permeability in drug discovery and development settings. *Adv. Drug Deliv. Rev.* **46**(1–3):3–26.

[15] Oprea TI, Davis AM, Teague SJ, Leeson PD (2001). Is there a difference between leads and drugs? A historical perspective. *J. Chem. Inf. Model.* **41**(5):1308–1315.

[16] Bleicher KH, Boehm H-J, Mueller K, Alanine AI (2003). Hit and lead generation: Beyond high-throughput screening. *Nat. Rev.* **2**:369–378.

[17] Spring DR (2003). Diversity-oriented synthesis: A challenge for synthetic chemists. *Org. Biomol. Chem.* **1**:3867–3870.

[18] Burke MD, Schreiber SL (2004). A planning strategy for diversity-oriented synthesis. *Angew. Chem. Int. Ed.* **43**:46–58.

[19] Lopez-Vallejo F, Giulianotti MA, Houghten RA, Medina-Franco JL (2012). Expanding the medicinally relevant chemical space with compound libraries. *Drug. Discov. Today* **17**(13/14):718–726.

[20] Kell DB (2013). Finding novel pharmaceuticals in the systems biology era using multiple effective drug targets, phenotypic screening and knowledge of transporters: Where drug discovery went wrong and how to fix it. *FEBS J.* **280**(23):5957–5980.

[21] Cordier C, Morton D, Murrison S, Nelson A, O'Leary-Steele C (2008). Natural products as an inspiration in the diversity-oriented synthesis of bioactive compound libraries. *Nat. Prod. Rep.* **25**:719–737.

[22] Kaiser M, Wetzel S, Kumar K, Waldmann H (2008). Biology-inspired synthesis of compound libraries. *Cell. Mol. Life Sci.* **65**:1186–1201.

[23] Over B, Wetzel S, Grütter C, Nakai Y, Renner S, Rauh D, Waldmann H (2013). Natural product-derived fragments for fragment-based ligand discovery. *Nat. Chem.* **5**:21–28.

[24] Trunzer M, Faller B, Zimmerlin A (2009). Metabolic soft spot identification and compound optimization in early discovery phases. *J. Med. Chem.* **52**:329–335.

[25] Kumar S, Sharma R, Roychowdhury A (2012). Modulation of cytochrome-P450 inhibition (CYP) in drug discovery: A medicinal chemistry perspective. *Curr. Med. Chem.* **19**(21): 3605–3621.

[26] Lipinski CA, Lombardo F, Dominy BW, Feeney PJ (1997). Experimental and computational approaches to estimate solubility and permeability in drug discovery and development settings. *Adv. Drug Deliv. Rev.* **23**:3–25.

[27] Wenlock MC, Austin RP, Barton P, Davies AM, Leeson PD (2003). A comparison of physicochemical property profiles of development and marketed oral drugs. *J. Med. Chem.* **46**:1250–1256.

[28] Bowes J, Brown AJ, Hamon J, Jarolimek W, Sridhar A, Waldron G, Whitebread S (2012). Reducing safety-related drug attrition: The use of in vitro pharmacological profiling. *Nat. Rev.* **11**:909–922.

[29] Sanguinetti MC, Jiang C, Curran ME, Keating MT (1995). A mechanistic link between an inherited and an acquired cardiac arrhythmia: HERG encodes the IKr potassium channel. *Cell* **81**:299–307.

[30] Redfern WS, Carlsson L, Davis AS, Lynch WG, MacKenzie I, Palethorpe S, Siegl PK, Strang I, Sullivan AT, Wallis R, Camm AJ, Hammond TG (2003). Relationships between preclinical cardiac electrophysiology, clinical QT interval prolongation and torsade de pointes for a broad range of drugs: Evidence for a provisional safety margin in drug development. *Cardiovasc. Res.* **58**:32–45.

[31] Baker M (2013). Fragment-based lead discovery grows up. *Nat. Rev. Drug Discov.* **12**:5–7.

[32] Jhoti H, Rees DC, Murray CW (2013). The "rule of three" for fragment-based drug discovery: Where are we now? *Nat. Rev. Drug Discov.* **12**:644.

[33] Wu B, Zhang Z, Noberini R, Barile E, Giulianotti M, Pinilla C, Houghten RA, Pasquale EB, Pellecchia M (2013). HTS by NMR of combinatorial libraries: A fragment-based approach to ligand discovery. *Chem. Biol.* **20**:19–33.

[34] Robinson JA, DeMarco S, Gombert F, Moehle K, Obrecht D (2008). The design, structures and therapeutic potential of protein epitope mimetics. *Drug Discov. Today* **13**:944–951.

[35] Serafimova IM, Pufall MA, Krishnan S, Duda K, Cohen MS, Maglathlin RL, McFarland JM, Miller RM, Frödin M, Taunton J (2013). Reversible targeting of noncatalytic cysteines with chemically tuned electrophiles. *Nat. Chem. Biol.* **8**(5):471–476.

[36] O'Shea R, Moser HE (2008). Physicochemical properties of antibacterial compounds: Implications for drug discovery. *J. Med. Chem.* **5**:2871–2878.

[37] Webb JL, Cullifer RE, Lee D (2011). Market trends of injectable drug delivery systems. *J. Med. Mark.* **11**(3):237–243.

[38] Herrero EP, Alonso MJ, Csaba N (2012). Polymer-based oral peptide nanomedicines. *Ther. Deliv.* **3**(5):657–668.

[39] Obrecht D, Chevalier E, Moehle K, Robinson JA (2012). ß-Hairpin protein epitope mimetic technology in drug discovery. *Drug Discov. Today Technol.* **9**(1):e63–e69.

[40] Coates WJ, Hunter DJ, MacLachlan WS (2000). Successful implementation of automation in medicinal chemistry. *Drug Discov. Today* **5**(11):521–527.

[41] Nettekoven M, Thomas AW (2002). Accelerating drug discovery by integrative implementation of laboratory automation in the work flow. *Curr. Med. Chem.* **9**:2179–2190.

[42] Weber A, von Roedern E, Stilz HU (2005). SynCar: An approach to automated synthesis. *J. Comb. Chem.* **7**:178–184.

[43] Koppitz M (2008). Maximizing efficiency in the production of compound libraries. *J. Comb. Chem.* **10**:573–579.

[44] Weller HN, Nirschl DS, Petrillo EW, Poss MA, Andres CJ, Cavallaro CL, Echols MM, Grant-Young KA, Houston JG, Miller AV, Swann RT (2006). Application of lean manufacturing concepts in drug discovery: Rapid analogue library synthesis. *J. Comb. Chem.* **8**:664–669.

[45] Djuric SW, Akritopoulou-Zanze I, Cox PB, Galasinski S (2010). Compound collection enhancement and paradigms for high-throughput screening—an update. *Annu. Rep. Med. Chem.* **45**:409–428.

[46] Feuston B, Chakravorty SJ, Conway JF, Culberson JC, Forbes J, Kraker B, Lennon PA, Lindsley C, McGaughey GB, Mosley R, Sheridan RP, Valenciano M, Kearsley SK (2005). Web enabling technology for the design, enumeration, optimization and tracking of compound libraries. *Curr. Top. Med. Chem.* **5**:773–783.

[47] Godfrey AG, Masquelin T, Hemmerle H (2013). A remote-controlled adaptive medchem lab: An innovative approach to enable drug discovery in the 21st century. *Drug Discov. Today* **18**:795–802.

[48] Kuroda N, Hird N, Cork DG (2006). Further development of a robust workup process for solution-phase high-throughput library synthesis to address environmental and sample tracking issues. *J. Comb. Chem.* **8**:505–512.

[49] Schaffrath M, von Roedern E, Hamley P, Stilz HU (2005). High-throughput purification of single compounds and libraries. *J. Comb. Chem.* **7**:546–553.

[50] Guth O, Krewer D, Freudenberg B, Paulitz C, Hauser M, Ilg K (2008). Automated modular preparative HPLC-MS purification laboratory with enhanced efficiency. *J. Comb. Chem.* **10**:875–882.

[51] Liu M, Chen K, Christian D, Fatima T, Pissarnitski N, Streckfuss E, Zhang C, Xia L, Borges S, Shi Z, Vachal P, Tata J, Athanasopoulos J (2012). High-throughput purification platform in support of drug discovery. *ACS Comb. Sci.* **14**:51–59.

[52] Weller HN, Nirschl DS, Paulson JL, Hoffman SL, Bullock WH (2012). Addressing the medicinal chemistry bottleneck: A lean approach to centralized purification. *ACS Comb. Sci.* **14**:520–526.

[53] Hochlowski JE, Searle PA, Tu NP, Pan JY, Spanton SG, Djuric SW (2011). An integrated synthesis-purification system to accelerate the generation of compounds in pharmaceutical discovery. *J. Flow Chem.* **2**:56–61.

[54] Sadler S, Moeller AR, Jones GB (2012). Microwave and continuous flow technologies in drug discovery. *Expert Opin. Drug Discov.* **7**(12):1107–1128.

[55] Bryan MC, Hein CD, Gao H, Xia X, Eastwood H, Bruenner BA, Louie SW, Doherty EM (2013). Disubstituted 1-aryl-4-aminopiperidine library synthesis using computational drug design and high-throughput batch and flow technologies. *ACS Comb. Sci.* **15**:503–511.

[56] Farrell WP, Aurigemma CM, Masters-Moore DF (2009). Advances in high throughput super-critical fluid chromatography. *J. Liq. Chromatogr. Relat. Technol.* **32**:1689–1710.

[57] Werner M, Kuratli C, Martin RE, Hochstrasser R, Wechsler D, Enderle T, Alanine AI, Vogel H (2014). Seamless integration of dose-response screening and flow chemistry: Efficient generation of structure–activity relationship data of beta-secretase (BACE1) inhibitors. *Angew. Chem. Int. Ed.* **53**:1704–1708.

[58] De la Hoz A., Loupy A, (eds.), *Microwaves in Organic Synthesis*. 3rd ed., Wiley-VCH, Weinheim, 2012.

[59] Larhed M, Olofsson K, (eds.), *Microwave Methods in Organic Synthesis, Topics in Current Chemistry*, Vol. **266**, Springer, Berlin, 2006.

[60] Kappe CO, Stadler A, Dallinger D, *Microwaves in Organic and Medicinal Chemistry*. 2rd ed., Wiley-VCH, Weinheim, 2012.

[61] Jacob J (2012). Microwave assisted reactions in organic chemistry: A review of recent advances. *Int. J. Chem.* **4**(6):29–43.

[62] Dallinger D, Kappe CO (2009). Controlled microwave heating in modern organic synthesis. Highlights from the 2004–2008 literature. *Mol. Divers.* **13**:71–193.

[63] Kremsner JM, Stadler A, Kappe CO (2007). High-throughput microwave-assisted organic synthesis: Moving from automated sequential to parallel library-generation formats in silicon carbide microtiter plates. *J. Comb. Chem.* **9**:285–291.

[64] Kappe CO, Damm M (2012). Parallel microwave chemistry in silicon carbide microtiter platforms: A review. *Mol. Divers.* **16**:5–25.

[65] Kappe CO, Pieber B, Dallinger D (2013). Microwave effects in organic synthesis: Myth or reality? *Angew. Chem. Int. Ed.* **52**:1088–1094.

[66] Anton Paar (2015). Microwave Reaction System: Multiwave PRO. Retrieved May 21, 2015, http://www.anton-paar.com/corp-en/products/details/microwave-reaction-system-multiwave-pro/microwave-synthesis/ (accessed May 28, 2015).

[67] Benali O, Davies G, Deal M, Farrant E, Guthrie D, Holden J, Wheeler R (2008). Fully automated open access platform for rapid, combined serial evaporation and sample reformatting. *J. Comb. Chem.* **10**:52–55.

[68] Klesper E, Corwin AH, Turner DA (1962). High pressure gas chromatography above critical temperatures. *J. Org. Chem.* **27**:700–701.

[69] Guiochon G, Tarafder A (2011). Fundamental challenges and opportunities for preparative supercritical fluid chromatography. *J. Chromatogr. A* **1218**:1037–1114.

[70] Rajendran A (2012). Design of preparative-supercritical fluid chromatography. *J. Chromatogr. A* **1250**:227–249.

[71] Smith RM (1999). Supercritical fluids in separation science—the dreams, the reality and the future. *J. Chromatogr. A* **856**:83–115.

[72] Tarafder A, Guiochon G (2011). Use of isopycnic plots in designing operations of supercritical fluid chromatography: II. The isopycnic plots and the selection of the operating pressure-temperature zone in supercritical fluid chromatography. *J. Chromatogr. A* **1218**:4576–4585.

[73] Ripka WC, Barker G, Krakover J (2001). High-throughput purification of compound libraries. *Drug Discov. Today* **6**:471–477.

[74] Hochlowski J, Olson J, Pan J, Sauer D, Searle P, Sowin T (2003). Purification of HTOS libraries by supercritical fluid chromatography. *J. Liq. Chromatogr. Relat. Technol.* **26**(3):333–354.

[75] Zhang X, Towle MH, Felice CE, Flamant JH, Goetzinger WK (2006). Development of a mass-directed preparative supercritical fluid chromatography purification system. *J. Comb. Chem.* **8**:705–714.

[76] Bozic A, inventor; Sanofi-Aventis Deutschland GMBH, applicant. Gas-liquid separator. WO patent 012962A1, January 23, 2014.

[77] Perrut M (1994). Advances in supercritical fluid chromatographic processes. *J. Chromatogr. A* **658**:293–313.

[78] Mason BP, Price KE, Steinbacher JL, Bogdan AR, McQuade DT (2007). Greener approaches to organic synthesis using microreactor technology. *Chem. Rev.* **107**:2300–2318.

[79] Wegner J, Ceylan S, Kirschning A (2011). Ten key issues in modern flow chemistry. *Chem. Commun.* **47**:4583–4592.

[80] Baumann M, Baxendale IR, Ley SV (2011). The flow synthesis of heterocycles for natural product and medicinal chemistry applications. *Mol. Divers.* **15**:613–630.

[81] McQuade DT, Seeberger PH (2013). Applying flow chemistry: Methods, materials, and multistep synthesis. *J. Org. Chem.* **78**:6384–6389.

[82] Carter CF, Baxendale IR, O'Brien M, Pavey JBJ, Ley SV (2009). Synthesis of acetal protected building blocks using flow chemistry with flow I.R. analysis: Preparation of butane-2,3-diacetal tartrates. *Org. Biomol. Chem.* **7**:4594–4597.

[83] Lei M, Hu R, Wang Y, Zhang H (2014). Synthesis of 1,4-disubstituted 1,2,3-triazoles via click reaction in micro flow reactor. *Heterocycles* **88**(2):1511–1517.

[84] Webb D, Jamison TF (2010). Continuous flow multi-step organic synthesis. *Chem. Sci.* **1**:675–680.

[85] Hopkin MD, Baxendale IR, Ley SV (2013). An expeditious synthesis of imatinib and analogues utilising flow chemistry methods. *Org. Biomol. Chem.* **11**:1822–1839.

[86] Wu J, Kozak JA, Simeon F, Hatton TA, Jamison TF (2014). Mechanism-guided design of flow systems for multicomponent reactions: Conversion of $CO_2$ and olefins to cyclic carbonates. *Chem. Sci.* **5**:1227–1231.

[87] Griffiths-Jones CM, Hopkin MD, Jönsson D, Ley SV, Tapolczay DJ, Vickerstaffe E, Ladlow M (2007). Fully automated flow-through synthesis of secondary sulfonamides in a binary reactor system. *J. Comb. Chem.* **9**:422–430.

[88] Watts P, Haswell SJ (2005). The application of micro reactors for organic synthesis. *Chem. Soc. Rev.* **34**:235–246.

[89] Malet-Sanz L, Susanne F (2012). Continuous flow synthesis. A pharma perspective. *J. Med. Chem.* **55**:4062–4098.

[90] Pastre JC, Browne DL, Ley SV (2013). Flow chemistry syntheses of natural products. *Chem. Soc. Rev.* **42**:8849–8869.

[91] Tsubogo T, Ishiwata T, Kobayashi S (2013). Asymmetric carbon–carbon bond formation under continuous-flow conditions with chiral heterogeneous catalysts. *Angew. Chem. Int. Ed.* **52**(26):6590–6604.

[92] Elvira KS, Casadevall i Solvas X, Wootton RCR, deMello AJ (2013). The past, present and potential for microfluidic reactor technology in chemical synthesis. *Nat. Chem.* **5**:905–915.

[93] Lange PP, James K (2012). Rapid access to compound libraries through flow technology: Fully automated synthesis of a 3-aminoindolizine library via orthogonal diversification. *ACS Comb. Sci.* **14**:570–578.

[94] Baxendale IR, Griffiths-Jones CM, Ley SV, Tranmer GK (2006). Preparation of the neolignan natural product grossamide by a continuous-flow process. Synlett (3):427–430.

[95] Wiles C, Watts P (2007). Parallel synthesis in an EOF-based micro reactor. Chem. Commun. (46):4928–4930.

[96] Ullah F, Zang Q, Javed S, Zhou A, Knudtson CA, Bi D, Hanson PR, Organ MG (2012). Multicapillary flow reactor: Synthesis of 1,2,5-thiadiazepane 1,1-dioxide library utilizing one-pot elimination and inter-/intramolecular double aza-Michael addition via microwave-assisted, continuous-flow organic synthesis (MACOS). *J. Flow Chem.* **2**(4):118–123.

[97] Jebrail MJ, Ng AHC, Rai V, Hili R, Yudin AK, Wheeler AR (2010). Synchronized synthesis of peptide-based macrocycles by digital microfluidics. *Angew. Chem. Int. Ed.* **49**:8625–8629.

[98] Martin RE, Lenz M, Alzieu T, Aebi JD, Forzy L (2013). Flow synthesis of annulated 5-aryl-substituted pyridines by tandem intramolecular inverse-electron-demand hetero-/retro-Diels-Alder reaction. *Tetrahedron Lett.* **54**:6703–6707.

[99] ThalesNano (2015). H-Cube Series. Retrieved May 21, 2015, http://thalesnano.com/h-cube-series (accessed May 28, 2015).

[100] Noel T, Hessel V (2013). Membrane microreactors: Gas–liquid reactions made easy. *ChemSusChem* **6**:405–407.

[101] O'Brian M, Baxendale IR, Ley SV (2010). Flow ozonolysis using a semipermeable Teflon AF-2400 membrane to effect gas-liquid contact. *Org. Lett.* **12**(7):1596–1598.

[102] Pastre JC, Browne DL, O'Brian M, Ley SV (2013). Scaling up of continuous flow processes with gases using a tube-in-tube reactor: Inline titrations and fanetizole synthesis with ammonia. *Org. Process Res. Dev.* **17**:1183–1191.

[103] O'Brian M, Taylor N, Polyzos A, Baxendale IR, Ley SV (2011). Hydrogenation in flow: Homogeneous and heterogeneous catalysis using Teflon AF-2400 to effect gas–liquid contact at elevated pressure. *Chem. Sci.* **2**:1250–1257.

[104] Petersen T, Polyzos A, O'Brien M, Ulven T, Baxendale IR, Ley SV (2012). The oxygen-mediated synthesis of 1,3-butadiynes in continuous flow: Using Teflon AF-2400 to effect gas/liquid contact. *ChemSusChem* **5**:274–277.

[105] Polyzos A, O'Brian M, Petersen TP, Baxendale IR, Ley SV (2011). The continuous-flow synthesis of carboxylic acids using $CO_2$ in a tube-in-tube gas permeable membrane reactor. *Angew. Chem. Int. Ed.* **50**:1190–1193.

[106] Koos P, Gross U, Polyzos A, O'Brian M, Baxendale IR, Ley SV (2011). Teflon AF-2400 mediated gas-liquid contact in continuous flow methoxycarbonylations and in-line FITR measurement of CO concentration. *Org. Biomol. Chem.* **9**:6903–6908.

[107] Cranwell PB, O'Brian M, Browne DL, Koos P, Polyzos A, Pena-López M, Ley SV (2012). Flow synthesis using gaseous ammonia in a Teflon AF-2400 tube-in-tube reactor: Paal-Knorr pyrrole formation and gas concentration measurement by inline flow titration. *Org. Biomol. Chem.* **10**:5774–5779.

[108] Mastronardi F, Gutmann B, Kappe CO (2013). Continuous flow generation and reactions of anhydrous diazomethane using a Teflon AF-2400 tube-in-tube reactor. *Org. Lett.* **15**(21):5590–5593.

[109] Pinho VD, Gutmann B, Miranda LSM, de Souza ROMA, Kappe CO (2014). Continuous flow synthesis of alpha-halo ketones: Essential building blocks of antiretroviral agents. *J. Org. Chem.* **79**(4):1555–1562.

# 13

# ADME PROFILING: AN INTRODUCTION FOR THE MEDICINAL CHEMIST

KATHARINA MERTSCH[1], MARTIN WILL[1], WERNGARD CZECHTIZKY[2], NIELS GRIESANG[1], ALEXANDER MARKER[1], and JACOB OLSEN[1]

[1] *Sanofi R&D, Frankfurt am Main, Germany*
[2] *Sanofi-Aventis Deutschland GmbH, Frankfurt am Main, Germany*

## 13.1 INTRODUCTION

Discovery and development of a new molecular entity (NME) cost more than 1 billion dollars; it currently takes 10–15 years to bring a drug to the market [5, 6]. Therefore, lowering of costs by reduction of attrition is one of the most important goals in pharmaceutical industry today [5, 6].

About 10–15 years ago, it was assumed that about 40% of drug candidates fail due to inappropriate pharmacokinetic (PK) parameters, while lack of clinical efficacy contributed to 29% of terminations of drug development in Britain [7]. In 2000, the attrition rate due to PK was reduced to about 10%, while failure due to lack of efficacy stayed constant at 30%, and failure caused by lack of drug safety increased to 30% [3]. One of the main reasons for the reduction of project failures due to PK issues is improved optimization strategies allowing for a multidimensional optimization of ADME properties including an improvement in the quality of *in vitro* ADME assays. ADME is an acronym for absorption, distribution, metabolism, and excretion of a compound; these four criteria influence drug levels and kinetics and hence the pharmacological effect of a compound *in vivo* [8]. As a consequence, a significantly better correlation of *in vitro–in vivo* parameters and an improved extrapolation from nonhuman to human PK parameters can be achieved [9, 10].

In contrast to former times, the optimization of physicochemical and ADME parameters is nowadays addressed very early in a project [11, 12]. It is generally accepted that

*Small Molecule Medicinal Chemistry: Strategies and Technologies*, First Edition.
Edited by Werngard Czechtizky and Peter Hamley.

multiparameter optimization as early as possible prevents time-consuming compound reoptimization in later project phases where structural changes have to be performed in an often already restricted chemical space [5]. Preferably, the optimization of physicochemical and ADME parameters starts right after screening with hit selection (a hit being a compound or compound series with reproducible biological activity in the screening assay and an overall physicochemical and ADME profile interesting enough to invest chemistry resources for optimization). Compound optimization of hits to clinical candidates depends on the required profile of the clinical candidate, the available resources, and the compound profiling process in place. In an iterative process, medicinal chemists synthesize new molecules in a series of synthesis cycles, in which the design of the compounds in a new synthesis cycle is guided by the compound profiles of the molecules prepared in the previous synthesis cycles. Parallel synthesis is frequently applied to generate focused compound libraries, typically consisting of 20–200 compounds with a compound quality comparable to compounds from classical medicinal chemistry, which are prepared in parallel to address specific questions on chemical space and are not accessible by automated derivatization.

A significant share of these compounds is, after being tested for biological activity, subsequently further profiled for physicochemical and ADME properties. The quality of profiling data and the knowledge of the structure–property relationships within a chemical series highly influence the number of synthesis cycles and compounds necessary to reach the compound quality of a clinical candidate.

## 13.2 COMPOUND PROFILING IN H2L OPTIMIZATION

The optimization of a hit series toward a clinical candidate is usually done in two phases. The goal of the first phase (the so-called hit-to-lead phase, H2L) is to identify strengths and weaknesses of a hit series identified after screening, to increase its biological potency, and to generate a structure–property relationship for critical optimization parameters in the most effective way. Parameters considered in this early phase (Fig. 13.1) are typically *in vitro* efficacy, selectivity versus antitargets [13] such as hERG [13], physicochemical properties such as thermodynamic solubility, log D, chemical stability, and ADME properties such as intestinal absorption [14–17], metabolic lability in microsomes and/or hepatocytes [15, 17, 18], and inhibition of cytochrome P450 enzymes such as CYP3A4 (see the following) [17–19].

Additionally, stability in plasma or gastrointestinal fluids [20] and binding to plasma proteins (in the so-called protein binding studies) [21] can also be included in H2L but can alternatively also be shifted to the lead optimization phase.

### 13.2.1 Intestinal Absorption

The absorption of drugs via the oral route is affected by drug properties and the physiology of the gastrointestinal tract [14–17]. The so-called rule of five [11] predicts that oral bioavailability will be poor for a drug if there are more than 5 H-bond donors and 10 H-bond acceptors, the log *P* value is greater than 5, and the molecular weight exceeds 500 [11]. In addition, oral absorption is influenced by solubility, drug dissolution from the dosage form, permeation across intestinal membranes, and irreversible drug removal by

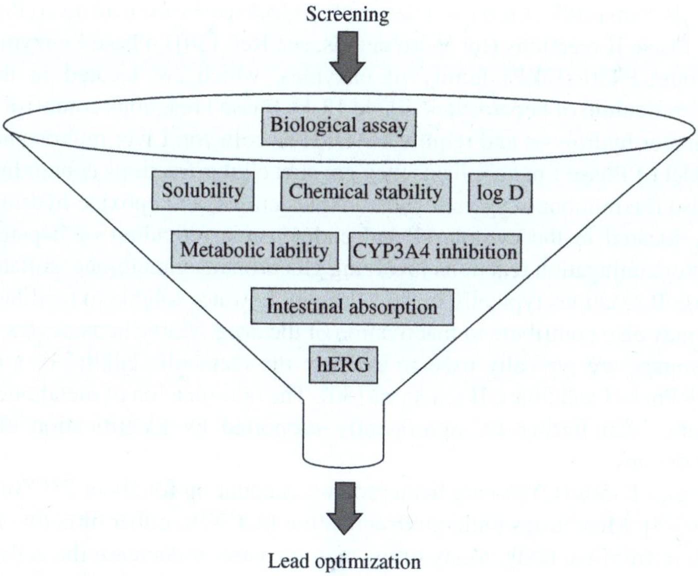


**FIGURE 13.1** Typical profiling tree for a small molecule hit-to-lead (H2L) optimization containing biological, physicochemical, and ADME profiling assays and an antitarget assay versus hERG blockade.

intestine and liver. Drug permeation across the intestinal membrane is described by the effective permeability coefficient, $P_{eff}$, which can be assessed from luminal perfusion studies by measurement of drug loss across the lumen of a given segment, where $C_{out,lumen}$ and $C_{in,lumen}$ are the concentrations of drug leaving and entering the lumen, respectively; $r$ is the radius; $l$ is the length; and $Q_{lumen}$ is the experimentally determined luminal flow rate [22, 23]:

$$P_{eff} = \frac{Q_{lumen}}{2\pi rl} \ln\left(\frac{C_{out,lumen}}{C_{in,lumen}}\right)$$

A typical *in vitro* test system uses the Caco-2 cell line, derived from human colon carcinoma cells [24]. In the permeability and transporter area, Caco-2 studies describe permeability, including concentration and pH effects, and efflux, as well as some transport phenomena related to efflux and/or uptake [22–26].

### 13.2.2 Drug Metabolism and Inhibition of CYP450 Enzymes

Drug metabolism is the biochemical modification of drugs by organisms, usually through specialized enzymatic systems converting lipophilic compounds into more readily excreted hydrophilic products [27, 28]. The rate of metabolism determines duration and quality of a drug's pharmacological action. Most drug metabolism occurs in the liver, but some processes also occur in the gut wall, lungs, and blood plasma; the compounds are usually eliminated by biotransformation and/or excretion into urine or bile [29].

Drug metabolism results from two classes of enzyme reactions in the liver, the so-called Phase I and Phase II reactions (for more details, see Ref. [30]). Phase I enzymes include the cytochrome P450 (CYP) family of enzymes, which are located in the smooth endoplasmic reticulum of hepatocytes (Table 13.1). Phase I reactions comprise oxidation, reduction, and/or hydrolysis and require NADPH as cofactor. Liver microsomes provide a useful model of Phase I metabolism; they are subcellular fractions containing not only CYPs but also flavin monooxygenases, carboxylesterases, and epoxide hydrolase. Phase II enzymes, located in the cytoplasm and endoplasmic reticulum of hepatocytes, are responsible for conjugation reactions involving glucuronate, glutathione, sulfate, and glutamine. Phase II reactions typically make a drug more water soluble to facilitate its elimination and may also contribute to inactivation of the drug. Viable hepatocytes, instead of liver microsomes, are typically used to evaluate the metabolic lability of a compound through both Phase I and Phase II reactions [30]. The optimization of metabolically labile chemical series can further be significantly supported by identification of typically formed metabolites.

Cytochromes P450s (CYPs) are hemeproteins accounting for about 75% of total drug metabolism [33]. Most drugs undergo deactivation by CYPs, either directly or by facilitated excretion from the body. Many drugs may increase or decrease the activity of various CYP isozymes either by inducing the biosynthesis of an isozyme (CYP induction) or by directly inhibiting the activity of the CYP (CYP inhibition). This is a major source of adverse drug interactions, since changes in CYP enzyme activity may affect the metabolism and clearance of drugs.

Table 13.1 shows an overview of major CYP isozymes of the CYP450 family.

In spite of many available protocols [29], CYP inhibition is usually tested *in vitro* by incubating various concentrations of a compound with liver microsomes, NADPH, and a substrate known to be metabolized to a certain probe metabolite in a CYP isoform-specific manner [31, 32, 34–36]. When formation of the probe metabolite is analyzed, the level of inhibition toward the CYP enzyme producing the probe metabolite can be measured in a concentration-dependent manner, enabling calculation of the IC50 (and Ki) value for inhibition [31, 32, 34–36]. While human liver microsomes have become the standard test system for *in vitro* CYP inhibition studies (FDA/EMA guidance), hepatocytes (cryopreserved or freshly prepared), a more liver-like *in vitro* test system, can also be relevant, for example, when the compound is significantly metabolized by Phase II enzymes (e.g., gemfibrozil; [37]). Some literature reports that hepatocyte CYP inhibition data are a better predictor of clinical outcome than data obtained with microsomes [38], while other authors report that data obtained in microsomes and hepatocytes are comparable [39].

Due to the fact that CYP3A4 is the major contributor to drug metabolism, some pharmaceutical companies only profile for CYP3A4 inhibition in liver microsomes in H2L, while others may already add additional isozymes such as CYP2D6 or CYP2C9 or use hepatocytes and liver microsomes in parallel [5, 19, 31, 32, 34–36].

### 13.2.3 Protein Binding

Drug protein binding is the reversible interaction of drugs with proteins in plasma such as albumin, alpha1-acid glycoprotein, or lipoproteins [21, 40]. However, drugs can also bind reversibly to red blood cells, tissue membranes, or other blood or tissue constituents. The

**TABLE 13.1 CYP450 Enzymes**

| CYP450 Enzyme | Selective Substrate (Concentration) | Positive Control (Concurrent Inhibition) | Positive Control (Time-Dependent Inhibition) | Marker Metabolite |
|---|---|---|---|---|
| CYP1A2 | Phenacetin | α-Naphthoflavone | Furafylline | 4-Acetamidophenol |
| CYP2B6 | Bupropion | Thio-TEPA[a] | Bergamottin | Hydroxybupropion |
| CYP2C8 | Amodiaquine | Quercetin | Gemfibrozil O-glucuronide | Desethylamodiaquine |
| CYP2C9 | Diclofenac | Sulfaphenazole | Tienilic acid | 4′-Hydroxydiclofenac |
| CYP2C19 | Mephenytoin | Tranylcypromine | Ticlopidine | 4′-Hydroxymephenytoin |
| CYP2D6 | Dextromethorphan | Quinidine | Paroxetine | Dextrorphan |
| CYP3A4/5 | Midazolam | Ketoconazole | Troleandomycin | 1′-Hydroxymidazolam |
| CYP3A4/5 | Testosterone | Ketoconazole | Troleandomycin | 6ß-Hydroxytestosterone |

Substrates and positive controls in protocols according to FDA/EMA requirements [31, 32].

[a] Thiotriethylenephosphoramide.

extent to which a drug is bound in plasma or blood is usually expressed as fraction unbound (fu), which is determined by the affinity of the drug to the protein, the concentration of the binding protein, and the relative concentrations of the drug versus protein. For most situations, the fraction unbound is a determinant of total steady-state drug concentration, but not of the pharmacologically active unbound drug concentration at steady state [41].

### 13.2.4 En Route to a Lead Series: *In Vivo* PK Studies

The goal of H2L optimization is a so-called lead series. Lead series usually exhibit an overall biological, physicochemical, and *in vitro* ADME profile sufficient to submit several representatives to *in vivo* pharmacological studies in one or several animal species. Prerequisite to approach an *in vivo* pharmacological efficacy study for a particular therapeutic readout in animals is usually the determination of a basic *in vivo* PK profile in the respective species.

Knowledge of the basic *in vivo* PK parameters of a drug is crucial for understanding how it is handled by the body and how it needs to be dosed to enable efficacy in a therapeutic setting. In order to plan the experiment, the following conditions need to be defined: route of administration (intravenous, oral, or others), dosing regimen, tissue distribution, sampling times, analytical methods to be used, and finally the animal species [42]. Blood (red blood cells, plasma, and serum), urine, and feces are the most commonly collected samples. In addition, a few representative organ and tissue samples are usually taken, such as the liver, kidney, brain, and other suspected target organs. Sampling times should depend on the substance being tested and the route of administration. In general, an equal number of blood samples should be taken in each phase of the concentration-versus-time curve. Intravenous (iv) studies usually require much shorter, and more frequent, sampling than is required for oral dosing. Time spacing of samples will depend on PK rate constants characterizing the change of drug concentration in specific reference regions [43]. PK rate constants characterize the speed at which a drug enters a compartment (absorption rate constant), distributes between central and peripheral compartments (distribution constants), and is eliminated (elimination constants) (for more details, see [43]). In a typical iv study, blood and tissue samples are taken in a "powers of 2" series, that is, samples at 2, 4, 8, 16, and 30 (32) min, 1, 2, 4, 8, and 16 h. Similar coverage can be obtained with only seven time points by using a "powers of 3" series: 3, 9, and 30 (27) min, 1, 3, 9, and 24 (27) h. Oral dosing studies extend to significantly longer periods. The sampling schedule for an oral dosing experiment might be 15 min, 30 min, 1, 2, 4, 8, 24, 48, and 72 h [44]. The PK parameters of compounds are usually defined by peak concentrations (Cmax), peak times (Tmax), terminal half-lives (t1/2), and areas under the curve (AUC). Tmax and Cmax may be directly obtained from the experimental observation of each subject. After oral administration, Cmax and Tmax are dependent on the extent and rate of drug absorption and its disposition profile [42]. The terminal half-life is a parameter used to describe the decay of the drug concentration in the terminal phase (when the semilog plot of the observed concentrations vs. time becomes linear). The AUC, very generally spoken, is a parameter that can either be used as an index of the drug exposure of the body (when referred to plasma drug levels) or as an index of the drug exposure of particular tissues (if referred to the drug levels in these tissues). It depends on the drug amount entering into systemic circulation and on the ability of the system to eliminate the drug (clearance). If the AUC is known from iv application and upon extravascular administration (e.g., oral), one can calculate the fraction absorbed (absolute bioavailability) of a drug [43].

AUC is calculated from time 0 (time of administration) to infinity after single drug administration. The AUC can further be used as a parameter delivering information on the drug elimination process. For more information on clearance, volume of distribution, and compartmental models, please consult respective review articles [42].

## 13.3 COMPOUND PROFILING IN LEAD OPTIMIZATION

In the second phase from hit to clinical candidate, the so-called lead optimization phase, the structure–activity relationship (SAR) of a lead series needs to be fully elaborated to select a so-called precandidate. This compound is subjected to full *in vivo* pharmacological profiling for the target indication and a tailored toxicological safety assessment to finally qualify as a clinical candidate.

In lead optimization, all previously mentioned physicochemical properties such as solubility, log D, chemical stability, ADME, and *in vivo* PK properties remain critical parameters and need to be continuously monitored. However, additional requirements arise to finally qualify a compound as a clinical candidate. The precandidate is therefore further profiled for inhibition of an extended panel of CYP enzymes, for mechanism-based inhibition (MBI) of CYP enzymes, and inhibition of transport proteins (see the following).

### 13.3.1 Extended CYP Inhibition Studies

Despite intensive early testing for CYP inhibition and drug–drug interaction (DDI) potential, there are thousands of adverse drug reactions every year. Withdrawals of terfenadine (Seldane®), mibefradil (Posicor®), astemizole (Hismanal®), cisapride (Propulsid®), and cerivastatin (Baycol®) resulted in extended inhibition testing of additional CYP enzymes such as CYP2D6, CYP2C9, CYP2C19, CYP1A2, and others ([19, 31, 32, 34–36]; see Table 13.1). Due to scientific consideration and improved LC–MS capacities, there is a trend to focus on clinically relevant substrates for reversible inhibition assays and time-dependent inhibition (TDI) [31, 34–36], while high-throughput assays using fluorogenic substrates and fluorescent plate readers were preferred in the past. Testing strategies in major companies follow FDA and EMA guidance's for industry [31, 32], not only for preclinical phase or later but also in discovery phases.

### 13.3.2 Mechanism-Based CYP Inhibition

A potential precandidate needs to be tested for the so-called MBI of CYP enzymes (profiled by IC50 shift assays and/or automated $K_i/K_{inact}$ determination) and for human hepatocyte clearance *resp.* hepatocyte induction [31, 32]. Studies for human hepatocyte clearance, human hepatocyte induction, and MBI evaluation of $K_i/K_{inact}$ have advanced during the last 10 years with advancing cell culture technologies. However, these models are being further developed using 3D cell cultures and combinations of hepatocyte–endothelial [45] or hepatocyte cocultures with Kupffer cells or stellate cells [46, 47]. TDI is mainly assessed *in vitro* using microsomes or hepatocytes and has been incorporated increasingly in the drug discovery process [35, 39]. While screening against MBI is performed with less sophisticated assays [35, 48–50], the determination of inhibition parameters (i.e., $k_{inact}$ and $K_I$) needed for predictions/risk assessment is usually only available for a handful of late-stage discovery compounds. In analogy to this, FDA/EMA guidance's and a recent PhRMA white paper [19] recommend $IC_{50}$ shift assays for initial evaluation of a compound's TDI (note: MBI accounts for the major share of TDI observed). A reasonable correlation between IC50 shift values [35, 39] and clinical data encourages many companies to switch to this assay design. However, a correlation between early discovery data and FDA compliant reports is expected, but not always easy to achieve (solubility issues, limited number of test concentrations resulting in IC50 determinations with certain variability, etc.).

### 13.3.3 Inhibition of Transport Proteins

A potential precandidate is further tested for being a substrate of physiological transport proteins [31, 32, 36, 51]. Caco-2 studies describe permeability, including concentration and pH effects, and efflux, as well as transport phenomena related to efflux and/or uptake. For an overview of transport proteins, see Table 13.2. For example, if a compound is an efflux substrate, Pgp, BCRP, and MRP2 are possible transporters involved in the efflux [36, 51]. They can be further investigated for efflux contribution in Caco-2, MDCK, and/or overexpressing cells. If a compound is inhibiting Pgp and risk assessment calculations result in a risk according to FDA recommendations, a possible clinical DDI study has to be performed later on.

For uptake transporters, it depends on the target and the envisaged patient population which transporters are characterized at which discovery stage. In diabetes and therapeutic

**TABLE 13.2 Recommended Transporters to be Tested According to Regulatory Demands [31, 32]**

| | Drug Transporter | FDA | EMA |
|---|---|---|---|
| Liver | OATP1B1 | X | X |
| | OATP1B3 | X | X |
| | OCT1 | | (X) |
| | Pgp | X | X |
| | BCRP | X | X |
| | BSEP | | (X) |
| | MATE1 | (X) | (X) |
| Kidney | OAT1 | X | X |
| | OAT3 | X | X |
| | OCT2 | | (X) |
| | Pgp | X | X |
| | MATE1 | (X) | (X) |
| | MATE2 | (X) | (X) |

**TABLE 13.3 The Biopharmaceutical Drug Disposition Classification System (BDDCS) Where Metabolism and Transporter Roles are Included [53]**

| | High Solubility | Low Solubility |
|---|---|---|
| Extensive metabolism | Class 1: transporter role minimal in gut/liver | Class 2: efflux transporters in gut might be involved as well as uptake/efflux transporter in liver |
| Poor metabolism | Class 3: absorptive transporters might be involved | Class 4: absorption and efflux transporters both could be important |

areas involving elderly patients, there is no doubt to include OATP1B1 and OCT2 transporters to exclude interactions with statins and metformin, respectively [36, 51]. Further uptake transporter studies can be performed if there are first suggestions from animal studies concerning preferred ways of drug elimination. If the liver is the predominant elimination pathway, OATP1B3, OCT1, and BSEP need investigation [31, 32, 36, 51]. If the predominant way of elimination is renal clearance, inhibition/substrate properties for OAT1, OAT3, and OCTs should be evaluated. For detailed decision trees, please refer to [31, 32, 51]. As a conclusion, there is no unique transporter strategy for early discovery. Transporters on the elimination pathway of a compound are rather very specific for the given compound, and early screening approaches are not useful except when one would like to avoid an interaction with a specific medication in a given therapeutic area.

### 13.3.4 Biopharmaceutical Classification of a Clinical Candidate (Classification of Potential Drugs into Biopharmaceutical Classification System or Biopharmaceutical Drug Disposition and Classification System)

The Biopharmaceutical Classification System (BCS) is based on permeability and solubility [52–54], and dissects drugs into four different categories: class 1 (high permeability and solubility), class 2 (low solubility, high permeability), class 3 (high solubility, low permeability), and class 4 (low solubility, low permeability). For details, see the FDA biowaiver documents [55] and Benet [54–56]. Nowadays, there is an interesting new twist to this that also considers solubility and metabolism according to Benets proposals: class 1 drugs (high solubility, extensive metabolism), class 2 drugs (low solubility, extensive metabolism), class 3 drugs (high solubility, poor metabolism, adding the proposal of renal and/or biliary elimination of unchanged drug), and class 4 drugs (low solubility, low metabolism, with additional consideration of renal and/or biliary elimination of unchanged drug) [53, 56] (Table 13.3). Benet proposes for class 1 drugs a minimal role for transporters; for class 2 drugs, there might be a role for efflux in the gut and both uptake and efflux transporters in the liver (eliminating parent and/or metabolites). For class 3 drugs, absorptive transporters might play a major role, while for class 4 drugs, both absorptive and efflux transporters could be important. The classification proposed by Benet can not only be used for drug development but also earlier in discovery, guiding characterization of a compound in preclinical and early clinical phase.

## 13.4 INTEGRATION OF MEDICINAL CHEMISTRY, BIOLOGY, PHYSICOCHEMICAL, AND ADME PROFILING: STRATEGIES TOWARD CYCLE TIME REDUCTIONS

In the modern drug discovery process, strategies for cycle time improvement during H2L and lead optimization until identification of a clinical candidate are in focus for continuous optimization. Historically, a sequential, less resource-intensive strategy for initial profiling including biological testing and physicochemical and early ADME profiling was typical.

Compounds were first filtered by *in vitro* biological potency, and only a few most potent compounds were then further followed up for physicochemical and ADME profiling.

This efficacy-biased, sequential profiling, however, bears the risk of sparse data sets for overall compound optimization. It often results in insufficient data sets for true multiparameter optimization beyond potency. Furthermore, this approach does not allow for making full use of the overall value of the whole compound set and its data. Knowing about good physicochemical and ADME parameters of less active compounds can nicely promote the optimization of highly potent, but, for example, insoluble or metabolically labile hits. It has become crucial for modern medicinal chemistry to avoid lengthy compound optimizations only focused on the generation of ever more potent but less bioavailable compounds [57–59].

The alternative to sequential compound profiling is a parallel compound profiling workflow.

Parallel profiling addresses the potency against the biological target as well as physicochemical and ADME properties. The right balance of resources between synthesis and profiling and operational excellence in the generation and interpretation of compound data are key to this approach, which clearly has several advantages. Right from the start, it enables a multidimensional compound optimization through availability of all critical optimization parameters at the time of designing the next cycle of compound synthesis. The comparison of Figures 13.2 and 13.3 further shows the reduction of overall cycle times through earlier start of profiling. At any given moment, all new compounds contribute to exploring critical parameters, creating an optimal SAR and structure–property matrix. In spite of the fact that profiling costs will become somewhat higher (see the following), a proper multiparameter optimization will lower the costs of attrition and particularly the efforts for the synthesis of ultimately useless compounds significantly.

The implementation of parallel profiling requires the setup of optimized *in vitro* assays with the right balance between sufficient data quality and acceptable resource needs and

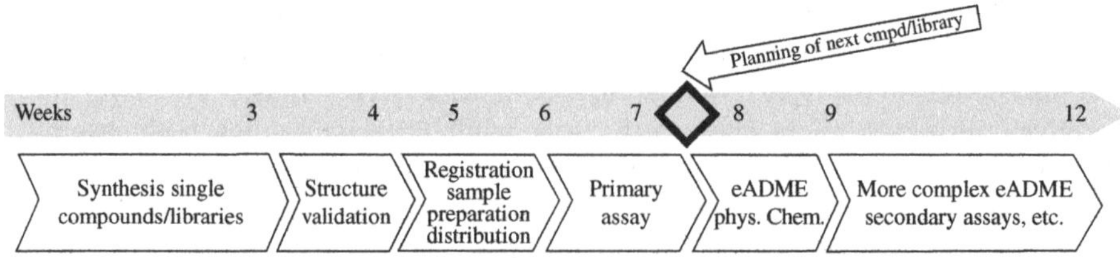


**FIGURE 13.2** Sequential profiling of lead compounds.

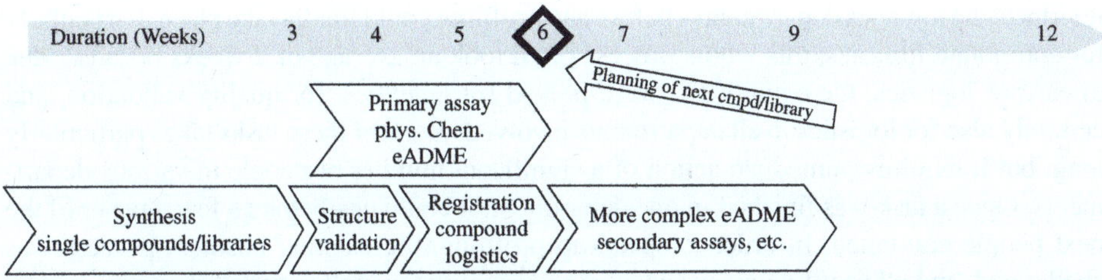


**FIGURE 13.3** Parallel profiling of lead compounds and effect on timelines.

costs. It should lead to a cycle time reduction of several weeks within one optimization loop of synthesis, analytics, and biological, physicochemical, and ADME profiling.

All data need to be stored in corporate databases immediately. They should become available for the whole project team to enable rapid data interpretation, model building, and design of the next synthesis cycle. Data quality, an optimal data visualization, and a qualified data interpretation are crucial for project progress. This requires a focused, integrated collaboration between medicinal chemists, drug designers, biologists, and respective scientists of analytics and PK departments responsible for physicochemical and ADME profiling.

Different companies use somewhat different compound optimization strategies; however, there seems to be a consensus on a basic set of assays used for the initial profiling of new molecules in drug discovery projects. It involves biological testing in a so-called primary biological assay; the assessment of purity, solubility, log D, and chemical stability; and ADME profiling including permeability testing (Caco2 [14–17], Pampa [60] assays), assessment of microsomal lability in human microsomes and/or other species [27, 28], and Cyp3A4 inhibition profiling [34–37].

The following section describes one possible option for a parallel profiling workflow used at Sanofi in Germany.

### 13.4.1 Planning Phase

At the planning stage of an integrated profiling process, the following parameters are considered to ultimately influence the final setup: number of compounds to profile, amounts and quality of the samples, physical state of the compounds, types of *in vitro* assays and physicochemical tests to be used, reproducibility of data, number of departmental interfaces and related logistics, expected workload, timelines to be achieved, and last, but not least, data quality, data management, data access, and visualization and interpretation tools.

During a process analysis of the Sanofi profiling workflow previously used, two major aspects leading to delays in compound profiling were identified. The first major aspect was availability and throughput of the existing protocols for physicochemical and *in vitro* ADME profiling assays (see the following). The second major point leading to time loss was a complex, sequential process to move newly prepared molecules through several departmental interfaces toward the final start of profiling. After preparation of the compounds by medicinal chemists, compound quality was first assessed by the analytical department. Only afterward, medicinal chemists validated the compounds in corporate databases and initiated the compound distribution to various assay labs—again via another interface, the logistics department—before any profiling could finally take place. Particularly for compound libraries, this whole process often took an average of 2 weeks because, due to ease of logistics, the compounds were pooled for analytics, for quality validation, and certainly also for logistics in all departments involved. None of these tasks takes particularly long, but it requires immediate action of a significant number of people in various departments. Once a task was finished in one department, it could easily take a few days until the next people continued. In order to speed up profiling, the existing business process was challenged, and all profiling assays adapted to ensure the overall required assay capacity.

### 13.4.2 Sample Preparation and Distribution

In pharmaceutical industry, analytical quality assessment, physicochemical profiling, biological assays, and ADME profiling are typically based on DMSO stock solutions. Some parts of the profiling process require defined concentrations, and some only a certain concentration range, the positive overall aspect being that the concentration range required for all assays is similar. The following scheme describes the new sample preparation and distribution process enabling a proper distribution of DMSO solutions for all parts of the profiling workflow (Fig. 13.4).

Medicinal chemists and chemists from library production deliver a defined amount of sample to the analytics labs—sufficient to supply analytics and all profiling assays. In this

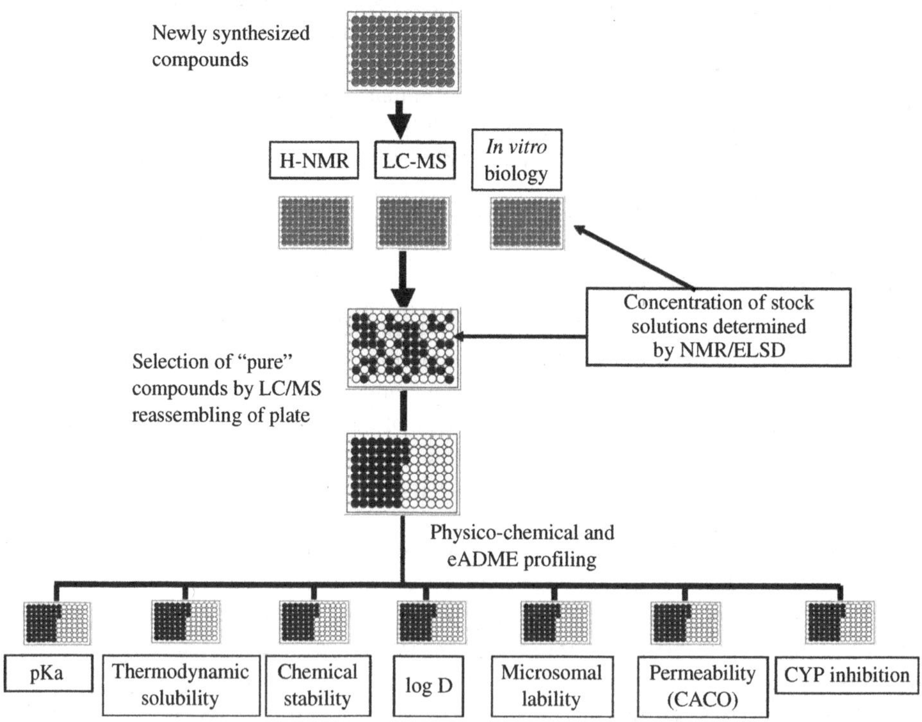


**FIGURE 13.4** Sample preparation and distribution scheme. Aliquots for *in vitro* biology are produced for all batches. Samples for physicochemical and ADME profiling are only produced if sample is pure by LC–UV/MS.

setup, *ca.* 3 mg of compound (depending on its molecular weight) is required. Samples with concentrations of 10 mmol are then prepared by dissolution of the solid sample with DMSO-d6 using a liquid handling platform.

This first solution is used to generate aliquots for 1H-NMR, LC–UV/MS, biological assays, and physicochemical resp. ADME profiling. Samples for biological assays are supplied without analytical structure validation (QC). This is due to several reasons. Primary biological assays are usually cheap and are run with sufficient throughput. In addition, automated LC–UV/MS quality control (QC) fails in *ca.* 5–10% to detect pure compounds because of their properties (poor MS response, poor or no UV response, and/or chromatographic behavior). In order not to lose time, the risk of testing some samples with improper quality is taken, particularly due to the fact that compounds (including their data) that are not positively validated by medicinal chemists after LC–UV/MS and 1H-NMR can be later erased from the corporate database. In the contrary, plates for physicochemical and ADME profiling only contain samples with sufficient LC–MS purity. Their selection is based on automated LC–UV/MS QC. This step includes the selection of pure samples (percentage of purity can be preselected by the operator), reassembling, and aliquoting for physicochemical and ADME profiling. The rationale behind this selection process is based on cost reduction and data quality maintenance.

The overall process further includes an experimental determination of concentrations, initially by 1H-NMR and then a second time by evaporative light scattering detection (ELSD) [61], before solubility testing. The concentration information is used for data evaluation, which is particularly valuable to keep control over sample concentrations for biological testing.

In this integrated profiling process, biological, physicochemical, and ADME samples are prepared and provided at the same time as the analytical samples, which saves a tremendous amount of time and reduces the number of required compound transfers between departments. Review of analytical data, final registration of compounds in corporate databases after QC by medicinal chemists, and preparation of stock solutions by compound logistics departments are off the critical path. Since the same DMSO stock solution is used for all initial profiling assays and NMR and LC–MS data are linked to all test samples, a direct proof of sample quality and concentration is immediately provided for all assays. No additional QC by biologists or ADME departments is required due to the fact that no additional reformatting is taking place.

Obviously, a process involving several disciplines, thousands of samples, and automated decisions requires a suitable IT infrastructure. It is critical to not only consider the physical flow of the samples but also the flow of data from the moment of compound synthesis to initiation of profiling requests, to profiling execution and documentation, and to data reporting and data accessibility. An optimal data transfer between the involved departments is crucial for the overall efficiency of such a process.

### 13.4.3 Compound QC

In a typical drug discovery workflow, compound synthesis is followed by QC, registration of compounds in corporate databases, and subsequent compound profiling. QC processes in research cover the assessment of compound identity and purity. The common QC process for a new compound is based on LC–MS, confirming the molecular mass and assessing the compound purity via UV response. Knowledge of the underlying reaction conditions leading to a compound supports the assumption to have the "right" compound when having the expected mass but is insufficient to prove the molecule's structure. 1D H-NMR is often used to have a second independent proof of molecular structure. 1D H-NMR spectra deliver structural information as well as information on impurities invisible in the LC–MS spectra. In case regiochemistry of synthetic transformations stays ambiguous, 2D NMR techniques [62, 63] can be applied. In order to minimize the risk that the desired compounds have the "wrong" structure, one should do a full NMR assignment of a few compounds in a new chemical series when starting to work on this series. The previously described QC process is well established; according to in-house evaluation, the risk to fail in determining the correct compound structure is less than 2%.

Whereas structure elucidation and purity assessments are easily accessible, there is no entirely satisfactory method for high-throughput sample concentration determination. Methods available for quantification are based on nitrogen detection [64] and ELSD [61], with both methods having their strengths and weaknesses. Nitrogen detection is not general because not all nitrogen-containing functional groups respond 100% and, in addition, the molecules of interest nowadays may not even contain nitrogen. ELSD is suitable for high-throughput testing, but concentration estimates are often not quantitative either [65].

As there is no satisfactory method to determine sample concentration, test solutions are usually prepared via weighing of samples and dissolution without experimental check of the finally reached concentration, assuming that molecules are completely dissolved in DMSO. However, this simple hypothesis often does not hold true. Several reports have been published on quality of DMSO stock solutions [66, 67]; they clearly show that the targeted 10 mmol DMSO stock solutions prepared for HTS and early *in vitro* assays have highly variable concentrations around an average of 7 mmol.

This issue is not only caused by errors in the weighing process, but can also be due to several other reasons listed below. Each individual factor mentioned in the following might not explain the very low concentrations of some samples, but all effects taken together may well:

- Limited inherent solubility of molecules in the used solvents. For example, 10% of drug-like samples are less soluble than 10 mmol in DMSO, and since DMSO is hygroscopic, its dissolution ability may decrease even further over time.
- Residual organic solvents and water in the samples are usually not considered at weighing; the solvent content often cannot be determined by routine LC–MS and is therefore ignored.
- Salt content is usually not systematically evaluated. As an example, hit/lead series often have weak basic or acidic moieties. Compounds are typically purified by reversed phase chromatography using TFA as a modifier. Depending on the nature of the molecule and the drying procedure, samples may contain stoichiometric or even nonstoichiometric amounts of TFA or no TFA. If TFA is not considered during weighing, the resulting solution of a molecule with, for example, a molecular weight of 400 can then reach a significantly lower true concentration than expected.
- Varying initial purities (e.g., one sample 98% pure, one only 81%) determined by LC–UV are usually not used to calibrate the amounts for weighing.

- Processing issues; e.g. in the numerous pipetting steps during high-throughput processing of compounds, sticky substances at the top of flasks may not be dissolved.
- Aging of stock solutions (water uptake, chemical degradation).

To fix the issues of sample content would require additional data from ion chromatography, water content, and elemental analysis. This would increase QC costs significantly. However, there is a proper way to circumvent this problem. As mentioned earlier, most organizations measure 1H-NMR spectra to confirm compound structure. 1H-NMR, however, is also a quantitative method [65, 68] and can therefore be used to determine the concentration of a sample. Due to the need of manual data evaluation, NMR is traditionally not widely used for quantification. However, new NMR hardware like Electronic Reference to Access *In Vivo* Concentrations (ERETIC) [69] and quantification software from vendors such as ACD/Labs, Mestrelab Research, or Bruker Corporation is now available [70]. Both developments enable concentration determination by 1H-NMR in high throughput.

### 13.4.4 Determination of Physicochemical Properties

Several physicochemical properties are generally measured in drug discovery—to predict bioavailability but to also estimate potential risks for development. Physicochemical properties in scope are log D [71–73], pKa [71], chemical stability, and compound solubility in relevant media. In this chapter, we only discuss solubility. For the other physicochemical properties, we refer to the literature [71–74]. Solubility in relevant media is the most important physicochemical compound property influencing a compound's PK parameters and its efficacy [75]. It is still difficult to provide exact solubility data for discovery compounds in high throughput. This is due to the dependence of solubility on the solid form (film, amorphous, crystalline) of a given molecule and due to the often incomplete characterization of compounds (see paragraph on compound QC). Crystalline compounds are usually much less soluble than their amorphous forms.

The historically common way of purification of compounds by recrystallization has been replaced by reversed phase chromatography and drying, which is predictable and time-saving. Chromatographic processes followed by evaporation or freeze-drying usually deliver amorphous solids or films. To support research teams in the assessment of both amorphous and crystalline forms of early research samples, we developed two solubility assays, one starting from amorphous and one from crystalline forms. As a requirement, the one starting from amorphous forms should enable a relative solubility ranking of compounds within one given chemical series.

To evaluate this, we compared solubility data determined from amorphous and crystalline solids of well-characterized compounds. Overall and visible in Figure 13.5, the solubility determined from both forms correlates reasonably well. As expected, the solubility of the amorphous solids is higher than the one determined from the crystalline solids, while poorly soluble compounds come out as poorly soluble in both assays.

It is important to mention that the solubility of individual compounds can still change dramatically when changing from an amorphous to a crystalline form. Therefore, important decisions like the selection of precandidates should always be based on solubility data from a crystalline form of the compound of interest.

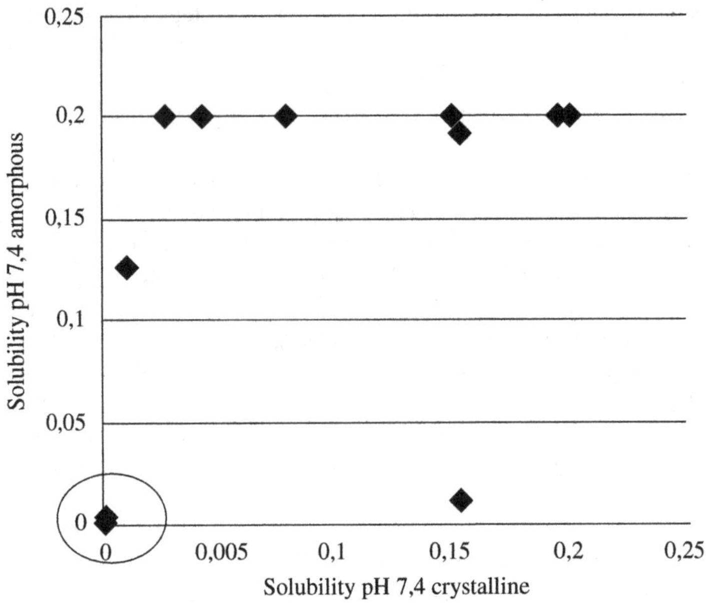


**FIGURE 13.5** Comparison of aqueous solubility at pH 7,4 and 25°C obtained from amorphous and crystalline solids. Assay target solubility is 0,2 mg/ml. (The diamonds in the circle represent individual compounds.)

There are different reasons why hits, leads, and even development candidates are often rather hydrophobic and of intrinsically low solubility (for a discussion, see chapter on drug-like properties). Among others, drivers toward poorly soluble compounds are the need to get potency via hydrophobic interactions, targets that allow only hydrophobic compounds to enter their binding site, complex multidimensional optimization problems not allowing optimization of all drug properties, but also the difficulty to deliver clear structure solubility relationships. If a high solubility is not achievable during compound optimization, a risk mitigation plan needs to be implemented to address the well-known issues associated with poorly soluble drugs:

- Most *in vitro* assays are run in aqueous solutions at compound concentrations in the nano- to millimolar range. It is well known that 40% of all discovery compounds do not reach a solubility of 1 μmol at pH 7,4 [76]. Targeted assay concentrations are therefore often not reached, and this is often not even obvious. As a consequence, the resulting SARs become unclear and can easily drive optimization in the wrong direction.
- Animal testing may be hampered by the fact that compounds cannot be dissolved in acceptable formulations tolerated by the animals. This strongly limits pharmacological and toxicological studies, if the required compound levels (exposures) cannot be reached.
- Finally, compounds often fail in the clinic due to insufficient exposure in humans, which leads to significant costs related to attrition.

So far, despite many attempts to design structure solubility models, there is no general recommendation on how to make insoluble compounds soluble beyond a few experimental standard rules such as increasing polarity of the molecule and interfering with the molecule's stacking options by reduction of flatness of the given molecule [77].

### 13.4.5 ADME Profiling: General Remarks

High-throughput assays are required to handle larger amounts of compounds in earlier phases of drug discovery projects. Usually, compounds are initially profiled for microsomal lability, permeability, and CYP inhibition. Early *in vitro* assays used by the majority of research organizations show a certain variability in assay design and performance, but finally, the majority determines stability in microsomal solutions after a certain defined time, permeability in a high-throughput mode, and CYP3A4 inhibition with midazolam (and sometimes testosterone) as a substrate. It is well accepted that the higher the data content, the lower usually the throughput of the corresponding test system. The metabolic lability of a compound, for example, is very well determined in plated human hepatocytes over 24 h including fraction metabolized (fm) for CYP3A4 or CYP2D6. But despite being of high scientific relevance and having excellent correlation to clinical data, hepatocyte clearance profiling rather fits the throughput and costs of late discovery phases; therefore, high-throughput assays like lability profiling in microsomes are still preferred over human hepatocyte clearance testing in early-phase discovery projects. LC–MS detection nowadays allows for screening of thousands of compounds per year in microsomes in a very cost-effective, sensitive mode. QC during assay performance assure consistent data output and reliable prediction of compound stability in microsomal fractions.

Permeability of compounds is tested by using either PAMPA [71] or CACO-2 cells (or MDCK cells) [14–17, 22–26]. Again, LC–MS capacity determines the throughput of a profiling facility, with each assay having advantages and disadvantages. It finally depends on the test strategy, modeling capacities, and the phase of compound optimization which assay is preferred. Cellular permeability data using, for example, CACO-2 cells have the advantage of including several biological events on cellular membranes like uptake and efflux transporters, certain metabolic activities, and a mechanistic potential to clarify issues of low permeability [23].

The implementation of a high-throughput profiling process requires new high-throughput assays and a reduction of the number of measurement repeats and controls by using standards, miniaturization, and improved data management. The new assays are validated against the EMA/FDA [31, 32] recommendations to ensure proper correlation, but with each new hit series, the correlation between data derived from the newly developed higher-throughput assays and data from lower-throughput, highest-quality assays needs to be reassessed. In addition, anytime hit series are moved to lead optimization, all higher-throughput profiling results of the best hit compounds are usually spot-checked with the lower-throughput, highest-quality assays.

### 13.4.6 Metabolic Lability Profiling

During the design of a higher-throughput metabolic lability assay [15, 17], it was decided to keep the same compound incubation conditions as in the previously used lower-throughput assay. Substrate and cofactor used, microsomal protein and final DMSO concentration, volume and duration of incubation, and pH value remain the same as compared to the previous lower-throughput assay. To achieve the desired higher throughput, the number of replicates was reduced, and the use of inhibitors for human microsomal incubations and the determination of the (−) NADPH value were omitted. The determination of lability not only in human but also in rat and mouse microsomes is retained, and the same plate layout used for all species. Furthermore, automation for the incubation and

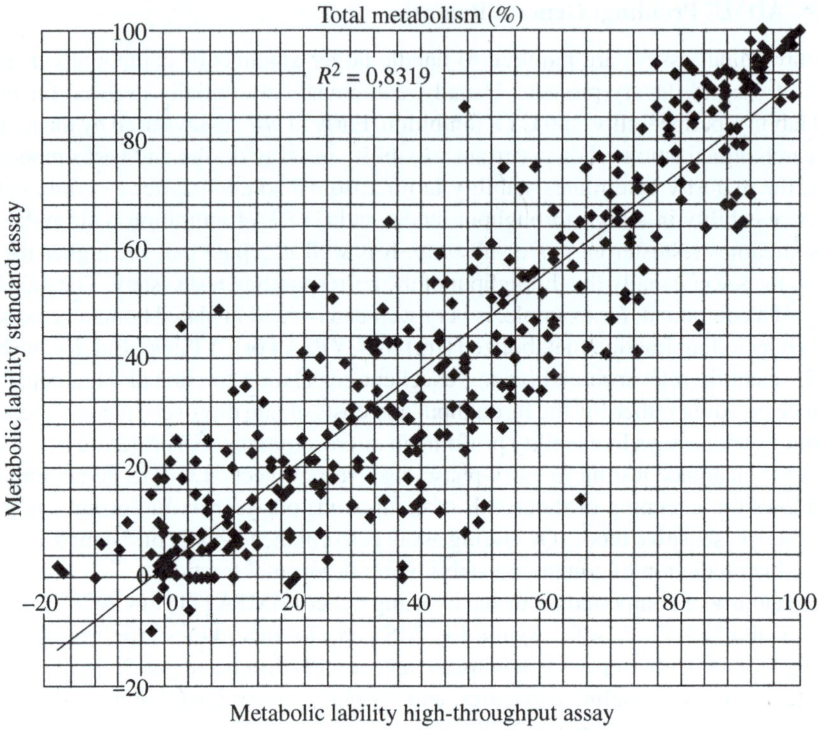


**FIGURE 13.6** Comparison of results for around 400 compounds derived from the newly developed higher-throughput assay and the standard protocol for the determination of metabolic lability. Values are shown in % total metabolism.

generation of plate maps of the incubation were improved. We further developed LC–MS/MS analytics using an automated method generation (software QuickQuan from Thermo [78]) and established efficient "workflow tools" for data processing to efficiently allow for review of results and final upload of data to the data warehouse. The workload per compound was reduced to about 50%, while the costs of consumables per test were reduced by about 40%.

The comparison of results retrieved either from the high-throughput or the previous lower-throughput assay for a set of *ca.* 400 well-profiled in-house Sanofi compounds at hand shows a good correlation (Fig. 13.6). The deviations are independent of the different assay conditions but are related to a different peak integration protocol used.

### 13.4.7 Permeability Testing

In line with a higher-throughput protocol for metabolic lability, a higher-throughput test for intestinal absorption [14, 15] using a CaCo2-derived cell line was developed. In a first step, the assay format was changed from 24- to 96-well plates—with the challenge that the comparability of results to those obtained with a 24-well setup had to be maintained. Parameters like the number of cells seeded, time of cultivation prior to the experiment, apical and basolateral buffer volumes, and duration of permeation time were optimized and adapted for three different types of transwell plates. Due to the decision to only incubate single replicates, an additional step of monolayer integrity testing using Lucifer yellow was introduced, and a threshold for barrier integrity defined. Once the assay conditions were finally determined, the assay layout and the pipetting workflow were optimized to maximum simplicity, speed, and highest automation feasible. The downstream LC–MS/MS analytics were optimized by introducing new software capable of automated compound tuning and analytics. Finally, the handling of raw data generated by LC–MS/MS was optimized by implementing new IS tools for data import and calculation of Papp values.

The comparison of Papp values determined by both the newly developed higher-throughput assay and the previous lower-throughput assay is shown in Table 13.4.

**TABLE 13.4 Comparison of Papp Values Measured in 24-Well and 96-Well Setup**

| Plate Size/Format | 24-Well | 96-Well |
|---|---|---|
| *Reference substance* | *Papp values [10E-7 cm/s]* | |
| Ranitidine | 1 | 2 |
| Mannitol | 3 | 1 |
| Fluvastatin | 25 | 43 |
| Minoxidil | 52 | 50 |
| Metoprolol | 84 | 57 |
| Verapamil | 141 | 75 |
| Propranolol | 163 | 111 |
| Carbamazepine | 254 | 196 |
| Antipyrine | 284 | 195 |
| Caffeine | 291 | 173 |

Regression analysis (Fig. 13.7) shows a high correlation ($R^2$=0,95) between the Papp values derived from 24-well to 96-well setup.

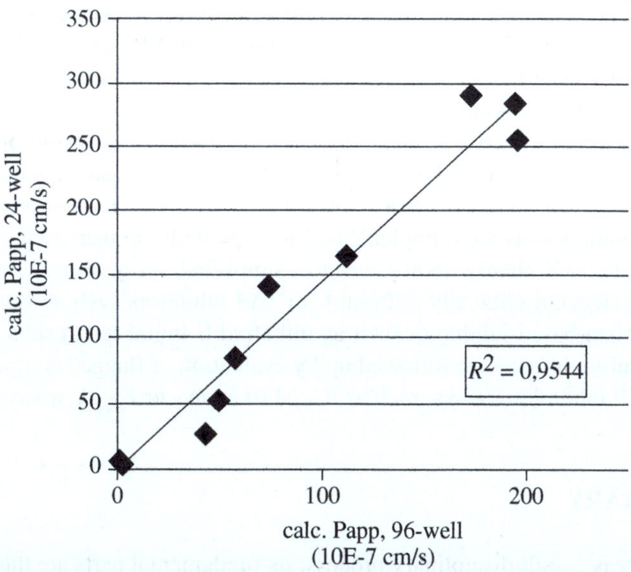


**FIGURE 13.7** Regression analysis—results obtained in 24-well and HTS 96-well setup.

**TABLE 13.5 Evaluation of Known CYP3A4 Inhibitors in the Competitive Higher-Throughput CYP3A4 Inhibition Assay Using Midazolam as Probe Substrate**

| Reference Substance | $IC_{50}$ Value (μM) | Rating $IC_{50}$ | Clinical DDI |
|---|---|---|---|
| Ketoconazole (control) | 0,013 | Red | Red |
| Itraconazole | <1 | Red | Red |
| Troleandomycin (MBI) | <1 | Red | Red |
| Nefazodone (MBI) | <1 | Red | Red |
| Mibefradil (MBI) | <1 | Red | Red |
| Voriconazole | 4,8 | Red | Red |
| Fluconazole | 8,2 | Orange | Red |
| Telithromycin (MBI) | 13 | Orange | Red |
| Clarithromycin (MBI) | >30 | Green | Red |
| Diltiazem (MBI) | >30 | Green | Red |
| Cimetidine | >30 | Green | Green |
| Fluoxetine | >30 | Green | Green |
| Azithromycin | >30 | Green | Green |

Test compound concentrations: from 1 to 30 μM. Simple $IC_{50}$ value ratings: high risk $IC_{50} < 5$ μM.

The compounds in Table 13.4 [16] are standard compounds used for such validation studies and recommended by authorities (FDA), which can further be extended to compare efflux ratio data or evaluate contribution of transporters to efflux [18].

### 13.4.8 CYP Inhibition Profiling

In the early discovery phase, the CYP3A4 inhibition potential of a new chemical series is conventionally assessed by a competitive assay in 96- or 384-well format [17]. The test compounds are incubated sequentially with the two FDA-compatible CYP3A4 probe substrates midazolam and testosterone and with human liver microsomes. Incubations are performed using a pipetting robot, and the amount of CYP3A4-specific metabolite is quantified by LC–MS/MS. An estimated IC50 value is determined by fitting the percent activity of control into a model.

In order to improve the assay throughput, the second substrate testosterone was omitted in the newly developed higher-throughput CYP3A4 inhibition assay. This reduction to the most clinically relevant substrate midazolam led to a *ca.* 50% increase in assay throughput. New data evaluation tools were implemented to cope with the increased amount of data generated. Table 13.5 clearly shows that the simplified assay allows for the flagging (traffic light system) of clinically relevant CYP3A4 inhibitors such as ketoconazole and potent mechanism-based inhibitors such as mibefradil. Initial test results in the higher-throughput inhibition assay are followed up by evaluation of flagged compounds in time-dependent CYP inhibition assays, such as $IC_{50}$ shift assays or $k_{inact}/K_I$ assays [19, 34–36].

## 13.5 SUMMARY

Drug discovery is a multidisciplinary process; its fundamental parts are the identification of suitable chemical starting points, so-called hits, and their optimization toward a clinical candidate that fulfills a given product profile. During optimization, compound properties such as biological and pharmacological efficacy, physicochemical properties, absorption, distribution, metabolism, excretion, selectivity for a specific target, PK/PD properties, and toxicological profile require permanent consideration. Additional parameters in scope are intellectual property, competitive situation, and cost of goods. This list clearly indicates that compound optimization is a multiparametric problem—to be solved by a multidisciplinary team of people.

This article summarizes state-of-the-art techniques to assess ADME and physicochemical properties—from screening methods to highly validated FDA/EMA compliant assays. Typically, screening assays are applied for the first profiling of new compounds, while before selection of a lead series, the predictivity of the screening assays needs to be reassessed by comparing initial data with data generated by more advanced assays. On the way to a clinical candidate, a further set of highly sophisticated tests is required to deliver a profiling package consistent with regulatory guidelines to gain approval for clinical testing. The consequent improvement of ADME assays and their application in drug discovery has led to a significant reduction of failure in clinical trials due to poor ADME properties.

Assays for the optimization of ADME and physicochemical properties are sufficiently advanced, but multidimensional optimization strategies utilizing all critical properties in parallel can still be improved to optimize hits toward leads and candidates. In order to generate sufficient data—while keeping costs and timelines under control—predictive screening assays and a well-organized process including sample and data management need to be in place. New workflows enable compound QC, direct determination of compound concentration for all samples profiled, and immediate profiling for ADME and physicochemical data to ensure rapid multiparametric optimization and shorter cycle times.

## REFERENCES


[1] Mullard, A., 2013 FDA drug approvals, *Nat Rev Drug Discov*, 2014, **13**, 85–9.

[2] Mullard, A., 2014 FDA drug approvals, *Nat Rev Drug Discov*, 2015, **14**, 77–81.

[3] Kola, I. Landis, J., Can the pharmaceutical industry reduce attrition rates? *Nat Rev Drug Discov*, 2004, **3**, 711–16.

[4] Smith, D. Schmid, E. Jones, B., Do drug metabolism and pharmacokinetic departments make any contribution to drug discovery? *Clin Pharmacokinet*, 2002, **41**(13), 1005–19.

[5] Hop, C.E. Cole, M.J. Davidson, R.E. Duignan, D.B. Federico, J. Janiszewski, J.S. Jenkins, K. Krueger, S. Lebowitz, R. Liston, T.E. Mitchell, W. Snyder, M. Steyn, S.J. Soglia, J.R. Taylor, C. Troutman, M.D. Umland, J. West, M. Whalen, K.M., Zelesky, V. Zhao, S.X., High throughput ADME screening: practical considerations, impact on the portfolio and enabler of in silico ADME models, *Curr Drug Metab*, 2008, **9**(9), 847–53.

[6] DiMasi, J.A. Hansen, R.W. Grabowski, H.G., The price of innovation: new estimates of drug development costs, *J Health Econ*, 2003, **22**(2), 151–85.

[7] Prentis, R.A. Lis, Y. Walker, S.R., Pharmaceutical innovation by the seven UK-owned pharmaceutical companies (1964–1985), *Br J Clin Pharmacol*, 1988, **25**(3), 387–96.

[8] Pellegatti, M., Preclinical *in vivo* ADME studies in drug development: a critical review, *Expert Opin Drug Metab Toxicol*, 2012, **8**(2), 161–72.

[9] Jones, H.M. Rowland-Yeo, K., Basic concepts in physiologically based pharmacokinetic modeling in drug discovery and development, *CPT Pharmacometrics Syst Pharmacol*, 2013, **2**, 1–7.

[10] Smith, D.A., Evolution of ADME science: where else can modeling and simulation contribute? *Mol Pharm*, 2013, **10**(4), 1162–70.

[11] Lipinski, C.A. Lombardo, F. Dominy, B.W. Feeny, P.J., Experimental and computational approaches to estimate solubility and permeability in drug discovery and development settings, *Adv Drug Delivery Rev*, 1997, **23**, 3–25.

[12] Kubinyi H., Drug research: myths, hype and reality, *Nat. Rev. Drug Discov*, 2003, **2**, 665–68.

[13] Price, D. Blagg, J. Jones, L. Greene, N. Wager, T., Physicochemical drug properties associated with *in vivo* toxicological outcomes: a review, *Expert Opin Drug Metab Toxicol*, 2009, **5**(8), 921–31.

[14] Deferme, S. Annaert, P. Augustijns P., *In vitro* screening models to assess intestinal drug absorption and metabolism. In: *Drug Absorption Studies*, Biotechnology: Pharmaceutical Aspects, Volume **VII**. C. Ehrhardt K.-J. Kim (eds), Springer, New York, 2008, 182–215.

[15] White, R.E., High-throughput screening in drug metabolism and pharmacokinetic support of drug discovery, *Annu Rev Pharmacol Toxicol*, 2000, **40**, 133–57.

[16] Marino, A.M. Yarde, M. Patel, H. Chong, S. Balimane, P.V., Validation of the 96 well Caco-2 cell culture model for high throughput permeability assessment of discovery compounds, *Int J Pharm*, 2005, **297**(1–2), 235–41.

[17] Vogel, H.G. Hock, F.J. Maas, J. Mayer, D., *Drug Discovery and Evaluation: Safety and Pharmacokinetic Assays*, Springer Verlag, Berlin/New York, 2006.

[18] Kerns, E. Di, L., *Drug-Like Properties: Concepts, Structure Design and Methods*, Elsevier Inc., Burlington, MA, 2008.

[19] Grimm, S.W. Einolf, H.J. Hall, S.D. He, K. Lim, H.K. Ling, K.-H. Lu, C. Nomeir, A.A. Seibert, E. Skordos, K.W. Tonn, G.R. Horn, R.V. Wang, R.W. Wong, Y.N. Yang, T.J. Obach, R.S., The conduct of *in vitro* studies to address time-dependent inhibition of drug-metabolizing enzymes: a perspective of the pharmaceutical research and manufacturers of America, *Drug Metab Dispos*, 2009, **37**, 1355–70.

[20] Augustijns P, Wuyts B, Hens B, Annaert P, Butler J, Brouwers J, A review of drug solubility in human intestinal fluids: implications for the prediction of oral absorption, *Eur J Pharm Sci*, 2014, **57**, 322–32.

[21] Smith, D.A. Di, L. Kerns, E.H., The effect of plasma protein binding on *in vivo* efficacy: misconceptions in drug discovery, *Nat Rev Drug Discov*, 2010, **9**, 929–39.

[22] Artursson, P. Karlsson, J., Correlation between oral drug absorption in humans and apparent drug permeability coefficients in human intestinal epithelial (Caco-2) cells, *Biochem Biophys Res Comm*, 1991, **175**(3), 880–5.

[23] Borchardt, R.T. Freidinger, R.M. Sawyer, T.K., *Integration of Pharmaceutical Discovery and Development*, Plenum Press, New York, 1998.

[24] Hidalgo, I.J. Raub, T.J. Borchardt, R.T., Characterization of the human colon carcinoma cell line (Caco-2) as a model system for intestinal epithelial permeability, *Gastroenterology*, 1989, **96**, 736–49.

[25] Artursson, P., Epithelial transport of drug in cell culture. I: a model for studying the passive diffusion of drugs over intestinal absorbtive (Caco-2) cells, *J Pharm Sci*, 1990, **79**(6), 476–82.

[26] Steffansen, B. Nielsen, C.U. Brodin, B. Eriksson, A.H. Andersen, R. Frokjaer, S., Intestinal solute carriers: an overview of trends and strategies for improving drug absorption, *Eur J Pharm Sci*, 2004, **21**(1), 3–16.

[27] Murphy, P.J., The development of drug metabolism research as expressed in the publications of ASPET (2008): Part 2, 1959–1983, *Drug Metab Dispos*, 2008, **36**(6), 981–5.

[28] Murphy, P.J., The development of drug metabolism research as expressed in the publications of ASPET: Part 3, 1984–2008. *Drug Metab Dispos*, 2008, **36**(10), 1977–82.

[29] Lin, J.H. Lu, A.Y.H., Role of pharmacokinetics and metabolism in drug discovery and development, *Pharmacol Rev*, 1997, **49**(4), 403–49.

[30] Dudda, A. Kuerzel, G.U., Metabolism studies *in vitro* and *in vivo*. Chapter II.H. In: *Drug Discovery and Evaluation: Safety and Pharmacokinetic Assays*. H.G. Vogel F.J. Hock J. Maas D. Mayer (eds), Springer Verlag, Berlin/New York, 2006, 493–520.

[31] U.S. Department of Health and Human Services. Food and Drug Administration. Center for Drug Evaluation and Research, Guidance for industry: drug interaction studies—study design, data analysis and implications for dosing and labeling recommendations, Draft Guidance 2012.

[32] (a) EMA, Guideline on the investigation of drug interactions, Draft Guideline 2010. (b) EMA, Guideline on the investigation of drug interactions, 2013.

[33] Guengerich, F.P., Cytochrome p450 and chemical toxicology, *Chem Res Toxicol*, 2008, **21**(1), 70–83.

[34] Obach, R.S. Walsky, R.L. Venkatakrishnan, K., The utility of *in vitro* cytochrome P450 inhibition data in the prediction of drug-drug interactions, *J Pharmacol Exp Ther*, 2006, **316**, 336–48.

[35] Obach, R.S. Robert, L.W. Venkatakrishnan, K., Mechanism-based inactivation of human cytochrome P450 enzymes and the prediction of drug-drug interactions, *Drug Metab Dispos*, 2007, **35**, 246–55.

[36] Zhang, L. Zhang, Y.D. Zhao, P. Shiew-Mei, H., Predicting drug–drug interactions: an FDA perspective, *AAPS J*, 2009, **11**(2), 300–6.

[37] Zhao, P., The use of hepatocytes in evaluating time-dependent inactivation of P450 *in vivo*, *Expert Opin Drug Metab Toxicol*, 2008, **4**(2), 151–64.

[38] Chen, Y. Liu, L. Monshouwer, M. Fretland, A.J., Determination of time-dependent inactivation of CYP3A4 in cryopreserved human hepatocytes and assessment of human drug-drug interactions, *Drug Metab Dispos*, 2011, **39**, 2085–92.

[39] Parkinson, A. Kazmi, F. Buckley, D.B. Yerino, P. Paris, B.L. Holsapple, J. Toren, P. Otradovec, S.M. Ogilvie, B.W., An evaluation of the dilution method for identifying metabolism-dependent inhibitors of cytochrome P450 enzymes, *Drug Metab Dispos*, 2011, **39**(8), 1370–87.

[40] Birkett, D.J., *Pharmacokinetics Made Easy*, McGraw-Hill, Sydney, 2002.

[41] Riedel, J., Distribution-*in vitro* tests-protein binding. Chapter II.F. In: *Drug Discovery and Evaluation: Safety and Pharmacokinetic Assays*. H.G. Vogel F.J. Hock J. Maas D. Mayer (eds), Springer Verlag, Berlin/New York, 2006, 473–86.

[42] Urso, R. Blardi, P. Giorgi, G., A short introduction to pharmacokinetics, *Eur Rev Med Pharmacol Sci*, 2002, **6**, 33–44.

[43] Ritschel, W.A. Kearns, G.L., *Handbook of Basic Pharmacokinetics*, 6th edition, Apha, Washington, DC, 2004.

[44] FDA, Additional studies part B. Metabolism and pharmacokinetic studies, 1993 Draft "Redbook II". *Guidance for Industry and Other Stakeholders Toxicological Principles for the Safety Assessment of Food Ingredients*. Redbook 2000 (Revised 2007), http://www.fda.gov/downloads/Food/GuidanceRegulation/UCM078741.pdf (accessed May 28, 2015).

[45] Kim, K. Ohashi, K. Utoh, R. Kano, K. Okano, T., Preserved liver-specific functions of hepatocytes in 3D co-culture with endothelial cell sheets, *Biomaterials*, 2012, **33**(5), 1406–13.

[46] Kostadinova, R. Boess, F. Applegate, D. Suter, L. Weiser, T. Singer, T. Naughton, B. Roth, A., A long-term three dimensional liver co-culture system for improved prediction of clinically relevant drug-induced hepatotoxicity, *Toxicol Appl Pharmacol*, 2013, **268**(1), 1–16.

[47] Mueller, D. Krämer, L. Hoffmann, E. Klein, S. Noor, F., 3D organotypic HepaRG cultures as *in vitro* model for acute repeated dose toxicity studies, *Toxicol In Vitro*, 2014, **28**(1), 104–12.

[48] Grime, K.H. Bird, J. Ferguson, D. Riley, R.J., Mechanism-based inhibition of cytochrome P450 enzymes: an evaluation of early decision making *in vitro* approaches and drug–drug interaction prediction methods, *Eur J Pharm Sci*, 2009, **36**, 175–91.

[49] Mukadam, S. Tay, S. Tran, D. Wang, L.M. Delarosa, E. Cyrus, K.S.S. Halladay, J.R. Kenny, J., Evaluation of time-dependent cytochrome P450 inhibition in a high-throughput, automated assay: introducing a novel area under the curve shift approach, *Drug Metab Lett*, 2012, **6**(1), 43–53.

[50] Ogilvie, B.W. Usuki, E. Yerino, P. Parkinson, A., *In vitro* approaches for studying the inhibition of drug-metabolizing enzymes responsible for the metabolism of drugs (reaction phenotyping) with emphasis on cytochrome P450. In: *Drug-Drug Interactions*, Drugs and the Pharmaceutical Sciences, 2nd edition, Volume **179**. A.D. Rodrigues (ed.), Informa Healthcare, New York, 2011, 231–358.

[51] Giacomini, K.M. Huang, S.M., Transporters in drug development and clinical pharmacology, *Clin Pharmacol Ther*, 2013, **94**(1), 3–9.

[52] Amidon, G.L. Lennernäs, H. Shah, V.P. Crison, J.R., A theoretical basis for a biopharmaceutics drug classification: the correlation of *in vitro* drug product dissolution and *in vivo* bioavailability, *Pharm Res*, 1995, **12**(3), 413–20.

[53] Wu, C.Y. Benet, L.Z., Predicting drug disposition via application of BCS: transport/absorption/elimination interplay and development of a biopharmaceutics drug disposition classification system, *Pharm Res*, 2005, **22**(1), 11–23.

[54] Benet, L.Z. Broccatelli, F. Oprea, T.I., BDDCS applied to over 900 drugs, *AAPS J*, 2011, **13**(4), 519–47.

[55] FDA, Guidance for industry: waiver of in vivo bioavailability and bioequivalence studies for immediate-release solid oral dosage forms based on a biopharmaceutics classification system, 2000.

[56] Benet, L.Z. Amidon, G.L. Barends, D.M. Lennernäs, H. Polli, J.E. Shah, V.P. Stavchansky, S.A. Yu, L.X. The use of BDDCS in classifying the permeability of marketed drugs, *Pharm Res*, 2008, **25**(3), 483–8.

[57] Hann, M.M. Keserü, G.M., Finding the sweet spot: the role of nature and nurture in medicinal chemistry, *Nat Rev Drug Discov*, 2012, **11**, 355–65.

[58] Gleeson, M.P. Hersey, A. Montanari, D. Overington, J., Probing the links between *in vitro* potency, ADMET and physicochemical parameters, *Nat Rev Drug Discov*, 2011, **10**, 197–208.

[59] Leeson, P.D. Springthorpe, B., The influence of drug-like concepts on decision-making in medicinal chemistry, *Nat Rev Drug Discov*, 2007, **6**, 881–90.

[60] Avdeef, A. Artursson, P. Neuhoff, S. Lazorova, L. Gråsjö, J. Tavelin, S., Caco-2 permeability of weakly basic drugs predicted with the double-sink PAMPA pKaflux method, *Eur J Pharm Sci*, 2005, **24**, 333–49.

[61] Bruns, A. Waldhoff, H. Winkle, W., Applications of HPLC with evaporative light scattering detection in fat and carbohydrate chemistry, *Chromatographia*, 1989, **27**, 340–2.

[62] Fuloria, N.J. Fuloria, S., Structural elucidation of small organic molecules by 1D, 2D and multi dimensional-solution NMR spectroscopy, J Anal Bioanal Tech, 2013, S11. doi: 10.4172/2155-9872.S11-001.

[63] Kessler, H. Gehrke, M. Griesinger, C., Two-dimensional NMR spectroscopy: background and overview of the experiments—new analytical methods (36), *Angew Chem Int Ed Engl*, 1988, **27**(4), 490–536.

[64] Fitch, W.L. Szardenings, A.K. Fujinari, E.M., Chemiluminescent nitrogen detection for HPLC: an important new tool in organic analytical chemistry, *Tetrahedron Lett*, 1997, **38**(10), 1689–92.

[65] Lane, S. Boughtflower, B. Mutton, I. Paterson, C. Farrant, D. Taylor, N. Blaxill, Z. Carmody, C. Borman, P., Toward single-calibrant quantification in HPLC. A comparison of three detection strategies: evaporative light scattering, chemiluminescent nitrogen, and proton NMR, *Anal Chem*, 2005, **77**(14), 4354–65.

[66] Popa-Burke, I.G. Issakova, O. Arroway, J.D. Bernasconi, P. Chen, M. Coudurier, L. Galasinski, S. Jadhav, A.P. Janzen, W.P. Lagasca, D. Liu, D. Lewis, R.S. Mohney, R.P. Sepetov, N. Sparkman, D.A. Hodge, C.N., Streamlined system for purifying and quantifying a diverse library of compounds and the effect of compound concentration measurements on the accurate interpretation of biological assay results, *Anal Chem*, 2004, **76**(24), 7278–87.

[67] Cheng, X. Hochlovski, J. Hepp, D. Beckner, C. Kantor, S. Schmitt, R., Studies on repository compound stability in DMSO under various conditions, *J Biomol Screen*, 2003, **8**, 292–304.

[68] Malz, F. Jancke, H., Validation of quantitative NMR, *J Pharm Biomed Anal*, 2005, **38**(5), 813–23.

[69] Akoka, S. Barantin, L. Trierweiler, M., Concentration measurement by proton NMR using the ERETIC method, *Anal Chem*, 1999, **71**(13), 2554–7.

[70] Pauli, G.F. Gödecke, T. Jaki, B.U. Lancin, D.C., Quantitative 1H NMR: development and potential of an analytical Method—an update, *J Nat Prod*, 2012, **75**(4), 834–51.

[71] Avdeef, A. Testa, B., Physicochemical profiling in drug research: a brief survey of the state-of-the-art of experimental techniques, *Cell Mol Life Sci*, 2002, **59**, 1681–9.

[72] Comer, J. Tam, K., Lipophilicity profiles: theory and measurement. In: *Pharmacokinetic Optimization in Drug Research: Biological, Physicochemical and Computational Strategies*. B. Testa H. van de Waterheemd G. Folkers R. Guy (eds), Verlag Helvetica Chimica Acta, Zürich, 2001, 275–304.

[73] Wilson, D.M. Wang, X.L. Walsh, E. Rourick, R.A., High throughput log D determination using liquid chromatography-mass spectrometry, *Comb Chem High Throughput Screen*, 2001, **4**, 511–19.

[74] Wan, H. Holmen, A.G., High throughput screening of physicochemical properties and *in vitro* ADME profiling in drug discovery, *Comb Chem High Throughput Screen*, 2009, **12**, 315–29.

[75] Savjani, K.T. Gajjar, A.K. Savjani, J.K., Drug solubility: importance and enhancement techniques, *ISRN Pharm*, 2012, **2012**, 195727.

[76] Williams, H.D. Trevaskis, N.L. Charman, S.A. Shanker, R.M. Charman, W.N. Pouton, C.W. Porter, C.J., Strategies to address low drug solubility in discovery and development, *Pharmacol Rev*, 2013, **65**(1), 315–499.

[77] Lipinski, C.A. Lombardo, F. Dominy, B.W. Feeney, P.J., Experimental and computational approaches to estimate solubility and permeability in drug discovery and development settings, *Adv Drug Deliv Rev*, 2001, **46**(1–3), 3–26.

[78] Duczak, N. Koers, J., Comparing manual compound optimizations against automated optimizations obtained by quickQuan on the TSQ quantum access. Thermo Fisher Scientific, San Jose, CA. Application note 401, https://www.thermoscientific.com/en/browse-results.html?searchWithinKW=Application+note+401&searchServiceType=browse&browseKeyword=Application+Notes&activeTab=DOC&docRefine=+11325+15581&continueSearch=true&changeRefine=true&initialRefine=11325%252525252B11280 (accessed May 28, 2015).

# PART V

# MEDICINAL CHEMISTRY BEYOND SMALL MOLECULES

# 14

# THE ROLE OF NATURAL PRODUCTS IN DRUG DISCOVERY: EXAMPLES OF MARKETED DRUGS

LARS OLE HAUSTEDT and KARSTEN SIEMS
*AnalytiCon Discovery GmbH, Potsdam, Germany*

## 14.1 NATURAL PRODUCTS AND NATURAL PRODUCT DERIVATIVES IN COMMERCIAL DRUGS

Natural product structures continue to play a significant role in the drug discovery and development process [1]. This review examines the distribution and origin of this important therapeutic class and examines several of the most medicinally important compounds in detail, highlighting the variety of key methods used to obtain quantities of the often complex molecules and the strategies used in their optimization.

Between 1980 and 2013, more than 1500 new drugs were introduced into the market including approximately 300 natural products and natural product derivatives (categories N and ND in the nomenclature of Cragg and Newman). Cragg and Newman also included, in the categories N and ND, drugs derived from polysaccharides and human peptides as lead structures, for example, heparin derivatives and gonadotropin-releasing hormone agonists (GnRH agonist, GnRH–A). If one only includes small molecule natural products, 262 drugs were launched between 1980 and 2013. Thirty of those drugs were unaltered natural products and 232 natural product derivatives.

A closer look at the 89 natural product scaffolds showed that all major classes of natural products including alkaloids, terpenoids, peptides, and polyketides are represented (Table 14.1, Fig. 14.1).

*Small Molecule Medicinal Chemistry: Strategies and Technologies*, First Edition.
Edited by Werngard Czechtizky and Peter Hamley.

**TABLE 14.1 Drugs Directly Derived from Natural Products Introduced Into Market between 1980 and 2013**

| Related Natural Product | Year of Natural Product Discovery | Disease Area | Generic Name | Trade Name | Year of Product Launch | Natural Product (N) or Derivative (ND) | Producing Organism |
|---|---|---|---|---|---|---|---|
| *Alkaloids* | | | | | | | |
| Camptothecin | 1966 | Anticancer | Irinotecan hydrochloride | Campto | 1994 | ND | P |
| | | Anticancer | Topotecan HCl | Hycamtin | 1996 | ND | P |
| | | Anticancer | Belotecan hydrochloride | Camtobell | 2004 | ND | P |
| Cephalotaxin | 2012 | Anticancer | Omacetaxine mepesuccinate | Synribo | 2012 | ND | P |
| Quinine | 1820 | Antiparasitic | Mefloquine HCI | Fansimef | 1985 | ND | P |
| Coffeine | 1820 | Bronchodilator | Doxofylline | Ansimar | 1985 | ND | P |
| | | Vasodilator; cerebral | Propentofylline propionate | Hextol | 1988 | ND | P |
| Elipticin | 1959 | Anticancer | Elliptinium acetate | Celiptium | 1983 | ND | P |
| Ergotamin | 1918 | Antihyperprolactinemia | Lisuride maleate | Dopergin | 1982 | ND | F |
| | | Antihyperprolactinemia | Metergoline | Liserdol | 1987 | ND | F |
| | | Antihyperprolactinemia | Cabergoline | Dostinex | 1993 | ND | F |
| | | Antihyperprolactinemia | Quinagolide hydrochloride | Norprolac | 1994 | ND | F |
| | | AntiParkinsonian | Pergolide mesylate | Permax | 1988 | ND | F |
| Hyoscyamin | Before 1900 | Antitussive | Flutropium bromide | Flubron | 1988 | ND | P |
| | | Bronchodilator | Oxitropium bromide | Oxivent | 1983 | ND | P |
| | | COPD | Tiotropium bromide | Spiriva | 2002 | ND | P |
| Cocaine | Before 1900 | Imaging drug Parkinson | Ioflupane i-123 | DaTscan | 2011 | ND | P |
| Morphin | Before 1900 | Antinarcotic | Naltrexone HCI | ReVia | 1984 | ND | P |
| | | Antinarcotic | Nalmefene HCl | Revex | 1995 | ND | P |
| | | Constipation | Methylnaltrexone bromide | Relistor | 2008 | ND | P |
| | | Uremic pruritis | Nalfurafine HCl | Remitch | 2009 | ND | P |

| | | | | | | | |
|---|---|---|---|---|---|---|---|
| Scopolamin | Before 1900 | Antispasmodic | Cimetropium bromide | Alginor | 1985 | ND | P |
| Ecteinascidin 743 | 1984 | Anticancer | Trabectedin | Yondelis | 2007 | N | M |
| Tubocurarin | 1948 | Muscle relaxant | Atracurium besylate | Nimbex | 1982 | ND | P |
| Vinblastin | 1959 | Anticancer | Vinorelbine | Navelbine | 1989 | ND | P |
| | | Anticancer | Vinflunine | Javlor | 2010 | ND | P |
| Vincamine | 1953 | Vasodilator; cerebral | Brovincamine fumarate | Sabromin | 1986 | ND | P |
| *Aminoacids and closely related compounds* | | | | | | | |
| Aminolevulinic acid | 1953 | Actinic keratoses | Aminolevulinic acid methyl ester | Metvix | 2001 | ND | H |
| | | Anticancer | Hexyl aminolevulinate | Hexvix | 2004 | ND | H |
| Carnitine | Before 1900 | Nootropic | Levacecarnine HCI | Nicetile | 1986 | ND | H |
| g-Aminobutyric acid | Before 1900 | Antiepileptic | Vigabatrin | Sabril | 1989 | ND | H |
| | | Antiepileptic | Pregabalin | Lyrica | 2004 | ND | H |
| N-Acetylglutamate | 1930 | Hyperammonemia | Carglumic acid | Carbaglu | 2003 | ND | H |
| Ornithine | Before 1900 | Antiparasitic | Eflornithine HCl | Ornidyl | 1990 | ND | H |
| 3-Hydroxy-3-methylglutaric acid | Before 1900 | Hypolipidemic | Meglutol | Lipoglutaren | 1983 | N | H |
| Phenylethylamine | Before 1900 | ADHD | Lisdexamfetamine mesilate | Vyvanse | 2007 | ND | P |
| Dopamine | Before 1900 | Antiparkinsonian | Melevodopa (CHF-1301) | Levomet | 1999 | ND | H |
| | | Cardiotonic | Ibopamine HCI | Inopamil/ Scandine | 1984 | ND | P |
| | | Cardiotonic | Docarpamine | Tanadopa | 1994 | ND | P |
| *Coenzymes* | | | | | | | |
| Biopterin | 1961 | Hyperphenylalaninemia | Sapropterin hydrochloride | Biopten | 1992 | ND | H |
| Coenzyme Q10 (ubiquinone) | 1962 | Nootropic | Idebenone | Avan | 1986 | ND | H |
| Folic acid | Before 1940 | Anticancer | Pralatrexate | Folotyn | 2009 | ND | H |

*(Continued)*

**TABLE 14.1 (*Continued*)**

| Related Natural Product | Year of Natural Product Discovery | Disease Area | Generic Name | Trade Name | Year of Product Launch | Natural Product (N) or Derivative (ND) | Producing Organism |
|---|---|---|---|---|---|---|---|
| Heme | Before 1930 | Anticancer | Talaporfin sodium | Laserphyrin | 2004 | ND | H |
| | | Porphyria | Heme arginate | Normosang | 1985 | ND | H |
| Vitamin D (cholecalciferol) | 1946 | Antipsoriatic | Calcipotriol | Daivonex | 1991 | ND | H |
| | | Calcium metabolism | Paricalcitol | Zemplar | 1998 | ND | H |
| | | Calcium metabolism | Doxercalciferol | Hectorol | 1999 | ND | H |
| | | Calcium metabolism | Maxacalcitol | Prezios | 2000 | ND | H |
| | | Sec. hyperthyroidism | Falecalcitriol | Horner/Fulstan | 2001 | ND | H |
| *Flavonoids and lignans* | | | | | | | |
| Genistein | Before 1900 | Calcium metabolism | Ipriflavone | Osten | 1989 | ND | P |
| Oligomeric procyanidin | Before 1950 | Diarrhea of HIV patients | Crofelemer | Fulyzaq | 2012 | N | P |
| Diosmin | 1938 | Cardiovascular disease | Hidrosmin | Venosmil | 1987 | ND | P |
| | | Gastroprotectant | Dosmalfate | Diotul | 2000 | ND | P |
| Sophoradin | 1969 | Antiulcer | Sofalcone | Solon | 1984 | ND | P |
| Nordihydroguaiaretic acid | 1944 | Anticancer | Masoprocol | Actinex | 1992 | N | P |
| Podophyllotoxin | 1932 | Anticancer | Etoposide phosphate | Etopophos | 1996 | ND | P |
| | | Genital warts | Podophyllotoxin | Condyline | 1985 | N | P |
| *Lipids* | | | | | | | |
| Lipstatin | 1987 | Antiobesity | Orlistat | Xenical | 1998 | ND | B |
| Myriocin | 1972 | Multiple sclerosis | Fingolimod HCl | Gilenya | 2010 | ND | F |

| | | | | | | | |
|---|---|---|---|---|---|---|---|
| *Macrolides* | | | | | | | |
| Epothilone | 1996 | Anticancer | Ixabepilone | Ixempra | 2007 | ND | **B** |
| Erythromycin | 1950 | Antibacterial | Roxithromycin | Rulid | 1987 | ND | B |
| | | Antibacterial | Azithromycin | Sumamed | 1988 | ND | B |
| | | Antibacterial | Erythromycin acistrate | Erasis | 1988 | ND | B |
| | | Antibacterial | Clarithromycin | Klaricid | 1990 | ND | B |
| | | Antibacterial | Dirithromycin | Nortron | 1993 | ND | B |
| | | Antibacterial | Flurithromycin ethylsuccinate | Ritro | 1997 | ND | B |
| | | Antibacterial | Telithromycin | Ketek | 2001 | ND | B |
| | 1975 | Antibacterial | Fidaxomicin | Dificid | 2011 | N | B |
| Fidaxomicin | | | | | | | |
| Ivermectin B1a | 1980 | Antiparasitic | Ivermectin | Mectizan | 1987 | N | B |
| Miocamycin | 1974 | Antibacterial | Miokamycin | Miocamycin | 1985 | N | B |
| Pristinamycin I | 1965 | Antibacterial | Quinupristin | Synercid | 1999 | ND | B |
| Pristinamycin II | 1965 | Antibacterial | Dalfopristin | Synercid | 1999 | ND | B |
| Rapamycin (sirolimus) | 1974 | Immunosuppressant | Sirolimus | Rapamune | 1999 | N | B |
| | | | | | | | B |
| | | Anticancer | Temsirolimus | Torisel | 2007 | ND | B |
| | | Immunosuppressant | Everolimus | Certican | 2004 | ND | B |
| | | | | Zortress | | | |
| | | | | Afinitor | | | |
| | | Restenosis | Zotarolimus | Endeavor | 2005 | ND | B |
| | | Restenosis | Umirolimus = biolimus a9 | BioMatrix | 2008 | ND | B |
| Rokitamycin | 1979 | Antibacterial | Rokitamycin | Ricamycin | 1986 | ND | B |
| | 1990 | Insecticide (head lice) | Spinosad | Natroba | 2011 | N | B |
| Spinosyn A | | | | | | | |
| Tacrolimus (FK 506) | 1986 | Antiallergic | Pimecrolimus (SDZ-ASM-981) | Elidel | 2002 | ND | B |
| | | Immunosuppressant | Tacrolimus | Prograf | 1993 | N | B |

(*Continued*)

**TABLE 14.1 (*Continued*)**

| Related Natural Product | Year of Natural Product Discovery | Disease Area | Generic Name | Trade Name | Year of Product Launch | Natural Product (N) or Derivative (ND) | Producing Organism |
|---|---|---|---|---|---|---|---|
| *Nucleosides* | | | | | | | |
| Mizoribine | 1973 | Immunosuppressant | Mizoribine | Bredinin | 1984 | N | F |
| *Peptides* | | | | | | | |
| Ciclosporin | 1976 | Immunosuppressant | Cyclosporine | Sandimmune | 1983 | N | B |
| Daptomycin | 1986 | Antibacterial | Daptomycin | Cubicin | 2003 | N | B |
| FR 901379 | 1991 | Antifungal | Caspofungin acetate | Cancidas | 2001 | ND | F |
| | | Antifungal | Micafungin sodium | Fungard | 2002 | ND | F |
| | | Antifungal | Anidulafungin | Eraxis | 2006 | ND | F |
| Muramyl dipeptide | 1974 | Anticancer | Mifamurtide | Junovan | 2010 | ND | B |
| Peplomycin | 1977 | Anticancer | Peplomycin | Pepleo | 1981 | N | B |
| Romidepsin (FR901228) | 1992 | Anticancer | Romidepsin | Istodax | 2010 | N | B |
| Teicoplanin | 1982 | Antibacterial | Teicoplanin | Targocid | 1988 | N | B |
| Thyroliberin | 1956 | CNS stimulant | Taltirelin | Ceredist | 2000 | ND | H |
| Ubenimex | 1976 | Immunostimulant | Ubenimex | Bestatin | 1987 | ND | B |
| Vancomycin | 1956 | Antibacterial | Telavancin HCl | Vibativ | 2009 | ND | B |
| *Polyketides* | | | | | | | |
| Aclarubicin (aclacinomycin A) | 1975 | Anticancer | Aclarubicin | Aclacin | 1981 | N | B |
| Doxorubicin | 1968 | Antibacterial | Tigecycline | Tygacil | 2005 | ND | B |
| | | Anticancer | Epirubicin HCI | Farmorubicin | 1984 | ND | F |
| | | Anticancer | Pirarubicin | Pinorubicin | 1988 | ND | B |
| | | Anticancer | Idarubicin hydrochloride | Zavedos | 1990 | ND | B |
| | | Anticancer | Valrubicin | Valstar | 1999 | ND | B |
| | | Anticancer | Amrubicin HCl | Calsed | 2002 | ND | B |

| | | | | | | | |
|---|---|---|---|---|---|---|---|
| Halichondrin B | 1986 | Anticancer | Eribulin | Halaven | 2010 | ND | M |
| Lovastatin | 1976 | Hypocholesterolemic | Lovastatin | Mevacor | 1987 | N | F |
| | | Hypocholesterolemic | Simvastatin | Zocor | 1988 | ND | F |
| | | Hypocholesterolemic | Pravastatin | Mevalotin | 1989 | ND | F |
| Mupirocin (pseudomonic acid A) | 1971 | Antibacterial | Mupirocin | Bactroban | 1985 | N | B |
| Mycophenolic acid | 1933 | Immunosuppressant | Mycophenolate mofetil | CellCept | 1995 | ND | F |
| | | Immunosuppressant | Mycophenolate sodium | Myfortic | 2003 | N | F |
| Rifamycin B | 1960 | Antibacterial | Rifaximin | Rifacol | 1985 | ND | B |
| | | Antibacterial | Rifapentine | Rifampin | 1988 | ND | B |
| | | Antibacterial | Rifabutin | Mycobutin | 1992 | ND | B |
| Calicheamicin | 1989 | Anticancer | Zinostatin stimalamer | Smancs | 1994 | ND | B |
| *Prostanes* | | | | | | | |
| Naturally occurring prostaglandins like prostaglandin E, prostacyclin etc. | 1962/1971 | Abortifacient | Sulprostone | Nalador | 1981 | ND | H |
| | | Abortifacient | Gemeprost | Cervagem | 1983 | ND | H |
| | | Antiglaucoma | Unoprostone isopropyl ester | Rescula | 1994 | ND | H |
| | | Antiglaucoma | Latanoprost | Xalatan | 1996 | ND | H |
| | | Antiglaucoma | Bimatoprost | Lumigan | 2001 | ND | H |
| | | Antiglaucoma | Travoprost | Travatan | 2001 | ND | H |
| | | Antiglaucoma | Tafluprost | Taflotan | 2008 | ND | H |
| | | Antihypertensive | Treprostinil sodium | Remodulin | 2002 | ND | H |
| | | Antithrombotic | Limaprost | Opalmon | 1988 | ND | H |
| | | Antiulcer | Enprostil | Gardrin | 1985 | ND | H |
| | | Antiulcer | Misoprostol | Cytotec | 1985 | ND | H |
| | | Antiulcer | Rosaprostol | Rosal | 1985 | ND | H |
| | | Antiulcer | Ornoprostil | Alloca | 1987 | ND | H |

(*Continued*)

**TABLE 14.1 (*Continued*)**

| Related Natural Product | Year of Natural Product Discovery | Disease Area | Generic Name | Trade Name | Year of Product Launch | Natural Product (N) or Derivative (ND) | Producing Organism |
|---|---|---|---|---|---|---|---|
| | | Chronic idiopathic constipation | Lubiprostone | Amitiza | 2006 | ND | H |
| | | Platelet aggreg. inhib. | Epoprostenol sodium | Flolan | 1983 | ND | H |
| | | Platelet aggreg. inhib. | Beraprost sodium | Dorner | 1992 | ND | H |
| | | Platelet aggreg. inhib. | Iloprost | Ilomedin | 1992 | ND | H |
| | | Antiglaucoma | Tafluprost | Zioptan | 2012 | ND | H |
| *β-Lactam* | | | | | | | |
| Cephalosporin C | 1961 | Antibacterial | Cefoperazone sodium | Cefobis | 1981 | ND | F |
| | | Antibacterial | Cefotiam HCl | Pansporin | 1981 | ND | F |
| | | Antibacterial | Cefsulodin sodium | Takesulin | 1981 | ND | F |
| | | Antibacterial | Apalcillin sodium | Lumota | 1982 | ND | F |
| | | Antibacterial | Ceftizoxime sodium | Epocelin | 1982 | ND | F |
| | | Antibacterial | Ceftriaxone sodium | Rocephin | 1982 | ND | F |
| | | Antibacterial | Moxalactam disodium | Shiomarin | 1982 | ND | F |
| | | Antibacterial | Cefmenoxime HCl | Tacef | 1983 | ND | F |
| | | Antibacterial | Ceftazidime | Fortam | 1983 | ND | F |
| | | Antibacterial | Aztreonam | Azactam | 1984 | ND | F |
| | | Antibacterial | Cefonicid sodium | Monocid | 1984 | ND | F |
| | | Antibacterial | Ceforanide | Precef | 1984 | ND | F |
| | | Antibacterial | Cefotetan disodium | Yamatetan | 1984 | ND | F |
| | | Antibacterial | Cefbuperazone sodium | Tomiporan | 1985 | ND | F |
| | | Antibacterial | Cefpiramide sodium | Sepatren | 1985 | ND | F |
| | | Antibacterial | Aspoxicillin | Doyle | 1987 | ND | F |
| | | Antibacterial | Cefixime | Cefspan | 1987 | ND | F |
| | | Antibacterial | Cefminox sodium | Meicelin | 1987 | ND | F |
| | | Antibacterial | Cefpimizole | Ajicef | 1987 | ND | F |

| | | | | | | |
|---|---|---|---|---|---|---|
| | | Antibacterial | Cefteram pivoxil | Tomiron | 1987 | ND | F |
| | | Antibacterial | Cefuroxime axetil | Zinnat | 1987 | ND | F |
| | | Antibacterial | Cefuzonam sodium | Cosmosin | 1987 | ND | F |
| | | Antibacterial | Flomoxef sodium | Flumarin | 1988 | ND | F |
| | | Antibacterial | Cefpodoxime proxetil | Banan | 1989 | ND | F |
| | | Antibacterial | Cefodizime sodium | Neucef | 1990 | ND | F |
| | | Antibacterial | Cefdinir | Cefzon | 1991 | ND | F |
| | | Antibacterial | Cefetamet pivoxil HCl | Globocef | 1992 | ND | F |
| | | Antibacterial | Cefpirome sulfate | Cefrom | 1992 | ND | F |
| | | Antibacterial | Cefprozil | Cefzil | 1992 | ND | F |
| | | Antibacterial | Ceftibuten | Seftem | 1992 | ND | F |
| | | Antibacterial | Loracarbef | Lorabid | 1992 | ND | F |
| | | Antibacterial | Cefepime | Maxipime | 1993 | ND | F |
| | | Antibacterial | Cefditoren pivoxil | Meiact | 1994 | ND | F |
| | | Antibacterial | Cefozopran HCl | Firstcin | 1995 | ND | F |
| | | Antibacterial | Cefcapene pivoxil | Flomox | 1997 | ND | F |
| | | Antibacterial | Cefoselis | Wincef | 1998 | ND | F |
| | | Antibacterial | Ceftobiprole medocaril | Zeftera | 2008 | ND | F |
| Nocardicin A | 1976 | Antibacterial | Carumonam | Amasulin | 1988 | ND | B |
| Penicillin G | 1928 | Antibacterial | Temocillin disodium | Temopen | 1984 | ND | F |
| | | Antibacterial | Lenampicillin HCI | Varacillin | 1987 | ND | F |
| | | Antibacterial | Sultamicillin tosylate | Unasyn | 1987 | ND | F |
| | | Antibacterial | Tazobactam sodium | Tazocillin | 1992 | ND | F |
| | | Beta-lactamase inhibitor | Sulbactam sodium | Sulperazone | 1986 | ND | F |
| Thienamycin | 1976 | Antibacterial | Imipenem/cilastatin | Zienam | 1985 | ND | B |
| | | Antibacterial | Meropenem | Merrem | 1994 | ND | B |
| | | Antibacterial | Panipenem/betamipron | Carbenin | 1994 | ND | B |
| | | Antibacterial | Biapenem | Omegacin | 2002 | ND | B |
| | | Antibacterial | Ertapenem sodium | Invanz | 2002 | ND | B |
| | | Antibacterial | Doripenem | Finibax | 2005 | ND | B |

*(Continued)*

**TABLE 14.1 (*Continued*)**

| Related Natural Product | Year of Natural Product Discovery | Disease Area | Generic Name | Trade Name | Year of Product Launch | Natural Product (N) or Derivative (ND) | Producing Organism |
|---|---|---|---|---|---|---|---|
| *Steroid* | | | | | | | |
| Naturally occurring steroids like hydrocortisone, estradiol etc. | 1954/1929 | 5-Alpha-reductase inhibitor | Finasteride | Proscar | 1992 | ND | H |
| | | Abortifacient | Mifepristone | Mifegyne | 1988 | ND | H |
| | | Anabolic metabolism | Tibolone | Livial | 1988 | ND | H |
| | | Antiallergic | Loteprednol etabonate | Lotemax | 1998 | ND | H |
| | | Antiallergic | Fluticasone furoate | Allermist | 2007 | ND | H |
| | | Antiasthmatic | Budesonide | Pulmicort | 1981 | ND | H |
| | | Antiasthmatic | Ciclesonide | Alvesco | 2005 | ND | H |
| | | Anticancer | Formestane | Lentaron | 1993 | ND | H |
| | | Anticancer | Exemestane | Aromasin | 1999 | ND | H |
| | | Anticancer | Fulvestrant | Faslodex | 2002 | ND | H |
| | | Antihypertensive | Eplerenone | Inspra | 2002 | ND | H |
| | | Anti-inflammatory | Halometasone | Sicorten | 1983 | ND | H |
| | | Anti-inflammatory | Hydrocortisone butyrate | Pandel | 1983 | ND | H |
| | | Anti-inflammatory | Alclometasone dipropionate | Vaderm | 1985 | ND | H |
| | | Anti-inflammatory | Deflazacort | Deflan | 1986 | ND | H |
| | | Anti-inflammatory | Prednicarbate | Dermatop | 1986 | ND | H |
| | | Anti-inflammatory | Mometasone furoate | Elocon | 1987 | ND | H |
| | | Antiinflammatory | Hydrocortisone aceponate | Retef | 1988 | ND | H |
| | | Anti-inflammatory | Fluticasone propionate | Flixonase | 1990 | ND | H |
| | | Anti-inflammatory | Halobetasol propionate | Ultravate | 1991 | ND | H |
| | | Anti-inflammatory | Deprodone propionate | Eclar | 1992 | ND | H |

| | | | | | | |
|---|---|---|---|---|---|---|
| | Anti-inflammatory | Betamethasone butyrate propionate | Antebate | 1994 | ND | H |
| | Anti-inflammatory | Rimexolone | Vexol | 1995 | ND | H |
| | Antiprogestogenic | Gestrinone | Dimetrose | 1986 | ND | H |
| | Benign prostatic hypertrophy | Dutasteride | Duagen | 2001 | ND | H |
| | Calcium metabolism | Secalciferol | Osteo D | 1988 | ND | H |
| | Contraception | Desogestrel/ethinylestradiol | Marvelon | 1981 | ND | H |
| | Contraception | Gestodene/ethinylestradiol | Femodene | 1987 | ND | H |
| | Contraception | Levonorgestrel/ethinyl estradiol | Preven | 1998 | ND | H |
| | Contraception | Drospirenone | Yasmin | 2000 | ND | H |
| | Contraception | Medroxyprogesterone/ estradiol | Lunelle | 2000 | ND | H |
| | Contraception | Norelgestromin/ethinyl estradiol | Ortho Evra | 2001 | ND | H |
| | Contraception | Etonogestrel/ ethinylestradiol | NuvaRing | 2002 | ND | H |
| | Contraception | Ulipristal acetate | ellaOne | 2009 | ND | H |
| | Endometriosis | Dienogest | Dinagest | 2008 | ND | H |
| | Hormone replacement therapy | Estrogen/ medroxyprogesterone | Premphase | 1995 | ND | H |
| | Hormone replacement therapy | Trimegestrone/estradiol | Totelle Sekvens | 2000 | ND | H |
| | Hormone replacement therapy | Drospirenone/estradiol | Angeliq | 2003 | ND | H |
| | Hormone replacement therapy | Estradiol/nomegestrol acetate | Naemis | 2003 | ND | H |
| | Hormone replacement therapy | Estradiol/levonorgestrel | Climara Pro | 2004 | ND | H |

(*Continued*)

**TABLE 14.1 (*Continued*)**

| Related Natural Product | Year of Natural Product Discovery | Disease Area | Generic Name | Trade Name | Year of Product Launch | Natural Product (N) or Derivative (ND) | Producing Organism |
|---|---|---|---|---|---|---|---|
| | | Hormone replacement therapy | Conjugated estrogens b | Enjuvia | 2005 | ND | H |
| | | Hypocholesterolemic | Chenodiol | Chenix | 1983 | N | H |
| | | Hypogonadism | Testosterone undecanoate | Restandol | 1981 | ND | H |
| | | Macular degeneration | Anecortave acetate | Retaane | 2006 | ND | H |
| | | Muscle relaxant | Vecuronium bromide | Norcuron | 1982 | ND | H |
| | | Muscle relaxant | Rocuronium bromide | Zemuron | 1994 | ND | H |
| | | Muscle relaxant | Rapacuronium bromide | Raplon | 1999 | ND | H |
| | | Ocular inflammation | Difluprednate | Durezol | 2008 | ND | H |
| | | Progestogen | Promegestone | Surgestone | 1983 | ND | H |
| | | Progestogen | Nomegestrol acetate | Lutenyl | 1986 | ND | H |
| | | Progestogen | Norgestimate | Cilest | 1986 | ND | H |
| | | Subarachnoid hemorrhage | Tirilazad mesylate | Freedox | 1995 | ND | H |
| | | Cancer | Abiraterone acetate | Zytiga | 2011 | ND | H |
| *Sugar* | | | | | | | |
| Acarbose | 1975 | Antidiabetic | Acarbose | Glucobay | 1990 | N | B |
| Nojirimycin | 1982 | Antidiabetic | Voglibose | Basen | 1994 | ND | B |
| | | Antidiabetic | Miglitol | Diastabol | 1998 | ND | B |
| | | Gaucher's disease | Miglustat | Zavesca | 2003 | ND | B |
| γ-Cyclodextrin | 1951 | Antidote (neuromuscular block) | Sugammadex sodium | Bridion | 2008 | ND | B |
| Kanamycin | 1957 | Antibacterial | Arbekacin | Habekacin | 1990 | ND | B |

| | | | | | | | |
|---|---|---|---|---|---|---|---|
| Neuraminic acid | 1941 | Antiviral | Oseltamivir | Tamiflu | 1999 | ND | H |
| | | Antiviral | Zanamivir | Relenza | 1999 | ND | H |
| | | Antiviral | Laninamivir octanoate | Inavir | 2010 | ND | H |
| *Terpenoid* | | | | | | | |
| Arglabin | 1982 | Anticancer | Arglabin | No trade name | 1999 | N | P |
| Artemisinin | 1977 | Antiparasitic | Artemether | Artemether | 1987 | ND | P |
| | | Antiparasitic | Artemisinin | Artemisin | 1987 | N | P |
| | | Antiparasitic | Artesunate | Arinate | 1987 | ND | P |
| | | Antiparasitic | Arteether | Artemotil | 2000 | ND | P |
| Chamazulene | 1926 | Antiulcer | Egualen sodium | Azuloxa | 2000 | ND | P |
| Ferruginol | 1939 | Antiulcer | Ecabet sodium | Gastrom | 1993 | ND | P |
| Forskolin | 1977 | Cardiotonic | Colforsin daropate HCl | Adele | 1999 | ND | P |
| Geranylgeraniol | 1939 | Antiulcer | Teprenone | Selbex | 1984 | ND | P |
| Ingenolmebutate | 1980 | Cancer | Ingenolmebutate | Picato | 2012 | N | P |
| Paclitaxel | 1971 | Anticancer | Paclitaxel | Taxol | 1993 | N | P |
| | | Anticancer | Docetaxel | Taxotere | 1995 | ND | P |
| | | Anticancer | Cabazitaxel | Jevtana | 2010 | ND | P |
| Plaunotol | 1977 | Antiulcer | Plaunotol | Kelnac | 1987 | N | P |
| Pleuromutilin | 1950 | Antibacterial | Retapamulin | Altabax | 2007 | ND | F |
| Retinoic acid | 1947 | Antiacne | Isotretinoin | Accutane | 1982 | N | H |
| | | Anticancer | Alitretinoin | Panretin | 1999 | ND | H |
| | | Antiulcer | Tretinoin tocoferil | Olcenon | 1993 | ND | H |
| | | Antipsoriatic | Acitretin | Neotigason | 1989 | ND | H |
| Tetrahydrocannabinol | 1940 | Antinauseant | Dronabinol | Marinol | 1986 | N | P |

B, bacterium; F, fungi; H, human metabolite; M, marine invertebrate; N, natural product; ND, natural product derivative; P, plant.

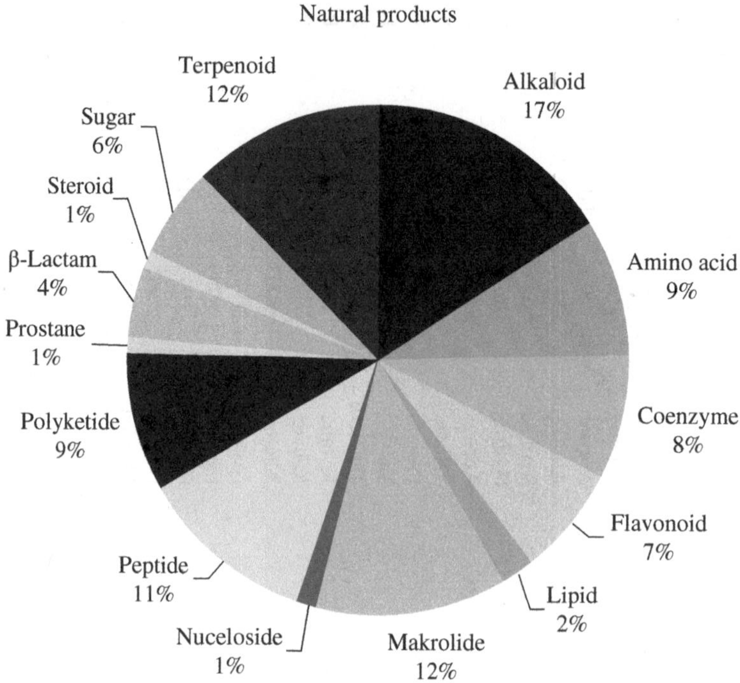


**FIGURE 14.1** Natural product scaffolds related to active ingredients in commercial drugs ($n=89$). (*See insert for color representation of the figure.*)

Forty-five percent of natural product-derived drugs are based on less than 10 natural products belonging to three smaller classes: steroids, prostanes, and β-lactams (Fig. 14.2).

Whereas most of the successful natural product scaffolds lead to only 1–5 different drugs on the market, at least 53 derivatives of steroids, 37 derivatives of cephalosporins and 17 derivatives of prostanes were launched between 1980 and 2013.

Eighty percent of the natural products leading to drugs on the market were isolated from bacteria, plants, fungi, and marine invertebrates (one sponge and one tunicate). Twenty percent are known as human metabolites, coenzymes, or vitamins (Fig. 14.3).

This distribution illustrates two general strategies in natural product-based drug discovery: a random screening approach and a more rational metabolomics approach.

The screening approach uses collections of pure natural products, extracts, or fractions together with suitable molecular or cellular screening systems to discover active compounds that are able to interact with a molecular target. Natural products are inherently suited to bind to proteins as they herald from nature's evolutionary process, synthesized by enzymes. Therefore, probably without exception, they are biologically active in some way. Nonetheless, they are not optimized to cure human diseases. Antibiotic natural products synthesized by fungi or bacteria to fight competing microorganisms or cytotoxic natural products in general (averting predators) being utilized in drug discovery as antibiotics or anticancer drugs might be deviations from this dogma. Screening of recently

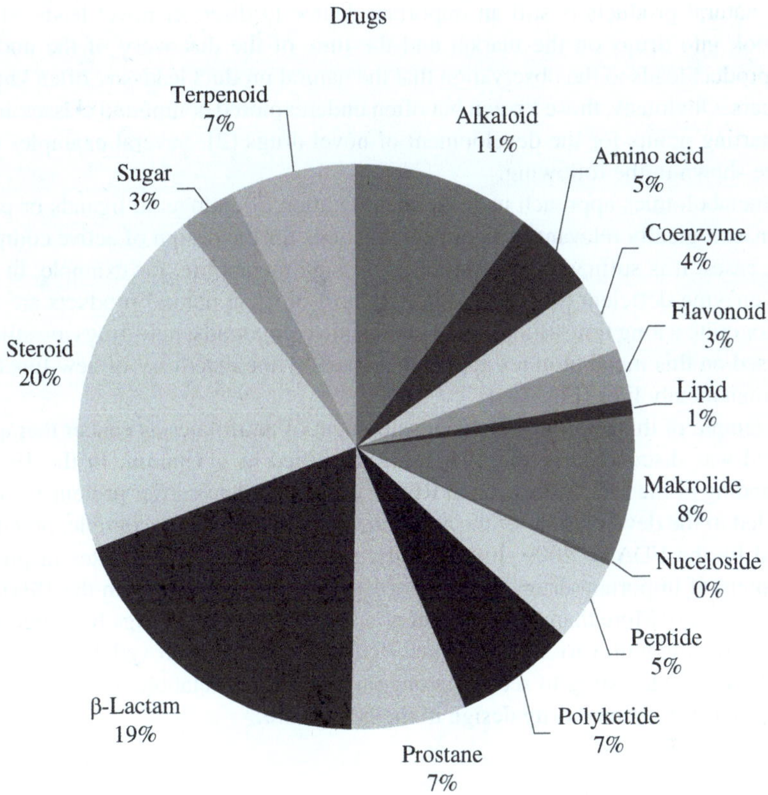


**FIGURE 14.2** 262 drugs based on natural products ($n=30$) or natural product derivatives ($n=232$). (*See insert for color representation of the figure.*)

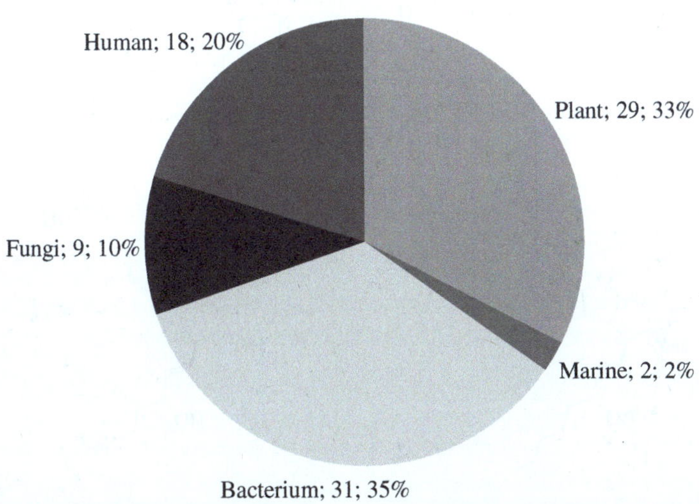


**FIGURE 14.3** Class of producing organism of corresponding natural product scaffolds ($n=89$). (*See insert for color representation of the figure.*)

isolated natural products is still an important source to discover novel leads. Having a closer look into drugs on the market and the time of the discovery of the underlying natural product leads to the observation that the natural product leads are often known for many years. Obviously, those known but often underexplored compound classes are interesting starting points for the development of novel drugs [2]. Several examples of such drugs are shown in the following.

The metabolomics approach uses the identification of endogenic ligands or products of pharmacologically relevant biochemical reactions for the design of active compounds. In some cases, it is sufficient to substitute a missing metabolite, for example, in case of specific enzyme-deficient diseases. Although most of these natural products are already known for quite a long time as important endogenic compounds, new drugs are still developed based on this metabolomics approach, driven by the discovery of new functions of such human metabolites (Fig. 14.4).

An example of this approach is the development of antifolates as cancer therapeutics. Folic acid was discovered in the 1940s and identified as a vitamin. In the 1980s, the understanding of the role of the protein RFC-1 (reduced folate carrier protein 1) in cancer biology led to the development of the antifolate class of drugs, for example, pralatrexate, approved by the FDA in 2009. Interestingly, the last human metabolites inspiring the development of important drugs were the prostaglandins, discovered in the 1960s, more than 50 years ago. More than 15 important prostaglandin-derived drugs for different indications interacting with various prostaglandin receptors were approved during the last 30 years. It will be interesting to see if the ongoing efforts in metabolomics will yield new starting points via rational drug design in the near future.

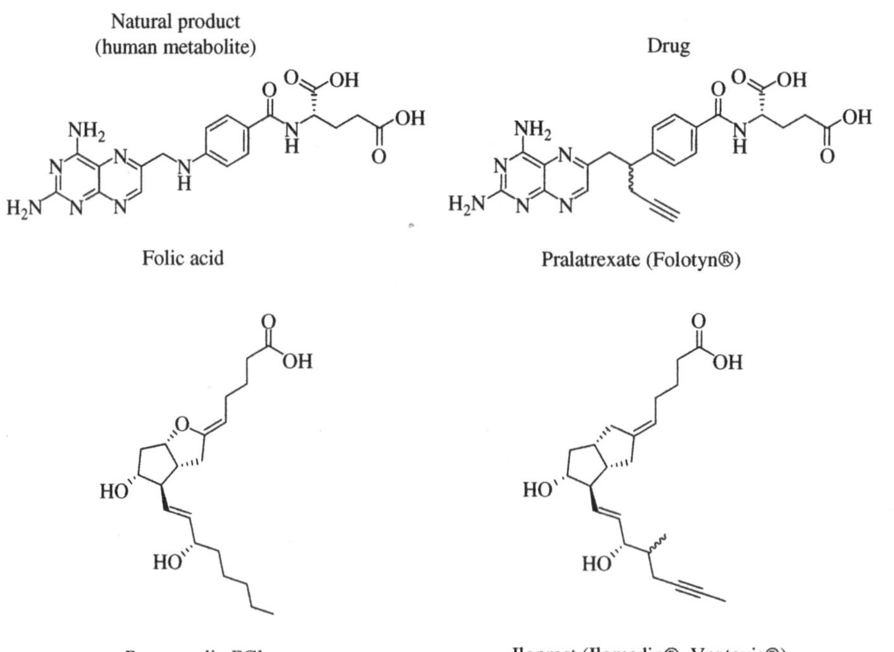


**FIGURE 14.4** Examples for drugs developed by the metabolomics approach.

## 14.2 HIT TO LEAD OPTIMIZATION OF NATURAL PRODUCT HITS

Most of the natural product hits leading to a drug and found by the screening approach are non-drug-like compounds. Notably, polyketides, macrolides, and terpenoids are often outside Lipinski space with a high number of stereocenters, functional groups (mostly hydroxyl and carbonyl), protein reactive groups such as epoxides (e.g., epothilone, scopolamine) and α,β-unsaturated carbonyls (e.g., rifamycin B), or potential metabolic liabilities such as β-lactams or oxetane (paclitaxel), endoperoxides (artemisinin) or esters (e.g., paclitaxel, lovastatin, scopolamine). Because each of these functional groups is a potential issue for unwanted side effects, the goal of hit-to-lead optimization programs is usually to reduce the complexity of the natural product and thus to enhance its stability and selectivity. In parallel, optimizing the PK, for example, solubility and stability, is usually necessary.

Interestingly enough, many of the aspects described earlier as being liabilities of natural products are at the same time critical for the observed activity. Warheads as β-lactams or endoperoxides are not generally introduced *de novo* into drug discovery programs. Such compounds are found in natural product screening campaigns or by careful observations of biological phenomena. These exceptional starting points are the unique selling point of natural products, and it is hard to imagine how they could be discovered without some sort of reverse engineering approach. Taking on the principles of nature can direct a drug discovery program in substantially different directions.

There are many different approaches to hit-to-lead optimization of natural products: using naturally occurring analogues, the generation of derivatives by semisynthesis or total synthesis, metabolic engineering, or even scaffold hopping.

## 14.3 CASE STUDY 1: TAXOL

While accounts exist in ancient literature describing the toxic properties of the European yew tree [3], its beneficial features were only discovered in the 1960s. Whereas the toxic properties are associated with the European yew tree (*Taxus baccata*), the American yew tree (*Taxus brevifolia*) does not contain these toxic compounds. The two substances responsible for the toxicity of the European yew tree were just recently identified to be taxine A and taxine B [4]. These two compounds seem to be responsible for livestock poisoning (Fig. 14.5).

As pointed out by Itokawa [5], if the biomaterial used in the discovery of Taxol would have been the European yew, the whole story would have been sadly different due to the toxic properties of these extracts.

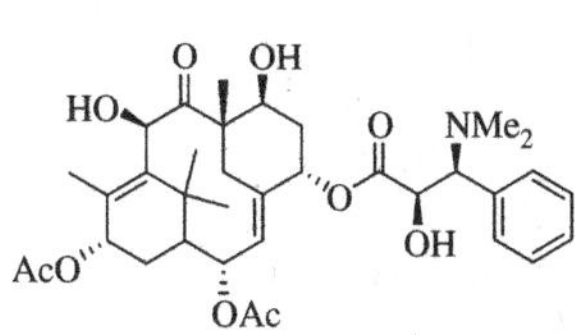


**8.1**

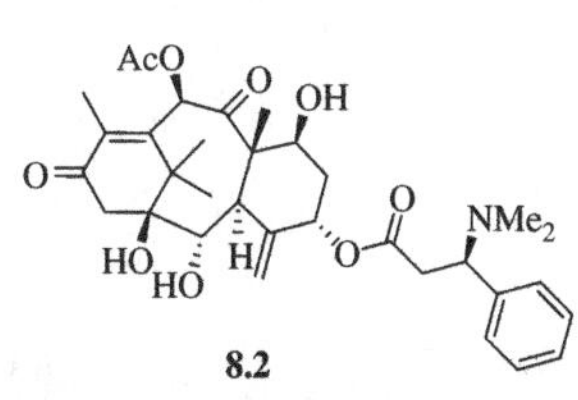


**8.2**

**FIGURE 14.5** Taxines A and B.

The stony path to reaping health benefits for millions of patients worldwide started even earlier in the late 1950s of the twentieth century. The US Department of Agriculture joined forces with the National Cancer Institute (NCI) to initiate a program that is even impressive in the era of high-throughput screening. From 1961 to 1981, 144,045 plant extracts and 16,000 extracts from animals were screened in multiple cellular assays. This program included a plant sample from the pacific yew tree, collected by Arthur S. Barclay in the Gifford Pinchot National Forrest Washington near the foot of Mt. St. Helens in August 1962 [6].

In Durham, North Carolina, Monroe Wall and his assistant, Mansukh Wani, were performing probably the first bioassay-guided fractionation process established on so-called 9kb cancer tissue cultures [7]. In December 1963, they isolated a sample and named it K172. K172 showed the broadest spectrum of activity ever noted in their screening on mouse leukemia cell lines. Four years later, after tedious structure elucidation by $^1$H-NMR and together with Coggon and McPhail from Duke University using X-ray, Wall and Wani published the structure, including the correct absolute stereochemistry of K172, which was named Taxol (Fig. 14.6).

Now, a second major obstacle was becoming obvious. In spite of being architecturally stunning, the molecular arrangement of functional groups around a unique, so far unprecedented, diterpene scaffold is a very demanding challenge in terms of synthetic organic chemistry. In fact, this challenging structure triggered one of the most difficult but competitive races in the field of total synthesis.

At this early stage, the issue of procurement became an increasingly pressing matter, although it was very clear that Taxol was the most promising lead the NCI was pursuing. The interest in Taxol was fuelled further in 1979 when Susan Horwitz published a so far unprecedented mechanism of action [8]. In contrast to the vinca alkaloids, Taxol was promoting *in vitro* microtubule assembly resulting in cell cycle arrest in the G2 and/or M phase. In 1984, the NCI compiled enough data to enter phase I clinical trials. At this point, 10 tons of Pacific yew bark were required. Promising results in phase I paved the way for the entry into clinical phase II trials in 1985. With the publication of phase II data in 1988, the public was becoming aware of the potential of Taxol in terms of curing ovarian cancer (results varied between 30 and 60% of patients displaying observed tumor shrinkage). The demand for Taxol exploded. To cure every US woman suffering from ovarian cancer, the NCI would have needed 240 pounds of pure Taxol—an equivalent of 360,000 lumbered trees. Normally left to rot after being felled, suddenly the Pacific yew tree caught so much attention that the public was getting concerned about the environmental impact [9]. Just as demand exploded, so did the costs invested by the NCI. By 1988, they had already spent US$25 million on the Taxol program. Due to this reason, in August 1989,

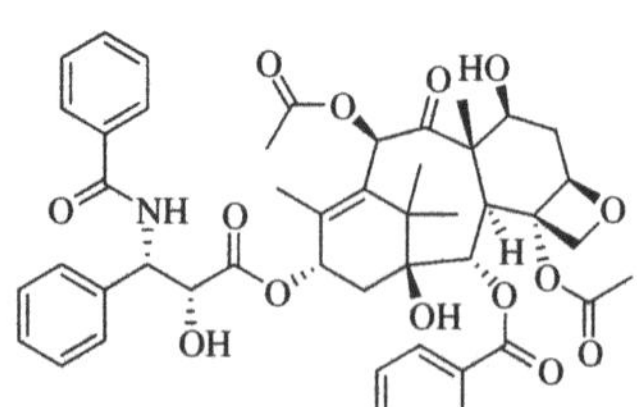

8.3

**FIGURE 14.6** Taxol.

the NCI advertised a Cooperative Research and Development Agreement (CRADA). In December 1989, Bristol-Myers signed an agreement with the NCI to obtain full access to the data the NCI had assembled so far. Furthermore, Bristol-Myers got the exclusive right to commercialize Taxol without being able to file a patent of matter.

As a starting material for the synthesis of Taxol, 10-deacetylbaccatin III was identified by Potier et al. (Fig. 14.7). They performed a tubulin assay by screening diverse parts of felled *T. baccata* L. (European species), which had to give way for a new approach on the Gif-sur-Yvette campus [10]. The extraction yields from the leaves of the European yew tree could be improved from 0.02 to 0.1%. Although 10-deacetylbaccatin III only showed weak bioactivity compared to Taxol, it was an excellent renewable starting material for the semisynthesis of Taxol [11].

As 10-deacetylbaccatin III is sensitive to acid (A-ring Wagner–Meerwein rearrangement and oxetane opening) and base (C2 benzoyl group hydrolysis, C4 acetyl group hydrolysis, and oxetane opening to tetrahydrofuran system), Potier et al. had to establish an orthogonal set of protecting groups that allowed the flexible introduction of the side chain at C13.

After the identification of an excellent starting material for a commercially viable synthesis of Taxol, the need for an efficient side chain synthesis had become more and more urgent. In 1986, Greene et al. published a synthetic approach to the natural Taxol side chain (Fig. 14.8) [12].

From here on, the way was paved for a renewable approach to Taxol. In 1988, Pierre Potier and Andrew Green published the first semisynthetic approach to Taxol (Fig. 14.9) [13].

This synthetic approach could further be used to synthesize side chain analogues. The *t*-butyl side chain analogue could also be synthesized by the efficient methodology published by Greene et al. (Fig. 14.10) [12].

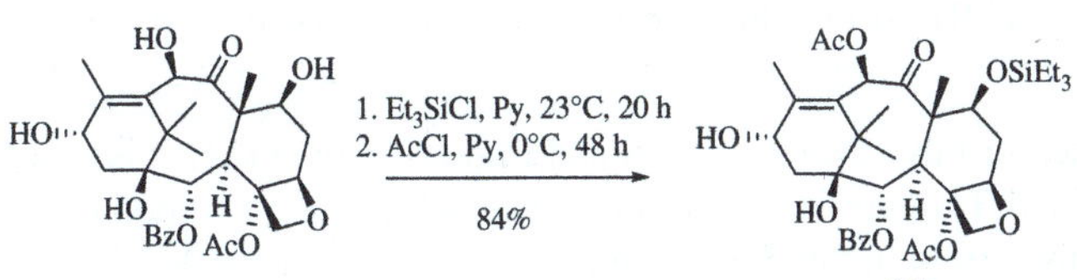


**FIGURE 14.7** Selective protection of deacetylbaccatin III by Poitier et al.

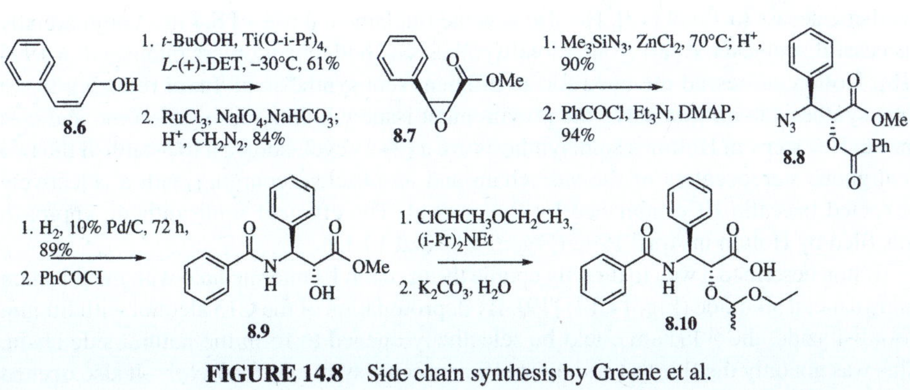


**FIGURE 14.8** Side chain synthesis by Greene et al.

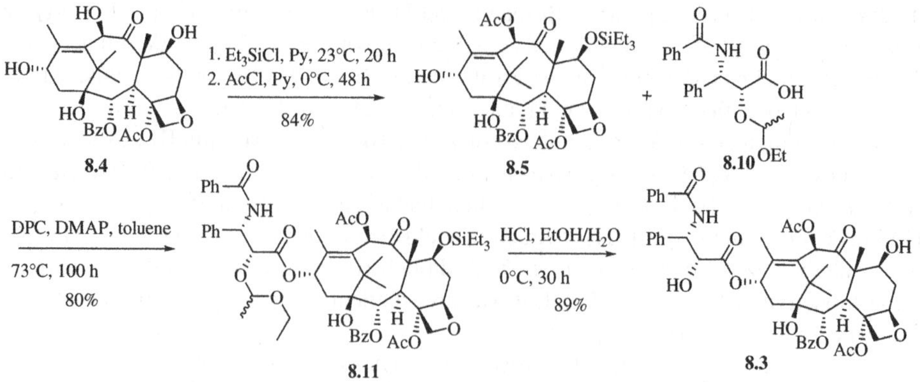


**FIGURE 14.9** Potier–Greene semisynthesis of Taxol.

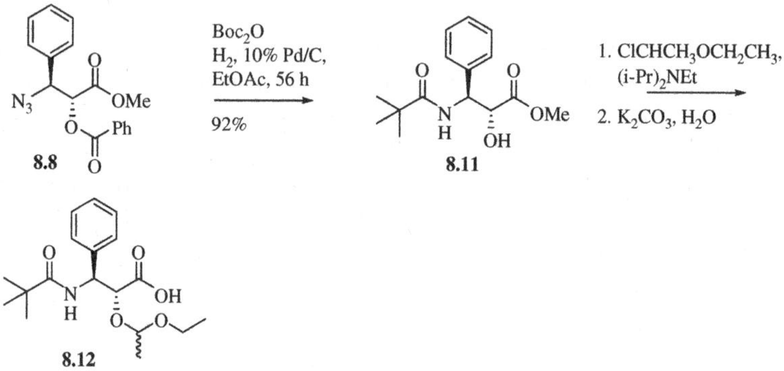


**FIGURE 14.10** Side chain synthesis of Taxotere by Greene et al.

A patent filed in July 1986 by CNRS and Rhône-Poulenc [14] comprised the structure later to be named Taxotere, now the most successful active compound from the taxane structural family. Found in tubulin binding assays, Taxotere showed slightly better activity on tubulin and also in cellular screenings [15]. The synthesis of Taxotere can be achieved in a similar manner to Taxol by using the same universal taxane precursor (Fig. 14.11) [16].

Robert Holton also appreciated the desperate need for an efficient and practical synthetic access to Taxol [17]. He also saw the fundamental role of **8.4** in a commercially successful synthesis. Apart from a total synthesis of both enantiomers published in 1994 [18], Holton addressed the need for an efficient semisynthesis of Taxol realizing that a total synthesis might not solve the procurement issue within a reasonable time and cost frame. Key steps in Holton's semisynthesis are a [2+2]-cycloaddition to establish the two contiguous stereocenters of the side chain and an efficient coupling with a selectively protected baccatin III established by Potier et al. The efficient semisynthetic approach was filed by Holton in April 1992 (Figs. 14.12 and 14.13).

Holton described a way to directly couple the baccatin III intermediate with the β-lactam using a metal alkoxide (Fig. 14.14) [19]. By deprotonation of the C13 alcohol with lithium hexadisilazide, the β-lactam could be selectively opened to form the natural side chain. This was not only the shortest and most efficient semi-synthesis for Taxol—it also opened

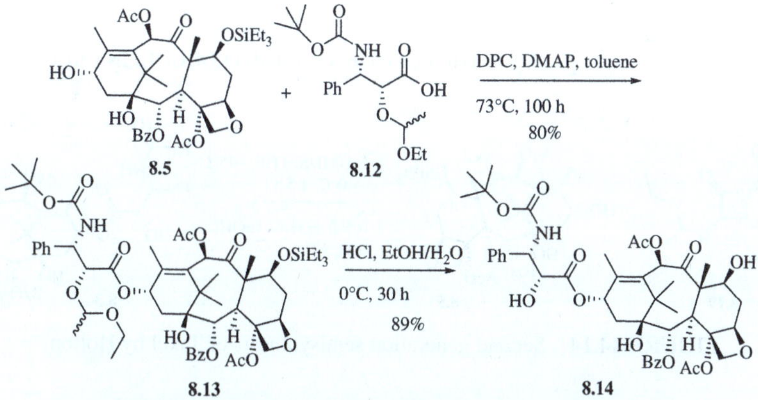

FIGURE 14.11 Semisynthesis of Taxotere by Potier et al.

FIGURE 14.12 10-Deacetylbaccatin III and *Taxus baccata* (background). (*See insert for color representation of the figure.*)

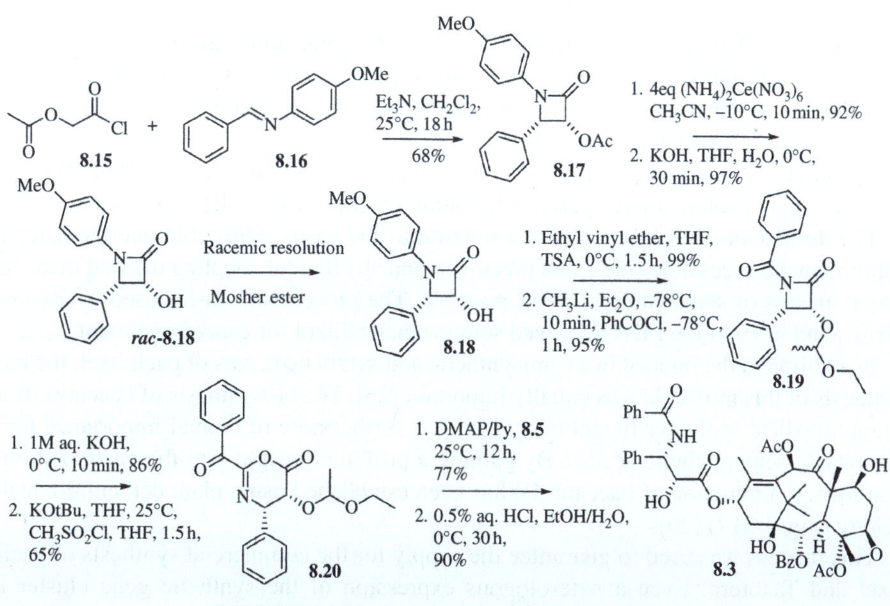

FIGURE 14.13 Holton semisynthesis of Taxol.

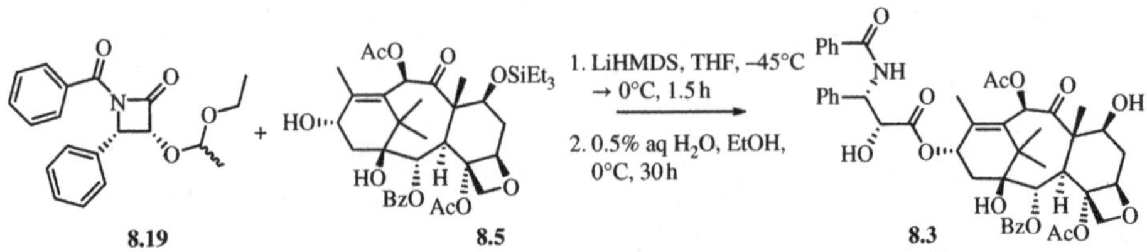

**FIGURE 14.14** Second generation semisynthesis of Taxol by Holton.

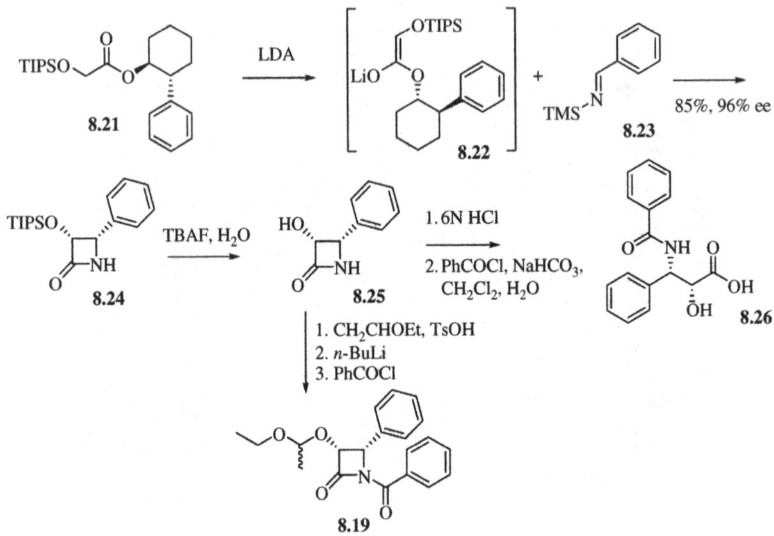

**FIGURE 14.15** Side chain synthesis by Ojima et al.

a way for side chain analogues. Holton published his total synthesis [18] after Nicolaou et al. in 1994 (other total syntheses were published by Danishefsky et al., Wender et al., and Mukaiyama et al.) [18, 20, 21–23] and pursued first the route having the highest impact for cancer patients worldwide.

Ojima et al. finally contributed a further improved enantioselective side chain synthesis using an enolate imine cyclocondensation reaction (Figs. 14.15 and 14.16) [24].

The three major breakthroughs of a renewable and easily adjustable intermediate, an unproblematic accessible side chain precursor, and an efficient coupling method made the major success of semisynthetic Taxol possible. The process was in-licensed by Bristol-Myers, and in 1994 the FDA approved semisynthetic Taxol for cancer treatment.

In addition to the interest in a semisynthetic and synthetic access of paclitaxel, the biosynthesis of this molecule was equally important [25]. The biosynthesis of baccatin III as an intermediate in the paclitaxel biosynthesis is furthermore of mutual importance for a sustainable semisynthetic access. By gaining a profound insight into the paclitaxel biosynthesis, a production of baccatin III has been established using plant cell culture technology (Fig. 14.17) [26].

This process was used to guarantee the supply for the commercial synthesis of paclitaxel and Taxotere. Even a heterologous expression of the synthetic gene cluster in

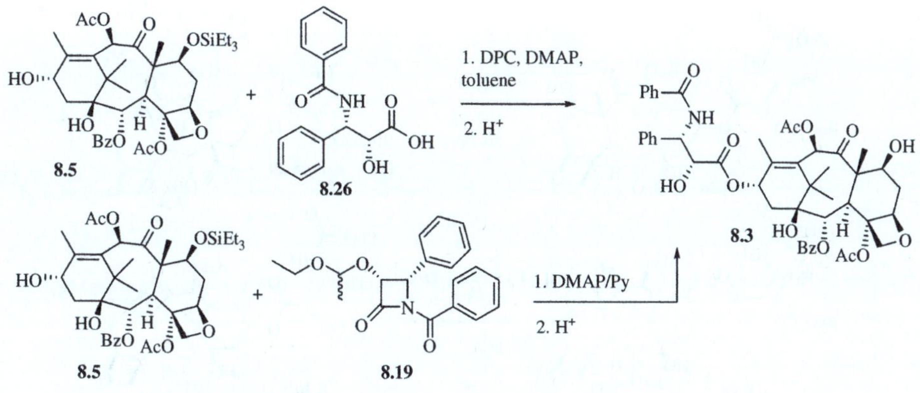

**FIGURE 14.16** Semisynthesis of Taxol by Ojima et al.

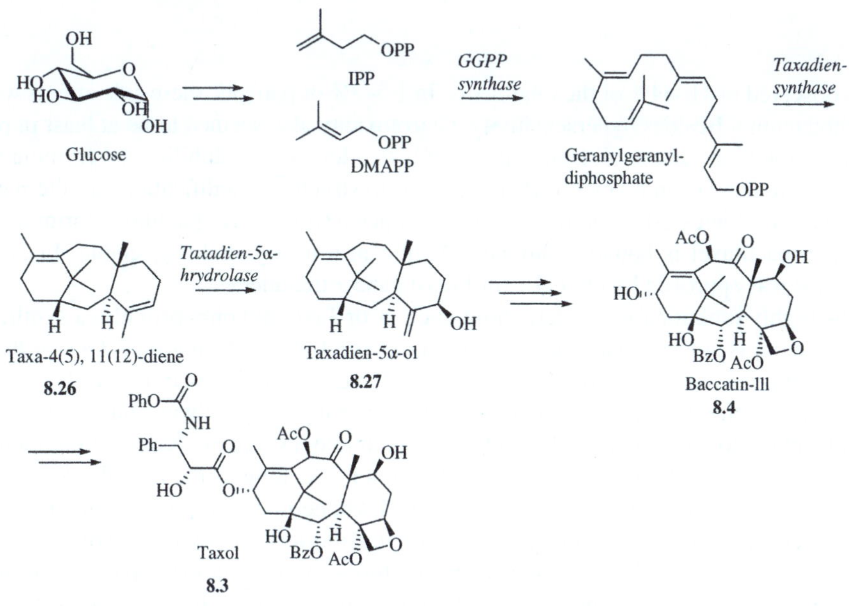

**FIGURE 14.17** Taxol biosynthesis

bacteria can be used for the production of the semisynthetic precursor, and excitingly, this will further lower the production costs for drugs belonging to the taxane family in the future [27].

Besides the accessibility of the drug, there was a second major drawback connected to Taxol. The compound was virtually insoluble in water. This problem was addressed by the NCI using a formulation called Cremophor EL made of castor oil. Cremophor EL is a widely used polyoxyethylated castor oil. With the solubility being a show stopper at this time, this was the only solution at hand. It later turned out to be a serious problem and a driving force for invention. Associated with the castor oil formulation is a hypersensitivity reaction in many patients, making a premedication with high doses of corticosteroids and H1 and H2 antagonists necessary. Even with this premedication, side reactions

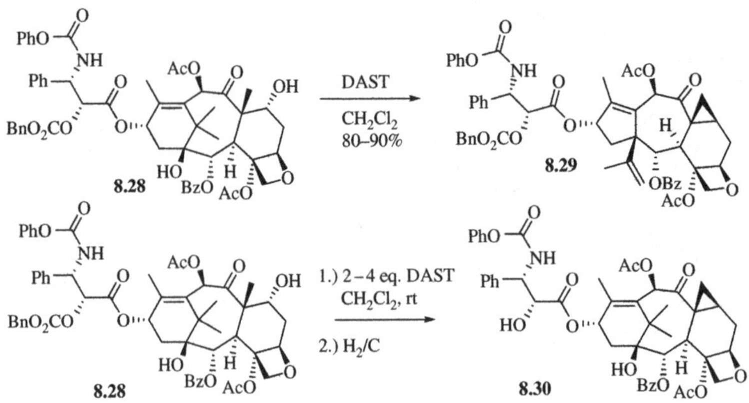

**FIGURE 14.18** Synthesis of a C7–C8 cyclopropyl analogue.

were observed in 41–44% of the treatments. In 1.5–3% of patients, these side effects were life threatening. Besides hypersensitivity, neurotoxicity also seemed to be at least in part associated with the castor oil formulation [28]. To address the solubility and formulation issues, there have been many efforts ranging from structural modifications to alternated formulation vehicles. Abraxis Bioscience developed a solvent-free paclitaxel formulation in which paclitaxel is bound to human albumin in a nanoparticle injectable. In 2005, Abraxane was approved by the FDA for breast cancer treatment.

The highly functionalized diterpenoid nucleus of Taxol not only provides a synthetic challenge but is subject to unusual and often unexpected transformations. One of these transformations found by serendipity even led to a compound now in phase III clinical trials. Unexpected transformation and skeleton rearrangements always have the potential advantage to alter the ligand backbone, which may eventually lead to a compound more efficient in multidrug-resistant (MDR) tumors. As long as the overall pharmacophore stays in place, subtle alterations of the backbone may lead to compounds being less effectively excluded by efflux pumps or transporters. A group at Bristol-Myers Squibb trying to establish a more thorough structure–activity relationship (SAR) around the C7 position of Taxol found an interesting transformation by chance: upon treatment of 7-epi-Taxol with diethylaminosulfur trifluoride (DAST), the C7–C8 cyclopropyl forms in conjunction with a ring contraction of the A-ring a five-membered ring (Fig. 14.18).

By lowering the reaction temperature and the equivalents of DAST, the A-ring contraction could be avoided [29]. After hydrogenation, the generated cyclopropyl analogue of Taxol showed very similar bioactivity as compared to the parent compound.

The initially formed 5-7-6 backbone showed remarkably diminished activity on cancer cell lines (KB cell culture) though compounds of the same backbone showed similar activities in tubulin depolymerization assays to Taxol itself [30].

Combining the Taxotere side chain with the cyclopropyl diterpenoid scaffold, Sanofi introduced a compound later named larotaxel into clinical trials. Larotaxel is currently in phase III clinical trials against metastatic breast cancer pretreated with taxanes [31]. Especially against P-gp-mediated efflux, larotaxel showed promising results as compared to the parent compounds Taxol and Taxotere (Fig. 14.19).

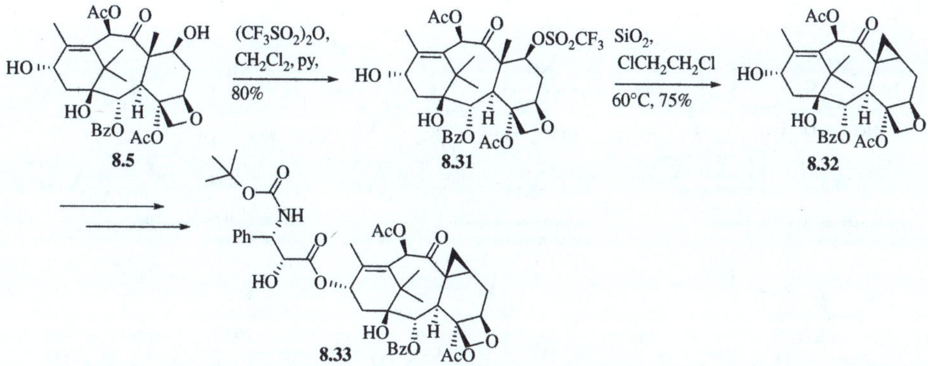

**FIGURE 14.19** Synthesis of larotaxel.

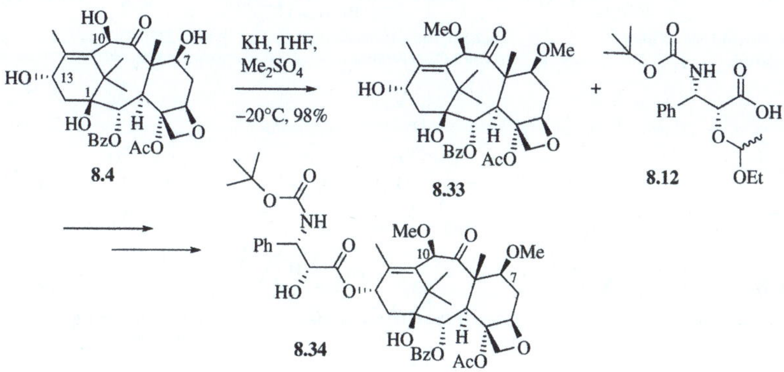

**FIGURE 14.20** Semisynthesis of cabazitaxel.

A C7, C10 bis-methoxy analogue of Taxotere known as cabazitaxel (Jevtana) from Sanofi recently received FDA approval for the treatment of hormone refractory prostate cancer in combination with prednisone. Even this very subtle structural change seems to influence the solubility and PK properties as compared to the parent compound. The elegant semi-synthesis starts from 10-deacetylbaccatin III. The 7 and 10 positions of the tetrol are selectively alkylated under low-temperature conditions (Fig. 14.20) [32].

As depicted in Figure 14.21, in addition to paclitaxel, Taxotere, cabazitaxel, and Abraxane, there are at least six further taxanes in clinical trials.

Larotaxel is in phase III clinical trials for pancreatic cancer, milataxel in phase II clinical trials for mesothelioma, ortataxel in phase II clinical trials for taxane-resistant tumors, and tesetaxel in phase II clinical trials for gastric cancer, colorectal cancer, and melanoma. BMS-275183 was in phase II clinical trials for non-small cell lung cancer (the clinical trial was terminated because of variable pharmacokinetics), and TPI-287 is in phase II clinical trials for prostate cancer and pediatric CNS cancer [33].

Alongside the compounds already in clinical trials or on the market, a tremendous amount of effort has been expended to develop even more selective analogues heralding from the taxane scaffold. While in its forties, it seems fair to say that Taxol has many fascinating sides still to be discovered. It is worth mentioning that, in some cases, nature does not uncover its secrets easily.

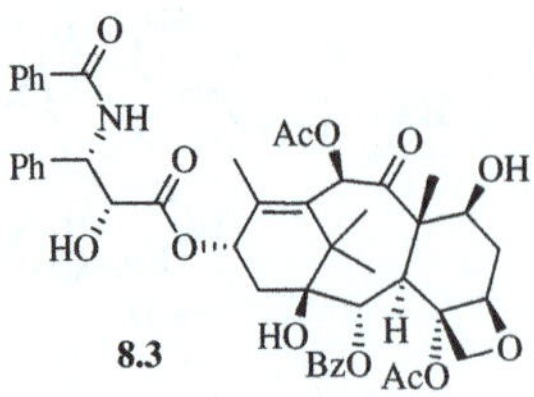

8.3

**Paclitaxel** (marketed by Bristol Meyer Squibb)

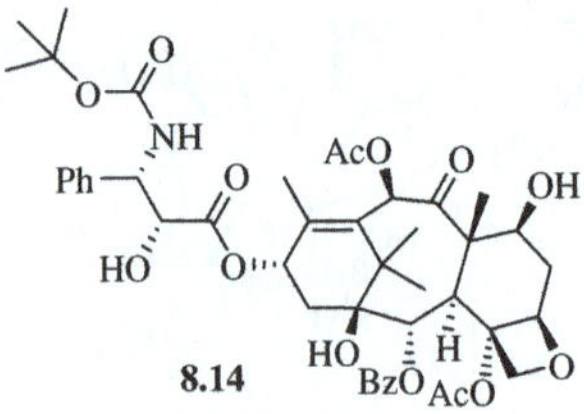

8.14

**Taxotere** (marketed by Sanofi Aventis)

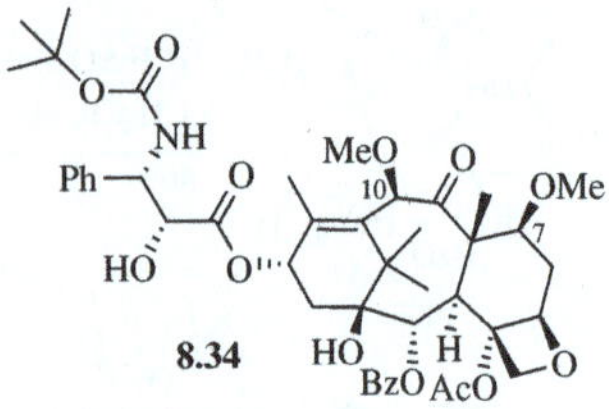

8.34

**Cabazitaxel** (marketed by Sanofi Aventis)

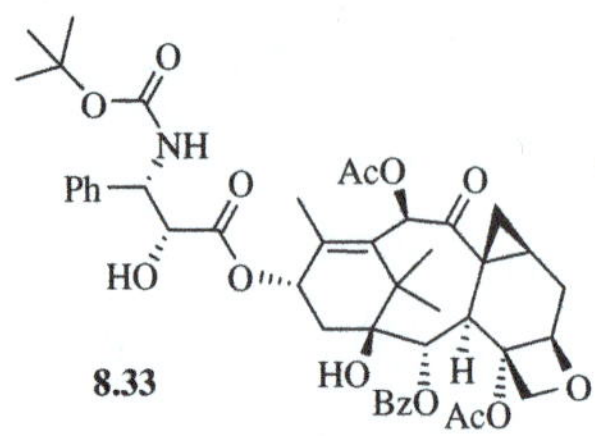

8.33

**Larotaxel** (Phase-III clinical trials, Sanofi Aventis)

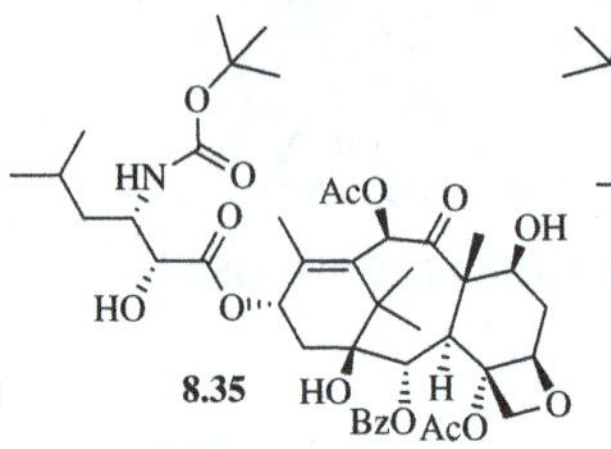

8.35

**Ortataxel** (Phase-II clinical trials, Spectrum Pharmaceuticals)

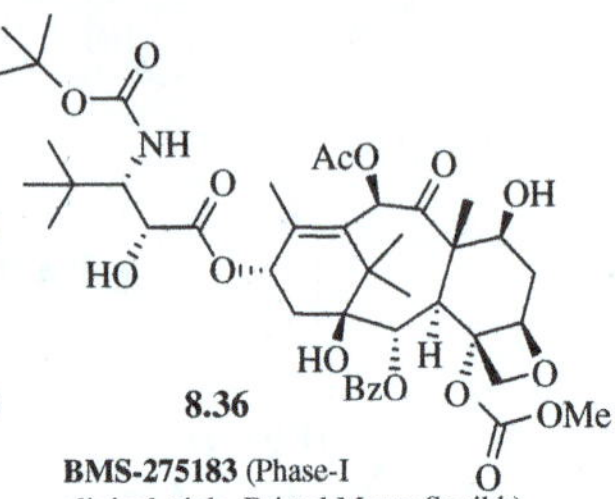

8.36

**BMS-275183** (Phase-I clinical trials, Bristol Meyer Squibb)

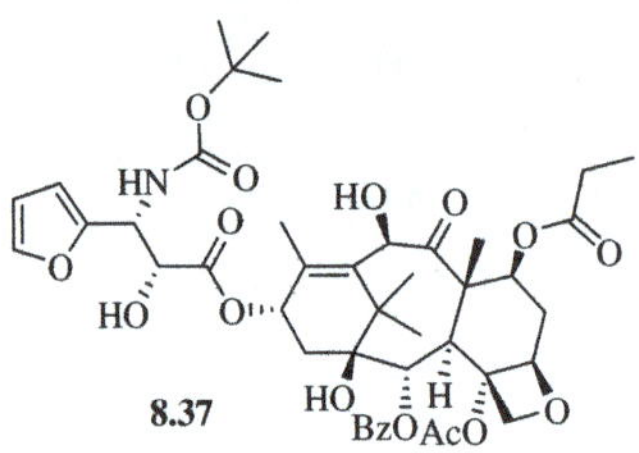

8.37

**Milataxel** (Phase-II clinical trials, Wyeth, Taxalog)

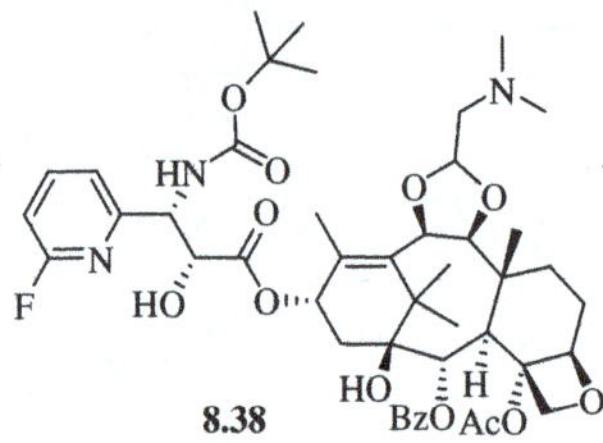

8.38

**Tesetaxel** (Phase-IIb clinical trials, Genta)

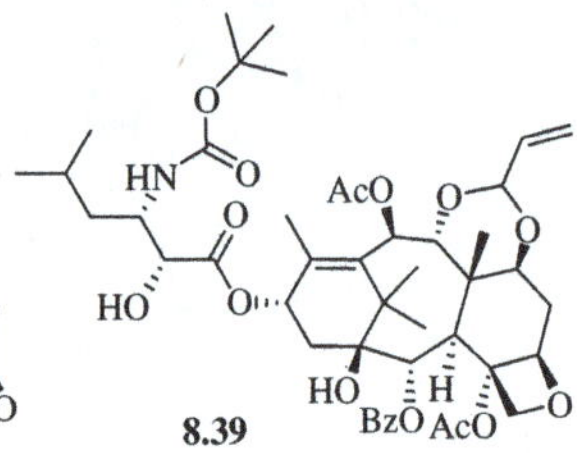

8.39

**TPI** (Phase-II clinical trials, Tapestry Pharmaceuticals)

**FIGURE 14.21** Taxanes on the market or in clinical trials.

## 14.4 CASE STUDY 2: EPOTHILONE

Epothilones were isolated from the myxobacterium *Sorangium cellulosum* and characterized as cytotoxic compounds in 1991 by Reichenbach and Höfle. The interesting activity including the microtubule-stabilizing mode of action similar to Taxol of the initially isolated epothilones A and B led to the isolation of at least 35 naturally occurring epothilones with the complete ring skeleton from different strains of *S. cellulosum*. They mainly differ in methylation, oxidation and double-bond patterns and in the presence of a thiazole or oxazole heterocycle in the side chain [34]. The natural product epothilone B itself (patupilone) was investigated in phase III clinical trials, but finally no significant improvement could be shown in ovarian cancer compared to treatment with doxorubicin [35].

Tang et al. (Kosan) cloned the complete gene cluster of the epothilone biosynthesis and expressed the genes heterologously in *Streptomyces coelicolor* [36] and *Myxococcus xanthus* [37]. Those strains grow much faster than *S. cellulosum*, but unfortunately with 0.1 mg/ml, the yields were 100-fold lower than in *S. cellulosum* [38]. Nevertheless this biochemical toolbox allowed the generation of several "unnatural natural products" [39]. Kosan selected epothilone D as clinical candidate (KOS-862), and it was investigated in

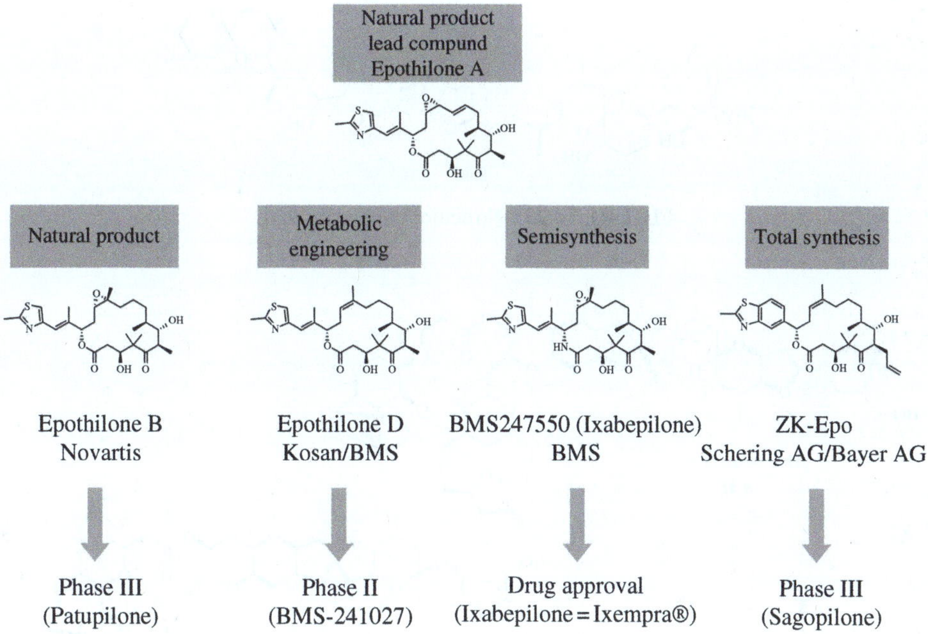


**FIGURE 14.22** Four different hit-to-lead strategies starting with natural products illustrated with anticancer compounds based on the lead epothilone A.

several clinical trials against different cancers but failed to reach the market. Currently, epothilone D is being evaluated for the reduction of Alzheimer-like pathology in aged tau transgenic mice [40].

Hunt et al. (BMS) explored the semisynthetic route with azaepothilone B (ixabepilone), a lactam metabolically more stable than the natural macrolactone. Although hundreds of semisynthetic derivatives were synthesized starting mainly from epothilones A and B [41, 42], the finally selected candidate ixabepilone is structurally quite similar to the natural product. The synthesis via a palladium-catalyzed ring opening reaction with azide as nucleophile, reduction of the azide, and closing of the lactam does not require any protecting groups [43].

Klar et al. (former Schering AG, now Bayer) developed a total synthesis of the epothilone skeleton, synthesized hundreds of derivatives, and finally selected sagopilone (ZK-EPO) as candidate for clinical development. Sagopilone differs from epothilone B by incorporating the double bond of the side chain into a benzothiazole and by replacement of a methyl group by an allyl group (Fig. 14.22) [44].

## 14.5 CASE STUDY 3: ERIBULIN

In 2011, Eisai obtained the FDA approval of Halaven (eribulin mesylate) [45]. This was the celebration of 25 years of hard work in labs both in academia and in industry [46]. The parent molecule of eribulin is halichondrin B, originally isolated from the marine sponge

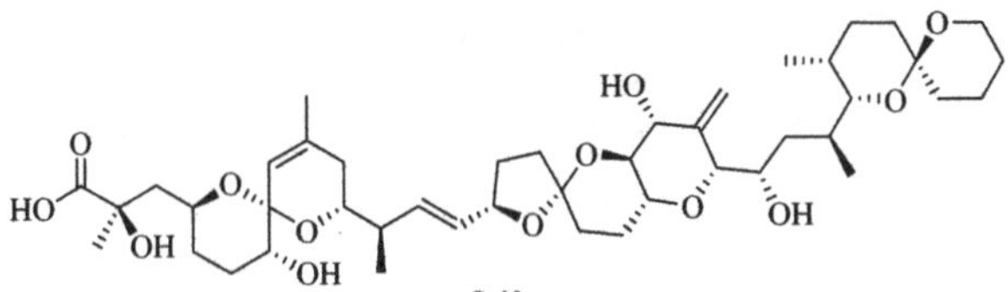

8.40

**FIGURE 14.23** Structure of okadaic acid.

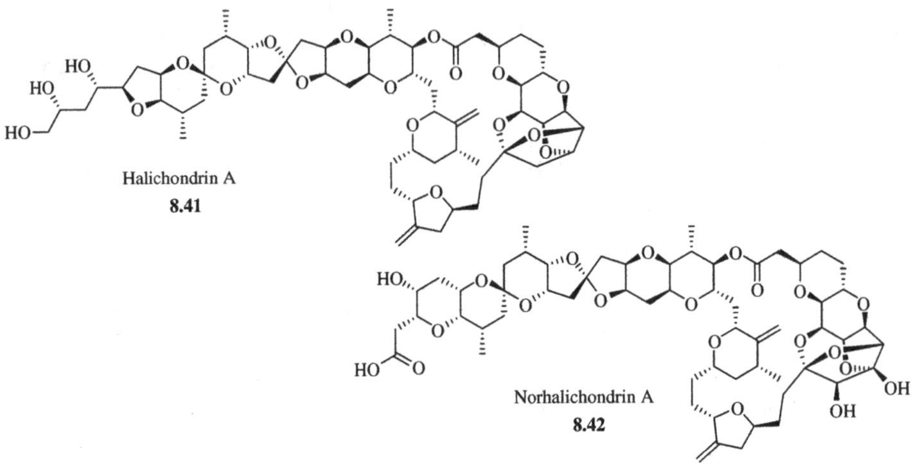

Halichondrin A
8.41

Norhalichondrin A
8.42

**FIGURE 14.24** Structures of halichondrin A and norhalichondrin A.

*Halichondria okadai* Kadota collected on the coast of Aburatsubo in the Miura Peninsula south of Tokyo—the same organism okadaic acid had been isolated from. Okadaic acid is remarkably toxic, unfortunately not showing any antitumor activity *in vivo* if dosed in subtoxic amounts (Fig. 14.23).

*Halichondria okadai* Kadota is a black-colored sponge growing in the mediolittoral zone (intertidal zone in particular the middle shore). Initially, 600 kg of material was collected [47, 48]. In a tedious procedure, 12.5 mg of halichondrin B was isolated from the original 600 kg of biomaterial by Hirata and Uemura using a bioassay-guided fractionation. Already in the first publication (together with NMR and X-ray structural data), very exciting *in vitro* (B-16 melanoma cells) and *in vivo* data were disclosed, and the structure of the related norhalichondrin A was unambiguously elucidated by X-ray crystallography of the corresponding *p*-bromophenacyl ester (Fig. 14.24).

Together with halichondrin B, a whole class (seven further compounds) of structurally related compounds were isolated and their structures elucidated. In 1988, Blunt and Munro isolated the halichondrins from an unrelated New Zealand sponge, the black shallow-water axinellid *Raspalia agminata* [49]. In addition, Munro and Blunt isolated isohomohalichondrin B from the bright yellow sponge *Lissodendoryx* sp. (family Myxillidae, order Poecilosclerida) collected from deep water off the Kaikoura Peninsula [50]. Pettit et al. isolated homohalichondrin B and halichondrin B from the Western Pacific marine sponge *Axinella* sp. collected in Palau (Fig. 14.25) [51].

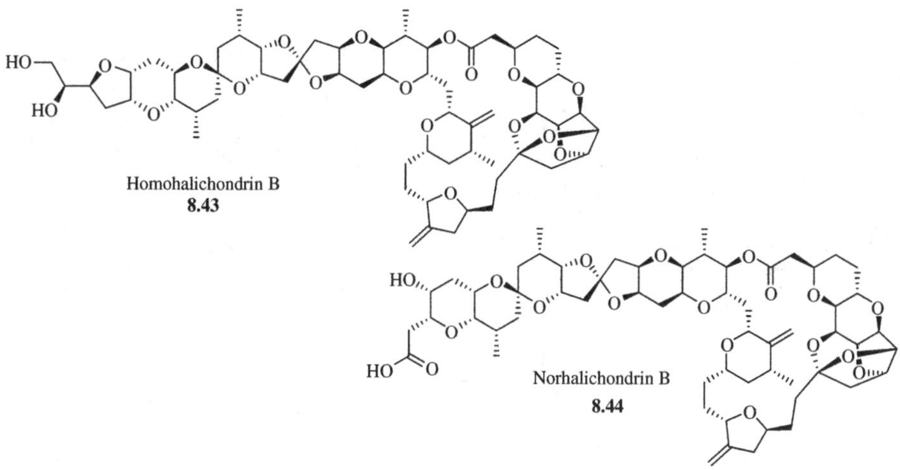


**FIGURE 14.25** Structures of homohalichondrin B and norhalichondrin B.

In spite of the diversity of locations and species, it was speculated if in fact the halichondrins were secondary metabolites by a symbiont of the sponge rather than metabolites of the sponge itself.

From all of the isolated compounds of the halichondrin structural family, halichondrin B seems to be the most active, displaying subnanomolar $IC_{50}$ values in cancer cell tissue cultures [52]. Halichondrin B binds to tubulin without interfering with colchicine. It binds to the vinca domain of tubulin, but not exactly to the vinca site. It inhibits vinca alkaloids (especially vinblastine) binding to tubulin in a noncompetitive manner. It has been shown *in vitro* that halichondrin B disrupts normal mitotic spindle architecture and induces cell cycle arrest in the G2/M phase. The *in vivo* data were very encouraging as well: in several human solid tumor xenograft models, halichondrin B showed exceptional results.

Due to its extraordinary bioactivity, halichondrin B was recommended by the NCI for preclinical trials in 1992. However, the amounts isolated from the original extraction by Hirata and Uemura were discouraging. A reisolation project by Blunt and Munro from *Lissodendoryx* sp. also didn't produce halichondrin B in sufficient amounts. From 200 kg of sponge, Blunt and Munro isolated 43 mg of halichondrin B [53]. A pilot study to cultivate the sponge in an aquaculture revealed an even decreased amount of isolated halichondrin B relative to the harvested biomaterial [54].

The only sustainable method for the procurement of halichondrin B emerged with Kishi's iconic hallmark total synthesis in 1992 (Fig. 14.26) [55].

The Kishi synthesis of halichondrin B is characterized by a highly convergent approach. It mastered the Ni/Cr Nozaki–Hiyama–Kishi cross-coupling reaction for bringing the convergent synthetic intermediates together. The readily accessible carbohydrate starting materials already made Kishi's initial synthesis highly efficient. When the success of halichondrin-derived compounds in the clinic was surfacing, the importance of an efficient synthetic access to the halichondrins became even more pressing. Accordingly, the Kishi group continuously improved the synthetic route to halichondrin B [56] and enabled access to unnatural analogues using the same synthetic protocol. Although there were numerous natural products showing similarly attractive bioactivity, they shared the same molecular complexity as halichondrin B.

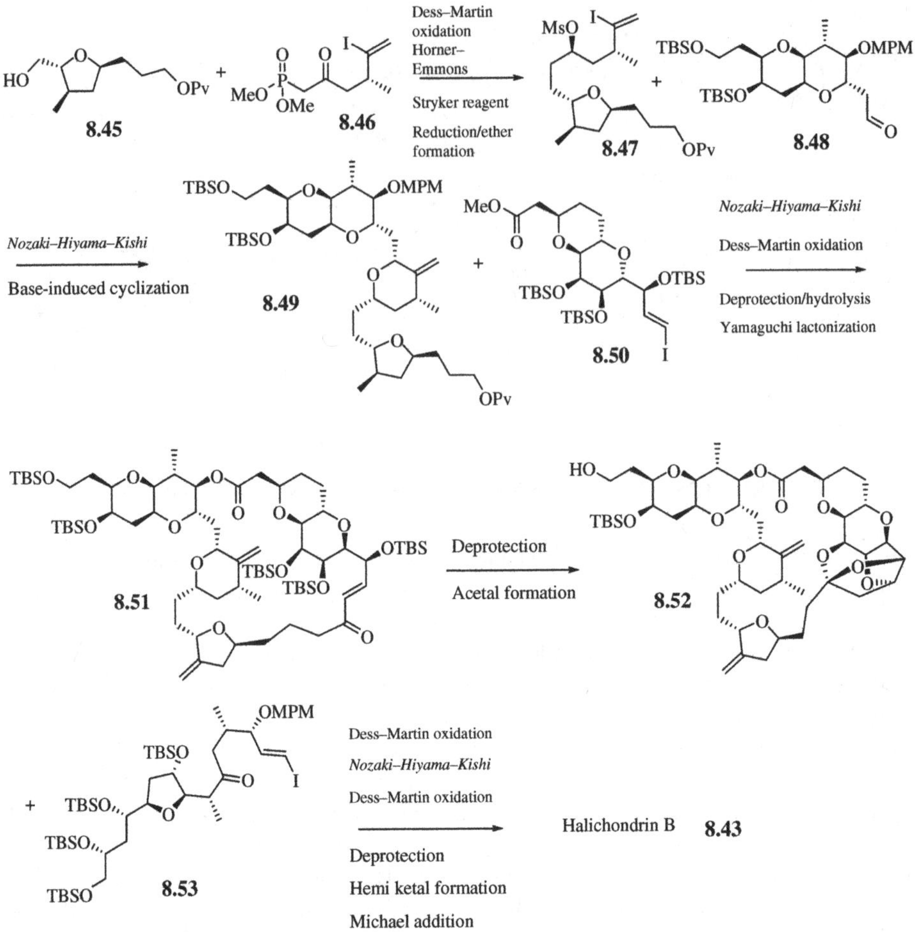


**FIGURE 14.26** Total synthesis of halichondrin B by Kishi et al.

In 1992, samples of the synthetic halichondrin B and several synthetic intermediates were sent to the Eisai Research Institute for evaluation of cell growth inhibitory activity. The most remarkable discovery was the activity of the macrocyclic ketone **8.56** comprising the C1–C38 fragment of halichondrin B [57, 58]. The *in vitro* results of compound **8.56** and halichondrin B were comparable within a magnitude, although **8.56** is a lot less complex in structure. With this observation, a synthetic access to halichondrin-related compounds became in reach of economic feasibility. Compound **8.54**, also prepared by the Kishi group, showed *in vivo* antitumor activity in the LOX human melanoma xenograft model (Fig. 14.27).

With these two structures in hand, a viable starting point for a drug discovery program was reached.

Although it is fair to say that Eisai's decision to start a lead optimization program from there was remarkably courageous, this status would have never been reached without the total synthesis that is a most impressive example of efficient contemporary organic synthesis in its impact on human health. One of the major hurdles in the development of

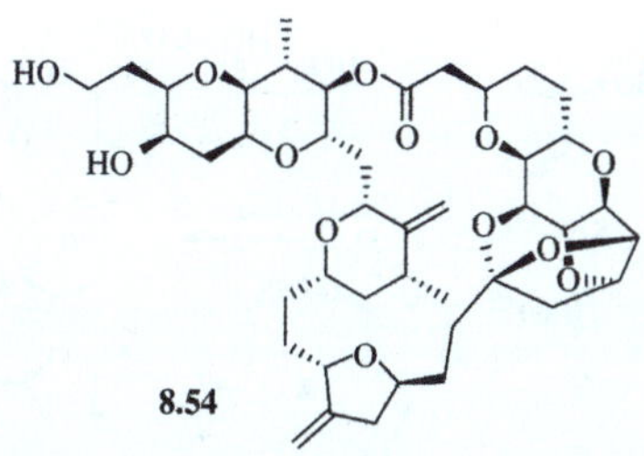

8.54

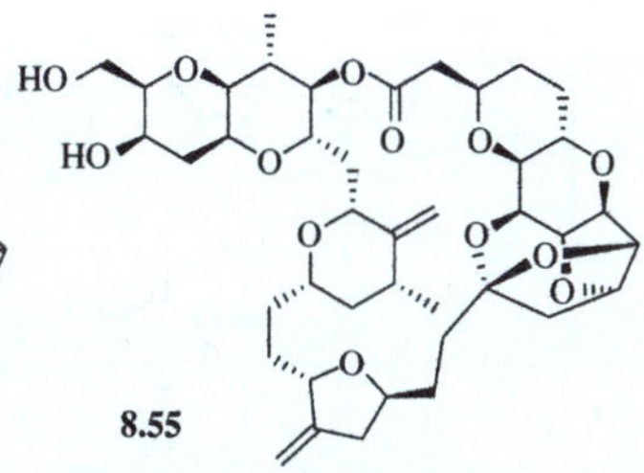

8.55

**FIGURE 14.27** Structure of the macrolactone diol part of halichondrin B.

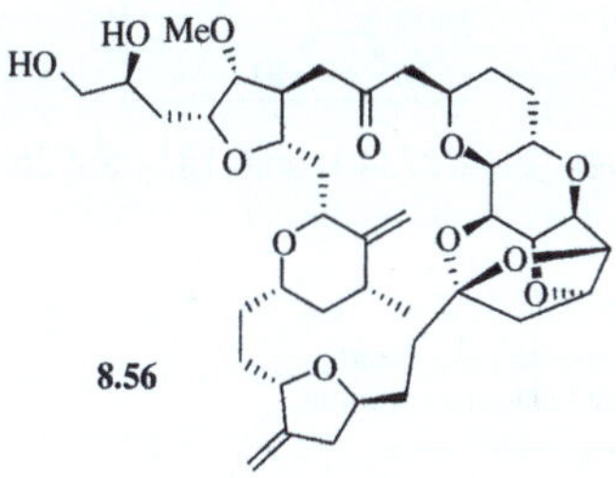

8.56

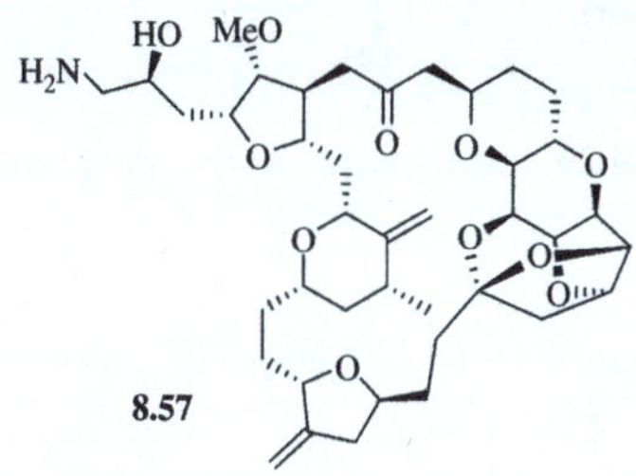

8.57

**FIGURE 14.28** Convergent key intermediates for the synthesis of **2.22.**

halichondrin-derived compounds is the instability of the macrocyclic lactone. At Eisai, more than 180 analogues of halichondrin B were generated and tested. From this tremendous effort, two macrocyclic ketone analogues were developed to clinical candidate status (Fig. 14.28) [59, 60].

The major differences between halichondrin B and the two candidates were the transformation of the C1 lactone to a ketone functionality, a substitution of the C31 methyl with a methoxy group, and the substitution of the C29–C38 tricyclic system with a single tetrahydrofuran. In contrast to the macrolactone, these macrocyclic ketones are stable in plasma for a sufficient period of time, underpinning the theory of a ketone being a more stable bioisostere to a lactone. ER-077349 and ER-086526 (ER-7389, eribulin) both showed a very exciting bioactivity *in vitro* (eribulin has an $IC_{50}$ of 0.07 nM on MDA-MB-435 human breast cancer cells as compared to 0.54 nM for vinblastine and 2.6 nM for paclitaxel) (Fig. 14.29).

Eribulin and ER-077349 were also active on several other human cancer cell lines, while they did not show significant toxicity on MR-90 human fibroblast cells, indicating excellent selectivity. The *in vitro* activities were on average comparable to the activities of halichondrin B. As with the total synthesis of halichondrin B, the Eisai synthesis of eribulin is highly convergent starting from carbohydrate synthons (Fig. 14.30) [61].

The three convergent fragments were coupled by a Nozaki–Hiyama–Kishi reaction followed by a sulfone aldehyde addition and a subsequent Nozaki–Hiyama–Kishi macrocyclization. The late stage intermediate diol was transformed into the amino alcohol **8.57** (eribulin) (Fig. 14.31).

Eribulin can subsequently be transformed into the mesylate salt that is the active ingredient in Eisai's Halaven—now on the market. After successful phase III clinical trials, Halaven received FDA approval in 2010 for metastatic breast cancer [62, 63].

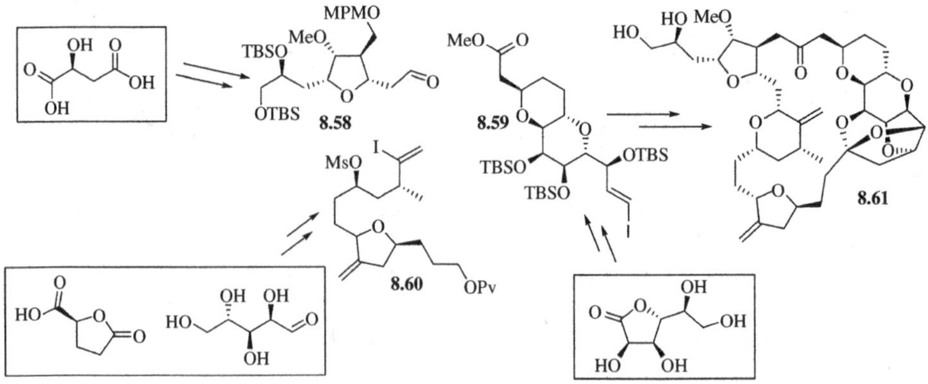


**FIGURE 14.29** Structures of the drug candidates ER-077349 and ER-086526 (ER-7389, eribulin).

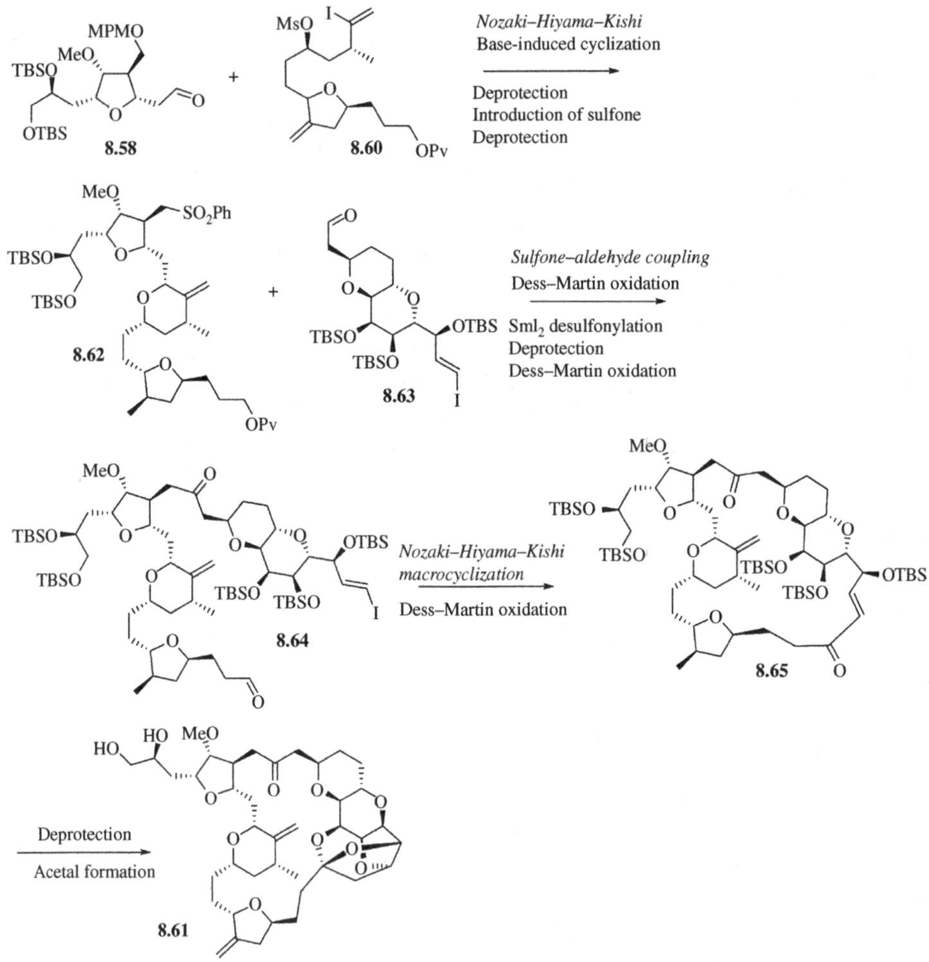


**FIGURE 14.30** Synthesis of late stage intermediate **2.22**.

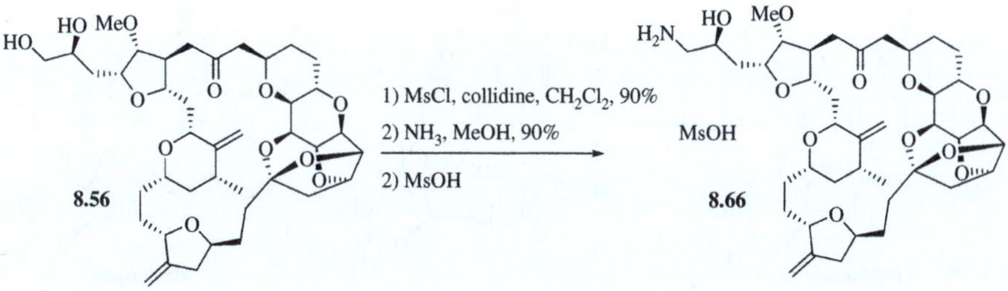


**FIGURE 14.31** Final steps in the synthesis of eribulin mesylate.

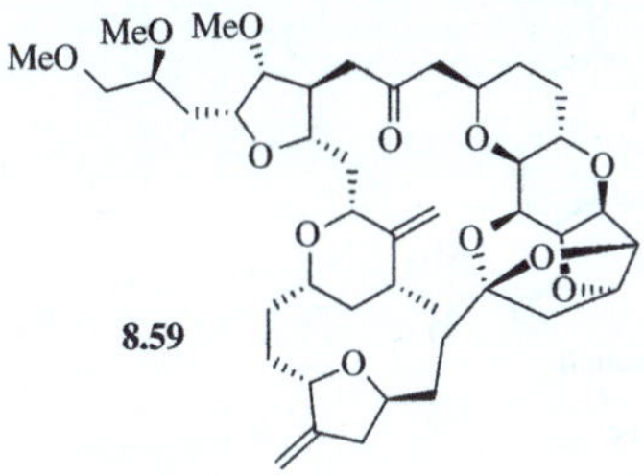

8.59

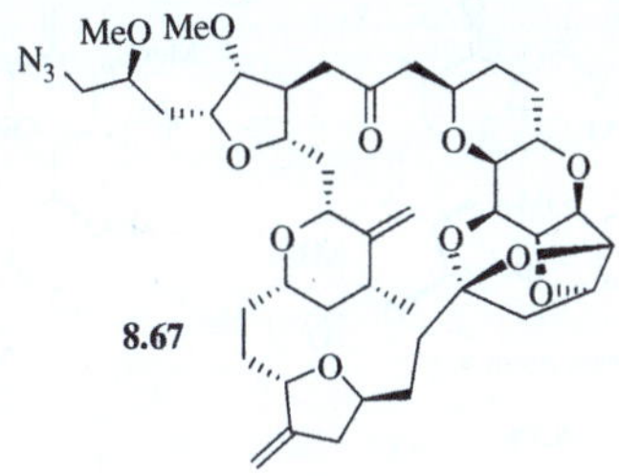

8.67

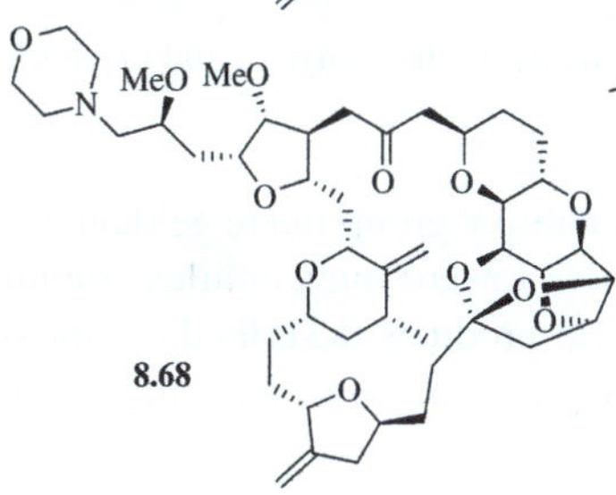

8.68

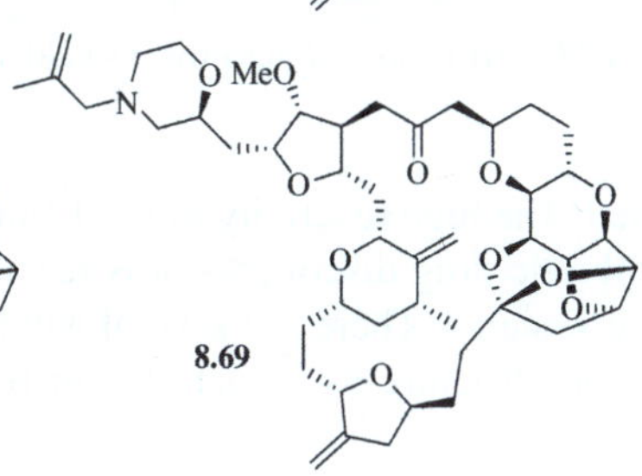

8.69

**FIGURE 14.32** Eribulin analogues showing improved MDR cell line activities or PK properties.

Apart from being a key synthetic intermediate and showing interesting activities in its own right, diol **8.56** can also function as a late stage intermediate for the generation of novel analogues with improved PK properties (Fig. 14.32).

MDR tumors are one major reason for shortfalls in clinical outcomes, especially in the case of refractory tumors. One mechanism by which cancer cells can develop MDR is the overexpression of efflux pumps such as P-gp. Eribulin is a substrate for P-gp, as are paclitaxel and doxorubicin. Diol **8.56**, showing similar *in vitro* and *in vivo* activities as compared to eribulin, is less susceptible to P-gp than eribulin.

## 14.6 CASE STUDY 4: GELDANAMYCIN

Geldanamycin is a benzoquinone ansamycin that binds to the heat shock protein Hsp90. It was first isolated by De Boer et al. from Upjohn by fermentation of *Streptomyces hygroscopicus* in 1970 [64]. The structure of geldanamycin was elucidated shortly after by NMR and IR spectroscopy [65]. Antiparasitic, antifungal, and antibacterial activities

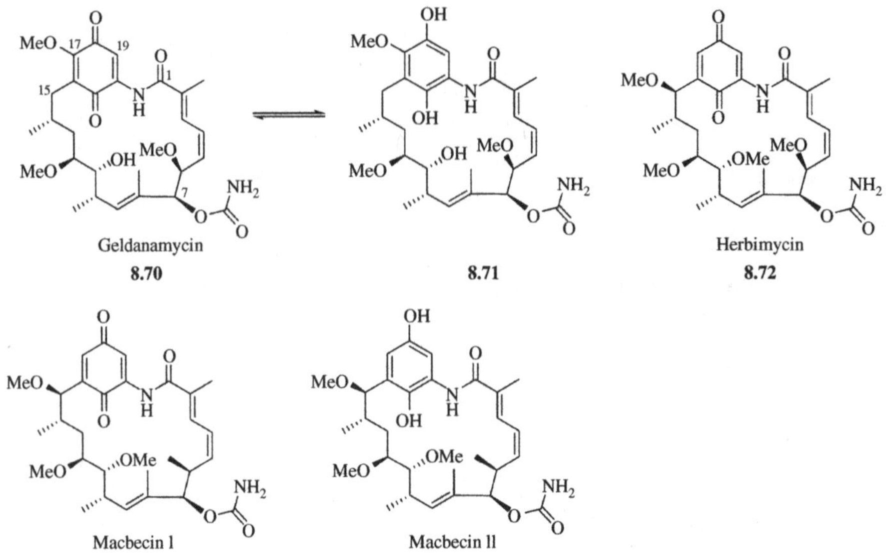


Geldanamycin **8.70** **8.71** Herbimycin **8.72** Macbecin I **8.74** Macbecin II **8.75**

**FIGURE 14.33** Structures of ansamycins geldanamycin, herbimycin, and macbecins I and II.

were reported. The high reactivity of the 17-methoxy group made geldanamycin unsuitable for antibiotic drug discovery—it is reactive toward nucleophiles commonly found in biological systems. There are several natural products described comprising similar bioactivity and structure as geldanamycin (e.g., herbimycin and macbecins I and II) (Fig. 14.33).

In 1995, several analogues of geldanamycin were reported blocking the expression of the ErbB2 oncoprotein in human breast cancer cell lines [66]. While geldanamycin had been postulated to have cell growth inhibition capabilities by kinase (v-Src, Raf-1, ErbB2) inhibition [67], inhibition of Hsp90 was subsequently shown *in vitro* [68]. Hsp90 is a 90 kDa member of a protein superfamily and exists in two isoforms. Hsp90α is the inducible form overexpressed in cancer cells, while Hsp90β is the constitutive form [69]. Hsp90 is one of the most abundant proteins in unstressed cells [70]. It is a ubiquitous molecular chaperone found in eubacteria and all branches of eukarya. Hsp90 is an elongated dimer with a dissociation constant in the low nanomolar range. It consists of three major domains: a highly conserved amino-terminal ATPase domain, a middle domain, and a carboxy-terminal dimerization domain. As a response to cellular stress, Hsp90 gets overexpressed, predominantly in cancer cells. Hsp90 helps its client proteins, many of them notorious oncogenes, to fold in the bioactive conformation. Some of these are validated cancer targets (HER-2/neu (Herceptin (trastuzumab)), Bcr–Abl (Gleevec (imatinib mesylate)), or VEGF (Sutent (sunitinib))). By inhibition of Hsp90, the client proteins undergo degradation as a result of misfolding—very effectively processed by a ubiquitylation proteasome pathway.

Herbimycin and geldanamycin were the first ligands described to bind to Hsp90. Likely due to the highly reactive 17-methoxy group, geldanamycin showed a high liver toxicity and was not suitable as a development candidate for drug discovery. However,

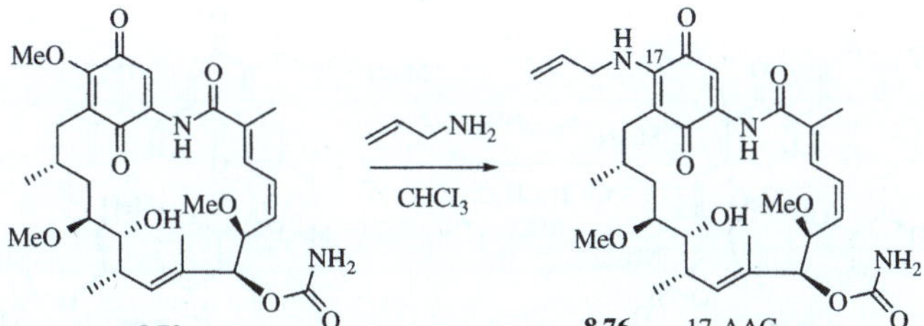


8.70 8.76

**FIGURE 14.34** Synthesis of 17-AAG (17-allylamino geldanamycin).

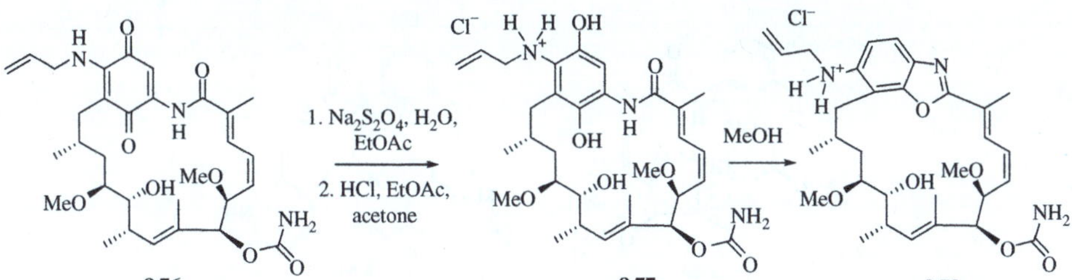


8.76 8.77 8.78

**FIGURE 14.35** Synthesis of the 17-AAG hydroquinone salt.

this high reactivity could be used for a replacement reaction by amines at room temperature. The *in vivo* anticancer activity and the synthesis of such derivatives were first described in a patent application from Kaken Chemical Company [71].

In 1994, 17-AAG was discovered to inhibit Hsp90 (Fig. 14.34). In 1999, 17-AAG entered clinical trials. Although to a lesser extent, 17-AAG still showed a certain hepatotoxicity.

Furthermore, 17-AAG is highly insoluble and difficult to formulate. As shown in Figure 14.35, geldanamycin can be converted *in vivo* into the hydroquinone form.

Infinity Pharmaceuticals reduced 17-AAG with sodium dithionite and used the HCl salt of the hydroquinone as a much more water-soluble compound as compared to the parent [72]. The hydroquinone 17-AAG entered clinical trials in 2005. Unfortunately, it is only stable at −20°C. Under acidic conditions, the benzoxazole analogue formed under these conditions shows diminished bioactivity [73].

By introducing an amino functionality into geldanamycin, Kosan finally increased the solubility of the parent compound [74]. More than 60 derivatives were synthesized and tested. From these, 17-DMAG showed the most promising *in vitro* and *in vivo* profile (Fig. 14.36).

By probing the 7-carbamate position for superior substituents, no residue with improved bioactivity could be identified [75]. It was shown that although many derivatives showed improved *in vitro* activity, only simple alkyl amine or alkoxy substituents showed improved *in vivo* activities. In 2006, 17-DMAG entered clinical trials, which were discontinued in 2008 as a result of focusing the resources on a 17-AAG formulation. Independently, the NCI demonstrated the efficacy, water solubility, and oral bioavailability of 17-DMAG and selected it for advanced studies. Currently, 17-DMAG is in phase I clinical trials for the treatment of relapsed chronic lymphocytic leukemia (CLL), small lymphocytic lymphoma (SLL), and B-cell prolymphocytic leukemia (B-PLL) [76].

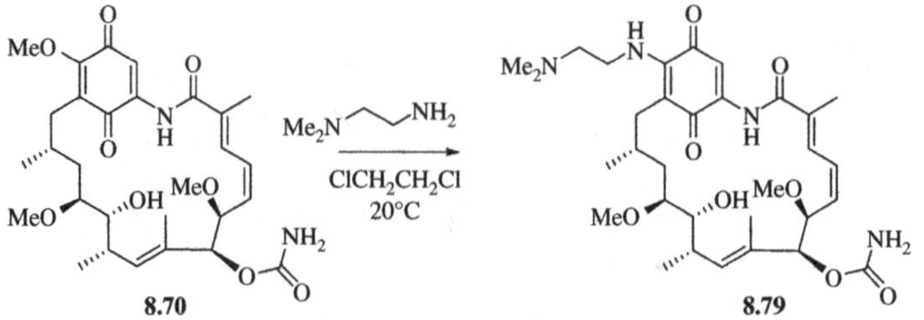

**FIGURE 14.36** Synthesis of 17-DMAG.

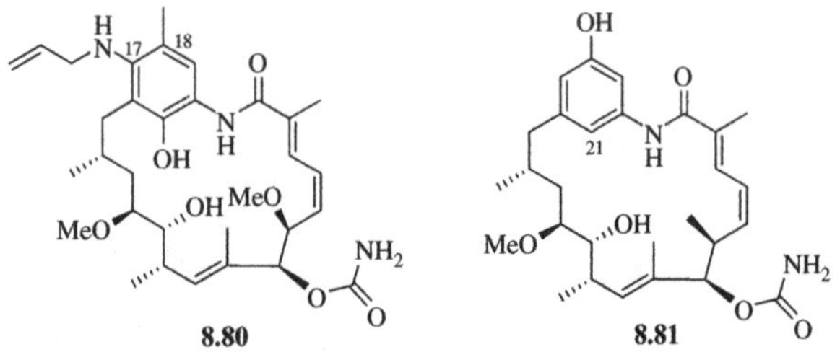

**FIGURE 14.37** Phenol analogues of geldanamycin and macbecin I.

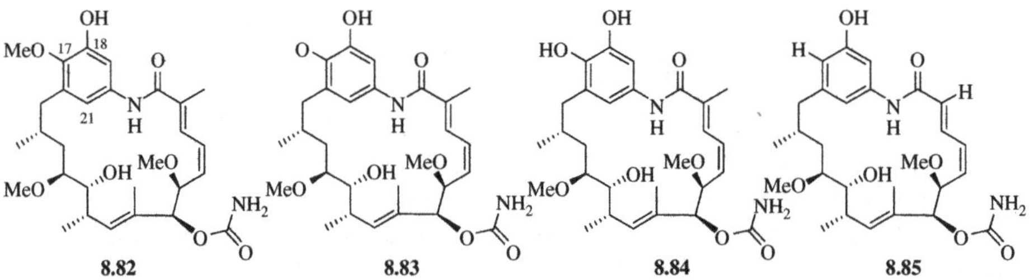

**FIGURE 14.38** Geldanamycin anlogues isolated by Kyowa Hakko.

Geldanamycin, 17-AAG, hydroxyquinone 17-AAG, and 17-DMAG share a common unfavorable feature, the benzoquinone toxicophore that undergoes redox cycles involving a NAD(P)H/quinine oxidoreductase (NQO1)-catalyzed 2-electron reduction of quinine to dihydroquinone. This leads to superoxide radical generation, which has been suggested as a component of their cytotoxicity [77, 78]. 17-AAG and analogues have been shown to undergo Michael addition with glutathione at C17 leading to cellular depletion. There are at least two compounds in the literature bearing a phenol substitute instead of the quinone moiety. Kyowa Hakko prepared geldanamycin derivatives devoid of quinones by reacting 17-AAG and analogues with methyl Grignard followed by reduction to obtain 4-methyl phenol derivatives **8.80** (Fig. 14.37) [79].

During the screening of extracts from fermentation of *Streptomyces* sp. from an Indonesian soil at Kyowa Hakko, four nonquinone Hsp90 inhibitors were discovered displaying interesting binding affinities to Hsp90 and *in vitro* activities against a variety of cancer cell lines (Fig. 14.38) [80].

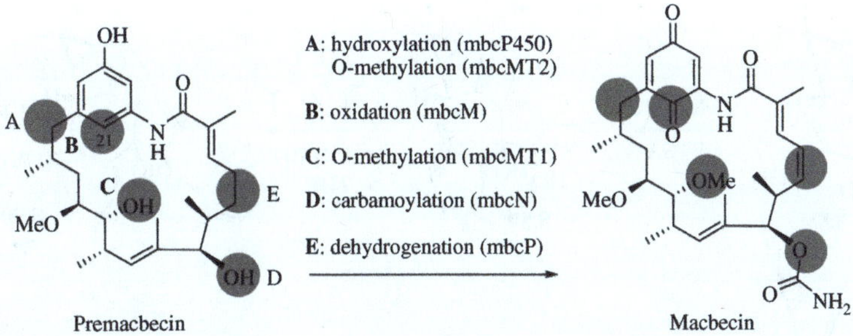


8.86 8.87

**FIGURE 14.39** Post-PKS modifications in the macbecin biosynthesis. (*See insert for color representation of the figure.*)

By biosynthetic engineering of the macbecin I synthase cluster, Biotica developed nonquinone Hsp90 inhibitors (Fig. 14.39) [81].

The macbecin biosynthetic gene cluster was identified and its sequence obtained. Biotica's strategy was based on a thorough understanding of the quinone biosynthetic pathway for ansamycin polyketides. The hydroxylation of a phenol precursor is followed by auto-oxidation with molecular oxygen. Molecular genetic manipulation of the producing organism to delete mbcM (responsible for introducing oxygen at C21) yielded a novel strain producing in the fermentation broth **8.81**. C21 oxidation is the second post-PKS step in the biosynthesis of macbecin I [82]. To produce **8.81** as a single product, the organism was further modified. Cultivation of the modified strain Biot-3982 led to the exclusive production of **8.81** in more than 200 mg/l fermentation titer. Mice dosed ip at 250 mg/kg/day for 5 days showed no apparent toxicity, whereas previous studies with macbecin had done (dosed at 10 or 20 mg/kg).

## 14.7 CASE STUDY 5: INGENOL MEBUTATE (PICATO)

Actinic keratosis is a premalignant lesion common in light-skinned populations worldwide. Effective treatments so far are based on cryosurgery. The generation of scars and the high reoccurrence rates are the main disadvantages of this procedure. Noninvasive self-treatments, comprising topical applications of, for example, fluorouracil, imiqimod, or diclofenac have to be applied over long periods of time, and the outcome is often less than favorable [83].

Ingenol mebutate has been shown *in vitro* to selectively activate PKCδ (protein kinase C, isoform delta) triggering immune response and the initiation of apoptosis [84]. *In vivo*, a neutrophil-mediated oxidative burst and clearance of tumors was shown after treatment with ingenol mebutate [85].

The anticancer and immunomodulating effects of ingenol mebutate correlate with the biological effects of many other terpenes from the tigliane and ingenane type. These privileged terpene scaffolds seem to have an important impact on many signaling pathways (PKC modulation playing a central role)—pivotal for cancer or HIV progression [86]. Ingenol mebutate is an activator of novel (δ, ε, η, θ) and classical ($\alpha$, $\beta_I$, $\beta_{II}$, $\gamma$) PKC isoforms [87].

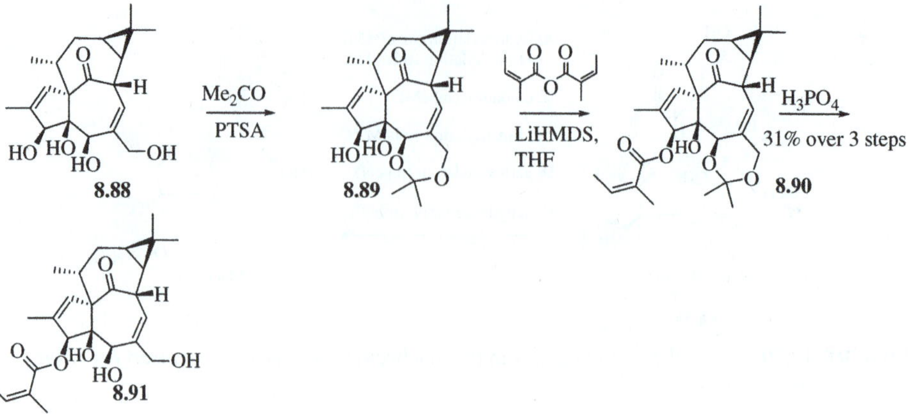

**FIGURE 14.40** Semisynthesis of ingenol-3-angelate.

Ingenol was approved by the FDA in 2012 for the topical first-in-class treatment of actinic keratosis. Furthermore, ingenol mebutate was successfully tested in a phase II clinical trial against basal cell carcinoma [88]. It has been isolated from *Euphorbia peplus* in very small amounts (1.1 mg/kg of dried biomaterial) [89]. *Euphorbia peplus* had traditionally been used as a remedy for common skin lesions [90]. Due to the poor extraction yields, ingenol mebutate can be usefully generated in three steps from ingenol [91]. Ingenol itself can be sourced from the dried seeds of *Euphorbia lathyris* [92], or other efficiently cultured euphorbia plants.

After protection of the C5–C20-diol functionality, the C3 position can be selectively addressed to form the angelate ester (Fig. 14.40). The angelate double bond is prone to isomerization to the thermodynamically more stable tiglate esters. Furthermore, acidic removal of the acetonide protecting group can equally cause isomerization. Therefore, Leo Pharma developed a semisynthetic sequence involving a nucleophilic attack of the 3-ingenol alcoholate onto a highly purified angelate anhydride. This was followed by a deprotection of the resulting angelate ester isopropylidene acetal under mild conditions. This three-step sequence, including a recrystallization after each step, could be established in 31% overall yield without any chromatographic steps. This sequence combines the excellent accessibility of ingenol from *E. lathyris* with excellent semisynthetic access. It is very suitable for the large-scale production of ingenol-3-angelate.

Although an efficient semisynthetic access has been established by Leo Pharma, the development of an efficient and sustainable *de novo* synthesis of ingenol mebutate is certainly a rewarding goal. The synthetically extremely challenging *in-out*-[4.4.1]-bicycloundecane core has been a motivation for three landmark total synthesis and one formal total syntheses prior to 2013 (Fig. 14.41).

Although all tremendous achievements, the synthetic routes range from 37 to 45 synthetic steps. They are neither feasible on a commercial scale nor suitable for an efficient generation of analogues. In 2002, the Winkler group established a 43-step total synthesis [93]. Starting from tetrahydropentalenone, the key synthetic steps involved a [2+2]-photocyclization (intramolecular De Mayo cycloaddition) followed by a retro-aldol (Fig. 14.42).

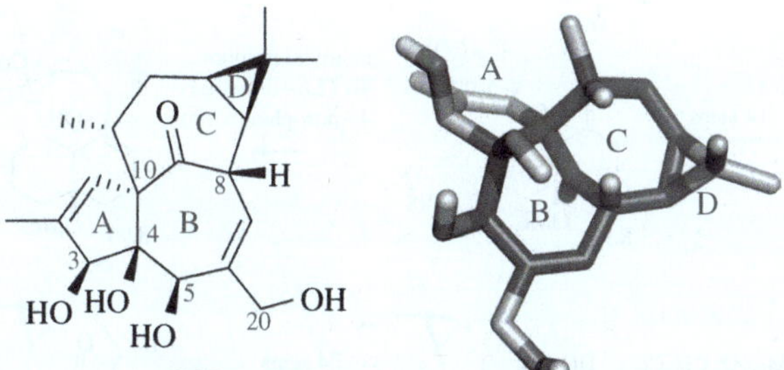

8.88

**FIGURE 14.41** (+)-Ingenol bearing an in-out-[4.4.1]-undecane ringsystem (blue). (*See insert for color representation of the figure.*)

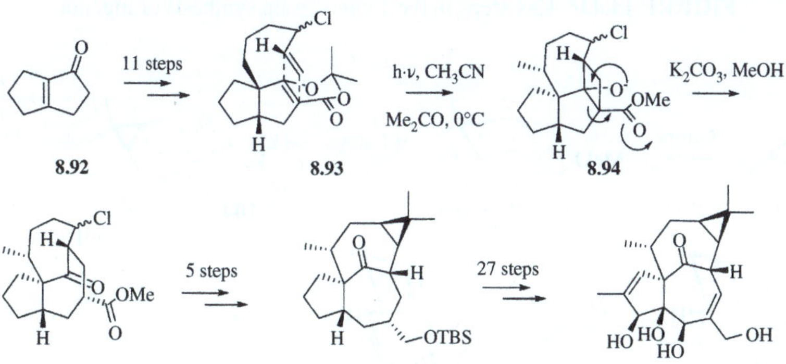

**FIGURE 14.42** Key steps in the Winkler group total synthesis of (+)-ingenol.

By this key reaction sequence, the *in-out*-[4.4.1]-undecane ring system was established. While the generation of the ingenane core scaffold was established in 18 steps, the stereoselective installation of the 3,4,5-triol and the introduction of one methyl group and two double bonds required 27 additional synthetic steps.

Starting from the commercially available dimethyl acetal of alpha-hydroxycyclohexanone, Tanino et al. developed a 45-step total synthesis to ingenol [94]. A *trans*-decaline ring system is suitably functionalized to generate a tricyclic intermediate formed in 15 steps. A di-Co–acetylene complex was used to stereoselectively introduce the construction of the cyclopropane D-ring system. After substrate-directed epoxidation of the bridgehead double bond, the stage was set for a pinacol rearrangement generating the ingenol core scaffold (Fig. 14.43).

Twenty-four additional steps introduced two C1 synthons together with redox steps necessary to implement stereoselectively the C3–C6 triol and two further double bonds.

In 2004, the Wood group published a 34-step synthesis of ingenol [95]. Wood et al. started from a precursor published by Funk et al., which is available in five steps from 3-carene. The seven-membered beta-keto ester was suitably functionalized in 11 steps to generate a ring-closing metathesis (RCM) precursor. After the RCM reaction forming the

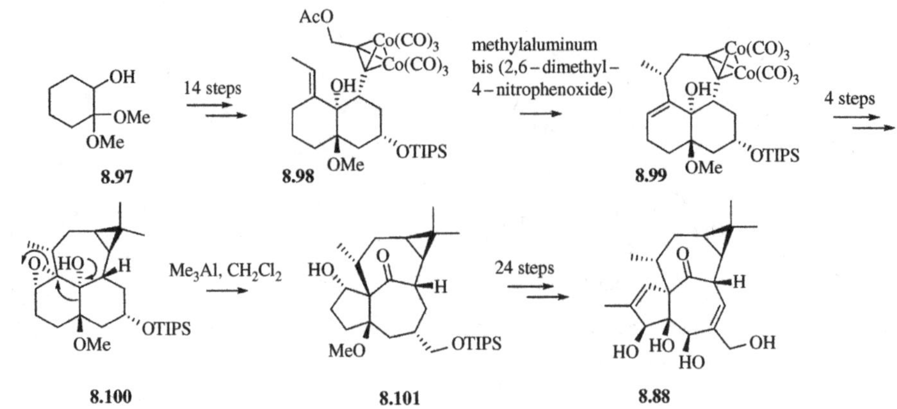


**FIGURE 14.43** Key steps in the Tanino group synthesis of ingenol.

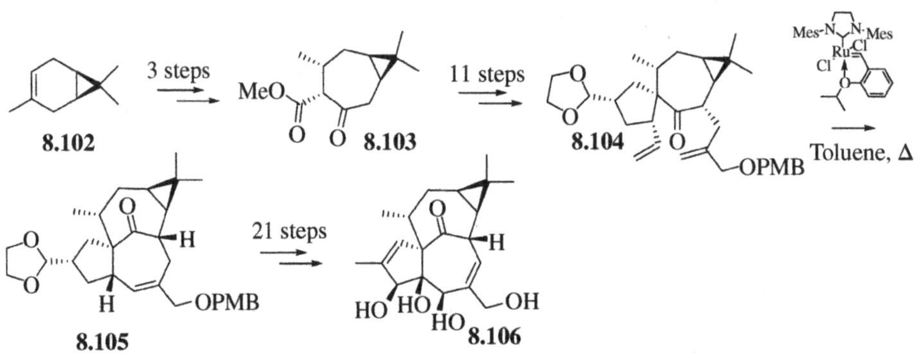


**FIGURE 14.44** Key steps of the Wood group total synthesis of ingenol.

ingenol core scaffold, 21 additional steps were required to install stereoselectively the ingenol oxidation pattern (Fig. 14.44).

The three total syntheses share an extensive sequence required for implementing the natural oxidation pattern. Hoffmann and Baran introduced the importance of redox economy as a key factor for an efficient synthesis of complex natural products [96]. By staying close to the biosynthetic pathway as a design principle, synthetic intermediates can be generated, which are spring loaded for oxidation [97]. Therefore, a more efficient synthesis in terms of redox economy was attempted.

Baran made extensive use of selective CH activation reactions. By sequential oxidation of allylic positions differentiated by the tuned liability of the allylic CH bond, a selective functionalization en route to Taxol could be established (Fig. 14.45) [98].

However, the biosynthesis of euphorbia diterpenes is still miraculous. A common unfunctionalized synthetic intermediate is diversified by various oxidation and rearrangement steps. It is likely that cyclase and oxidase steps are intertwined.

The concept of a defined cyclase phase (seven steps) and an oxidase phase (seven steps) was stringently implemented in the synthesis of ingenol allowing a high level of redox economy. The biosynthetic pathway and Baran's total synthesis of ingenol share a

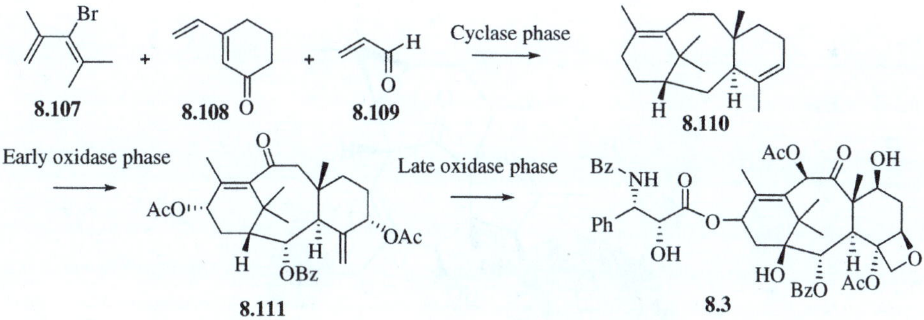


**FIGURE 14.45** En route to Taxol—cyclase and oxidase phase.

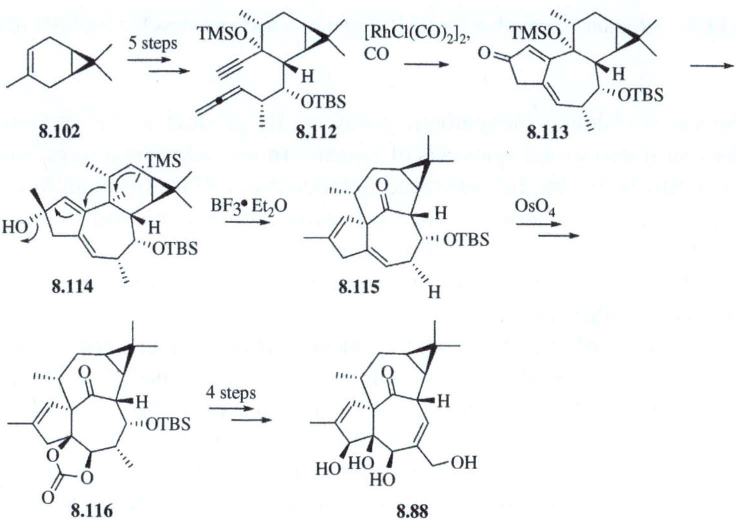


**FIGURE 14.46** Baran group total synthesis of (+)-ingenol.

defined assembly phase, a cyclase, and an oxidase phase—as already exemplified by synthetic efforts in the Taxol case study.

Baran et al. designed key intermediates suitable for selective oxidation. The precursor for selective CH activation reactions must be fine-tuned in terms of the electronic and steric nature to be selectively addressable. The retrosynthetic design by Baran et al. would furthermore allow a late stage diversification and therefore access to a comprehensive SAR study.

Baran et al. [99] developed a 14-step synthesis to ingenol starting from (+)-3-carene (Fig. 14.46).

In five steps, an allenyl/alkynyl precursor suitable for a Pauson–Khand reaction starting from (+)-3-carene was generated. The Pauson–Khand reaction sets in place four contiguous stereocenters and forms three carbon–carbon bonds. Thereby, the complete stereocontrol over seven stereocenters was achieved by a combination of substrate and reagent control. In three steps, the precursor for a pivotal vinylogous pinacol rearrangement was generated. The vinylogous pinacol rearrangement is a synthetic step

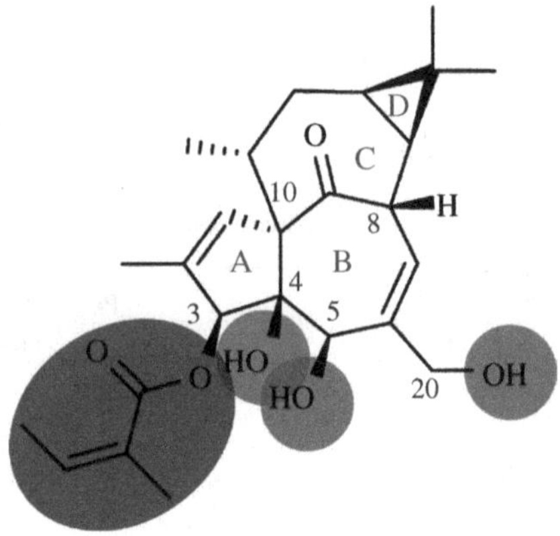


**FIGURE 14.47** Regions accessible for SAR (lightgrey) and unaccessible for SAR (darkgrey).

inspired by the envisioned biosynthetic pathway. Its product is the endpoint of the cyclase phase in Baran's total synthesis of ingenol. In four additional steps, the ingenol synthesis was finished using full substrate stereocontrol [100]. Although this synthetic sequence is exceptional in terms of step and redox economy, the three stoichiometric oxidation steps (an $OsO_4$-mediated dihydroxylation and two subsequent $SeO_2$ allylic oxidations) will probably require the substitution of less toxic reagents to be feasible for a commercial synthesis of ingenol.

The Baran group further established a synthetic access to the ingenol scaffold, which can be used for extensive SAR exploration. In a first series of papers, Leo Pharma shared their insight into the established SAR around the ingenol scaffold (Fig. 14.47) [101].

Using a cascade of distinct assays, the different ingenol regions could be probed for biological activity. PKCδ activation, ROS activity, and cytokine release (IL-8 and TNF-α) as well as stability were explored in specific settings. Stability—being a key factor for bioactivity—had to be carefully monitored. Ingenol mebutate is most stable at pH 2.0–4.5. One major factor accounting for the instability is the acyl migration of the C3 ester side chain. Under thermodynamic conditions an equilibrium is formed between the positions C3, C5, and C20. With the C5 and C20 hydroxy groups being essential for biological activity, even subtle changes in the ingenol scaffold generate inactive derivatives. Sterically demanding esters are mandatory for the stability of the C3 esters.

Further insight into the SAR (especially concerning the complex A,B,C,D-ring system) are expected in the near future. Simplifications of the diterpene scaffold by conserving the pharmacophoric features could influence the commercial success of fully synthetic ingenol mebutate followers.

## 14.8 SUMMARY

The previous examples clearly show the importance of different orthogonal access routes to potential new natural product-derived drugs. Being able to access biologically active natural products by microbial fermentation or extraction from collected biomaterial such as plants and marine invertebrates in parallel to synthetic access via semisynthesis or total synthesis can tremendously accelerate drug discovery efforts, while new synthetic or biosynthetic routes can give access to unexpected and important SAR observations.

Slower to be realized than the expectation of 20 years ago, genetic engineering (formerly often called "combinatorial biosynthesis") opens the door for the production of analogues—usually quite close to the initial natural product, but often difficult to reach by semi- or total synthetic approaches. Understanding biosynthetic pathways and utilizing these to build biomimetic synthetic routes often enable investigation of SARs. A deep understanding of pharmacophores of natural product scaffolds has led to the development of generations of drugs originating from one common ancestor (e.g., β-lactam antibiotics or prostaglandins).

Efficient drug discovery from natural product starting points is seldom the exclusive domain of one technique over others but rather the synergistic usage of synthetic chemistry, semisynthetic chemistry, molecular biology, and molecular pharmacology.

## REFERENCES

[1] Newman, D J; Cragg, G M (2012) Natural products as sources of new drugs over the 30 years from 1981 to 2010 *J. Nat. Prod.* **75**, 311–335.

[2] von Nussbaum, F; Brands, M; Hinzen, B; Weigand, S; Häbich, D (2006) Antibacterial natural products in medicinal chemistry—exodus or revival? *Angew. Chem. Int. Ed.* **45**, 5072.

[3] (a)Mann, J (1994) Steps to a successful synthesis *Nature* **367**, 594;(b)Caesar, G J (1965) *Commentarii de Bello Gallico, Book VI*, **32**, Weidmann, Berlin.

[4] Wilson, C R; Sauer, J-M; Hooser, S B (2001) Taxines: a review of the mechanism and toxicity of yew (*Taxus* spp.) alkaloids *Toxicon* **39**, 175.

[5] Itokawa, H; Lee, K H (2003) *Taxus: The Genus Taxus*, Taylor and Francis, London/New York, pp 1–18.

[6] Stephenson, F (2002) A tale of taxol Florida State University Research in Review **12** (3). http://www.rinr.fsu.edu/fall2002/taxol.html. Accessed August 26, 2015.

[7] Leighton, J; Kalla, R L (1958) Comparison of biological qualities of "transformed" cells of normal and cancerous origin *Ann. N. Y. Acad. Sci.* **76**, 513.

[8] Schiff, P B; Fant, J; Horwitz, S B (1979) Promotion of microtubule assembly in vitro by taxol *Nature* **277**, 665.

[9] *The New York Times* (May 2, 1987) Demand for yew tree concerns environmentalists, p 29.

[10] Guénard, D; Guèritte-Voegelein, F; Potier, P (1993) Taxol and taxotere: discovery, chemistry, and structure-activity relationships *Acc. Chem. Res.* **26**, 160.

[11] Sènilh, V; Blechert, S; Colin, M; Guènard, D; Picot, F; Potier, P; Varenne, P (1984) Mise en Évidence de Nouveaux Analogues du Taxol Extraits de *Taxus baccata J. Nat. Prod.* **47**, 131.

[12] Denis, J-N; Greene, A E; Serra, A A; Luche, M-J (1986) An efficient, enantioselective synthesis of the taxol side chain *J. Org. Chem.* **51**, 46.

[13] Denis, J-N; Greene, A E; Guénard, D; Guéritte-Voegelein, F; Mangatal, L; Potier, P (1988) Highly efficient, practical approach to natural taxol *J. Am. Chem. Soc.* **110**, 5917.

[14] Colin, M; Guènard, D; Guèritte-Voegelein, F; Potier, P (1987) Procédé de préparation du taxol et du désacétyl-10 taxol FR 8610401.

[15] Guèritte-Voegelein, F; Guènard, D; Lavelle, F; Le Goff, M-T; Mangatal, L; Potier, P (1991) Relationships between the structure of taxol analogs and their antimitotic activity *J. Med. Chem.* **34**, 992.

[16] Collin, M; Guènard, D; Guèritte-Voegelein, F; Potier, P (1987) Taxal-Derivate, deren Herstellung und diese enthaltende pharmazeutische Zusammenstellungen EP 0253738.

[17] (a) Holton, R A (1989) Method for preparation of taxol using an oxazinone US 5,015,744;(b) Holton, R A (1989) Reacting with alcohol in presence of altivating agent US 5,175,315.

[18] (a) Holton, R A; Somoza, C; Kim, H-B; Liang, F; Biediger, R J; Boatman, D; Shindo, M; Smith, C C; Kim, S; Nadizadeh, H; Suzuki, Y; Tao, C; Vu, P; Tang, S; Zhang, P; Murthi, K K; Gentile, L N; Liu, J H (1994) First total synthesis of taxol. 1. Functionalization of the B ring *J. Am. Chem. Soc.* **116**, 1597;(b) Holton, R A; Kim, H-B; Somoza, C; Liang, F; Diediger, R J; Boatman, P D; Shindo, M; Smith, C C; Kim, S; Nadizadeh, H; Suszuki, Y; Tao, C; Vu, P; Tang, S; Zhang, P; Murthi, K K; Gentile, L N; Liu, J H (1994) First total synthesis of taxol. 2. Completion of the C and D rings *J. Am. Chem. Soc.* **116**, 1599.

[19] Holton, R A (1992) Taxol derivatives US 5,274,124.

[20] Nicolaou, K C; Yang, Z; Liu, J J; Ueno, H; Natermet, P G; Guy, R K; Claiborne, C F; Renaud, J; Couladouros, E A; Paulvannan, K; Sorensen, E J (1994) Total synthesis of taxol *Nature* **367**, 630.

[21] Danishefsky, S J; Masters, J J; Young, W B; Link, J T; Snyder, L B; Magee, T V; Jung, D K; Isaacs, R C A; Bornmann, W G; Alaimo, C A; Coburn, C A; Di Grandi, M J (1996) Total synthesis of baccatin III and taxol *J. Am. Chem. Soc.* **118**, 2843.

[22] Wender, P A; Badham, N F; Conway, S P; Floreancig, P E; Glass, T E; Houze, J B; Krauss, N E; Lee, D; Marquess, D G; McGrane, P L; Meng, W; Natchus, M G; Shuker, A J; Sutton, J C; Taylor, R E (1997) The pinene path to taxanes. 6. A concise stereocontrolled synthesis of taxol *J. Am. Chem. Soc.* **119**, 2757.

[23] Mukaiyama, T; Shiina, S; Iwadare, H; Saitoh, M; Nishimura, T; Ohkawa, N; Sakoh, H; Nishimura, K; Tani, Y-I; Hasegawa, M; Yamada, K; Saitoh, K (1999) Asymmetric total synthesis of taxol® *Chem. Eur. J.* **5**, 121.

[24] Ojima, I; Habus, I; Zhao, M; Georg, G I; Jayasinghe, L R (1991) Efficient and practical asymmetric synthesis of the taxol C-13 side chain, *N*-benzoyl-(2R,3S)-3-phenylisoserine, and its analogs via chiral 3-hydroxy-4-aryl-.beta.-lactams through chiral ester enolate-imine cyclocondensation *J. Org. Chem.* **56**, 1681.

[25] Croteau, R; Ketchum, R E B; Long, R M; Kaspera, R; Wildung, M R (2006) Taxol biosynthesis and molecular genetics *Phytochem. Rev.* 2006, **5**, 75.

[26] Yukimune, Y; Tabata, H; Higashi, Y; Hara, Y (1996) Methyl jasmonate-induced overproduction of paclitaxel and baccatin III in Taxus cell suspension cultures *Nat. Biotechnol.* **14**, 1129.

[27] Ajikumar, P K; Xiao, W-H; Tyo, K E J; Wang, Y; Simeon, F; Leonard, E; Mucha, O; Phon, T H; Pfeifer, B; Stephanopoulos, G (2010) Isoprenoid pathway optimization for Taxol precursor overproduction in *Escherichia coli Science* **330**, 70.

[28] Gelderblom, H; Verweij, J; Nooter, K; Spareboom, A (2001) Cremophor EL: the drawbacks and advantages of vehicle selection for drug formulation *Eur. J. Cancer* **37**, 1590.

[29] Chen, S-H; Huang, S; Wie, J; Farina, V (1993) Serendipitous synthesis of a cyclopropane-containing taxol analog via anchimeric participation of an unactivated angular methyl group *J. Org. Chem.* **58**, 4520.

[30] Samaranayake, G; Magri, N F; Jitrangsri, C; Kingston, D G I (1991) Modified taxols. 5. Reaction of taxol with electrophilic reagents and preparation of a rearranged taxol derivative with tubulin assembly activity *J. Org. Chem.* **56**, 5114.

[31] Diéras, V; Limentani, S; Romieu, G; Tubiana-Hulin, M; Lortholary, A; Kaufman, P; Girre, V; Besenval, M; Valero, V (2008) Phase II multicenter study of larotaxel (XRP9881), a novel taxoid, in patients with metastatic breast cancer who previously received taxane-based therapy *Ann. Oncol.* **19**, 1255.

[32] (a) Didier, E; Oddon, G; Panze, D; Leion, P; Piquet, D (1999) Process for production of trioxane US 5,962,702;(b) Bissery, M-C (2000) Use of taxoid derivatives US 6,403,634.

[33] Dumontet, C; Jordan, M A (2010) Microtubule-binding agents: a dynamic field of cancer therapeutics *Nat. Rev. Drug Discov.* **9**, 790.

[34] Hardt, I H; Steinmetz, H; Gerth, K; Sasse, F; Reichenbach, H; Höfle, G (2001) New natural epothilones from *Sorangium cellulosum*, strains So ce90/B2 and So ce90/D13: isolation, structure elucidation, and SAR studies *J. Nat. Prod.* **64**, 847–856.

[35] Colombo, N; Kutarska, E; Dimopoulos, M; Bae, D S; Rzepka-Gorska, I; Bidzinski, M; Scambia, G; Engelholm, S A; Joly, F; Weber, D; El-Hashimy, M; Li, J; Souami, F; Wing, P; Engelholm, S; Bamias, A; Schwartz, P (2012) Randomized, open-label, phase III study comparing patupilone (EPO906) with pegylated liposomal doxorubicin in platinum-refractory or -resistant patients with recurrent epithelial ovarian, primary fallopian tube, or primary peritoneal cancer *J. Clin. Oncol.* **30**, 3841–3847.

[36] Tang, L; Shah, S; Chung, L; Carney, J; Katz, L; Khosla, C; Julien, B (2000) Cloning and heterologous expression of the epothilone gene cluster *Science* **287**, 640–642.

[37] Julien, B; Shah, S (2002) Heterologous expression of epothilone biosynthetic genes in *Myxococcus xanthus Antimicrob. Agents Chemother.* **46**, 2772–2778.

[38] Ongley, S E; Bian, X; Neilan, B A; Müller, R (2013) Recent advances in the heterologous expression of microbial natural product biosynthetic pathways *Nat. Prod. Rep.* **30**, 1121–1138.

[39] Starks, C M; Zhou, Y; Liu, F; Licari, P J (2003) Isolation and characterization of new epothilone analogues from recombinant *Myxococcus xanthus* fermentations *J. Nat. Prod.* **66**, 1313–1317.

[40] Zhang, B; Caroll, J; Trojanowski, J Q; Yao, Y; Iba, M; Potuzak, J S; Hogan, A-M L; Xie, S X; Ballatore, C; Smith, A B III; Lee, V M-Y; Brunden, K R (2012) The microtuble-stabilizing agent, epothilone D, reduces axonal dysfunction, neurotoxicity, cognitive deficits and Alzheimer-like pathology in an interventional study with aged tau transgenic mice *J. Neurosci.* **32**, 3601–3611.

[41] Altmann, K H; Gaugaz, F Z; Schiess, R (2011) Diversity through semisynthesis: the chemistry and biological activity of semisynthetic epothilone derivatives *Mol. Divers.* **15**, 383–399.

[42] Hunt, J T (2009) Discovery of ixabepilone *Mol. Cancer Ther.* **8**, 275–281.

[43] Borzilleri, R M; Zheng, X; Schmidt, R J; Johnson, J A; Kim, S H; DiMarco, J D; Fairchild, C R; Gougoutas, J C; Lee, F Y F; Long, B H; Vite, G D (2000) A novel application of a Pd(0)-catalyzed nucleophilic substitution reaction to the regio- and stereoselective synthesis of lactam analogues of the epothilone natural products *J. Am. Chem. Soc.* **122**, 8890–8897.

[44] Klar, U; Buchmann, B; Schwede, W; Skuballa, W; Hoffmann, J; Lichtner, R B (2006) Total synthesis and antitumor activity of ZK-EPO: the first fully synthetic epothilone in clinical development *Angew. Chem. Int. Ed.* **45**, 7942–7948.

[45] Ledford, H (2010) Complex synthesis yields breast-cancer therapy *Nature* **468**, 608.

[46] Jackson, K L; Henderson, J A; Phillips, A J (2009) The halichondrins and E7389 *Chem. Rev.* **109**, 3044.

[47] Uemura, D; Takahashi, K; Yamamoto, T; Katayama, C; Tanaka, J; Okumura, Y; Hirata, Y (1985) Norhalichondrin A: an antitumor polyether macrolide from a marine sponge *J. Am. Chem. Soc.* **107**, 4796.

[48] Hirata, Y; Uemura, D (1986) Halichondrins—antitumor polyether macrolides from a marine sponge *Pure Appl. Chem.* **58**, 701.

[49] Lake, R J (1988) Internal research report. Post-Doctoral Fellow, Marine Chemistry Group, University of Canterbury.

[50] Litaudon, M; Hart, J B; Blunt, J W; Lake, R J; Munro, M H G (1994) Isohomohalichondrin B, a new antitumor polyether macrolide from the New Zealand deep-water sponge *Lissodendoryx* sp. *Tetrahedron Lett.* **35**, 9435.

[51] Pettit, G; Herald, C L; Boyd, M R; Leet, J E; Dufresne, C; Doubek, D L; Schmidt, J M; Cerney, R L; Hooper, J N A; Rützler, K C (1991) Antineoplastic agents. 219. Isolation and structure of the cell growth inhibitory constituents from the western Pacific marine sponge *Axinella* sp. *J. Med. Chem.* **34**, 3339.

[52] Bai, R; Pauli, K D; Herald, C L; Malspeis, L; Pettit, G R; Hamel, E (1991) Halichondrin B and homohalichondrin B, marine natural products binding in the vinca domain of tubulin. Discovery of tubulin-based mechanism of action by analysis of differential cytotoxicity data *J. Biol. Chem.* **266**, 15882.

[53] Litaudon, M; Hickford, S J H; Lill, R E; Lake, R J; Blunt, J W; Munro, M H G (1997) Antitumor polyether macrolides: new and hemisynthetic halichondrins from the New Zealand deep-water sponge *Lissodendoryx* sp. *J. Org. Chem.* **62**, 1868.

[54] Munro, M G H; Blunt, J W; Dumdei, E J; Hickford, S J H; Lill, R E; Battershill, C N; Duckworth, A R (1999) The discovery and development of marine compounds with pharmaceutical potential *Prog. Ind. Microbiol.* **35**, 155.

[55] Aicher, T D; Buszek, K R; Fang, F G; Forsyth, C J; Jung, S H; Kishi, Y; Matelich, M C; Scola, P M; Spero, D M; Yoon, S K (1992) Total synthesis of halichondrin B and norhalichondrin B *J. Am. Chem. Soc.* **114**, 3162.

[56] (a) Duan, J J W; Kishi, Y (1996) Synthetic studies on halichondrins: a new practical synthesis of the C.1–C.12 segment *Tetrahedron Lett.* **34**, 7541;(b) Stamos, D P; Kishi, Y (1996) Synthetic studies on halichondrins: a practical synthesis of the C.1 C.13 segment *Tetrahedron Lett.* **37**, 8643;(c) Stamos, D P; Sean, S C; Kishi, Y (1997) New synthetic route to the C.14–C.38 segment of halichondrins *J. Org. Chem.* **62**, 7552;(d) Guo, H; Dong, C-G; Kim, D-S; Urabe, D; Wang, J; Kim, J T; Liu, X; Sasaki, T; Kishi, Y (2009) Toolbox approach to the search for effective ligands for catalytic asymmetric Cr-mediated coupling reactions *J. Am. Chem. Soc.* **131**, 15387;(e) Kim, D-S; Dong, C-G; Kim, J T; Guo, H; Huang, J; Tiseni, P S; Kishi, Y (2009) New syntheses of E7389 C14–C35 and halichondrin C14–C38 building blocks: double-inversion approach *J. Am. Chem. Soc.* **131**, 15636;(f) Dong, C-G; Henderson, J A; Kaburagi, Y; Sasaki, T; Kim, J T; Urabe, D; Guo, H; Kishi, Y (2009) New syntheses of E7389 C14–C35 and halichondrin C14–C38 building blocks: reductive cyclization and oxy-Michael cyclization approaches *J. Am. Chem. Soc.* **131**, 15642.

[57] Wang, Y; Habgood, G J; Christ, W J; Kishi, Y; Littlefield, B A; Yu, M (2000) Structure–activity relationships of halichondrin B analogues: modifications at C.30–C.38 *J. Bioorg. Med. Chem. Lett.* **10**, 1029.

[58] Kishi, Y; Fang, F G; Forsyth, C J; Scola, P M; Yoon, S K (1995) Halichondrins and related compounds US 5,436,238.

[59] Zheng, W; Seletsky, B M; Palme, M H; Lydon, P J; Singer, L A; Chase, C E; Lemelin, C A; Shen, Y; Davis, H; Tremblay, L; Towle, M J; Salvato, K A; Wels, B F; Aalfs, K K; Kishi, Y; Littlefield, B A; Yu, M J (2004) Structurally simplified macrolactone analogues of halichondrin B *Bioorg. Med. Chem. Lett.* **14**, 5551.

[60] Littlefield, B A; Palme, M; Seletsky, B M; Towle, M J; Yu, M J (1999) Macrocyclic analogs and methods of their use and preparation WO 99/65894.

[61] (a) Smith, A B III; Chen, S S-Y; Nelson, F C; Reichert, J M; Salvatore, B A (1995) Total syntheses of (+)-acutiphycin and (+)-*trans*-20,21-didehydroacutiphycin *J. Am. Chem. Soc.* **117**, 12013;(b) Stamos, D P; Kishi, Y (1996) Synthetic studies on halichondrins: a practical synthesis of the C.1 C.13 segment *Tetrahedron Lett.* **37**, 8643; and Refs. 56, 58–60.

[62] Towle, M J; Salvato, K A; Budrow, J; Wels, B F; Kuznetsov, G; Aalfs, K K; Welsh, S; Zheng, W; Seletsky, B M; Palme, M H; Habgood, G J; Singer, L A; DiPietro, L V; Wang, Y; Chen, J J; Quincy, D A; Davis, A; Yoshimatsu, K; Kishi, Y; Yu, M J; Littlefield, B A (2001) In vitro and in vivo anticancer activities of synthetic macrocyclic ketone analogues of halichondrin B1 *Cancer Res.* **61**, 1013.

[63] (a)Cortes, J; O'Shaughnessy, J; Loesch, D; Blum, J L; Vahdat, L T; Petrakova, K; Chollet, P; Manikas, A; Diéras, V; Delozier, T; Vladimirov, V; Cardoso, F; Koh, H; Bougnoux, P; Dutcus, C E; Seegobin, S; Mir, D; Meneses, N; Wanders, J; Twelves, C (2011) Eribulin monotherapy versus treatment of physician's choice in patients with metastatic breast cancer (EMBRACE): a phase 3 open-label randomised study *Lancet* **377**, 914;(b) Twelves, C; Cortes, J; Vahdat, L T; Wanders, J; Akerele, C; Kaufman, P A (2010) Eribulin mesylate in the treatment of metastatic breast cancer *Clin. Breast Cancer* **10**, 160;(c) Cortes, J; Vahdat, L; Blum, J L; Twelves, C; Campone, M; Roche, H; Bacelot, T; Awada, A; Paridaens, R; Goncalves, A; Shuster, D E; Wanders, J; Fang, F; Gurnani, R; Richmond, E; Cole, P E; Ashworth, S; Allison, M A (2010) Phase II study of the halichondrin B analog eribulin mesylate in patients with locally advanced or metastatic breast cancer previously treated with an anthracycline, a taxane, and capecitabine *J. Clin. Oncol.* **28**, 3922.

[64] De Boer, C; Meulman, P A; Wnuk, R J; Peterson, D H (1970) Geldanamycin, a new antibiotic *J. Antibiot.* **23**, 442.

[65] Sasaki, K; Rinehart, K L; Slomp, G; Grostic, M F; Olson, E C (1970) Geldanamycin. I. Structure assignment *J. Am. Chem. Soc.* **92**, 7591.

[66] (a) Schnur, R C; Corman, M L; Gallaschun, R J; Cooper, B A; Dee, M F; Doty, M L; Muzzi, M L; Moyer, J D; DiOrio, C I; Barbacci, E G; Miller, P E; O'Brien, A T; Morin, M J; Foster, B A; Pollack, V A; Savage, D M; Sloan, D E; Pustilnik, L R; Moyer, M P (1995) Inhibition of the oncogene product p185erbB-2 in vitro and in vivo by geldanamycin and dihydrogeldanamycin derivatives *J. Med. Chem.* **38**, 3806;(b)Schnur, R C; Corman, M L; Gallaschun, R J; Cooper, B A; Dee, M F; Doty, J L; Muzi, M L; DiOrio, C I; Barbacci, E G; Miller, P E; Pollack, V A; Savage, D M; Pustilnik, L R; Moyer, J D; Moyer, M P (1995) erbB-2 oncogene inhibition by geldanamycin derivatives: synthesis, mechanism of action, and structure-activity relationships *J. Med. Chem.* **38**, 3813.

[67] Fukasawa, H; Li, P-M; Yamamoto, C; Murakami, Y; Mizuno, S; Uehara, Y (1991) Specific inhibition of cytoplasmic protein tyrosine kinases by herbimycin A *in vitro Biochem. Pharmacol.* **42**, 1661.

[68] Whitesell, L; Mimnaugh, E G; De Costa, B; Myers, C E; Neckers, L M (1994) Inhibition of heat shock protein Hsp90-pp60v-src heteroprotein complex formation by benzoquinone ansamycins: essential role for stress proteins in oncogenic transformation *Proc. Natl. Acad. Sci. U. S. A.* **91**, 8324.

[69] Biamonte, M A; Van de Water, R; Arndt, J W; Scannevin, R H; Perret, D; Lee, W-C (2010) Heat shock protein 90: inhibitors in clinical trials *J. Med. Chem.* **53**, 3.

[70] Wegele, H; Müller, L; Buchner, J (2004) Hsp70 and Hsp90—a relay team for protein folding *Rev. Physiol. Biochem. Pharmacol.* **151**, 1.

[71] Sasaki, K; Inoue, Y (1980) Geldanamycin derivatives and anti-tumor agent containing the same DE 3006097.

[72] Adams, J; Gao, Y; Georges Evangelinos, A T; Grenier, L; Pak, R; Porter, J R; Wright, J L (2005) Analogs of benzoquinone-containing ansamycins for the treatment of cancer WO2005063714.

[73] Ge, J; Normant, E; Porter, J R; Ali, J A; Dembski, M S; Gao, Y; Georges, A T; Grenier, L; Pak, R H; Patterson, J; Sydor, J R; Tibbitts, T T; Tong, J K; Adams, J; Palombella, V J (2006) Design, synthesis, and biological evaluation of hydroquinone derivatives of 17-amino-17-demethoxygeldanamycin as potent, water-soluble inhibitors of Hsp90 *J. Med. Chem.* **49**, 4606.

[74] Tian, Z-Q; Liu, Y; Zhang, D; Wang, Z; Dong, S D; Cerreras, C W; Zhou, Y; Rastelli, G; Santi, D V; Myles, D C (2004) Synthesis and biological activities of novel 17-aminogeldanamycin derivatives *Bioorg. Med. Chem.* **12**, 5317.

[75] Rastelli, G; Tian, Z-Q; Wang, Z; Myles, D C; Liu, Y (2005) Structure-based design of 7-carbamate analogs of geldanamycin *Bioorg. Med. Chem. Lett.* **15**, 5016.

[76] ClinicalTrials.gov (2010) A phase I study of the Hsp90 inhibitor 17-DMAG (alvespimycin) in patients with relapsed chronic lymphocytic leukemia (CLL)/small lymphocytic lymphoma (SLL), and B-cell prolymphocytic leukemia (B-PLL). http://clinicaltrials.gov/ct/show/NCT01126502. Accessed June 3, 2015.

[77] Kelland, L R; Sharp, S Y; Rogers, P M; Myers, T G; Workman, P (1999) DT-Diaphorase expression and tumor cell sensitivity to 17-allylamino,17-demethoxygeldanamycin, an inhibitor of heat shock protein 90 *J. Natl. Cancer Inst.* **91**, 1940.

[78] Dikalov, S; Landmesser, U; Harrison, D G (2002) Geldanamycin leads to superoxide formation by enzymatic and non-enzymatic redox cycling. Implications for studies of Hsp90 and endothelial cell nitric-oxide synthase *J. Biol. Chem.* **277**, 25480.

[79] Yamaguchi, S; Nakashima, T; Kanda, Y (2007) Benzenoid ansamycin derivative WO 2007/001049.

[80] Onodera, H; Kaneko, M; Takahashi, Y; Uochi, Y; Funahashi, J; Nakashima, T; Soga, S; Suzuki, M; Ikeda, S; Yamashita, Y; Rahayu, E S; Kanda, Y; Ichimura, M (2008) Conformational significance of EH21A1-A4, phenolic derivatives of geldanamycin, for Hsp90 inhibitory activity *Bioorg. Med. Chem. Lett.* **18**, 1588.

[81] Zhang, M-Q; Gaisser, S; Nur-E-Alam, M; Sheehan, L S; Vousden, W A; Gaitatzis, N; Peck, G; Coates, N J; Moss, S J; Radzom, M; Foster, T A; Sheridan, R M; Gregory, M A; Roe, S M; Prodromou, C; Pearl, L; Boyd, S M; Wilkinson, B; Martin, C J (2008) Optimizing natural products by biosynthetic engineering: discovery of nonquinone Hsp90 inhibitors *J. Med. Chem.* **51**, 5494.

[82] Martin, C; Zhang, M; Gaisser, S; Coates, N; Wilkinson, B (2011) 4,5-dihydromacbecin derivatives and their use in the treatment of cancer or B-cell malignancies US 2011/00991452.

[83] Lebwohl, M; Swanson, N; Anderson, L L; Melgaard, A; Xu, Z; Berman, B (2012) Ingenol mebutate gel for actinic keratosis *N. Engl. J. Med.* **366**, 1010.

[84] Rosen, R H; Gupta, A K; Tyring, S K (2011) Dual mechanism of action of ingenol mebutate gel for topical treatment of actinic keratoses: rapid lesion necrosis followed by lesion-specific immune response *J. Am. Acad. Dermatol.* **66**, 486–493.

[85] Challacombe, J M; Suhrbier, A; Parsons, P G; Jones, B; Hampson, P; Kavanagh, D; Rainger, G E; Morris, M; Lord, J M; Le, T T; Hoang-Le, D; Ogbourne, S M (2006) Neutrophils are a key component of the antitumor efficacy of topical chemotherapy with ingenol-3-angelate *J. Immunol.* **177**, 8123.

[86] Wender, P A; Kee, J M; Warrington, J M (2008) Practical synthesis of prostratin, DPP, and their analogs, adjuvant leads against latent HIV *Science* **320**, 649–652.

[87] Kedei, N; Lundberg, D J; Toth, A; Welburn, P; Garfield, S H; Blumberg, P M (2004) Characterization of the interaction of ingenol 3-angelate with protein kinase C *Cancer Res.* **64**, 3243.

[88] Siller, G; Rosen, R; Freeman, M; Welburn, P; Katsamas, J; Ogbourne, S M (2010) PEP005 (ingenol mebutate) gel for the topical treatment of superficial basal cell carcinoma: results of a randomized phase IIa trial *Australas. J. Dermatol.* **51**, 99.

[89] Hohmann, J; Evanics, F; Berta, L; Bartók, T (2000) Diterpenoids from *Euphorbia peplus* *Planta Med.* **66**, 291.

[90] (a) Fallen, R S; Gooderham, M (2012) Ingenol mebutate: an introduction *Skin Therapy Lett.* **17**, 1;(b) Ramsay, J R; Suhrbier, A; Aylward, J H; Ogbourne, S; Cozzi, S J; Poulsen, M G; Baumann, K C; Welburn, P; Redlich, G L; Parsons, P G (2011) The sap from *Euphorbia peplus* is effective against human nonmelanoma skin cancers *Br. J. Dermatol.* **164**, 633.

[91] Liang, X; Grue-Sørensen, G; Petersen, A K; Högberg, T (2012) Semisynthesis of ingenol 3-angelate (PEP005): efficient stereoconservative angeloylation of alcohols *Synlett* **23**, 2647; WO2012/010172.

[92] Appendino, G; Tron, G C; Cravatto, G; Palmisano, G; Jakupovic, J (1999) An expeditious procedure for the isolation of ingenol from the seeds of *Euphorbia lathyris J. Nat. Prod.* **62**, 76–79.

[93] Winkler, J D; Rouse, M B; Greaney, M F; Harrison, S J; Yeon, Y T (2002) The first total synthesis of (±)-ingenol *J. Am. Chem. Soc.* **124**, 9726–9728.

[94] Tanino, K; Onuki, K; Asano, K; Miyashita, M; Nakamura, T; Takahashi, Y; Kuwajima, I (2003) Total synthesis of ingenol *J. Am. Chem. Soc.* **125**, 1498–1500.

[95] Nickel, A; Maruyama, T; Tang, H; Murphy, P D; Greene, B; Yusuff, N; Wood, J L (2004) Total synthesis of ingenol *J. Am. Chem. Soc.* **126**, 16300–16301.

[96] Burns, N Z; Baran, P S; Hoffmann, R W (2009) Redox economy in organic synthesis *Angew. Chem. Int. Ed.* **48**, 2854–2867.

[97] Mendoza, A; Ishihara, Y; Baran, P S (2012) Scalable enantioselective total synthesis of taxanes *Nat. Chem.* **4**, 21–25.

[98] Wilde, N C; Isomura, M; Mendoza, A; Baran, P S (2014) Two-phase synthesis of (−)-taxuyunnanine D *J. Am. Chem. Soc.* **136**, 4909–4912.

[99] Jørgensen, L; McKerral, S J; Kuttruff, C A; Ungeheuer, F; Felding, J; Baran, P S (2013) 14-step synthesis of (+)-ingenol from (+)-3-carene *Science* **341**, 878–882.

[100] McKerrall, S J; Jørgensen, L; Kuttruff, C A; Ungeheuer, F; Baran, P S (2014) Development of a concise synthesis of (+)-ingenol *J. Am. Chem. Soc.* **136**, 5799–5810.

[101] (a) Liang, X; Grue-Sørensen, G; Månsson, K; Vedsø, P; Soor, A; Stahlhut, M; Bertelsen, M; Engell, K M; Högberg, T (2013) Synthesis, biological evaluation and SAR of ingenol mebutate analogues for treatment of actinic keratosis and non-melanoma skin cancer *Bioorg. Med. Chem. Lett.* **23**, 5624–5629;(b) Grue-Sørensen, G; Liang, X; Månsson, K; Vedsø, P; Sørensen, M D; Soor, A; Stahlhut, M; Bertelsen, M; Engell, K M; Högberg, T (2014) Synthesis, biological evaluation and SAR of 3-benzoates of ingenol for treatment of actinic keratosis and non-melanoma skin cancer *Bioorg. Med. Chem. Lett.* **24**, 54–60.

# 15

# PEPTIDOMIMETICS OF α-HELICAL AND β-STRAND PROTEIN BINDING EPITOPES

NINA BIONDA and RUDI FASAN
*University of Rochester, Rochester, NY, USA*

## 15.1 PROTEIN–PROTEIN INTERACTIONS AS THERAPEUTIC TARGETS

Protein–protein interactions (PPIs) are central in the regulation and maintenance of virtually all cellular functions, from signal transduction to gene regulation, proliferation, and apoptosis. The complexity and physiological relevance of these interactions is apparent by the fact that from 300,000 to 650,000 distinct pairwise PPIs are estimated to occur in a human cell [1, 2]. Aberrant, ill-regulated, or inappropriate PPIs due to gene inactivation, gene overexpression, or bacterial and viral infections are associated to a host of pathologies, which make them attractive targets for therapeutic intervention [3–5]. However, the development of small-molecule agents capable of modulating PPIs has proven to be very challenging, in particular when compared to more established pharmacological targets such as enzymes and G-protein-coupled receptors (GPCR) [3, 5, 6]. A problematic aspect associated with targeting PPIs relates to the typically shallow topology and extended contact area of protein–protein interfaces, which are typically 3- to 10-fold larger (1200–4000 Å$^2$) [7, 8] than those of protein–small molecule complexes (300–800 Å$^2$) [9]. Furthermore, in terms of physicochemical properties (i.e., atom packing density, solvent accessibility, polar/nonpolar character), protein–protein recognition sites often do not differ significantly from regions of the protein surface not involved in binding [10]. Finally, protein interfaces are characterized by a high degree of plasticity enabling grooves and cavities to form upon interaction with a binding partner, which are not observed in the crystal structure of the protein in unbound form or in complex with other molecules [6, 11, 12].

*Small Molecule Medicinal Chemistry: Strategies and Technologies*, First Edition.
Edited by Werngard Czechtizky and Peter Hamley.

While these features have challenged both the rational design of small-molecule PPI inhibitors as well as their identification through high-throughput screening approaches [13, 14], a breakthrough in this field was the recognition that a few interfacial amino acid residues typically contribute most of the free energy of binding in protein complexes [11, 15, 16]. These so-called hot spot residues can be identified via alanine scanning mutagenesis [15, 17–19] and have been defined as sites whose substitution (with alanine) causes an increase in $\Delta\Delta G_{binding}$ of at least 2.0 kcal/mol (i.e., >30-fold increase in $K_D$). Analysis of PPI databases indicates that hot spot regions occupy only a small fraction of the entire contact interface (10–15%) [16] and are characterized by high shape complementarity [20] and exclusion from bulk solvent upon complex formation [19, 21]. Furthermore, these regions often exhibit cooperativity with respect to their contribution to the binding free energy [22, 23] and can mediate binding to different binding partners, showing high functional and structural adaptivity [17, 24]. The hot spot residue theory has had profound implications for PPI drug discovery, the major being that small molecules capable of interacting with these hot spot regions can provide, at least in principle, an effective means for disrupting a target protein–protein complex [5]. This theory has also spurred the development of computational methods for predicting hot spot residues [25, 26], thus providing a tool, complementary to mutagenesis, to assist in the identification of promising target sites for blocking PPIs [27].

In support of the potential feasibility of disrupting PPIs with small-molecule agents, there has been successful development of selective and potent small-molecule inhibitors for a few PPIs [3], namely, the interaction between the oncoprotein human protein double minute 2 (HDM2) and tumor suppressor p53 [28, 29], between the antiapoptotic members of the Bcl-2 family and proapoptotic proteins BAK/BAD [30], between cytokine interleukin-2 (IL-2) and its receptor IL-2Rα [31], and between the human papillomavirus (HPV) transcription factor E2 and the viral helicase E1 [32]. Notably, compound similarity analyses showed that these compounds share low similarity with members of chemical libraries employed against known pharmacological targets (i.e., enzymes or GPCR) [3]. Furthermore, analysis of the relationship between the molecular size of the aforementioned PPI inhibitors and their binding affinity for the respective protein (via plots of $\Delta G_{binding}$ vs. number of nonhydrogen atoms) revealed an approximately linear correlation (correlation coefficient: 0.7) and a rather uniform "ligand efficiency" (0.2–0.25 cal/mol/nonhydrogen atom) [3]. These trends indicate that, in general, the inhibitory potency of these compounds scales with molecular size and that molecular masses of 650–750 Da are expected to be required to grant high affinity for the target protein ($K_D$: 1–10 nM) [3]. Altogether, these case studies suggest that new chemotypes whose molecular features extend beyond the limits set by Lipinski's "rule of five" are required for effective targeting of PPIs.

A promising and complementary approach toward the development of PPI inhibitors has involved the design and development of mimics of α-helical and β-sheet protein binding motifs. At the basis of these strategies lies the recognition that PPIs are often mediated by well-defined (super)secondary structural elements, such as α-helices [33], β-structures [34] (i.e., β-sheet and β-hairpins), and turns [35]. Importantly, these protein recognition motifs are often encompassed by relatively short peptide sequences (8–12 residues), which make them amenable to structural and/or functional mimicry by means of compact peptidic or nonpeptidic scaffolds. Two main approaches have been explored in this area. A first one has focused on the development of peptide-based compounds that closely reproduce the topographical features of the target structural motif (a.k.a. "type I" peptidomimetics). A second one focuses on type III peptidomimetics, which are nonpeptidic, oligomeric scaffolds designed to direct side-chain functionalities in a spatial orientation analogous to that of the target structural motif. This chapter places emphasis on contemporary strategies in the field of type I peptidomimetics, in particular those involving peptide macrocyclization. An overview of current strategies in the field of type III peptidomimetics will also be provided. Furthermore, our discussion will focus on peptidomimetics of α-helical and β-strand structural motifs, while the reader is referred to other sources for a survey on the design and applications of turn mimics [36].

## 15.2 PEPTIDOMIMETICS OF α-HELICAL PROTEIN BINDING EPITOPES

### 15.2.1 α-Helix-Mediated PPIs

The α-helix (or $3_{14}$-helix) represents the most common secondary structure in polypeptides, comprising over 30% of the secondary structural elements found in structurally characterized proteins. The α-helix is characterized by 3.4 residues per turn, a rise of 5.4 Å, and $\varphi$ and $\psi$ backbone dihedral angles of −60° and −45°, respectively, and is stabilized by a hydrogen-bond network involving the C=O of residue $i$ and the NH group of the $i+4$ residue. In α-helices, the amino acid side chains are positioned on top of each other every three to four residues along the peptide chain; as a result, α-helices are considered to display three "faces."

Due to their prevalence in protein structures, α-helical motifs are frequently involved in mediating PPIs. A recent survey of the Protein Data Bank (PDB) (August 2009) revealed that 62% of the structurally characterized protein–protein complexes feature an α-helical motif at the interface [33]. Furthermore, it was found that most of these α-helix-mediated PPIs (60%) involve a one-face α-helix interaction, with two-face (30%) and three-face (10%) interactions being significantly less abundant. Most common hot spot residues occurring at these interfaces comprise aromatic (Phe, Tyr, Trp) and hydrophobic (Leu) amino acids, which is consistent with the finding that PPIs are largely driven by hydrophobic interactions [10].

A number of α-helix-mediated PPIs have been identified as promising therapeutic targets due to their implication in the development and progression of cancer and other diseases (Fig. 15.1) [37, 38]. For example, a relatively short (2.5 turns) α-helical motif from the N-terminal transcriptional activation domain of the tumor suppressor p53 protein mediates its interaction with the negative regulators, HDM2 and HDMX (Fig. 15.1a) [39]. In this case, a triad of cofacial $i$/i + 4/$i$ + 7 residues ($Phe^{19}$, $Trp^{23}$, $Leu^{26}$) in the p53 helix insert into a hydrophobic cleft of the HDM2/X proteins [39]. Inhibition of the p53/HDM2 interaction has been recognized as a viable therapeutic strategy for activating p53-dependent apoptosis in cancer cells containing a functional, wild-type p53 protein [29]. A similar one-face α-helix interaction mode, involving hydrophobic residues at the $i$, $i+4$, $i+7$, and $i+11$ positions, drives the interaction between the prosurvival BH (Bcl-2 homology) domain proteins (e.g., Bcl-xL and Mcl-1) and the proapoptotic BH3 proteins (e.g., BID, BAD, and BIM) (Fig. 15.1b) [40, 41]. Disruption of these interactions has also proven to be a promising approach for cancer therapy [42]. The activities of several transcription factors are also regulated by α-helical protein binding motifs [38]. For example, estrogen receptors (ER) are activated by an α-helical peptide coactivator, in which three

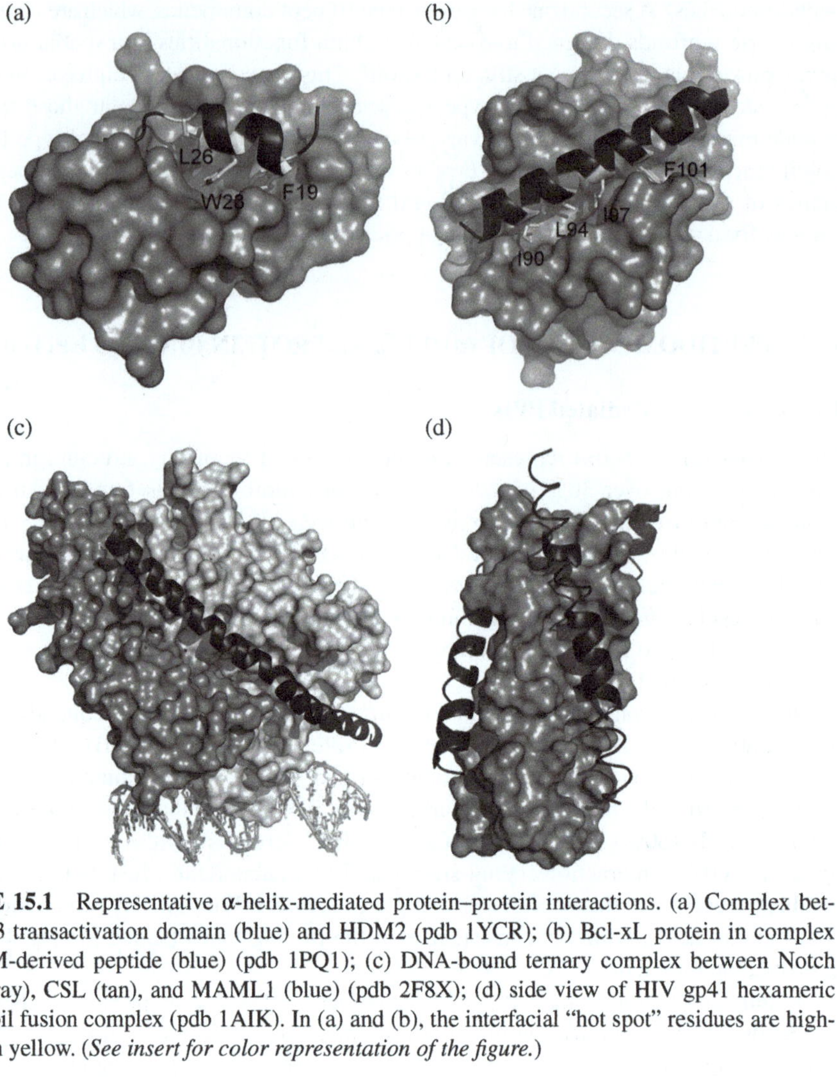


**FIGURE 15.1** Representative α-helix-mediated protein–protein interactions. (a) Complex between p53 transactivation domain (blue) and HDM2 (pdb 1YCR); (b) Bcl-xL protein in complex with BIM-derived peptide (blue) (pdb 1PQ1); (c) DNA-bound ternary complex between Notch ICN1 (gray), CSL (tan), and MAML1 (blue) (pdb 2F8X); (d) side view of HIV gp41 hexameric coiled-coil fusion complex (pdb 1AIK). In (a) and (b), the interfacial "hot spot" residues are highlighted in yellow. (*See insert for color representation of the figure.*)

Leu residues at the $i$, $i+3$, and $i+4$ positions bury into the ER pocket via a two-face α-helix binding mode [43]. Among the α-helix-mediated PPIs occurring in signaling pathways implicated in tumor progression, there is that involving an α-helical motif from the transcription factor TCF4 and one from BCL9 that bind to two distinct grooves in the β-catenin protein [44], a key mediator of intracellular signal transduction in the Wnt pathway [45]. In the Notch signaling pathway, an extended 60 amino acid-long α-helical domain from the MAML transcriptional coactivator protein binds in a cleft at the interface of CSL and Notch intracellular domain (Fig. 15.1c) [46], a process critical for mediating transcription of Notch-regulated genes [47]. Finally, therapeutically relevant α-helix-mediated PPIs involved in other pathologies include the gp41 (glycoprotein 41) hexameric coiled-coil assembly (Fig. 15.1d) [48], which plays a key role in cell invasion by the human immunodeficiency virus (HIV-1).

Excised from the protein context, short peptides spanning α-helical motifs (<15–20 residues) rarely adopt a stable α-helical structure in solution. In addition, linear peptides are readily degraded by proteases and display poor cell penetrability and pharmacokinetics. To overcome these issues, a number of strategies have been explored to stabilize the α-helical conformation in peptides [37, 49], in particular through the installation of covalent linkages in between the side chains and the backbone of these molecules. Conformational rigidification of these molecules has proven beneficial toward improving their binding affinity and specificity to the target protein, stability against proteolytic degradation, and ability to cross biological membranes.

### 15.2.2 Side-Chain Cross-Linked α-Helices

***15.2.2.1 Disulfide- and Lactam-Constrained α-Helices*** Given the predictable arrangement of $i$, $i+4$, $i+7$, and $i+11$ residues along the same face of the helix, a widely investigated strategy for α-helix stabilization has involved introduction of covalent bonds between the $i$ and the $i+4$ or $i+7$ side-chain groups (Fig. 15.2).

Some of the earliest efforts in this direction have employed amino acids capable of forming intrachain disulfide bridges. Schultz et al. reported significant α-helical stabilization (20 → 99% α-helicity) of a model peptide via disulfide formation between a D- and L-2-amino-6-mercaptohexanoic acid residue at the $i$ and $i+7$ positions, respectively [50].

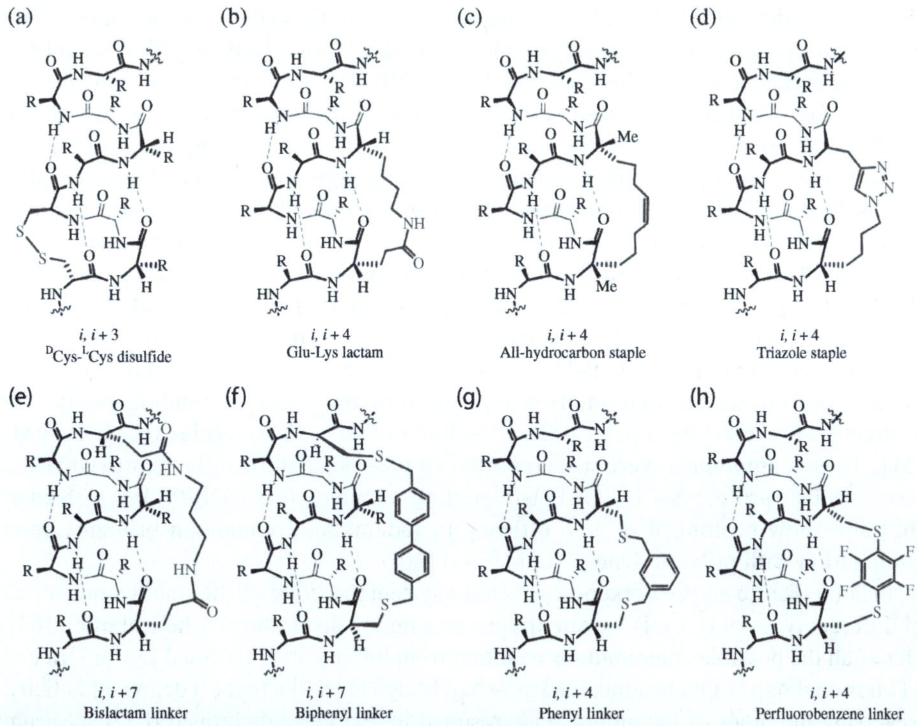


**FIGURE 15.2** Representative side-chain cross-linking strategies for α-helix stabilization, involving interside-chain bridges (a–d) and non-peptidic linkers (e–h).

The L-,L- and L-,D- analogs show minimal α-helical content, highlighting the critical role of the D-,L- arrangement for optimal stabilization of the α-helical structure. For the same purpose, other groups have investigated the use of disulfide bridges spanning a single turn of the helix via oxidation of two cysteine residues at the $i$ and $i+3$ positions (Fig. 15.2a). Mierke et al. reported a ($i$, $i+3$) disulfide-bridged peptide derived from the N-terminal parathyroid hormone-related protein (PTHrP) [51], whereas Spatola and coworkers applied this approach to develop inhibitors of the ER, ERα, and ERβ, based on the LXXLL-containing nuclear receptor box pentapeptide [52]. In the latter work, the most potent ($i$, $i+3$) disulfide-bridged peptide showed a $K_i$ of 70 pM for ERα and 350 pM for ERβ, exhibiting superior potency (10-fold lower $K_i$) compared to a ($i$, $i+4$) lactam-bridged analog [53]. Also in this case, a D-,L- arrangement (i.e., $^{D}$Cys at $i$ and $^{L}$Cys at $i+3$) was found to be optimal for α-helical stabilization (Fig. 15.4a; see following text) as well as biological activity owing to a best match of the distance between the cysteine side-chain thiol groups with the length of the disulfide bridge [52].

Due to the chemical instability of disulfide bonds, nonreducible interside-chain linkages have been explored for α-helix stabilization. A widely investigated approach has involved the use of lactam bridges formed by amide coupling of Asp/Glu and Lys residues (Fig. 15.2b). One of the earliest examples was the generation of cyclic analogs of the growth hormone-releasing factor (GRF) via a ($i$, $i+4$) $Asp^{8}\rightarrow Lys^{12}$ lactamization by scientists at Roche [56]. These compounds were reported to exhibit high potency as well as prolonged *in vivo* stability compared to the natural hormone. Using an analogous strategy, Chorev et al. developed an analog of the PTHrP stabilized by a ($i$, $i+4$) lactam bridge between $Lys^{13}$ and $Asp^{17}$ [57]. This modification led to a 10-fold greater affinity for the parathyroid hormone (PTH) receptor as compared to the linear peptide, showing a $K_i$ of 17 nM in a PTH-stimulated adenylate cyclase assay with human bone-derived B10 cells. Interestingly, inverting the stereochemistry of either or both residues had detrimental effect on binding. A similar design principle was applied to constrain other osteogenic peptides with varying success [58].

Ahn et al. used a positional cyclization scanning approach to identify more stable lactam-bridged analogs of glucagon [59]. Although all of the constrained analogs showed higher helical content compared to glucagon, only macrolactamization between $Lys^{5}/Glu^{9}$ and $Lys^{17}/Glu^{21}$ resulted in an increase in binding affinity (sixfold) for the glucagon receptor ($IC_{50}$: 240 nM) [59]. In another work, Miranda et al. [60] and Murage et al. [61, 62] designed a series of conformationally constrained glucagon-like peptide-1 (GLP-1) derivatives bearing $i$, $i$+4/5/7 lactam linkages. Several derivatives showed improved potency in the receptor binding assays, in particular those carrying a ($i$, $i+4$) $Lys^{30}\rightarrow Glu^{34}$ or $Glu^{22}\rightarrow Lys^{26}$ bridge ($IC_{50}$ ~ 1.5 nM) [61]. These compounds were also reported to possess significantly improved resistance against proteolysis by GLP-1-degrading enzymes (e.g., DPPV-IV; $t_{1/2}$ > 24 h) [62] and show prolonged *in vivo* efficacy in rodents and nonhuman primates upon conjugation with polyethylene glycol (PEG) [60].

In 2005, Fairlie and coworkers conducted a systematic study on the stabilization effect of different types of ($i$, $i+4$) lactam bridges on a minimalist (1 turn) α-helical motif [63]. Out of all the possible combinations resulting from linking an N-terminal Lys or Orn and a C-terminal Asp or Glu residue, the Lys→Asp bridge led to the highest degree of helicity. Reversing the order of the amino acids resulted in a circular dichroism (CD) spectrum consistent with a $3_{10}$-helix, whereas shortening the amide linkage by one methylene unit (Lys to Orn substitution) resulted in complete loss of helicity. This stabilization strategy was then applied to generate peptidomimetics of the respiratory syncytial virus fusion protein, leading to the identification of a constrained 13-residue peptide with high helicity in water and potent antiviral activity ($IC_{50}$: 36 nM) [64]. Other successful applications of this design principle were reported in the context of mimetics of the bacterial quorum-sensing pheromone, competence-stimulating peptide (CSP-1), and the GPCR "superagonist" ORL-1. In the latter case, the constrained peptide showed a ninefold improvement as a nociceptin agonist ($EC_{50}$: 40 pM) as compared to unconstrained counterpart ($EC_{50}$: 360 pM) [65]. The possibility to achieve further stabilization of α-helical peptides via the introduction of multiple lactam bridges has also been explored [66, 67] although these studies have remained for the most part limited to structural analyses.

***15.2.2.2 All-Hydrocarbon Stapled α-Helices*** Peptide "stapling" via interside-chain hydrocarbon linkages (e.g., Fig. 15.2c) has emerged as one of the most promising approaches for the generation of potent and stable type I peptidomimetics of α-helical protein binding motifs [68, 69]. This approach, developed in the Verdine lab [70] and currently serving as a drug discovery platform at Aileron Therapeutics, relies on two design principles for α-helix stabilization: (i) modification of the target peptide sequence with α,α-disubstituted amino acids (e.g., α-methyl-α-alkenyl amino acids), which can act as strong α-helix inducers [71], and (ii) formation of an interside-chain all-hydrocarbon cross-link ("staple") via ruthenium-catalyzed olefin metathesis, a reaction that was previously found to be compatible with macrocyclization of peptide sequences by Blackwell and Grubbs [72].

Systematic analysis of the stereochemistry, length of the alkenyl side chain, and relative location of the olefin-containing unnatural amino acids has enabled the identification of optimal parameters for α-helix stabilization via all-hydrocarbon stapling [68]. Typically, optimal stability for $(i, i+4)$ stapled peptides is achieved through an 8-carbon hydrocarbon linkage formed by ring-closing metathesis (RCM) between two α-methyl-α-pentenyl glycine residues with an (*S*) configuration (Fig. 15.3) [73]. A similar linkage yet between residues with opposite configuration (i.e., (*R*) at *i*, (*S*) at $i+3$) was found optimal to favor α-helical structures in $(i, i+3)$ stapled peptides [74].

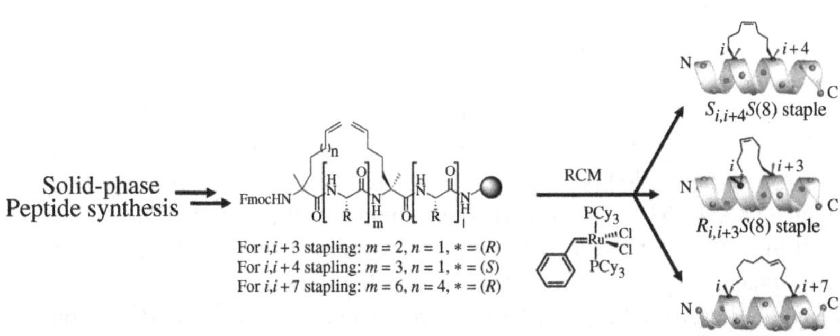


**FIGURE 15.3** All-hydrocarbon stapled peptides. The cross-linking α-methyl-α-alkenyl glycine amino acids are incorporated during solid-phase peptide synthesis, followed by in-solution or on-resin cyclization via ring-closing metathesis (RCM) with Grubb's first-generation catalyst. Three types of stapled peptides are schematically illustrated along with the preferred size of the linker and stereochemistry of the α-methylated amino acids.

Finally, in (*i*, *i* + 7) stapled peptides, in which the staple encompasses two consecutive α-helical turns (as opposed to about one turn in the *i*, *i* + 4 and *i*, *i* + 3 stapled counterparts), a 11-carbon tether between two α-methylated residues with an (*R*) configuration at the *i* site and an (*S*) configuration at the *i* + 7 site was determined to be most effective toward α-helical stabilization (Fig. 15.3) [70].

In a number of studies by the Verdine's, Walensky's, and other groups, the installation of one or more hydrocarbon staples into α-helical protein recognition motifs was found to lead to peptidomimetics with enhanced protein binding affinity, reduced sensitivity to proteolytic degradation, and increased cell permeability [68, 69]. Because of these favorable features, some of these peptidomimetics have proven effective to achieve specific inhibition of the corresponding α-helix-mediated PPI *in vitro*, in cells, and, in some instances, also *in vivo*. In a key study by Walensky et al., this approach was applied to generate a stapled analog of a 23-mer amphipathic α-helical motif derived from the BH3 domain of the proapoptotic protein BID, resulting in a potent inhibitor ($SAHB_A$, $IC_{50}$ ~ 40 nM) of antiapoptotic members of the Bcl-2 protein family [75]. The stapled peptide showed increased helical content (87% vs. 16%) as well as a 10-fold longer half-life in serum as compared to the linear peptide. Notably, this compound was found to be capable of entering cells (via an energy-dependent endocytotic mechanism), activating the apoptotic pathway, and reducing the viability of various leukemia cell lines [75]. Furthermore, *in vivo* experiments with mice bearing human leukemia xenografts demonstrated the compound's ability to prolong the life of the animals by reducing progression of the disease [75]. Using a similar approach, (*i*, *i* + 4) stapled peptides based on other BH3 helices (BAD [76], BIM [77], NOXA [54], PUMA [54], and Mcl-1 [54]) were developed for targeting antiapoptotic Bcl-2 proteins. Screening of a panel of these compounds enabled the identification of a potent ($K_D$: 10 nM) and selective Mcl-1 inhibitor, which could sensitize leukemia cells to caspase-dependent apoptosis. This stapled peptide (MCL-1 $SAHB_D$) was found to engage MCL-1 at the canonical BH3-binding groove [54]. In contrast, a BIM BH3-derived (*i*, *i* + 4) stapled peptide, designed to target and activate proapoptotic BAX, was found to interact with a site on BAX surface that is distinct from the canonical BH3 binding site [77].

Peptide all-hydrocarbon stapling has been also successfully applied to develop inhibitors of the p53/HDM2 interaction [78], which is misregulated in a variety of tumors [29]. In this case, a (*i*, *i* + 7) stapling approach was found to be most effective to obtain peptidomimetics of the α-helical HDM2-binding domain in p53 (Fig. 15.1a) with increased stability (30–85% vs. 11% α-helicity for the wild-type p53 peptide) and higher binding affinity for HDM2 ($K_D$: 0.8–100 nM vs. 410 nM) [78]. Optimization of the sequence to increase the overall charge of the stapled peptide from −2 to +1 resulted in cell-permeable compounds, one of which (SAH-p53-8) was found able to induce cell death in HDM2-overexpressing cancer cell lines [78]. Since the wild-type p53 peptide or other HDM2-targeting peptides isolated by phage display can also bind to the homologous protein HDMX, optimized stapled peptides based on these sequences were found capable of inducing apoptosis in both HDM2- and HDMX-overexpressing tumor cells [79, 80], an opportunity not offered by small-molecule HDM2 inhibitors (e.g., nutlin-3a, RG7112) [29] that show negligible inhibition against HDMX. Finally, researchers at Aileron Therapeutics recently disclosed the development of an *i*, *i* + 7 stapled dual HDM2/X inhibitor peptide (ATSP-7041) capable of inducing robust p53-dependent tumor growth in HDM2/HDMX-overexpressing xenograft cancer models [81]. This compound, which was obtained via extensive iterative optimization of the peptidomimetic HDM2/X binding properties ($K_D$, $k_{on}$, $k_{off}$), cell penetration, and pharmacokinetic/pharmacodynamic properties, was shown to exhibit excellent pharmaceutical properties with broad tissue distribution, extended half-life (18 h in monkey), and more sustained activation of the p53 pathway (up to 48 h) than that achieved with a small-molecule HDM2 antagonist drug candidate (RG7112 [29]). ATSP-7041 has recently entered phase I clinical trials.

In addition to the aforementioned examples, stapled peptide inhibitors for a number of other α-helix-mediated PPIs have been reported over the past few years. For example, ($i$, $i+4$) stapled peptides based on mastermind-like (MAML1) coactivator protein could be used to disrupt the ternary complex between MAML, NOTCH intracellular domain ICN-1, and CSL transcription factor, resulting in the repression of the NOTCH pathway and antiproliferative effect on leukemia cells and T-ALL animal model [82]. Similarly, an α-helical β-catenin-binding motif in axin provided a starting point for the generation of ($i$, $i+4$) and ($i$, $i+7$) stapled analogs capable of blocking β-catenin interaction with the transcriptional regulator T-cell factor (TCF) [83]. Compared to the wild-type peptide, the most promising compound (fStAx-3) showed improved α-helicity (51% vs. 15%) and affinity to β-catenin ($K_D$: 60 nM vs. 5 μM) and could be further optimized to effectively inhibit β-catenin-mediated transcriptional activities and reduce the viability of Wnt-dependent colorectal cancer cell lines [83]. Other successful examples of the use of peptide stapling in α-helix peptidomimetic design include the development of inhibitors of the estrogen receptors ERα and ERβ [84] and the HIV-1 integrase [85], disruptors of the Rab8a/effector [86], and protein kinase A/AKAP interactions [87]. In addition, a double stapling strategy was applied to a peptide sequence related to the 36-mer HIV-1 fusion inhibitor enfuvirtide [88]. The clinical usefulness of this peptidic drug is limited by poor pharmacokinetics and oral bioavailability, both of which could be improved via the installation of two ($i$, $i+4$) hydrocarbon linkages along the peptide sequence [88].

The structure of various stapled peptide PPI inhibitors in complex with their respective target proteins has been recently elucidated by X-ray crystallography [54, 81, 83, 84, 89] or NMR [90]. These studies confirmed the α-helical structure of the stapled peptide inhibitors but also revealed that in a number of cases, the hydrocarbon linkage interacts with the protein surface, thereby contributing to the stability of the protein–inhibitor complex (e.g., in MCL-1 $SAHB_D$/MCL-1 [54] (Fig. 15.4b), SAH-p53-8/HDM2 [89], and SP1/ERβ [84] complex).

***15.2.2.3 Thioether-, Triazole-, and Oxime-Cross-Linked α-Helices*** A number of other types of covalent linkages between side chains have been investigated for α-helix stabilization. Using a 10-mer helical epitope from the HIV transmembrane envelope glycoprotein gp41 as a model peptide, Dawson and coworkers reported the preparation of constrained peptidomimetics featuring a ($i$, $i+4$) and ($i$, $i+3$) thioether linkage formed between a *N*-bromoacetyl-L-ornithine residue at the $i$ site and a downstream cysteine residue [91]. The cyclic analogs showed increased α-helicity as compared to the linear counterparts and to a ($i$, $i+3$) lactam-constrained analog.

Focusing on α-helical motifs from the PTHrP, Chorev and coworkers investigated the stabilizing effect of interside-chain 1,4-disubstituted [1,2,3]triazole linkages, which were generated via copper-catalyzed 1,3-dipolar Huisgen cycloaddition between an alkyne- and an azide-bearing unnatural amino acid (e.g., (L)-ε-azidonorleucine and (L)-propargylglycine) (Fig. 15.2d). All of the cyclic analogs in which the triazolyl moiety is flanked by a total of five to six methylene ($-CH_2-$) units were found to be compatible with the

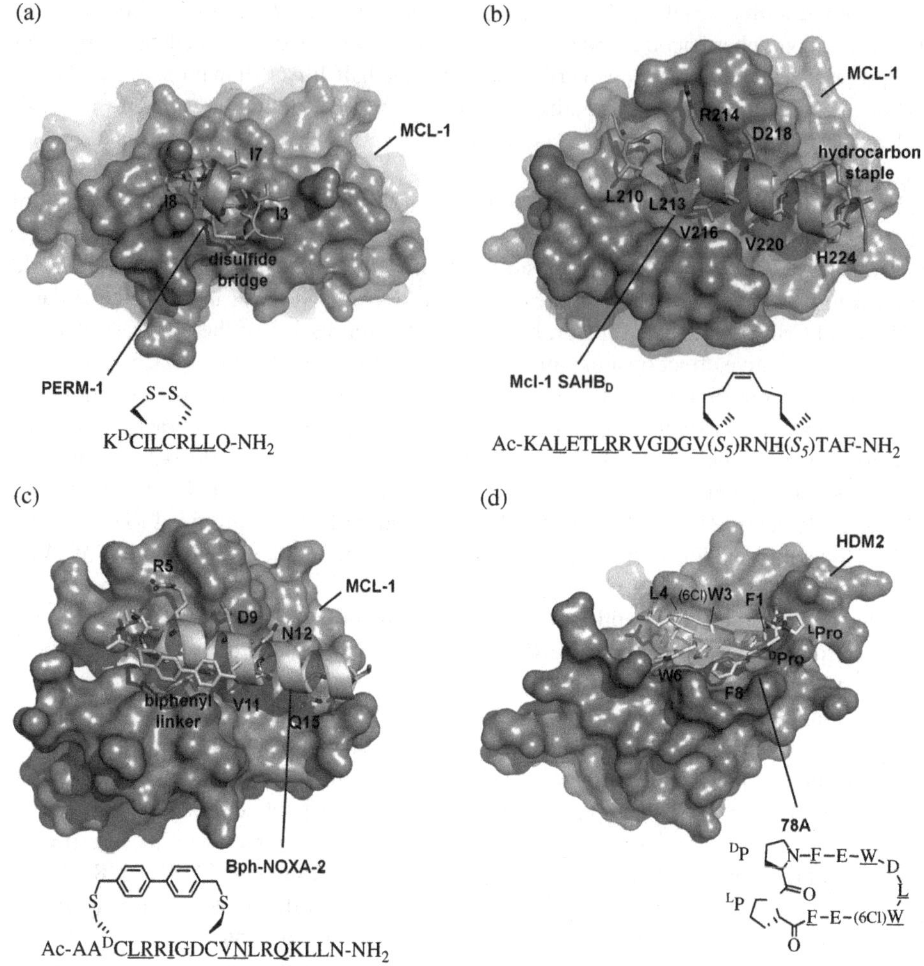


**FIGURE 15.4** Crystal structure of α-helix peptidomimetics bound to their target proteins. (a) Crystal structure of estrogen receptor α (gray) in complex with disulfide cross-linked PERM-1 (purple, PDB 1pcg); (c–d) structure of Mcl-1 (gray) in complex with (c) stapled peptide Mcl-1 $SAHB_D$ (orange, PDB 3mk8) [54] and (d) biphenyl-bridged peptide Bph-NOXA-2 (cyan, PDB 4g35) [55]; (d) crystal structure of complex between HDM2 (gray) and beta-hairpin inhibitor 78A (yellow, PDB 2axi) [12]. The residues in closest contact with the protein surface are displayed as stick models and highlighted in the corresponding structures. (*See insert for color representation of the figure.*)

α-helical structure [92, 93]. Expanding on this approach, Wang and coworkers more recently applied this strategy to generate triazole-stapled analogs of the β-catenin-binding α-helical motif from the B-cell CLL/lymphoma 9 (BCL9) protein [94]. The most promising compound was obtained by cyclization via (L)-ε-azidonorleucine and the D-stereoisomer of propargylglycine at the $i$ and $i+4$ positions, respectively. Compared to the uncyclized peptide, this compound showed increased α-helicity (90% vs. 44%), an approximately fivefold longer half-life in the presence of proteases, and fivefold lower $K_i$ (130 nM) in disrupting the β-catenin/BCL9 interaction *in vitro* [94]. Alternative approaches recently investigated to stabilize helical structures include the use of interside-chain oxime- [95] and pyrazoline-based linkages [96]. Although the latter was reported to favor the stabilization of $3_{10}$-helices, it could be applied to generate viable cyclic peptide inhibitors of the p53/HDM2(X) interaction ($IC_{50}$: 5–500 nM), some of which were found to activate p53 in cell-based luciferase assays [97].

***15.2.2.4 Side-Chain Cross-Linked α-Helices with Nonpeptidic Linkers*** While the methods described previously (Sections 15.2.2.1–15.2.2.3) rely on the formation of covalent linkages between amino acid side chains, other groups have focused on identifying nonpeptidic cross-linking moieties for stabilizing α-helical protein binding epitopes. In early studies, two Glu residues in an $i$, $i+7$ arrangement were linked through α,ω-diamino alkanes (Fig. 15.2e), and most effectively via 1,3-diaminopropane and 1,5-diaminopentane, to generate stabilized analogs of C-terminal α-helical motifs from apamin, RNAse A, and HIV-1 gp41 proteins [98, 99]. Using a related strategy, optimal dicarboxylic acid-based linkers were identified to stabilize a model α-helical peptide via connecting two lysine residues at $i$ and $i+4$, $i+7$, or $i+11$ positions [100].

Symmetrical cross-linking agents for achieving α-helical stabilization via linking of the side-chain sulfhydryl groups of two cysteine residues were also investigated. Wooley and coworkers reported the design of a "distance-matching" diaryl-chloroacetamido reagent that greatly enhanced the α-helical content of a model peptide upon *S*-alkylation of two cysteines at ($i$, $i+11$) positions [101]. A similar approach, but involving alkylation of ($i$, $i+7$) cysteines with *bis*-bromomethyl-biphenyl/bipyridine linkers (Fig. 15.2f), was utilized by Lin and coworkers to generate cyclic analogs of a phage-displayed peptide inhibitor of the p53-HDM2/X interaction [102] and of a NoxaB-derived peptide capable of binding to Mcl-1 with high selectivity over other Bcl-2 family members [55]. In the latter study, the most promising derivative, which featured a $^{D}$Cys at $i$ and a $^{L}$Cys at $i+7$ cross-linked with a 4,4′-*bis*-bromomethyl-biphenyl moiety, was reported to inhibit Mcl-1 with a $K_i$ of 30 nM and to reduce the viability of a Mcl-1-overexpressing cancer cell line (35%) with an $IC_{50}$ of 13 μM. Interestingly, higher activity in cells was obtained only upon sequence optimization to obtain a charge-neutral peptide and backbone *N*-methylation at the N-terminal amino acid positions not involved in intrahelical hydrogen bonding [55]. The crystal structure of the complex confirmed the α-helical structure of the peptidomimetic molecule (Fig. 15.4c). In another recent study, Greenbaum and coworkers screened a panel of cysteine-reactive bivalent reagents to stabilize an α-helical motif from calpastatin, a natural inhibitor of the calcium-regulated calpain proteases [103]. Here, the largest stabilization was obtained by alkylating two L-cysteines installed at solvent-exposed positions of the helix with α,α′-dibromo-*m*-xylene (Fig. 15.2g). The resulting 10-mer cyclopeptide, which shows a $K_i$ of 10 μM against calpain-1, could then be coupled to an epoxide warhead to generate an affinity-based probe that is selective for calpain-1 and calpain-2 over other cysteine proteases [103]. Finally, Pentelute and coworkers recently reported the stapling of cysteine-containing peptides via thiol-mediated nucleophilic aromatic substitution of perfluorinated phenyl linkers (Fig. 15.2h) [104].

The relative ability of the different α-helix stabilization strategies described in Sections 15.2.2.1–15.2.2.4 were recently compared side by side by Fairlie et al. using a model peptide, Ac-*cyclo*[$X_1$-$(Ala)_3$-$X_5$]-$NH_2$, where $X_1$ and $X_5$ are the cross-linked amino acid residues [105]. This study showed that the $Lys^1 \rightarrow Asp^5$ lactam linker was most effective in inducing α-helicity in water, followed by the all-hydrocarbon and triazole cross-links.

In comparison, the peptides constrained by a (*i*, *i*+4) *m*-xylene or perfluorophenyl thioether cross-links showed much lower α-helical content.

### 15.2.3 Hydrogen-Bond Surrogate-Stabilized α-Helices

According to the Zimm–Bragg model for α-helix formation [106], the formation of the first turn of the α-helix (nucleation) is associated with the largest energy barrier, after which propagation of the α-helical structure along the peptide chain occurs more rapidly. Based on these premises, the replacement of a backbone N–H···O=C hydrogen bond with a covalent "hydrogen-bond surrogate" (HBS) has provided an alternative approach for favoring α-helix nucleation and thus stabilizing α-helical structures in short peptides (Fig. 15.5) [107].

As compared to the interside-chain stapling strategies described previously (Section 15.2.2), an inherent advantage of the HBS-based approach in the context of α-helix peptidomimetic design is that the covalent cross-link is placed on the inside of the α-helix, that is, without involving side-chain functionalities, which can thus remain available for molecular recognition and interaction with the target protein. Over the past decade, a number of different HBS have been investigated for α-helix stabilization (Fig. 15.5).

The first example of HBS was reported in 1999 by Cabezas and Satterthwait, who replaced the main-chain hydrogen bond (C=O···NH) at the N-terminus of a 10-mer model peptide with a hydrazone link (Fig. 15.5b) [108]. The synthesis of the hydrazone-linked peptide was achieved by Fmoc solid-phase peptide synthesis (SPPS) by introducing a protected glycine hydrazide derivative ($(CH_3)_2C{=}N{-}NHCH_2CO{-}$) at the $i+4$ position and coupling 5,5-dimethoxy-1-oxopentanoic acid to the N-terminus of the peptide. Upon acid-catalyzed deprotection of the imine and acetal group and subsequent formation of the hydrazone linkage, the cross-linked peptide could be isolated in good yields (40%) and was determined by NMR to adopt a full-length α-helix in water. The hydrazone HBS-based approach was applied by Calvo et al. to generate a conformationally constrained mimic of an α-helical epitope from human papillomavirus type 16 E7 (HPV-16 E7) oncoprotein, which could be used to analyze sera of patients with cervical carcinoma [109].

Using a similar design principle, Arora and coworkers developed a strategy for the preparation of peptide constrained by a hydrocarbon HBS linkage (Fig. 15.5c), which

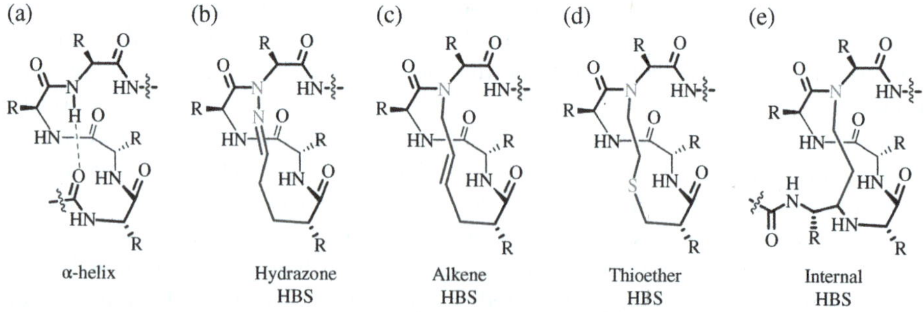


**FIGURE 15.5** Hydrogen-bond surrogates for α-helix nucleation and stabilization. The N-terminal (b–d) or internal (e) backbone hydrogen bond (C=O···H–N) between the *i*th and *i*+4th residue of an α-helix (a) is replaced with a covalent bond.

is installed by a ruthenium-catalyzed RCM reaction between an $N^{\alpha}$-allyl amino acid residue at position $i+4$ and an N-terminal 4-pentenoyl moiety [110]. Unlike the hydrazone strategy, the RCM-based approach affords a stable and irreversible intramain-chain bond, and a resulting HBS-stabilized model peptide (8-mer) was found to be fully α-helical in solution and to retain most of the helical structure (>85%) under thermal (95°C) or chemical (4 M guanidinium chloride) denaturing conditions. Further studies showed that this methodology is compatible with a variety of sequences, except when β-branched residues such as *N*-allyl-threonine, *N*-allyl-valine, and *N*-allyl-isoleucine are used at the $i+4$ site [111]. Under optimal catalytic conditions for the RCM step, the desired *trans*-alkene is obtained as the major product (*trans/cis* ratio of ~3:1) [111], but, depending on the sequence, its isolation by chromatographic means can be problematic [112]. However, saturation of the olefinic group to give a 13-membered macrocycle was found to retain the α-helix stabilizing effect of the unsaturated HBS counterpart [113].

Conformational stabilization via all-hydrocarbon HBS has been applied to a number of medicinally relevant α-helical protein binding domains. In a first study, a HBS-stabilized 16-mer peptide inhibitor of the oncoprotein Bcl-xL ($K_D$: 69 nM) was obtained based on an α-helical motif from the BAK BH3 domain [114]. In addition to increased α-helicity (65% vs. 20%) and twofold higher binding affinity for Bcl-xL, the constrained peptide displayed 60-fold higher stability against proteolytic degradation as compared to the unconstrained counterpart. HBS-containing peptides for targeting other α-helix-mediated PPIs such as those involved in p53/HDM2 complex formation ($K_D$: 160 nM) [115], Ras/SOS interaction ($K_D$: 28 nM for nucleotide-free Ras) [116], and HIV-1 gp41 six-helix bundle formation [117] were also reported, the latter being able to inhibit HIV-1 fusion in cell-based assays with an $EC_{50}$ of 42 μM. In more recent studies, HBS-stabilized helices from two distinct α-helical domains of the hypoxia-inducible factor 1α (HIF-1α) were developed to antagonize its interaction with p300/CBP coactivator protein and thus HIF-1α-mediated transcription of hypoxia-inducible genes [118, 119], a process involved in cancer growth and metastasis [120]. While types of HBS peptides could downregulate the expression of HIF-1α-inducible genes in cell-based assays, those based on HIF-1α C-terminal α-helical motif were also found to be able to induce a 50% tumor volume reduction in murine tumor xenografts [119].

Due to synthetic challenges associated with the preparation of HBS peptide libraries via RCM, alternative HBS have been explored, which include the use of thioether (Fig. 15.5d) [121], disulfide [122], and "internal" HBS linkages (Fig. 15.5e) [123]. Among them, the thioether and "internal" HBS cross-links were found to exhibit comparable α-helix stabilizing effects to that of hydrocarbon HBS based on CD and NMR studies, although the effectiveness of these peptidomimetics toward blocking PPIs *in vivo* awaits further investigation.

### 15.2.4 Other Type I α-Helix Peptidomimetics

***15.2.4.1 α-Helical Macrocyclic Organo-Peptide Hybrids*** Fasan and coworkers recently investigated macrocyclic organo-peptide hybrids (MOrPHs) [124, 125] as a molecular scaffold to accommodate α-helical protein recognition motifs. These compounds are constructed by side-chain-to-C-end cyclization of short (4–12-mer) peptide sequences via a variable nonpeptidic moiety through a dual orthogonal ligation strategy [124, 125]. Using the p53-HDM2/X interaction as a model for α-helix-mediated PPI, a series of designer MOrPHs were prepared by constraining a linear HDM2-binding peptide previously isolated by phage display by means of different nonpeptidic linkers that would match the distance between the side chain of a solvent-exposed residue (*i*) and the C-terminal carboxy group of a residue at the $i+6$ or $i+10$ position [126]. The most potent compound was found to act as a dual inhibitor for HDM2 and HDMX oncoproteins with an $IC_{50}$ of 110 and 340 nM, respectively, showing greatly (10-fold) enhanced half-life in the presence of proteases as compared to the acyclic counterpart. Interestingly, the nonpeptidic moiety was found to have a significant influence in modulating the conformational properties and HDM2/X isoform selectivity of the macrocycles [126].

***15.2.4.2 β-Hairpin Mimics of α-Helical Protein Epitopes*** Upon recognition that the distance between the C(α) atoms of the *i* and $i+4$ residues in an α-helix is close to that between the *i* and $i+2$ C(α) atoms along one strand of a β-hairpin, Robinson and coworkers designed β-hairpin peptidomimetics of p53 α-helical motif involved in the interaction with HDM2 protein [127]. Adoption of a β-hairpin structure in these compounds was enforced by head-to-tail cyclization with a $^{D}$Pro–$^{L}$Pro dipeptide template (*vide infra*), and upon sequence optimization, a potent β-hairpin inhibitor of the p53/HDM2 interaction ($IC_{50}$: 125 nM [127]; $K_D$: 25 nM [128]) was obtained. Structural elucidation of the β-hairpin inhibitor/HDM2 complex revealed that the $i/i+4/i+6$ triad of amino acid residues ($Phe^{19}/Trp^{23}/Leu^{26}$) in p53 directly involved in HDM2 binding was reproduced by the *i* (Phe)/$i+2$ (6-Cl-Trp)/$i+3$ (Leu) counterparts arranged on the β-hairpin scaffold (Fig. 15.4d) [12]. Furthermore, cofacial side chains from the second strand of the β-hairpin were found to establish energetically relevant interactions with the protein surface, which were not exploited by the p53 α-helical motif.

Using a similar design principle, β-hairpin "protein epitope mimetics" (PEMs) were successfully developed to target other α-helix-mediated protein–protein and protein–nucleic acid interactions [129, 130]. For example, this approach was applied to generate mimics of an α-helical domain of HIV-1 Rev protein, which exerts a crucial role in viral replication upon binding to a stem loop of HIV-1 Rev responsive element (RRE) mRNA [131]. The most potent derivatives were found to inhibit Rev/RRE mRNA interaction with a $K_D$ of 1–10 nM, that is, 10–100-fold lower than that of a Rev-derived α-helical peptide ($Rev_{34-50}$). In another work, another series of Arg-rich β-hairpin inhibitors targeted against the interaction between HIV-1 transactivator protein Tat and transactivation response (TAR) RNA were reported [132]. These peptidomimetics were found to effectively mimic Tat α-helix-mediated interaction with a major groove on TAR molecule, disrupting this complex with low nanomolar *in vitro* potency [133]. In subsequent studies, these compounds were found to inhibit Tat-dependent transcription in cell-based reporter assays and to inhibit the replication of diverse HIV-1 strains, including both CXCR44-tropic and CCR5-tropic HIV-1 isolates, with an $IC_{50}$ of 0.25–1 μM [134]. Finally, a β-hairpin compound containing two sulfated tyrosine residues in positions *i* and $i+2$ was designed to mimic the helical epitope from the cellular receptor CCR5 implicated in mediating HIV-1 cellular invasion via interaction with the viral coat protein gp120 [135]. The cyclic peptidomimetic was found to act as a low micromolar inhibitor of gp120/CD4 interaction and to be able to block HIV-1 infection of a luciferase reporter cell with moderate potency.

### 15.2.5 Type III α-Helix Peptidomimetics

This class of peptidomimetics comprises nonpeptidic molecules that replicate the spatial orientation of key residues displayed on the α-helical peptide structure (Fig. 15.6) [37, 136]. The main idea behind these designs is to utilize a template that could exhibit more "drug-like" properties than peptide-based molecules and potentially allow for a modular approach toward targeting different PPIs by matching the side chains to the target α-helical motif.

A prototypical example is the trisubstituted 3,2′,2″-terphenyl scaffold introduced by Hamilton and coworkers, in which the *ortho*-substituents are arranged in a spatial arrangement similar to the *i*, *i* + 3(4), and *i* + 7 residues of an α-helix (Fig. 15.6b) [137]. Differently substituted terphenyl derivatives were designed to inhibit calmodulin/phosphodiesterase [137] and Bcl-xL/Bak [138] interactions, as well as gp41 assembly [139] with $IC_{50}$ values in the submicromolar range (110–800 nM). The selectivity of these compounds for different α-helix-binding proteins (e.g., p53/HDM2 vs. Bcl-xL/Bak interaction) could be tuned by exchanging the substituents connected to the terphenyl core structure [140]. Subsequent modification of this prototypical scaffold in order to

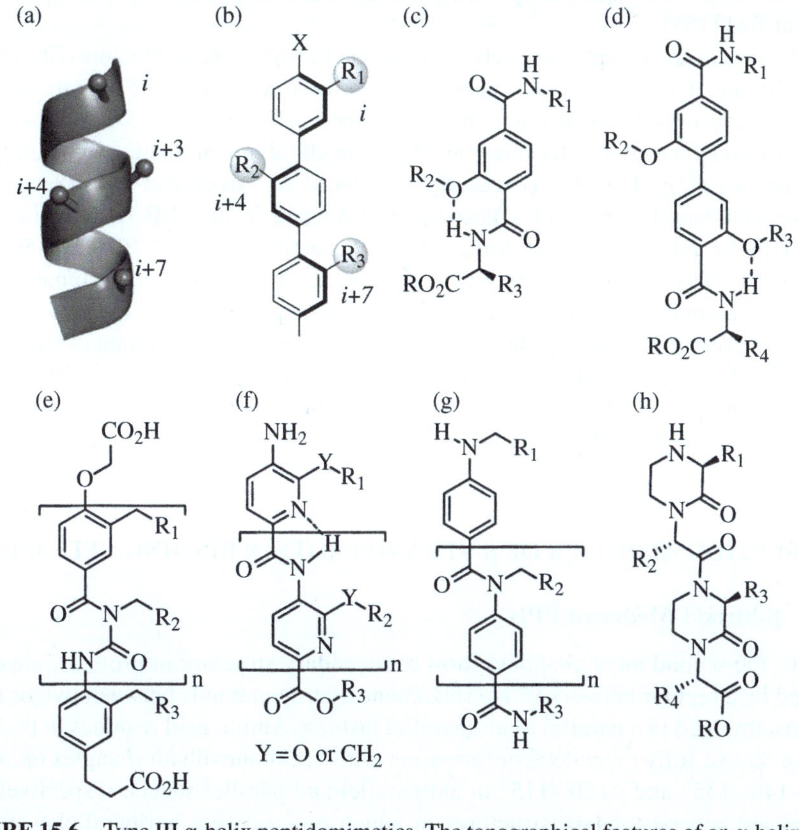


**FIGURE 15.6** Type III α-helix peptidomimetics. The topographical features of an α-helix (a) are mimicked using nonpeptidic scaffolds based on (b) 3,2′,2″-terphenyl, (c) terephthalamide, (d) biphenyl-4,4-dicarboxamide, (e) oligobenzoylurea, (f) oligopyridylcarboxamide, (g) *N*-alkylated oligobenzamide, and (h) oxopiperazine.

simplify the synthesis and increase the water solubility of the corresponding derivatives led to the development of terephthalamide (Fig. 15.6c) [141], biphenyl-4,4-dicarboxamide (Fig. 15.6d) [142], and benzoylurea (Fig. 15.6e) [143, 144] templates, which resulted in inhibitors of the Bcl-xL/Bak interaction with activities comparable to those of the terphenyl-based compounds ($K_i$: 0.8–2.4 μM).

Scaffolds based on aromatic oligoamides [145] such as oligopyridylcarboxamides (Fig. 15.6f) [146], oligophenylcarboxamides (Fig. 15.6g) [146], and regioisomers thereof [37] have attracted significant interest due to their synthetic accessibility and modular structure, which have in turn facilitated the preparation of large peptidomimetic libraries (e.g., >8000 members) [147] by solution-phase [147, 148] or solid-phase [149] parallel synthesis. Compounds based on these scaffolds have furnished low micromolar inhibitors against islet amyloid polypeptide (IAPP) aggregation ($IC_{50}$: 8 μM) [150], the p53/HDM2 interaction ($K_i$: 3–12 μM) [148, 151, 152], and Kaposi's sarcoma-associated herpesvirus protease ($IC_{50}$: 3.1 μM) [153]. A potent oligophenylcarboxamide-based mimetic developed by Yap et al. was found to inhibit Bak/Bcl-xL with a $K_i$ of 94 nM, showing antiproliferative activity against Bcl-xL-overexpressing cancer cell lines ($IC_{50}$: 1–4 μM) [154]. In another study, an oligoamide peptidomimetic capable of inhibiting gp41 assembly ($K_i$: 0.6–1 μM) and showing activity in cell–cell fusion assay ($IC_{50}$: 5–8 μM) was identified [155].

While the classes of peptidomimetics reproduce the topographical features of one face of the helix (i.e., $i$, $i+4$, $i+7$ side chains), some of these scaffolds were later modified to enable mimicry of side-chain functionalities at the $i+3$ [156] or $i+5$ positions [155]. Finally, a novel class of α-helix mimetics based on chiral, nonaromatic oligooxopiperazine structures (Fig. 15.6h) was recently introduced by Arora and coworkers [157]. Oxopiperazine-based mimics of a key α-helical domain in the HIF-1α were found to inhibit HIF-1α interaction with transcriptional coactivator p300/CBP ($K_D$ ~ 500 nM) *in vitro*, repress transcription of HIF-1α-regulated genes in cells, and reduce tumor burden (~50%) in experimental animal models [158]. In a subsequent study by the same group, a Rosetta-based computational method was implemented to guide the design of oxopiperazine α-helix mimics, enabling further optimization of the aforementioned HIF-1α mimics ($K_D$: 30–160 nM) as well as the development of submicromolar inhibitors of the p53/HDM2 interaction ($K_D$: 300 nM) [159].

## 15.3 PEPTIDOMIMETICS OF β-STRAND PROTEIN BINDING EPITOPES

### 15.3.1 β-Strand-Mediated PPIs

β-Sheets, the second most abundant form of secondary structure in proteins, are characterized by a regular network of interbackbone hydrogen bonds between two or more β-strands arranged in a parallel or antiparallel fashion. Amino acid residues in β-sheets are in an almost fully extended conformation with backbone dihedral angles (ϕ, ψ) of about −140°/135° and −120°/115° in antiparallel and parallel sheets, respectively. β-Hairpins are supersecondary structures in which two adjacent antiparallel β-strands are connected through a loop region (e.g., type I, II, I′, II′, or VI turn). β-Strand structures (i.e., β-sheets and β-hairpins) constitute important functional elements in PPIs and protein recognition (Fig. 15.7) [34].

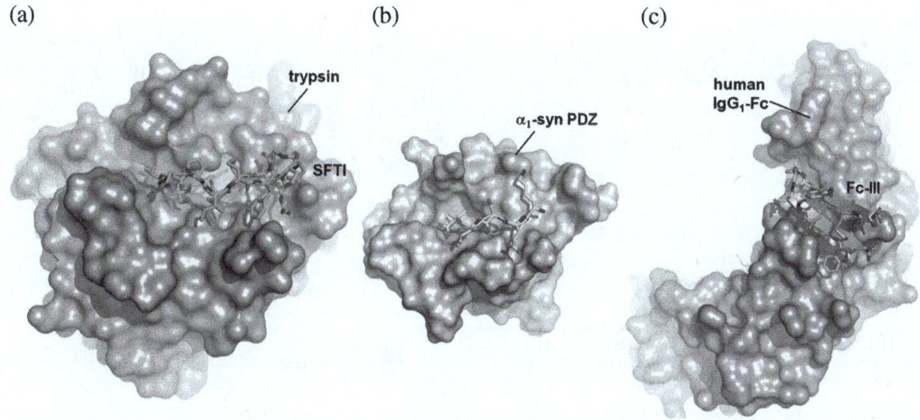


**FIGURE 15.7** Representative protein complexes involving β-strand binding epitopes. (a) Crystal structure of bovine trypsin (gray) in complex with β-hairpin cyclic peptide inhibitor from sunflower seeds (SFTI, cyan) (PDB 1SFI); (b) structure of $\alpha_1$-syntrophin PDZ domain in complex with single β-strand heptapeptide recognition motif (yellow, PDB 2PDZ); (c) structure of human IgG Fc region (gray) in complex with disulfide-constrained β-hairpin peptide binder (Fc III, purple) isolated by phage display (PDB 1DN2). (*See insert for color representation of the figure.*)

A recent survey identified about 15,000 interfacial β-strands making energetically significant contributions to protein complexes in the PDB [34]. This study also highlighted the inherent diversity of interaction modalities established by β-strands as compared to α-helices, which stems from their ability to interact with a protein partner as single entity or as part of β-sheet/β-hairpin structures, and via the side-chain functionalities exclusively or also through backbone hydrogen bonding (e.g., Fig. 15.7b). Aromatic amino acids, histidine, and arginine were found to be the most common energetically important residues within these structural motifs, with hot spots residues being localized predominantly on one face (e.g., Fig. 15.7c) or on both faces of the strand (e.g., Fig. 15.7a), in 62% and 32% of the cases, respectively.

Strategies for generating type I and type III peptidomimetics of β-hairpin and β-sheet structures have attracted significant interest in recent years, owing to their relevance in protein folding studies [160], in drug discovery [130, 161], and as chemical agents for studying and potentially interfering with processes associated with β-structure aggregation and fibril formation [162], which are implicated in a variety of amyloid diseases (e.g., Alzheimer's, Parkinson's, Huntington's disease) [163, 164].

### 15.3.2 Type I β-Strand Peptidomimetics

With a few exceptions, natural peptides encompassing β-hairpins or other β-structures found in proteins are devoid of native structure in aqueous solution or form aggregates. Although cross-strand noncovalent interactions (e.g., salt bridges, π–π stacking, π–cation interactions) can stabilize β-hairpin conformations in linear peptides [160], a more effective approach to achieve this goal has involved the use of organic templates to nucleate a β-structure conformation in the appended peptide chain (Fig. 15.8a). For this purpose, a variety of structures have been investigated (Fig. 15.8) [165], which share the common

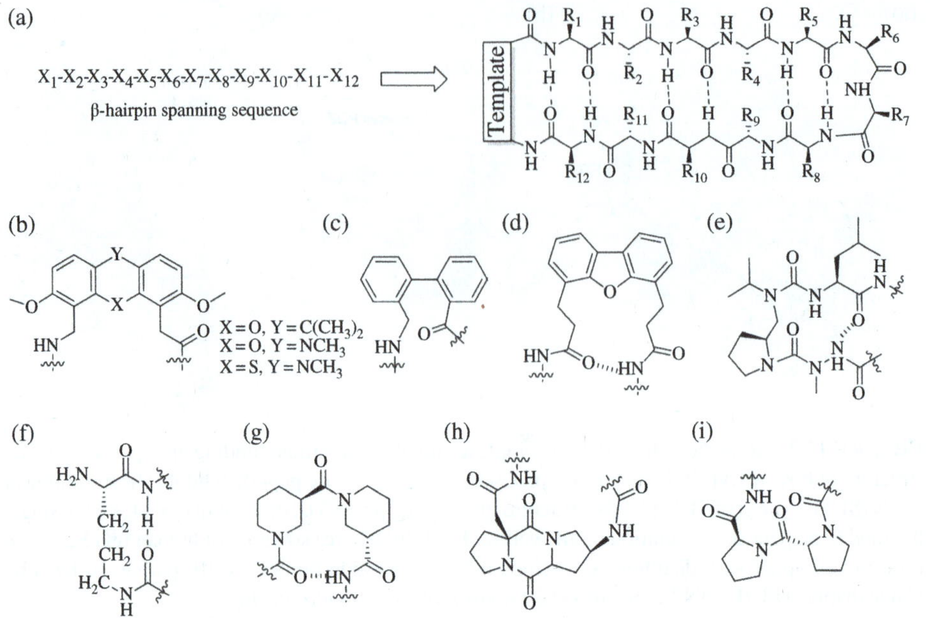


**FIGURE 15.8** Template-stabilized β-hairpin peptidomimetics. (a) General description of the design strategy for stabilizing β-hairpin (or β-sheet) motifs via the use of turn-mimicking templates; (b–i) selected templates used for β-hairpin/β-sheet stabilization of β-hairpin and β-sheet conformations based on phenoxathiin ring (b), 2,2′-substituted biphenyl (c), dibenzofuran (d), peptide/oligourea/azapeptide hybrid (e), (L)-ornithine (f), heterochiral nipecotic acid dimer (g), bicyclic proline derivative (h), and ᴰPro–ᴸPro unit (i).

feature of providing a relatively rigid skeleton to orient the amino and carboxyl functional groups at a distance comparable to that between opposite β-hairpin strands (5–9 Å).

In pioneering studies, Feigel and colleagues introduced aromatic scaffolds such as the phenoxathiin rings (Fig. 15.8b) [166] and 2,2′-substituted biphenyls (Fig. 15.8c) [167] to stabilize a β-structure in peptide loops. Other examples of aromatic scaffolds include the dibenzofuran turn units (Fig. 15.8d) investigated by Schneider and Kelly, whose efficiency in stabilizing β-hairpin structures was found to depend on the formation of a hydrophobic cluster between the template and the side chains of the flanking residues [168].

Alternative, nonaromatic scaffolds include the peptide/oligourea/azapeptide hybrid oligomer (Fig. 15.8e), reported by Soth and Nowick to adopt a hairpin turn in chloroform [169], and the heterochiral nipecotic acid-based structure, reported by Gellman and coworkers to induce sheet structures in the attached β-peptide strands (Fig. 15.8g) [170]. Significant interest has also been associated with templates derived from proline and structurally related lactams (Fig. 15.8h and i), due to their polar character, good β-turn mimetic properties, and synthetic accessibility [165]. Mono- and bicyclic proline-based heterocycles have been incorporated into receptor antagonists or agonists and in vaccine candidates [165].

A particularly successful strategy toward the development of β-hairpin peptidomimetics has involved the incorporation of a ᴰPro–ᴸPro dipeptide template into head-to-tail peptide cyclic peptides (Fig. 15.8i), an approach developed by Robinson and coworkers and used as drug discovery platform by Polyphor [130, 171]. This heterochiral dipeptide adopts a rigid type II′ β-turn [172], which is ideal to initiate antiparallel strands held together by registered cross-strand hydrogen bonds. For example, upon transferring the reactive loop (7 or 11 residues) from a member of the Bowman–Birk (BB) family of serine protease inhibitors onto a $^{D}$Pro–$^{L}$Pro template, structurally well-defined β-hairpin mimetics with low nanomolar $K_i$ values for inhibition of bovine trypsin were obtained [173]. This cyclic peptide was later utilized as a starting point for developing potent inhibitors against a number of pharmaceutically relevant serine proteases [130], such as cathepsin G (POL5196, $K_i$: 25 nM), human neutrophil elastase (HNE) (POL5975, $K_i$: 1.1 nM), and tryptase (POL9428, $K_i$: 8 nM), which are implicated in pulmonary diseases such as cystic fibrosis, acute lung injury, chronic obstructive pulmonary disease (COPD), and asthma [174–176].

Numerous naturally occurring antimicrobial peptides (AMPs) adopt a β-hairpin fold [177]. By designing $^{D}$Pro–$^{L}$Pro-templated β-hairpin mimetics based on the host-defense peptide protegrin 1 (e.g., L27-11; Fig. 15.9a) [178], potent antimicrobial macrocyclic compounds with MICs in the nanomolar range against multiple-drug-resistant *Pseudomonas* strains were obtained [180, 181]. The antimicrobial activity of optimized, nonmembrane-lytic derivatives of these compounds was found to depend upon their ability to target the β-barrel membrane protein LptD, which plays a critical role in the biogenesis of the bacterial outer membrane [182]. One of these peptides, POL7080, which shows optimal target selectivity, plasma stability properties, and toxicological profile, has recently entered phase I clinical trials for the treatment of Gram-negative bacterial infection [130].

In another work, the natural β-hairpin peptide polyphemusin II [183] served as the starting point for the development of CXCR4 antagonists [184]. CXCR4 is a transmembrane GPCR involved in many physiopathological processes, from mediating CD4-dependent HIV-1 infection of human T cells to regulating trafficking of normal and cancer stem cells [185]. CXCR4 inhibitors are thus of interest for the treatment of HIV and inflammation and as agents for mobilization of hematopoietic stem cells. The key residues from polyphemusin II were first transferred onto the $^{D}$Pro–$^{L}$Pro template, followed by further constraining of the peptide through cyclization. Using parallel synthesis, the potency and ADME properties of the initial peptidomimetic were then optimized in iterative cycles, resulting in potent and selective CXCR4 inhibitors ($IC_{50}$ ~ 1–2 nM) with prolonged plasma stability ($t_{1/2}$ > 5 h) [184]. One of these compounds (POL6326) successfully moved into a phase II clinical trial for autologous stem cell transplantation in newly diagnosed multiple myeloma patients [130].

In addition to aforementioned cases, the strategy based on $^{D}$Pro–$^{L}$Pro-templated macrocyclic peptides has proven useful to obtain structural mimetics of other biologically relevant protein epitopes, including α-helical motifs (see Section 15.2.4.2), turn structures [186], and engineered β-hairpin peptides isolated by phage display (Fig. 15.7c) [187].

### 15.3.3 Type III β-Strand Peptidomimetics

Similarly to the α-helix-mimicking counterparts, type III β-strand peptidomimetics consist of nonpeptidic (or hybrid nonpeptidic/peptidic) scaffolds designed to recreate the topological features of β-strand structures (Fig. 15.10).

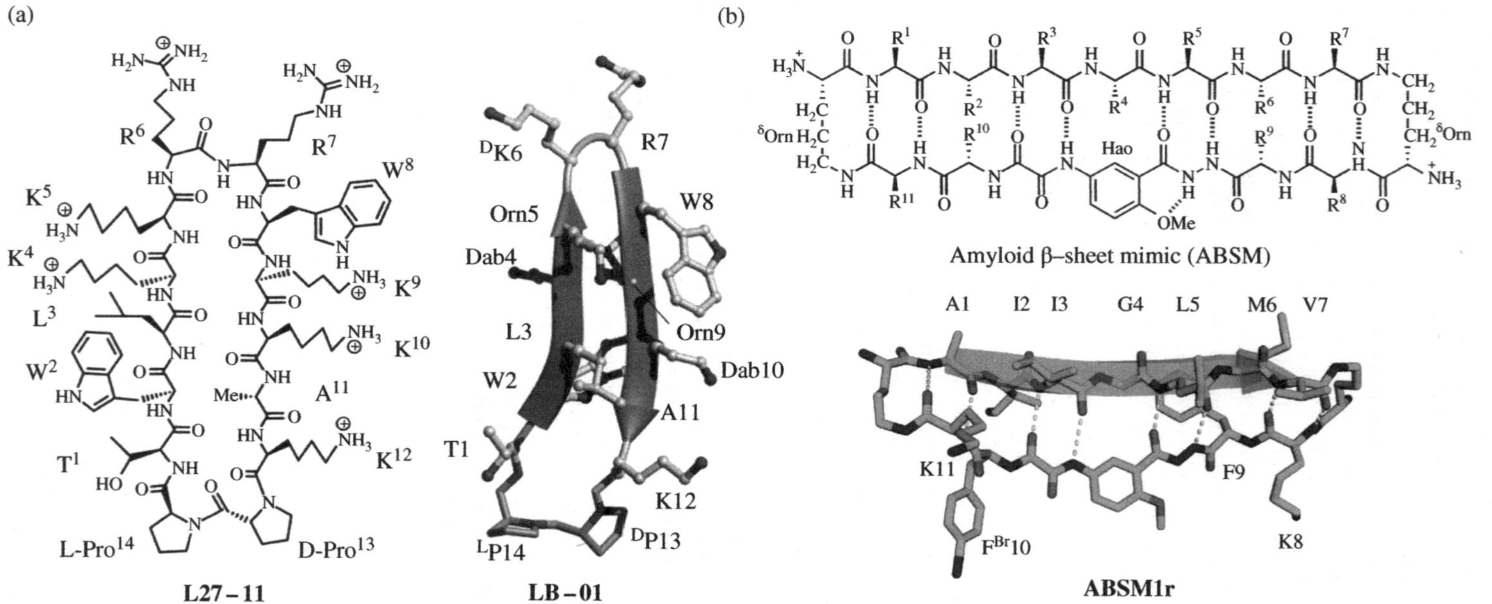


**FIGURE 15.9** β-hairpin peptidomimetics. (a) Chemical structure of protegrin I-derived cyclic peptidomimetic L27-11 [178] and NMR structure of its analog LB-01 [179]; (b) general structure of amyloid β-sheet mimics (ABSMs) and X-ray crystal structure of ABSM1r, an ABSM incorporating the heptapeptide sequence AIIGLMV from the amyloid-β peptide, Aβ [162]. (*See insert for color representation of the figure.*)

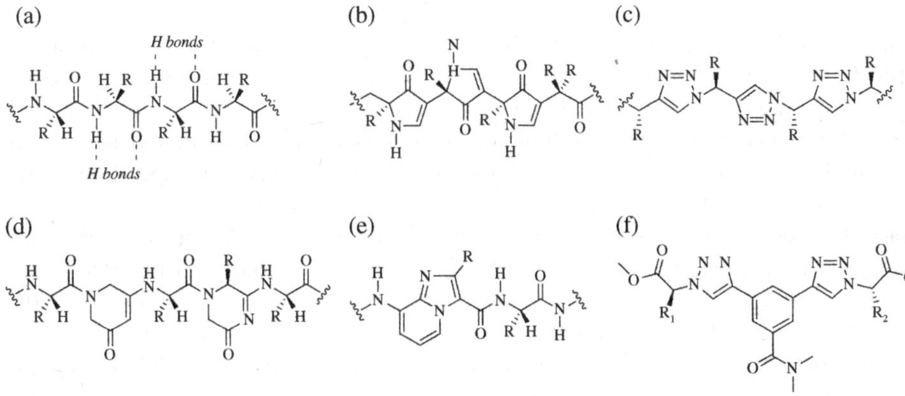

**FIGURE 15.10** (a) Representative type III peptidomimetics of peptide β-strand motifs, (b) oligopyrrolinones, (c) triazolamers, (d) dihydropyridinone/dihydropyrazinone hybrids, (e) imidazo[1,2-a]pyridine, and (f) "kinked" bis-triazoles.

One of the first examples of single β-strand mimics was reported by Smith and Hirschmann who designed a substituted pyrrolinone scaffold (Fig. 15.10b) to mimic both the backbone hydrogen-bond donors/acceptors and side-chain functionalities of an extended peptide β-strand [188]. These scaffolds were later utilized to develop inhibitors of HIV-1 protease ($IC_{50}$ ~ 2 nM) and other proteolytic enzymes (renin, matrix metalloproteases) [189]. More recently, an alternative class of β-strand mimics based on 1,3-substituted triazole oligomers (called "triazolamers") was introduced by Arora and coworkers (Fig. 15.10c) [190, 191]. NMR analysis revealed that these compounds mainly adopt a zigzag structure reminiscent of that of Hirschmann's oligopyrrolinones [188], with their overall conformation being dictated by the dipole–dipole interactions between adjacent rings. Although this backbone lacks the hydrogen-bond network of a typical β-strand, it contains hydrogen-bond capabilities for potential interaction with a protein [192]. The triazolamer scaffold was utilized to obtain peptidomimetics capable of inhibiting HIV-1 protease with moderate activity ($K_i$ ~ 25 μM) [193].

While the aforementioned cases are examples of fully nonpeptidic β-strand peptidomimetics, a parallel approach in β-strand mimicry has involved the incorporation of organic moieties into peptidic structures to enforce adoption of a β-hairpin (or β-sheet) conformation. For this purpose, Bartlett and coworkers introduced azacyclohexenone units such as 1,2-dihydro-3(*6H*)-pyridinone and 4,5-dihydro-2(*3H*)-pyrazinone (Fig. 15.10d) [194] into linear peptides to mimic a single β-strand interaction involved in the recognition of the cognate peptide substrate by the PDZ domain of $\alpha_1$-syntrophin (Fig. 15.7b) [195]. The resulting compounds were found to inhibit this protein with a $K_i$ of 320 nM, a binding affinity 20-fold higher than that of the linear peptide. Nowick and coworkers identified in the unnatural amino acid 5-amino-2-methoxybenzoic acid hydrazide (Hao; Fig. 15.9b), a well-suited nonpeptidic unit capable of duplicating the hydrogen-bonding functionality of one edge of a tripeptide β-strand [196]. Peptides incorporating the Hao moiety and connected through an L-ornithine-based β-turn mimic (Fig. 15.8f) [197] were found to be able to form artificial β-hairpin and β-sheet structures [198]. More recently, these scaffolds were utilized to display a variety of heptapeptide sequences from different amyloid proteins (Fig. 15.9b), including, among others, the Aβ-peptide associated with Alzheimer's disease, β2-microglobulin associated with dialysis-related amyloidosis, and α-synuclein associated with Parkinson's disease [162]. These so-called amyloid β-sheet mimics (ABSMs) were able to antagonize the aggregation of the corresponding amyloid proteins, thereby reducing the toxicity of amyloid aggregates.

Using a related approach, Hamilton and coworkers inserted a *p*-aminobenzoic acid (Abz) moiety, as a dipeptide β-strand mimic, into a linear peptide corresponding to the single β-strand epitope involved in the activation of Akt kinase by glycogen synthase kinase 3 (GSK3β). The resulting GSK3β-derived mimics were found to inhibit Akt kinase activity with an $IC_{50}$ of about 30 μM [199]. The same group later reported series of other β-strand peptidomimetics incorporating disubstituted indolinone- [200], imidazolinone- [201], and diphenylacetylene-based moieties [202], all of which exhibited structural characteristics of β-strands in the context of model peptide sequences. A series of Akt1 inhibitors with submicromolar activity ($IC_{50}$ ~ 0.5 μM) were prepared by Del Valle and coworkers by modifying a short Akt consensus peptide substrate with an imidazo[1,2-*a*]pyridine-based dipeptide mimic (Fig. 15.10e) [203, 204].

Finally, a series of potentially "universal" peptidomimetics have been recently reported by the Burgess' group [205]. These scaffolds, which are based on various heterocyclic core structures (e.g., 1,3,4-oxadiazoles, 1,3-butadienes, *bis*-triazoles, diketopiperazines), were found suitable for displaying pairs of functionalities in a spatial arrangement similar to that of amino acid side chains in common protein secondary structures, including β-strands (e.g., "kinked" *bis*-triazoles; Fig. 15.10f). Biological screening of these compounds led to the identification of micromolar inhibitors for a variety of protein targets, including the Bcl-2 family proteins Mcl-1 and Bid and the MEK5/MEK kinase 2 interaction, suggesting that these types of minimalistic peptidomimetics could constitute a potentially useful source of lead structures for further medicinal chemistry efforts [205].

## 15.4 CONCLUSION

The recognition that a large fraction of biomolecular interactions are mediated by well-defined protein secondary structural elements has spurred the development of different yet complementary approaches for the generation of peptidic and nonpeptidic mimics of α-helical and β-strand protein binding epitopes. As outlined in this chapter, these strategies have provided a route for the identification of inhibitors of a number of therapeutically relevant PPIs, which have previously eluded or remained inaccessible to modulation via well-established classes of therapeutics such as small molecules or antibodies. Conformationally constrained peptides have provided a viable, and currently the most effective, means for achieving potent and selective inhibition of a variety of α-helix- or β-hairpin-mediated PPIs. Importantly, structural rigidification of these molecules has proven useful toward overcoming notorious shortcomings of peptide-based agents, such as low metabolic stability and poor cell penetrability, in addition to having beneficial effects on potency and specificity. While the successful application of type I peptidomimetics across a number of different protein targets is encouraging, the process of turning these designs into viable drug candidates is still far from being a straightforward one, requiring a considerable amount of empirical optimization. In this regard, a major challenge is, for example, associated with the currently limited understanding of how and to what extent the different types of structural constraints employed in these strategies affect the cell permeability properties and intracellular fate (i.e., endosomal escape vs. intraendosomal trapping) of the resulting peptidomimetics. Progress in this direction is expected to greatly accelerate the development of cell-permeable agents that can physically access and effectively interfere with intracellular PPIs. In the context of β-strand-mediated interactions, another challenge is associated with the inherent heterogeneity of this subclass of PPIs, which is unlikely to be addressed by a unified peptidomimetic approach. One the other hand, the nonpeptidic nature of type III peptidomimetics confers them with potentially advantageous drug-like properties, but open challenges in this field remain that of defining modular strategies for controlling their molecular recognition properties as well as that of achieving high potency and selectivity toward a given protein target. These challenges notwithstanding, the recent advancement of stapled α-helical peptides as well as of β-hairpin PEMs to clinical testing suggests that these and other classes of peptidomimetics have the realistic potential to provide a valuable new source of therapeutics in the near future.

## REFERENCES


[1] Stumpf, M. P., Thorne, T., de Silva, E., Stewart, R., An, H. J., Lappe, M., and Wiuf, C. (2008) Estimating the size of the human interactome, *Proc. Natl. Acad. Sci. U. S. A.* **105**, 6959–6964.

[2] De Las Rivas, J., and Fontanillo, C. (2010) Protein-protein interactions essentials: key concepts to building and analyzing interactome networks, *PLoS Comput. Biol.* **6**, e1000807.

[3] Wells, J. A., and McClendon, C. L. (2007) Reaching for high-hanging fruit in drug discovery at protein-protein interfaces, *Nature* **450**, 1001–1009.

[4] Zinzalla, G., and Thurston, D. E. (2009) Targeting protein-protein interactions for therapeutic intervention: a challenge for the future, *Future Med. Chem.* **1**, 65–93.

[5] Milroy, L. G., Grossmann, T. N., Hennig, S., Brunsveld, L., and Ottmann, C. (2014) Modulators of protein-protein interactions, *Chem. Rev.* **114**, 4695–4748.

[6] Surade, S., and Blundell, T. L. (2012) Structural biology and drug discovery of difficult targets: the limits of ligandability, *Chem. Biol.* **19**, 42–50.

[7] Lo Conte, L., Chothia, C., and Janin, J. (1999) The atomic structure of protein-protein recognition sites, *J. Mol. Biol.* **285**, 2177–2198.

[8] Chakrabarti, P., and Janin, J. (2002) Dissecting protein-protein recognition sites, *Proteins* **47**, 334–343.

[9] Cheng, A. C., Coleman, R. G., Smyth, K. T., Cao, Q., Soulard, P., Caffrey, D. R., Salzberg, A. C., and Huang, E. S. (2007) Structure-based maximal affinity model predicts small-molecule druggability, *Nat. Biotechnol.* **25**, 71–75.

[10] Keskin, O., Gursoy, A., Ma, B., and Nussinov, R. (2008) Principles of protein-protein interactions: what are the preferred ways for proteins to interact? *Chem. Rev.* **108**, 1225–1244.

[11] Thanos, C. D., DeLano, W. L., and Wells, J. A. (2006) Hot-spot mimicry of a cytokine receptor by a small molecule, *Proc. Natl. Acad. Sci. U. S. A.* **103**, 15422–15427.

[12] Fasan, R., Dias, R. L., Moehle, K., Zerbe, O., Obrecht, D., Mittl, P. R., Grutter, M. G., and Robinson, J. A. (2006) Structure-activity studies in a family of beta-hairpin protein epitope mimetic inhibitors of the p53-HDM2 protein-protein interaction, *Chembiochem* **7**, 515–526.

[13] Spencer, R. W. (1998) High-throughput screening of historic collections: observations on file size, biological targets, and file diversity, *Biotechnol. Bioeng.* **61**, 61–67.

[14] Cochran, A. G. (2000) Antagonists of protein-protein interactions, *Chem. Biol.* **7**, R85–R94.

[15] Clackson, T., and Wells, J. A. (1995) A hot spot of binding energy in a hormone-receptor interface, *Science* **267**, 383–386.

[16] Moreira, I. S., Fernandes, P. A., and Ramos, M. J. (2007) Hot spots—a review of the protein-protein interface determinant amino-acid residues, *Proteins* **68**, 803–812.

[17] DeLano, W. L., Ultsch, M. H., de Vos, A. M., and Wells, J. A. (2000) Convergent solutions to binding at a protein-protein interface, *Science* **287**, 1279–1283.

[18] Thorn, K. S., and Bogan, A. A. (2001) ASEdb: a database of alanine mutations and their effects on the free energy of binding in protein interactions, *Bioinformatics* **17**, 284–285.

[19] DeLano, W. L. (2002) Unraveling hot spots in binding interfaces: progress and challenges, *Curr. Opin. Chem. Biol.* **12**, 14–20.

[20] Binkowski, T. A., Adamian, L., and Liang, J. (2003) Inferring functional relationships of proteins from local sequence and spatial surface patterns, *J. Mol. Biol.* **332**, 505–526.

[21] Bogan, A. A., and Thorn, K. S. (1998) Anatomy of hot spots in protein interfaces, *J. Mol. Biol.* **280**, 1–9.

[22] Reichmann, D., Rahat, O., Albeck, S., Meged, R., Dym, O., and Schreiber, G. (2005) The modular architecture of protein-protein binding interfaces, *Proc. Natl. Acad. Sci. U. S. A.* **102**, 57–62.

[23] Reichmann, D., Cohen, M., Abramovich, R., Dym, O., Lim, D., Strynadka, N. C., and Schreiber, G. (2007) Binding hot spots in the TEM1-BLIP interface in light of its modular architecture, *J. Mol. Biol.* **365**, 663–679.

[24] Ma, B., Shatsky, M., Wolfson, H. J., and Nussinov, R. (2002) Multiple diverse ligands binding at a single protein site: a matter of pre-existing populations, *Protein Sci.* **11**, 184–197.

[25] Geppert, T., Hoy, B., Wessler, S., and Schneider, G. (2011) Context-based identification of protein-protein interfaces and "hot-spot" residues, *Chem. Biol.* **18**, 344–353.

[26] Thangudu, R. R., Bryant, S. H., Panchenko, A. R., and Madej, T. (2012) Modulating protein-protein interactions with small molecules: the importance of binding hotspots, *J. Mol. Biol.* **415**, 443–453.

[27] Golden, M. S., Cote, S. M., Sayeg, M., Zerbe, B. S., Villar, E. A., Beglov, D., Sazinsky, S. L., Georgiadis, R. M., Vajda, S., Kozakov, D., and Whitty, A. (2013) Comprehensive experimental and computational analysis of binding energy hot spots at the NF-kappa B essential modulator/IKK beta protein-protein interface, *J. Am. Chem. Soc.* **135**, 6242–6256.

[28] Wang, W., and Hu, Y. (2012) Small molecule agents targeting the p53-MDM2 pathway for cancer therapy, *Med. Res. Rev.* **32**, 1159–1196.

[29] Khoo, K. H., Verma, C. S., and Lane, D. P. (2014) Drugging the p53 pathway: understanding the route to clinical efficacy, *Nat. Rev. Drug Discov.* **13**, 217–236.

[30] Thomas, S., Quinn, B. A., Das, S. K., Dash, R., Emdad, L., Dasgupta, S., Wang, X. Y., Dent, P., Reed, J. C., Pellecchia, M., Sarkar, D., and Fisher, P. B. (2013) Targeting the Bcl-2 family for cancer therapy, *Expert Opin. Ther. Targets* **17**, 61–75.

[31] Braisted, A. C., Oslob, J. D., Delano, W. L., Hyde, J., McDowell, R. S., Waal, N., Yu, C., Arkin, M. R., and Raimundo, B. C. (2003) Discovery of a potent small molecule IL-2 inhibitor through fragment assembly, *J. Am. Chem. Soc.* **125**, 3714–3715.

[32] White, P. W., Titolo, S., Brault, K., Thauvette, L., Pelletier, A., Welchner, E., Bourgon, L., Doyon, L., Ogilvie, W. W., Yoakim, C., Cordingley, M. G., and Archambault, J. (2003) Inhibition of human papillomavirus DNA replication by small molecule antagonists of the E1-E2 protein interaction, *J. Biol. Chem.* **278**, 26765–26772.

[33] Bullock, B. N., Jochim, A. L., and Arora, P. S. (2011) Assessing helical protein interfaces for inhibitor design, *J. Am. Chem. Soc.* **133**, 14220–14223.

[34] Watkins, A. M., and Arora, P. S. (2014) Anatomy of beta-strands at protein-protein interfaces, *ACS Chem. Biol.* **9**, 1747–1754.

[35] Gavenonis, J., Sheneman, B. A., Siegert, T. R., Eshelman, M. R., and Kritzer, J. A. (2014) Comprehensive analysis of loops at protein-protein interfaces for macrocycle design, *Nat. Chem. Biol.* **10**, 716–722.

[36] Kee, K. S., and Jois, S. D. (2003) Design of beta-turn based therapeutic agents, *Curr. Pharm. Des.* **9**, 1209–1224.

[37] Azzarito, V., Long, K., Murphy, N. S., and Wilson, A. J. (2013) Inhibition of alpha-helix-mediated protein-protein interactions using designed molecules, *Nat. Chem.* **5**, 161–173.

[38] Guarracino, D. A., Bullock, B. N., and Arora, P. S. (2011) Mini review: protein-protein interactions in transcription: a fertile ground for helix mimetics, *Biopolymers* **95**, 1–7.

[39] Kussie, P. H., Gorina, S., Marechal, V., Elenbaas, B., Moreau, J., Levine, A. J., and Pavletich, N. P. (1996) Structure of the MDM2 oncoprotein bound to the p53 tumor suppressor transactivation domain, *Science* **274**, 948–953.

[40] Czabotar, P. E., Lee, E. F., van Delft, M. F., Day, C. L., Smith, B. J., Huang, D. C., Fairlie, W. D., Hinds, M. G., and Colman, P. M. (2007) Structural insights into the degradation of Mcl-1 induced by BH3 domains, *Proc. Natl. Acad. Sci. U. S. A.* **104**, 6217–6222.

[41] Liu, X., Dai, S., Zhu, Y., Marrack, P., and Kappler, J. W. (2003) The structure of a Bcl-xL/Bim fragment complex: implications for Bim function, *Immunity* **19**, 341–352.

[42] Billard, C. (2013) BH3 mimetics: status of the field and new developments, *Mol. Cancer Ther.* **12**, 1691–1700.

[43] Shiau, A. K., Barstad, D., Loria, P. M., Cheng, L., Kushner, P. J., Agard, D. A., and Greene, G. L. (1998) The structural basis of estrogen receptor/coactivator recognition and the antagonism of this interaction by tamoxifen, *Cell* **95**, 927–937.

[44] Xu, W., and Kimelman, D. (2007) Mechanistic insights from structural studies of beta-catenin and its binding partners, *J. Cell Sci.* **120**, 3337–3344.

[45] Anastas, J. N., and Moon, R. T. (2013) WNT signalling pathways as therapeutic targets in cancer, *Nat. Rev. Cancer* **13**, 11–26.

[46] Wilson, J. J., and Kovall, R. A. (2006) Crystal structure of the CSL-Notch-Mastermind ternary complex bound to DNA, *Cell* **124**, 985–996.

[47] Andersson, E. R., and Lendahl, U. (2014) Therapeutic modulation of Notch signalling—are we there yet? *Nat. Rev. Drug Discov.* **13**, 357–378.

[48] Chan, D. C., Fass, D., Berger, J. M., and Kim, P. S. (1997) Core structure of gp41 from the HIV envelope glycoprotein, *Cell* **89**, 263–273.

[49] Andrews, M. J. I., and Tabor, A. B. (1999) Forming stable helical peptides using natural and artificial amino acids, *Tetrahedron* **55**, 11711–11743.

[50] Jackson, D. Y., King, D. S., Chmielewski, J., Singh, S., and Schultz, P. G. (1991) General approach to the synthesis of short .alpha.-helical peptides, *J. Am. Chem. Soc.* **113**, 9391–9392.

[51] Pellegrini, M., Royo, M., Chorev, M., and Mierke, D. F. (1997) Conformational consequences of i, i + 3 cystine linkages: nucleation for alpha-helicity? *J. Pept. Res.* **49**, 404–414.

[52] Galande, A. K., Bramlett, K. S., Trent, J. O., Burris, T. P., Wittliff, J. L., and Spatola, A. F. (2005) Potent inhibitors of LXXLL-based protein-protein interactions, *Chembiochem* **6**, 1991–1998.

[53] Leduc, A.-M., Trent, J. O., Wittliff, J. L., Bramlett, K. S., Briggs, S. L., Chirgadze, N. Y., Wang, Y., Burris, T. P., and Spatola, A. F. (2003) Helix-stabilized cyclic peptides as selective inhibitors of steroid receptor–coactivator interactions, *Proc. Natl. Acad. Sci. U. S. A.* **100**, 11273–11278.

[54] Stewart, M. L., Fire, E., Keating, A. E., and Walensky, L. D. (2010) The MCL-1 BH3 helix is an exclusive MCL-1 inhibitor and apoptosis sensitizer, *Nat. Chem. Biol.* **6**, 595–601.

[55] Muppidi, A., Doi, K., Edwardraja, S., Drake, E. J., Gulick, A. M., Wang, H. G., and Lin, Q. (2012) Rational design of proteolytically stable, cell-permeable peptide-based selective Mcl-1 inhibitors, *J. Am. Chem. Soc.* **134**, 14734–14737.

[56] Felix, A. M., Heimer, E. P., Wang, C. T., Lambros, T. J., Fournier, A., Mowles, T. F., Maines, S., Campbell, R. M., Wegrzynski, B. B., Toome, V., Fry, D., and Madison, V. S. (1988) Synthesis, biological activity and conformational analysis of cyclic GRF analogs, *Int. J. Pept. Protein Res.* **32**, 441–454.

[57] Chorev, M., Roubini, E., McKee, R. L., Gibbons, S. W., Goldman, M. E., Caulfield, M. P., and Rosenblatt, M. (1991) Cyclic parathyroid hormone related protein antagonists: lysine 13 to aspartic acid 17 [i to (i + 4)] side chain to side chain lactamization, *Biochemistry* **30**, 5968–5974.

[58] Willick, G. E., Morley, P., and Whitfield, J. F. (2004) Constrained analogs of osteogenic peptides, *Curr. Med. Chem.* **11**, 2867–2881.

[59] Ahn, J.-M., Gitu, P. M., Medeiros, M., Swift, J. R., Trivedi, D., and Hruby, V. J. (2001) A new approach to search for the bioactive conformation of glucagon: positional cyclization scanning, *J. Med. Chem.* **44**, 3109–3116.

[60] Miranda, L. P., Winters, K. A., Gegg, C. V., Patel, A., Aral, J., Long, J., Zhang, J., Diamond, S., Guido, M., Stanislaus, S., Ma, M., Li, H., Rose, M. J., Poppe, L., and Veniant, M. M. (2008) Design and synthesis of conformationally constrained glucagon-like peptide-1 derivatives with increased plasma stability and prolonged in vivo activity, *J. Med. Chem.* **51**, 2758–2765.

[61] Murage, E. N., Schroeder, J. C., Beinborn, M., and Ahn, J. M. (2008) Search for alpha-helical propensity in the receptor-bound conformation of glucagon-like peptide-1, *Bioorg. Med. Chem.* **16**, 10106–10112.

[62] Murage, E. N., Gao, G., Bisello, A., and Ahn, J. M. (2010) Development of potent glucagon-like peptide-1 agonists with high enzyme stability via introduction of multiple lactam bridges, *J. Med. Chem.* **53**, 6412–6420.

[63] Shepherd, N. E., Hoang, H. N., Abbenante, G., and Fairlie, D. P. (2005) Single turn peptide alpha helices with exceptional stability in water, *J. Am. Chem. Soc.* **127**, 2974–2983.

[64] Shepherd, N. E., Hoang, H. N., Desai, V. S., Letouze, E., Young, P. R., and Fairlie, D. P. (2006) Modular α-helical mimetics with antiviral activity against respiratory syncitial virus, *J. Am. Chem. Soc.* **128**, 13284–13289.

[65] Harrison, R. S., Shepherd, N. E., Hoang, H. N., Ruiz-Gómez, G., Hill, T. A., Driver, R. W., Desai, V. S., Young, P. R., Abbenante, G., and Fairlie, D. P. (2010) Downsizing human, bacterial, and viral proteins to short water-stable alpha helices that maintain biological potency, *Proc. Natl. Acad. Sci. U. S. A.* **107**, 11686–11691.

[66] Osapay, G., and Taylor, J. W. (1992) Multicyclic polypeptide model compounds. 2. Synthesis and conformational properties of a highly .alpha.-helical uncosapeptide constrained by three side-chain to side-chain lactam bridges, *J. Am. Chem. Soc.* **114**, 6966–6973.

[67] Bracken, C., Gulyas, J., Taylor, J. W., and Baum, J. (1994) Synthesis and nuclear magnetic resonance structure determination of an .alpha.-helical, bicyclic, lactam-bridged hexapeptide, *J. Am. Chem. Soc.* **116**, 6431–6432.

[68] Verdine, G. L., and Hilinski, G. J. (2012) Stapled peptides for intracellular drug targets, *Methods Enzymol.* **503**, 3–33.

[69] Walensky, L. D., and Bird, G. H. (2014) Hydrocarbon-stapled peptides: principles, practice, and progress, *J. Med. Chem.* **57**, 6275–6288.

[70] Schafmeister, C. E., Po, J., and Verdine, G. L. (2000) An all-hydrocarbon cross-linking system for enhancing the helicity and metabolic stability of peptides, *J. Am. Chem. Soc.* **122**, 5891–5892.

[71] Karle, I. L., Flippen-Anderson, J. L., Uma, K., and Balaram, P. (1990) Apolar peptide models for conformational heterogeneity, hydration, and packing of polypeptide helices: crystal structure of hepta- and octapeptides containing alpha-aminoisobutyric acid, *Proteins* **7**, 62–73.

[72] Blackwell, H. E., and Grubbs, R. H. (1998) Highly efficient synthesis of covalently cross-linked peptide helices by ring-closing metathesis, *Angew. Chem. Int. Ed.* **37**, 3281–3284.

[73] Kim, Y. W., and Verdine, G. L. (2009) Stereochemical effects of all-hydrocarbon tethers in i,i+4 stapled peptides, *Bioorg. Med. Chem. Lett.* **19**, 2533–2536.

[74] Kim, Y. W., Kutchukian, P. S., and Verdine, G. L. (2010) Introduction of all-hydrocarbon i,i+3 staples into alpha-helices via ring-closing olefin metathesis, *Org. Lett.* **12**, 3046–3049.

[75] Walensky, L. D., Kung, A. L., Escher, I., Malia, T. J., Barbuto, S., Wright, R. D., Wagner, G., Verdine, G. L., and Korsmeyer, S. J. (2004) Activation of apoptosis in vivo by a hydrocarbon-stapled BH3 helix, *Science* **305**, 1466–1470.

[76] Braun, C. R., Mintseris, J., Gavathiotis, E., Bird, G. H., Gygi, S. P., and Walensky, L. D. (2010) Photoreactive stapled BH3 peptides to dissect the BCL-2 family interactome, *Chem. Biol.* **17**, 1325–1333.

[77] Gavathiotis, E., Suzuki, M., Davis, M. L., Pitter, K., Bird, G. H., Katz, S. G., Tu, H. C., Kim, H., Cheng, E. H., Tjandra, N., and Walensky, L. D. (2008) BAX activation is initiated at a novel interaction site, *Nature* **455**, 1076–1081.

[78] Bernal, F., Tyler, A. F., Korsmeyer, S. J., Walensky, L. D., and Verdine, G. L. (2007) Reactivation of the p53 tumor suppressor pathway by a stapled p53 peptide, *J. Am. Chem. Soc.* **129**, 2456–2457.

[79] Bernal, F., Wade, M., Godes, M., Davis, T. N., Whitehead, D. G., Kung, A. L., Wahl, G. M., and Walensky, L. D. (2010) A stapled p53 helix overcomes HDMX-mediated suppression of p53, *Cancer Cell* **18**, 411–422.

[80] Brown, C. J., Quah, S. T., Jong, J., Goh, A. M., Chiam, P. C., Khoo, K. H., Choong, M. L., Lee, M. A., Yurlova, L., Zolghadr, K., Joseph, T. L., Verma, C. S., and Lane, D. P. (2013) Stapled peptides with improved potency and specificity that activate p53, *ACS Chem. Biol.* **8**, 506–512.

[81] Chang, Y. S., Graves, B., Guerlavais, V., Tovar, C., Packman, K., To, K. H., Olson, K. A., Kesavan, K., Gangurde, P., Mukherjee, A., Baker, T., Darlak, K., Elkin, C., Filipovic, Z., Qureshi, F. Z., Cai, H., Berry, P., Feyfant, E., Shi, X. E., Horstick, J., Annis, D. A., Manning, A. M., Fotouhi, N., Nash, H., Vassilev, L. T., and Sawyer, T. K. (2013) Stapled alpha-helical peptide drug development: a potent dual inhibitor of MDM2 and MDMX for p53-dependent cancer therapy, *Proc. Natl. Acad. Sci. U. S. A.* **110**, E3445–E3454.

[82] Moellering, R. E., Cornejo, M., Davis, T. N., Del Bianco, C., Aster, J. C., Blacklow, S. C., Kung, A. L., Gilliland, D. G., Verdine, G. L., and Bradner, J. E. (2009) Direct inhibition of the NOTCH transcription factor complex, *Nature* **462**, 182–188.

[83] Grossmann, T. N., Yeh, J. T., Bowman, B. R., Chu, Q., Moellering, R. E., and Verdine, G. L. (2012) Inhibition of oncogenic Wnt signaling through direct targeting of beta-catenin, *Proc. Natl. Acad. Sci. U. S. A.* **109**, 17942–17947.

[84] Phillips, C., Roberts, L. R., Schade, M., Bazin, R., Bent, A., Davies, N. L., Moore, R., Pannifer, A. D., Pickford, A. R., Prior, S. H., Read, C. M., Scott, A., Brown, D. G., Xu, B., and Irving, S. L. (2011) Design and structure of stapled peptides binding to estrogen receptors, *J. Am. Chem. Soc.* **133**, 9696–9699.

[85] Long, Y. Q., Huang, S. X., Zawahir, Z., Xu, Z. L., Li, H., Sanchez, T. W., Zhi, Y., De Houwer, S., Christ, F., Debyser, Z., and Neamati, N. (2013) Design of cell-permeable stapled peptides as HIV-1 integrase inhibitors, *J. Med. Chem.* **56**, 5601–5612.

[86] Spiegel, J., Cromm, P. M., Itzen, A., Goody, R. S., Grossmann, T. N., and Waldmann, H. (2014) Direct targeting of Rab-GTPase-effector interactions, *Angew. Chem. Int. Ed.* **53**, 2498–2503.

[87] Wang, Y., Ho, T. G., Bertinetti, D., Neddermann, M., Franz, E., Mo, G. C., Schendowich, L. P., Sukhu, A., Spelts, R. C., Zhang, J., Herberg, F. W., and Kennedy, E. J. (2014) Isoform-selective disruption of AKAP-localized PKA using hydrocarbon stapled peptides, *ACS Chem. Biol.* **9**, 635–642.

[88] Bird, G. H., Madani, N., Perry, A. F., Princiotto, A. M., Supko, J. G., He, X., Gavathiotis, E., Sodroski, J. G., and Walensky, L. D. (2010) Hydrocarbon double-stapling remedies the proteolytic instability of a lengthy peptide therapeutic, *Proc. Natl. Acad. Sci. U. S. A.* **107**, 14093–14098.

[89] Baek, S., Kutchukian, P. S., Verdine, G. L., Huber, R., Holak, T. A., Lee, K. W., and Popowicz, G. M. (2012) Structure of the stapled p53 peptide bound to Mdm2, *J. Am. Chem. Soc.* **134**, 103–106.

[90] Moldoveanu, T., Grace, C. R., Llambi, F., Nourse, A., Fitzgerald, P., Gehring, K., Kriwacki, R. W., and Green, D. R. (2013) BID-induced structural changes in BAK promote apoptosis, *Nat. Struct. Mol. Biol.* **20**, 589–597.

[91] Brunel, F. M., and Dawson, P. E. (2005) Synthesis of constrained helical peptides by thioether ligation: application to analogs of gp41, Chem. Commun. (Camb.) 20, 2552–2554.

[92] Cantel, S., Isaad Ale, C., Scrima, M., Levy, J. J., DiMarchi, R. D., Rovero, P., Halperin, J. A., D'Ursi, A. M., Papini, A. M., and Chorev, M. (2008) Synthesis and conformational analysis of a cyclic peptide obtained via i to i+4 intramolecular side-chain to side-chain azide-alkyne 1,3-dipolar cycloaddition, *J. Org. Chem.* **73**, 5663–5674.

[93] Scrima, M., Le Chevalier-Isaad, A., Rovero, P., Papini, A. M., Chorev, M., and D'Ursi, A. M. (2010) Cu-I-catalyzed azide-alkyne intramolecular i-to-(i+4) side-chain-to-side-chain cyclization promotes the formation of helix-like secondary structures, *Eur. J. Org. Chem.* **2010**, 446–457.

[94] Kawamoto, S. A., Coleska, A., Ran, X., Yi, H., Yang, C. Y., and Wang, S. (2012) Design of triazole-stapled BCL9 alpha-helical peptides to target the beta-catenin/B-cell CLL/lymphoma 9 (BCL9) protein-protein interaction, *J. Med. Chem.* **55**, 1137–1146.

[95] Haney, C. M., Loch, M. T., and Horne, W. S. (2011) Promoting peptide [small alpha]-helix formation with dynamic covalent oxime side-chain cross-links, *Chem. Commun. (Camb.)* **47**, 10915–10917.

[96] Madden, M. M., Rivera Vera, C. I., Song, W., and Lin, Q. (2009) Facile synthesis of stapled, structurally reinforced peptide helices via a photoinduced intramolecular 1,3-dipolar cycloaddition reaction, *Chem. Commun. (Camb.)* **37**, 5588–5590.

[97] Madden, M. M., Muppidi, A., Li, Z., Li, X., Chen, J., and Lin, Q. (2011) Synthesis of cell-permeable stapled peptide dual inhibitors of the p53-Mdm2/Mdmx interactions via photoinduced cycloaddition, *Bioorg. Med. Chem. Lett.* **21**, 1472–1475.

[98] Phelan, J. C., Skelton, N. J., Braisted, A. C., and McDowell, R. S. (1997) A general method for constraining short peptides to an alpha-helical conformation, *J. Am. Chem. Soc.* **119**, 455–460.

[99] Sia, S. K., Carr, P. A., Cochran, A. G., Malashkevich, V. N., and Kim, P. S. (2002) Short constrained peptides that inhibit HIV-1 entry, *Proc. Natl. Acad. Sci. U. S. A.* **99**, 14664–14669.

[100] Fujimoto, K., Kajino, M., and Inouye, M. (2008) Development of a series of cross-linking agents that effectively stabilize alpha-helical structures in various short peptides, *Chemistry* **14**, 857–863.

[101] Zhang, F., Sadovski, O., Xin, S. J., and Woolley, G. A. (2007) Stabilization of folded peptide and protein structures via distance matching with a long, rigid cross-linker, *J. Am. Chem. Soc.* **129**, 14154–14155.

[102] Muppidi, A., Wang, Z., Li, X., Chen, J., and Lin, Q. (2011) Achieving cell penetration with distance-matching cysteine cross-linkers: a facile route to cell-permeable peptide dual inhibitors of Mdm2/Mdmx, *Chem. Commun. (Camb.)* **47**, 9396–9398.

[103] Jo, H., Meinhardt, N., Wu, Y., Kulkarni, S., Hu, X., Low, K. E., Davies, P. L., DeGrado, W. F., and Greenbaum, D. C. (2012) Development of alpha-helical calpain probes by mimicking a natural protein-protein interaction, *J. Am. Chem. Soc.* **134**, 17704–17713.

[104] Spokoyny, A. M., Zou, Y., Ling, J. J., Yu, H., Lin, Y. S., and Pentelute, B. L. (2013) A perfluoroaryl-cysteine S(N)Ar chemistry approach to unprotected peptide stapling, *J. Am. Chem. Soc.* **135**, 5946–5949.

[105] de Araujo, A. D., Hoang, H. N., Kok, W. M., Diness, F., Gupta, P., Hill, T. A., Driver, R. W., Price, D. A., Liras, S., and Fairlie, D. P. (2014) Comparative alpha-helicity of cyclic pentapeptides in water, *Angew. Chem. Int. Ed.* **53**, 6965–6969.

[106] Zimm, B. H., and Bragg, J. K. (1959) Theory of the phase transition between helix and random coil in polypeptide chains, *J. Chem. Phys.* **31**, 526–535.

[107] Patgiri, A., Jochim, A. L., and Arora, P. S. (2008) A hydrogen bond surrogate approach for stabilization of short peptide sequences in alpha-helical conformation, *Acc. Chem. Res.* **41**, 1289–1300.

[108] Cabezas, E., and Satterthwait, A. C. (1999) The hydrogen bond mimic approach: solid-phase synthesis of a peptide stabilized as an α-helix with a hydrazone link, *J. Am. Chem. Soc.* **121**, 3862–3875.

[109] Calvo, J. C., Choconta, K. C., Diaz, D., Orozco, O., Bravo, M. M., Espejo, F., Salazar, L. M., Guzman, F., and Patarroyo, M. E. (2003) An alpha helix conformationally restricted peptide is recognized by cervical carcinoma patients' sera, *J. Med. Chem.* **46**, 5389–5394.

[110] Chapman, R. N., Dimartino, G., and Arora, P. S. (2004) A highly stable short α-helix constrained by a main-chain hydrogen-bond surrogate, *J. Am. Chem. Soc.* **126**, 12252–12253.

[111] Dimartino, G., Wang, D., Chapman, R. N., and Arora, P. S. (2005) Solid-phase synthesis of hydrogen-bond surrogate-derived α-helices, *Org. Lett.* **7**, 2389–2392.

[112] Wang, D., Chen, K., Kulp, J. L., and Arora, P. S. (2006) Evaluation of biologically relevant short α-helices stabilized by a main-chain hydrogen-bond surrogate, *J. Am. Chem. Soc.* **128**, 9248–9256.

[113] Wang, D., Chen, K., Dimartino, G., and Arora, P. S. (2006) Nucleation and stability of hydrogen-bond surrogate-based α-helices, *Org. Biomol. Chem.* **4**, 4074–4081.

[114] Wang, D., Liao, W., and Arora, P. S. (2005) Enhanced metabolic stability and protein-binding properties of artificial alpha helices derived from a hydrogen-bond surrogate: application to Bcl-xL, *Angew. Chem. Int. Ed.* **44**, 6525–6529.

[115] Henchey, L. K., Porter, J. R., Ghosh, I., and Arora, P. S. (2010) High specificity in protein recognition by hydrogen-bond-surrogate alpha-helices: selective inhibition of the p53/MDM2 complex, *Chembiochem* **11**, 2104–2107.

[116] Patgiri, A., Yadav, K. K., Arora, P. S., and Bar-Sagi, D. (2011) An orthosteric inhibitor of the Ras-Sos interaction, *Nat. Chem. Biol.* **7**, 585–587.

[117] Wang, D., Lu, M., and Arora, P. S. (2008) Inhibition of HIV-1 fusion by hydrogen-bond-surrogate-based alpha helices, *Angew. Chem. Int. Ed.* **47**, 1879–1882.

[118] Henchey, L. K., Kushal, S., Dubey, R., Chapman, R. N., Olenyuk, B. Z., and Arora, P. S. (2010) Inhibition of hypoxia inducible factor 1-transcription coactivator interaction by a hydrogen bond surrogate alpha-helix, *J. Am. Chem. Soc.* **132**, 941–943.

[119] Kushal, S., Lao, B. B., Henchey, L. K., Dubey, R., Mesallati, H., Traaseth, N. J., Olenyuk, B. Z., and Arora, P. S. (2013) Protein domain mimetics as in vivo modulators of hypoxia-inducible factor signaling, *Proc. Natl. Acad. Sci. U. S. A.* **110**, 15602–15607.

[120] Semenza, G. L. (2003) Targeting HIF-1 for cancer therapy, *Nat. Rev. Cancer* **3**, 721–732.

[121] Mahon, A. B., and Arora, P. S. (2012) Design, synthesis and protein-targeting properties of thioether-linked hydrogen bond surrogate helices, *Chem. Commun. (Camb.)* **48**, 1416–1418.

[122] Miller, S. E., Kallenbach, N. R., and Arora, P. S. (2012) Reversible α-helix formation controlled by a hydrogen bond surrogate, *Tetrahedron* **68**, 4434–4437.

[123] Vernall, A. J., Cassidy, P., and Alewood, P. F. (2009) A single alpha-helical turn stabilized by replacement of an internal hydrogen bond with a covalent ethylene bridge, *Angew. Chem. Int. Ed.* **48**, 5675–5678.

[124] Smith, J. M., Vitali, F., Archer, S. A., and Fasan, R. (2011) Modular assembly of macrocyclic organo-peptide hybrids using synthetic and genetically encoded precursors, *Angew. Chem. Int. Ed.* **50**, 5075–5080.

[125] Frost, J. R., Vitali, F., Jacob, N. T., Brown, M. D., and Fasan, R. (2013) Macrocyclization of organo-peptide hybrids through a dual bio-orthogonal ligation: insights from structure-reactivity studies, *Chembiochem* **14**, 147–160.

[126] Smith, J. M., Frost, J. R., and Fasan, R. (2014) Designer macrocyclic organo-peptide hybrids inhibit the interaction between p53 and HDM2/X by accommodating a functional α-helix, *Chem. Commun. (Camb.)* **50**, 5027–5030.

[127] Fasan, R., Dias, R. L. A., Moehle, K., Zerbe, O., Vrijbloed, J. W., Obrecht, D., and Robinson, J. A. (2004) Using a β-hairpin to mimic an α-helix: cyclic peptidomimetic inhibitors of the p53–hdm2 protein–protein interaction, *Angew. Chem. Int. Ed.* **43**, 2109–2112.

[128] Grasslin, A., Amoreira, C., Baldridge, K. K., and Robinson, J. A. (2009) Thermodynamic and computational studies on the binding of p53-derived peptides and peptidomimetic inhibitors to HDM2, *Chembiochem* **10**, 1360–1368.

[129] Robinson, J. A., Demarco, S., Gombert, F., Moehle, K., and Obrecht, D. (2008) The design, structures and therapeutic potential of protein epitope mimetics, *Drug Discov. Today* **13**, 944–951.

[130] Obrecht, D., Chevalier, E., Moehle, K., and Robinson, J. A. (2012) β-Hairpin protein epitope mimetic technology in drug discovery, *Drug Discov. Today Technol.* **9**, e63–e69.

[131] Moehle, K., Athanassiou, Z., Patora, K., Davidson, A., Varani, G., and Robinson, J. A. (2007) Design of beta-hairpin peptidomimetics that inhibit binding of alpha-helical HIV-1 Rev protein to the rev response element RNA, *Angew. Chem. Int. Ed.* **46**, 9101–9104.

[132] Athanassiou, Z., Dias, R. L., Moehle, K., Dobson, N., Varani, G., and Robinson, J. A. (2004) Structural mimicry of retroviral tat proteins by constrained beta-hairpin peptidomimetics: ligands with high affinity and selectivity for viral TAR RNA regulatory elements, *J. Am. Chem. Soc.* **126**, 6906–6913.

[133] Davidson, A., Leeper, T. C., Athanassiou, Z., Patora-Komisarska, K., Karn, J., Robinson, J. A., and Varani, G. (2009) Simultaneous recognition of HIV-1 TAR RNA bulge and loop sequences by cyclic peptide mimics of Tat protein, *Proc. Natl. Acad. Sci. U. S. A.* **106**, 11931–11936.

[134] Lalonde, M. S., Lobritz, M. A., Ratcliff, A., Chamanian, M., Athanassiou, Z., Tyagi, M., Wong, J., Robinson, J. A., Karn, J., Varani, G., and Arts, E. J. (2011) Inhibition of both HIV-1 reverse transcription and gene expression by a cyclic peptide that binds the Tat-transactivating response element (TAR) RNA, *PLoS Pathog.* **7**, e1002038.

[135] Seitz, M., Rusert, P., Moehle, K., Trkola, A., and Robinson, J. A. (2010) Peptidomimetic inhibitors targeting the CCR5-binding site on the human immunodeficiency virus type-1 gp120 glycoprotein complexed to CD4, *Chem. Commun. (Camb.)* **46**, 7754–7756.

[136] Yin, H., and Hamilton, A. D. (2005) Strategies for targeting protein-protein interactions with synthetic agents, *Angew. Chem. Int. Ed.* **44**, 4130–4163.

[137] Orner, B. P., Ernst, J. T., and Hamilton, A. D. (2001) Toward proteomimetics: terphenyl derivatives as structural and functional mimics of extended regions of an α-helix, *J. Am. Chem. Soc.* **123**, 5382–5383.

[138] Kutzki, O., Park, H. S., Ernst, J. T., Orner, B. P., Yin, H., and Hamilton, A. D. (2002) Development of a potent Bcl-xL antagonist based on α-helix mimicry, *J. Am. Chem. Soc.* **124**, 11838–11839.

[139] Ernst, J. T., Kutzki, O., Debnath, A. K., Jiang, S., Lu, H., and Hamilton, A. D. (2002) Design of a protein surface antagonist based on α-helix mimicry: inhibition of gp41 assembly and viral fusion, *Angew. Chem. Int. Ed.* **41**, 278–281.

[140] Yin, H., Lee, G.-I., Sedey, K. A., Kutzki, O., Park, H. S., Orner, B. P., Ernst, J. T., Wang, H.-G., Sebti, S. M., and Hamilton, A. D. (2005) Terphenyl-based Bak BH3 α-helical proteomimetics as low-molecular-weight antagonists of Bcl-xL, *J. Am. Chem. Soc.* **127**, 10191–10196.

[141] Yin, H., Lee, G.-I., Sedey, K. A., Rodriguez, J. M., Wang, H.-G., Sebti, S. M., and Hamilton, A. D. (2005) Terephthalamide derivatives as mimetics of helical peptides: disruption of the Bcl-xL/Bak interaction, *J. Am. Chem. Soc.* **127**, 5463–5468.

[142] Rodriguez, J. M., Nevola, L., Ross, N. T., Lee, G.-I., and Hamilton, A. D. (2009) Synthetic inhibitors of extended helix–protein interactions based on a biphenyl 4,4′-dicarboxamide scaffold, *Chembiochem* **10**, 829–833.

[143] Rodriguez, J. M., Ross, N. T., Katt, W. P., Dhar, D., Lee, G.-I., and Hamilton, A. D. (2009) Structure and function of benzoylurea-derived α-helix mimetics targeting the Bcl-xL/Bak binding interface, *ChemMedChem* **4**, 649–656.

[144] Rodriguez, J. M., and Hamilton, A. D. (2007) Benzoylurea oligomers: synthetic foldamers that mimic extended alpha helices, *Angew. Chem. Int. Ed.* **46**, 8614–8617.

[145] Ernst, J. T., Becerril, J., Park, H. S., Yin, H., and Hamilton, A. D. (2003) Design and application of an α-helix-mimetic scaffold based on an oligoamide-foldamer strategy: antagonism of the bak BH3/Bcl-xL complex, *Angew. Chem. Int. Ed.* **42**, 535–539.

[146] Saraogi, I., Incarvito, C. D., and Hamilton, A. D. (2008) Controlling curvature in a family of oligoamide alpha-helix mimetics, *Angew. Chem. Int. Ed.* **47**, 9691–9694.

[147] Shaginian, A., Whitby, L. R., Hong, S., Hwang, I., Farooqi, B., Searcey, M., Chen, J., Vogt, P. K., and Boger, D. L. (2009) Design, synthesis, and evaluation of an alpha-helix mimetic library targeting protein-protein interactions, *J. Am. Chem. Soc.* **131**, 5564–5572.

[148] Lu, F., Chi, S. W., Kim, D. H., Han, K. H., Kuntz, I. D., and Guy, R. K. (2006) Proteomimetic libraries: design, synthesis, and evaluation of p53-MDM2 interaction inhibitors, *J. Comb. Chem.* **8**, 315.

[149] Campbell, F., Plante, J. P., Edwards, T. A., Warriner, S. L., and Wilson, A. J. (2010) N-alkylated oligoamide alpha-helical proteomimetics, *Org. Biomol. Chem.* **8**, 2344–2351.

[150] Saraogi, I., Hebda, J. A., Becerril, J., Estroff, L. A., Miranker, A. D., and Hamilton, A. D. (2010) Synthetic alpha-helix mimetics as agonists and antagonists of islet amyloid polypeptide aggregation, *Angew. Chem. Int. Ed.* **49**, 736–739.

[151] Plante, J. P., Burnley, T., Malkova, B., Webb, M. E., Warriner, S. L., Edwards, T. A., and Wilson, A. J. (2009) Oligobenzamide proteomimetic inhibitors of the p53-hDM2 protein-protein interaction, *Chem. Commun. (Camb.)* **34**, 5091–5093.

[152] Azzarito, V., Prabhakaran, P., Bartlett, A. I., Murphy, N. S., Hardie, M. J., Kilner, C. A., Edwards, T. A., Warriner, S. L., and Wilson, A. J. (2012) 2-O-alkylated para-benzamide alpha-helix mimetics: the role of scaffold curvature, *Org. Biomol. Chem.* **10**, 6469–6472.

[153] Shahian, T., Lee, G. M., Lazic, A., Arnold, L. A., Velusamy, P., Roels, C. M., Guy, R. K., and Craik, C. S. (2009) Inhibition of a viral enzyme by a small-molecule dimer disruptor, *Nat. Chem. Biol.* **5**, 640–646.

[154] Yap, J. L., Cao, X., Vanommeslaeghe, K., Jung, K. Y., Peddaboina, C., Wilder, P. T., Nan, A., MacKerell, A. D., Jr., Smythe, W. R., and Fletcher, S. (2012) Relaxation of the rigid backbone of an oligoamide-foldamer-based alpha-helix mimetic: identification of potent Bcl-xL inhibitors, *Org. Biomol. Chem.* **10**, 2928–2933.

[155] Whitby, L. R., Boyle, K. E., Cai, L., Yu, X., Gochin, M., and Boger, D. L. (2012) Discovery of HIV fusion inhibitors targeting gp41 using a comprehensive α-helix mimetic library, *Bioorg. Med. Chem. Lett.* **22**, 2861–2865.

[156] Kim, I. C., and Hamilton, A. D. (2006) Diphenylindane-based proteomimetics reproduce the projection of the i, i+3, i+4, and i+7 residues on an alpha-helix, *Org. Lett.* **8**, 1751–1754.

[157] Tošovská, P., and Arora, P. S. (2010) Oligooxopiperazines as nonpeptidic α-helix mimetics, *Org. Lett.* **12**, 1588–1591.

[158] Lao, B. B., Grishagin, I., Mesallati, H., Brewer, T. F., Olenyuk, B. Z., and Arora, P. S. (2014) In vivo modulation of hypoxia-inducible signaling by topographical helix mimetics, *Proc. Natl. Acad. Sci. U. S. A.* **111**, 7531–7536.

[159] Lao, B. B., Drew, K., Guarracino, D. A., Brewer, T. F., Heindel, D. W., Bonneau, R., and Arora, P. S. (2014) Rational design of topographical helix mimics as potent inhibitors of protein-protein interactions, *J. Am. Chem. Soc.* **136**, 7877–7888.

[160] Stotz, C. E., and Topp, E. M. (2004) Applications of model beta-hairpin peptides, *J. Pharm. Sci.* **93**, 2881–2894.

[161] Loughlin, W. A., Tyndall, J. D., Glenn, M. P., and Fairlie, D. P. (2004) Beta-strand mimetics, *Chem. Rev.* **104**, 6085–6117.

[162] Cheng, P. N., Liu, C., Zhao, M., Eisenberg, D., and Nowick, J. S. (2012) Amyloid beta-sheet mimics that antagonize protein aggregation and reduce amyloid toxicity, *Nat. Chem.* **4**, 927–933.

[163] Ross, C. A., and Poirier, M. A. (2004) Protein aggregation and neurodegenerative disease, *Nat. Med.* **10** Suppl, S10–S17.

[164] Aguzzi, A., and O'Connor, T. (2010) Protein aggregation diseases: pathogenicity and therapeutic perspectives, *Nat. Rev. Drug Discov.* **9**, 237–248.

[165] Robinson, J. A. (2000) The design, synthesis and conformation of some new beta-hairpin mimetics: novel reagents for drug and vaccine discovery, *Synlett* **4**, 429–441.

[166] Feigel, M. (1986) 2,8-Dimethyl-4-(carboxymethyl)-6-(aminomethyl)phenoxathiin S-dioxide—an organic substitute for the beta-turn in peptides, *J. Am. Chem. Soc.* **108**, 181–182.

[167] Brandmeier, V., and Feigel, M. (1989) A macrocycle containing 2 biphenyl and 2 alanine subunits, synthesis and conformation in solution, *Tetrahedron* **45**, 1365–1376.

[168] Tsang, K. Y., Diaz, H., Graciani, N., and Kelly, J. W. (1994) Hydrophobic cluster formation is necessary for dibenzofuran-based amino-acids to function as beta-sheet nucleators, *J. Am. Chem. Soc.* **116**, 3988–4005.

[169] Soth, M. J., and Nowick, J. S. (1999) A peptide/oligourea/azapeptide hybrid that adopts a hairpin turn, *J. Org. Chem.* **64**, 276–281.

[170] Chung, Y. J., Christianson, L. A., Stanger, H. E., Powell, D. R., and Gellman, S. H. (1998) A beta-peptide reverse turn that promotes hairpin formation, *J. Am. Chem. Soc.* **120**, 10555–10556.

[171] Robinson, J. A. (2008) Beta-hairpin peptidomimetics: design, structures and biological activities, *Acc. Chem. Res.* **41**, 1278–1288.

[172] Nair, C. M., Vijayan, M., Venkatachalapathi, Y. V., and Balaram, P. (1979) X-ray crystal-structure of pivaloyl-D-Pro-L-Pro-L-Ala-N-methylamide—observation of a consecutive beta-turn conformation, J. Chem. Soc. Chem. Commun. 24, 1183–1184.

[173] Descours, A., Moehle, K., Renard, A., and Robinson, J. A. (2002) A new family of β-hairpin mimetics based on a trypsin inhibitor from sunflower seeds, *Chembiochem* **3**, 318–323.

[174] Maryanoff, B. E. (2004) Inhibitors of serine proteases as potential therapeutic agents: the road from thrombin to tryptase to cathepsin G, *J. Med. Chem.* **47**, 769–787.

[175] Ohbayashi, H. (2002) Neutrophil elastase inhibitors as treatment for COPD, *Expert Opin. Investig. Drugs* **11**, 965–980.

[176] Fiorucci, L., and Ascoli, F. (2004) Mast cell tryptase, a still enigmatic enzyme, *Cell. Mol. Life Sci.* **61**, 1278–1295.

[177] Hancock, R. E. W., and Sahl, H. G. (2006) Antimicrobial and host-defense peptides as new anti-infective therapeutic strategies, *Nat. Biotechnol.* **24**, 1551–1557.

[178] Brogden, K. A. (2005) Antimicrobial peptides: pore formers or metabolic inhibitors in bacteria? *Nat. Rev. Microbiol.* **3**, 238–250.

[179] Schmidt, J., Patora-Komisarska, K., Moehle, K., Obrecht, D., and Robinson, J. A. (2013) Structural studies of beta-hairpin peptidomimetic antibiotics that target LptD in *Pseudomonas* sp, *Bioorg. Med. Chem.* **21**, 5806–5810.

[180] Shankaramma, S. C., Athanassiou, Z., Zerbe, O., Moehle, K., Mouton, C., Bernardini, F., Vrijbloed, J. W., Obrecht, D., and Robinson, J. A. (2002) Macrocyclic hairpin mimetics of the cationic antimicrobial peptide protegrin I: a new family of broad-spectrum antibiotics, *Chembiochem* **3**, 1126–1133.

[181] Robinson, J. A., Shankaramma, S. C., Jettera, P., Kienzl, U., Schwendener, R. A., Vrijbloed, J. W., and Obrecht, D. (2005) Properties and structure-activity studies of cyclic beta-hairpin peptidomimetics based on the cationic antimicrobial peptide protegrin I, *Bioorg. Med. Chem.* **13**, 2055–2064.

[182] Srinivas, N., Jetter, P., Ueberbacher, B. J., Werneburg, M., Zerbe, K., Steinmann, J., Van der Meijden, B., Bernardini, F., Lederer, A., Dias, R. L., Misson, P. E., Henze, H., Zumbrunn, J., Gombert, F. O., Obrecht, D., Hunziker, P., Schauer, S., Ziegler, U., Kach, A., Eberl, L., Riedel, K., DeMarco, S. J., and Robinson, J. A. (2010) Peptidomimetic antibiotics target outer-membrane biogenesis in *Pseudomonas aeruginosa*, *Science* **327**, 1010–1013.

[183] Miyata, T., Tokunaga, F., Yoneya, T., Yoshikawa, K., Iwanaga, S., Niwa, M., Takao, T., and Shimonishi, Y. (1989) Antimicrobial peptides, isolated from horseshoe crab hemocytes, tachyplesin II, and polyphemusins I and II: chemical structures and biological activity, *J. Biochem.* **106**, 663–668.

[184] DeMarco, S. J., Henze, H., Lederer, A., Moehle, K., Mukherjee, R., Romagnoli, B., Robinson, J. A., Brianza, F., Gombert, F. O., Lociuro, S., Ludin, C., Vrijbloed, J. W., Zumbrunn, J., Obrecht, J. P., Obrecht, D., Brondani, V., Hamy, F., and Klimkait, T. (2006) Discovery of novel, highly potent and selective beta-hairpin mimetic CXCR4 inhibitors with excellent anti-HIV activity and pharmacokinetic profiles, *Bioorg. Med. Chem.* **14**, 8396–8404.

[185] Burger, J. A., and Burkle, A. (2007) The CXCR4 chemokine receptor in acute and chronic leukaemia: a marrow homing receptor and potential therapeutic target, *Br. J. Haematol.* **137**, 288–296.

[186] Favre, M., Moehle, K., Jiang, L., Pfeiffer, B., and Robinson, J. A. (1999) Structural mimicry of canonical conformations in antibody hypervariable loops using cyclic peptides containing a heterochiral diproline template, *J. Am. Chem. Soc.* **121**, 2679–2685.

[187] Dias, R. L., Fasan, R., Moehle, K., Renard, A., Obrecht, D., and Robinson, J. A. (2006) Protein ligand design: from phage display to synthetic protein epitope mimetics in human antibody Fc-binding peptidomimetics, *J. Am. Chem. Soc.* **128**, 2726–2732.

[188] Smith, A. B., Keenan, T. P., Holcomb, R. C., Sprengeler, P. A., Guzman, M. C., Wood, J. L., Carroll, P. J., and Hirschmann, R. (1992) Design, synthesis, and crystal structure of a pyrrolinone-based peptidomimetic possessing the conformation of a .beta.-strand: potential application to the design of novel inhibitors of proteolytic enzymes, *J. Am. Chem. Soc.* **114**, 10672–10674.

[189] Smith, A. B., Charnley, A. K., and Hirschmann, R. (2010) Pyrrolinone-based peptidomimetics. "let the enzyme or receptor be the judge", *Acc. Chem. Res.* **44**, 180–193.

[190] Angelo, N. G., and Arora, P. S. (2005) Nonpeptidic foldamers from amino acids: synthesis and characterization of 1,3-substituted triazole oligomers, *J. Am. Chem. Soc.* **127**, 17134–17135.

[191] Angelo, N. G., and Arora, P. S. (2007) Solution- and solid-phase synthesis of triazole oligomers that display protein-like functionality, *J. Org. Chem.* **72**, 7963–7967.

[192] Brik, A., Alexandratos, J., Lin, Y. C., Elder, J. H., Olson, A. J., Wlodawer, A., Goodsell, D. S., and Wong, C. H. (2005) 1,2,3-triazole as a peptide surrogate in the rapid synthesis of HIV-1 protease inhibitors, *Chembiochem* **6**, 1167–1169.

[193] Jochim, A. L., Miller, S. E., Angelo, N. G., and Arora, P. S. (2009) Evaluation of triazolamers as active site inhibitors of HIV-1 protease, *Bioorg. Med. Chem. Lett.* **19**, 6023–6026.

[194] Phillips, S. T., Rezac, M., Abel, U., Kossenjans, M., and Bartlett, P. A. (2001) "@-Tides": the 1,2-dihydro-3(6H)-pyridinone unit as a β-strand mimic, *J. Am. Chem. Soc.* **124**, 58–66.

[195] Hammond, M. C., Harris, B. Z., Lim, W. A., and Bartlett, P. A. (2006) β Strand peptidomimetics as potent PDZ domain ligands, *Chem. Biol.* **13**, 1247–1251.

[196] Nowick, J. S., Chung, D. M., Maitra, K., Maitra, S., Stigers, K. D., and Sun, Y. (2000) An unnatural amino acid that mimics a tripeptide β-strand and forms β-sheetlike hydrogen-bonded dimers, *J. Am. Chem. Soc.* **122**, 7654–7661.

[197] Nowick, J. S., and Brower, J. O. (2003) A new turn structure for the formation of beta-hairpins in peptides, *J. Am. Chem. Soc.* **125**, 876–877.

[198] Nowick, J. S. (2008) Exploring β-sheet structure and interactions with chemical model systems, *Acc. Chem. Res.* **41**, 1319–1330.

[199] Kayser, K. J., Glenn, M. P., Sebti, S. M., Cheng, J. Q., and Hamilton, A. D. (2007) Modifications of the GSK3β substrate sequence to produce substrate-mimetic inhibitors of Akt as potential anti-cancer therapeutics, *Bioorg. Med. Chem. Lett.* **17**, 2068–2073.

[200] Wyrembak, P. N., and Hamilton, A. D. (2009) Alkyne-linked 2,2-disubstituted-indolin-3-one oligomers as extended β-strand mimetics, *J. Am. Chem. Soc.* **131**, 4566–4567.

[201] Sutherell, C. L., Thompson, S., Scott, R. T. W., and Hamilton, A. D. (2012) Aryl-linked imidazolidin-2-ones as non-peptidic [small beta]-strand mimetics, *Chem. Commun. (Camb.)* **48**, 9834–9836.

[202] Lingard, H., Han, J. T., Thompson, A. L., Leung, I. K., Scott, R. T., Thompson, S., and Hamilton, A. D. (2014) Diphenylacetylene-linked peptide strands induce bidirectional beta-sheet formation, *Angew. Chem. Int. Ed.* **53**, 3650–3653.

[203] Kang, C. W., Sun, Y., and Del Valle, J. R. (2012) Substituted imidazo[1,2-a]pyridines as beta-strand peptidomimetics, *Org. Lett.* **14**, 6162–6165.

[204] Kim, Y. B., Kang, C. W., Ranatunga, S., Yang, H., Sebti, S. M., and Del Valle, J. R. (2014) Imidazo[1,2-a]pyridine-based peptidomimetics as inhibitors of Akt, *Bioorg. Med. Chem. Lett.* **24**, 4650–4653.

[205] Ko, E., Liu, J., Perez, L. M., Lu, G., Schaefer, A., and Burgess, K. (2010) Universal peptidomimetics, *J. Am. Chem. Soc.* **133**, 462–477.

# 16

# *IN VIVO* IMAGING OF DRUG ACTION

OLIVER PLETTENBURG and MATTHIAS LÖHN
*Sanofi Deutschland GmbH, Frankfurt am Main, Germany*

## 16.1 INTRODUCTION

Despite the rise of modern technologies in drug discovery, the success rates of bringing new compounds to the market have been unsustainably low [1]. In order to improve on this, more predictive translation of results obtained in animal studies to the clinical situation in humans has to be achieved [2].

*In vivo* imaging has evolved to be one of the most powerful methods to assist modern drug development, to enable monitoring of disease progression in longitudinal studies by noninvasive methods in living animals, and to demonstrate treatment efficacy. A plethora of methods for following development of disease pathogenesis, as well as directly or indirectly observing target engagement *in vivo*, have been developed over the last years. To provide a comprehensive overview of this rapidly evolving research area has been the subject of many excellent monographs [3–5] and is clearly exceeding the scope of this chapter. The purpose of this chapter is to provide an introduction to general principles and concepts of *in vivo* imaging and probe design, used to visualize the distribution behavior and effects of drug treatment. We will exemplify these concepts by providing some selected case studies from recent literature in the areas of oncology, CNS, and diabetes. Also, some examples for translating these results into clinical practice will be discussed.

Although many starting points for probes are originating from compounds derived from medicinal chemistry efforts, their desired overall profiles may differ substantially, and the desired overall compound profile has to be carefully adjusted to the specific imaging question. For instance, for most therapeutic compounds, oral bioavailability and long half-lives are highly desirable to allow convenient once daily dosing. Prolonged circulation times for probes, however, frequently lead to strong background signal and

*Small Molecule Medicinal Chemistry: Strategies and Technologies*, First Edition.
Edited by Werngard Czechtizky and Peter Hamley.

prevent clear observation of the desired effects. In addition, most probes are injected. Thus, ligand structure, metabolic degradation, uptake and excretion pathways, and, in case of radionuclides, half-life have to be carefully balanced and considered when devising the optimization strategy. Furthermore, specific characteristics of the chosen imaging method have to be taken into account.

## 16.2 OVERVIEW OF IMAGING METHODS

Modern *in vivo* imaging techniques are indispensable and important tools for the anatomical or functional evaluation of an organ or tissue and for the evaluation of the pharmacokinetic and pharmacodynamic properties of a specific compound on different organs or tissues and may also be useful for the detection and evaluation of novel biomarkers. Per se, *in vivo* imaging techniques allow noninvasive serial monitoring of the compound's effect or the organ's function. Among such imaging techniques are different methods, such as magnetic resonance imaging (MRI), X-ray computer tomography (CT), ultrasound (US), single-photon emission computed tomography (SPECT), positron emission tomography (PET), photoacoustic tomography (PAT), thermoacoustic imaging (TI), electrical impedance tomography (EIT), and fluorescence- or luminescence-based technologies. In recent years, fluorescence-based imaging has received great attention due to the increasing availability of different fluorescent proteins, dyes, and probes. Tremendous efforts have been spent on the development of high quantum yield and stable fluorophores to facilitate noninvasive studies of protein function, gene expression, and other cellular processes [6]. Furthermore, great progress has been made with the adaptation of existing imaging technologies in the clinical sector to the needs of preclinical research, such as smaller devices compatible with the smaller sizes of the animals, requiring improved resolution and quality of the image [7].

### 16.2.1 Fluorescence-Based Methods

Fluorescence- and bioluminescence-based imaging are widely used versatile *in vivo* imaging techniques. Both methods are relatively robust, reliable, and inexpensive and even allow moderate throughput screening. However, both methods are only semiquantitative, since the emitted fluorescence and bioluminescence light is scattered and attenuated by the surrounding tissue which precludes precise quantification. An additional problem of conventional fluorescence microscopy is the limited penetration depth into tissue for imaging purposes.

The utilization of bioluminescence is even more complex since it requires the incorporation of the luciferase gene into cells or tissue in living animals. Therefore, use of bioluminescence is limited to validation of targets in transgenic animals Scheme 16.1, where it has shown great value [8–11].

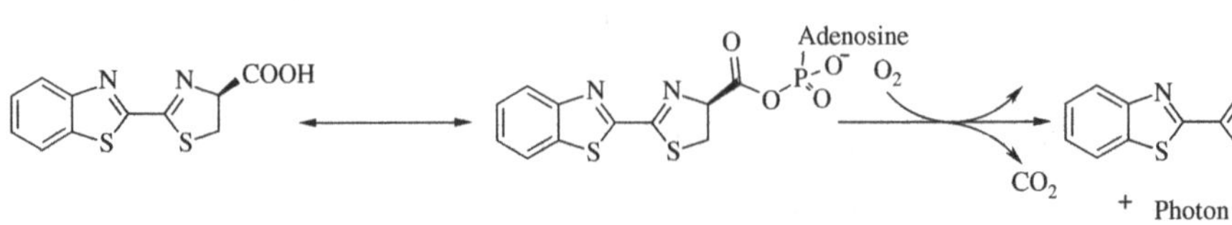


**SCHEME 16.1** Luciferase reaction.

Many excellent references describe methods for biomolecular fluorescence labeling, as well as reagents and hands-on protocols for detection, most of them though with a major focus on cellular imaging [12, 13].

Some representative structures of the commonly used class of cyanine dyes are given in Figure 16.1.

This frequently used class of highly fluorescent fluorophores [14] consists of two 2,2-dimethylindoline elements connected by an unsaturated carbon chain, the number of methine units defining the nomenclature (i.e., Cy3 contains three linking methine units). Cy5.5 differs from Cy5 by presence of fused rings at the indoline moieties, increasing the emission maximum by about 30 nm. Each (CH=CH)-unit increases the maximum by about 100 nm. When choosing the reporter moiety for imaging applications, it needs to be considered that many dyes are hydrophobic and can interact with membranes; careful selection and employment of proper controls is therefore required to avoid misinterpretations [15, 16]. Therefore, sulfonated cyanine dyes represent a valuable alternative, further enriching the palette of available dyes. Influence of the sulfonate groups on the emission properties of the respective dyes is usually negligible. Recently, a new class of symmetrical cyanine dyes was reported, named CyAL, which is easily available via a one-pot reaction [17]. It is interesting that two structurally different dyes with almost identical fluorescence properties corresponding to the frequently used Cy5.5 system could be

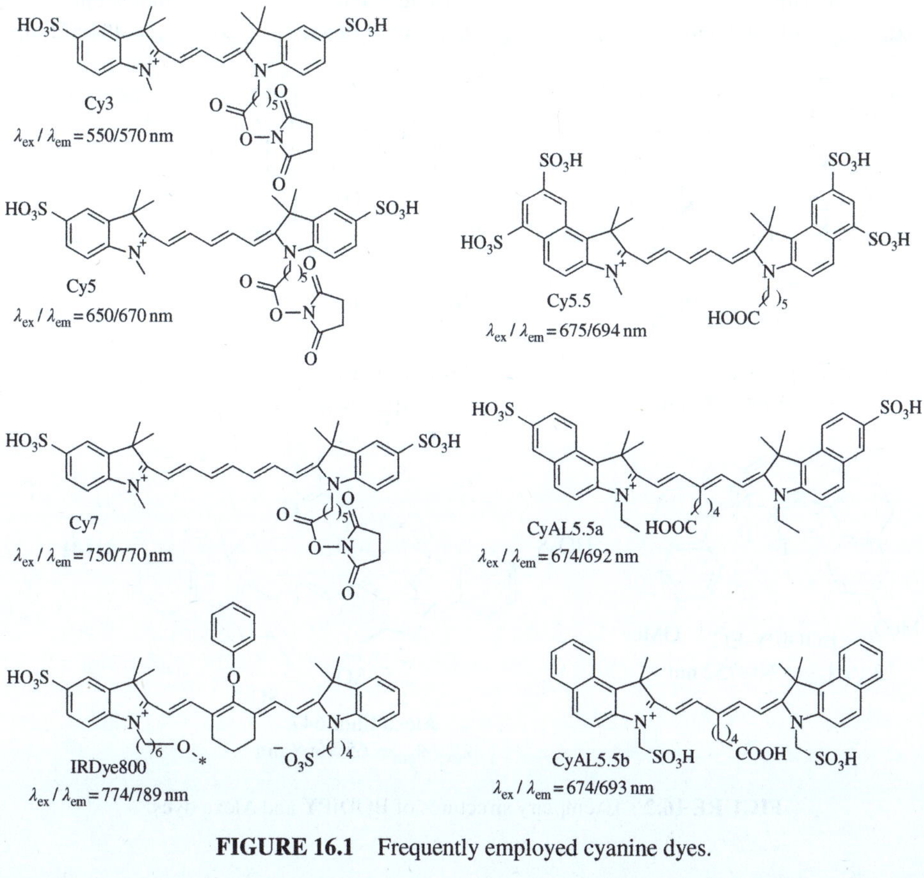


**FIGURE 16.1** Frequently employed cyanine dyes.

obtained, either bearing the negative charge on the annulated ring system or tethered to the indoline nitrogen.

Examples for alternative fluorescent scaffolds are shown in Figure 16.2. Nondecorated boron dipyrromethene (BODIPY) [18] systems display a maximal emission at 503 nm; these systems are characterized by favorable photostability and strong absorption and emission properties. BODIPY dyes are relatively nonpolar, and the chromophore is electrically neutral, leading to good cellular permeability. Efficient syntheses to conformationally restricted Aza-BODIPY derivatives with emission in the near infrared had been developed [19], but their utility is hampered by insufficient water solubility.

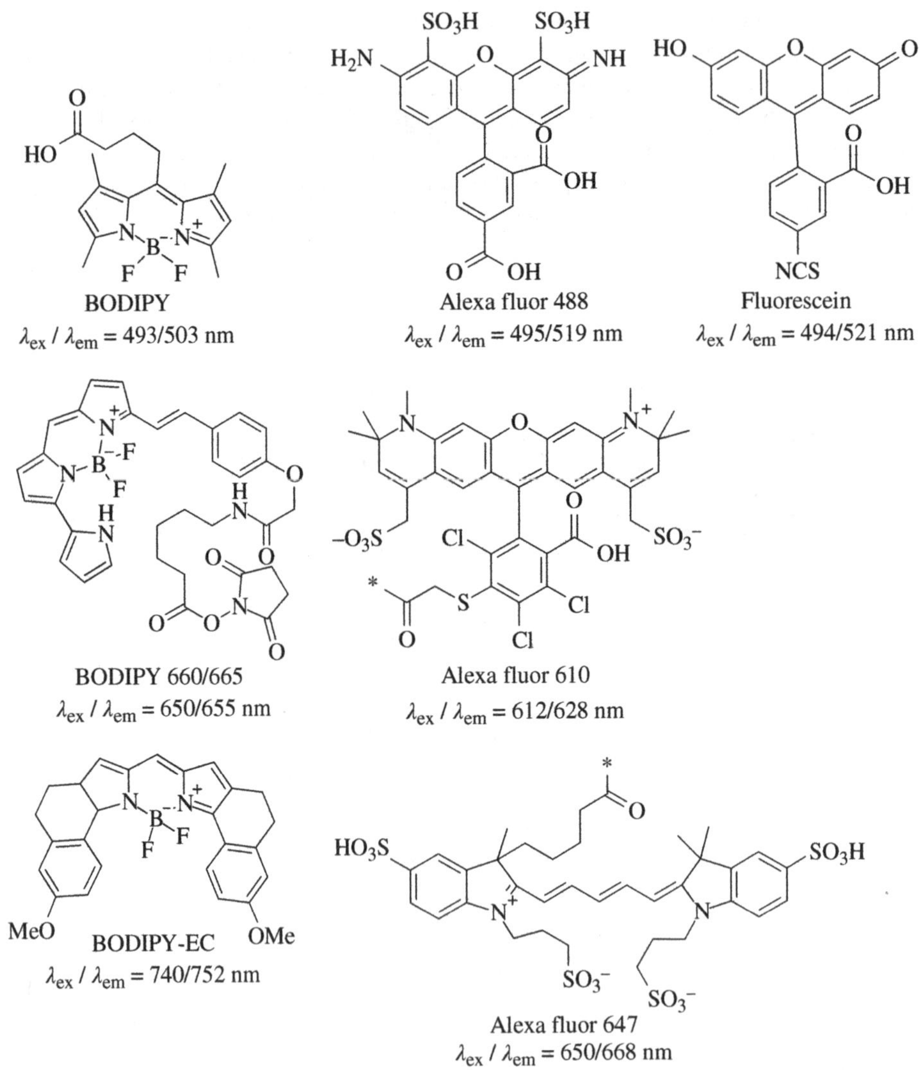


**FIGURE 16.2** Exemplary structures of BODIPY and Alexa dyes.

Another broadly applied class of compounds is the Alexa Fluor series, mostly based on scaffolds related to fluorescein, but showing superior bleaching behavior and spanning a variety of wavelengths. The structure of the near-infrared dye Alexa Fluor 750 is not disclosed yet, but it is interesting to see that the closest reported analogue, Alexa 647, is a member of the cyanine dye family. There are many other structural types of near-infrared dyes published; a thorough review on structures and respective properties has been provided by Ptaszek [20].

For fluorescence-based imaging, a dye is connected via a linker to a probe, for example, an antibody, protein, peptide, or a chemical compound to achieve targeted delivery of the fluorescent dye to the tissue or target of interest. The optimal combination of these three components has to be chosen with care, balancing overall lipophilicity and respective size to minimize nonspecific interactions.

The use of near-infrared fluorescence (NIRF) imaging takes advantage of the optical properties of the skin, tissue, and bone, which are mostly transparent to NIRF light (Fig. 16.3) between 680 and 1300 nm.

Several reviews have been published discussing structures, chemistry, and underlying principles of fluorescence-based imaging, as well as future directions [20–26]. Great effort has been undertaken to develop efficient dyes to directly monitor various analytes like pH, different ion concentrations, or reactive oxygen species in an *in vivo* environment [26, 27].

Further improvement of NIRF imaging probes has been achieved using a "prodrug" approach to the NIRF probe that generates a strong NIRF signal only upon tissue penetration and proteolysis of the linker-quenched NIRF probe. Thereby, the NIRF probe becomes unquenched and highly fluorescent. This approach significantly reduced background fluorescence and consequently improved signal-to-noise ratio leading to a high quantum yield. It potentially allows the early detection of small primary tumors, like breast and prostate cancer [21, 28].

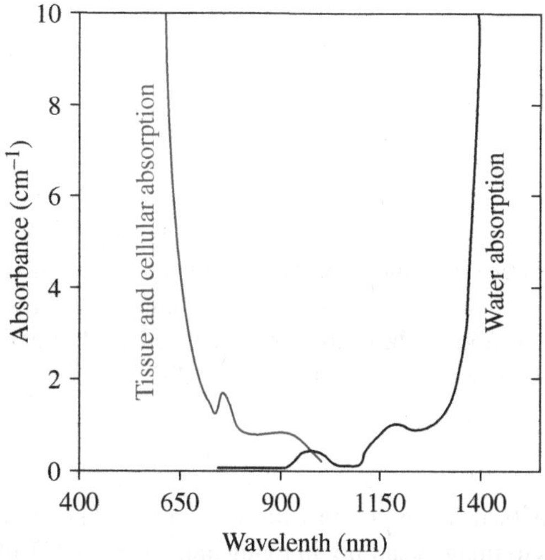


**FIGURE 16.3** To allow fluorescence imaging *in vivo*, a wavelength window between 680 and 1300 nm could be envisaged.

A further significant progress of fluorescence-based imaging technology development was made by the introduction of the two-photon excitation microscopy by Denk, Strickler, and Webb in 1990. In contrast to conventional fluorescence microscopy, where one "high-energy" photon is required to excite fluorescence, described by the characteristic Stokes shift of a dye, the two-photon technology is based on the concept that two "low-energy" photons can also excite a fluorescent dye. As a consequence, "low-energy photons" have a greater tissue penetration due to reduced scattering in tissue, which results in improved optical resolution. Using this technology, it is possible to obtain high-resolution images of living tissue up to 1 mm in depth compared to the penetration depth of 50–80 μm using conventional confocal technology. Further advantages of the two-photon microscopy are more efficient light detection, lower background, and significantly reduced phototoxicity [29–33].

### 16.2.2 MRI

MRI is a powerful technology allowing high-resolution images with excellent contrast between tissues without the need of applying radiopharmaceuticals. The significance of this technology was acknowledged by awarding two Nobel Prizes to its inventors: the Nobel Prize in physics to Purcell [34] and Bloch [35] for the discovery of the principles of nuclear magnetic resonance (1952) and the Nobel Prize in medicine to Lauterbur [36] and Mansfield [37] for their discoveries concerning resonance magnetic imaging (2003). It is based on the principle that in the presence of a magnetic field a local net macroscopic magnetization is induced, which can be perturbed by a radio frequency pulse. The return process to the equilibrium state is characterized by two independent processes, called $T_1$ (spin–lattice) and $T_2$ (spin–spin) relaxation. MRI contrast agents can either increase the $T_1$ signal (mainly paramagnetic compounds like Gd(III) complexes) or decrease the signal on $T_2$-weighted images. The latter class mainly consists of superparamagnetic compounds like superparamagnetic iron oxide nanoparticles, so-called SPIOs. Gadolinium(III) bears the maximal number of seven unpaired electrons, resulting in high magnetism and symmetry.

MRI requires a very strong and uniform magnetic field around the body to detect nuclear magnetic resonance emissions of hydrogen atoms of water molecules. The majority of clinically used MRI systems to date operate at 1.5 T, but machines with up to 9.4 T are in clinical use and greater than 15 T in preclinical use. To induce nuclear magnetic resonance emission, the strong magnetic field is rapidly switched on and off. The rate at which excited atoms return to the equilibrium state depends on the local environment and thus generates the contrast between different tissues. A broad variety of MRI systems are available. Usually, MRI is recommended over CT if the same diagnostic information can be obtained by the safer MRI method. MRI delivers excellent optical resolution but requires high contrast-enhancing compound concentration to be present. To increase local concentration, targeting approaches have been devised [38]; furthermore, efforts have been undertaken to enhance the sensitivity in MRI [39].

### 16.2.3 CT

X-ray CT and its miniaturized variation to allow scanning of small animals, μCT, are frequently used methods to image anatomy in living animals [40]. CT utilizes external X-ray sources to generate images from different angles around the body to avoid shielding effects of very dense tissues on other tissues. Subsequently, the obtained data are processed by a computer in order to provide cross-sectional images of the tissue, organs, or body. 3D reconstructions of the body can be obtained. Since the penetration of X-rays depends on the density of the respective tissue, lung, bone, and soft tissue can be easily distinguished. In particular, bone imaging is frequently used due to the bone's high density; the reconstructed whole body image is often integrated in multimodal images to provide anatomical information, as well as calculation of attenuation coefficients for SPECT and PET image reconstructions. For high-resolution images, the animal usually has to be anesthetized to avoid disturbing motion.

### 16.2.4 PET/SPECT

PET and SPECT are both emission tomography methods. In both cases, the subject is injected with a radioisotope. In case of PET, the radioisotope decays via emission of a positron ($\beta^+$), which annihilates quickly with an electron. This neutralization process liberates energy in form of two γ-rays with an energy of 511 keV each [41]. Since the pair of gamma rays is emitted in opposite direction to each other, the PET system is equipped with multiple detectors for γ-photons in a circular order around the subject's body. Upon detection of two coincidentally arriving photons on two opposing detectors, a signal is generated. Cross-sectional images of the organs or of the body can be generated using the spatiotemporal characteristics of these particular signals. Similar to the CT technology, images obtained by a PET system are processed afterward by a computer in order to provide cross-sectional images of the tissue, organs, or body. To enhance the resolution and to provide three-dimensional images of the radio-pharmacon's distribution, modern PET systems are frequently combined with CT scanners. This tandem PET/CT system is able to generate three-dimensional imaging of the patient's body or organ during the same scanning procedure. In contrast to PET, the technical platform of SPECT is more related to conventional scintigraphy. A γ-emitting radioisotope is injected into the patient, and a two-dimensional snapshot is taken using a γ ray-sensitive camera. For the generation of three-dimensional images, multiple planar snapshots are taken by a single or multiple cameras rotating around the patient's body. In general, it is possible to take a "static" snapshot of the radionuclide's distribution in the patient's body or to perform multiple snapshots, so-called dynamic SPECT analyses in a minute-, hour-, or day-based intervals to evaluate time-dependent distribution of the radionuclide.

PET displays a slightly improved spatial resolution and higher sensitivity compared to SPECT (up to $10^{-12}$ M for PET compared to $10^{-10}$–$10^{-11}$ M for SPECT). These differences are connected to the nature of $\beta^+$ emission. Achievable resolution is dependent on the so-called positron range (i.e., the distance the positron travels before annihilation); thus, low-energy $\beta^+$ emission leads to improved resolution. Also, the requirement of coincidence detection of two simultaneous γ-rays leads to high sensitivity. Advantages of SPECT over PET are improved quantitative accuracy in deeper tissue areas, relatively easy access to the most commonly used radionuclides, and in particular the option to detect and image multiple radionuclides simultaneously when these emit different energy photons.

The most frequently applied radionuclides for SPECT are $^{99m}$Tc (available by benchtop generators from $^{99}$Mo, $t_{1/2}$, 6h) and $^{111}$In with a much longer half-life of 2.8 days.

Most of radionuclides used for PET like $^{18}$F have to be generated by cyclotrons. However, often this implies transportation over long distances from the location of the

**TABLE 16.1 Properties of Different Radionuclides Used for Imaging**

| Radionuclide | Emission | Energy (keV) | Abundance (%) | Half-Life | Radius (nm) |
|---|---|---|---|---|---|
| $^{11}C$ | $\beta^+$ | 511 | 100 | 20 min | |
| $^{68}Ga^{3+}$ | $\beta^+$ | 511 | 89 | 68 min | 76 |
| $^{18}F$ | $\beta^+$ | 511 | 100 | 110 min | |
| $^{64}Cu^{2+}$ | $\beta^+$ | 511 | 18 | 12.7 h | 87 |
| $^{89}Zr^{2+}$ | $\beta^+$ | 511 | 23 | 3.3 days | 86 |
| $^{111}In^{3+}$ | $\gamma$ | 172, 245 | 100 | 2.8 days | 94 |
| $^{99m}Tc^{7+}$ | $\gamma$ | 140 | 88 | 6 h | 56 |
| $Eu^{3+}$ | None | | — | Stable | 109 |
| $Gd^{3+}$ | None | | — | Stable | 108 |

cyclotron to the PET facility. Because of the longer half-life time of $^{18}F$ compared to $^{11}C$, fluorine is used in more than 90% of all PET scans. Properties of frequently used radionuclides are listed below (Table 16.1) [42–44]. Gallium with a half-life of 68 minutes can be derived from $^{68}$Germanium by readily available benchtop generators. The availability of bioorthogonal click chemistries for effective attachment of chelating moieties makes it a useful alternative for labeling of biomolecules like peptides, proteins, and antibodies.

In order to improve contrast and enrich the imaging agent in specific tissues of interest, targeting moieties have to be attached to the respective probe. For employing cationic metal radioisotopes, a binding moiety needs to be installed to keep the metal attached to the targeting molecule. To avoid toxic effects of leaching radiotracers or gadolinium [45], tight complexation has to be ensured. Some representative structures are given in Figure 16.4.

Derived from ethylenediamine (EDTA), several acyclic (e.g., DTPA) and cyclic chelators (e.g., DOTA consisting of four and NOTA consisting of three ethylenediamine moieties) are frequently used [46, 47]. In case of NODAGA, a position for attachment of biomolecules is present; also in DOTA derivatives, often one carboxylic acid is replaced by an amide to attach a tether for conjugation. In all cases, the cyclic core in combination with the attached amide/carboxylic acid residues forms a cavity in which the metal is deeply buried. In case of gadolinium, only one remaining coordination site for water remains open to generate the MRI signal (Figure 16.5).

The binding strength is dependent on the electronic properties, as well as on the diameter of the complexed cation; for instance, NOTA is a very strong chelator for the smaller $Ga^{3+}$, while the bigger $Gd^{3+}$ is less effectively complexed by the smaller NOTA. Also, attachment of amide groups can result in lowered complexation efficacy; the gadolinium complex of tetra-glycine amide of DOTA (also named 1,4,7,10-tetraazacyclododecane 1,4,7,10-tetraacetic acid amide (DOTAM)), for example, displays a dissociation constant of only 14.5 compared to 24.7 for the corresponding free DOTA complex.

It needs to be considered that, as most of these methods rely on the application of radiotracers, their inherent toxicity needs to be kept in mind, in particular when performing chronic studies; Funk [48] reported that in many published SPECT experiments applied radiation doses already reach a significant fraction of the dose causing 50% mortality within 30 days, which may, for instance, result in altered gene expression (and thus possibly altered phenotype) as a direct consequence of administration of a harmful dose of the radiolabelled agent. This may compromise results of longitudinal studies. Also, doses necessary to obtain high-quality images may have no pharmacological effect in human patients, but may very

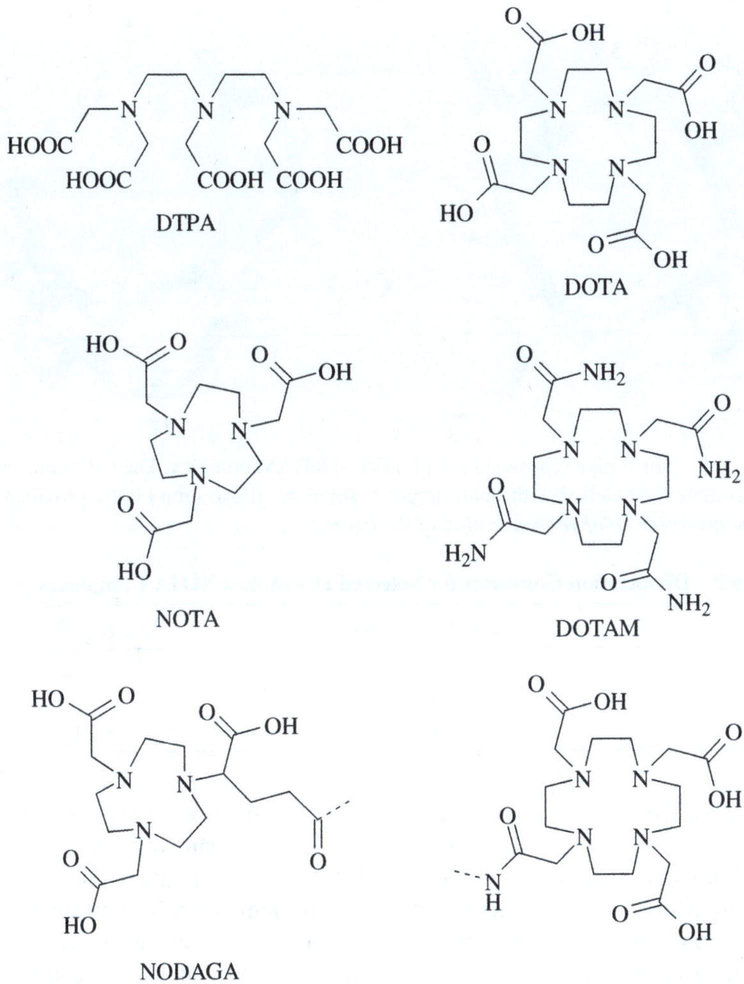


**FIGURE 16.4** Structures of frequently used chelators for MRI and PET.

well do so in small animals like mice [49], either by direct action of the targeting agent or by effects caused by radiation. Table 16.2 summarizes some characteristics of cations commonly used for imaging. $Eu^{3+}$ (fluorescence imaging) and $Gd^{3+}$ (MRI), albeit no radioisotopes, are included for comparison. The half-life of the respective isotopes has to be taken into consideration for both synthetic access to the probe and for designing the optimal imaging experiment. It has to be ensured that at time of image acquisition sufficient tracer is present in the target tissue. The periphery, though, should be essentially cleared from probe. Thus, the pharmacokinetic properties of the probe and the half-life of the incorporated isotope have to be combined in an optimal fashion. For cations, the radius has to be considered to choose the optimal chelating moiety, as discussed previously.

Design and synthesis of PET and SPECT tracers require careful consideration of multiple factors, for example, choice of optimal isotope and chelating moiety in case of radiometals, ideal synthetic access, envisaged dose, and duration of treatment and targeting mechanism of choice [7, 50–56].

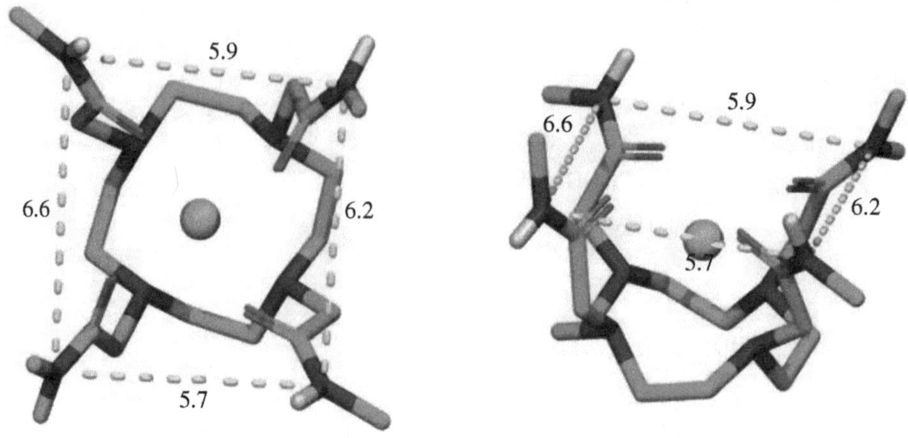


**FIGURE 16.5** Illustrative representation of a $Gd^{3+}$-DOTAM complex. The $Gd^{3+}$ ion is buried in a cage-like complex. Modelled with Schrödinger™ software, illustration kindly provided by L.-H. Wang. (*See insert for color representation of the figure.*)

**TABLE 16.2 Dissociation Constants for Selected DOTA and NOTA Complexes**

| Log $K_{ML}$ | Log $K_{ML}$ (DOTA) | Log $K_{ML}$ (NOTA) |
|---|---|---|
| $Ga^{3+}$ | 21.3 | 31.0 |
| $Cu^{2+}$ | 22.3 | 21.6 |
| $Gd^{3+}$ | 24.7 | 14.3 |

Optimal PET tracers for evaluation of therapeutic effects of a given drug should not alter its structure to exclude potentially misleading results caused by introduction of the modification itself. In case of utilization of $^{18}F$ for PET, this requires fluorine to already be incorporated in the drug. As fluorine is a versatile element to block metabolic sites and fine-tune overall physicochemical properties of a molecule, many examples of marketed drugs contain fluorine [57]. It is understood that for most molecules completely new synthetic routes have to be developed, as chemical sources and methods for $^{18}F$ incorporation are limited. Furthermore, as a general rule, syntheses with radioactive molecules should not require more time than three half-lives of the respective isotope. For example that would leave in the case of fluorine 5.5 h, in the case of $^{11}C$ even only 1 h for completion of the synthesis, which calls for synthetic schemes allowing incorporation of the respective radioisotope in the last step of the synthetic sequence and avoiding the need of time-consuming workup and purification.

However, the synthetic repertoire for incorporation of $^{18}F$ is steadily increasing [58]. Two recent examples of potentially very versatile methods developed by the Gouvernor and Ritter groups are depicted in Scheme 16.2 [59–62].

One option to overcome the limitations of individual imaging technologies and to increase the collected information is utilization of multimodal imaging [63]. For instance, clear determination of a physiological state may require detection of more than one molecular marker or inclusion of anatomical imaging. It is of high importance to take the different sensitivity ranges of imaging techniques into account when combining different imaging modalities in one probe; for sequential application of individual tracers, it is important to consider their pharmacokinetic and tissue distribution behavior, as well as to

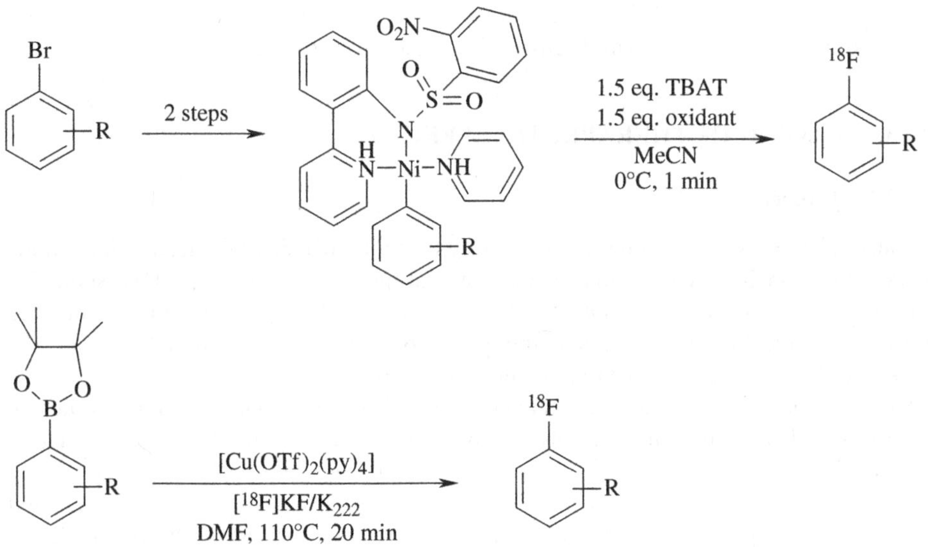


**SCHEME 16.2** Novel rapid methods for incorporation of $^{18}$F into aryl and heteroaryl systems.

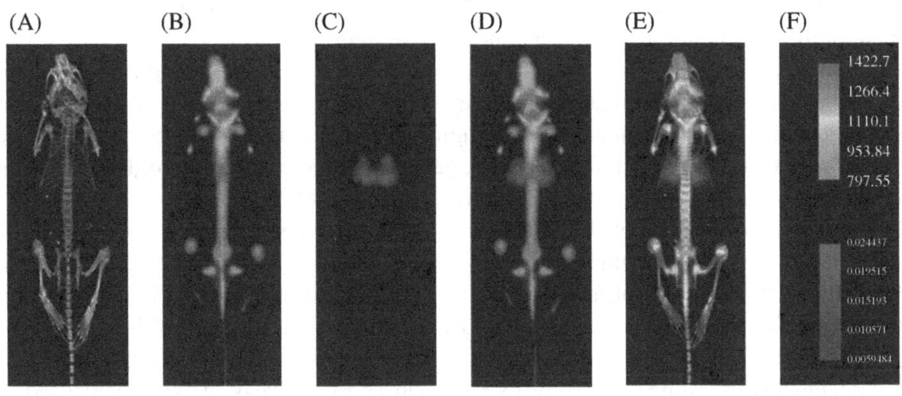


**FIGURE 16.6** Images from Chapman [64]. Example of dual *in vivo* $^{99m}$Tc-pentetate and Na$^{18}$F tracer imaging showing individual modality displays A, CT; B, PET; C, SPECT; D, a merged dual modality SPECT/PET display; E, a merged trimodal PET/SPECT/CT display; F, unit scale bars for PET (kBq/cc) and SPECT (cps/voxel, voxel size = 250 μm isotropic). Reproduced with permission from Chapman [64]. (*See insert for color representation of the figure.*)

ensure unaltered orientation of the animal. While PET and CT are coapplied routinely, simultaneous SPECT/PET imaging is hampered by interference with 511 keV photons, preventing obtainment of useful SPECT information. By careful optimization of sequence and protocol, it is now possible though to obtain this kind of information as recently demonstrated by Chapman et al. [64]. In this example, trimodal imaging was performed in rat, applying a combination of the lung-specific SPECT tracer $^{99m}$Tc-pentetate, the PET tracer Na$^{18}$F, which is showing enrichment in bone and anatomical CT imaging.

Figure 16.6 shows examples of dual *in vivo* $^{99m}$Tc-pentetate and Na$^{18}$F tracer imaging with individual modality displays (A. CT, B. PET, C. SPECT, D. a merged dual modality SPECT/PET display, and E. a merged trimodal PET/SPECT/CT display).

The methods discussed earlier have found numerous applications in drug discovery; on the following pages, we will discuss some illustrative examples.

## 16.3 IMAGING OF THERAPEUTIC EFFECTS

### 16.3.1 Cancer

Cancer is a leading cause of death, accounting for more than 580,000 deaths alone in the United States in 2013. Estimations by the National Institutes of Health (NIH) assume that the overall costs of cancer in 2008 were $201.5 billion with roughly one-third accounting for direct medical costs and the remaining $124 billion for indirect mortality costs, that is, cost of lost productivity due to premature death [65].

Traditionally, assessment of treatment efficacy by imaging methods strongly relies on monitoring changes in tumor size, as, for instance, defined by the Response Evaluation Criteria in Solid Tumors (RECIST) guidelines [66]. However, the RECIST criteria are controversially discussed as they may, for instance, underestimate the effect of targeted therapy. Biopsies provide an extremely valuable source of information, allowing to define cancer subtypes and to stratify patients according to presence of certain target mutations. However, due to their invasive nature, they cannot be frequently repeated, and target expression may not be homogeneous within the tumor, as well as it may change in the course of therapy; therefore, additional methods are required.

Applications for imaging in oncology mainly comprise detection of primary tumors and metastases and determination of drug distribution behavior and assessment of treatment efficacy. One particularly challenging question that needs to be addressed by imaging is detection of small metastases. A general feature of rapidly proliferating cells is their high metabolic turnover. Therefore, 2-deoxy-2-($^{18}$F)-fluorodeoxyglucose (18-FDG) can be used as an effective PET tracer. FDG is trapped within the metabolizing cell as it is not a substrate for glucose-6-phosphate isomerase [67]. An alternative reagent correlating with general proliferative activity is ($^{18}$F)-fluorothymidine (18-FLT). Upon cellular uptake, the compound gets rapidly phosphorylated by cytosolic thymidine kinase 1 (TK1). The activity of this enzyme is very high in proliferating cells, peaking in S-phase, but low in resting ones. Both 18-FDG and 18-FLT have been extensively used to predict overall and progression-free survival. Meta-analyses of published data on various cancers indicate that $^{18}$FDG-PET/CT may not be a suitable primary diagnostic procedure due to its low sensitivity for small (≤10 mm) tumors; but the authors stress its potential to be useful for detection of distant metastases [68]. Also, application of these probes represents an easy and generally applicable method for monitoring initial response to chemotherapy, comparing FDG or FLT uptake after a defined period of time to the uptake observed before initiation of treatment. Exemplary recent clinical applications predict clinical outcome after 2 days of gefitinib treatment [69], seek to identify patients benefitting most from erlotinib treatment [70], or discuss target inhibition of an AKT inhibitor [71]. An analysis of nine different human xenograft models in mice demonstrated general suitability of most of these models for pharmacodynamic studies applying either FDG or FLT [72].

One efficient way to target cancer is the use of monoclonal antibodies [73, 74]. More than a dozen therapeutics are available for cancer treatment to date. Treatment is either performed using a naked antibody, often in combination with chemo- or radiotherapy, or using antibody–drug conjugates [75–79]. In this concept, the antibody is used as a targeting moiety to deliver a toxic payload, which can either consist of radioisotopes or extremely toxic chemicals, to the tumor. Currently approved and marketed are gemtuzumab ozogamicin (loaded with calicheamicin), brentuximab vedotin (loaded with auristatin), and ado-trastuzumab emtansine (loaded with mertansine). Usually, one to three toxic moieties are bound to one antibody carrier molecule. In the case of toxic conjugates, optimal tumor targeting is an essential requirement to reduce toxic side effects. Several techniques to improve tumor targeting have been developed over time, many of them either relying on attachment of homing moieties [80–83] or exploiting the fenestration of the tumor vasculature, making it more permeable to particle-conjugated drugs [84, 85].

PET imaging is an ideal tool to accompany drug development of these molecules, for example, to effectively assess optimal dosing studies and tumor targeting efficacy in first in man studies, to follow drug accumulation in tumor and nontarget tissues over time, and to measure response to therapy.

The large size of monoclonal antibodies allows effective labeling with radiotracers without compromising their biological activity. However, the usually observed very long half-lives of antibodies have to be taken into account when choosing the positron emitter. $^{89}Zr$ ($t_{½}$ = 78.4 h) and $^{124}I$ ($t_{½}$ = 100.3 h) are frequently used.

In order to produce material useful for clinical profiling of radiolabeled conjugates, effective and highly reproducible methods had to be developed [86–88].

One issue associated with profiling of antibody drugs in rodent models is the limited cross-reactivity between humans and rodents, limiting selectivity profiling to human xenograft models. However, important information on general biodistribution can be effectively obtained. In this context, $^{89}Zr$-PET is particularly useful due to its good sensitivity and translatability from mouse to human [89].

***16.3.1.1 Tumor imaging*** Several efforts have been undertaken to combine different imaging modalities. Pomper [90] describes a DOTA-based agent bearing a conjugated IRDye800CW moiety and small lysine–glutamate urea residues for effective targeting of the prostate-specific membrane antigen (PSMA). Sequential SPECT/CT and optical imaging were performed in SCID mice, bearing PSMA-positive and PSMA-negative tumors. Sensitive NIRF-based visualization could be achieved at doses as low as 0.1 nmol, effectively distinguishing between PSMA-positive and PSMA-negative tumor tissue (Fig. 16.7).

In order to allow multimodal imaging, different reporter groups need to be attached to a single molecule, often requiring incorporation of multiple copies of the individual residues for increased sensitivity. Nanoparticles offer good opportunities for attaching multiple moieties and span a wide variety of sizes and physicochemical properties, making them ideal carrier systems for multimodal imaging. Many inorganic (e.g., quantum dots, gold, iron oxide based) and organic nanoparticle platforms (e.g., liposomes, micelles, dendrimeric structures) have been established, and their binding chemistries, biodistribution, and targeting behavior have been thoroughly studied [91–95]. Potential toxicity of nanoparticles and in particular of quantum dots due to their loading with toxic metals has to be considered; therefore, incorporation of active targeting mechanisms proves to be

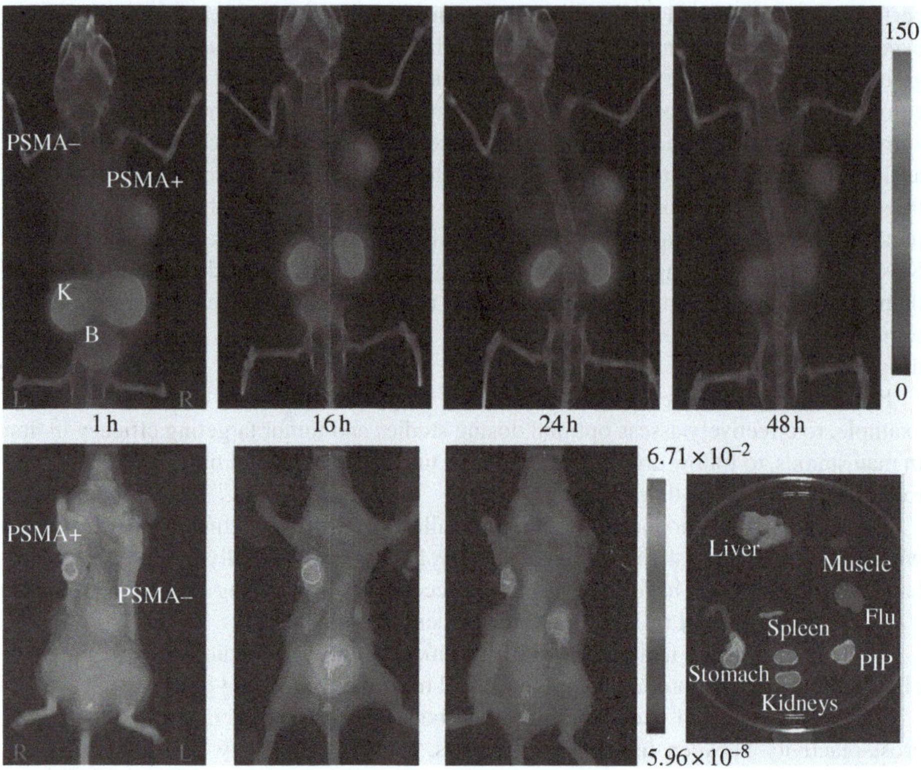


**FIGURE 16.7** Images from Pomper [90]. Sequential SPECT/CT (top row) and optical (bottom row) imaging of dual SPECT–NIRF reagent using PSMA-positive (PIP) and PSMA-negative (flu) tumors in a male SCID mouse. Reproduced with permission from Banerjee et al. [90]. (*See insert for color representation of the figure.*)

very valuable to increase local concentration at tissue of interest and thus being able to reduce overall applied dose.

It is another intriguing concept to use labeling of cancer cell-specific surface proteins during surgery to efficiently resect tumor tissue. Three different questions have to be resolved to bring fluorescence-guided surgery (FGS) [96] to the clinics: firstly, efficient and specific tumor labeling needs to be accomplished; secondly, the availability of predictive patient-like animal models to allow effective development needs to be ensured; and lastly, portable and practicable systems to excite the used fluorophores during surgery have to be developed.

In a recent study, five clinically used monoclonal antibodies were linked to the near-infrared fluorescence dye IRDye800. The probes were intravenously injected into nude mice bearing human breast cancer. Tumors were surgically removed using two different imaging systems. Both imaging platforms were able to detect tumors as small as 0.5 mg; the EGFR targeting drugs cetuximab and panitumumab proved to be the best for imaging in this particular model [97].

To effectively stain tumor tissue, specific antigens need to be identified. Carcino-embryonic antigen (CEA) is a glycosylphosphatidylinositol (GPI) cell surface-anchored glycoprotein. Normally produced in gastrointestinal tissue during fetal development, expression in adults is restricted to cancer cells, particularly in adenocarcinomas, for example, in colon or pancreatic cancers. This makes CEA a promising cell marker for FGS.

In recent work by Hiroshima et al. [98], patient-derived colon tumors were transplanted into nude mice and allowed to grow for eight weeks. Mice were i.v. injected with a monoclonal anti-CEA antibody conjugated to Alexa Fluor 488, twenty-four hours later, the tumor was surgically removed using a handheld fluorescence imaging system. All animals had complete resection and did survive for at least 6 months without reoccurrence of tumor.

In a similar study [99] on pancreatic cancer, efficacy of conventional bright-field and FGS was compared. Complete tumor resection was achieved in 92% of mice in the FGS group as compared to 46% in the control group, cure rates improved from 5 to 40%, and 1-year postoperative survival rates increased from 0 to 28% with FGS, underlining the potential utility of this method to improve outcome of surgical treatment of cancer.

The increased understanding of underlying signal transduction processes has led to an enormous effort to develop kinase inhibitors. As of April 2015, 28 small-molecule kinase inhibitors have been approved by the FDA for the treatment of cancer, a majority of them being tyrosine kinase inhibitors (TKIs). In contrast to antibodies, kinase inhibitors are usually small molecules, thus displaying typical physicochemical and pharmacokinetic properties like being able to penetrate cells, being orally absorbable, and showing increased protein binding. This has to be taken into consideration when choosing the optimal radiolabeling or imaging strategy. In contrast to antibodies, kinases are reasonably well conserved among species, allowing the assessment of inhibitor selectivity and potential side effects in rodents and not depending on studies in non-human primates. The vast majority of inhibitors under development target the ATP domain of kinases. As ATP is the universal substrate of all kinases and its binding domain is highly conserved, obtaining specific inhibition of the target kinase remains challenging. Nonselective interference with signal transduction can lead to serious side effects. Although most kinase inhibitors have undergone significant efforts to optimize their selectivity profiles, other targets are frequently hit. This can be associated with toxicities or lead to beneficial effects if the involved targets are involved in proliferative effects. Knowledge of the target profiles of kinases in clinical development is essential to understand their pharmacological properties [100–102], but solely relying on *in vitro* results is insufficient. Another important point to consider is the occurrence of resistant mutations, either within different cancer subtypes or by clonal selection as a consequence of chemotherapy [103, 104].

To provide a thorough understanding of the *in vivo* properties of these compounds and in particular to shed light on their interaction with their targets, PET using specifically radiolabeled inhibitors is of great value. Synthesis of the radiolabeled derivatives and preclinical work has recently been comprehensively summarized by Slobbe [105].

For the design of the radiolabeled molecule, it is essential that the original structure of the inhibitor is ideally not altered to avoid misinterpretation due to altered distribution, which limits labeling in most cases to $^{11}C$ and $^{18}F$. Introduction of these isotopes will usually require establishment of specifically designed synthetic routes due to their short half-lives. A second point to consider is metabolism to avoid cleavage of the radiolabel, also potentially leading to misinterpretation. To avoid this, different positions of the molecule of interest should be radiolabeled and subsequently studied in biodistribution and metabolism studies before selecting a radiotracer for clinical development.

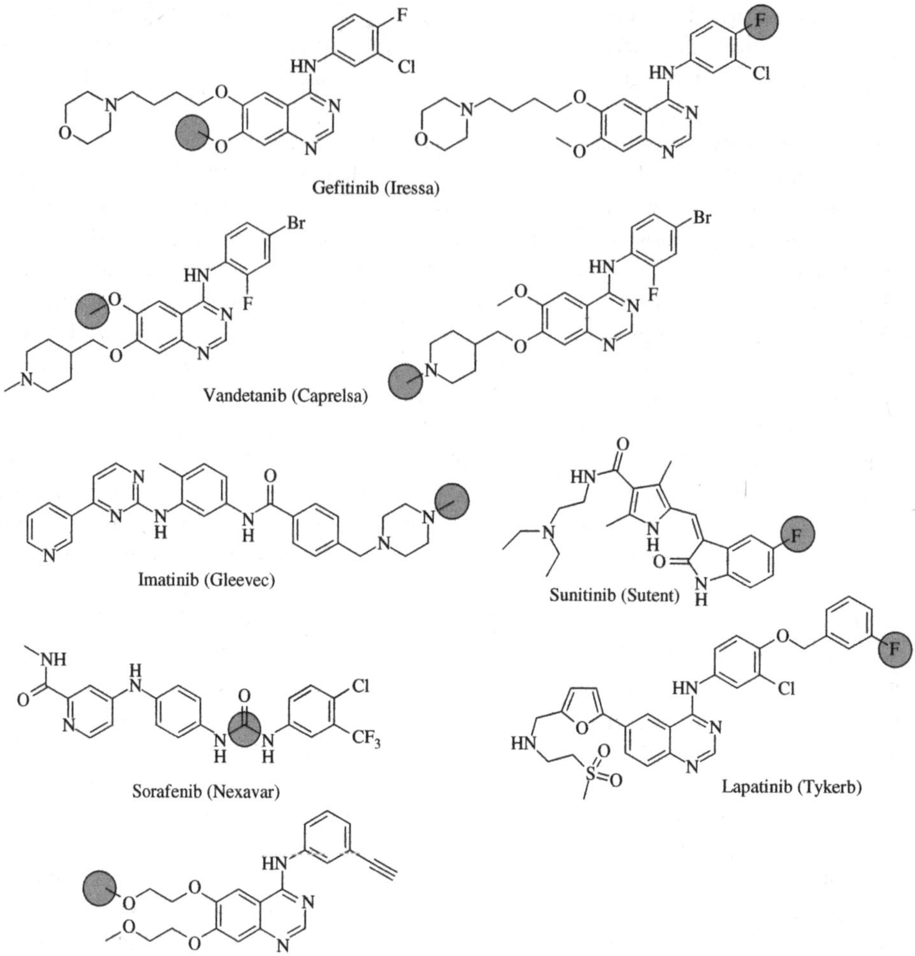


**FIGURE 16.8** Position of introduced radiolabels is indicated by gray circles.

A summary of published TKI radiotracers is given in Figure 16.8.

Studying appropriately labeled compounds in patients will open up new venues for personalized therapy [106–108]. Erlotinib (Tarceva™, Roche) was optimized to be a potent ATP competitive inhibitor of the EGFR kinase domain. However, it only shows efficacy in 10–30% of treated patients, which calls for methods to preselect the fraction of actually benefitting patients. [$^{11}$C]erlotinib had been studied in preclinical [109] and several clinical studies [110]. It could be shown that [$^{11}$C]erlotinib binds specifically to xenografts bearing activating mutations in the kinase domain, but not in models containing activating mutations in the extracellular domain of EGFR [111]. In clinical trials, a significantly higher uptake in patients bearing exon19-deletions could be observed, indicating improved response rates in preselected patients [107, 112].

Mining genomic and expression data in defined disease states can help to generate hypotheses on the involvement of certain targets in cellular processes. However, the complex regulation underlying many cellular processes, for example, the activation of phosphorylation, translocation events, or formation of protein complexes, is not captured by these methods. Also, the significance of low-abundance proteins with high activity may be underestimated. Thus, the determination of the specific activity of a given enzyme can be an invaluable measure for determination of distinct cellular states and dynamics of pathological processes. Therefore, the development of activity-based probes to monitor the activity of a target is an area of growing scientific attention [113]. Of particular interest is the imaging of enzyme activity in an *in vivo* context [114], reflecting disease state and progression. For kinases, this approach is still in its infancy, due to difficulties of reaching the intracellular location with a probe in sufficient quantities. Reported probes to date are limited to *in vitro* experiments using cell lysates [115], but it will be highly interesting to observe the development of this area in the future.

Another class of therapeutically relevant enzymes that have been extensively examined in terms of development of activity-based probes are proteases. Proteases are key players in many important cellular processes, particularly in cancer and inflammation [116–118]. In particular for cysteine proteases, promising activity-based probes based on fluorophore–quencher pairs [119] are available today [120, 121], as the high nucleophilicity of the thiol group in the catalytic center renders them particularly susceptible to binding to a scavenger. The underlying principle of designing activity-based protease probes is illustrated in Figure 16.9.

In general, a warhead is connected to a peptide sequence specifically recognized by the target protease. The warhead is allowing nucleophilic attack by the catalytic cysteine of the protease, binding irreversibly to the catalytic center of the protease. This will result in inactivation of the protease and simultaneous cleavage of the quenching moiety, resulting in generation of a fluorescence signal. Acyloxymethylketones (AOMKs) are frequently used for this purpose; the general mechanism for enzyme inactivation by this class of mechanism-based inhibitors is given in Scheme 16.3.

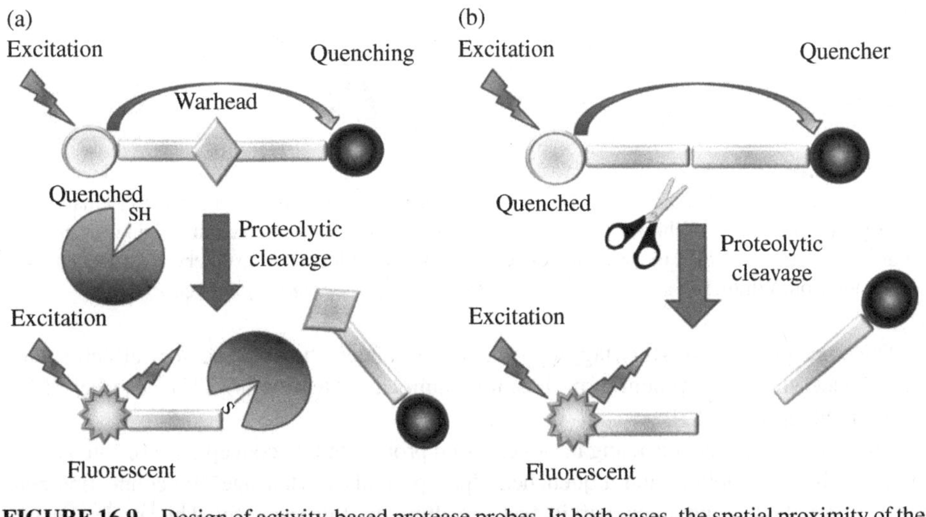


**FIGURE 16.9** Design of activity-based protease probes. In both cases, the spatial proximity of the fluorophore–quencher pair is abrogated, either by (a) irreversibly tagging and inactivating the target protease or (b) selective cleavage of the connecting linker using turnover-based protease probes, preserving the activity of the target enzyme. (*See insert for color representation of the figure.*)

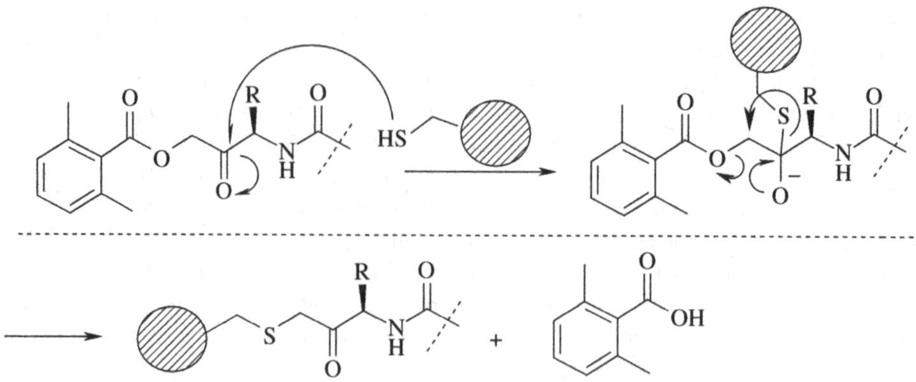

**SCHEME 16.3** Protease inactivation by acyloxymethylketones (AOMKs).

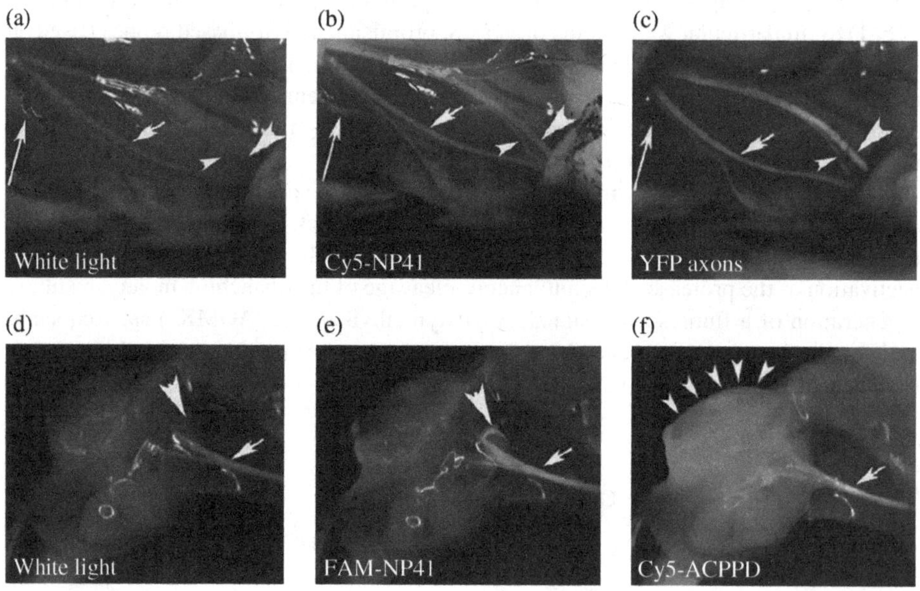


**FIGURE 16.10** (a–c) Right facial nerve and its arborizations in a thy1-YFP mouse treated with Cy5-NP41. (d–f) Left sciatic nerve (arrow) and its arborization in a mouse with a syngeneic 8119 mammary tumor graft. Picture from Whitney [133]. Simultaneous tumor (green) and neural staining (cyan, near infrared, and yellow (YFP)) to assist nerve-sparing tumor resection during surgery. Reproduced with permission from Whitney et al. [133]. (*See insert for color representation of the figure.*)

The concept of irreversibly tagging proteases has been shown to be very effective and led to successful development of probes, for example, for legumain [122], caspases [123, 124], or cathepsins [125].

An alternative concept is using turnover-based probes. In this concept, a substrate is decorated with a fluorophore and a quencher. Upon proteolytic cleavage, the connection between fluorophores and quencher is lost, resulting in fluorescence. This probe design preserves the catalytic activity of the target protease, thereby being particularly useful for detection of low-abundance enzymes. A recent example describes probes for monitoring of cathepsin B and MMP activity [126]. Identification of peptidic sequences being selectively cleaved by the target protease can represent a veritable challenge. An alternative approach for probe design, the so-called *reverse design* concept, relies on utilizing thoroughly optimized inhibitors resulting from medicinal chemistry efforts. By turning the warhead of these inhibitors into a scissile bond, it is likely that the overall favorable attributes of the scaffold in terms of selectivity and physicochemical properties can be conserved. Decoration with a fluorophore and quencher leads to highly useful probes; however, it should be noted that nature and connection site of the dyes will impact selectivity and enzymatic turnover and therefore require careful optimization. Examples include probes for cathepsin S [127, 128] and caspase 1 [129]. A drawback of this method is that the probes, after cleavage, are not located to a specific cellular environment, leading to potential dilution of the signal. In a recent work, lipidation could be successfully applied to locate the fluorophores to the tumor environment [130], resulting in fluorescence signal with excellent tumor to tissue ratios.

Tsien described utilization of matrix metalloprotease probes in combination with cell-penetrating peptides to improve cellular uptake [131]. The concept was utilized to assist tumor resection [132]. Simultaneous staining with probes containing a peptide, which was identified by phage display to specifically bind to nerves, significantly improved outcome by allowing optimal tumor resection but preserving nervous tissue [133–135] (Fig. 16.10).

### 16.3.2 Diabetes

An important hallmark of diabetes is the progressive loss of insulin-producing beta cells, either due to autoimmune destruction (type 1) or as a consequence of insulin resistance and gluco- and lipotoxicity (type 2), leading to constantly increased insulin secretion and subsequent beta-cell death. To date, the status of disease progression can only be assessed indirectly by recording the physiological response to a glucose challenge or determination of fasting plasma glucose values. In the course of disease during early stages of beta-cell apoptosis, remaining beta cells will try to compensate by increased insulin production. Noninvasive imaging of beta-cell mass in animals and ultimately in humans has been a long-standing goal in diabetes research. Correlation of response to glucose tolerance tests with functional beta-cell mass would allow a much more precise classification, in particular of early disease states [136, 137]. Another highly interesting area for noninvasive beta-cell imaging is the monitoring of survival of transplanted islets [138–140]. Transplantation either of human donor material or of encapsulated pig islets represents the only curative approach for treatment of type 1 diabetes available to date, albeit autoimmune destruction of the implanted islets will eventually result in loss of implanted cells over time [141, 142].

Beta cells are exclusively located in islets of Langerhans in the pancreas, which is located in the deep abdomen behind the stomach and in close proximity to the liver. This close proximity imposes particular challenges to molecular imaging in terms of tissue selectivity. While for small animal species like mouse NIRF methods can be used in principle, for larger animals or human use, MRI and PET can be envisaged.

Little work has been performed using unspecific agents, although recently $Mn^{2+}$ has been examined *in vitro* and *in vivo* in various rodent models of diabetes [143] and even has been administered to healthy and type 2 diabetic subjects, followed by manganese-enhanced MRI [144]. From these studies, it was evident that $Mn^{2+}$ is taken up in healthy human islets, resulting in an improved signal-to-noise ratio. This effect was less pronounced in diabetic patients, resulting in reduction of signal enhancement and allowing to discriminate diabetic and nondiabetic patients. The authors concluded that in principle high-resolution MRI experiments could be used in the future to assess β-cell mass, although concerns, for example, on safety of $Mn^{2+}$ infusions remain, and this approach as currently described is clearly not suitable for determining beta-cell mass *in vivo*.

In terms of small molecule agents, the SUR1-binding insulin releaser glibenclamide has been intensively studied [145]. Biodistribution studies of the $^{18}$F-labeled derivative (**1**) revealed a quick uptake and slow clearance of the compound (Fig. 16.11). Liver uptake was about 100-fold higher as compared to pancreas. In an exploratory *in vivo* PET study in healthy volunteers, time activity curves showed a fast uptake and very slow washout of tracer. The high background and the strong liver uptake (about 100-fold higher than pancreas) did not allow effective imaging of the pancreas. An attempt to increase the overall polarity of the identified binders by glycosylation has also been described [146]. However, the increased polarity (logD = 0.43) of (**2**) was accompanied by a decreased *in vitro* and *in vivo* activity, and this molecule has not been further converted into a labeled probe. It has also been described by Schneider [145] that a $^{99}$Tc-labeled glibenclamide derivative (**3**) did show altered tissue distribution properties, displaying less uptake in the liver. In a recent publication by Oh [147], an alternative $^{99}$Tc-labeled SUR1 binder, a derivative of glipizide (**4**), is described in more detail. The overall increased polarity of the complex led to a favorable muscle to pancreas ratio and allowed a clear visualization of the pancreas in mice; in a control experiment, mice were treated with streptozotocin, which induces effective β-cell destruction. In these animals, the signal from the pancreas was not observable.

As the pancreas is highly innervated with sympathetic and parasympathetic neurons, several imaging probes originally developed for CNS applications have been examined for their utility as beta-cell imaging agents.

An intensively studied class of probes is derived from dihydrotetrabenazine (DBTZ), targeting the islet β-cell vesicular monoamine transporter 2 (VMAT2) [148, 149].

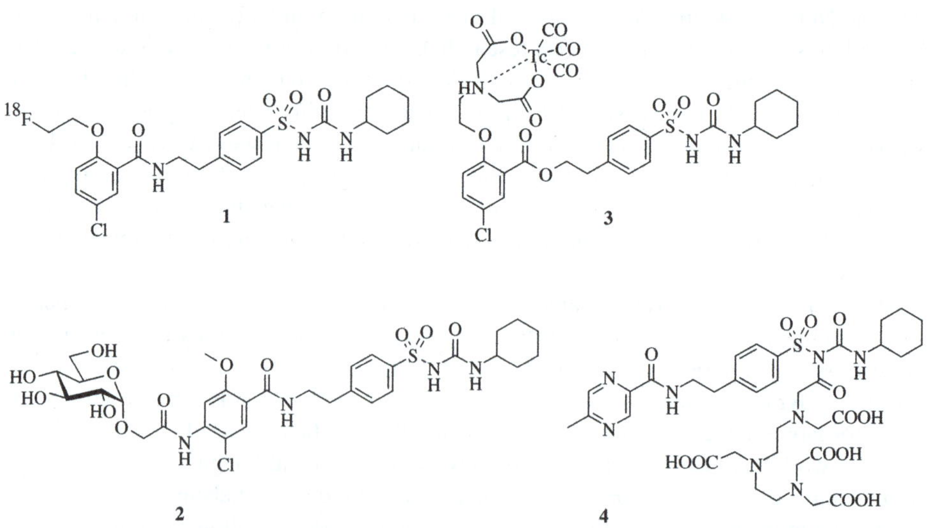

**FIGURE 16.11** Representative examples of SUR1 targeting moieties.

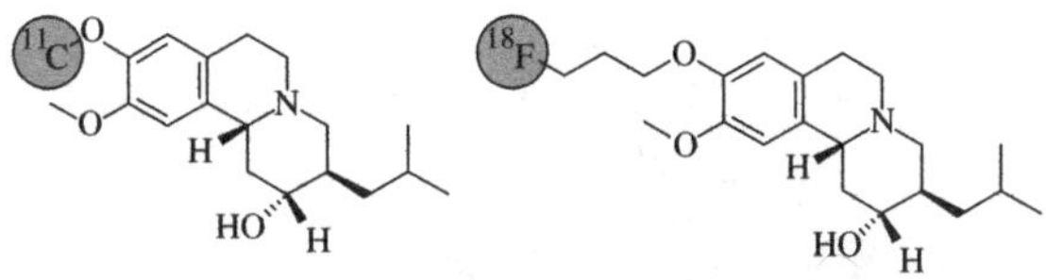

**FIGURE 16.12** DTBZ-derived radioligands targeting VMAT2.

$^{11}$C-DTBZ has been clinically used as a specific PET ligand for brain imaging (Fig. 16.12). However, the utilization of these probes for β-cell imaging has been controversially discussed [150].

Small molecule-based imaging approaches have been hampered by quick penetration of the probe into various tissues, leading to high background. Furthermore, the high protein binding of several agents results in long plasma residence, which renders these moieties useless for PET experiments based on $^{18}$F.

Rudin and coworkers described the generation of an antibody directed against Tmem-27, a surface N-glycoprotein that is highly expressed in β cells [151]. First, *in vivo* examinations were published describing imaging experiments in transgenic animals. High cellular uptake of the generated antibodies was observed. Unfortunately, the antibody does not seem to display cross-reactivity to the mouse protein, so studying the utility of this approach for beta-cell imaging in models of diabetes is yet to be demonstrated.

So far, the best studied class of β-cell imaging agents are based on GLP-1 agonists. Involvement of this class B GPCR in glucose metabolism and disease pathogenesis of diabetes has been extensively studied [152]. The receptor displays high expression on beta cells, and the medicinal chemistry underlying GLP-1 analogues is well understood [153, 154].

Weissleder reported the synthesis of a NIR-fluorescent exendin analogue, the fluorophore being attached to position 12 of the peptidic chain [155, 156]. In a subsequent work, different dyes, attachment positions (12, 27, and 40), and strategies have been examined for the binding properties of the resulting derivatives [157]. Attachment at position 40 and 12 displays a comparable affinity, while attachment to position 27 proves to be inferior. Positions for attachment of reporters to Exendin are depicted in Fig. 16.13. It also gets clear from the data that the overall nature of the used dye influences binding properties to the receptor, and optimal probe design has to be performed with great care. The general approach has also been transformed into a $^{18}$F-labeling strategy to allow PET imaging [158]. Alternative PET labeling strategies mainly exploit exendin-Cys-40 derivatives [159].

Wang described the synthesis of a $^{18}$F derivative of the inactive Ex(9–36), labeled via reaction of [$^{18}$F]-4-fluorobenzaldehyde with a 6-hydrazinonicotinyl group on the ε-amine of Lys27 as a potential ligand useful for beta-cell mass imaging, but this approach did not result in specific binding [160]. Boermann reported the synthesis of $^{111}$In- and $^{68}$Ga-labeled derivatives of exendin-3 with the chelating moiety attached via Cys-40 [161]. Specific uptake of these probes into subcutaneously implanted INS1 tumors in mice could be observed; an almost complete blocking of uptake by coadministration of unlabeled compound was demonstrated. Also, specific pancreatic uptake was observed, and optimal targeting was achieved at very low doses (≤0.1 μg). In another study, copper- and

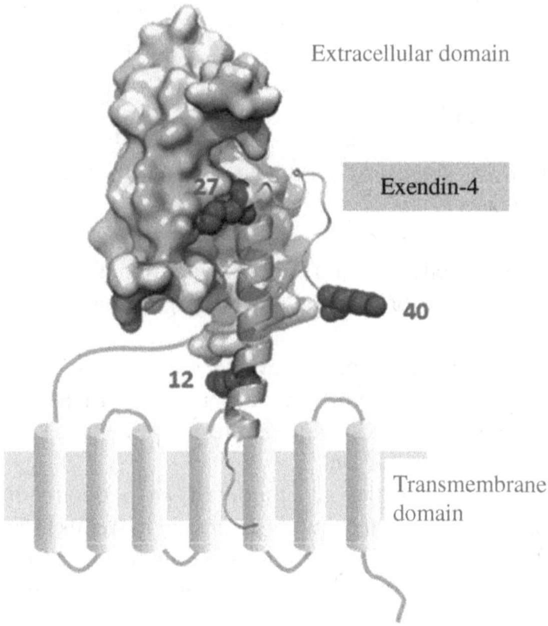


**FIGURE 16.13** Points of modification for currently explored exendin derivatives. Figure kindly provided by A. Evers. (*See insert for color representation of the figure.*)

in a first study in human, the $^{111}$In derivative described previously did show a heterogeneous pancreatic uptake, which is in accordance with available autopsy data. In a cohort of type 1 diabetic patients, uptake was significantly reduced [162].

Only recently, a first bimodal GLP-1R agonist has been described [163], in which $^{64}$Cu is complexed to a sarcophagine chelator, which is connected to a Cy5 dye. Subsequent attachment of the imaging moiety via click chemistry to position 12 of an alkyne-modified exendin-4 leads to a derivative with only slightly reduced binding affinity. Its application in *in vivo* imaging of insulinoma-bearing mice revealed that at the doses applied, good islet visualization could be achieved for both imaging modalities, the fluorescence signal being more closely associated with the control beta-cell marker.

From all these efforts, it can be concluded that although noninvasive β-cell imaging is a highly attractive target and tremendous progress has been made over the last years, there is still significant work to be done. The close proximity of liver and kidney represents a particular challenge, but recently published data suggests that by careful optimization of probe and imaging protocol, the long-lived dream of noninvasive determination of human beta-cell mass may come true eventually.

### 16.3.3 CNS Disorders

Global life expectancy constantly increased over the last 200 years almost all over the world, first driven by significantly improved hygiene conditions and education, then by the development of vaccines and antibiotics in the nineteenth century and the subsequent continuous improvement of the public healthcare system. In Germany, for example, the average life expectancy rose by more than 10 years between 1960 and 2010.

This inevitably led to a strong increase in the occurrence of central nervous system diseases, ranging from degenerative brain disorders, such as Alzheimer's and Parkinson's diseases, to vascular problems such as strokes. The need for cost-effective diagnosis and in particular for improved therapy is obvious, but the search for new treatment options remains extremely challenging. The identification of the underlying pathophysiological mechanisms of CNS diseases is hampered by their heterogeneous appearance and the limited predictivity of current animal models. Although there is clearly no shortage of putative targets for treatment, their validation remains cumbersome. Furthermore, the blood–brain barrier effectively limits penetration of substances into the brain.

To select a molecule for further clinical profiling, it is key to demonstrate that the compound actually reaches the target and to link target engagement to a pharmacological response, for example, a behavioral change.

From an imaging point of view, the brain represents a well-insulated physiological compartment, which can be monitored without overlapping signals from clearing organs like liver or kidney.

To assess biodistribution of CNS drug candidates, PET imaging of radiolabeled drugs is frequently applied [164]. In order to assess target engagement, highly specific ligands are required to discriminate between areas with high target expression and regions lacking the target receptor. To assess general utility of a ligand, several orientating criteria have been introduced. For effective PET imaging, a signal-to-noise ratio (i.e., the ratio of specific to nonspecific binding) of at least 1.4 or greater is usually required. As the affinity of the ligand to its target also plays an important role, as another rule of thumb, a value of greater than or equal to 10 for the ratio of tracer concentration in the target tissue divided by the receptor affinity ($B_{max}/K_d$) should be achieved, four being the lower limit. For many CNS receptors, selective PET tracers meeting these requirements are available [164].

A frequently used term to describe the interaction of a drug and its target is the so-called receptor occupancy, describing what percentage of the population of given target receptors is actually occupied by the experimental drug at a given dose. This information is important to determine the optimal dose of a drug, as well as to link the observed effects to the pharmacodynamic behavior of the drug. For the experimental determination of the receptor occupancy, a drug is administered at a given dose, and the reduction of uptake of a second specific tracer due to receptor binding of the drug to its target is measured. This requires two PET scans per subject, one before drug administration to determine the baseline tracer uptake and one after drug administration to determine the altered uptake of the radiolabeled tracer. Target occupancy studies at different doses can reveal important information on CNS permeability and can help to improve the design of phase II and III clinical studies, but require careful design to deliver valuable information [165–167].

For MRK-409, an α3-preferential partial $GABA_A$ agonist [168], receptor occupancy studies were performed using [$^3$H]flumazenil as radioligand (labeled at the methyl position). Minimal effective doses in preclinical models corresponded to 35–65% of receptor occupancy, depending on the examined model. As a side effect, a modest reduction in response rate hinting toward a slight sedation was observed at doses corresponding to a target occupancy of greater than 90%. However, subsequent human studies revealed a pronounced sedation already at doses of 2 mg, limiting the tolerated dose to only 1 mg. In a human PET study, [$^{11}$C] flumazenil uptake following a dose of 1 mg MRK-409 was not

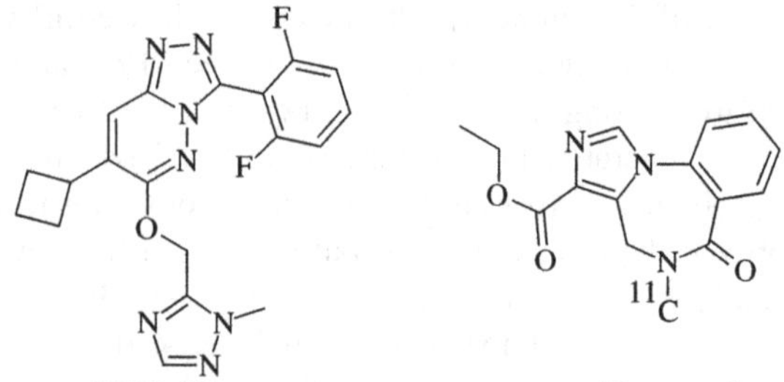


**FIGURE 16.14** Structure of MRK-409 and $^{11}$C-flumazenil.

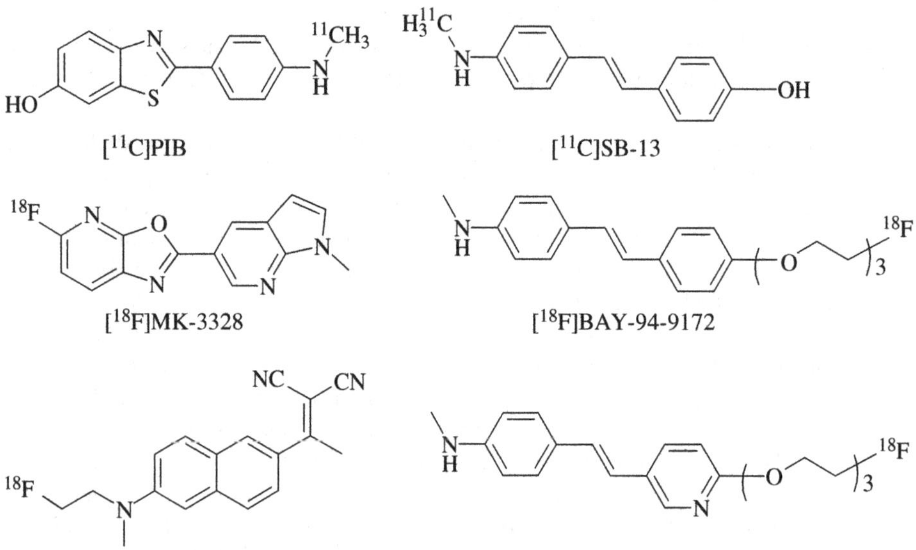


**FIGURE 16.15** Structures of some successfully applied PET amyloid imaging agents.

altered compared to placebo, suggesting inadequate receptor occupancy at this dose. Consequently, the further development of MRK-409 (Fig. 16.14) was stopped.

Although drug discovery for CNS diseases is clearly not short of challenges, the development of therapeutics for treatment of Alzheimer's disease has been particularly devastating. Massive effort has been undertaken (and still is) to develop viable therapeutic strategies to combat this pandemic disease. Key to success is an early, reliable diagnosis, capable of discriminating between different cognitive disorders and allowing clear assessment of the actual disease stage, thus opening up opportunities for therapeutic intervention already at an early point in time [169].

Several radiolabeled plaque imaging reagents, binding to amyloid beta (Aβ), have been examined in the clinics. Exemplary structures are shown in Figure 16.15.

One issue associated with imaging of amyloid β is that although the correlation between increased Aβ levels and Alzheimer's disease is well established, it is not yet fully clear, if

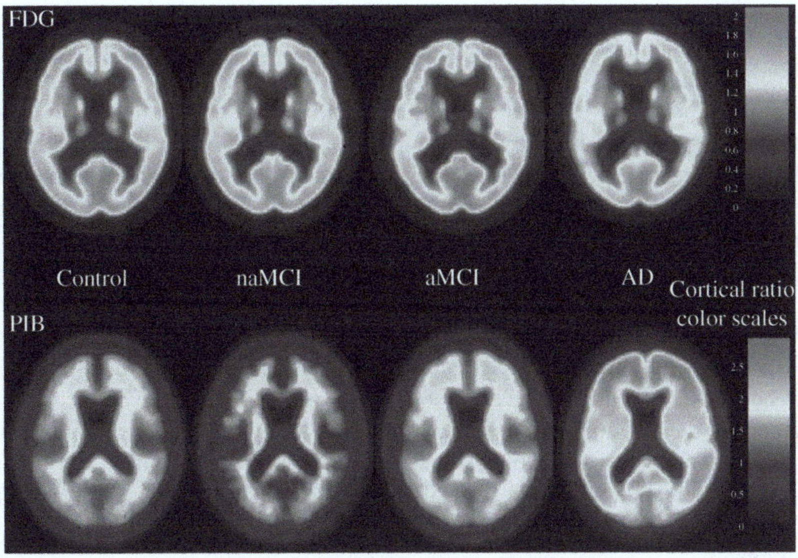


**FIGURE 16.16** Representative figure comparing $^{18}$F-FDG and PIB imaging in control subjects, patients with nonamnestic mild cognitive impairment (naMCI), amnestic mild cognitive impairment (aMCI), and Alzheimer's disease (AD). Reproduced with permission from Lowe et al. [172]. (*See insert for color representation of the figure.*)

a change in the observed signal can be correlated with response to treatment and possible influence on disease progression. An alternative surrogate biomarker to detection of Aβ levels is the use of [$^{18}$F]FDG-PET scans for determination of brain activity. The brains of Alzheimer's patients frequently show areas of severe hypometabolism. This phenotype worsens in parallel with disease progression. FDG-PET imaging can be used to distinguish Alzheimer's disease from other forms of dementia, as different types of dementias show specific regional patterns of metabolic reduction [170]. Several studies demonstrate good inverse correlation between Aβ imaging reagents and glucose metabolism [171].

In the study displayed in Figure 16.16, a small, but consistent decrease in signal of $^{18}$F-FDG can be observed, correlating inversely with increasingly impaired memory function. In this particular study, the correlating increase in PIB (Pittsburgh Compound-B) signal was even slightly more pronounced. The two imaging methods may very well be complimentary to each other.

Another highly valuable technique for assessing the therapeutic potential of CNS drugs is functional magnetic resonance imaging (fMRI) [173]. It allows production of brain activity maps displaying which regions of the brain are involved in a particular mental process. This is achieved by monitoring changes in blood flow and oxygenation, occurring in response to neural activity. Neuronal activity triggers an increase in blood flow to the surrounding capillary beds to ensure sufficient oxygen and glucose supply. This leads to an increase of oxyhemoglobin concentration, while concentration of deoxyhemoglobin, being more paramagnetic than oxyhemoglobin, decreases. Because oxyhemoglobin is less paramagnetic than deoxyhemoglobin, the MR signal intensities in the local brain environment change. The resulting fMRI signal changes are in the range of 0.5–5%.

In the past, fMRI studies in human subjects required comparison to a reference state; thus, only "task-based" imaging was possible. This limited its utility for translational research as for preclinical studies many animals require to be anesthetized or sedated during the experiment. Recent advances in the field led to an improved understanding of intrinsic brain networks [174] displaying the reactivity and interconnectivity of different brain regions. They are present in different states like rest and anesthesia, and many of these networks are broadly conserved over species. Analysis of these networks does not require performance of specific actions, thus opening up opportunities to interrogate them task independently. These networks are significantly altered in diseases of the central nervous system like Alzheimer's disease [175].

Molecular imaging has significantly improved our knowledge of the time course of events leading to Alzheimer's disease and allowed distinguishing AD from other forms of dementias. Amyloid imaging has shown high accumulation of Aβ already in a very early state of disease, which frequently precedes cognitive impairment and functional changes that can be measured via FDG-PET. Although to date FDG-PET may be the most effective method for the diagnosis of Alzheimer's disease, the combination with other imaging techniques, as well as newly evolving ones like fMRI [176, 177], may help to further improve the understanding of pathological changes in the course of disease, as well as to identify well-defined but early time points that would allow for disease-modifying intervention.

## 16.4 CONCLUSION AND OUTLOOK

As can be seen from the examples earlier, already today a rich arsenal of imaging tools is available. The pace of technical improvement for individual imaging techniques is stunning and will lead to drastically improved spatial resolution, shorter time intervals for measurement, and increased sensitivities. By combination of multiple different imaging modalities into one machine, the individual strengths of the selected methods can be efficiently combined and furthermore allow simultaneous observation of several different parameters in one living animal or patient. For probes, the development of efficient targeting mechanisms combined with new chelators, fluorophores, or carrier systems, carefully optimized to address specific imaging questions, led to significantly improved sensitivities and signal-to background ratios.

Molecular imaging is providing significant insights into mechanism of action and efficacy of exploratory drugs. But despite the tremendous advances made over the last years, many imaging approaches are still exploratory. In order to support decision making in the drug discovery process, thorough validation by properly designed animal studies in a variety of disease models and employment of well-characterized positive and negative controls is essential. Careful histological and immunohistochemical analyses need to be correlated with the obtained imaging data and pharmacological outcome to assess scope and limitations of the examined visualization method. As ultimately the goal of many of these efforts is to use the designed imaging reagents for clinical validation of new therapeutic concepts, thorough clinical validation of the method is required, which is very costly and time consuming. For this reason, many of the efforts are undertaken in public–private consortia, joining forces of public funding, academic centers, and industry partners. So far, many of these agents will only be useful as surrogate endpoints, which may help internal selection processes or support go/no-go decisions in preclinical or early clinical development. For demonstration of therapeutic efficacy though, large clinical trials will still be required to demonstrate positive impact on hard endpoints, which are accepted by regulatory agencies. Thus, cost of phase III studies, which account for the lion share of development costs of new drugs, will not be significantly reduced by the advent of new imaging methods any time soon. However, it is likely that the choice which compounds to actually progress into clinical trials will be made on a significantly improved state of knowledge, which can ultimately lead to improved success rates.

## REFERENCES

[1] Hay, M., Thomas, D.W., Craighead, J.L., Economides, C., Rosenthal, J. (2014) Clinical development success rates for investigational drugs. *Nat Biotechnol* **32**, 40–51.

[2] Wehling, M. (ed.) *Principles of Translational Science in Medicine – From Bench to Bedside*, (Cambridge University Press Cambridge, 2010).

[3] Weissleder, R., Ross, B.D., Rehemtulla, A., Gambhir, S.S. *Molecular Imaging*, (Pmph-USA Shelton, CT, 2010).

[4] Rudin, M. *Molecular Imaging: Basic Principles and Applications in Biomedical Research*, (Imperial College Press London, 2013).

[5] Braddock, M. *Biomedical Imaging: The Chemistry of Labels, Probes and Contrast Agents*, (RCS Publishing Cambridge, 2012).

[6] Ntziachristos, V. (2006) Fluorescence molecular imaging. *Annu Rev Biomed Eng* **8**, 1–33.

[7] Jelicks, L.A., Tanowitz, H.B., Albanese, C. (2013) Small animal imaging of human disease: from bench to bedside and back. *Am J Pathol* **182**, 294–295.

[8] O'Neill, K., Lyons, S.K., Gallagher, W.M., Curran, K.M. ,Byrne, A.T. (2010) Bioluminescent imaging: a critical tool in pre-clinical oncology research. *J Pathol* **220**, 317–327.

[9] Kocher, B., Piwnica-Worms, D. (2013) Illuminating cancer systems with genetically engineered mouse models and coupled luciferase reporters in vivo. *Cancer Discov* **3**, 616–629.

[10] Dothager, R.S., Flentie, K., Moss, B., Pan, M.H., Kesarwala, A., Piwnica-Worms, D. (2009) Advances in bioluminescence imaging of live animal models. *Curr Opin Biotechnol* **20**, 45–53.

[11] Fraga, H. (2008) Firefly luminescence: a historical perspective and recent developments. *Photochem Photobiol Sci* **7**, 146–158.

[12] Johnson, I., Spence, M.T.Z., *Molecular Probes Handbook: A Guide to Fluorescent Probes and Labeling Technologies* (Life Technologies, Carlsbad, CA, 2011).

[13] Goldman, R.D., Spector, D.L., Swedlow, J.R. *Live Cell Imaging: A Laboratory Manual* (Cold Spring Harbor Laboratory Press Cold Spring Harbor, NY, 2009).

[14] Lin, Y., Weissleder, R.,Tung, C.H. (2002) Novel near-infrared cyanine fluorochromes: synthesis, properties, and bioconjugation. *Bioconjug Chem* **13**, 605–610.

[15] Hughes, L.D., Rawle, R.J., Boxer, S.G. (2014) Choose your label wisely: water-soluble fluorophores often interact with lipid bilayers. *PLoS One* **9**, e87649.

[16] Zanetti-Domingues, L.C., Tynan, C.J., Rolfe, D.J., Clarke, D.T., Martin-Fernandez, M. (2013) Hydrophobic fluorescent probes introduce artifacts into single molecule tracking experiments due to non-specific binding. *PLoS One* **8**, e74200.

[17] Shao, F., Yuan, H., Josephson, L., Weissleder, R., Hilderbrand, S.A. (2011) Facile synthesis of monofunctional pentamethine carbocyanine fluorophores. *Dyes Pigments* **90**, 119–122.

[18] Loudet, A., Burgess, K. (2007) BODIPY dyes and their derivatives: syntheses and spectroscopic properties. *Chem Rev* **107**, 4891–4932.

[19] Zhao, W., Carreira, E.M. (2006) Conformationally restricted aza-BODIPY: highly fluorescent, stable near-infrared absorbing dyes. *Chemistry* **12**, 7254–7263.

[20] Ptaszek, M. (2013) Rational design of fluorophores for in vivo applications. *Prog Mol Biol Transl Sci* **113**, 59–108.

[21] Frangioni, J.V. (2003) In vivo near-infrared fluorescence imaging. *Curr Opin Chem Biol* **7**, 626–634.

[22] Escobedo, J.O., Rusin, O., Lim, S., Strongin, R.M. (2010) NIR dyes for bioimaging applications. *Curr Opin Chem Biol* **14**, 64–70.

[23] Hilderbrand, S.A. (2010) Labels and probes for live cell imaging: overview and selection guide. *Methods Mol Biol* **591**, 17–45.

[24] Hilderbrand, S.A., Weissleder, R. (2010) Near-infrared fluorescence: application to in vivo molecular imaging. *Curr Opin Chem Biol* **14**, 71–79.

[25] Kobayashi, H., Ogawa, M., Alford, R., Choyke, P.L., Urano, Y. (2010) New strategies for fluorescent probe design in medical diagnostic imaging. *Chem Rev* **110**, 2620–2640.

[26] Guo, Z., Park, S., Yoon, J., Shin, I. (2014) Recent progress in the development of near-infrared fluorescent probes for bioimaging applications. *Chem Soc Rev* **43**, 16–29.

[27] Yuan, L., Lin, W., Zheng, K., He, L., Huang, W. (2013) Far-red to near infrared analyte-responsive fluorescent probes based on organic fluorophore platforms for fluorescence imaging. *Chem Soc Rev* **42**, 622–661.

[28] Weissleder, R., Tung, C.H., Mahmood, U., Bogdanov, A., Jr. (1999) In vivo imaging of tumors with protease-activated near-infrared fluorescent probes. *Nat Biotechnol* **17**, 375–378.

[29] Denk, W., Strickler, J.H., Webb, W.W. (1990) Two-photon laser scanning fluorescence microscopy. *Science* **248**, 73–76.

[30] Niesner, R., Andresen, V., Neumann, J., Spiecker, H., Gunzer, M. (2007) The power of single and multibeam two-photon microscopy for high-resolution and high-speed deep tissue and intravital imaging. *Biophys J* **93**, 2519–2529.

[31] Rubart, M. (2004) Two-photon microscopy of cells and tissue. *Circ Res* **95**, 1154–1166.

[32] Bort, G., Gallavardin, T., Ogden, D., Dalko, P.I. (2013) From one-photon to two-photon probes: "caged" compounds, actuators, and photoswitches. *Angew Chem Int Ed Engl* **52**, 4526–4537.

[33] Helmchen, F., Denk, W. (2005) Deep tissue two-photon microscopy. *Nat Methods* **2**, 932–940.

[34] Purcell, E.M. (1952) "E. M. Purcell – Nobel Lecture: Research in Nuclear Magnetism". Nobelprize.org. Nobel Media AB 2014. Web. August 14, 2014. http://www.nobelprize.org/nobel_prizes/physics/laureates/1952/purcell-lecture.html (accessed on May 12, 2015).

[35] Bloch, F. (1952) "Felix Bloch – Nobel Lecture: The Principle of Nuclear Induction". Nobelprize.org. Nobel Media AB 2014. Web. August 14, 2014. http://www.nobelprize.org/nobel_prizes/physics/laureates/1952/bloch-lecture.html (accessed on May 12, 2015).

[36] Lauterbur, P.C. (2003) "Paul C. Lauterbur – Nobel Lecture: All Science is Interdisciplinary – from Magnetic Moments to Molecules to Men". Nobelprize.org. Nobel Media AB 2014. Web. August 14, 2014. http://www.nobelprize.org/nobel_prizes/medicine/laureates/2003/lauterbur-lecture.html (accessed on May 12, 2015).

[37] Mansfield, P. (2003) "Sir Peter Mansfield – Nobel Lecture: Snap-Shot MRI". Nobelprize.org. Nobel Media AB 2014. Web. August 14, 2014. http://www.nobelprize.org/nobel_prizes/medicine/laureates/2003/mansfield-lecture.html (accessed on May 12, 2015).

[38] Caravan, P. (2009) Protein-targeted gadolinium-based magnetic resonance imaging (MRI) contrast agents: design and mechanism of action. *Acc Chem Res* **42**, 851–862.

[39] Aime, S., Castelli, D.D., Crich, S.G., Gianolio, E., Terreno, E. (2009) Pushing the sensitivity envelope of lanthanide-based magnetic resonance imaging (MRI) contrast agents for molecular imaging applications. *Acc Chem Res* **42**, 822–831.

[40] Goldman, L.W. (2007) Principles of CT and CT technology. *J Nucl Med Technol* **35**, 115–128; quiz 129–130.

[41] Ramogida, C.F., Orvig, C. (2013) Tumour targeting with radiometals for diagnosis and therapy. *Chem Commun (Camb)* **49**, 4720–4739.

[42] Olberg, D.E., Hjelstuen, O.K. (2010) Labeling strategies of peptides with (1)(8)F for positron emission tomography. *Curr Top Med Chem* **10**, 1669–1679.

[43] Herzog, H. (2007) Methods and applications of positron-based medical imaging. *Radiat Phys Chem* **76**, 337–342.

[44] Lundqvist, H., Tolmachev, V. (2002) Targeting peptides and positron emission tomography. *Biopolymers* **66**, 381–392.

[45] Penfield, J.G., Reilly, R.F., Jr. (2007) What nephrologists need to know about gadolinium. *Nat Clin Pract Nephrol* **3**, 654–668.

[46] Wadas, T.J., Wong, E.H., Weisman, G.R., Anderson, C.J. (2010) Coordinating radiometals of copper, gallium, indium, yttrium, and zirconium for PET and SPECT imaging of disease. *Chem Rev* **110**, 2858–2902.

[47] Cai, Z., Anderson, C.J. (2014) Chelators for copper radionuclides in positron emission tomography radiopharmaceuticals. *J Label Compd Radiopharm* **57**, 224–230.

[48] Funk, T., Sun, M., Hasegawa, B.H. (2004) Radiation dose estimate in small animal SPECT and PET. *Med Phys* **31**, 2680–2686.

[49] Kung, M.P., Kung, H.F. (2005) Mass effect of injected dose in small rodent imaging by SPECT and PET. *Nucl Med Biol* **32**, 673–678.

[50] Chakraborty, S., Liu, S. (2010) (99m)Tc and (111)In-labeling of small biomolecules: bifunctional chelators and related coordination chemistry. *Curr Top Med Chem* **10**, 1113–1134.

[51] Heffern, M.C., Matosziuk, L.M., Meade, T.J. (2014) Lanthanide probes for bioresponsive imaging. *Chem Rev* **114**, 4496–4539.

[52] Koba, W., Jelicks, L.A., Fine, E.J. (2013) MicroPET/SPECT/CT imaging of small animal models of disease. *Am J Pathol* **182**, 319–324.

[53] Lopci, E., Nanni, C., Castellucci, P., Montini, G.C., Allegri, V., Rubello, D., Chierichetti, F., Ambrosini, V., Fanti, S. (2010) Imaging with non-FDG PET tracers: outlook for current clinical applications. *Insights Imaging* **1**, 373–385.

[54] Pimlott, S.L., Sutherland, A. (2011) Molecular tracers for the PET and SPECT imaging of disease. *Chem Soc Rev* **40**, 149–162.

[55] Mach, R.H., Schwarz, S.W. (2010) Challenges for developing PET tracers: isotopes, chemistry and regulatory aspects. *PET Clin* **5**, 131–153.

[56] Hutchins, G.D., Miller, M.A., Soon, V.C., Receveur, T. (2008) Small animal PET imaging. *ILAR J* **49**, 54–65.

[57] Wang, J., Sanchez-Rosello, M., Acena, J.L., del Pozo, C., Sorochinsky, A.E., Fustero, S., Soloshonok, V.A., Liu, H. (2014) Fluorine in pharmaceutical industry: fluorine-containing drugs introduced to the market in the last decade (2001–2011). *Chem Rev* **114**, 2432–2506.

[58] Tredwell, M., Gouverneur, V. (2012) 18F labeling of arenes. *Angew Chem Int Ed Engl* **51**, 11426–11437.

[59] Lee, E., Kamlet, A.S., Powers, D.C., Neumann, C.N., Boursalian, G.B., Furuya, T., Choi, D.C., Hooker, J.M., Ritter, T. (2011) A fluoride-derived electrophilic late-stage fluorination reagent for PET imaging. *Science* **334**, 639–642.

[60] Lee, E., Hooker, J.M., Ritter, T. (2012) Nickel-mediated oxidative fluorination for PET with aqueous [18F] fluoride. *J Am Chem Soc* **134**, 17456–17458.

[61] Kamlet, A.S., Neumann, C.N., Lee, E., Carlin, S.M., Moseley, C.K., Stephenson, N., Hooker, J.M., Ritter, T. (2013) Application of palladium-mediated (18)F-fluorination to PET radiotracer development: overcoming hurdles to translation. *PLoS One* **8**, e59187.

[62] Tredwell, M., Preshlock, S.M., Taylor, N.J., Gruber, S., Huiban, M., Passchier, J., Mercier, J., Genicot, C., Gouverneur, V. (2014) A general copper-mediated nucleophilic F Fluorination of arenes. *Angew Chem Int Ed Engl* **126**, 7885–7889.

[63] Kobayashi, H., Longmire, M.R., Ogawa, M., Choyke, P.L. (2011) Rational chemical design of the next generation of molecular imaging probes based on physics and biology: mixing modalities, colors and signals. *Chem Soc Rev* **40**, 4626–4648.

[64] Chapman, S.E., Diener, J.M., Sasser, T.A., Correcher, C., Gonzalez, A.J., Avermaete, T.V., Leevy, W.M. (2012) Dual tracer imaging of SPECT and PET probes in living mice using a sequential protocol. *Am J Nucl Med Mol Imaging* **2**, 405–414.

[65] American Cancer Society. *Cancer Facts & Figures 2013*, (American Cancer Society Atlanta, GA, 2013). http://www.cancer.org/acs/groups/content/@epidemiologysurveilance/documents/document/acspc-036845.pdf (accessed on May 28, 2015).

[66] Eisenhauer, E.A., Therasse, P., Bogaerts, J., Schwartz, L.H., Sargent, D., Ford, R., Dancey, J., Arbuck, S., Gwyther, S., Mooney, M., Rubinstein, L., Shankar, L., Dodd, L., Kaplan, R., Lacombe, D., Verweij, J. (2009) New response evaluation criteria in solid tumours: revised RECIST guideline (version 1.1). *Eur J Cancer* **45**, 228–247.

[67] Buck, A.K., Reske, S.N. (2004) Cellular origin and molecular mechanisms of 18F-FDG uptake: is there a contribution of the endothelium? *J Nucl Med* **45**, 461–463.

[68] Belhocine, T., Spaepen, K., Dusart, M., Castaigne, C., Muylle, K., Bourgeois, P., Bourgeois, D., Dierickx, L., Flamen, P. (2006) 18FDG PET in oncology: the best and the worst (Review). *Int J Oncol* **28**, 1249–1261.

[69] Takahashi, R., Hirata, H., Tachibana, I., Shimosegawa, E., Inoue, A., Nagatomo, I., Takeda, Y., Kida, H., Goya, S., Kijima, T., Yoshida, M., Kumagai, T., Kumanogoh, A., Okumura, M., Hatazawa, J., Kawase, I. (2012) Early [18F]fluorodeoxyglucose positron emission tomography at two days of gefitinib treatment predicts clinical outcome in patients with adenocarcinoma of the lung. *Clin Cancer Res* **18**, 220–228.

[70] Zander, T., Scheffler, M., Nogova, L., Kobe, C., Engel-Riedel, W., Hellmich, M., Papachristou, I., Toepelt, K., Draube, A., Heukamp, L., Buettner, R., Ko, Y.D., Ullrich, R.T., Smit, E., Boellaard, R., Lammertsma, A.A., Hallek, M., Jacobs, A.H., Schlesinger, A., Schulte, K., Querings, S., Stoelben, E., Neumaier, B., Thomas, R.K., Dietlein, M., Wolf, J. (2011) Early prediction of nonprogression in advanced non-small-cell lung cancer treated with erlotinib by using [(18)F]fluorodeoxyglucose and [(18)F]fluorothymidine positron emission tomography. *J Clin Oncol* **29**, 1701–1708.

[71] Maynard, J., Ricketts, S.A., Gendrin, C., Dudley, P., Davies, B.R. (2013) 2-Deoxy-2-[18F]fluoro-D-glucose positron emission tomography demonstrates target inhibition with the potential to predict anti-tumour activity following treatment with the AKT inhibitor AZD5363. *Mol Imaging Biol* **15**, 476–485.

[72] Keen, H., Pichler, B., Kukuk, D., Duchamp, O., Raguin, O., Shannon, A., Whalley, N., Jacobs, V., Bales, J., Gingles, N., Ricketts, S.A., Wedge, S.R. (2012) An evaluation of 2-deoxy-2-[18F]fluoro-D-glucose and 3′-deoxy-3′-[18F]-fluorothymidine uptake in human tumor xenograft models. *Mol Imaging Biol* **14**, 355–365.

[73] Shuptrine, C.W., Surana, R., Weiner, L.M. (2012) Monoclonal antibodies for the treatment of cancer. *Semin Cancer Biol* **22**, 3–13.

[74] Duebel, S. *Handbook of Therapeutic Antibodies*, (Wiley-Blackwell, Hoboken, NJ, 2010).

[75] Carter, P.J., Senter, P.D. (2008) Antibody-drug conjugates for cancer therapy. *Cancer J* **14**, 154–169.

[76] Senter, P.D. (2009) Potent antibody drug conjugates for cancer therapy. *Curr Opin Chem Biol* **13**, 235–244.

[77] Sievers, E.L., Senter, P.D. (2013) Antibody-drug conjugates in cancer therapy. *Annu Rev Med* **64**, 15–29.

[78] Panowksi, S., Bhakta, S., Raab, H., Polakis, P., Junutula, J.R. (2014) Site-specific antibody drug conjugates for cancer therapy. *MAbs* **6**, 34–45.

[79] Chari, R.V., Miller, M.L., Widdison, W.C. (2014) Antibody-drug conjugates: an emerging concept in cancer therapy. *Angew Chem Int Ed Engl* **53**, 3796–3827.

[80] Boohaker, R.J., Lee, M.W., Vishnubhotla, P., Perez, J.M., Khaled, A.R. (2012) The use of therapeutic peptides to target and to kill cancer cells. *Curr Med Chem* **19**, 3794–3804.

[81] Zhang, Y., Hong, H., Cai, W. (2011) Tumor-targeted drug delivery with aptamers. *Curr Med Chem* **18**, 4185–4194.

[82] Gautam, A., Kapoor, P., Chaudhary, K., Kumar, R., Raghava, G.P. (2014) Tumor homing peptides as molecular probes for cancer therapeutics, diagnostics, and theranostics. *Curr Med Chem* **21**, 2367–2391.

[83] Kapoor, P., Singh, H., Gautam, A., Chaudhary, K., Kumar, R., Raghava, G.P. (2012) TumorHoPe: a database of tumor homing peptides. *PLoS One* **7**, e35187.

[84] Maeda, H. (2001) The enhanced permeability and retention (EPR) effect in tumor vasculature: the key role of tumor-selective macromolecular drug targeting. *Adv Enzyme Regul* **41**, 189–207.

[85] Ogawara, K., Yoshizawa, Y., Un, K., Araki, T., Kimura, T., Higaki, K. (2013) Nanoparticle-based passive drug targeting to tumors: considerations and implications for optimization. *Biol Pharm Bull* **36**, 698–702.

[86] Holland, J.P., Sheh, Y., Lewis, J.S. (2009) Standardized methods for the production of high specific-activity zirconium-89. *Nucl Med Biol* **36**, 729–739.

[87] Verel, I., Visser, G.W., Boellaard, R., Stigter-van Walsum, M., Snow, G.B., van Dongen, G.A. (2003) 89Zr immuno-PET: comprehensive procedures for the production of 89Zr-labeled monoclonal antibodies. *J Nucl Med* **44**, 1271–1281.

[88] Vosjan, M.J., Perk, L.R., Visser, G.W., Budde, M., Jurek, P., Kiefer, G.E., van Dongen, G.A. (2010) Conjugation and radiolabeling of monoclonal antibodies with zirconium-89 for PET imaging using the bifunctional chelate p-isothiocyanatobenzyl-desferrioxamine. *Nat Protoc* **5**, 739–743.

[89] Williams, S.P. (2012) Tissue distribution studies of protein therapeutics using molecular probes: molecular imaging. *AAPS J* **14**, 389–399.

[90] Banerjee, S.R., Pullambhatla, M., Byun, Y., Nimmagadda, S., Foss, C.A., Green, G., Fox, J.J., Lupold, S.E., Mease, R.C., Pomper, M.G. (2011) Sequential SPECT and optical imaging of experimental models of prostate cancer with a dual modality inhibitor of the prostate-specific membrane antigen. *Angew Chem Int Ed Engl* **50**, 9167–9170.

[91] Xing, Y., Zhao, J., Conti, P.S., Chen, K. (2014) Radiolabeled nanoparticles for multimodality tumor imaging. *Theranostics* **4**, 290–306.

[92] Astruc, D., Boisselier, E., Ornelas, C. (2010) Dendrimers designed for functions: from physical, photophysical, and supramolecular properties to applications in sensing, catalysis, molecular electronics, photonics, and nanomedicine. *Chem Rev* **110**, 1857–1959.

[93] Zhu, C., Liu, L., Yang, Q., Lv, F., Wang, S. (2012) Water-soluble conjugated polymers for imaging, diagnosis, and therapy. *Chem Rev* **112**, 4687–4735.

[94] Cassette, E., Helle, M., Bezdetnaya, L., Marchal, F., Dubertret, B., Pons, T. (2013) Design of new quantum dot materials for deep tissue infrared imaging. *Adv Drug Deliv Rev* **65**, 719–731.

[95] Reddy, L.H., Arias, J.L., Nicolas, J., Couvreur, P. (2012) Magnetic nanoparticles: design and characterization, toxicity and biocompatibility, pharmaceutical and biomedical applications. *Chem Rev* **112**, 5818–5878.

[96] Gibbs, S.L. (2012) Near infrared fluorescence for image-guided surgery. *Quant Imaging Med Surg* **2**, 177–187.

[97] Korb, M.L., Hartman, Y.E., Kovar, J., Zinn, K.R., Bland, K.I., Rosenthal, E.L. (2014) Use of monoclonal antibody-IRDye800CW bioconjugates in the resection of breast cancer. *J Surg Res* **188**, 119–128.

[98] Hiroshima, Y., Maawy, A., Metildi, C.A., Zhang, Y., Uehara, F., Miwa, S., Yano, S., Sato, S., Murakami, T., Momiyama, M., Chishima, T., Tanaka, K., Bouvet, M., Endo, I., Hoffman, R.M. (2014) Successful fluorescence-guided surgery on human colon cancer patient-derived orthotopic xenograft mouse models using a fluorophore-conjugated anti-CEA antibody and a portable imaging system. *J Laparoendosc Adv Surg Tech A* **24**, 241–247.

[99] Metildi, C.A., Kaushal, S., Pu, M., Messer, K.A., Luiken, G.A., Moossa, A.R., Hoffman, R.M., Bouvet, M. (2014) Fluorescence-guided Surgery with a fluorophore-conjugated antibody to carcinoembryonic antigen (CEA), that highlights the tumor, improves surgical resection and increases survival in orthotopic mouse models of human pancreatic cancer. *Ann Surg Oncol* **21**, 1405–1411.

[100] Kitagawa, D., Yokota, K., Gouda, M., Narumi, Y., Ohmoto, H., Nishiwaki, E., Akita, K., Kirii, Y. (2013) Activity-based kinase profiling of approved tyrosine kinase inhibitors. *Genes Cells* **18**, 110–122.

[101] Uitdehaag, J.C., de Roos, J.A., van Doornmalen, A.M., Prinsen, M.B., de Man, J., Tanizawa, Y., Kawase, Y., Yoshino, K., Buijsman, R.C., Zaman, G.J. (2014) Comparison of the cancer gene targeting and biochemical selectivities of all targeted kinase inhibitors approved for clinical use. *PLoS One* **9**, e92146.

[102] Davis, M.I., Hunt, J.P., Herrgard, S., Ciceri, P., Wodicka, L.M., Pallares, G., Hocker, M., Treiber, D.K., Zarrinkar, P.P. (2011) Comprehensive analysis of kinase inhibitor selectivity. *Nat Biotechnol* **29**, 1046–1051.

[103] Sawyers, C.L. (2001) Research on resistance to cancer drug Gleevec. *Science* **294**, 1834.

[104] Daley, G.Q. (2003) Gleevec resistance: lessons for target-directed drug development. *Cell Cycle* **2**, 190–191.

[105] Slobbe, P., Poot, A.J., Windhorst, A.D., van Dongen, G.A. (2012) PET imaging with small-molecule tyrosine kinase inhibitors: TKI-PET. *Drug Discov Today* **17**, 1175–1187.

[106] Tolmachev, V., Stone-Elander, S., Orlova, A. (2010) Radiolabelled receptor-tyrosine-kinase targeting drugs for patient stratification and monitoring of therapy response: prospects and pitfalls. *Lancet Oncol* **11**, 992–1000.

[107] Poot, A.J., Slobbe, P., Hendrikse, N.H., Windhorst, A.D., van Dongen, G.A. (2013) Imaging of TKI-target interactions for personalized cancer therapy. *Clin Pharmacol Ther* **93**, 239–241.

[108] Altai, M., Orlova, A., Tolmachev, V. (2013) Radiolabeled probes targeting tyrosine-kinase receptors for personalized medicine. *Curr Pharm Des* **20**, 2275–2292.

[109] Memon, A.A., Jakobsen, S., Dagnaes-Hansen, F., Sorensen, B.S., Keiding, S., Nexo, E. (2009) Positron emission tomography (PET) imaging with [11C]-labeled erlotinib: a micro-PET study on mice with lung tumor xenografts. *Cancer Res* **69**, 873–878.

[110] Memon, A.A., Weber, B., Winterdahl, M., Jakobsen, S., Meldgaard, P., Madsen, H.H., Keiding, S., Nexo, E., Sorensen, B.S. (2011) PET imaging of patients with non-small cell lung cancer employing an EGF receptor targeting drug as tracer. *Br J Cancer* **105**, 1850–1855.

[111] Petrulli, J.R., Sullivan, J.M., Zheng, M.Q., Bennett, D.C., Charest, J., Huang, Y., Morris, E.D., Contessa, J.N. (2013) Quantitative analysis of [11C]-erlotinib PET demonstrates specific binding for activating mutations of the EGFR kinase domain. *Neoplasia* **15**, 1347–1353.

[112] Bahce, I., Smit, E.F., Lubberink, M., van der Veldt, A.A., Yaqub, M., Windhorst, A.D., Schuit, R.C., Thunnissen, E., Heideman, D.A., Postmus, P.E., Lammertsma, A.A., Hendrikse, N.H. (2013) Development of [(11)C]erlotinib positron emission tomography for in vivo evaluation of EGF receptor mutational status. *Clin Cancer Res* **19**, 183–193.

[113] Nodwell, M.B., Sieber, S.A. (2012) ABPP methodology: introduction and overview. *Top Curr Chem* **324**, 1–41.

[114] Razgulin, A., Ma, N., Rao, J. (2011) Strategies for in vivo imaging of enzyme activity: an overview and recent advances. *Chem Soc Rev* **40**, 4186–4216.

[115] Pazos, E., Vazquez, M.E. (2014) Advances in lanthanide-based luminescent peptide probes for monitoring the activity of kinase and phosphatase. *Biotechnol J* **9**, 241–252.

[116] Turk, B. (2006) Targeting proteases: successes, failures and future prospects. *Nat Rev Drug Discov* **5**, 785–799.

[117] Lopez-Otin, C., Bond, J.S. (2008) Proteases: multifunctional enzymes in life and disease. *J Biol Chem* **283**, 30433–30437.

[118] Ossovskaya, V.S., Bunnett, N.W. (2004) Protease-activated receptors: contribution to physiology and disease. *Physiol Rev* **84**, 579–621.

[119] Marras, S.A. (2006) Selection of fluorophore and quencher pairs for fluorescent nucleic acid hybridization probes. *Methods Mol Biol* **335**, 3–16.

[120] Edgington, L.E., Bogyo, M. (2013) In vivo imaging and biochemical characterization of protease function using fluorescent activity-based probes. *Curr Protoc Chem Biol* **5**, 25–44.

[121] Edgington, L.E., Verdoes, M., Bogyo, M. (2011) Functional imaging of proteases: recent advances in the design and application of substrate-based and activity-based probes. *Curr Opin Chem Biol* **15**, 798–805.

[122] Edgington, L.E., Verdoes, M., Ortega, A., Withana, N.P., Lee, J., Syed, S., Bachmann, M.H., Blum, G., Bogyo, M. (2013) Functional imaging of legumain in cancer using a new quenched activity-based probe. *J Am Chem Soc* **135**, 174–182.

[123] Edgington, L.E., van Raam, B.J., Verdoes, M., Wierschem, C., Salvesen, G.S., Bogyo, M. (2012) An optimized activity-based probe for the study of caspase-6 activation. *Chem Biol* **19**, 340–352.

[124] Edgington, L.E., Berger, A.B., Blum, G., Albrow, V.E., Paulick, M.G., Lineberry, N., Bogyo, M. (2009) Noninvasive optical imaging of apoptosis by caspase-targeted activity-based probes. *Nat Med* **15**, 967–973.

[125] Verdoes, M., Oresic Bender, K., Segal, E., van der Linden, W.A., Syed, S., Withana, N.P., Sanman, L.E., Bogyo, M. (2013) Improved quenched fluorescent probe for imaging of cysteine cathepsin activity. *J Am Chem Soc* **135**, 14726–14730.

[126] Yhee, J.Y., Kim, S.A., Koo, H., Son, S., Ryu, J.H., Youn, I.C., Choi, K., Kwon, I.C., Kim, K. (2012) Optical imaging of cancer-related proteases using near-infrared fluorescence matrix metalloproteinase-sensitive and cathepsin B-sensitive probes. *Theranostics* **2**, 179–189.

[127] Caglic, D., Globisch, A., Kindermann, M., Lim, N.H., Jeske, V., Juretschke, H.P., Bartnik, E., Weithmann, K.U., Nagase, H., Turk, B., Wendt, K.U. (2011) Functional in vivo imaging of cysteine cathepsin activity in murine model of inflammation. *Bioorg Med Chem* **19**, 1055–1061.

[128] Watzke, A., Kosec, G., Kindermann, M., Jeske, V., Nestler, H.P., Turk, V., Turk, B., Wendt, K.U. (2008) Selective activity-based probes for cysteine cathepsins. *Angew Chem Int Ed Engl* **47**, 406–409.

[129] Kindermann, M., Roschitzki-Voser, H., Caglic, D., Repnik, U., Miniejew, C., Mittl, P.R., Kosec, G., Grutter, M.G., Turk, B., Wendt, K.U. (2010) Selective and sensitive monitoring of caspase-1 activity by a novel bioluminescent activity-based probe. *Chem Biol* **17**, 999–1007.

[130] Hu, H.Y., Vats, D., Vizovisek, M., Kramer, L., Germanier, C., Wendt, K.U., Rudin, M., Turk, B., Plettenburg, O., Schultz, C. (2014) In vivo imaging of mouse tumors by a lipidated cathepsin s substrate. *Angew Chem Int Ed Engl* **53**, 7669–7673.

[131] Olson, E.S., Jiang, T., Aguilera, T.A., Nguyen, Q.T., Ellies, L.G., Scadeng, M., Tsien, R.Y. (2010) Activatable cell penetrating peptides linked to nanoparticles as dual probes for in vivo fluorescence and MR imaging of proteases. *Proc Natl Acad Sci U S A* **107**, 4311–4316.

[132] Nguyen, Q.T., Olson, E.S., Aguilera, T.A., Jiang, T., Scadeng, M., Ellies, L.G., Tsien, R.Y. (2010) Surgery with molecular fluorescence imaging using activatable cell-penetrating peptides decreases residual cancer and improves survival. *Proc Natl Acad Sci U S A* **107**, 4317–4322.

[133] Whitney, M.A., Crisp, J.L., Nguyen, L.T., Friedman, B., Gross, L.A., Steinbach, P., Tsien, R.Y., Nguyen, Q.T. (2011) Fluorescent peptides highlight peripheral nerves during surgery in mice. *Nat Biotechnol* **29**, 352–356.

[134] Nguyen, Q.T., Tsien, R.Y. (2013) Fluorescence-guided surgery with live molecular navigation–a new cutting edge. *Nat Rev Cancer* **13**, 653–662.

[135] Savariar, E.N., Felsen, C.N., Nashi, N., Jiang, T., Ellies, L.G., Steinbach, P., Tsien, R.Y., Nguyen, Q.T. (2013) Real-time in vivo molecular detection of primary tumors and metastases with ratiometric activatable cell-penetrating peptides. *Cancer Res* **73**, 855–864.

[136] Gotthardt, M. (2010) Beta cell imaging–why we need it and what has been achieved. *Curr Pharm Des* **16**, 1545–1546.

[137] Blomberg, B.A., Codreanu, I., Cheng, G., Werner, T.J., Alavi, A. (2013) Beta-cell imaging: call for evidence-based and scientific approach. *Mol Imaging Biol* **15**, 123–130.

[138] Wang, P., Medarova, Z., Moore, A. (2011) Molecular imaging: a promising tool to monitor islet transplantation. *J Transplant* **2011**, 202915.

[139] Borot, S., Crowe, L.A., Toso, C., Vallee, J.P., Berney, T. (2011) Noninvasive imaging techniques in islet transplantation. *Curr Diab Rep* **11**, 375–383.

[140] Hwang, J.H., Noh, Y.W., Choi, J.H., Noh, J.R., Kim, Y.H., Gang, G.T., Kim, K.S., Park, H.S., Lim, Y.T., Moon, H., Hong, K.S., Lee, H.G., Chung, B.H., Lee, C.H. (2013) In vivo imaging of islet transplantation using PLGA nanoparticles containing iron oxide and indocyanine green. *Magn Reson Med* **71**, 1054–1063.

[141] Barton, F.B., Rickels, M.R., Alejandro, R., Hering, B.J., Wease, S., Naziruddin, B., Oberholzer, J., Odorico, J.S., Garfinkel, M.R., Levy, M., Pattou, F., Berney, T., Secchi, A., Messinger, S., Senior, P.A., Maffi, P., Posselt, A., Stock, P.G., Kaufman, D.B., Luo, X., Kandeel, F., Cagliero, E., Turgeon, N.A., Witkowski, P., Naji, A., O'Connell, P.J., Greenbaum, C., Kudva, Y.C., Brayman, K.L., Aull, M.J., Larsen, C., Kay, T.W., Fernandez, L.A., Vantyghem, M.C., Bellin, M., Shapiro, A.M. (2012) Improvement in outcomes of clinical islet transplantation: 1999–2010. *Diabetes Care* **35**, 1436–1445.

[142] Low, G., Hussein, N., Owen, R.J., Toso, C., Patel, V.H., Bhargava, R., Shapiro, A.M. (2010) Role of imaging in clinical islet transplantation. *Radiographics* **30**, 353–366.

[143] Lamprianou, S., Immonen, R., Nabuurs, C., Gjinovci, A., Vinet, L., Montet, X.C., Gruetter, R., Meda, P. (2011) High-resolution magnetic resonance imaging quantitatively detects individual pancreatic islets. *Diabetes* **60**, 2853–2860.

[144] Botsikas, D., Terraz, S., Vinet, L., Lamprianou, S., Becker, C.D., Bosco, D., Meda, P., Montet, X. (2012) Pancreatic magnetic resonance imaging after manganese injection distinguishes type 2 diabetic and normoglycemic patients. *Islets* **4**, 243–248.

[145] Schneider, S., Feilen, P.J., Schreckenberger, M., Schwanstecher, M., Schwanstecher, C., Buchholz, H.G., Thews, O., Oberholzer, K., Korobeynikov, A., Bauman, A., Comagic, S., Piel, M., Schirrmacher, E., Shiue, C.Y., Alavi, A.A., Bartenstein, P., Rosch, F., Weber, M.M., Klein, H.H., Schirrmacher, R. (2005) In vitro and in vivo evaluation of novel glibenclamide derivatives as imaging agents for the non-invasive assessment of the pancreatic islet cell mass in animals and humans. *Exp Clin Endocrinol Diabetes* **113**, 388–395.

[146] Schneider, S., Ueberberg, S., Korobeynikov, A., Schechinger, W., Schwanstecher, C., Schwanstecher, M., Klein, H.H., Schirrmacher, E. (2007) Synthesis and evaluation of a glibenclamide glucose-conjugate: a potential new lead compound for substituted glibenclamide derivatives as islet imaging agents. *Regul Pept* **139**, 122–127.

[147] Oh, C.S., Kohanim, S., Kong, F.L., Song, H.C., Huynh, N., Mendez, R., Chanda, M., Edmund Kim, E., Yang, D.J. (2012) Sulfonylurea receptor as a target for molecular imaging of pancreas beta cells with (99m)Tc-DTPA-glipizide. *Ann Nucl Med* **26**, 253–261.

[148] Singhal, T., Ding, Y.S., Weinzimmer, D., Normandin, M.D., Labaree, D., Ropchan, J., Nabulsi, N., Lin, S.F., Skaddan, M.B., Soeller, W.C., Huang, Y., Carson, R.E., Treadway, J.L., Cline, G.W. (2011) Pancreatic beta cell mass PET imaging and quantification with [11C]DTBZ and [18F]FP-(+)-DTBZ in rodent models of diabetes. *Mol Imaging Biol* **13**, 973–984.

[149] Normandin, M.D., Petersen, K.F., Ding, Y.S., Lin, S.F., Naik, S., Fowles, K., Skovronsky, D.M., Herold, K.C., McCarthy, T.J., Calle, R.A., Carson, R.E., Treadway, J.L., Cline, G.W. (2012) In vivo imaging of endogenous pancreatic beta-cell mass in healthy and type 1 diabetic subjects using 18F-fluoropropyl-dihydrotetrabenazine and PET. *J Nucl Med* **53**, 908–916.

[150] Blomberg, B.A., Eriksson, O., Saboury, B., Alavi, A. (2013) Beta-cell mass imaging with DTBZ positron emission tomography: is it possible? *Mol Imaging Biol* **15**, 1–2.

[151] Vats, D., Wang, H., Esterhazy, D., Dikaiou, K., Danzer, C., Honer, M., Stuker, F., Matile, H., Migliorini, C., Fischer, E., Ripoll, J., Keist, R., Krek, W., Schibli, R., Stoffel, M., Rudin, M. (2012) Multimodal imaging of pancreatic beta cells in vivo by targeting transmembrane protein 27 (TMEM27). *Diabetologia* **55**, 2407–2416.

[152] Spellman, C.W. (2011) Pharmacology of GLP-1 agonists: describing the therapeutic potential to patients. *J Am Osteopath Assoc* **111**, eS10–14.

[153] Lorenz, M., Evers, A., Wagner, M. (2013) Recent progress and future options in the development of GLP-1 receptor agonists for the treatment of diabesity. *Bioorg Med Chem Lett* **23**, 4011–4018.

[154] Jones, L.H., Price, D.A. (2013) Medicinal chemistry of glucagon-like peptide receptor agonists. *Prog Med Chem* **52**, 45–96.

[155] Reiner, T., Thurber, G., Gaglia, J., Vinegoni, C., Liew, C.W., Upadhyay, R., Kohler, R.H., Li, L., Kulkarni, R.N., Benoist, C., Mathis, D., Weissleder, R. (2011) Accurate measurement of pancreatic islet beta-cell mass using a second-generation fluorescent exendin-4 analog. *Proc Natl Acad Sci U S A* **108**, 12815–12820.

[156] Reiner, T., Kohler, R.H., Liew, C.W., Hill, J.A., Gaglia, J., Kulkarni, R.N., Weissleder, R. (2010) Near-infrared fluorescent probe for imaging of pancreatic beta cells. *Bioconjug Chem* **21**, 1362–1368.

[157] Clardy, S.M., Keliher, E.J., Mohan, J.F., Sebas, M., Benoist, C., Mathis, D., Weissleder, R. (2013) Fluorescent exendin-4 derivatives for pancreatic beta-cell analysis. *Bioconjug Chem* **21**, 171–177.

[158] Keliher, E.J., Reiner, T., Thurber, G.M., Upadhyay, R., Weissleder, R. (2012) Efficient F-Labeling of synthetic exendin-4 analogues for imaging beta cells. *ChemistryOpen* **1**, 177–183.

[159] Brom, M., Andralojc, K., Oyen, W.J., Boerman, O.C., Gotthardt, M. (2010) Development of radiotracers for the determination of the beta-cell mass in vivo. *Curr Pharm Des* **16**, 1561–1567.

[160] Wang, Y., Lim, K., Normandin, M., Zhao, X., Cline, G.W., Ding, Y.S. (2012) Synthesis and evaluation of [18F] exendin (9–39) as a potential biomarker to measure pancreatic beta-cell mass. *Nucl Med Biol* **39**, 167–176.

[161] Brom, M., Oyen, W.J., Joosten, L., Gotthardt, M., Boerman, O.C. (2010) 68Ga-labelled exendin-3, a new agent for the detection of insulinomas with PET. *Eur J Nucl Med Mol Imaging* **37**, 1345–1355.

[162] Brom M., Woliner-van der Weg W., Joosten L., Frielink C., Bouckenooghe T., Rijken P., Andralojc K., Göke B.J., de Jong M., Eizirik D.L., Béhé M., Lahoutte T., Oyen W.J., Tack C.J., Janssen M., Boerman O.C., Gotthardt M. (2014) Non-invasive quantification of the beta cell mass by SPECT with $^{111}$In-labelled exendin. *Diabetologia* **57**, 950–959.

[163] Brand, C., Abdel-Atti, D., Zhang, Y., Carlin, S., Clardy, S.M., Keliher, E.J., Weber, W.A., Lewis, J.S., Reiner, T. (2014) In vivo imaging of GLP-1R with a targeted bimodal PET/fluorescence imaging agent. *Bioconjug Chem* **25**, 1323–1330.

[164] Jones, T., Rabiner, E.A. (2012) The development, past achievements, and future directions of brain PET. *J Cereb Blood Flow Metab* **32**, 1426–1454.

[165] Passchier, J., Gee, A., Willemsen, A., Vaalburg, W., van Waarde, A. (2002) Measuring drug-related receptor occupancy with positron emission tomography. *Methods* **27**, 278–286.

[166] Zhang, Y., Fox, G.B. (2012) PET imaging for receptor occupancy: meditations on calculation and simplification. *J Biomed Res* **26**, 69–76.

[167] Abanades, S., van der Aart, J., Barletta, J.A., Marzano, C., Searle, G.E., Salinas, C.A., Ahmad, J.J., Reiley, R.R., Pampols-Maso, S., Zamuner, S., Cunningham, V.J., Rabiner, E.A., Laruelle, M.A., Gunn, R.N. (2011) Prediction of repeat-dose occupancy from single-dose data: characterisation of the relationship between plasma pharmacokinetics and brain target occupancy. *J Cereb Blood Flow Metab* **31**, 944–952.

[168] Atack, J.R., Wafford, K.A., Street, L.J., Dawson, G.R., Tye, S., Van Laere, K., Bormans, G., Sanabria-Bohorquez, S.M., De Lepeleire, I., de Hoon, J.N., Van Hecken, A., Burns, H.D., McKernan, R.M., Murphy, M.G., Hargreaves, R.J. (2011) MRK-409 (MK-0343), a GABAA receptor subtype-selective partial agonist, is a non-sedating anxiolytic in preclinical species but causes sedation in humans. *J Psychopharmacol* **25**, 314–328.

[169] Johnson, K.A., Fox, N.C., Sperling, R.A., Klunk, W.E. (2012) Brain imaging in Alzheimer disease. *Cold Spring Harb Perspect Med* **2**, a006213.

[170] Ishii, K. (2013) PET approaches for diagnosis of dementia. *Am J Neuroradiol* **35**, 2030–2038.

[171] Forsberg, A., Engler, H., Blomquist, G., Langstrom, B., Nordberg, A. (2012) The use of PIB-PET as a dual pathological and functional biomarker in AD. *Biochim Biophys Acta* **1822**, 380–385.

[172] Lowe, V.J., Kemp, B.J., Jack, C.R., Jr., Senjem, M., Weigand, S., Shiung, M., Smith, G., Knopman, D., Boeve, B., Mullan, B., Petersen, R.C. (2009) Comparison of 18F-FDG and PiB PET in cognitive impairment. *J Nucl Med* **50**, 878–886.

[173] Borsook, D., Becerra, L., Hargreaves, R. (2006) A role for fMRI in optimizing CNS drug development. *Nat Rev Drug Discov* **5**, 411–424.

[174] Smucny, J., Wylie, K.P., Tregellas, J.R. (2014) Functional magnetic resonance imaging of intrinsic brain networks for translational drug discovery. *Trends Pharmacol Sci* **35**, 397–403.

[175] Liu, Y., Wang, K., Yu, C., He, Y., Zhou, Y., Liang, M., Wang, L., Jiang, T. (2008) Regional homogeneity, functional connectivity and imaging markers of Alzheimer's disease: a review of resting-state fMRI studies. *Neuropsychologia* **46**, 1648–1656.

[176] Vemuri, P., Jones, D.T., Jack, C.R., Jr. (2012) Resting state functional MRI in Alzheimer's disease. *Alzheimers Res Ther* **4**, 2.

[177] Damoiseaux, J.S. (2012) Resting-state fMRI as a biomarker for Alzheimer's disease? *Alzheimers Res Ther* **4**, 8.

# INDEX



*Small Molecule Medicinal Chemistry: Strategies and Technologies*, First Edition.
Edited by Werngard Czechtizky and Peter Hamley.

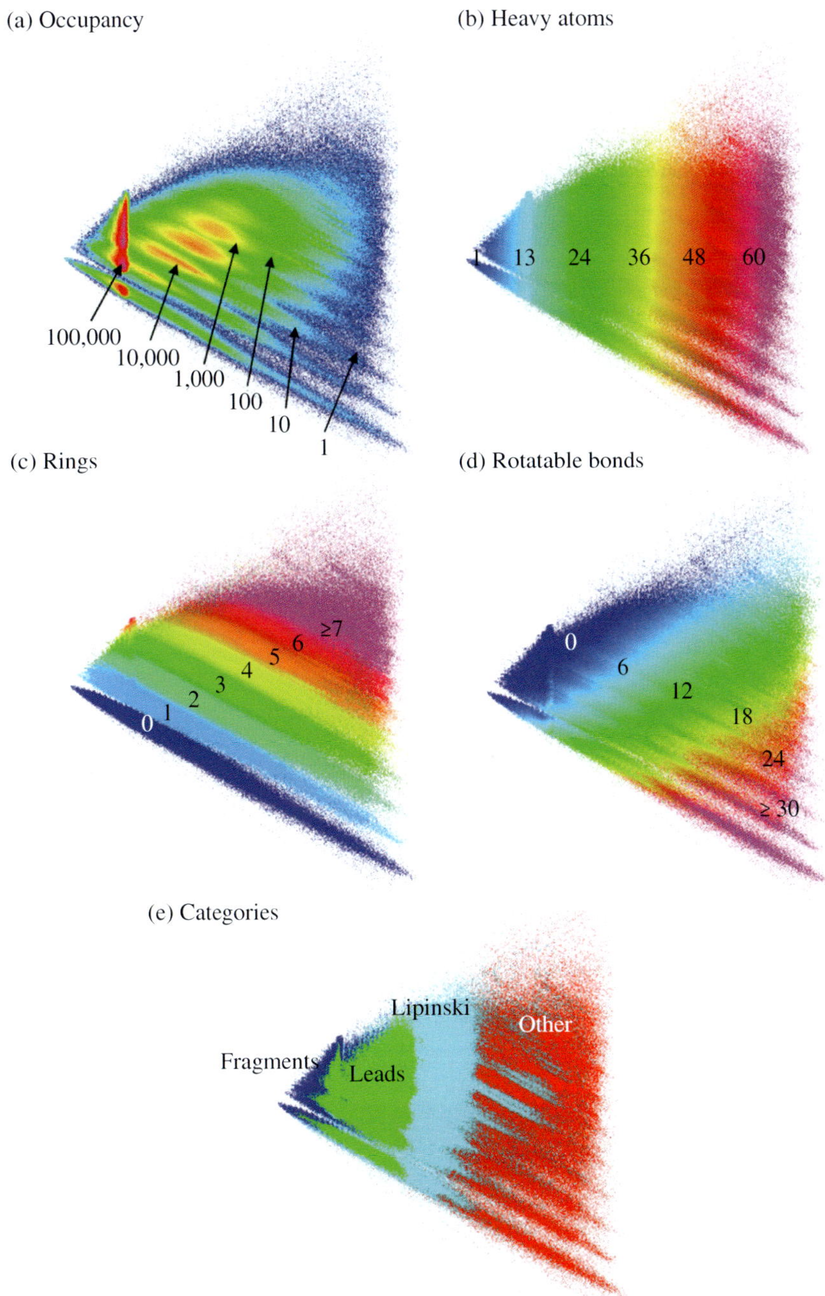


**FIGURE 1.1** See text page 15 for figure caption.

*Small Molecule Medicinal Chemistry: Strategies and Technologies*, First Edition.
Edited by Werngard Czechtizky and Peter Hamley.

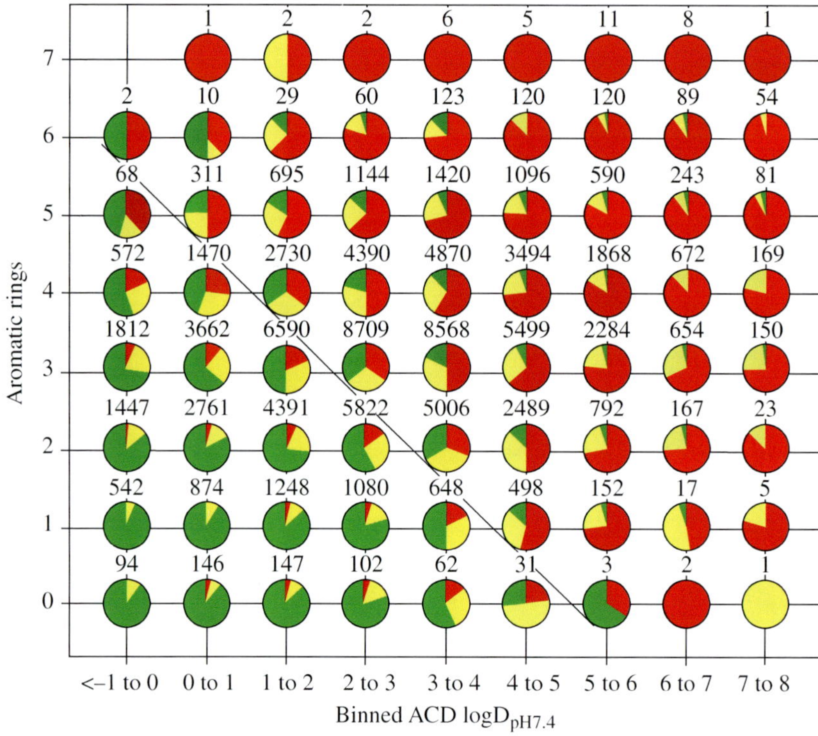


**FIGURE 1.3** See text page 21 for figure caption.

**FIGURE 2.1** See text page 41 for figure caption.

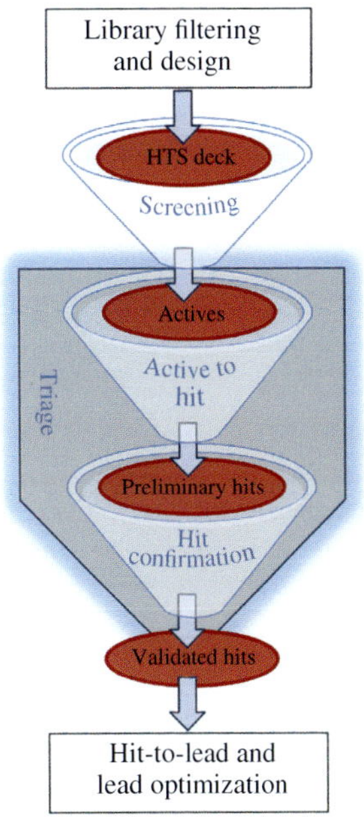


| | Triage (Soft decisions) | Lead-op (Hard decisions) |
|---|---|---|
| Potency cutoff | Weak (up to 10 mM) | <100 nM |
| Properties | Broad ranges; Failures acceptable | Drug like; Limit failures |
| Solubility | Moderate | High |
| ADMET | Liabilities acceptable optimization expected | No liabilities |
| Off-targets | Acceptable; Generally optimizable | Minimize |
| SAR | Not necessary | Necessary not flat |
| Reactive groups | Not acceptable | |

**FIGURE 7.2** See text page 193 for figure caption.

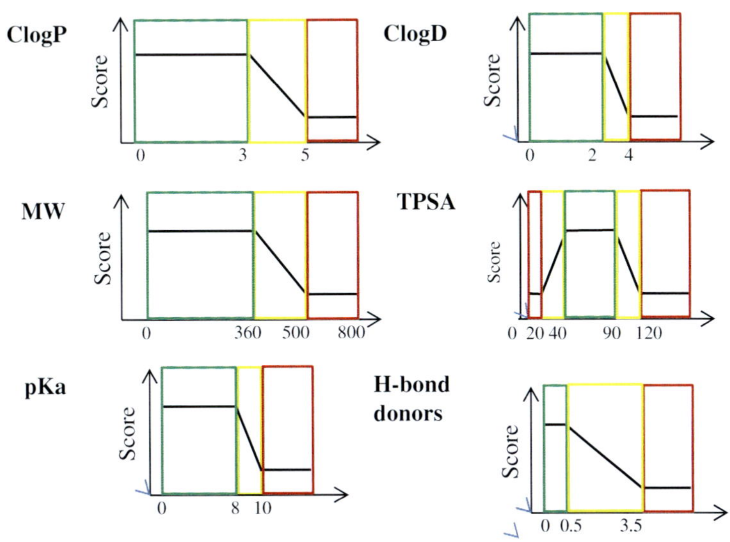


CNS MPO
Green: fully within acceptable limits
Yellow: partially inside limits
Red: beyond acceptable limits

**MPO score** $= \sum_{\{X\}} \frac{fx}{fx_{\max}}$

$$f_x = w \quad x < y_0$$

$$f_x = \frac{w(x - y_o)}{(y_{\max} - y_0)}$$

$$f_x = 0 \; x > y_{\max}$$

**FIGURE 7.6** See text page 207 for figure caption.

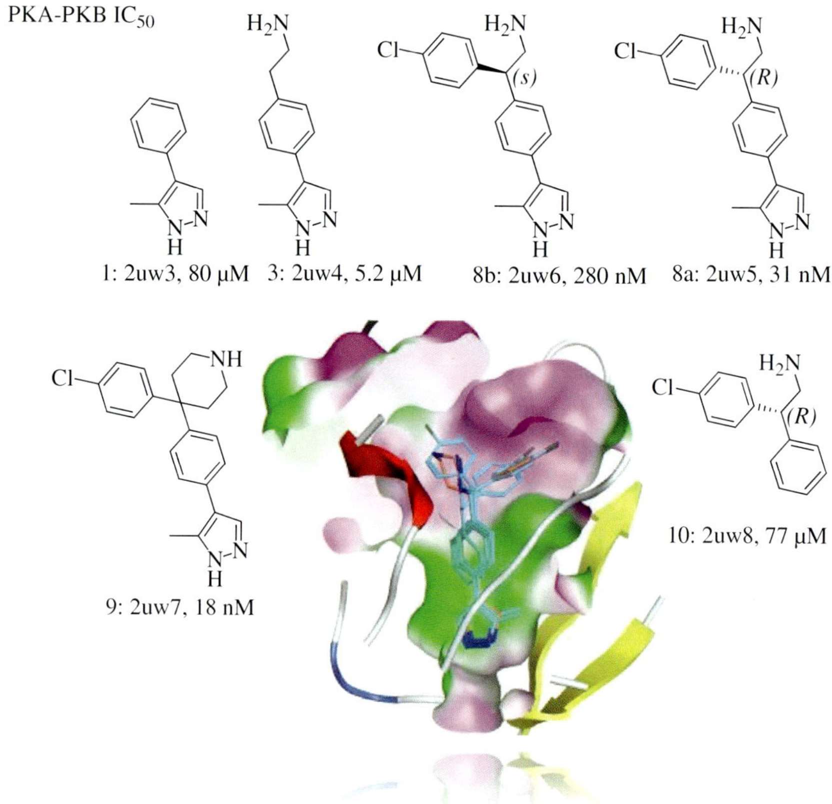


**FIGURE 7.9** See text page 213 for figure caption.

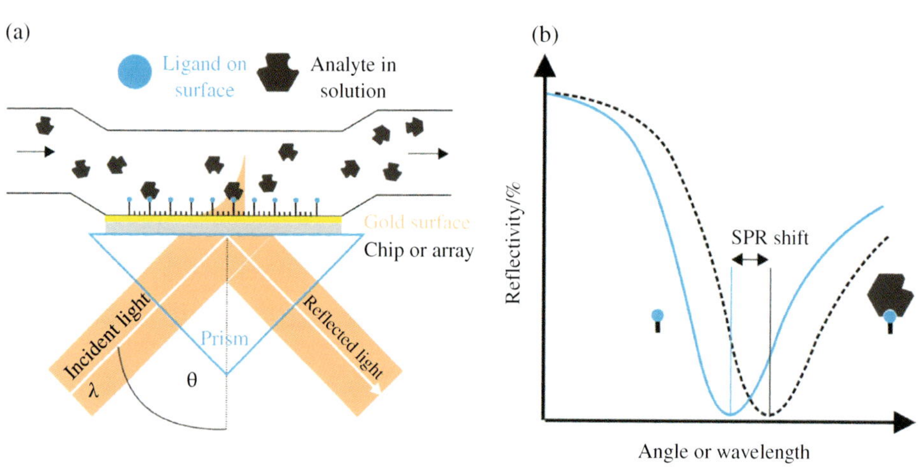


**FIGURE 8.1** See text page 226 for figure caption.

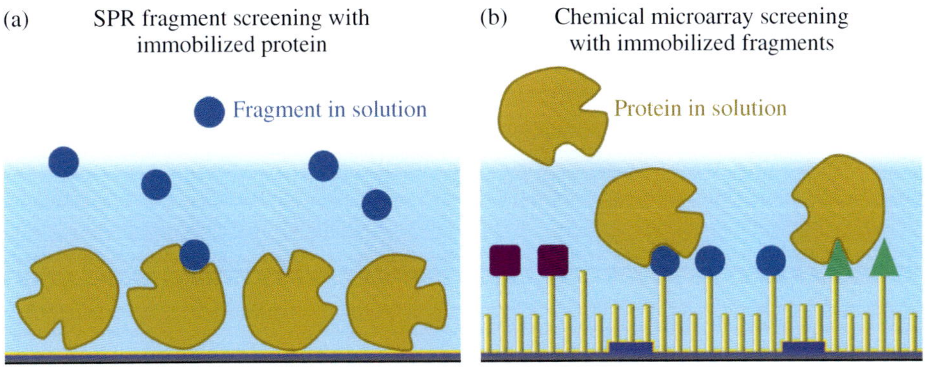


**FIGURE 8.2** See text page 227 for figure caption.

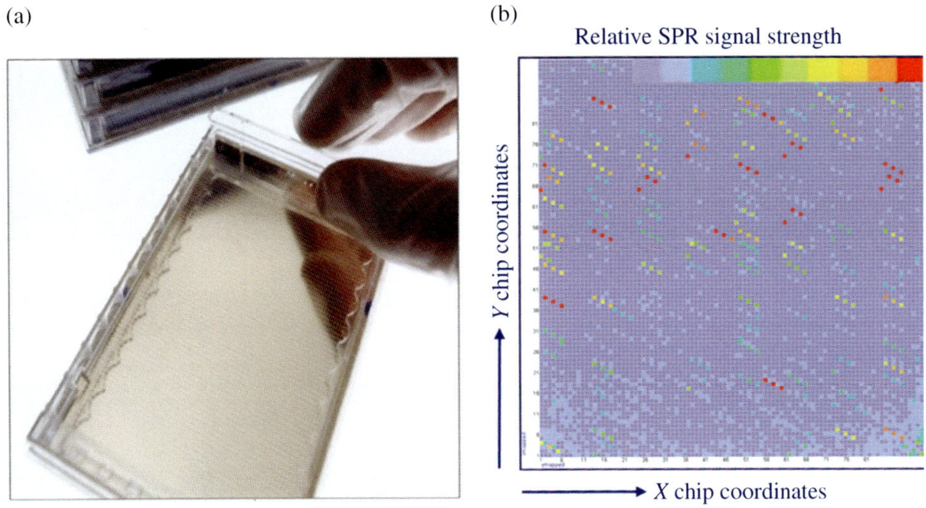


**FIGURE 8.4** See text page 230 for figure caption.

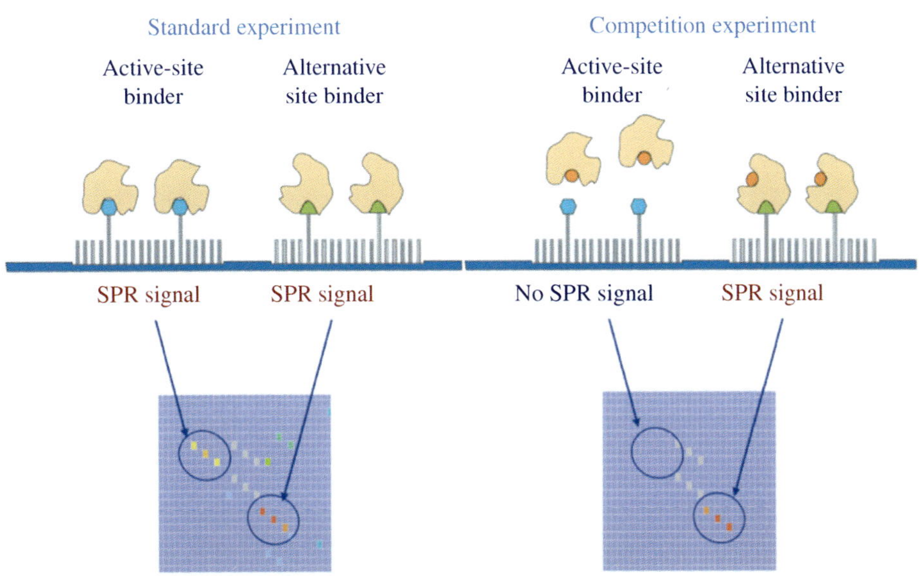


**FIGURE 8.5** See text page 231 for figure caption.

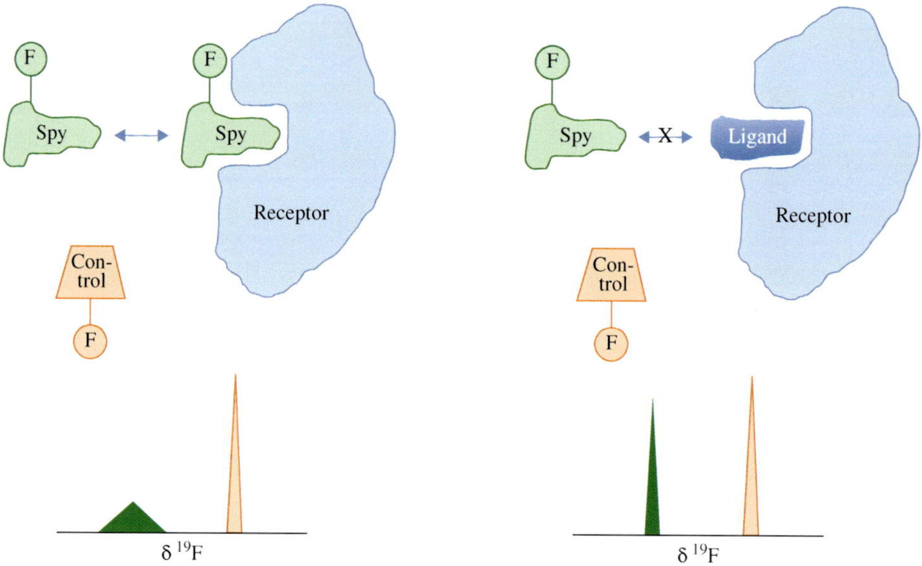


**FIGURE 8.6** See text page 233 for figure caption.

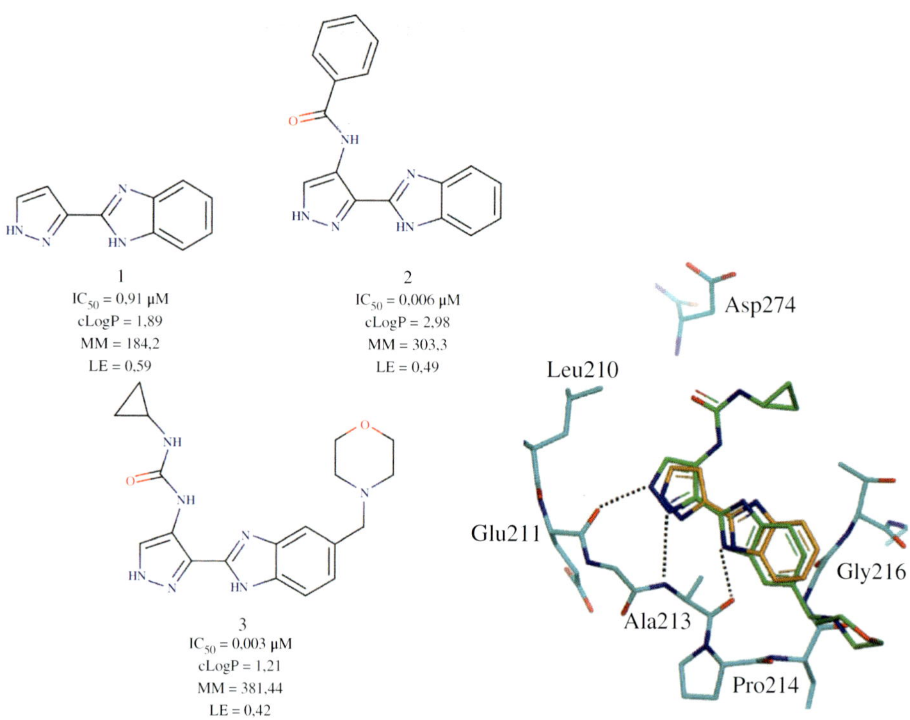


**FIGURE 8.7** See text page 241 for figure caption.

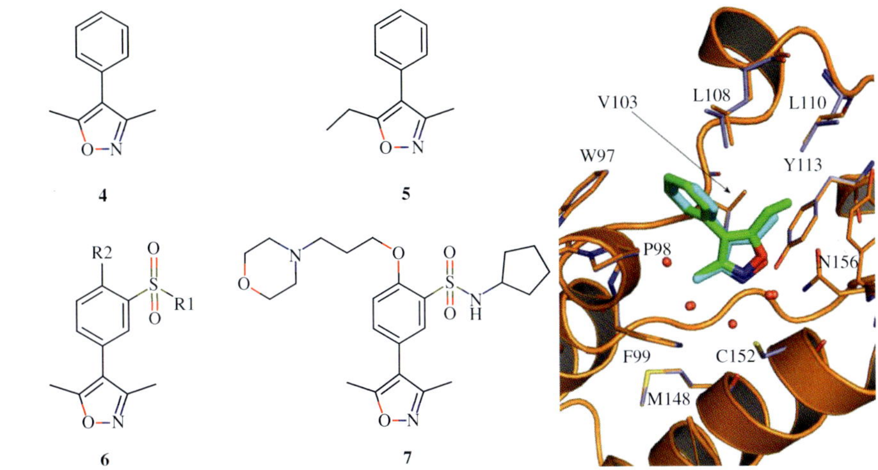


FIGURE 8.8 See text page 242 for figure caption.

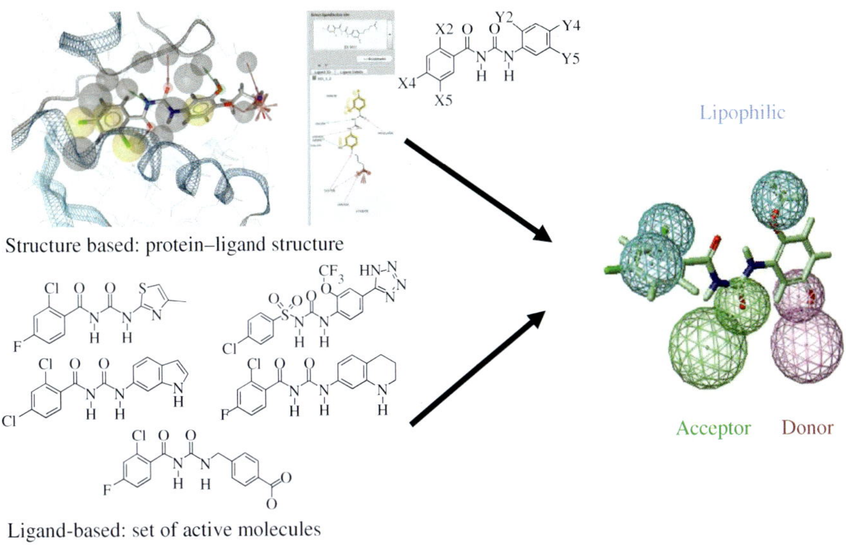


FIGURE 9.5 See text page 262 for figure caption.

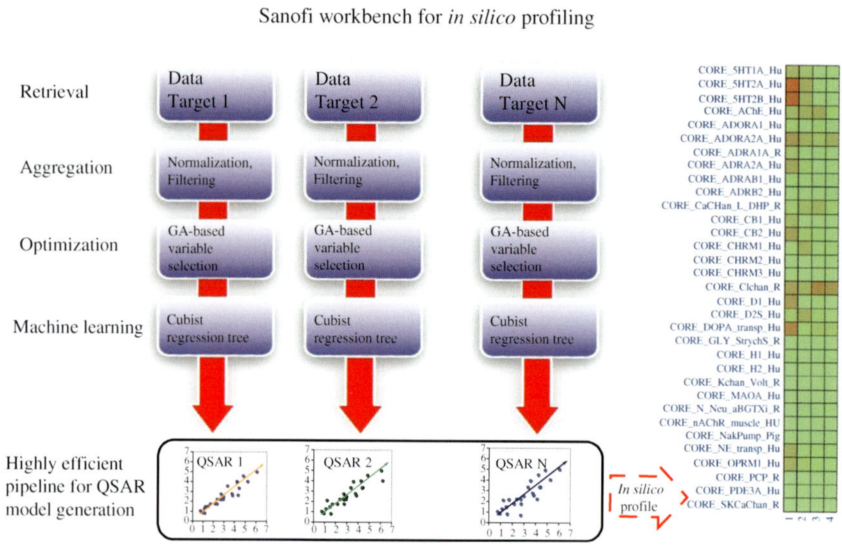


**FIGURE 9.6** See text page 268 for figure caption.

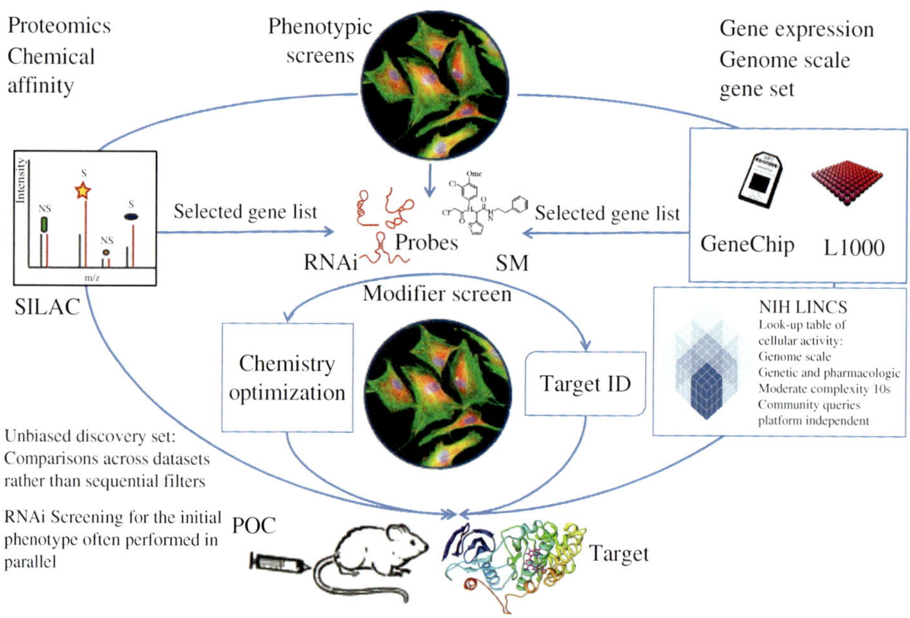


**FIGURE 10.3** See text page 293 for figure caption.

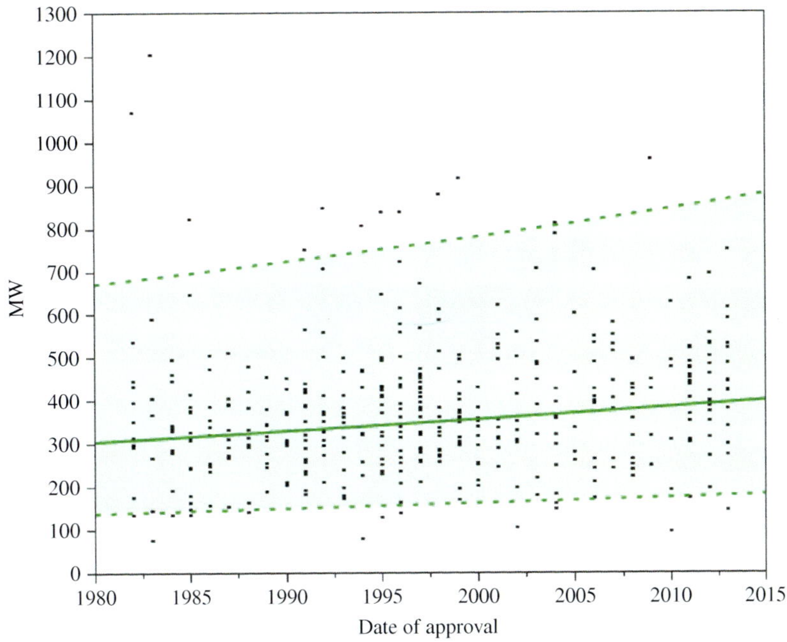


**FIGURE 11.1** See text page 309 for figure caption.

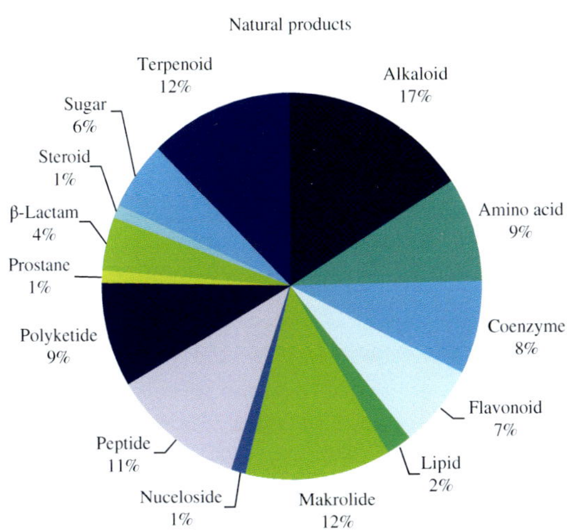


**FIGURE 14.1** See text page 394 for figure caption.

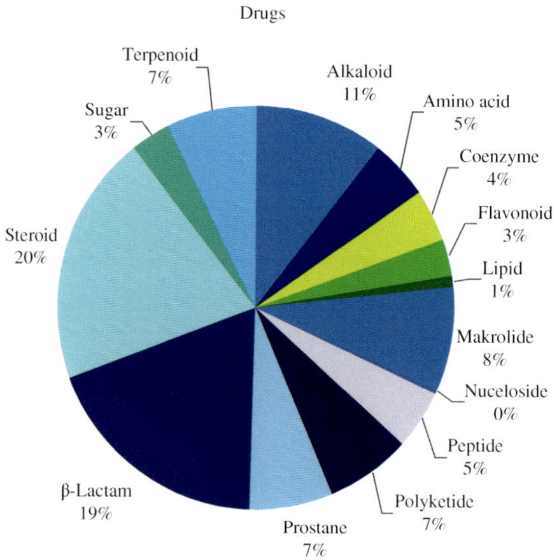


**FIGURE 14.2** See text page 395 for figure caption.

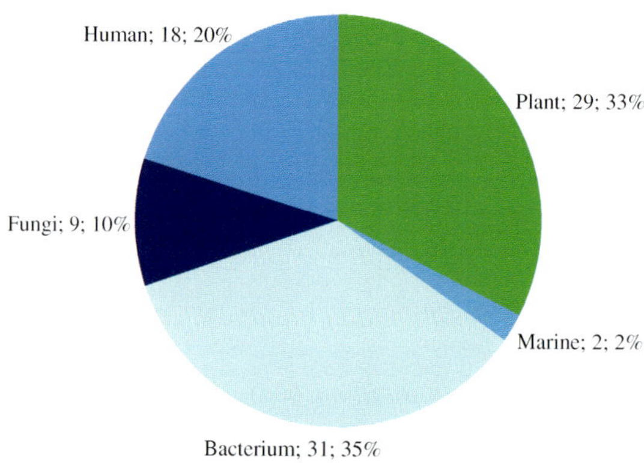


**FIGURE 14.3** See text page 395 for figure caption.

**FIGURE 14.12** See text page 401 for figure caption.

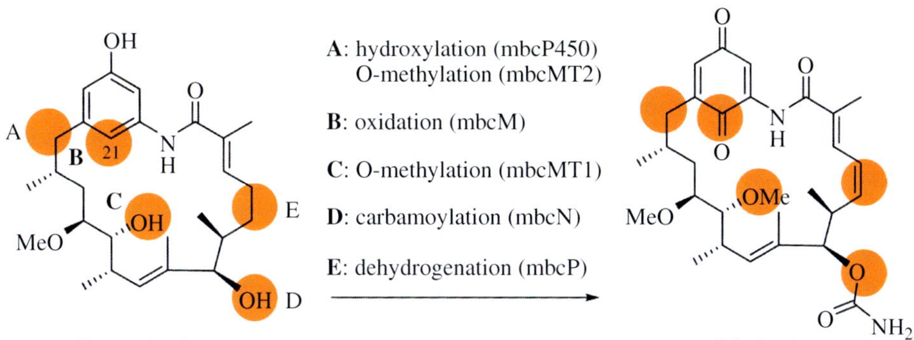


Premacbecin

**8.86**

Macbecin

**8.87**

**FIGURE 14.39** See text page 417 for figure caption.

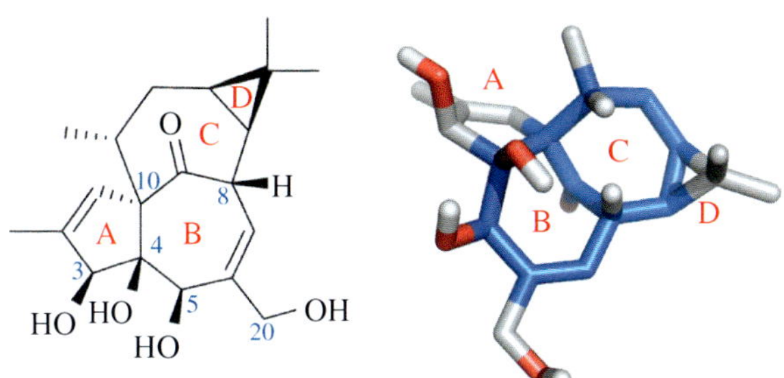


**8.88**

**FIGURE 14.41** See text page 419 for figure caption.

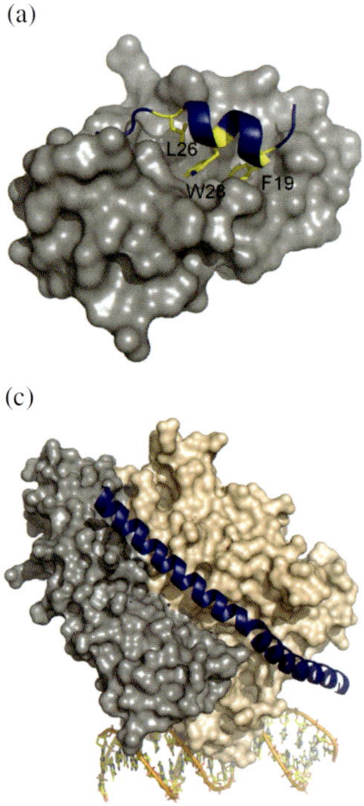


**FIGURE 15.1** See text page 434 for figure caption.

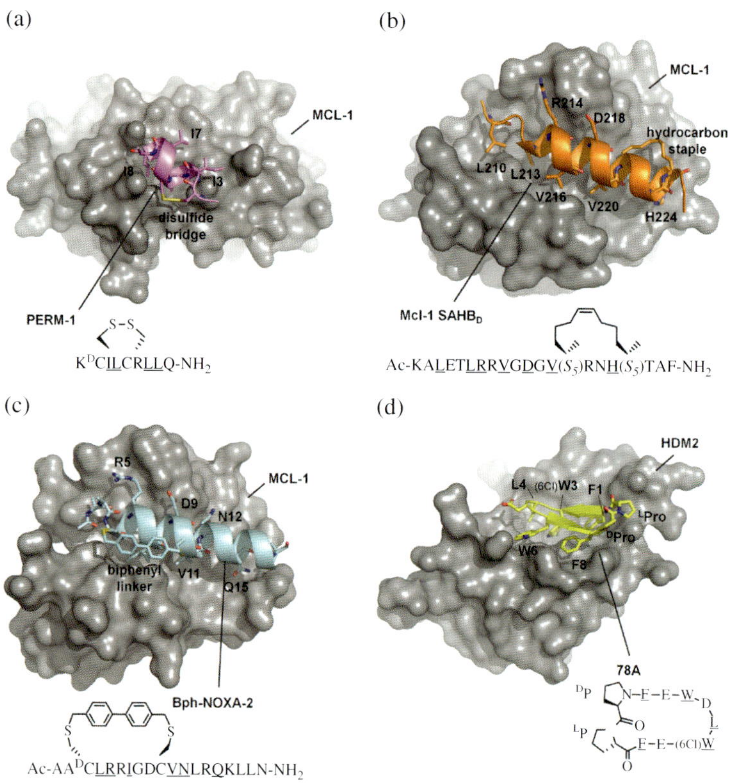


**FIGURE 15.4** See text page 440 for figure caption.

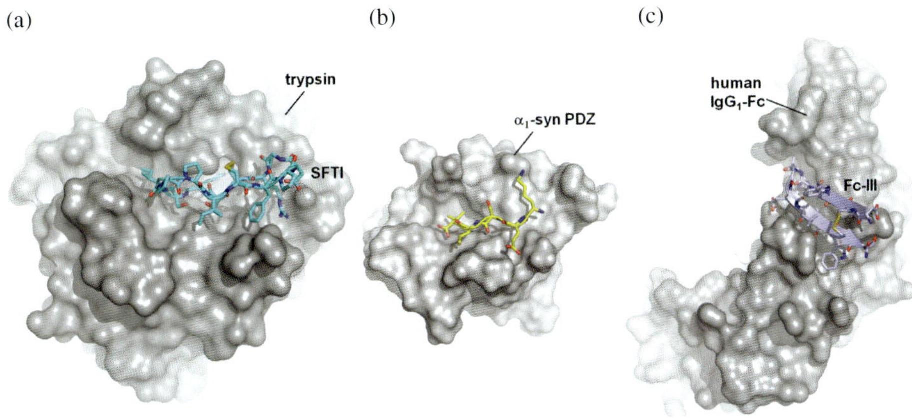


**FIGURE 15.7** See text page 447 for figure caption.

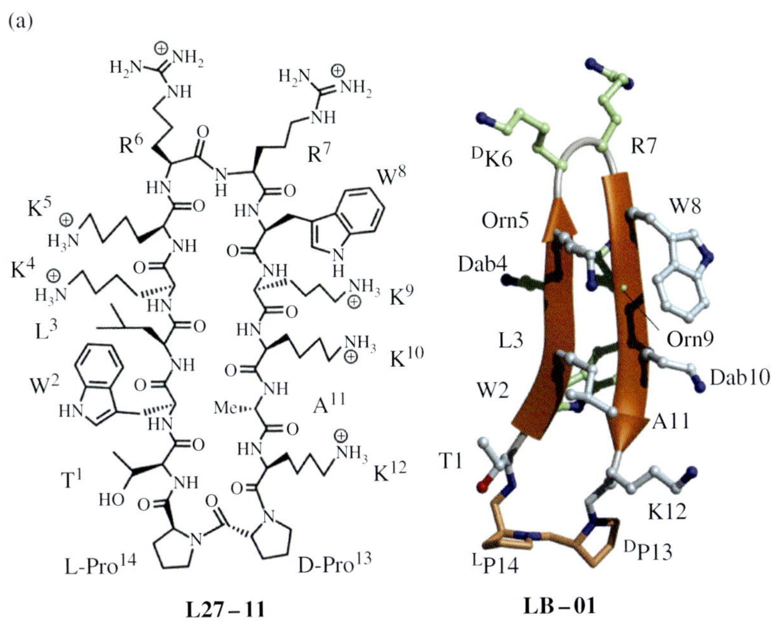


**FIGURE 15.9(a)** See text page 450 for figure caption.

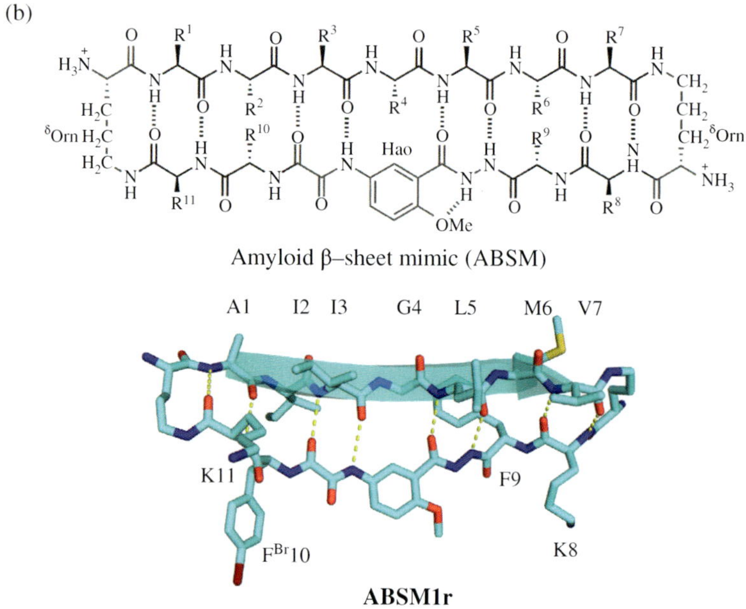


**FIGURE 15.9(b)** See text page 450 for figure caption.

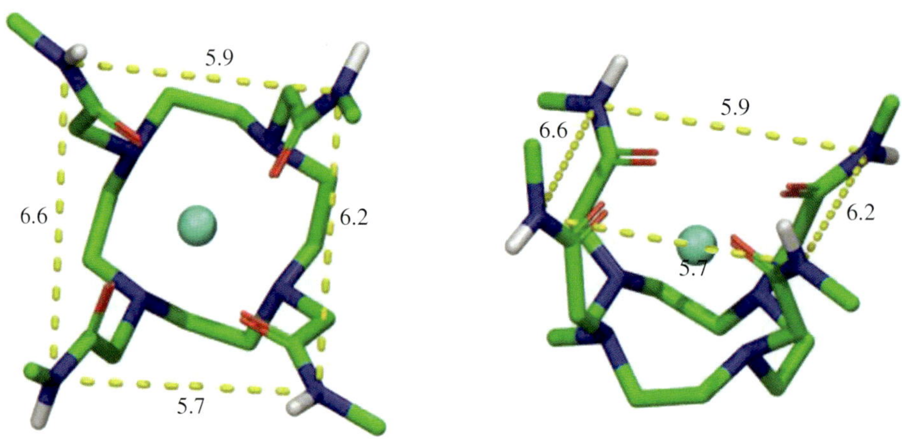


**FIGURE 16.5** See text page 474 for figure caption.

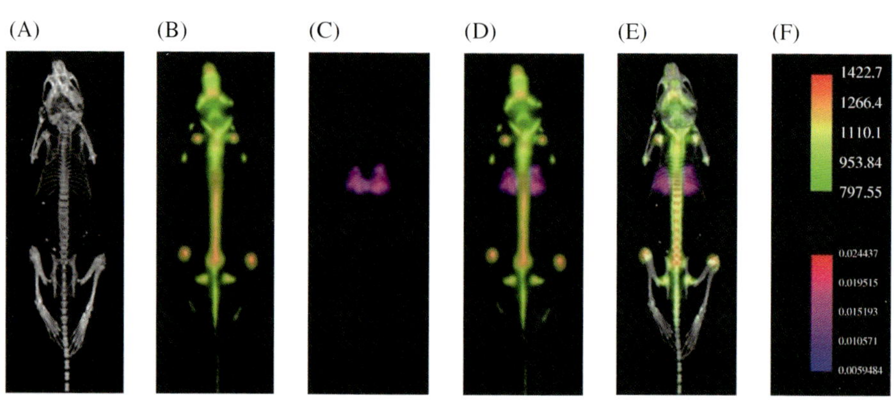


**FIGURE 16.6** See text page 475 for figure caption.

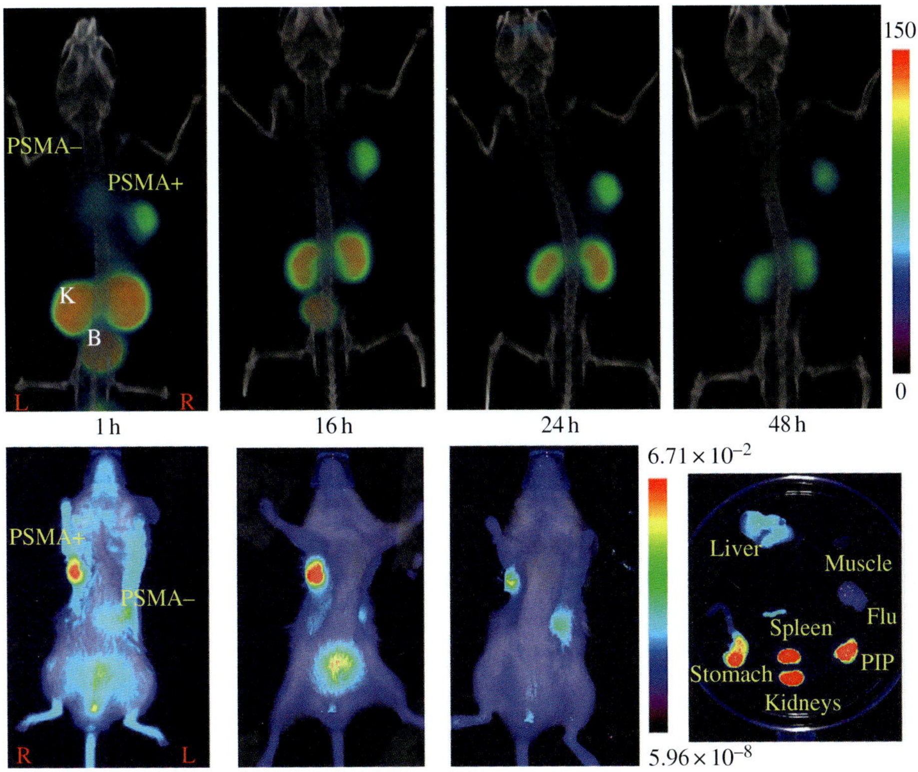


**FIGURE 16.7** See text page 478 for figure caption.

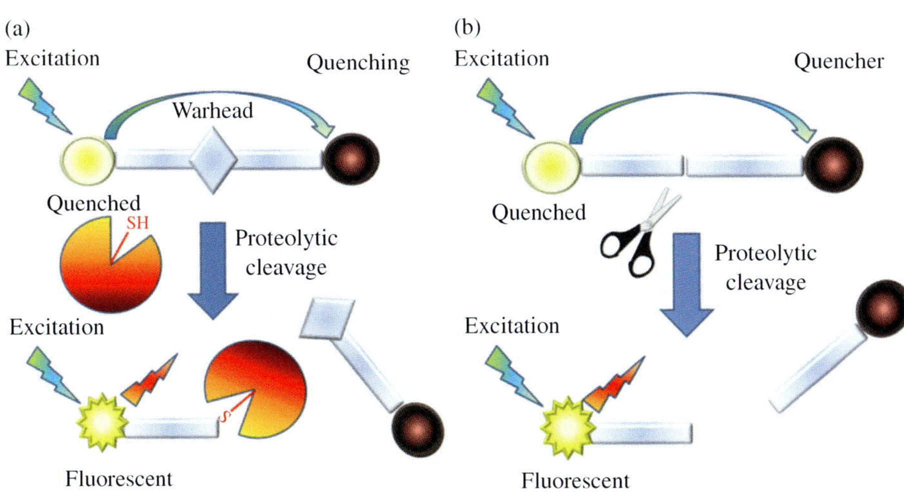


**FIGURE 16.9** See text page 481 for figure caption.

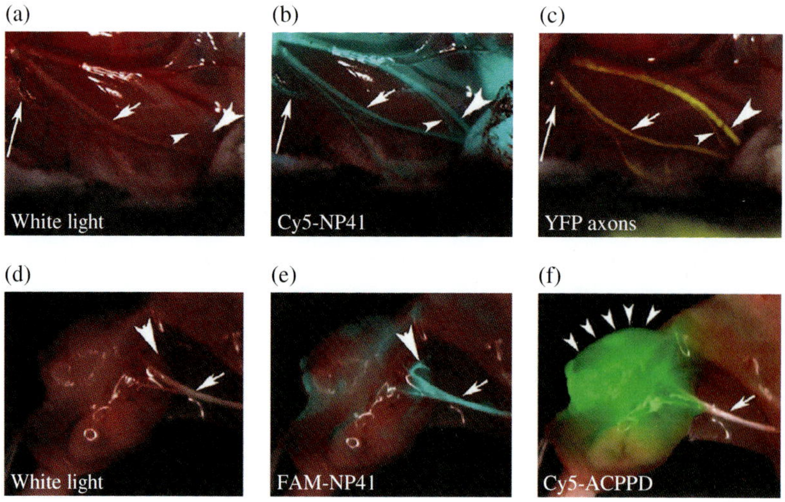


**FIGURE 16.10** See text page 482 for figure caption.

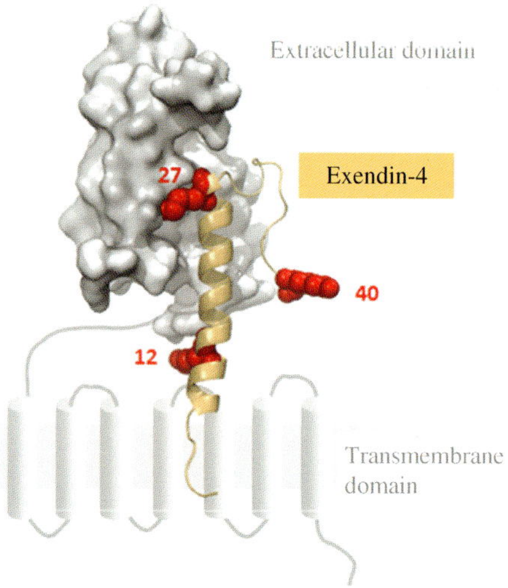


**FIGURE 16.13** See text page 486 for figure caption.

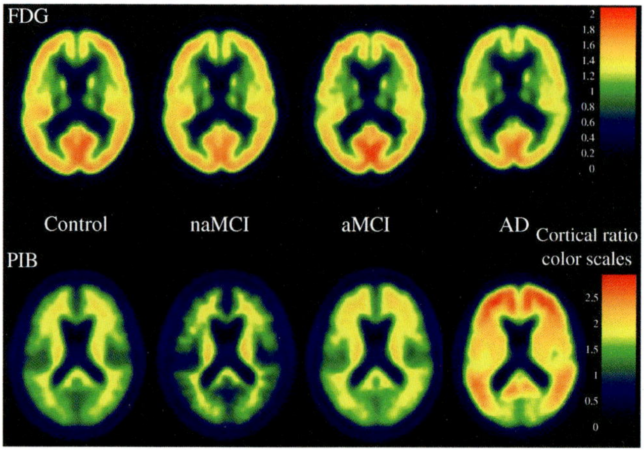


**FIGURE 16.16** See text page 489 for figure caption.

Printed in the USA/Agawam, MA
June 30, 2026

907143.012

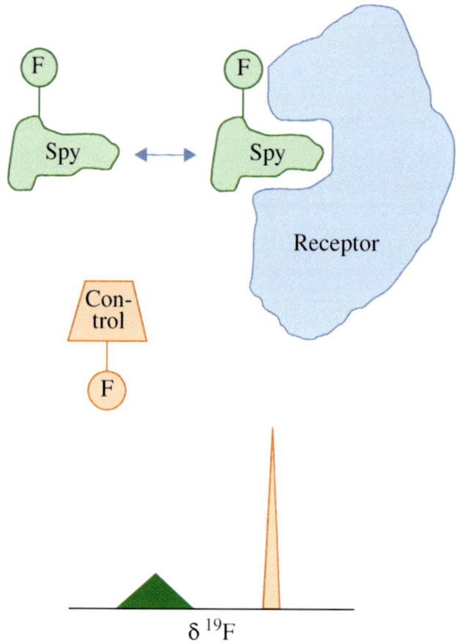


**FIGURE 8.6** See text page 233 for figure caption.

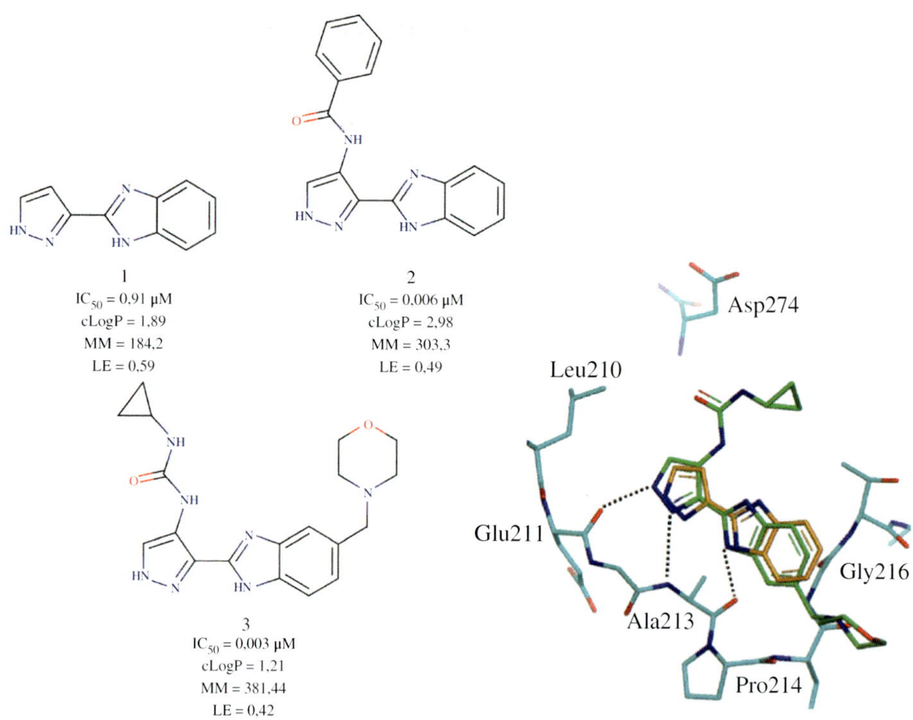


**FIGURE 8.7** See text page 241 for figure caption.